Fermions and Anomalies in Quantum Field Theories

Theoretical and Mathematical Physics

This series, founded in 1975 and formerly entitled (until 2005) *Texts and Monographs in Physics* (TMP), publishes high-level monographs in theoretical and mathematical physics. The change of title to *Theoretical and Mathematical Physics* (TMP) signals that the series is a suitable publication platform for both the mathematical and the theoretical physicist. The wider scope of the series is reflected by the composition of the editorial board, comprising both physicists and mathematicians.

The books, written in a didactic style and containing a certain amount of elementary background material, bridge the gap between advanced textbooks and research monographs. They can thus serve as a basis for advanced studies, not only for lectures and seminars at graduate level, but also for scientists entering a field of research.

Loriano Bonora

Fermions and Anomalies in Quantum Field Theories

Springer

Loriano Bonora
Theoretical Particle Physics
International School for Advanced Studies
Trieste, Italy

ISSN 1864-5879 ISSN 1864-5887 (electronic)
Theoretical and Mathematical Physics
ISBN 978-3-031-21927-6 ISBN 978-3-031-21928-3 (eBook)
https://doi.org/10.1007/978-3-031-21928-3

This Springer imprint is published by the registered company Springer Nature Switzerland AG
The registered company address is: Gewerbestrasse 11, 6330 Cham, Switzerland

For Anita and Caterina

Preface

Quantum field theory is a theory under construction. It is not an axiom-based theory, where everything can be derived, at least potentially, from first principles (although attempts in this direction have been made). It is rather a mixture of heuristic intuitions and pieces of mathematics, sometimes of very rigorous mathematics; fantastically successful, it must be stressed, in describing quantum physics. The path integral, which is at the basis of modern quantum field theory, is the typical example in this sense: it is not (yet) mathematically rigorous, but it is nevertheless an extremely efficient and powerful tool. And it is an incredible alloy of mathematics and physical intuition. There is a sector of quantum field theory, where this composite texture is particularly visible: anomalies. The anomaly problem is in principle very simple: there are classically conserved (divergenceless, traceless) quantities that are not conserved anymore upon quantization. But, in practice, this simplicity quickly disappears when faced with the ambiguities and variety of approaches the (still) incomplete nature of the quantum field theory formulation requires. The main focus of this book is on how an acceptable degree of confidence about our knowledge of anomalies can be gathered by comparing various methods and techniques of calculation.

The main objects of study are the consistent chiral gauge and trace anomalies of various local symmetries (gauge, diffeomorphisms, conformal). They are produced in theories of elementary fields (fermions, but, occasionally, also other fields) minimally interacting with a gauge potential and/or a metric. In order to deal with them, three different approaches are illustrated: perturbative (mostly based on Feynman diagrams), non-perturbative (which, for simplicity, are referred to as heat kernel methods) and family's index theorem methods. None of them, alone, is sufficient to disentangle all the problems the anomaly calculation poses. However, all together they provide a complex of results that dispel any doubt about the detailed knowledge of anomalies, their meaning and their implications. For this reason, a considerable effort throughout is spent in order to compare the various methods and results, and attention is focused on spelling out the ambiguities, which are, unfortunately, scattered all through the derivation of anomalies.

The central issue for chiral consistent anomalies is related in all cases to the existence of the fermion propagator, that is of the inverse of the kinetic Dirac operator for fermions. In the case of a Weyl fermion theory, the naive propagator does not exist. This is a catastrophe for perturbative methods, because the fermion propagator is an essential ingredient in any type of calculation. The usual trick one resorts to in order to bypass this difficulty is to add a *free* fermion of opposite chirality. This works well, except when there are (chiral consistent) anomalies. Anomalies show up as violations of conservation laws and disruptions of the BRST Ward identities, which are basic pillars for unitarity and renormalizability. In a sense, this means that repairing the fermion propagator by adding a free fermion of opposite chirality is illusory; inverting the Dirac-Weyl operator is simply impossible.

The heat kernel methods are not essentially different from the perturbative approach. They use the same kind of trick to bypass the non-invertibility of the Dirac-Weyl operator, except that they start from the full kinetic operator. Their relevance is of course due to their providing complete results instead of the first few terms of a series expansion, like in the case of perturbative methods. The approach with the family's index theorem is instead not only methodologically but also conceptually different, because its aim is studying the obstructions to the existence of the fermion propagator. All approaches lead to the same results: (consistent chiral) anomalies of perturbative and non-perturbative methods show up when the index theorem signals the existence of obstructions and vice-versa. The interesting and unifying aspect is that anomalies and obstructions have a common origin in the cohomology of the classifying space.

Trace anomalies are, instead, divided into two groups, even and odd parity anomalies. Odd parity anomalies are akin to chiral anomalies, and they are related to the existence of the fermion propagator and appear as obstructions to inverting the Dirac-Weyl operator. Even trace anomalies, instead, although they signal a violation of Weyl invariance, do not correspond to obstructions to the existence of propagators, they have nothing to do with the geometry of the classifying space and do not represent in general any threat to unitarity and renormalizability (with the notable exception of 2d string theory).

The purpose of this book is not a complete historical review of all the methods used to analyze anomalies and their results, nor a description of all the anomalies discovered in the last fifty or so years. This would be impossible in a single volume due to the vastness of the relevant literature. Our purpose is rather to provide a coherent view (as opposed to a sparse set of notions) of the anomaly problem. This implies a selective organization of the material (so, for instance, not all the regularization prescriptions in the perturbative and non-perturbative approaches will be analyzed, and not all the interpretations of the index theorem that have appeared in the literature will be considered). Selective choices of material accompanied by detailed and explicit calculations and an accurate comparison of the three above-mentioned methods (perturbative, heat kernel like and family's index theorem) are organized in order to produce a coherent and logical design.

It is opportune, at this stage, to dispel some terminological confusion in the literature. The term 'chiral anomaly' is rather universally used but may have different meanings; the same is the case for Adler-Bell-Jackiw (ABJ) anomaly. Our basic distinction is between *consistent* and *covariant* anomalies, and this is our preferred terminology. Covariant anomalies in gauge theories may be referred to as ABJ. Same for diffeomorphism anomalies in gravitational theories. At times, the simple term chiral anomaly will be used to include both (odd parity) of them. We use also the term 'split' anomalies for those that have opposite sign for opposite chiralities. Trace anomalies may also be split and consistent, and even and odd-parity.

This said, it is worth spelling out a few characteristic features of the layout of this book. Although for practical and historical reasons, the distinction between consistent and covariant anomalies is maintained, it is shown that the latter are a particular limit of the former (this limit accounts not only for their form but also for their coefficient). Another distinctive feature to be stressed is the rigid link between ABJ and odd parity trace anomalies which is unveiled here. An important aspect to be highlighted is that no symmetry is considered in isolation, but the anomaly problem is analyzed by taking into consideration all the symmetries in the game, for instance, gauge plus diffeomorphisms, conformal symmetry plus diffeomorphisms, vector plus axial symmetries and so on (which, of course, is the only correct way to derive anomalies; it does not make sense, for instance, to study trace anomalies while ignoring diffeomorphisms). Next, as pointed out above, the family's index theorem is an essential ingredient for our comprehension of anomalies. It requires a geometrical interpretation of the latter, of the BRST algebra and the descent equations. Such an interpretation is presented here via the evaluation map, which allows to connect the spacetime where anomalies live with the classifying space. The classifying space is a universal (i.e. it stands there for any spacetime with given gauge group) object, from which, on one side, the local nature of anomalies is inherited, and in which, on the other side, the family's index theorem is rooted. The interpretation of a non-trivial family's index as an obstruction to the existence of the fermion propagator is perhaps well-known; the choice here is to stress it as the pivot element for interpreting anomalies. One final distinctive feature of the present book concerns the presentation of global anomalies by means of differential characters. The latter are the mathematical objects that globally extend the notion of anomaly. They may carry torsion, in which case an indeterminacy of the path integral may originate. This indeterminacy is what is called global anomaly.

Anomalies are divided into two large families: even and odd parity anomalies. In this book, the stress is largely on the latter. Odd parity anomalies may appear both in specific current divergences or in the trace of the energy-momentum tensor, and the consistent ones are non-trivial (they cannot be eliminated by subtracting local counterterms from the effective action). Even parity anomalies are non-trivial only as trace anomalies (they may appear in intermediate stages of the calculation of a current divergence, but in that case they can always be eliminated by a local subtraction).

The upshot of the above choices and results is that the present book, apart from standard material, is largely complementary to the existing textbooks on anomalies, which, however, does not mean that it covers all that is not present in the latter. It has programmatic limitations, the emphasis being more on methodology than on applications, and the list of topics which are not treated is still rather large: phenomenological applications of the ABJ anomalies to particle physics, to quantum

Hall effect, non-renormalization theorems, anomalies and supersymmetry, anomalies on lattice gauge theories, anomalies and Weyl semi-metals, anomalies in spacetimes with boundaries, anomalies on orbifolds, anomalies and CFT/AdS correspondence, anomalies for higher spin fields, anomalies in noncommutative field theories, the cobordism approach to global anomalies, Wess-Zumino terms for trace anomalies, trace anomalies and related entanglement calculations. In particular, we do not consider anomalies produced by integrating out fields with spin higher than $\frac{1}{2}$, with the exception of the zero modes of the superstrings in the last chapter.

Now a short description of the contents. The book is divided into six parts. Part I is preparatory. Chapter 1 is an introduction to fermions. It is particularly focused on distinguishing among Dirac, Weyl and Majorana fermions in four dimensions. Chapter 2 is devoted to a description of the classical symmetries (gauge, diffeomorphisms and local Lorentz) and their BRST extension. Chapter 3 contains a short introduction to conformal symmetry, in two and higher dimensions.

Part II of the book is devoted to the relation between anomalies and cohomology. In Chap. 4, effective actions are introduced and the consistency conditions satisfied by anomalies are derived. Then in Chap. 5, it is shown that the latter subtend a cohomology problem, and the BRST (or local) cohomology of field theory is defined. Finally, it is shown how to construct non-trivial solutions of the consistency conditions, the non-trivial cocycles, for all the symmetries involved.

Part III deals with perturbative methods to compute anomalies. Chapter 6 is devoted in particular to Weyl fermions coupled to a vector potential, and to the problems they pose from the perturbative point of view. A rather detailed derivation of their consistent chiral anomalies is given. Then the same is done for a Dirac fermion coupled to a vector and an axial potential. In Chap. 7, it is the turn of trace anomalies. To start with, a long discussion is devoted to the definition of trace anomaly. Then trace and diffeomorphism anomalies in fermion theories in 2d are calculated. The 2d example is in fact a useful playground in preparation to the more complicated 4d cases. Next come the computations of the odd parity trace and diffeomorphism anomaly due to a gauge field and to a background metric in 4d. In all these derivations, the dimensional regularization is (mostly) used as the fittest one, and, in this regard, a legend about dimensional regularization 'having problems with γ_5' should be contradicted. None of such problems will be met. This leads to the conviction that the latter are to be rather attributed to unresolved ambiguities which have nothing to do with dimensional regularization.

In Part IV, nonperturbtive heat kernel-like methods are treated. Chapter 8 is devoted to various Dirac operators and square thereof, after which the Schwinger-DeWitt and the Seeley-DeWitt approaches (denoted henceforth generically by SDW) are introduced. In Chap. 9, these methods are used for explicit calculations. In particular, Bardeen's anomaly, due to the coupling of a Dirac fermion with a vector and an axial gauge fields, is derived in various ways. Then an explicit derivation is given of the consistent chiral anomaly of Weyl fermions in interaction with a gauge field. Chapter 10 starts with an introduction to the metric-axial-tensor (MAT) background and to the relevant formalism. Then the trace anomaly generated by such a background in interaction with a Dirac fermion is calculated with two methods, from which one can derive in particular the odd trace anomaly for a Weyl fermion in an ordinary metric background, the ABJ-like trace anomaly and the gauge-induced odd trace anomaly.

Part V is devoted to the family's index theorem and its application to chiral anomalies. It starts with a preliminary introduction to the geometry of anomalies (Chap. 11), that is to a geometrical interpretation of the BRST symmetry. This allows us not only to geometrize the algebraic treatment of the first part of the book, but especially to gain a new and unifying understanding of anomalies in field theory: local anomalies do not have to do with the topology of spacetime, but with the geometry of the classifying space, they are 'universal'. Chapter 12 deals with the index theorem, to which a long mathematical introduction is devoted. Then the family's index theorem is applied to anomalies to discover that in this context anomalies do not appear as violations of classical conservation laws, but as obstructions to the existence of the fermion propagator (or the fermion determinant). This approach to anomalies has the advantage that it gives the obstructions in a very general and, at the same time, detailed form (which includes their coefficients), so that the problem of anomaly cancelation can be analyzed in a very efficient way also in higher dimensions. Finally Chap. 13 deals with global anomalies, analyzed with the mathematical tool of differential characters, and the criteria for their absence.

Part VI is devoted to a few special topics. Chapter 14 contains a complete calculation of trace and diffeomorphism anomalies of a Dirac fermion in a MAT brackground in 2d, both with perturbative and non-perturbative methods. Chapter 15 deals with Wess-Zumino terms in field theories. In particular, the Wess-Zumino term that connect diffeomorphisms and local Lorentz anomalies is explicitly calculated without and with background connection. Chapter 16 is devoted to sigma models and their local anomalies, to the role of Wess-Zumino terms in sigma models and, finally, to global sigma model anomalies. Chapter 17 analyzes the worldsheet, sigma model and target space anomalies of string and superstring theories.

Note on the Prerequisites In order to follow the techniques and arguments used in this book, familiarity with some basic theories is necessary: elementary quantum field theory and gravity theory, on the physical side, differential geometry including bundle geometry as well as some rudimentary knowledge of homotopy, homology and cohomology, on the mathematical side. When more advanced mathematics is needed (for the family's index theorem and global anomalies, for instance), it is briefly introduced in specific (hopefully helpful) sections.

Note on Bibliography A full bibliography on anomalies consists of thousands of papers and books and is impractical for editorial reasons. Therefore I have made the choice of a short bibliography for each chapter containing papers and books which are a direct source of each chapter or have been basic for my understanding of fermions and anomalies. A more complete Bibliography is collected in an online file, whose reference can be found in Chap. 18.

Trieste, Italy Loriano Bonora

Acknowledgments

It is a duty and a pleasure for me to thank and express my gratitude to all my collaborators at various stages of my forty-year research activity on anomalies: Alexander Andrianov, Mauro Bregola, Chong-Sun Chu, Paolo Cotta-Ramusino, Maro Cvitan, Predrag Dominis Prester, Antonio Duarte Pereira, Ricardo Gamboa-Saravi, Bruno Lima de Souza, Silvio Pallua, Paolo Pasti, Mateo Paulisić, Cesare Reina, Maurizio Rinaldi, Alexander Sorin, Martin Schnabl, Ivica Smolić, Jim Stasheff, Tamara Štemberga, Alessandro Tomasiello, Mario Tonin and Stav Zalel. I would like to thank in particular Roberto Soldati for his collaboration and encouragements during the gestation of this book and for allowing me to use his lecture notes for Sects. 1.1 and 1.5.

I feel the obligation to acknowledge and thank the hospitality and support of several institutions: the Physics Department of Padua University, SISSA, INFN in primis, but also the Center for Theoretical Physics at MIT, the Theory Division at CERN, the Yukawa Institute at Kyoto and the KEK Theory Center at Tsukuba, where I found stimulating environments for my research.

My acknowledgment and gratitude goes also to the ArXiv repository, to the INSPIRE database and finally to Wikipedia, which have been of invaluable help especially in time of COVID.

Concerning the publication phase of this book, I would like to thank Masud Chaichian for his warm interest in it and his friendly advice. Finally I would like to acknowledge Ramon Khanna, Executive Editor of Springer Physics & Astronomy Books for his constant assistance and the personnel of Books Production of Springer Nature.

Trieste, Italy — Loriano Bonora

Contents

Part I Basic Tools

Part V Nonperturbative Methods. (B) Index Theorem

Part I
Basic Tools

Chapter 1
Fermions

In the first chapter, we introduce the formalism for fermions. We start by defining fermions in 4d Minkowski spacetime. This is motivated by the fact that in the first part of the book, in particular in the perturbative approach to anomalies, we will be mostly concerned with four-dimensional problems but also by the fact that the properties of fermions we will need in other dimensions do not differ much from the four-dimensional ones. The main characters in this book are massless fermions: Dirac and Majorana, but especially Weyl. For this reason, a sizable portion of this initial chapter is spent in order to introduce and discuss the properties of massless fermions. We start with Dirac fermions, which transform according to a reducible representation of the Lorentz group. Next, we focus on the two distinct irreducible representations the Dirac fermions decompose into: the Weyl and Majorana fermions. Since this distinction is crucial for anomalies (Majorana fermions cannot carry consistent chiral anomalies, while Weyl fermions can), we will devote a section to explain the properties that distinguish them, and will try to ward off the reader from the most frequent pitfalls in this regard.

Non-perturbative calculations of anomalies, in particular the index theorem approach, are usually formulated in a generic dimension. In the fifth and sixth part of the book, we will need fermions in any (even) dimensions. Fermions are representations of spin groups, which are strictly related to Clifford algebras (the algebra of gamma matrices). Some of the relevant theorems for anomalies are better formulated in terms of Clifford algebras, corresponding spin groups and representations. For this reason, the next topic in this chapter is a concise introduction to Clifford algebras, spin groups and representations, both in even dimensional Minkowski and Euclidean spacetimes. After that, we deem it appropriate to insert a comment about the Wick rotation, an operation used in quantum field theory to pass from a Minkowski background metric to a corresponding Euclidean one, and assign in this way a precise meaning to (ordinary and path) integrals. This technique, which is the backbone of the computational power in quantum field theory, will be used throughout the book.

L. Bonora, *Fermions and Anomalies in Quantum Field Theories*, Theoretical and Mathematical Physics, https://doi.org/10.1007/978-3-031-21928-3_1

Notation. In this chapter and until further notice, we use a flat Minkowski metric $\eta_{\mu\nu}$ with mostly $(-)$ signature. The gamma matrices satisfy $\{\gamma^\mu, \gamma^\nu\} = 2\eta^{\mu\nu}$ and

$$\gamma_\mu^\dagger = \gamma_0\gamma_\mu\gamma_0.$$

At times we use also the α matrices, defined by $\alpha_\mu = \gamma_0\gamma_\mu$.

The charge conjugation operator C is defined to satisfy

$$\gamma_\mu^T = -C^{-1}\gamma_\mu C, \quad CC^* = -1, \quad CC^\dagger = 1. \tag{1.1}$$

For example, $C = C^\dagger = C^{-1} = \gamma^0\gamma^2 = \alpha^2$ does satisfy all the above requirements. This representation of C will be used below, but it must be clear that it holds only in some γ-matrix representations, such as the Dirac and Weyl ones, not in the Majorana. The chiral matrix $\gamma_5 = i\gamma^0\gamma^1\gamma^2\gamma^3$ has the properties

$$\gamma_5^\dagger = \gamma_5, \quad (\gamma_5)^2 = 1, \quad C^{-1}\gamma_5 C = \gamma_5^T.$$

The generators of the Lorentz group are $\Sigma_{\mu\nu} = \frac{i}{4}[\gamma_\mu, \gamma_\nu]$. We use also the notation $\Sigma_i = \varepsilon_{ijk}\Sigma_{jk}$ for $i, j, k = 1, 2, 3$. When dealing with anomalies it is sometimes more convenient to use the 'anti-Hermitean' version of $\Sigma_{\mu\nu}$ without the i. To avoid misunderstanding we will always specify the definition we use.

1.1 Massless Spinors

In this section, we review the basic and fundamental properties of the massless spinors, both from the classical and quantum field theoretical points of view. Our treatment follows the transition from the classical to the canonical quantum theory and is quite complete and explicit. Continuous and discrete symmetries are carefully analyzed and discussed for all the three kinds of massless spinors: namely Weyl, Dirac and Majorana. The purpose of the present section is also to fill a surprising gap in the textbooks devoted to this subject. While massive Dirac fermions are introduced and discussed in all existing textbooks, the formalism for massless fermions is often understood.

1.1.1 Massless Dirac and Weyl Spinors

Consider a massless Dirac bispinor, i.e. any complex and Grassmann valued bispinor on the Minkowski space, which is a solution of the differential equation

$$\partial\!\!\!/\, \psi(x) = 0 \quad \psi(x) = \begin{bmatrix} \psi_L(x) \\ \psi_R(x) \end{bmatrix}$$

Hereafter and until further notice, we adopt the Weyl representation of γ matrices. In order to obtain the most general solution of the above equation, one has to somewhat move away from the conventional approach and formalism that has been developed for the celebrated Dirac equation

$$i\partial\!\!\!/\,\psi(x) = M\,\psi(x)$$

In the chiral representation for the gamma matrices, we can build up a very convenient set of spin-states as follows. Consider the matrix $\alpha^\nu \equiv \gamma^0\gamma^\nu$ with matrix elements

$$\alpha^\nu = \begin{bmatrix} \sigma^\nu & 0 \\ 0 & \bar{\sigma}^\nu \end{bmatrix}$$

where we have set $\sigma^\nu = (\mathbb{I}, -\sigma_k)$ with $k = 1, 2, 3$, while $\bar{\sigma}^\nu = (\mathbb{I}, \sigma_k)$, and σ_k are the Pauli matrices, which satisfy $\sigma_i\sigma_j = \delta_{ij} + i\varepsilon_{ijk}\sigma_k$. Then the massless Dirac equation can also be cast in the form

$$\alpha\cdot\partial\,\psi(x) = 0 \quad \Longleftrightarrow \quad \begin{cases} \sigma\cdot\partial\,\psi_L(x) = 0 \\ \bar{\sigma}\cdot\partial\,\psi_R(x) = 0 \end{cases} \tag{1.2}$$

in such a manner that it can be rewritten as a Schrödinger equation,

$$i\hbar\,\frac{\partial\psi}{\partial t} = H_0\,\psi,$$

where the massless Dirac Hamiltonian operator, in physical units, reads

$$H_0 = c\,\alpha^k\,(-\,i\hbar\nabla_k) = c\,\alpha^k\hat{p}^k$$

To solve the massless Dirac equation, let us first consider the plane wave stationary solutions

$$\psi_p(x) = \Gamma(p)\,\exp\{-\,i\,p\cdot x\},$$

where the spinor $\Gamma(p)$ fulfills the algebraic equations

$$(\,\gamma^0 p_0 + \gamma^k p_k\,)\,\Gamma(p) = 0 \quad \Longleftrightarrow \quad \vec{\alpha}\cdot\vec{p}\,\Gamma(p) = p_0\,\Gamma(p), \tag{1.3}$$

The latter admit non-trivial solutions iff

$$\det\,\|\,\vec{\alpha}\cdot\vec{p} - p_0\,\| = 0$$

This amounts to the biquadratic algebraic equation $(p_0^2 - \mathbf{p}^2)^2 = 0$. It follows that there are two pairs of eigenvectors of the Hamiltonian H_0, such that $(\mathbf{p} \equiv \vec{p})$

$$\begin{cases} H_0 u_r(\mathbf{p}) = |\,\mathbf{p}\,| u_r(\mathbf{p}) \\ H_0 v_r(\mathbf{p}) = -\,|\,\mathbf{p}\,| v_r(\mathbf{p}) \end{cases} \quad (r = 1, 2) \tag{1.4}$$

Moreover, since $[\,H_0, \gamma_5\,] = 0$, it is evident that the energy eigenstates will also be eigenstates of the chiral matrix, i.e. they will exhibit definite chirality.

A positive frequency solution of (1.3) $\Gamma(p) = \Gamma_+(\wp, \mathbf{p})$, with $p_0 = \wp$, satisfies

$$\vec{\alpha} \cdot \vec{p}\, \Gamma_+(\wp, \mathbf{p}) = \wp\, \Gamma_+(\wp, \mathbf{p}) \quad (\wp = |\,\mathbf{p}\,|),$$

while a negative frequency solution will satisfy

$$\vec{\alpha} \cdot \vec{p}\, \Gamma_-(-\,\wp, \mathbf{p}) = -\,\wp\, \Gamma_-(-\,\wp, \mathbf{p}) \quad (\wp = |\,\mathbf{p}\,|)$$

If we consider the parity transformed state $\Gamma'_+(\wp, -\,\mathbf{p}) = \gamma^0\, \Gamma_+(\wp, \mathbf{p})$ we can readily check that

$$(\,\vec{\alpha} \cdot \vec{p} + \wp)\, \gamma^0\, \Gamma_+(\wp, \mathbf{p}) = \gamma^0\, (\,-\,\vec{\alpha} \cdot \vec{p} + \wp)\, \Gamma_+(\wp, \mathbf{p}) = 0 \tag{1.5}$$

Hence, we are allowed to make the identification

$$\Gamma_-(-\,\wp, \mathbf{p}) = \gamma^0\, \Gamma_+(\wp, \mathbf{p}), \tag{1.6}$$

which is also a consequence of $\{H_0, \gamma^0\} = 0$.

Spin-States

Our next purpose is to define a complete and orthogonal quartet of spin-states for the 4d massless bispinor space (i.e. solutions of (1.3)). To this end, we have to search among states of opposite chirality and frequency. The construction goes as follows. Explicit evaluation of the relevant spin-matrices yields for $p_0 = |\,\vec{p}\,| = \wp$

$$\sigma_\mu\, p^\mu = \wp\, \mathbb{I} + \sigma_k\, p^k = \wp + \vec{\sigma} \cdot \vec{p} = \begin{pmatrix} \wp + p_z & p_x - i p_y \\ p_x + i p_y & \wp - p_z \end{pmatrix},$$

$$\bar{\sigma}_\mu\, p^\mu = \wp\, \mathbb{I} - \sigma_k\, p^k = \wp - \vec{\sigma} \cdot \vec{p} = \begin{pmatrix} \wp - p_z & -\,p_x + i p_y \\ -\,p_x - i p_y & \wp + p_z \end{pmatrix},$$

so that we find

$$\sigma \cdot p\; \bar{\sigma} \cdot p = \bar{\sigma} \cdot p\; \sigma \cdot p = p^2 = \wp^2 - \vec{p}^{\,2} = p\!\!\!/^{\,2} = \tilde{p\!\!\!/}^{\,2} = 0. \tag{1.7}$$

Notice that $\tilde{p}^{\nu} = (\wp, -\vec{p}\,) = p_{\nu}$ is the dual light-cone four-vector that satisfies

$$\tilde{p}^{2} = p^{2} = 0 \quad p \cdot \tilde{p} = 2\vec{p}^{\,2}$$

Moreover we get

$$\Pi \equiv \frac{\not{p}\,\tilde{\not{p}}}{4\wp^2} = \frac{1}{2}\left(1 + \frac{\vec{\alpha}\cdot\vec{p}}{\wp}\right) = \frac{1}{2\wp}\begin{bmatrix} \sigma\cdot p & 0 \\ 0 & \bar{\sigma}\cdot p \end{bmatrix} \tag{1.8}$$

$$\overline{\Pi} \equiv \frac{\tilde{\not{p}}\,\not{p}}{4\wp^2} = \frac{1}{2}\left(1 - \frac{\vec{\alpha}\cdot\vec{p}}{\wp}\right) = \frac{1}{2\wp}\begin{bmatrix} \bar{\sigma}\cdot p & 0 \\ 0 & \sigma\cdot p \end{bmatrix} \tag{1.9}$$

which satisfy by construction

$$\Pi = \Pi^{\dagger} \quad \Pi^{2} = \Pi \quad \mathrm{Tr}\,\Pi = 2 \tag{1.10}$$

$$\overline{\Pi} = \overline{\Pi}^{\dagger} \quad \overline{\Pi}^{2} = \Pi \quad \mathrm{Tr}\,\overline{\Pi} = 2 \tag{1.11}$$

$$\Pi\,\overline{\Pi} = 0 = \overline{\Pi}\,\Pi \quad [\,\gamma_5,\,\Pi\,] = [\,\overline{\Pi},\,\gamma_5\,] = 0 \tag{1.12}$$

$$\gamma_0\,\Pi\,\gamma_0 = \overline{\Pi} \quad \gamma_0\,\overline{\Pi}\,\gamma_0 = \Pi \tag{1.13}$$

This means that Π and $\overline{\Pi}$ are chirality preserving, though parity exchanging, projectors onto two-dimensional orthogonal spaces. Hence, we can set, with $p^{\nu} = (\wp, \mathbf{p}) = (\wp, \vec{p}\,)$,

$$u_{-}(\mathbf{p}) = \frac{1}{\sqrt{\wp - p_z}}\,\gamma^{\nu} p_{\nu}\,\xi_{+} = \begin{bmatrix} u_L(\mathbf{p}) \\ 0 \\ 0 \end{bmatrix} = \frac{1}{\sqrt{\wp - p_z}}\begin{bmatrix} \wp - p_z \\ -p_x - ip_y \\ 0 \\ 0 \end{bmatrix} \tag{1.14}$$

$$\text{where}\quad \xi_{+} = \begin{bmatrix} 0 \\ 0 \\ 1 \\ 0 \end{bmatrix} \quad \text{and} \quad u_L(\mathbf{p}) = \frac{\bar{\sigma}\cdot p}{\sqrt{\wp - p_z}}\begin{vmatrix} 1 \\ 0 \end{vmatrix}$$

The above spin-state is a positive frequency solution of equation (1.3) with negative chirality:

$$\not{p}\;u_{-}(\mathbf{p}) = 0, \quad (\gamma_5 + 1)\,u_{-}(\mathbf{p}) = 0, \tag{1.15}$$

$$\Pi\,u_{-}(\mathbf{p}) = 0, \quad \overline{\Pi}\,u_{-}(\mathbf{p}) = u_{-}(\mathbf{p}). \tag{1.16}$$

Moreover, the 2-component left-handed Weyl spinor $u_L(\mathbf{p})$ satisfies

$$u_L^\dagger(\mathbf{p})\, u_L(\mathbf{p}) = 2\wp, \quad u_L(\mathbf{p}) \otimes u_L^\dagger(\mathbf{p}) = \bar{\sigma} \cdot p \quad (p_0 = \wp) \tag{1.17}$$

Notice that the rank-two square matrix $\mathsf{P}_L \equiv \frac{\bar{\sigma} \cdot p}{2\wp}$ satisfies, for $p_0 = \wp$,

$$\sigma \cdot p\, \mathsf{P}_L = 0, \quad \mathsf{P}_L^\dagger = \mathsf{P}_L, \quad \mathsf{P}_L^2 = \mathsf{P}_L, \quad \mathrm{Tr}\, \mathsf{P}_L = 1,$$

so that it corresponds to the one-dimensional projector on positive energy and left-handed spin states. Next we set

$$u_+(\mathbf{p}) = -\gamma^0 u_-(-\mathbf{p}) = \frac{-1}{\sqrt{\wp + p_z}} \begin{pmatrix} 0 \\ 0 \\ \wp + p_z \\ p_x + ip_y \end{pmatrix} = (-1) \begin{pmatrix} 0 \\ u_R(\mathbf{p}) \end{pmatrix} \tag{1.18}$$

$$\text{where} \qquad u_R(\mathbf{p}) = \frac{\sigma \cdot p}{\sqrt{\wp + p_z}} \begin{vmatrix} 1 \\ 0 \end{vmatrix},$$

which satisfies by construction

$$\begin{aligned} &\not{p}\, u_+(\mathbf{p}) = 0, \;\; (\gamma_5 - 1)\, u_+(\mathbf{p}) = 0, \\ &\Pi\, u_+(\mathbf{p}) = 0, \;\; \overline{\Pi}\, u_+(\mathbf{p}) = u_+(\mathbf{p}), \end{aligned} \tag{1.19}$$

$$u_R^\dagger(\mathbf{p})\, u_R(\mathbf{p}) = 2\wp, \;\; u_R(\mathbf{p}) \otimes u_R^\dagger(\mathbf{p}) = \sigma \cdot p \tag{1.20}$$

Therefore $\mathsf{P}_R = (\sigma \cdot p)/2\wp\ (p_0 = \wp)$ represents the 1d projector on right-handed spin states of positive energy. Thus, we see that we can view $\overline{\Pi}$ as the projector on the two-dimensional space of spin states with momentum $\vec{p}$, positive energy $p_0 = \wp$ and both chiralities. The corresponding progressive plane waves will thereby be $u_\mp(\vec{p}\,) \exp\{-i\wp t + i\vec{p} \cdot \vec{x}\}\ (\wp = |\,\vec{p}\,|)$.

Quite analogously, we can build as well a pair of negative frequency spin-states of momentum **p** but opposite chirality:

$$\tilde{\not{p}}\, v_\mp(\mathbf{p}) = 0 \quad (\gamma_5 \pm 1)\, v_\mp(\mathbf{p}) = 0 \tag{1.21}$$

$v_\mp$ are constructed as follows

$$v_-(\mathbf{p}) = (\wp - p_z)^{-\frac{1}{2}}\,\gamma^\nu \tilde{p}_\nu\,\eta_+ = \begin{pmatrix} v_L(\mathbf{p}) \\ 0 \\ 0 \end{pmatrix} = \frac{1}{\sqrt{\wp - p_z}} \begin{pmatrix} p_x - ip_y \\ \wp - p_z \\ 0 \\ 0 \end{pmatrix},$$

$$\text{where}\quad \eta_+ = \begin{pmatrix} 0 \\ 0 \\ 0 \\ 1 \end{pmatrix} \qquad v_L(\mathbf{p}) = \frac{\sigma \cdot p}{\sqrt{\wp - p_z}} \begin{vmatrix} 0 \\ 1 \end{vmatrix},$$

which evidently satisfies

$$\overline{\Pi}\, v_-(\mathbf{p}) = 0, \quad \Pi\, v_-(\mathbf{p}) = v_-(\mathbf{p}), \tag{1.22}$$

$$v_L^\dagger(\mathbf{p})\, v_L(\mathbf{p}) = 2\wp, \quad v_L(\mathbf{p}) \otimes v_L^\dagger(\mathbf{p}) = \sigma \cdot p, \quad (p_0 = \wp\,). \tag{1.23}$$

Likewise

$$v_+(\mathbf{p}) = \gamma^0\, v_-(-\,\mathbf{p}) = \frac{1}{\sqrt{\wp + p_z}} \begin{pmatrix} 0 \\ 0 \\ -p_x + ip_y \\ \wp + p_z \end{pmatrix} = \begin{pmatrix} 0 \\ v_R(\mathbf{p}) \end{pmatrix} \tag{1.24}$$

$$\text{where}\quad v_R(\mathbf{p}) = \frac{\bar{\sigma} \cdot p}{\sqrt{\wp + p_z}} \begin{vmatrix} 0 \\ 1 \end{vmatrix},$$

which is such that

$$\tilde{\not{p}}\, v_+(\mathbf{p}) = 0, \quad (\gamma_5 - 1)\, v_+(\mathbf{p}) = 0, \quad (p_0 = \wp\,), \tag{1.25}$$

$$v_R^\dagger(\mathbf{p})\, v_R(\mathbf{p}) = 2\wp, \quad v_R(\mathbf{p}) \otimes v_R^\dagger(\mathbf{p}) = \bar{\sigma} \cdot p, \quad (p_0 = \wp\,). \tag{1.26}$$

From the obvious equality $\sigma \cdot \tilde{p} = \bar{\sigma} \cdot p$, it follows that we can also understand

$$(2\wp)^{-1} v_L(-\mathbf{p}) \otimes v_L^\dagger(-\mathbf{p}) = \frac{\bar{\sigma} \cdot p}{2\wp} = \mathsf{P}_L \quad (p_0 = \wp\,)$$

as the one-dimensional projector on left-handed spin states of negative frequency, while

$$(2\wp)^{-1} v_R(-\mathbf{p}) \otimes v_R^\dagger(-\mathbf{p}) = \frac{\sigma \cdot p}{2\wp} = \mathsf{P}_R \quad (p_0 = \wp\,)$$

can be interpreted as the one-dimensional projector on the negative frequency right-handed spin-states. In sum, we have the following resolution of the rank-two unit matrix: $\mathbb{I} = \mathsf{P}_R + \mathsf{P}_L$.

The above defined quartet of massless and chiral bispinor spin-states fulfills orthogonality and closure relations. On the one hand, spin-states of opposite chirality and/or opposite frequency are orthogonal

$$\left.\begin{array}{c} u_{\pm}^{\dagger}(\mathbf{p})\, u_{\mp}(\mathbf{p}) = 0 = v_{\pm}^{\dagger}(\mathbf{p})\, v_{\mp}(\mathbf{p}) \\ \\ u_{\pm}^{\dagger}(\mathbf{p})\, v_{\mp}(\mathbf{p}) = 0 \\ \\ u_{\pm}^{\dagger}(\mathbf{p})\, v_{\pm}(\mathbf{p}) = 0 = v_{\mp}^{\dagger}(\mathbf{p})\, u_{\mp}(\mathbf{p}) \end{array}\right\} \tag{1.27}$$

On the other hand, we have

$$u_{\pm}^{\dagger}(\mathbf{p})\, u_{\pm}(\mathbf{p}) = v_{\pm}^{\dagger}(\mathbf{p})\, v_{\pm}(\mathbf{p}) = 2\wp \tag{1.28}$$

which exhausts the orthogonality relations. Furthermore, we get

$$(2\wp)^{-1}[\, u_{\pm}(\mathbf{p}) \otimes u_{\pm}^{\dagger}(\mathbf{p}) + v_{\pm}(\mathbf{p}) \otimes v_{\pm}^{\dagger}(\mathbf{p})\,] = \mathbb{P}_{\pm} \tag{1.29}$$

with $\mathbb{P}_{\pm} = \frac{1}{2}(\mathbb{I} \pm \gamma_5)$, which represent the closure or completeness relations for the basis quartet of massless and chiral bispinor spin-states. The corresponding plane wave functions, which are normal mode solutions of the massless Dirac equation, read

$$\left\{\begin{array}{l} u_{\pm,\,\mathbf{p}}(x) = [\,(2\pi)^3 2\wp\,]^{-\frac{1}{2}} u_{\pm}(\mathbf{p})\, \mathrm{e}^{-i\wp t + i\mathbf{p}\cdot\mathbf{x}} \\ \\ v_{\pm,\,\mathbf{p}}(x) = [\,(2\pi)^3 2\wp\,]^{-\frac{1}{2}}\, v_{\pm}(-\,\mathbf{p})\, \mathrm{e}^{i\wp t - i\mathbf{p}\cdot\mathbf{x}} \end{array}\right. \tag{1.30}$$

and satisfy in turn orthonormality and closure relations with respect to the usual Poincaré invariant inner product: for instance

$$\int \mathrm{d}\mathbf{x}\, u_{\imath,\mathbf{p}}^{\dagger}(x) v_{\jmath,\mathbf{p}'}(x) = 0, \quad \int \mathrm{d}\mathbf{x}\, u_{\imath,\mathbf{p}}^{\dagger}(x) u_{\jmath,\mathbf{p}'}(x) = \delta_{\imath\jmath}\delta(\mathbf{p} - \mathbf{p}')$$
$$= \int \mathrm{d}\mathbf{x}\, v_{\imath,\mathbf{p}}^{\dagger}(x) v_{\jmath,\mathbf{p}'}(x),\ (\forall \imath,\, \jmath = +, - \vee \mathbf{p}, \mathbf{p}' \in \mathbb{R}^3),\ etc.$$

For a better understanding of the sequel, it is crucial to observe that, for any given frequency and chirality, the spin-states and wave functions of particle and antiparticle have opposite wave vectors, i.e. opposite helicity. Moreover, for later convenience, we realize that the normal modes of the massless Dirac spinor satisfy by direct inspection the following relationships:

$$i\gamma^2 u_{\pm,\,\mathbf{p}}^{*}(x) = v_{\mp,\,\mathbf{p}}(x) \quad i\gamma^2 v_{\pm,\,\mathbf{p}}^{*}(x) = u_{\mp,\,\mathbf{p}}(x) \tag{1.31}$$

With less general but more colorful terminology we will adopt names of the standard model (SM) and call $u_{-}(\mathbf{p})$, $u_{-,\,\mathbf{p}}(x)$ the left-handed SM neutrino spin-state and normal modes, respectively, while $v_{-}(\mathbf{p})$, $v_{-,\,\mathbf{p}}(x)$ will refer to the corresponding negative frequency and chirality spin-state and normal modes, which will be related to the neutrino antiparticles of opposite helicity, inasmuch as their masses can be neglected[1]. The remaining pair of chiral spin-states and normal modes correspond to hypothetical right-handed neutrinos and related negative frequency spin-states and normal modes, that have never been observed so far. Their existence represents one of the most intriguing and exciting actual possibilities beyond the standard model.

Symmetries

The classical Lagrange density for the massless Dirac field can be written in different forms: namely,

$$\mathcal{L}_0 = \psi^\dagger(x)\, i\alpha \cdot \partial\, \psi(x) \doteq \tfrac{1}{2}\, \psi^\dagger(x)\, i\alpha \cdot \overleftrightarrow{\partial}\, \psi(x) \tag{1.32}$$

$$\doteq \tfrac{1}{2}\, \psi_L^\dagger(x)\, i\sigma \cdot \overleftrightarrow{\partial}\, \psi_L(x) + \tfrac{1}{2}\, \psi_R^\dagger(x)\, i\bar{\sigma} \cdot \overleftrightarrow{\partial}\, \psi_R(x), \tag{1.33}$$

up to irrelevant four divergences, where the spinor fields are understood to be Grassmann valued functions in a Minkowski spacetime, and

$$\psi(x) = \begin{bmatrix} \psi_L(x) \\ \psi_R(x) \end{bmatrix} \qquad \alpha^\nu = \begin{bmatrix} \sigma^\nu & 0 \\ 0 & \bar{\sigma}^\nu \end{bmatrix}$$

The Euler-Lagrange field equations are

$$\alpha \cdot \partial\, \psi(x) = 0 \quad \Leftrightarrow \quad \sigma \cdot \partial\, \psi_L(x) = \bar{\sigma} \cdot \partial\, \psi_R(x) = 0,$$

whose general plane wave solutions have been obtained and discussed above. The canonical energy-momentum tensor reads

$$T_{\mu\nu} = \psi^\dagger(x)\, \alpha_\mu\, i\partial_\nu \psi(x) = \psi_L^\dagger(x)\, \sigma_\mu\, i\partial_\nu \psi_L(x) + \psi_R^\dagger(x)\, \bar{\sigma}_\mu\, i\partial_\nu \psi_R(x), \tag{1.34}$$

It is not symmetric and defines the conserved energy-momentum four-vector

$$P_0 = \int \mathrm{d}\mathbf{x}\, \psi^\dagger(x)\, i\partial_0 \psi(x) = \int \mathrm{d}\mathbf{x}\, \psi^\dagger(x) H_0 \psi(x) \tag{1.35}$$

$$= \int \mathrm{d}\mathbf{x} \left[\psi_L^\dagger(x) H_L \psi_L(x) + \psi_R^\dagger(x) H_R \psi_R(x) \right] \tag{1.36}$$

$$\mathbf{P} = \int \mathrm{d}\mathbf{x}\, \psi^\dagger(x)(-i\hbar\nabla)\psi(x) \tag{1.37}$$

[1] We recall that the neutrino and/or antineutrino mass has not yet been directly measured and the present experimental limit is < 2 eV.

The energy-momentum tensor is traceless $g^{\mu\nu}T_{\mu\nu} = 0$, on the eom shell. The canonical total angular momentum density tensor can always be written in the form

$$M^{\lambda\mu\nu} = x^{\mu}T^{\lambda\nu} - x^{\nu}T^{\lambda\mu} + S^{\lambda\mu\nu} \tag{1.38}$$

$$S^{\lambda\mu\nu} = \frac{1}{2}\psi^{\dagger}(x)\,\gamma^{0}\{\gamma^{\lambda},\ \Sigma^{\mu\nu}\}\,\psi(x), \tag{1.39}$$

where $\Sigma_{\mu\nu} = \frac{i}{4}[\gamma_{\mu}, \gamma_{\nu}]$. So, from the Noether theorem related to the Lorentz symmetry, we get

$$T_{\nu\mu} - T_{\mu\nu} = \partial^{\lambda}S_{\lambda\mu\nu},$$

which is known to entail the helicity conservation. In order to determine the helicity assignments, we can proceed as follows. The generators of the rotation group for both irreducible Weyl spinor representations are $S_{jk} = \frac{1}{2}\,\varepsilon_{jk\ell}\,\sigma_{\ell}$. It follows that the corresponding Pauli-Lubanski operator reads

$$W_0 = \mathbf{P}\cdot\mathbf{S} = \vec{p}\cdot\frac{1}{2}\,\hbar\,\vec{\sigma}$$

$$W_j = \frac{1}{2}\,\wp\sigma_j - \frac{1}{2}\,i\varepsilon_{jk\ell}\,p_k\,\sigma_{\ell},$$

which entails

$$\begin{aligned} g^{\mu\nu}\,W_{\mu}W_{\nu} &= \frac{1}{4}\,p_j\,p_k\,\sigma_j\sigma_k - \frac{1}{4}\left(\wp\sigma_j - i\varepsilon_{jk\ell}\,p_k\,\sigma_{\ell}\right)\left(\wp\sigma_j - i\varepsilon_{jrs}\,p_r\,\sigma_s\right) \\ &= -\frac{1}{2}\,\wp^2 + \frac{1}{4}\left(\delta_{kr}\delta_{\ell s} - \delta_{ks}\delta_{\ell r}\right)p_k\,p_r\,\sigma_{\ell}\,\sigma_s = 0 \end{aligned}$$

where the 2×2 identity matrix has been understood. Then, thanks to the light-like nature of the Pauli-Lubanski operator, we can identify the *helicity operator* with

$$\mathrm{h} \equiv \frac{W_0}{\wp} = \frac{|\mathbf{W}|}{\wp} = \frac{1}{2}\,\hbar\,\hat{n}\cdot\vec{\sigma} \qquad\qquad \hat{n} = \vec{p}/\wp$$

and define the Weyl's Hamiltonian operators via

$$H_0 = \begin{bmatrix} H_L & 0 \\ 0 & H_R \end{bmatrix}, \qquad H_R = c\,\hbar\,\vec{p}\cdot\vec{\sigma} = -\,H_L$$

It follows that we have $H_L = -\,\hbar c\wp h$, $H_R = \hbar c\wp h$ and, consequently,

$$\begin{aligned} &H_L\,u_L(\mathbf{p}) = \hbar c\wp\,u_L(\mathbf{p}) \quad \mathrm{h}u_L(\mathbf{p}) = -\,\frac{1}{2}\,u_L(\mathbf{p}) \\ &H_R\,u_R(\mathbf{p}) = \hbar c\wp\,u_R(\mathbf{p}) \quad \mathrm{h}u_R(\mathbf{p}) = \frac{1}{2}\,u_R(\mathbf{p}) \\ &H_L\,v_L(\mathbf{p}) = -\,\hbar c\wp\,v_L(\mathbf{p}) \quad \mathrm{h}v_L(\mathbf{p}) = \frac{1}{2}\,v_L(\mathbf{p}) \\ &H_R\,v_R(\mathbf{p}) = -\,\hbar c\wp\,v_R(\mathbf{p}) \quad \mathrm{h}v_R(\mathbf{p}) = -\,\frac{1}{2}\,v_R(\mathbf{p}) \end{aligned}$$

This means that:

- Neutrino spin-states carry negative helicity, while the antineutrino spin-states carry positive helicity, both of them being of negative chirality, i.e. $\gamma_5 u_-(\mathbf{p}) = \gamma_5 v_-(\mathbf{p}) = -1$.
- The situation is reversed for positive chirality spin-states. Thus, positive helicity is exhibited by positive frequency particles—the would-be right-handed neutrinos—while negative helicity by negative frequency antiparticles, both of negative chirality, i.e. $\gamma_5 u_+(\mathbf{p}) = \gamma_5 v_+(\mathbf{p}) = 1$.

The Weyl action integrals

$$S_L = \int \mathrm{d}^4x\, \tfrac{1}{2}\, \psi_L^\dagger(x)\, i\sigma\cdot \overleftrightarrow{\partial}\, \psi_L(x) \quad \text{and} \quad S_R = \int \mathrm{d}^4x\, \tfrac{1}{2}\, \psi_R^\dagger(x)\, i\bar{\sigma}\cdot \overleftrightarrow{\partial}\, \psi_R(x) \tag{1.40}$$

and the massless Dirac action integral, which is the sum of the left- and right-handed Weyl action integrals, are invariant under scale or dilatation symmetry. The latter acts upon any inertial coordinate system and spinor field variables on the Minkowski space according to

$$x^\lambda \quad \longmapsto \quad x'^{\lambda} = \mathrm{e}^{-\varrho} x^\lambda \quad (\varrho \in \mathbb{R}) \tag{1.41}$$

$$\Psi(x) \quad \longmapsto \quad \Psi'(x) = \mathrm{e}^{\frac{3}{2}\varrho}\, \Psi(\mathrm{e}^{\varrho} x) \quad (\Psi = \psi, \psi_L, \psi_R) \tag{1.42}$$

To obtain the infinitesimal generators of the scale transformations, we consider an infinitesimal dilatation/contraction, $x'^{\mu} = (1 - \delta\varrho + \cdots)\, x^\mu$ with $|\,\delta\varrho\,| \ll 1$

$$\Psi'(x) = \Psi(x) + \tfrac{3}{2}\delta\varrho\, \Psi(x) + x\cdot\partial\, \Psi(x)\, \delta\varrho + \cdots$$

and thereby

$$\delta\Psi(x) = \delta\varrho\, \left[\, \tfrac{3}{2} + x\cdot\partial\,\right]\, \Psi(x)$$

Hence, we can identify

$$\mathrm{D}_\Psi \equiv \tfrac{3}{2} + x\cdot\partial \tag{1.43}$$

as the generator of the local scale transformations on classical spinor fields. Now, since $\Psi'(x') = \mathrm{e}^{\frac{3}{2}\varrho}\, \Psi(x)$, the Lagrangian (1.32) satisfies

$$\mathcal{L}_0'(x') = \mathrm{e}^{4\varrho}\mathcal{L}_0(x)$$

so that the action integral for the massless Dirac spinor field is invariant, *viz.*,

$$S = \int \mathrm{d}^4x'\, \mathcal{L}_0'(x') = \int \mathrm{d}^4x\, \mathcal{L}_0(x). \tag{1.44}$$

The dilatation four-current can be readily obtained from the Noether theorem and can be written in different, though equivalent, forms, *viz.*,

$$\mathcal{D}^{\mu}(x) = \frac{\delta \mathcal{L}_0}{\delta \partial_\mu \psi(x)} \, \mathrm{D}_\psi \, \psi(x) - \mathcal{L}_0(x)\, x^\mu = \tfrac{3}{2}\, \psi^\dagger(x) i \alpha^\mu \psi(x) + x_\nu \, T^{\mu\nu}(x),$$

so that, on shell,

$$\partial \cdot \mathcal{D} = g^{\mu\nu} T_{\mu\nu} = 0 \tag{1.45}$$

Finally, the massless Dirac action integral turns out to be invariant under the U(1) groups of the ordinary and chiral phase changes, *viz.*,

$$\psi'(x) = \mathrm{e}^{-i\theta}\psi(x), \quad \psi'(x) = \mathrm{e}^{-i\gamma_5\theta}\psi(x) \quad (0 \le \theta < 2\pi) \tag{1.46}$$

or equivalently

$$\psi_L'(x) = \mathrm{e}^{-i\theta}\psi_L(x), \quad \psi_L'(x) = \mathrm{e}^{i\theta}\psi_L(x), \quad (0 \le \theta < 2\pi) \tag{1.47}$$

$$\psi_R'(x) = \mathrm{e}^{-i\theta}\psi_R(x), \quad \psi_R'(x) = \mathrm{e}^{-i\theta}\psi_R(x) \quad (0 \le \theta < 2\pi) \tag{1.48}$$

The Noether theorem leads to the conserved currents

$$J^\mu(x) = \psi^\dagger(x)\alpha^\mu\psi(x) \quad J_5^\mu(x) = \psi^\dagger(x)\alpha^\mu\gamma_5\psi(x) \tag{1.49}$$

The corresponding charges

$$\mathcal{Q} = \int \mathrm{d}\mathbf{x} \left[\psi_L^\dagger(x)\,\psi_L(x) + \psi_R^\dagger(x)\,\psi_R(x) \right] \tag{1.50}$$

$$\mathcal{Q}_5 = \int \mathrm{d}\mathbf{x} \left[-\psi_L^\dagger(x)\,\psi_L(x) + \psi_R^\dagger(x)\,\psi_R(x) \right] \tag{1.51}$$

are constant in time $\dot{\mathcal{Q}} = \dot{\mathcal{Q}}_5 = 0$.

To sum up, for a massless Dirac field, we have seen until now the occurrence of a 13-parameter Lie group of spacetime and internal symmetries leading to thirteen conserved charges. Actually, as we shall see later on, under certain general conditions, scale invariance implies in fact the invariance under a larger 15-parameter symmetry group, the conformal group, which includes the ten-dimensional Poincaré group. Scale invariance implies conformal invariance if and only if there exists in the considered theory a symmetric traceless conserved energy-momentum tensor $\Theta_{\mu\nu}$, called the improved energy-momentum tensor (which may not coincide with the canonical energy-momentum tensor), such that $\partial \cdot \mathcal{D} = g^{\mu\nu}\Theta_{\mu\nu} = 0$. It turns out that this is true for all power counting renormalizable massless models involving fields of spin ≤ 1. Hence, the massless Dirac action integral is actually invariant under a 17-parameter Lie group involving spacetime as well as internal symmetries.

Concerning discrete symmetries, it turns out that the action integral $S = \int \mathrm{d}^4x\mathcal{L}_0(x)$ respects $\mathcal{C}$, $\mathcal{P}$ and $\mathcal{T}$ transformations separately. The charge conjugation $\mathcal{C}$ is an internal discrete symmetry which acts on the classical Dirac spinors according to (here we make a particular choice of the matrix C, see notation above)

$$\psi(x) \quad \longmapsto \quad \psi^c(x) = \gamma^2 \psi^*(x)$$

while a parity $\mathcal{P}$ transform is defined by $\psi'(t, -\mathbf{x}) = \gamma^0 \psi(x)$, and time reversal $\mathcal{T}$ by $\psi'(-t, \mathbf{x}) = \Theta\, \psi(x)$, $\Theta = \gamma^3\gamma^1$. It is straightforward to verify by direct inspection that the action integral is invariant.

The effect of these discrete transformations is different for chiral spinors, for they mix the chiral components and drive them out of the specific chiral spinor spaces specified by one of the two irreducible non-equivalent Weyl representations of the Lorentz group $D(\frac{1}{2}, 0)$ and $D(0, \frac{1}{2})$, see below.

Quantum Theory

The massless Dirac quantum field is the operator valued tempered distribution defined by $\psi(x) = \psi_-(x) + \psi_+(x)$ where

$$\begin{aligned} \psi^{(-)}(x) &= \int \mathbf{dp}\, [\,(2\pi)^3\, 2\wp\,]^{-\frac{1}{2}} \left[\, c_{-,\mathbf{p}}\, u_-(\mathbf{p}) + c_{+,\mathbf{p}}\, u_+(\mathbf{p})\,\right] \mathrm{e}^{\,i\mathbf{p}\cdot\mathbf{x}-i\wp t} \\ \psi^{(+)}(x) &= \int \mathbf{dp}\, [\,(2\pi)^3\, 2\wp\,]^{-\frac{1}{2}} \left[\, d^{\dagger}_{-,\mathbf{p}}\, v_-(-\,\mathbf{p}) + d^{\dagger}_{+,\mathbf{p}}\, v_+(-\,\mathbf{p})\,\right] \mathrm{e}^{\,i\wp t-i\mathbf{p}\cdot\mathbf{x}} \end{aligned} \tag{1.52}$$

We define in addition

$$\begin{aligned} \psi_{\mp}(x) = \mathbb{P}_{\mp}\, \psi(x) &= \int \mathbf{dp}\, \left[\, c_{\mp,\mathbf{p}}\, u_{\mp,\mathbf{p}}(x) + d^{\dagger}_{\mp,\mathbf{p}}\, v_{\mp,\mathbf{p}}(x)\,\right], \\ \psi^{\dagger}_{\mp}(y) = \psi^{\dagger}(y)\, \mathbb{P}_{\mp} &= \int \mathbf{dp}'\, \left[\, c^{\dagger}_{\mp,\mathbf{p}'}\, u^{\top *}_{\mp,\mathbf{p}'}(y) + d_{\mp,\mathbf{p}'}\, v^{\top *}_{\mp,\mathbf{p}'}(y)\,\right], \end{aligned} \tag{1.53}$$

which corresponds to the most general solution of the Weyl-Dirac equations (1.2). The creation and annihilation operators satisfy the canonical anticommutation relations

$$\{c_{\mp,\mathbf{p}}, c^{\dagger}_{\mp,\mathbf{p}'}\} = \{d_{\mp,\mathbf{p}}, d^{\dagger}_{\mp,\mathbf{p}'}\} = \delta(\mathbf{p} - \mathbf{p}'), \tag{1.54}$$

all the other anticommutators being null. It follows that, for example,

$$\begin{aligned} \{\psi_-(x), \psi^{\dagger}_-(y)\}_{x_0=y_0} &= \int \mathbf{dp} \int \mathbf{dp}' \\ \times\Big[\{c_{-,\mathbf{p}}, c^{\dagger}_{-,\mathbf{p}'}\}\, u_{-,\mathbf{p}}(x) u^{\dagger}_{-,\mathbf{p}'}(y) &+ \{d^{\dagger}_{-,\mathbf{p}}, d_{-,\mathbf{p}'}\}\, v_{-,\mathbf{p}}(x) v^{\dagger}_{-,\mathbf{p}'}(y)\Big]_{x_0=y_0} \\ = \int \frac{\mathbf{dp}}{(2\pi)^3 2\wp} \left[\, u_-(\mathbf{p})\, u^{\dagger}_-(\mathbf{p}) + v_-(\mathbf{p})\, v^{\dagger}_-(\mathbf{p})\,\right] &\mathrm{e}^{\,i\mathbf{p}\cdot(\mathbf{x}-\mathbf{y})} = \mathbb{P}_L\, \delta(\mathbf{x} - \mathbf{y}), \end{aligned}$$

where $\mathbb{P}_L = \frac{\mathbb{P}_-}{2\wp}$, and, analogously,

$$\{\psi_+(x), \psi_+^\dagger(y)\}_{x_0=y_0} = \mathbb{P}_R\, \delta(\mathbf{x}-\mathbf{y}), \quad \mathbb{P}_R = \frac{\mathbb{P}_+}{2\wp},$$

while all the remaining anticommutators vanish at arbitrary times, e.g. $\{\psi_-(x), \psi_+^\dagger(y)\} = 0$ etc.

From the previously obtained expressions, one can readily derive the normal modes expansions of the Weyl quantum fields in the two-component formalism

$$\psi_L(x) = \int \mathrm{d}\mathbf{p}\, [\,(2\pi)^3\, 2\wp\,]^{-\frac{1}{2}} \left[c_{-,\,\mathbf{p}}\, u_L(\mathbf{p})\, \mathrm{e}^{-\,ip\cdot x} + d_{-,\,\mathbf{p}}^\dagger\, v_L(-\,\mathbf{p})\, \mathrm{e}^{\,ip\cdot x} \right]_{p_0=\wp} \tag{1.55}$$

$$\text{where } u_L(\mathbf{p}) = \frac{\bar\sigma\cdot p}{\sqrt{\wp - p_z}} \begin{vmatrix} 1 \\ 0 \end{vmatrix}, \quad v_L(-\,\mathbf{p}) = \frac{\bar\sigma\cdot p}{\sqrt{\wp + p_z}} \begin{vmatrix} 0 \\ 1 \end{vmatrix}, \quad (p_0 = \wp\,)$$

$$\psi_R(x) = \int \mathrm{d}\mathbf{p}\, [\,(2\pi)^3\, 2\wp\,]^{-\frac{1}{2}} \left[c_{+,\,\mathbf{p}}\, u_R(\mathbf{p})\, \mathrm{e}^{-\,ip\cdot x} + d_{+,\,\mathbf{p}}^\dagger\, v_R(-\,\mathbf{p})\, \mathrm{e}^{\,ip\cdot x} \right]_{p_0=\wp} \tag{1.56}$$

$$\text{where } u_R(\mathbf{p}) = \frac{\sigma\cdot p}{\sqrt{\wp + p_z}} \begin{vmatrix} 1 \\ 0 \end{vmatrix}, \quad v_R(-\,\mathbf{p}) = \frac{\sigma\cdot p}{\sqrt{\wp - p_z}} \begin{vmatrix} 0 \\ 1 \end{vmatrix}, \quad (p_0 = \wp\,),$$

with canonical anticommutation relations

$$\{\psi_L(x), \psi_L^\dagger(y)\}_{x_0=y_0} = \delta(\mathbf{x}-\mathbf{y}) = \{\psi_R(x), \psi_R^\dagger(y)\}_{x_0=y_0}, \tag{1.57}$$

all the other anticommutators being null at arbitrary times as expected, because the Fock spaces of the left (negative) and right (positive) chirality are evidently orthogonal to each other: for instance $\{\psi_L(x), \psi_R^\dagger(y)\} \equiv 0$ *etc.*

Let us come now to the evaluation of the invariants and conserved charges—in the massless spinor quantum theory. For example, since we have

$$\psi_\pm^\dagger(y) i\partial_\mu \psi_\pm(x) = :\psi_\pm^\dagger(y) i\partial_\mu \psi_\pm(x): \; - \int \mathrm{d}\mathbf{p}\, p_\mu\, v_{\pm,\,\mathbf{p}}^{\top *}(y)\, v_{\pm,\,\mathbf{p}}(x)$$

with $p_0 = \wp$, one can easily obtain

$$P^\mu = \sum_{\iota=-,+} \int \mathrm{d}\mathbf{p}\, p^\mu \left[c_{\iota,\,\mathbf{p}}^\dagger\, c_{\iota,\,\mathbf{p}} + d_{\iota,\,\mathbf{p}}^\dagger\, d_{\iota,\,\mathbf{p}} \right]_{p_0=\wp} = P_L^\mu + P_R^\mu \tag{1.58}$$

In order to construct the spin operator, consider for instance a left Weyl field $\psi_-(t, z)$ traveling along the $Oz-$axis, so that $T_-^{12}(t, z) = T_-^{21}(t, z) = 0 = \partial_\nu S_-^{\nu 21}(t, z)$. It follows that

$$\Sigma_- = \int_{-\infty}^{\infty} \mathrm{d}z : S_-^{021}(t, z) : \qquad \dot\Sigma_- = 0$$

where ($\Sigma_3 \equiv \Sigma_{12}$)

$$\Sigma_- = -\frac{1}{2}\int_{-\infty}^{\infty} \mathrm{d}z : \psi_-^\dagger(0,z)\,\Sigma_3\,\psi_-(0,z) : = -\frac{1}{2}\int_{-\infty}^{\infty} \mathrm{d}z : \psi_L^\dagger(0,z)\,\sigma_3\,\psi_L(0,z) :$$

Now, since for a one-dimensional motion we have for $\wp = p_z\,\mathrm{sgn}(p_z)$,

$$\psi_L(t,z) = \int_{-\infty}^{\infty} \mathrm{d}p_z\,[\,4\pi\wp\,]^{-\frac{1}{2}}\left[c_{-,\,p_z}\,u_L(p_z)\,\mathrm{e}^{-i\wp t + izp_z} + d_{-,\,p_z}^{\,\dagger}\,v_L(-\,p_z)\,\mathrm{e}^{\,i\wp t - izp_z}\right]$$

$$\text{with}\qquad u_L(p_z) = \frac{\wp - \sigma_3\,p_z}{\sqrt{\wp - p_z}}\begin{vmatrix}1\\0\end{vmatrix}\qquad v_L(-\,p_z) = \frac{\wp - \sigma_3\,p_z}{\sqrt{\wp + p_z}}\begin{vmatrix}0\\1\end{vmatrix}$$

and, thereby,

$$\sigma_3\,u_L(p_z) = u_L(p_z)\qquad \sigma_3\,v_L(-\,p_z) = -\,v_L(-\,p_z)$$

$$\sigma_3\,\psi_L(t,z) = \int_{-\infty}^{\infty} \mathrm{d}p_z\,[\,4\pi\wp\,]^{-\frac{1}{2}}\left[c_{-,\,p_z}\,u_L(p_z)\,\mathrm{e}^{-\,i\wp t + izp_z} - d_{-,\,p_z}^{\,\dagger}\,v_L(-\,p_z)\,\mathrm{e}^{\,i\wp t - izp_z}\right].$$

This entails that the general form of the spin operator for a one-dimensional motion along, e.g. the $Oz-$axis becomes

$$\Sigma_z = -\frac{1}{2}\int_{-\infty}^{\infty} \mathrm{d}p_z\left[c_{-,\,p_z}^{\,\dagger}\,c_{-,\,p_z} + d_{-,\,p_z}^{\,\dagger}\,d_{-,\,p_z} - c_{+,\,p_z}^{\,\dagger}\,c_{+,\,p_z} - d_{+,\,p_z}^{\,\dagger}\,d_{+,\,p_z}\right] \tag{1.59}$$

which confirms the fact that neutrino particles are of negative helicity, while the antineutrinos exhibit positive helicity, owing to the opposite direction of the wave vector, the situation being reversed for the hypothetical right-handed neutrinos of positive chirality. As a matter of fact, the projection of the spin along the direction of motion is $-\frac{1}{2}$ for both particles and antiparticles. Now, since the spin-state are $u_L(p_z)$ and $v_L(-p_z)$ respectively, it clearly follows that particles of negative chirality share negative helicity, for the spin waves are progressive, while antiparticles exhibit positive helicity, spin waves being regressive. The opposite assignments hold for positive chirality quanta.

Turning to the internal symmetries and charges, we find

$$\mathcal{Q} = \int \mathrm{d}\mathbf{x} : \psi_-^\dagger(x)\,\psi_-(x) + \psi_+^\dagger(x)\,\psi_+(x) : \equiv \mathcal{Q}_- + \mathcal{Q}_+ \tag{1.60}$$

$$\mathcal{Q}_\mp = \int \mathrm{d}\mathbf{p}\left[c_{\mp,\,\mathbf{p}}^{\dagger}\,c_{\mp,\,\mathbf{p}} - d_{\mp,\,\mathbf{p}}^{\dagger}\,d_{\mp,\,\mathbf{p}}\right] \tag{1.61}$$

$$\mathcal{Q}_5 = \mathcal{Q}_+ - \mathcal{Q}_- \tag{1.62}$$

Finally, the dilatation charge reads

$$\mathcal{D} = \int \mathrm{d}\mathbf{x} : \frac{3}{2} i\psi^\dagger(x)\psi(x) + x_\nu T^{0\nu}(x) := x_0 P_0 + i \left\{ \tfrac{3}{2}\,\mathcal{Q} - \int \mathrm{d}\mathbf{x}\, x^k : \psi^\dagger(x)\nabla_k\psi(x) : \right\}. \tag{1.63}$$

We find

$$\begin{aligned}\partial_0\mathcal{D} &= P_0 + \int \mathrm{d}\mathbf{x}\, x^k : [\, i\partial_0\psi(x)\,]^\dagger\, \nabla_k\psi(x) : - \int \mathrm{d}\mathbf{x}\, x^k : \psi^\dagger(x)\nabla_k i\partial_0\psi(x) : \\ &= P_0 + \int \mathrm{d}\mathbf{x}\, x^k : \psi^\dagger(x) H_0 \nabla_k\psi(x) : - \int \mathrm{d}\mathbf{x}\, x^k : \psi^\dagger(x)\nabla_k H_0\psi(x) := P_0\end{aligned}$$

Since we have the well-known relation $\partial_0\mathcal{D} = i\,[\,\mathcal{D}, P_0\,] = P_0$, it follows that, as expected,

$$\dot{\mathcal{D}} = \frac{\mathrm{d}\mathcal{D}}{\mathrm{d}t} = \partial_0\mathcal{D} + i\,[\,P_0, \mathcal{D}\,] = 0,$$

the dilatation charge is globally time independent.

From the above analysis and correspondences, it follows that the 1-particle states $c_\mp^\dagger(\mathbf{p})|0\rangle$ of the Fock space of the massless Dirac spinor field represent the particles with the following quantum number assignments:

(i) Opposite chirality;
(ii) The same energy-momentum $p^\nu = (\wp, \mathbf{p})$;
(iii) Equal internal charge, called *lepton number*, $\ell = +1$;
(iv) Opposite spin-helicity.

Notice that spin-helicity and chirality quantum numbers do not coincide at all.

The antimatter 1-particle states $d_\mp^\dagger(\mathbf{p})|0\rangle$ exhibit opposite charge and spin-helicity, with respect to particles of the same energy-momentum and chirality.

Discrete Symmetries

Turning to the discrete symmetries, we start with the charge conjugation unitary transformation of the quantum theory for a massless Dirac spinor field, which amounts to the simultaneous exchange between chirality and particle-antiparticle operators, up to a phase factor, *viz.*,

$$\mathcal{C}\, c_{\mp,\,\mathbf{p}}\, \mathcal{C}^{-1} = \mathrm{e}^{i\theta} d_{\pm,\,\mathbf{p}} \quad \mathcal{C}\, d_{\mp,\,\mathbf{p}}\, \mathcal{C}^{-1} = \mathrm{e}^{-i\theta} c_{\pm,\,\mathbf{p}} \tag{1.64}$$

$$\mathcal{C}^2 = \mathbb{I} \quad \mathcal{C} = \mathcal{C}^\dagger = \mathcal{C}^{-1} \quad 0 \le \theta < 2\pi, \tag{1.65}$$

which yields

$$\mathcal{C}\,\psi_\mp(x)\,\mathcal{C} = \mathrm{e}^{i\theta} \int \mathrm{d}\mathbf{p}\, \left[\, d_{\pm,\,\mathbf{p}}\, u_{\mp,\,\mathbf{p}}(x) + c_{\pm,\,\mathbf{p}}^\dagger\, v_{\mp,\,\mathbf{p}}(x) \,\right].$$

Now, from the above-mentioned relations (1.31), we can write

$$\begin{aligned}\mathcal{C}\,\psi_{\mp}(x)\,\mathcal{C} &= \mathrm{e}^{i\theta}\int \mathbf{dp}\,\Big[\,d_{\pm,\,\mathbf{p}}\,u_{\mp,\,\mathbf{p}}(x) + c_{\pm,\,\mathbf{p}}^{\dagger}\,v_{\mp,\,\mathbf{p}}(x)\,\Big]\\ &= i\gamma^2\,\mathrm{e}^{i\theta}\int \mathbf{dp}\,\Big[\,c_{\pm,\,\mathbf{p}}^{\dagger}\,u_{\pm,\,\mathbf{p}}^{*}(x) + d_{\pm,\,\mathbf{p}}\,v_{\pm,\,\mathbf{p}}^{*}(x)\,\Big]\\ &= \gamma^2\left(\psi_{\pm}^{\dagger}(x)\right)^{\top} \qquad \text{for } \theta = -\,\pi/2\end{aligned}$$

Now, since we have $\psi(x) = \psi_{-}(x) + \psi_{+}(x)$, we come to the result

$$\psi^{c}(x) = \mathcal{C}\,\psi(x)\,\mathcal{C} = \gamma^2\left(\psi^{\dagger}(x)\right)^{\top} \tag{1.66}$$

which is nothing but the quantum generalization of the charge conjugation rule for classical spinor fields, i.e. $\psi^{c}(x) = \gamma^2\psi^{*}(x)$, as expected. It is important to remark that the internal and discrete charge conjugation is a chirality preserving transformation: for example

$$\gamma_5\,\psi_{-}^{c}(x) = -\,\gamma^2\int \mathbf{dp}\,\Big[\,c_{+,\,\mathbf{p}}^{\dagger}\,\gamma_5\,u_{+,\,\mathbf{p}}^{*}(x) + d_{+,\,\mathbf{p}}\,\gamma_5\,v_{+,\,\mathbf{p}}^{*}(x)\,\Big] = -\psi_{-}^{c}(x)$$

because γ_5 is real in the Weyl representation of the Clifford algebra. Thus, the charge conjugated of a ν particle of negative chirality, negative spin-helicity and positive lepton number is a $\bar{\nu}$ antiparticle of still negative chirality, though positive spin-helicity and negative lepton number. Nonetheless, it turns out that the charge conjugation transformation necessarily involves both chiralities, in order to be implemented. Hence, the $\mathcal{C}$ transform is well-defined only for four-component bispinors and not for two-component Weyl spinors belonging to the two irreducible two-dimensional representations.

The parity transformation is a discrete spacetime symmetry of the action integral $S = \int \mathrm{d}^4x\,\mathcal{L}_0(x)$ that can be implemented at the quantum field theory level as

$$\begin{aligned}\mathcal{P}\,c_{\mp,\,\mathbf{p}}\,\mathcal{P}^{\dagger} &= \mathrm{e}^{i\eta_{\mp}}\,c_{\pm,-\,\mathbf{p}} \quad \mathcal{P}\,d_{\mp,\,\mathbf{p}}\,\mathcal{P}^{\dagger} = \mathrm{e}^{i\theta_{\mp}}\,d_{\pm,-\,\mathbf{p}}\\ &\qquad 0 \le \eta_{\mp} < 2\pi \quad 0 \le \theta_{\mp} < 2\pi\end{aligned} \tag{1.67}$$

Then, after insertion of the normal mode expansion of the massless Dirac quantum field, with $x'^{\mu} = \tilde{x}^{\mu} = (x_0, -\,\vec{x}\,) = x_{\mu}$, we find

$$\begin{aligned}\psi'(x') = \mathcal{P}\,\psi(\tilde{x})\,\mathcal{P}^{\dagger} &= \int \mathbf{dp}\,\Big[\,\mathrm{e}^{i\eta_{\mp}}\,c_{\pm,-\,\mathbf{p}}\,u_{\mp,\,\mathbf{p}}(\tilde{x}) + \mathrm{e}^{-\,i\theta_{\mp}}\,d_{\pm,-\,\mathbf{p}}^{\dagger}\,v_{\mp,\,\mathbf{p}}(\tilde{x})\,\Big]\\ &= \int \mathbf{dp}\,\Big[\,\mathrm{e}^{i\eta_{\mp}}\,c_{\pm,\,\mathbf{p}}\,u_{\mp,-\,\mathbf{p}}(\tilde{x}) + \mathrm{e}^{-\,i\theta_{\mp}}\,d_{\pm,\,\mathbf{p}}^{\dagger}\,v_{\mp,-\,\mathbf{p}}(\tilde{x})\,\Big]\,.\end{aligned}$$

Now we have

$$u_{\mp,-\mathbf{p}}(\tilde{x}) = [\,(2\pi)^3 2\wp\,]^{-\frac{1}{2}}\, u_{\mp}(-\mathbf{p})\, \mathrm{e}^{-i\wp t + i\mathbf{p}\cdot\mathbf{x}} = -\,\gamma^0\, u_{\pm,\,\mathbf{p}}(x),$$
$$v_{\mp,-\mathbf{p}}(\tilde{x}) = [\,(2\pi)^3 2\wp\,]^{-\frac{1}{2}}\, v_{\mp}(\mathbf{p})\, \mathrm{e}^{i\wp t - i\mathbf{p}\cdot\mathbf{x}} = \gamma^0\, v_{\pm,\,\mathbf{p}}(x),$$

so that, after choosing

$$\eta_{\mp} = \mp k\pi \quad \theta_{\mp} = \mp 2k\pi \quad (k \in \mathbb{Z}),$$

we finally obtain the usual transformation law

$$\psi'_{\mp}(x') = \mathcal{P}\,\psi_{\mp}(\tilde{x})\,\mathcal{P} = \gamma^0\,\psi_{\pm}(x) \tag{1.68}$$
$$\psi'(x') = \mathcal{P}\,\psi(\tilde{x})\,\mathcal{P} = \gamma^0\,\psi(x) \tag{1.69}$$
$$\mathcal{P}^2 = \mathbb{I} \quad \mathcal{P} = \mathcal{P}^{\dagger} = \mathcal{P}^{-1} \tag{1.70}$$

Once again, it is worth remarking that even the parity transform is not a well-defined operation for a two-component Weyl fermion, because it necessarily involves both chiralities, i.e. both irreducible two-dimensional representations of the proper Lorentz group.

In quantum field theory, the time reversal operator $\mathcal{T}$ corresponds to an antilinear and antiunitary transformation [1] with the following properties:

$$[\,P_0,\ \mathcal{T}\,] = 0 \quad \{\vec{P},\ \mathcal{T}\,\} = 0 \quad \{M^{\rho\sigma},\ \mathcal{T}\,\} = 0 \tag{1.71}$$

where, as usual, $(\,P^{\mu},\ M^{\rho\sigma}\,)$ are the ten generators of the Poincaré group. It follows therefrom that the antilinear and antiunitary time reversal operation reverses the signs of all particle momenta and spin angular momenta, both for particles and antiparticles. In this regard, we need to explain what is meant by spin-flip [2]. Consider, for example, the spin state

$$u_{-}(\mathbf{p}) = \begin{pmatrix} u_L(\mathbf{p}) \\ 0 \\ 0 \end{pmatrix} = \frac{1}{\sqrt{\wp - p_z}} \begin{pmatrix} \wp - p_z \\ -p_x - ip_y \\ 0 \\ 0 \end{pmatrix} \tag{1.72}$$
$$u_L(\mathbf{p}) = \frac{\bar{\sigma}\cdot p}{\sqrt{\wp - p_z}} \begin{vmatrix} 1 \\ 0 \end{vmatrix}$$

This positive energy, left-handed spin state of negative chirality, involves the spin-up constant spinor $\begin{vmatrix} 1 \\ 0 \end{vmatrix}$, and for our purpose, it will be better labeled as $u_{-}(\mathbf{p},\uparrow)$. Its corresponding spin reversed state $u_{-}(\mathbf{p},\downarrow)$ has the very same structure, but for the constant spinor which will be the spin-down $\begin{vmatrix} 0 \\ 1 \end{vmatrix}$. Thus, we can denote the pair of normalized spin flip states as

$$u_-(\mathbf{p},\uparrow) = \frac{1}{\sqrt{\wp - p_z}} \begin{bmatrix} \wp - p_z \\ -p_x - ip_y \\ 0 \\ 0 \end{bmatrix}, \quad u_-^*(-\mathbf{p},\downarrow) = \frac{1}{\sqrt{\wp - p_z}} \begin{bmatrix} p_x + ip_y \\ \wp - p_z \\ 0 \\ 0 \end{bmatrix} \tag{1.73}$$

with analogous relations for the remaining spin states. Then, we have

$$\mathcal{T}\, c_{\mp,\mathbf{p}}\, \mathcal{T}^{-1} = \mathrm{e}^{i\eta_\mp}\, c_{\mp,-\mathbf{p}}, \quad \mathcal{T}\, d_{\mp,\mathbf{p}}\, \mathcal{T}^{-1} = \mathrm{e}^{i\theta_\mp}\, d_{\mp,-\mathbf{p}}$$

$$\begin{aligned}\psi'_\mp(x') = \mathcal{T}\,\psi_\mp(-x_0,\vec{x})\,\mathcal{T}^{-1} &= \int \mathbf{dp}\left[\mathrm{e}^{i\eta_\mp}\, c_{\mp,-\mathbf{p}}\, u^*_{\mp,\mathbf{p}}(-\tilde{x}) + \mathrm{e}^{-i\theta_\mp}\, d^\dagger_{\mp,-\mathbf{p}}\, v^*_{\mp,\mathbf{p}}(-\tilde{x})\right] \\ &= \int \mathbf{dp}\left[\mathrm{e}^{i\eta_\mp}\, c_{\mp,\mathbf{p}}\, u^*_{\mp,-\mathbf{p}}(-\tilde{x}) + \mathrm{e}^{-i\theta_\mp}\, d^\dagger_{\mp,\mathbf{p}}\, v^*_{\mp,-\mathbf{p}}(-\tilde{x})\right]\end{aligned}$$

the complex conjugation being due to the antilinear and antiunitary nature of the $\mathcal{T}-$ operation, while the spin-flip operation in the spinor plane wave functions is understood. Consider now, for instance,

$$\mathcal{T}\psi^{(-)}(-t,\mathbf{x})\mathcal{T}^{-1} = \int \frac{\mathbf{dp}}{\sqrt{(2\pi)^3 2\wp}}\left[\mathrm{e}^{i\eta_-}\, c_{-,\mathbf{p}}\, u_-^*(-\mathbf{p}\downarrow) + \mathrm{e}^{i\eta_+}\, c_{+,\mathbf{p}}\, u_+^*(-\mathbf{p}\downarrow)\right]\mathrm{e}^{-i\wp t + i\mathbf{p}\cdot\mathbf{x}}$$

It can be readily seen that the right-hand side of the above equation becomes a local expression for the quantum field at the instant t, notably $\Theta\,\psi^{(-)}(t,\mathbf{x})$, if a 4×4 matrix can be found such that

$$\mathrm{e}^{i\eta_\mp}\, u^*_\mp(-\mathbf{p}\downarrow) = \Theta\, u_\mp(\mathbf{p}\uparrow). \tag{1.74}$$

From the explicit form (1.73), it turns out that the above relations hold true for $\eta_\mp = 0$. The orthogonality and closure relations for the spin-states require the unitarity property $\Theta\,\Theta^\dagger = \mathbb{I} = \Theta^\dagger\Theta$. Moreover, from the spin-state equations, we obtain

$$\begin{aligned} &\vec{\alpha}\cdot\vec{p}\, u_\mp(\vec{p}) = \wp u_\mp(\vec{p}), \quad -\vec{\alpha}^*\cdot\vec{p}\, u^*_\mp(-\vec{p}) = \wp u^*_\mp(-\vec{p}), \\ &-\vec{\alpha}^*\cdot\vec{p}\,\Theta u_\mp(\vec{p}) = \wp\Theta u_\mp(\vec{p}) \qquad \text{i. e.} \quad -\Theta^\dagger\vec{\alpha}^*\Theta = \vec{\alpha}\end{aligned}$$

The unitary solution of the last equation is unique up to an irrelevant phase factor. In the Weyl representation of the Clifford algebra, the real matrix

$$\Theta = -\gamma^1\gamma^3 = -i\,\Sigma_2 \equiv -i\begin{pmatrix}\sigma_2 & 0 \\ 0 & \sigma_2\end{pmatrix} \tag{1.75}$$

is a solution with the important property $\Theta^*\Theta + \mathbb{I} = 0$, which does not depend upon the specific representation of the Dirac matrices. Thus we finally find

$$\psi'_{\pm}(x') = \mathcal{T}\,\psi_{\pm}(-t,\mathbf{x})\,\mathcal{T}^{-1} = \Theta\,\psi_{\pm}(t,\mathbf{x}) \quad (x' = -\tilde{x}) \tag{1.76}$$

which shows that the time reversal is a chirality preserving transformation, just like the charge conjugation, as already stressed. Moreover, time reversal is well-defined even for two-component Weyl spinors, since the spin-flip operation does not mix left and right components of a massless Dirac bispinor. This means that also the $\mathcal{CP}$ and $\mathcal{CPT}$ transformations are well-defined for two-component Weyl spinors. In summary, the quantum theory which arises from the classical action integral

$$S_0 \equiv \int \mathrm{d}^4x\,\mathcal{L}_0(x) = \int \mathrm{d}^4x\,\psi^{\dagger}(x)\,i\alpha\cdot\partial\,\psi(x)$$

is invariant under the discrete $\mathcal{C}, \mathcal{P}, \mathcal{T}$ transformation separately, while the Weyl action integrals

$$S_L = \int \mathrm{d}^4x\,\tfrac{1}{2}\,\psi_L^{\dagger}(x)\,i\sigma\cdot\overleftrightarrow{\partial}\,\psi_L(x), \quad S_R = \int \mathrm{d}^4x\,\tfrac{1}{2}\,\psi_R^{\dagger}(x)\,i\bar{\sigma}\cdot\overleftrightarrow{\partial}\,\psi_R(x)$$

and related quantum theories are invariant under $\mathcal{T}$ and $\mathcal{CP}$ transforms. I.e. the latter are well-defined symmetries for left and right-handed Weyl spinors.

Causal Green's Functions

The causal Green's function, or Feynman propagator, of a theory is identified with the inverse of the kinetic operator in the Lagrangian. Here, we are interested in the theory of a Dirac fermion (1.32) and of a Weyl fermion, represented by one of the two pieces in (1.33). While for a Dirac fermion no obstacle is foreseen in the construction of the Green's function, for a Weyl fermion we expect trouble, because in the four-component formalism of (1.33), the kinetic operator contains a chiral projector. We can, nevertheless, proceed like in the construction of the massless Dirac Green's function hoping we can find a way out. The following paragraphs are intended to illustrate the obstacles met in this procedure.

Like in the Dirac case, consider the rank-two square matrix of the vacuum expectation value of the time ordered product of a pair of, e.g. right-handed quantum Weyl fields: namely

$$S_c(x-y) \equiv \langle 0|T\,\psi_R(x)\,\psi_R^{\dagger}(y)|0\rangle \tag{1.77}$$

After inserting the normal mode expansion (1.56) and using (1.20, 1.26), we get

$$\begin{aligned} S_c(x-y) &= \theta(x_0-y_0)\langle\psi_R(x)\,\psi_R^{\dagger}(y)\rangle_0 - \theta(y_0-x_0)\langle\psi_R^{\dagger}(y)\,\psi_R(x)\rangle_0 \\ &= \theta(x_0-y_0)\,i\sigma\cdot\partial\int \mathrm{d}\mathbf{p}\,[\,(2\pi)^3\,2\wp\,]^{-1}\,\exp\{-i\wp\,(x_0-y_0)+i\mathbf{p}\cdot(\mathbf{x}-\mathbf{y})\} \\ &+ \theta(y_0-x_0)\,i\sigma\cdot\partial\int \mathrm{d}\mathbf{p}\,[\,(2\pi)^3\,2\wp\,]^{-1}\,\exp\{\,i\wp\,(x_0-y_0)-i\mathbf{p}\cdot(\mathbf{x}-\mathbf{y})\} \end{aligned}$$

Now, if we recall the definitions of the massless Wightman distributions [3]

$$\begin{aligned} iD_0^{(\pm)}(x-y) &\equiv \pm \int Dp \exp\{\pm ip\cdot x\} \rfloor_{p_0=\wp} \\ &= \pm \int \frac{\mathbf{dp}}{(2\pi)^3 2\wp} \exp\{\pm i\wp(x_0-y_0) \mp i\mathbf{p}\cdot(\mathbf{x}-\mathbf{y})\}, \end{aligned}$$

we obtain

$$\begin{aligned} S_c(x-y) &= \theta(x_0-y_0)\, i\sigma\cdot\partial\,(-i)D_0^{(-)}(x-y) + \theta(y_0-x_0)\, i\sigma\cdot\partial\, iD_0^{(+)}(x-y) \\ &= i\sigma\cdot\partial[\,\theta(y_0-x_0)\, iD_0^{(+)}(x-y) - \theta(x_0-y_0)\, iD_0^{(-)}(x-y)\,] \\ &- \delta(x_0-y_0)[\,D_0^{(-)}(x-y) + D_0^{(+)}(x-y)\,] \\ &= i\sigma\cdot\partial_x\, D_0^c(x-y), \end{aligned}$$

where $D_0(\xi) = D_0^{(-)}(\xi) + D_0^{(+)}(\xi)$ is the massless Pauli-Jordan distribution [3], which fulfills $D_0(0,\vec{\xi}\,) = 0$, while

$$D_0^c(x-y) \equiv \theta(y_0-x_0)\, iD_0^{(+)}(x-y) - \theta(x_0-y_0)\, iD_0^{(-)}(x-y)$$

is the Feynman propagator, or causal Green's function, of a massless scalar field. Its Fourier representation is

$$D_0^c(x-y) = \frac{i}{(2\pi)^4}\int \mathrm{d}^4p\, \frac{\mathrm{e}^{-ip\cdot(x-y)}}{p^2+i\epsilon}$$

so that $iD_0^c(\xi)$ acts as the unique fully causal inverse of the d'Alembert differential operator.

As a consequence, we eventually find

$$\bar{\sigma}\cdot\partial_x\, S_c(x-y) = \delta^{(4)}(x-y) \tag{1.78}$$

which means that the rank-two matrix-valued causal Green's function S_c is the fully causal inverse of the right-handed differential operator $\bar{\sigma}\cdot\partial$. By following step by step, the above derivation *mutatis mutandis*, it is not difficult to prove that after setting

$$\bar{S}_c(x-y) \equiv \langle 0|T\,\psi_L(x)\,\psi_L^\dagger(y)|0\rangle \tag{1.79}$$

we get

$$\bar{S}_c(x-y) = i\bar{\sigma}\cdot\partial_x\, D_0^c(x-y) \qquad \sigma\cdot\partial_x\,\bar{S}_c(x-y) = \delta^{(4)}(x-y). \tag{1.80}$$

Consider now a change of the inertial reference frame, i.e. $x'^\mu = \Lambda^\mu_\nu(x+a)^\nu$. We have

$$D_0^{c\,\prime}(x'-y') = D_0^c(x-y)$$

$$\Lambda_L^{-1}\bar{\sigma}^\mu\Lambda_R\partial'_\mu = \Lambda_R^\dagger\bar{\sigma}^\mu\Lambda_R\partial'_\mu = \Lambda^\mu_\kappa\bar{\sigma}^\kappa\Lambda^\rho_\mu\partial_\rho = \bar{\sigma}^\mu\partial_\mu$$

and thereby

$$\bar{\sigma}^\mu\,\partial'_\mu = \Lambda_L\,\bar{\sigma}^\mu\,\partial_\mu\,\Lambda_R^{-1}$$

It follows that the transformation law of the Weyl causal Green's functions,

$$\bar{S}'_c(x'-y') = \Lambda_L\,\bar{S}_c(x-y)\,\Lambda_R^{-1}, \quad S'_c(x'-y') = \Lambda_R\,S_c(x-y)\,\Lambda_L^{-1},$$

does necessarily involve the $SL(2,\mathbb{C})$ matrices of both irreps $D(0,\frac{1}{2})$ and $D(\frac{1}{2},0)$. To better grasp this subtle point, let us rewrite e.g. the causal Green's function in its Fourier integral representation in the most suggestive form, viz.,

$$
\begin{aligned}
\bar{S}_c(x-y) &= \frac{i}{(2\pi)^4}\int \mathrm{d}^4p\,\frac{p_0-\vec{\sigma}\cdot\vec{p}}{p_0^2-\vec{p}^{\,2}+i\epsilon}\,\mathrm{e}^{-ip\cdot(x-y)} \\
&= \frac{i}{(2\pi)^4}\int \mathrm{d}^4p\,\mathrm{e}^{-ip\cdot(x-y)}\,\frac{p_0-\vec{\sigma}\cdot\vec{p}}{(p_0-\wp+i\epsilon)(p_0+\wp-i\epsilon)} \\
&= \int\frac{\mathrm{d}^4p}{(2\pi)^4}\,\mathrm{e}^{-ip\cdot(x-y)}\left(\frac{i}{p_0-\wp+i\epsilon}-\frac{i}{p_0+\wp-i\epsilon}\right)\frac{p_0-\vec{\sigma}\cdot\vec{p}}{2\wp}
\end{aligned}
$$

It turns out that in the above off-shell expression, the rank-two square matrix

$$
\mathbb{W}=\frac{p_0-\vec{\sigma}\cdot\vec{p}}{2\wp}
$$

is no longer a projector on left-handed two-component spin states because

$$
\mathbb{W}^2=\frac{p_0^2-2p_0\vec{\sigma}\cdot\vec{p}+\vec{p}^{\,2}}{4\vec{p}^{\,2}}\neq\mathbb{W}\qquad\qquad \mathrm{Tr}\,\mathbb{W}=\frac{p_0}{\wp}\in\mathbb{R}
$$

as it was true for the corresponding on shell matrix with $p_0=\wp$. *Hence, in the process of inversion for the Weyl differential operators, leading to the Weyl causal Green's functions, the very notion of chirality is definitely and unavoidably lost, just owing to the nature of the time ordered products of the Weyl quantum fields.*

In order to get a *bona fide* fully causal inversion for the kinetic differential operator of a massless spinor, we have to turn back to Dirac bispinors. Consider in fact the quantity

$$
\begin{aligned}
S_0(x-y) &= \langle 0|T\psi(x)\,\bar{\psi}(y)|0\rangle=\begin{bmatrix}0 & \bar{S}_c(x-y)\\ S_c(x-y) & 0\end{bmatrix} \\
&= \begin{bmatrix}0 & i\partial\cdot\bar{\sigma}\\ i\partial\cdot\sigma & 0\end{bmatrix} D_0^c(x-y)
\end{aligned}
$$

so that

$$
i\not{\partial}S_0(x-y)=(-1)\begin{bmatrix}\Box & 0\\ 0 & \Box\end{bmatrix} D_0^c(x-y)=i\delta^{(4)}(x-y), \tag{1.81}
$$

which shows, as expected, that the massless Schwinger propagator $S_c(x-y)$ is the unique inverse of the massless Dirac differential operator $\not{\partial}$, that fulfills causality forward and backward in time, its Fourier representation being

$$
S_c(x-y)=\frac{i}{(2\pi)^4}\int \mathrm{d}^4p\,\frac{\not{p}}{p^2+i\epsilon}\,\mathrm{e}^{-\,ip\cdot(x-y)} \tag{1.82}
$$

The problem of the propagator for a Weyl fermion cannot be solved in the above simple-minded way. It is in fact going to be the central problem in this book and will be taken up anew in Chap. 6.

1.1.2 Massive and Massless Majorana Spinors

The second part of the present section is devoted to Majorana spinors.

One can write a relativistic invariant field equation for a massive 2-component spinor field: to this end, let us start from a left-handed Weyl spinor $\psi_L \in D(\frac{1}{2}, 0)$ that transforms according to the $SL(2, \mathbb{C})$ matrix Λ_L. Call such a 2-component spinor field $\chi_a(x)$ $(a = 1, 2)$.

Definitions and Lagrangians

Let us consider the Weyl spinor wave field as a classical anticommuting field, i.e. a *Grassmann valued* left Weyl spinor field over the Minkowski space which satisfies

$$\{\chi_a(x),\ \chi_b(y)\} = 0 \quad (x, y \in \mathcal{M} \mid a, b = 1, 2)$$

together with the complex conjugation rule

$$(\chi_1\chi_2)^* = \chi_2^*\chi_1^* = -\chi_1^*\chi_2^* \tag{1.83}$$

so as to imitate the Hermitean conjugation of quantum fields. *A Majorana classical spinor field is a self-conjugated bispinor*, which can be constructed, for example, out of the left-handed spinor $\chi_a(x)$ $(a = 1, 2)$ as follows

$$\chi_M(x) = \begin{bmatrix} \chi(x) \\ -\sigma_2\chi^*(x) \end{bmatrix} = \chi_M^c(x)$$

the charge conjugation rule for any classical bispinors ψ being defined by the general relationship

$$\psi^c(x) = \mathrm{e}^{i\theta}\,\gamma^2\,\psi^*(x) \quad (0 \le \theta < 2\pi)$$

which is the discrete internal—i.e. spacetime point independent—symmetry transformation introduced before. In the sequel, we shall conveniently choose $\theta = 0$. The Majorana bispinor has a right-handed lower Weyl spinor component $-\sigma_2\chi^* \in D(0, \frac{1}{2})$, albeit functional dependent, due to the charge self-conjugation constraint, so that χ_M possesses both chiralities and polarizations, at variance with its left-handed Weyl component spinor $\chi(x)$. There is another kind of self-conjugated Majorana bispinor, which can be constructed out of a right-handed Weyl building spinor $\varphi \in D(0, \frac{1}{2})$

$$\varphi_M(x) = \begin{bmatrix} \sigma_2\varphi^*(x) \\ \varphi(x) \end{bmatrix} = \varphi_M^c(x)$$

From the Majorana self-conjugated bispinors, one can readily construct the most general Poincaré invariant and power counting renormalizable Lagrangian. For instance, by starting from the bispinor $\chi_M(x)$ we have

$$\mathcal{L}_M = \tfrac{1}{4}\,\overline{\chi}_M(x)\,\gamma^\mu\, i\,\overleftrightarrow{\partial}_\mu \chi_M^c(x) - \tfrac{1}{2}\, m\overline{\chi}_M(x)\chi_M^c(x)$$

where $\alpha^\nu = \gamma_0\gamma^\nu$, while the employed notation reminds us that the upper and lower components of a Majorana bispinor *can never be treated as functionally independent, even formally, due to the presence of the self-conjugation constraint.* It follows that *the Majorana mass term* can be written in the two equivalent forms

$$\begin{aligned}\mathcal{L}_M^m &= \tfrac{1}{2}\, m\left[\,\chi^\top(x)\sigma_2\,\chi(x) + \chi^\dagger(x)\sigma_2\,\chi^*(x)\,\right]\\ &= -\, im\left[\,\chi_1(x)\chi_2(x) + \chi_1^*(x)\chi_2^*(x)\,\right]\\ &= im\left[\,\chi_2^*(x)\chi_1^*(x) + \chi_2(x)\chi_1(x)\,\right] = \left(\mathcal{L}_M^m\right)^*,\end{aligned}$$

whence it is clear that the corresponding integral $\int \mathrm{d}^4x\, \mathcal{L}_M^m(x)$ is not invariant under the internal $U(1)$ phase transformation

$$\chi(x) \quad\mapsto\quad \chi'(x) = \chi(x)\,\mathrm{e}^{\,i\theta} \quad (0 \le \theta < 2\pi)$$

Concerning the kinetic term, from the relations

$$\chi_M^\dagger = \left[\chi^\dagger,\quad -\chi^\top\sigma_2\right], \quad \chi^\dagger = (\chi^*)^\top,$$

together with

$$\alpha^\nu = \gamma_0\gamma^\nu = \begin{pmatrix}\sigma^\nu & 0\\ 0 & \bar{\sigma}^\nu\end{pmatrix}, \quad \sigma_\mu = \bar{\sigma}^\mu = (1, \vec{\sigma}\,), \quad \alpha^\nu\gamma^2 = \begin{pmatrix}0 & \sigma^\nu\sigma_2\\ -\bar{\sigma}^\nu\sigma_2 & 0\end{pmatrix}$$

one can obtain the Majorana kinetic term in the 2-component formalism

$$\begin{aligned}&\tfrac{1}{4}\,\chi_M^\dagger(x)\,\alpha^\mu\gamma^2\, i\,\overleftrightarrow{\partial}_\mu\chi_M^*(x) = \tfrac{1}{4}\,\chi^\dagger(x)\,\sigma^\mu\, i\,\overleftrightarrow{\partial}_\mu\chi(x) + \tfrac{1}{4}\,\chi^\top(x)\,\sigma_2\,\bar{\sigma}^\mu\,\sigma_2\, i\,\overleftrightarrow{\partial}_\mu\chi^*(x)\\ &= \tfrac{1}{4}\chi^\dagger(x)\sigma^\mu i\,\overleftrightarrow{\partial}_\mu\chi(x) + \tfrac{1}{4}\chi^\top(x)(\bar{\sigma}^\mu)^* i\,\overleftrightarrow{\partial}_\mu\chi^*(x) = \chi^\dagger(x)\sigma^\mu i\partial_\mu\chi(x)\\ &-\frac{1}{2}i\partial_\mu\!\left[\chi^\dagger(x)\sigma^\mu\,\chi(x)\right],\end{aligned}$$

which is *twice* the Lagrangian for a left-handed Weyl spinor, as expected. Here, the above complex conjugation rule (1.83) for Grassmann valued field has been used.

It turns out that even the kinetic action integral

$$\tfrac{1}{4}\int \mathrm{d}^4x\,\chi_M^\dagger(x)\,\alpha^\mu\gamma^2\, i\,\overleftrightarrow{\partial}_\mu\chi_M^*(x)$$

is not invariant under the overall phase transformation $\chi_M'(x) = \mathrm{e}^{\,i\theta}\chi_M(x)$ of the Majorana bispinor, just like the previously discussed mass term. Hence, as it will be further confirmed after the transition to the Majorana representation of the Dirac matrices, *there is no invariant scalar charge for a Majorana spinor, it is a genuinely neutral spin $\frac{1}{2}$ field.* As it will be discussed and clarified in the sequel, *there is a relic continuous U(1) symmetry only for Majorana massless spinors*, which leads to the existence of a conserved pseudo-scalar charge, the meaning of which will be better explained further on.

The full Majorana Lagrangian in the 2-component formalism reads

$$\widehat{\mathcal{L}}_M = \tfrac{1}{2}\,\chi^\dagger(x)\,\sigma^\mu\, i\partial_\mu\chi(x) + \tfrac{1}{2}\,m\chi^\top(x)\,\sigma_2\,\chi(x) + \text{c.c.} \tag{1.84}$$

so that the Euler-Lagrange field equation becomes

$$i\,\sigma^\mu\,\partial_\mu\chi(x) + m\,\sigma_2\,\chi^*(x) = 0 \tag{1.85}$$

which is known as the Majorana field equation. Multiplying from the left by σ_2 and taking complex conjugation yields

$$i\sigma_2\,\partial_0\chi^*(x) - i\sigma_2\,\sigma_k^*\,\partial_k\chi^*(x) + m\chi(x) = 0.$$

Remembering that we have $\sigma_2\,\vec{\sigma}\,\sigma_2 = -\,\vec{\sigma}$ and that $\bar{\sigma}^\mu = (1, \vec{\sigma}\,)$ we come to the equivalent form of the Majorana wave equation

$$i\bar{\sigma}^\mu\,\sigma_2\,\partial_\mu\chi^*(x) + m\chi(x) = 0 \tag{1.86}$$

Now, if we act from the left with the operator $i\bar{\sigma}^\nu\partial_\nu$ on equation (1.85) and use equation (1.86), we find

$$\bar{\sigma}^\nu\,\sigma^\mu\,\partial_\nu\partial_\mu\chi(x) + m^2\chi(x) = (\Box + m^2\,)\chi(x) = 0$$

which means that the left-handed spinor $\chi \in D(\frac{1}{2}, 0)$, i.e. the building block of the self-conjugated Majorana bispinor, is actually solution of the Klein-Gordon wave equation. Moreover, by understanding the Majorana bispinor to be defined by the self-conjugation constraint $\chi_M(x) = \chi_M^c(x)$, it is easy to check that the pair of equivalent wave equations (1.85) and (1.86) is equivalent to the single bispinor wave equation

$$(\,\alpha^\nu i\partial_\nu - \beta m\,)\chi_M(x) = 0 \tag{1.87}$$

where use has been made of the Dirac notation $\beta = \gamma_0$, while the Majorana Lagrangian can be recast in a further 4-component form

$$\mathcal{L}_M = \tfrac{1}{4}\,\chi_M^\dagger(x)\,\alpha^\mu\, i\,\overleftrightarrow{\partial}_\mu\chi_M^c(x) - \tfrac{1}{2}\,m\,\chi_M^\dagger(x)\,\beta\,\chi_M^c(x) \tag{1.88}$$

It is immediate to verify that the bispinor form (1.87) of the field equations coincides with the pair of functionally dependent and equivalent forms (1.85) and (1.86) of the

Majorana spinor wave equation. Notice that the Majorana self-conjugated bispinor transforms under the Poincaré group according to

$$\chi'_M(x') = \Lambda_{\frac{1}{2}}\,\chi^c_M(x) \quad \Lambda_{\frac{1}{2}} = \begin{bmatrix} \Lambda_L & 0 \\ 0 & \Lambda_R \end{bmatrix} \tag{1.89}$$

with $x' = \Lambda(x + \mathrm{a})$.

The Majorana Real Representation

By definition, the Majorana bispinor $\chi_M(x) = \chi^c_M(x)$ must fulfill the self-conjugation constraint, which linearly relates the lower spinor component to the complex conjugate of the upper spinor component. Then, a representation must exist which makes the Majorana bispinor real, in such a way that the previously introduced pair of complex variables $\chi_a \in \mathbb{C}$ $(a = 1, 2)$ can be replaced by the four real variables $\psi_{M,\alpha} \in \mathbb{R}$ $(\alpha = 1, 2, 3, 4)$. To obtain this real representation, we note that

$$\chi_M = \begin{bmatrix} \chi \\ -\sigma_2\,\chi^* \end{bmatrix} \quad \chi^*_M = \begin{bmatrix} 0 & -\sigma_2 \\ \sigma_2 & 0 \end{bmatrix} \chi_M = -\gamma^2\chi_M$$

A transformation to *real* bispinor fields $\psi_M = \psi^*_M$ can be made by writing

$$\chi_M = S\,\psi_M\,, \quad \chi^*_M = S^*\,\psi_M = S^*\,S^{-1}\,\chi_M\,, \quad \text{whence} \quad S^* = \begin{bmatrix} 0 & -\sigma_2 \\ \sigma_2 & 0 \end{bmatrix} S\,.$$

Now, let us set $\begin{bmatrix} 0 & i\sigma_2 \\ -i\sigma_2 & 0 \end{bmatrix} = i\gamma^2 \equiv \rho, \quad \rho = \rho^\dagger, \quad \rho^2 = \mathbb{I}$, so that

$$\exp\{i\rho\,\theta\} = \mathbb{I}\,\cos\theta + i\rho\,\sin\theta$$

Then the solution for the above relation is the unitary matrix

$$S = \exp\left\{-\tfrac{1}{4}\,\pi i\rho\right\} = \frac{\sqrt{2}}{2}\,(1 - i\rho)$$

i.e. $S = \frac{\sqrt{2}}{2}\begin{bmatrix} 1 & \sigma_2 \\ -\sigma_2 & 1 \end{bmatrix} = \frac{\sqrt{2}}{2}(1 + \gamma^2)$, or, even more explicitly,

$$S = \frac{\sqrt{2}}{2}\begin{bmatrix} 1 & 0 & 0 & -i \\ 0 & 1 & i & 0 \\ 0 & i & 1 & 0 \\ -i & 0 & 0 & 1 \end{bmatrix}.$$

From the above relation $\psi_M = S^{-1}\chi_M = \psi^*_M$, one can immediately obtain the correspondence rule between the complex and real forms of the self-conjugated Majorana bispinor, namely

$$\psi_{M1} = \sqrt{2}\,\Re\mathrm{e}\,\chi_1 \quad \psi_{M2} = \sqrt{2}\,\Re\mathrm{e}\,\chi_2$$
$$\psi_{M3} = -\sqrt{2}\,\Im m\,\chi_2 \; \psi_{M4} = \sqrt{2}\,\Im m\,\chi_1$$

Thus, we write down the *Majorana representation for the Clifford algebra*, given by the similarity transformation acting on the γ–matrices in the Weyl representation, *viz.*,

$$\gamma_M^\mu \equiv S^\dagger \gamma^\mu S, \quad \text{i.e.}: \quad \gamma_M^0 = \begin{pmatrix} 0 & i & 0 & 0 \\ -i & 0 & 0 & 0 \\ 0 & 0 & 0 & -i \\ 0 & 0 & i & 0 \end{pmatrix}, \quad \gamma_M^1 = \begin{pmatrix} -i & 0 & 0 & 0 \\ 0 & i & 0 & 0 \\ 0 & 0 & -i & 0 \\ 0 & 0 & 0 & i \end{pmatrix}$$

$$\gamma_M^2 = \begin{pmatrix} 0 & 0 & 0 & -i \\ 0 & 0 & i & 0 \\ 0 & i & 0 & 0 \\ -i & 0 & 0 & 0 \end{pmatrix}, \quad \gamma_M^3 = \begin{pmatrix} 0 & i & 0 & 0 \\ i & 0 & 0 & 0 \\ 0 & 0 & 0 & i \\ 0 & 0 & i & 0 \end{pmatrix}, \quad \gamma_M^5 = \begin{pmatrix} 0 & 0 & 0 & i \\ 0 & 0 & -i & 0 \\ 0 & i & 0 & 0 \\ -i & 0 & 0 & 0 \end{pmatrix},$$

These satisfy by direct inspection

$$\{\gamma_M^\mu, \gamma_M^\nu\} = 2\,g^{\mu\nu} \quad \{\gamma_M^\nu, \gamma_M^5\} = 0$$

$$\gamma_M^0 = \beta_M^\dagger \quad \gamma_M^k = -\gamma_M^{k\,\dagger} \quad \gamma_M^5 = \gamma_M^{5\,\dagger}$$

$$\gamma_M^\nu \; = \; -\,\gamma_M^{\nu\,*} \quad \gamma_M^5 \; = \; -\,\gamma_M^{5\,*}$$

The result is that, instead of a complex self-conjugated bispinor, which has been constructed out of a left-handed Weyl spinor, one can simply employ a real Majorana bispinor:

$$\chi_M(x) = \chi_M^c(x) \quad \leftrightarrow \quad \psi_M(x) = S^\dagger \chi_M(x) = \psi_M^*(x)$$

An analogous construction holds starting from a right-handed Weyl spinor $\varphi \in D(0, \frac{1}{2})$. Then, the Majorana Lagrangian and the ensuing Majorana wave field equation take the manifestly real forms

$$\mathcal{L}_M = \tfrac{1}{4}\,\psi_M^\top(x)\,\alpha_M^\nu\, i\,\overleftrightarrow{\partial}_\nu\,\psi_M(x) - \tfrac{1}{2}\,m\,\psi_M^\top(x)\,\beta_M\,\psi_M(x)$$

$$(i\partial\!\!\!/_M - m)\psi_M(x) = 0 \quad \psi_M(x) = \psi_M^*(x)$$

$$\alpha_M^\nu = \gamma_M^0\,\gamma_M^\nu \quad \alpha_M^0 = \mathbb{I} \quad \beta_M \equiv \gamma_M^0$$

with $\vec{\alpha}_M = \vec{\alpha}_M^* = \vec{\alpha}_M^\top$. It turns out that, from the manifestly real form of the Majorana Lagrangian, *the only relic internal symmetry of the Majorana's action integral—except in the massless case—is the discrete* $\mathbb{Z}_2$ *symmetry, i.e.* $\psi_M(x) \longmapsto -\psi_M(x)$.

Singular Massless Limit

Actually, it turns out that *only in the massless case* there exists a further accidental invariance of the action integral under the U(1) internal symmetry group

$$\psi_M(x) \quad \mapsto \quad \psi'_M(x) = \exp\left\{\pm i\theta\,\gamma_M^5\right\}\psi_M(x) \quad (0 \le \theta < 2\pi)$$

the imaginary unit being needed in order to preserve reality of the transformed Majorana bispinor. From the Noether theorem, we get the corresponding real four-current, which satisfies the continuity equation

$$J_5^{\mu}(x) = \tfrac{1}{2}\,\psi_M^{\top}(x)\,\alpha_M^{\mu}\,i\gamma_M^5\,\psi_M(x) \qquad\qquad \partial\cdot J_5(x) = 0$$

as well as the ensuing conserved pseudo-scalar charge

$$\pm\,Q_5 = \pm\frac{1}{2}\int \mathrm{d}\mathbf{x}\,\psi_M^{\top}(t,\mathbf{x})\,i\gamma_M^5\psi_M(t,\mathbf{x}) \qquad\qquad \dot{Q}_5 = 0$$

the overall $\pm$ sign being conventional and irrelevant. In order to better appreciate the meaning of this quantity we notice that, if we turn to the complex form $\chi_M(x)$ of the self-conjugated bispinor, we can immediately realize that the above U(1) internal symmetry of the massless case is nothing but the ordinary phase transformation for the independent building spinor $\chi(x)$. As a matter of fact, the phase transformation $\chi'(x) = \mathrm{e}^{-i\theta}\chi(x)$ just induces the chiral transformation $\chi'_M(x) = \mathrm{e}^{\,i\theta\gamma_5}\chi_M(x)$ on the complex form of the Majorana bispinor.

Spin States

The Majorana Hamiltonian reads

$$H_M = \alpha_M^k\,\hat{p}^k + m\,\beta_M \quad (\hat{p}^k = -i\hbar\nabla_k)$$

To solve the Majorana wave equation, we set

$$\psi_M(x) = \int \frac{\mathrm{d}^4 p}{(2\pi)^{3/2}}\,\widetilde{\psi}_M(p)\,\exp\{-\,ip\cdot x\}$$

with the reality condition

$$\widetilde{\psi}_M^{*}(p) = \widetilde{\psi}_M(-p)$$

so that

$$(\not{p}_M - m)\,\widetilde{\psi}_M(p) = 0 \quad \not{p}_M \equiv p_\nu\gamma_M^\nu$$

which implies

$$\widetilde{\psi}_{M,\alpha}(p) = (\not{p}_M + m)_{\alpha\beta}\,\widetilde{\phi}_\beta(p), \quad \left(p^2 - m^2\right)\widetilde{\phi}_\alpha(p) = 0, \quad \widetilde{\phi}_\alpha(p) = \delta\left(p^2 - m^2\right) f_\alpha(p),$$

for $\alpha = 1, 2, 3, 4$. Furthermore, from the reality condition on the Majorana spinor field

$$\widetilde{\psi}^{*}_{M,\alpha}(p) = \widetilde{\psi}_{M,\alpha}(-p),$$

we find

$$\widetilde{\psi}^{*}_{M,\alpha}(p) = (\not{p}^{*}_{M} + m)_{\alpha\beta}\,\widetilde{\phi}^{*}_{\beta}(p) = (m - \not{p}_{M})_{\alpha\beta}\,\widetilde{\phi}^{*}_{\beta}(p) = \widetilde{\psi}_{M,\alpha}(-p) \quad \Longleftrightarrow \quad f^{*}_{\beta}(p) = f_{\beta}(-p)$$

thanks to the circumstance that the $\gamma-$matrices in the Majorana representation are purely imaginary. Then we can write

$$\begin{aligned}
\psi_M(x) &= \int_{-\infty}^{\infty} \mathrm{d}p_0 \int \mathrm{d}\mathbf{p}\, \frac{\theta(p_0)}{(2\pi)^{3/2}\, 2\omega_{\mathbf{p}}}\, (\not{p}_M + m)_{\alpha\beta} f_\beta(p)\, \delta\left(p_0 - \omega_{\mathbf{p}}\right) \exp\{-\,ip\cdot x\} \\
&+ \int_{-\infty}^{\infty} \mathrm{d}p_0 \int \mathrm{d}\mathbf{p}\, \frac{\theta(-\,p_0)}{(2\pi)^{3/2}\, 2\omega_{\mathbf{p}}} (\not{p}_M + m)_{\alpha\beta} f_\beta(p)\, \delta\left(p_0 + \omega_{\mathbf{p}}\right) \exp\{-\,ip\cdot x\} \\
&= 2m \int \mathrm{d}\mathbf{p}\, \frac{1}{(2\pi)^{3/2}\, 2\omega_{\mathbf{p}}} \left[\mathcal{E}^{+}_{M}(p) f(p)\, \mathrm{e}^{-\,ip\cdot x} + \mathcal{E}^{-}_{M}(p) f^{\,*}(p)\, \mathrm{e}^{\,ip\cdot x}\right] \\
&\stackrel{\text{def}}{=} \sum_{\mathbf{p}} \left[\, \mathcal{E}^{+}_{M}(p)\, f_{\mathbf{p}}\, \mathrm{e}^{-\,ip\cdot x} + \mathcal{E}^{-}_{M}(p)\, f^{\,*}_{\mathbf{p}}\, \mathrm{e}^{\,ip\cdot x} \,\right] = \psi^{*}_{M}(x),
\end{aligned}$$

where $p_0 = \omega_{\mathbf{p}}$, whereas $f_{\mathbf{p}} = 2mf(p)/(2\pi)^{\frac{3}{2}}\, 2\omega_{\mathbf{p}}$. The projectors onto the spin states are

$$\mathcal{E}^{\pm}_{M}(p) = \left(m \pm \not{p}_{M}\right)/2m \quad \left(p_0 = \omega_{\mathbf{p}}\right)$$

with ($\tilde{p}^{\mu} = p_{\mu}$)

$$\begin{aligned}
&[\mathcal{E}^{\pm}_{M}(p)]^{*} = \mathcal{E}^{\mp}_{M}(p), \quad [\mathcal{E}^{\pm}_{M}(p)]^{\dagger} = \mathcal{E}^{\pm}_{M}(\tilde{p}), \quad [\mathcal{E}^{\pm}_{M}(p)]^{2} = \mathcal{E}^{\pm}_{M}(p), \\
&\mathcal{E}^{\pm}_{M}(p)\mathcal{E}^{\mp}_{M}(p) = 0, \quad \operatorname{tr}\mathcal{E}^{\pm}_{M}(p) = 2, \quad \mathcal{E}^{+}_{M}(p) + \mathcal{E}^{-}_{M}(p) = \mathbb{I}.
\end{aligned}$$

Now, in order to construct the spin states of the Majorana real spinor field, let us start from the spin matrices in the Majorana representation

$$\Sigma_{M,1} = i\gamma^{2}_{M}\,\gamma^{3}_{M} = \begin{bmatrix} 0 & i\sigma_3 \\ -\,i\sigma_3 & 0 \end{bmatrix}, \qquad \Sigma_{M,2} = i\gamma^{3}_{M}\,\gamma^{1}_{M} = \begin{bmatrix} \sigma_2 & 0 \\ 0 & \sigma_2 \end{bmatrix},$$
$$\Sigma_{M,3} = \gamma^{1}_{M}\,\gamma^{2}_{M} = \begin{bmatrix} 0 & -\,i\sigma_1 \\ i\sigma_1 & 0 \end{bmatrix}$$

and from the common eigenvectors of the matrix β_M and of the diagonal spin matrix $\Sigma_{M,2}$, which are, *e.g.*,

$$\xi_+ \equiv \begin{bmatrix} 0 \\ 0 \\ 1 \\ i \end{bmatrix}, \quad \xi_- \equiv \begin{bmatrix} i \\ 1 \\ 0 \\ 0 \end{bmatrix}, \quad \eta_+ \equiv \begin{bmatrix} 1 \\ i \\ 0 \\ 0 \end{bmatrix} \quad \eta_- \equiv \begin{bmatrix} 0 \\ 0 \\ i \\ 1 \end{bmatrix},$$

where

$$\gamma_M^0\,\xi_\pm = \xi_\pm, \quad \gamma_M^0\,\eta_\pm = -\,\eta_\pm$$

They do indeed satisfy by direct inspection

$$\gamma_M^0\,\xi_\pm = \xi_\pm, \quad \xi_\mp^\dagger\,\xi_\pm = 0, \quad \left(\Sigma_{M,2} \mp 1\right)\xi_\pm = 0$$

so that we have by construction

$$\xi_r^\dagger\,\xi_s = 2\,\delta_{rs}, \quad \xi_r^\dagger\,\gamma_M^k\,\xi_s = 0, \quad \forall\, r, s = \pm \vee k = 1, 2, 3$$

The spin states ξ_r ($r = \pm$) are two degenerate eigenstates of the Majorana Hamiltonian in the massive neutral spinor particle rest frame $\mathbf{p} = 0$ with positive eigenvalue $p_0 = m$ and with opposite spin projections on the OY axis, *viz.*,

$$\sigma_M^{31}\,\xi_\pm \equiv \tfrac{1}{4}\,i\,[\,\gamma_M^3,\,\gamma_M^1\,]\,\xi_\pm = \tfrac{1}{2}\,\Sigma_{M,2}\,\xi_\pm = \pm\,\tfrac{1}{2}\,\xi_\pm$$

Then we define the Majorana *complex spin-states* to be

$$w_r(\mathbf{p}) \equiv 2m(2\omega_{\mathbf{p}} + 2m)^{-\frac{1}{2}}\,\mathcal{E}_M^+(p)\,\xi_r \quad (r = \pm \vee p_0 = \omega_{\mathbf{p}}),$$

which are the two eigenstates of the positive energy projector

$$\mathcal{E}_M^+(p)\,w_r(\mathbf{p}) = w_r(\mathbf{p}) \quad (r = +, - \vee p_0 = \omega_{\mathbf{p}}),$$

with

$$w_r^\dagger(\mathbf{p})\,w_s(\mathbf{p}) = 2\omega_{\mathbf{p}}\,\delta_{rs}$$

In fact we have for instance

$$\begin{aligned} w_r^\dagger(\mathbf{p})\,w_s(\mathbf{p}) &= (2\omega_{\mathbf{p}} + 2m)^{-1}\,\xi_r^\dagger(m + \tilde{\not p})(m + \not p_M)\,\xi_s \\ &= (2\omega_{\mathbf{p}} + 2m)^{-1}\,\xi_r^\dagger\left(2\omega_{\mathbf{p}}^2 + 2m\omega_{\mathbf{p}}\right)\xi_s \\ &= 2\omega_{\mathbf{p}}\,\tfrac{1}{2}\,\xi_r^\dagger\,\xi_s = 2\omega_{\mathbf{p}}\,\delta_{rs} \end{aligned}$$

where use has been made of the property $\gamma_M^0\,\xi_r = \xi_r$ $(r = \pm)$. Moreover we can easily verify that $\overline{w}_r(\mathbf{p})\,w_s(\mathbf{p}) = 2m\,\delta_{rs}$, so that we can write

$$\sum_{r=+,-} w_r(\mathbf{p}) \otimes \overline{w}_r(\mathbf{p}) = m + \not p_M$$

It turns out that the pair of orthogonal negative frequency spin-states are the complex conjugates of the former ones

$$w_r^*(\mathbf{p}) = 2m(2\omega_{\mathbf{p}} + 2m)^{-\frac{1}{2}}\, \mathcal{E}_M^-(p)\, \xi_r^* = 2m(2\omega_{\mathbf{p}} + 2m)^{-\frac{1}{2}}\, \mathcal{E}_M^+(-p)\, \xi_r^*,$$

which evidently fulfill

$$\begin{aligned}
\mathcal{E}_M^-(p)\, w_r^*(\mathbf{p}) &= w_r^*(\mathbf{p}) \quad (r = +, - \ \vee \ p_0 = \omega_{\mathbf{p}}) \\
w_r^\dagger(\mathbf{p}) w_s^*(-\mathbf{p}) &= w_r^\top(-\mathbf{p}) w_s(\mathbf{p}) = 0 \quad (\forall\, r, s = +, -) \\
\overline{w}_r^*(\mathbf{p})\, w_r^*(\mathbf{p}) &= 2m\, \delta_{rs} = -\, w_r^\top(\mathbf{p})\, \gamma_M^0\, w_s^*(\mathbf{p}) \\
\sum_{r=+,-} w_r^*(\mathbf{p}) \otimes \overline{w}_r^*(\mathbf{p}) &= m - \not{p}_M = m + \not{p}_M^* \quad (p_0 = \omega_{\mathbf{p}}),
\end{aligned}$$

as implied by the Majorana representation of the Clifford algebra. Of course, an absolutely equivalent construction can be made, had we started from the other degenerate pair of constant eigenspinors η_r $(r = +, -)$ of the matrix β_M. In conclusion, the most general normal mode decomposition of the Grassmann valued classical Majorana real spinor wave field becomes

$$\psi_M(x) = \sum_{\mathbf{p},\, r} \left[a_{\mathbf{p},\, r}\, w_{\mathbf{p},\, r}(x) + a_{\mathbf{p},\, r}^*\, w_{\mathbf{p},\, r}^*(x) \right] = \psi_M^*(x),$$

with

$$w_{\mathbf{p},\, r}(x) \equiv [\, (2\pi)^3 2\omega_{\mathbf{p}}\,]^{\frac{1}{2}}\; w_r(\mathbf{p})\; \exp\{-\, i\omega_{\mathbf{p}} t + i\mathbf{p}\cdot\mathbf{x}\}$$

$$\{a_{\mathbf{p},\, r},\, a_{\mathbf{q},\, s}\} = \{a_{\mathbf{p},\, r},\, a_{\mathbf{q},\, s}^*\} = \{a_{\mathbf{p},\, r}^*,\, a_{\mathbf{q},\, s}^*\} = 0$$
$$\forall\, \mathbf{p},\, \mathbf{q} \in \mathbb{R}^3 \quad \forall\, r,\, s = +, -$$

It is important to realize that the real Majorana spinor wave field exhibits two opposite helicity states.

Quantum Theory

The transition to the quantum theory is performed as usual by means of the creation and annihilation operators $a_{\mathbf{p},\, r}$ and $a_{\mathbf{p},\, s}^\dagger$ which satisfy the canonical anticommutation relations

$$\{a_{\mathbf{p},\, r},\, a_{\mathbf{q},\, s}\} = 0 = \{a_{\mathbf{p},\, r}^\dagger,\, a_{\mathbf{q},\, s}^\dagger\}$$

$$\{a_{\mathbf{p},\, r},\, a_{\mathbf{q},\, s}^\dagger\} \;=\; \delta_{rs}\, \delta\,(\mathbf{p} - \mathbf{q})$$

$$\forall\, \mathbf{p},\, \mathbf{q} \in \mathbb{R}^3 \quad \forall\, r, s = +, -$$

so that

$$\psi_M(x) = \sum_{\mathbf{p},\, r} \left[a_{\mathbf{p},\, r}\, w_{\mathbf{p},\, r}(x) + a^\dagger_{\mathbf{p},\, r}\, w^*_{\mathbf{p},\, r}(x) \right] = \psi^c_M(x)$$

where the Hermitean conjugation refers to the creation and annihilation operators acting on the Fock space. Moreover, we readily get the related and useful normal mode expansions

$$\psi^\top_M(x) = \sum_{\mathbf{p},\, r} \left[a_{\mathbf{p},\, r}\, w^\top_{\mathbf{p},\, r}(x) + a^\dagger_{\mathbf{p},\, r}\, w^{\top *}_{\mathbf{p},\, r}(x) \right]$$

$$\overline{\psi}_M(x) = \sum_{\mathbf{q},\, s} \left[a^\dagger_{\mathbf{q},\, s}\, \overline{w}_{\mathbf{q},\, s}(x) + a_{\mathbf{q},\, s}\, \overline{w}^*_{\mathbf{q},\, s}(x) \right]$$

in which

$$\overline{w}_{\mathbf{p},\, r}(x) \equiv [\, (2\pi)^3 2\omega_{\mathbf{p}}\,]^{-\frac{1}{2}}\, [\, w^\top_r(\mathbf{p})\,]^*\, \beta_M\, \exp\{\, i\omega_{\mathbf{p}} t - i\mathbf{p}\cdot\mathbf{x}\}$$

From the orthogonality relations

$$\left(w_{\mathbf{p},\, r},\, w_{\mathbf{q},\, s} \right) = \int \mathrm{d}\mathbf{x}\, \overline{w}_{\mathbf{p},\, r}(x)\, \beta_M\, w_{\mathbf{q},\, s}(x) = \delta_{rs}\, \delta(\mathbf{p}-\mathbf{q}),$$

$$\left(w_{\mathbf{p},\, r},\, w^*_{\mathbf{q},\, s} \right) = \int \mathrm{d}\mathbf{x}\, \overline{w}_{\mathbf{p},\, r}(x)\, \beta_M\, w^*_{\mathbf{q},\, s}(x) = 0,$$

$$\left(w^*_{\mathbf{p},\, r},\, w^*_{\mathbf{q},\, s} \right) = \int \mathrm{d}\mathbf{x}\, w^\top_{\mathbf{p},\, r}(x)\, w^*_{\mathbf{q},\, s}(x) = -\, \delta_{rs}\, \delta(\mathbf{p}-\mathbf{q}),$$

one can easily obtain all the observable quantities involving the Majorana spinor field. For example, the energy-momentum four-vector takes the form

$$P_\mu = \frac{i}{4} \int \mathrm{d}\mathbf{x}\, :\overline{\psi}_M(x)\, \beta_M\, \overleftrightarrow{\partial}_\mu\, \psi_M(x) := \frac{1}{2} \sum_{\mathbf{p},\, r} p_\mu \left[a^\dagger_{\mathbf{p},\, r} a_{\mathbf{p},\, r} - a_{\mathbf{p},\, r} a^\dagger_{\mathbf{p},\, r} \right]$$

$$= \sum_{\mathbf{p},\, r} p_\mu\, a^\dagger_{\mathbf{p},\, r} a_{\mathbf{p},\, r} \quad (p_0 = \omega_{\mathbf{p}})$$

Moreover, for example, if $\partial_x \psi_M = \partial_z \psi_M = 0$ so that $\psi_M(t, y) = \psi_M(t, 0, y, 0)$, then we obtain for the energy-momentum tensor

$$2i\, T^{13}(t, y) = \overline{\psi}_M(t, y)\, \gamma^1_M\, \overleftrightarrow{\partial}_z\, \psi_M(t, y) = 0,$$

$$2i\, T^{31}(t, z) = \overline{\psi}_M(t, y)\, \gamma^3_M\, \overleftrightarrow{\partial}_x\, \psi_M(t, y) = 0,$$

and thereby

$$\partial_\mu M^{\mu}_{13} = \partial_\mu S^{\mu}_{13} = 0,$$

whence it follows that the helicity is conserved in time. After insertion of the normal modes expansion, one gets

$$\begin{aligned}
\mathrm{h} &= \int_{-\infty}^{\infty} \mathrm{d}y \; : \tfrac{1}{2}\, \psi_M(t,y)\, \Sigma_{M,2}\, \psi_M(t,y) : \\
&= \int_{-\infty}^{\infty} \mathrm{d}y \; : \sum_{p,r} \Big[a_{p,r}\, w_{p,r}(t,y) + a^{\dagger}_{p,r}\, w^{*}_{p,r}(t,y) \Big] \\
&\times \tfrac{1}{2}\, \Sigma_{M,2} \sum_{q,s} \Big[a^{\dagger}_{q,s}\, w^{*}_{q,s}(t,y) + a_{q,s}\, w_{q,s}(t,y) \Big],
\end{aligned}$$

where we have set

$$\mathbf{p} = (0, p, 0) \quad \mathbf{q} = (0, q, 0) \quad \omega_p = \sqrt{p^2 + m^2}$$

$$w_{p,r}(t,y) = [\,4\pi\omega_p\,]^{-1/2}\, w_r(p)\, \exp\{ipy - it\omega_p\} \quad (r = +, -).$$

The normalization is now consistent with the occurrence that the spinor plane waves are independent of the transverse spatial coordinates $x_\perp = (x^1, x^3)$. From the commutation relation

$$[\,\omega_p \gamma^0_M - p\,\gamma^2_M,\ \Sigma_{M,2}\,] = 0,$$

together with the definition

$$w_\pm(p) \equiv (2\omega_p + 2m)^{-1/2}\, \left(m + \omega_p \gamma^0_M - p\,\gamma^2_M\right) \xi_\pm,$$

it can be readily derived that

$$(\,\Sigma_{M,2} \mp 1\,)\,\xi_\pm = 0 \quad \Rightarrow \quad (\,\Sigma_{M,2} \mp 1\,)\, w_\pm(p) = 0,$$

which yields in turn

$$\mathrm{h} = \tfrac{1}{2} \int_{-\infty}^{\infty} \mathrm{d}p \left[a^{\dagger}_{p,+}\, a_{p,+} - a^{\dagger}_{p,-}\, a_{p,-} \right].$$

The 1-particle states $a^{\dagger}_{\mathbf{p},\pm}\,|0\rangle$ represent spin $\frac{1}{2}$ neutral Majorana massive particles with energy-momentum $p^\mu = (\omega_{\mathbf{p}}, \mathbf{p})$ and positive/negative helicity.

From the Majorana Lagrangian and related wave equation, one can immediately realize that the special distributions for the Majorana neutral spinor field are the same as for a Dirac charged quantum field, namely,

$$\left\{\psi_M(x),\, \overline{\psi}_M(y)\right\} = S(x-y;\, m) \quad \text{and} \quad \langle\, 0\,|\, T\,\psi_M(x)\,\overline{\psi}_M(y)\,|\, 0\,\rangle = S^F(x-y;\, m),$$

the massless limit being smooth.

1.2 Dirac, Majorana and Weyl Fermions in 4d

This section is a summary of the main properties and formulas for massless Dirac, Weyl and Majorana fermions, derived in the previous section. The purpose is to highlight those that will play a major role in the rest of the book.

A Dirac fermion ψ is a four-component complex spinor. Under a Lorentz transformation, it transforms as

$$\psi(x) \to \psi'(x') = \exp\left[-\frac{1}{2}\lambda^{\mu\nu}\Sigma_{\mu\nu}\right]\psi(x), \tag{1.90}$$

for $x'^{\mu} = \Lambda^{\mu}{}_{\nu}\, x^{\nu}$. Here $\lambda^{\mu\nu} + \lambda^{\nu\mu} = 0$ are six real canonical coordinates for the Lorentz group, $\Sigma_{\mu\nu} = \frac{1}{4}[\gamma_\mu, \gamma_\nu]$ are the generators in the 4d representation of Dirac bispinors, while $\Lambda^{\mu}{}_{\nu}$ are the Lorentz matrices in the fundamental vector representation. The invariant kinetic Lagrangian for a free Dirac field is

$$i\bar{\psi}\gamma^\mu\partial_\mu\psi \tag{1.91}$$

where $\bar{\psi} = \psi^\dagger\gamma_0$. A Dirac fermion admits a Lorentz invariant mass term $m\bar{\psi}\psi$.

Since γ_5 commutes with $\Sigma_{\mu\nu}$, a Dirac bispinor is a reducible representation of the Lorentz group, and it can be seen as the direct sum of two Weyl spinor representations

$$\psi_L = P_L\psi, \quad \psi_R = P_R\psi, \quad \text{where} \quad P_L = \frac{1-\gamma_5}{2}, \quad P_R = \frac{1+\gamma_5}{2}$$

with opposite chiralities

$$\gamma_5\psi_L = -\psi_L, \quad \gamma_5\psi_R = \psi_R.$$

A left-handed Weyl fermion admits a Lagrangian kinetic term

$$i\overline{\psi}_L\gamma^\mu\partial_\mu\psi_L \tag{1.92}$$

but not a mass term, because $(\overline{\psi}_L\psi_L) = 0$, since $\gamma_5\gamma^0 + \gamma^0\gamma_5 = 0$. So a Weyl fermion is massless and this property is protected by the chirality conservation.

For Majorana fermions, we need the notion of *Lorentz covariant conjugate* spinor, for which we use the more convenient notation $\hat{\psi}$[2] :

$$\hat{\psi} = \gamma_0 C \psi^*. \tag{1.93}$$

Notice that $\gamma_5 \widehat{\psi_R} = -\widehat{\psi_R}$ and $\gamma_5 \widehat{\psi_L} = \widehat{\psi_L}$.

It is not hard to show that if ψ transforms like (1.90), then

$$\hat{\psi}(x) \to \hat{\psi}'(x') = \exp\left[-\frac{1}{2}\lambda^{\mu\nu}\Sigma_{\mu\nu}\right]\hat{\psi}(x). \tag{1.94}$$

Therefore, one can impose on ψ the condition

$$\psi = \hat{\psi} \tag{1.95}$$

because both sides transform in the same way. By definition, a spinor satisfying (1.95) is a Majorana spinor. It admits both kinetic and mass term.

To state the above more precisely, Weyl representations are irreducible representations of the group $SL(2, C)$, which is the covering group of the *proper ortochronous* Lorentz group. They are usually denoted $(\frac{1}{2}, 0)$ and $(0, \frac{1}{2})$ in the $SU(2) \times SU(2)$ notation of the $SL(2, C)$ irreducible representations. Lorentz transformations commute also with the charge conjugation operation

$$\mathcal{C}\psi\mathcal{C}^{-1} = \eta_C \gamma_0 C \psi^* \tag{1.96}$$

where η_C is a phase which, for simplicity, in the sequel we set equal to 1. This also implies that Dirac spinors are reducible and suggests another possible reduction: by imposing (1.95) we single out another irreducible representation, the Majorana one. The Majorana representation is the minimal irreducible representation of a (one out of eight) covering of the *complete* Lorentz group. It is evident, and well-known, that Majorana and Weyl representations in 4d are incompatible.

Let us consider next charge conjugation and parity, and recall the relevant properties of a Weyl fermion. We have

$$\mathcal{C}\psi_L\mathcal{C}^{-1} = P_L\mathcal{C}\psi\mathcal{C}^{-1} = P_L\hat{\psi} = \hat{\psi}_L. \tag{1.97}$$

The charge conjugate of a Majorana field is, by definition, itself. While a Majorana field is invariant under charge conjugation, for a Weyl fermion, we have

$$\mathcal{C}\left(\int i\overline{\psi_L}\gamma^\mu\partial_\mu\psi_L\right)\mathcal{C}^{-1} = \int i\overline{\hat{\psi}_L}\gamma^\mu\partial_\mu\hat{\psi}_L = \int i\overline{\psi_R}\gamma^\mu\partial_\mu\psi_R \tag{1.98}$$

i.e. a Weyl fermion is, so to say, maximally non-invariant.

---

[2] Here and in the sequel we use $\overline{\psi}$ and $\overline{\psi}$, $\hat{\psi}$ and $\widehat{\psi}$ in a precise sense, which is hopefully unambiguous. For instance, $\overline{\psi} = \psi^\dagger\gamma_0$, $\overline{\psi_R} = (\psi_R)^\dagger\gamma_0$, $\widehat{\psi_R} = \gamma_0 C(\psi_R)^*$, $\hat{\psi}_R = P_R\hat{\psi}$, and so on.

The parity operation is defined by

$$\mathcal{P}\psi_L(t,\vec{x})\mathcal{P}^{-1} = \eta_P \gamma_0 \psi_R(t,-\vec{x}) \tag{1.99}$$

where η_P is a phase, which in the sequel we set to 1. In terms of the action, we have

$$\mathcal{P}\left(\int \overline{\psi}_L \gamma^\mu \partial_\mu \psi_L\right)\mathcal{P}^{-1} = \int \overline{\psi}_R \gamma^\mu \partial_\mu \psi_R. \tag{1.100}$$

For a Majorana fermion the action is invariant under parity.

This also suggests a useful representation for a Majorana fermion. Let $\psi_R = P_R\psi$ be a generic Weyl fermion. We have $P_R\psi_R = \psi_R$ and it is easy to prove that $P_L\widehat{\psi_R} = \widehat{\psi_R}$, i.e. $\widehat{\psi_R}$ is left-handed. Therefore, the sum $\psi_M = \psi_R + \widehat{\psi_R}$ is a Majorana fermion because it satisfies (1.95). And any Majorana fermion can be represented in this way. This representation is instrumental in the calculus of anomalies, see below.

Considering next CP, from the above it follows that the action of a Majorana fermion is obviously invariant under it. On the other hand for a Weyl fermion, we have

$$\mathcal{CP}\psi_L(t,\vec{x})(\mathcal{CP})^{-1} = \gamma_0 P_R \hat{\psi}(t,-\vec{x}) = \gamma_0 \hat{\psi}_R(t,-\vec{x}). \tag{1.101}$$

Applying, now, CP to the action for a Weyl fermion, one gets

$$\begin{aligned}&\mathcal{CP}\left(\int i\overline{\psi_L}\gamma^\mu \partial_\mu \psi_L\right)(\mathcal{CP})^{-1} \\ &= \int i\overline{\hat{\psi}_R}(t,-\vec{x})\gamma^{\mu\dagger}\partial_\mu \hat{\psi}_R(t,-\vec{x}) = \int i\overline{\hat{\psi}_R}(t,\vec{x})\gamma^\mu \partial_\mu \hat{\psi}_R(t,\vec{x}).\end{aligned} \tag{1.102}$$

But one can prove as well that

$$\int i\overline{\hat{\psi}_R}(t,\vec{x})\gamma^\mu \partial_\mu \hat{\psi}_R(t,\vec{x}) = \int i\overline{\psi_L}(x)\gamma^\mu \partial_\mu \psi_L(x). \tag{1.103}$$

Therefore, the action for a Weyl fermion is CP invariant. As we have seen it is also, separately, T invariant, and, so, CPT invariant. The transformation properties of the Weyl and Majorana spinor fields are summarized in the following table:

$$\begin{array}{lll} & \text{Majorana} & \text{Weyl} \\ P: & \mathcal{P}\psi(t,\vec{x})\mathcal{P}^{-1} = \gamma_0\psi(t,-\vec{x}) & \mathcal{P}\psi_L(t,\vec{x})\mathcal{P}^{-1} = \gamma_0\psi_R(t,-\vec{x}) \\ C: & \mathcal{C}\psi\mathcal{C}^{-1} = \gamma_0 C\psi^* = \psi & \mathcal{C}\psi_L\mathcal{C}^{-1} = P_L\hat{\psi} = \hat{\psi}_L \\ \text{CP}: & \mathcal{CP}\psi(t,\vec{x})(\mathcal{CP})^{-1} = \psi(t,-\vec{x}) & \mathcal{CP}\psi_L(t,\vec{x})(\mathcal{CP})^{-1} = \gamma_0\hat{\psi}_R(t,-\vec{x}) \end{array} \tag{1.104}$$

The quantum interpretation of the field ψ_L starts from the plane wave expansion

$$\psi_L(x) = \int d^4p \left(a(p)u_-(p)\mathrm{e}^{-ipx} + b^\dagger(p)v_+(p)\mathrm{e}^{ipx}\right) \tag{1.105}$$

where u_-, v_+ are fixed and independent left-handed spinors (there are only two of them). Such spin states have been explicitly constructed above. We interpret (1.105) as follows: $b^\dagger(p)$ creates a left-handed particle while $a(p)$ destroys a left-handed particle with negative helicity (because of the opposite momentum). However Eqs. (1.101,1.103) force us to identify the latter with a right-handed antiparticle: C maps particles to antiparticles, while P invert helicities, so CP maps left-handed particles to right-handed antiparticles. One needs not stress that in this game right-handed particles or left-handed antiparticles are absent.

Let us focus now on the difference between Weyl fermions and massless Majorana fermions. The question whether a massless Majorana fermion is or is not the same as a Weyl fermion at both the classical and the quantum level is often a source of confusion. In relation to anomalies, it is of utmost importance to clarify this point: if the two concepts were equivalent, the problem of anomalies would be drastically different, since Majorana fermions, as we shall see, cannot carry consistent anomalies. The following note is devoted to clarify this fundamental distinction.

Note on Weyl and massless Majorana fermions. Let us consider the simplest case in which there is no additional quantum number appended to a fermion. To start with, let us recall the obvious differences between the two. The first, and most obvious, has already been mentioned: they belong to two different irreducible representations of the Lorentz group (in 4d there cannot exist a spinor that is simultaneously Majorana and Weyl, unlike in 2 and 10 d). Another important difference is that the helicity for a Weyl fermion is well defined and linked to its chirality, while for a Majorana fermion chirality is undefined, so that the relation with its helicity is also undefined. Next, a parity operation maps the Majorana action into itself, while it maps the Weyl action (1.92) into the same action for the opposite chirality. The same holds for the charge conjugation operator. Moreover, a classical Majorana spinor is a self-conjugated bispinor that can always be chosen to be real and always contains both chiralities in terms of four real independent component functions. From a physical point of view, they describe neutral spin 1/2 objects—not yet detected in Nature—and consequently, there is no phase transformation (U(1) continuous symmetry) involving self-conjugated Majorana spinors, independently of the presence or not of a mass term. Hence, e.g. its particle states do not admit antiparticles of opposite charge, simply because a charge does not exist at all for charge-self-conjugate spinors. The general solution of the wave field equations for a free Majorana spinor always entails the presence of two polarization states with opposite helicity. On the contrary, it is well-known that a chiral Weyl spinor, describing for instance massless neutrinos in the standard model, admits only one polarization or helicity state, it always involves antiparticles of opposite helicity, and it always carries a conserved internal quantum number such as the lepton number, which is opposite for particles and antiparticles.

In conclusion, there is no room for confusing massless Majorana spinors with chiral Weyl spinors. Why they are sometimes considered the same objects may be due to the fact that we can establish a one-to-one correspondence between the components of a Weyl spinor and those of a Majorana spinor in such a way that the Lagrangian, in two-component notation, looks the same. If, in the chiral representation, we write ψ_L as $\begin{pmatrix}\omega\\0\end{pmatrix}$, where ω is a two-component spinor, then (1.92) above becomes

$$i\omega^\dagger\sigma^\mu\partial_\mu\omega \tag{1.106}$$

which has the same form (up to an overall factor) as a massless Majorana action, which in the two-component formalism takes the form

$$\widehat{\mathcal{L}}_M = \frac{1}{2}\,\chi^\dagger(x)\,\sigma^\mu\, i\partial_\mu\chi(x) + c.c. \tag{1.107}$$

see (1.84) and (1.85). But the action is not everything in a theory, it must be accompanied by a set of specifications. Even though numerically the two actions coincide, the way they respond to a variation of the Weyl and Majorana fields is different. One leads to the Weyl equation of motion, the other to the Majorana equation of motion. The delicate issue is precisely this: when we take the variation of an action with respect to a field in order to extract the equations of motion, we must make sure that the variation respects the symmetries and the properties that are expected in the equations of motion. Specifically in the present case, if we wish the equation of motion to preserve chirality we must use variations that preserve chirality, i.e. variations that are eigenfunctions of γ_5. If instead we wish the equation of motion to transform in the Majorana representation we have to use variations that transform suitably, i.e. variations that are eigenfunctions of the charge conjugation operator. If we do so, we obtain two different results which are irreducible to each other no matter which action we use.

Finally, and most importantly, in the quantum theory, a crucial role is played by the functional measure, which is different for Weyl and Majorana fermions. This point will be discussed later on. Before ending this section, we would like to clarify the issue of the already mentioned U(1) continuous symmetry of Weyl fermions, which is sometime confused with the axial $\mathbb{R}$ symmetry of massless Majorana fermions and assumed to justify the identification of Weyl and massless Majorana fermions. We have already discussed above this point. Let us see it from another point of view. To start with let us consider a free massless Dirac fermion ψ. Its free action is clearly invariant under the transformation $\delta\psi = i(\alpha + \gamma_5\beta)\psi$, where α and β are real numbers. This symmetry can be gauged by minimally coupling ψ to a vector potential V_μ and an axial potential A_μ, in the combination $V_\mu + \gamma_5 A_\mu$, so that α and β become arbitrary real functions. For convenience, let us choose the Majorana representation for gamma matrices (see previous section), so that all of them, including γ_5, are imaginary. If we now impose ψ to be a Majorana fermion, its four-component can be chosen to be real and only the symmetry parametrized by β makes sense in the action (let us call it β symmetry). If instead we impose ψ to be Weyl, say $\psi = \psi_L$, then, since $\gamma_5\psi_L = -\psi_L$, the symmetry transformation will be $\delta\psi_L = i(\alpha - \beta)\psi_L$. This fact may be the origin of the confusion, because it looks like we can merge the two parameters α and β into a single parameter identified with the β of the Majorana axial β symmetry. However this is not correct because for a right-handed Weyl fermion the symmetry transformation is $\delta\psi_R = i(\alpha + \beta)\psi_R$. Forgetting β, the Majorana fermion does not transform. Forgetting α, both Weyl and Majorana fermions do transform, but Weyl fermions transform with opposite signs for opposite chiralities. This distinction will become crucial in the computation of anomalies (see below).

1.3 Clifford Algebras and Spinors

At the basis of all the elaborations on spinor fields in quantum field theories is the mathematical concept of Clifford algebra and its representations. Spinors are in fact representations both of a Clifford algebra and of a spin group, which is in turn a subgroup of the Clifford algebra. This section is a short introduction to this subject and to its connection with spin and spinor fields (for more complete accounts see, for instance, [4, 5]).

Given a vector space V with a quadratic form $q(v)$, consider the tensor algebra $\mathcal{T}(V) = \sum_{r=0}^{\infty} \otimes^r V$. The quotient by the ideal $\mathcal{I}_q(V)$ formed by all the elements of the type $v \otimes v - q(v)1$ defines the Clifford algebra

$$C\ell(V, q) = \mathcal{T}(V)/\mathcal{I}_q(V)$$

It can be defined alternatively as the algebra generated by V (and 1), subject to the relation

$$v \cdot w + w \cdot v = 2\, g(v, w) \tag{1.108}$$

for $u, w \in V$, where $2g(v, w) = q(v + w) - q(v) - q(w)$. The dot in (1.108) is the product

$$v{\cdot}w = v \wedge w + g(v, w) \tag{1.109}$$

If the quadratic form is the one defined on the basis elements $\{e_i\}$, $i = 1, \ldots n$ of V by $g(e_i, e_j) = \eta_{ij}$, where η_{ij} is the diagonal matrix with r entries equal to 1 and s entries equal to -1, with $r+s = n$, the relation (1.108) becomes

$$e_i \cdot e_j + e_j \cdot e_i = 2\eta_{ij} \tag{1.110}$$

One easily recognizes the form of the γ-matrix algebra (although the definition (1.108) is more general): in fact, the field theory γ-matrices are a representation of the Clifford algebra. When we refer specifically to the canonical form (1.110), the Clifford algebra will be denoted $C\ell_{r,s}$. These algebras have been classified in terms of fields and real or complex 2×2 matrix rings, $\mathcal{M}(2, \mathbb{R})$ and $\mathcal{M}(2, \mathbb{C})$, respectively. For instance,

$$C\ell_{3,0} = \mathcal{M}(2, \mathbb{C}), \quad C\ell_{3,1} = \mathcal{M}(2, \mathbb{R}) \otimes \mathcal{M}(2, \mathbb{R}), \quad \text{etc.}$$

So far we have considered real Clifford algebras. Complex Clifford algebra (which are the complexification of real ones) are denoted $C\ell_{\mathbb{C}}(n)$. Their classification is rather simple

$$C\ell_{\mathbb{C}}(2k) = \mathcal{M}(2^k, \mathbb{C}), \quad C\ell_{\mathbb{C}}(2k+1) = \mathcal{M}(2^k, \mathbb{C}) \oplus \mathcal{M}(2^k, \mathbb{C})$$

The automorphism of $C\ell(V, q)$ induced by $v \to \tilde{v} = -v$ for $v \in V$ defines a decomposition into even and odd subspaces

$$C\ell(V, q) = C\ell^0(V, q) \oplus C\ell^1(V, q) \tag{1.111}$$

and gives $C\ell(V, q)$ a $\mathbb{Z}_2$-graded algebra structure: $C\ell^i(V, q) \cdot C\ell^j(V, q) \subseteq C\ell^{i+j}(V, q)$, $i, j = 0, 1$.

Inside a Clifford algebra, one can define various groups. The largest one is the group of invertible elements

$$C\ell^{\times}_{r,s} = \{a \in C\ell_{r,s} | \exists\, a^{-1}\} \tag{1.112}$$

An important subgroup is the Clifford-Lipschitz group

$$\mathcal{C}_{r,s} = \{a \in C\ell^{\times}_{r,s} | \, a\, v\, a^{-1} \in V, \;\; \forall v \in V = \mathbb{R}^{r,s}\} \tag{1.113}$$

This definition is based on the adjoint action in V. We can extend it to the full Clifford algebra

$$\sigma_a(x) = a\, x\, a^{-1}, \quad \forall a \in \mathcal{C}_{r,s}, \; \forall x \in C\ell_{r,s} \tag{1.114}$$

σ_a is an automorphism of $\mathcal{C}_{r,s}$. A more instrumental operation is the twisted adjoint

$$\widetilde{\sigma}_a(x) = \tilde{a}\, x\, a^{-1} \tag{1.115}$$

where $\tilde{a}$ has been defined before.

Let us define the group

$$O(V, q) = \{t \in GL(V) : t^* q = q\}$$

where $t^* q$ is the pullback of q by $t : V \to V$. I.e. $O(V, q)$ is the group of linear invertible transformations of V that leave the quadratic form q unchanged. And let us define the subgroup

$$SO(V, q) = \{t \in O(V, q) : \det(t) = 1\}$$

As it turns out, we have the correspondences

$$\widetilde{\sigma}(\mathcal{C}_{r,s}) = O(r, s), \quad \widetilde{\sigma}(\mathcal{C}^0_{r,s}) = SO(r, s) \tag{1.116}$$

where $\mathcal{C}^0_{r,s} = \mathcal{C}_{r,s} \cap C\ell^0_{r,s}$ is the even Clifford subalgebra.

Now let us come to the spin groups. We can define a norm $N(a)$ for any element $a \in C\ell(V, q)$, such that $N(v) = q(v)$ for any $v \in V$. Then the spin group of V endowed with a quadratic form q is

$$\mathrm{Spin}(V, q) = \{a \in \mathcal{C}^0_{r,s} | N(a) = \pm 1\} \tag{1.117}$$

The restricted spin group $\mathrm{Spin}^+(V, q)$ is identified by the condition $N(a) = 1$.

One can prove that the following sequence is exact

$$0 \longrightarrow \mathbb{Z}_2 \longrightarrow \mathrm{Spin}(V, q) \xrightarrow{\widetilde{\sigma}} SO(V, q) \longrightarrow 0$$

where the twisted adjoint map $\widetilde{\sigma}$, for any $v, w \in V$, can also be represented as

$$\widetilde{\sigma}_v(w) = w - 2\frac{q(v, w)}{q(v)}v \tag{1.118}$$

By the above exact sequence $Spin(V, q)$ is a double covering of $SO(V, q)$.

A $\mathbb{K}$ ($\mathbb{K}$=$\mathbb{C}$ or $\mathbb{R}$) representation of $C\ell(V, q)$ is a $\mathbb{K}$-algebra homomorphism

$$C\ell(V, q) \longrightarrow Hom_{\mathbb{K}}(W, W)$$

of $C\ell(V, q)$ into the linear transformations of a finite dimensional vector space W. W is called a $\mathbb{K}$-*module* of $C\ell(V, q)$. Clearly, this defines also representations of $\mathrm{Spin}(V, q)$ and $SO(V, q)$ as groups contained in $C\ell(V, q)$. In particular $C\ell(V, q)$ can be itself a representation space for the adjoint or twisted adjoint action defined above. Of course, this will be a representation only of $C\ell^{\times}(V, q)$ or of its subgroups $\mathrm{Spin}(V, q)$ and $SO(V, q)$.

We are now ready to define spinors. Among various definitions that can be met in the literature, we choose the following one:

- *Given a vector space $V = \mathbb{R}^{r,s}$ with canonical quadratic form q and the associated Clifford algebra $C\ell(V, q)$, a spinor is any element of an irreducible representation space of the associated spin group* $\mathrm{Spin}(V, q)$.

A classification of spin representations can be found in the literature. Since, in the anomaly problems considered in this book, we are interested in complex representations, we limit ourselves to them. They are as follows

- $n = 2k, \quad \mathbb{C}^{2^{k-1}} \oplus \mathbb{C}^{2^{k-1}}$
- $n = 2k + 1, \quad \mathbb{C}^{2^k}$

($\mathbb{C}$ is the field of complex numbers).

The meaning of the decomposition in two irreducible representations when $n = 2k$ is related to the volume element. In the complex case, the latter is

$$\omega_{\mathbb{C}} = i^k e_1 \dots e_{2k}, \tag{1.119}$$

where e_i is an orthonormal basis of $V = \mathbb{R}^{r,s}$. It has the property $\omega_{\mathbb{C}}^2 = 1, \forall k$. The two irreducible representations correspond to the eigenvalues ± 1 of $\omega_{\mathbb{C}}$ and are identified by the projectors $\pi_{\pm} = \frac{1}{2}(1 \pm \omega_{\mathbb{C}})$.

The element $\omega_{\mathbb{C}}$ is evidently an alias of the chirality operator in quantum field theory.

1.3.1 Spinor Representations in Even Dimension

In this book, the main interest is in spacetimes of even dimensions. The properties of spinors in such spaces are summarized in Table 1.1:

Table 1.1 Spinors in even dimension

Metric sign.	Weyl, $\mathbb{C}$		Conjugacy	Dirac, $\mathbb{C}$	Majorana-Weyl, $\mathbb{R}$		Majorana, $\mathbb{R}$
(2,0)	1L	1R	Mutual	2	–	–	2
(1,1)	1L	1R	Self	2	1L	1R	2
(4,0)	2L	2R	Self	4	–	–	–
(3,1)	2L	2R	Mutual	4	–	–	4
(6,0)	4L	4R	Mutual	8	–	–	8
(5,1)	4L	4R	Self	8	–	–	–
(8,0)	8L	8R	Self	16	8L	8R	16
(7,1)	8L	8R	Mutual	16	–	–	–
(10,0)	16L	16R	Mutual	32	–	–	32
(9,1)	16L	16R	Self	32	16L	16R	32

where L, R denotes left-handed or right-handed spinors; the numbers in the columns with $\mathbb{C}$ or $\mathbb{R}$ tag denote the number of complex or real components according to whether they are in column $\mathbb{C}$ or $\mathbb{R}$, respectively. 'Self' means self-conjugate and 'mutual' means that the L and R representations are conjugate to each other.

1.4 Minkowski Versus Euclidean

We are now ready to speak about the Wick rotation, an omnipresent technique in quantum field theory. Notwithstanding the results of Osterwalder and Schrader and others, [6–10], if there is an aspect of quantum field theory that may sound unattractive or even hardly acceptable to an external observer, and conveys a sense of incompleteness to any conscious practitioner, this is the frequent, if not continuous, recourse to the Wick rotation, i.e. to the passage from the original Minkowski spacetime, with Lorentzian background metric, to the Euclidean counterpart, and vice-versa. Field theories of physical interest (except those describing stationary statistical phenomena) are formulated in a Minkowski spacetime, and involve hyperbolic quantities, equations and integrals, of which we do not have yet full control. In order to assign an unambiguous meaning to the path integral and the Feynman integrals, we resort to analytic continuation: we complexify coordinates and momenta, for instance, by rotating the time axis by ninety degrees in the anticlockwise direction, counting on the (generally assumed) absence of singularities in the swept quadrants. This usually produces mathematically well-defined objects, which are assumed by definition to be 'the right thing'. The evaluation of anomalies is not exempt of such heuristic acrobatics, because it involves quantum corrections. There is, as yet, no universally recognized procedure to make a Wick rotation when fermions are involved, there are in fact several different recipes (see the bibliography below). In this book, we would like to be as practical as possible. Hereafter, we explain the minimalistic practical recipe we will be using throughout.

1.4.1 Wick Rotation and Fermions

We use a metric $\eta_{\mu\nu}$ with mostly $(-)$ signature. The gamma matrices satisfy $\{\gamma^\mu, \gamma^\nu\} = 2\eta^{\mu\nu}$ and

$$\gamma_\mu^\dagger = \gamma_0\gamma_\mu\gamma_0.$$

The generators of the Lorentz group are $\Sigma_{\mu\nu} = \frac{1}{4}[\gamma_\mu, \gamma_\nu]$.

The chiral matrix $\gamma_5 = i\gamma^0\gamma^1\gamma^2\gamma^3$ has the properties

$$\gamma_5^\dagger = \gamma_5, \quad (\gamma_5)^2 = 1, \quad C^{-1}\gamma_5 C = \gamma_5^T.$$

and

$$\mathrm{tr}(\gamma_5\gamma_\mu\gamma_\nu\gamma_\lambda\gamma_\rho) = -4i\varepsilon_{\mu\nu\lambda\rho} \tag{1.120}$$

A Wick rotation means[3] : $x^0 \to \tilde{x}^0 = ix^0$, $k^0 \to \tilde{k}^0 = ik^0$ and $\gamma^0 \to \tilde{\gamma}^0 = i\gamma^0$, while x^i, k_i, γ^i remain unchanged. From now on a tilde represents a Euclidean object. We have in particular

$$\{\tilde{\gamma}^\mu, \tilde{\gamma}^\nu\} = 2\tilde{\eta}^{\mu\nu} = -2\delta^{\mu\nu} \tag{1.121}$$

and

$$\tilde{\gamma}_\mu^\dagger = -\tilde{\gamma}_\mu \tag{1.122}$$

A naive analytic continuation gives also $\tilde{\gamma}_5 = i\gamma_5$. Therefore,

$$\tilde{\gamma}_5^\dagger = -\tilde{\gamma}_5, \quad \tilde{\gamma}_5^2 = -1, \quad \mathrm{tr}(\tilde{\gamma}_5\tilde{\gamma}_\mu\tilde{\gamma}_\nu\tilde{\gamma}_\lambda\tilde{\gamma}_\rho) = -4i\varepsilon_{\mu\nu\lambda\rho} \tag{1.123}$$

A consequence of the above is, however, that, while $P_R = \frac{1+\gamma_5}{2}$ is a projector: $P_R^2 = P_R$, its 'analytic continuation' $\frac{1+\tilde{\gamma}_5}{2}$ is not! Therefore, a Weyl spinor in a 4d Euclidean spacetime cannot be defined by means of this simple-minded Wick rotation starting from $\psi_R = \frac{1+\gamma_5}{2}\psi$, where ψ is Dirac fermion. This can be argued also from general properties of Weyl fermions in even dimensions $\mathsf{d} = 2k$: the Weyl fermions in dimension $(2k-1, 1)$ have a different nature than the Weyl fermions in $(2k, 0)$ dimensions, although they have the same number of complex components (see table

---

[3] This is one of the possible recipes for Wick rotations. It is also possible to rotate the space directions, as we will do in the next subsection.

above). While in (3, 1) dimensions, the two Weyl spinors are conjugate to each other; in (4, 0) dimensions, they are self-conjugate. The above rules, no doubt, need a refinement. More refined recipes for Wick rotation have been proposed which prescribe, for instance, also a rotation of the spinor fields themselves. We would like to avoid (for most of the book at least) these complications and introduce a minimum of modifications.

It is clear that if we wish to Wick rotate an expression that contains γ matrices, we can proceed straightforwardly as above for γ^μ, i.e. $\gamma^\mu \to \tilde{\gamma}^\mu$, but for γ_5 we have to replace it with $\widehat{\widetilde{\gamma}}_5 = i\tilde{\gamma}_5 = -\gamma_5$, which, of course, has the properties

$$\widehat{\widetilde{\gamma}}_5^\dagger = \widehat{\widetilde{\gamma}}_5, \quad \widehat{\widetilde{\gamma}}_5^2 = 1, \quad \mathrm{tr}(\widehat{\widetilde{\gamma}}_5\tilde{\gamma}_\mu\tilde{\gamma}_\nu\tilde{\gamma}_\lambda\tilde{\gamma}_\rho) = 4\varepsilon_{\mu\nu\lambda\rho} \tag{1.124}$$

We remark that the algebra properties

$$\{\gamma_5, \gamma^\mu\} = 0, \quad \{\widehat{\widetilde{\gamma}}_5, \tilde{\gamma}^\mu\} = 0 \tag{1.125}$$

in any case, do not change. In other words, a sensible Wick rotation must be such that, while the γ_μ are analytically continued as above, γ_5 remain unchanged. A Wick rotation of this kind can be understood as an additional rotation in the axial direction in the axial-complex enlargement of spacetime considered in Chap. 10. In practice, however, we will avoid a direct use of the prescription (1.124), as explained hereafter.

As we shall see, a Wick rotation is always necessary when using Feynman diagrams in order to assign a meaning to the loop integrals. It is also necessary with non-perturbative methods based on the SDW approach, where an expression like $\int_0^\infty \frac{ds}{is}\mathrm{Tr}\left(e^{i\mathcal{F}s}\right)$, accompanied by the ϵ-prescription, makes sense only if $\mathcal{F}$ is a positive quadratic operator. And, of course, it is needed when applying the family's index theorem, because the latter has been proved in Riemannian spaces.

Our attitude throughout the book in regard to Wick rotation is defined by a minimal set of rules. We have mostly in mind Minkowski field theories and we wish to derive results in a Minkowski background, so the purpose of Wick rotations is not to emigrate to corresponding full-fledged Euclidean field theories and carry out the calculations there. Our purpose is to give a definite mathematical meaning to our manipulations. So, in the case of Feynman diagrams, we limit ourselves to operate the Wick rotation only on the relevant loop integrals so as to transform them into convergent ones; in the case of heat kernel-like methods, we perform the Wick rotation on the relevant squared kinetic operators, the basic objects in that trade, so as to define well-behaved positive elliptic operators. Once this is done, and a set of rules for calculations is defined via the Euclidean operator, we shall replace it with the Minkowski one. By our experience, this is the safest way to avoid miscalculations. Finally in the family's index theorem, we Wick rotate only the Dirac operator itself, because the theorem deals with the operator rather than with the fermionic action. In all these cases, we do not pretend to answer the question: what are the Euclidean field theories subtended by these Wick rotations. The question what Euclidean field theory can be formulated with these Wick rotations is not relevant to us (with the exception of Sect. 13.6). This 'halfway' attitude may not be mathematically satisfactory, but it is heuristically the simplest thing to do. The results we obtain are consistent among themselves and with the best pre-existing literature. We leave to a future more advanced stage of the field theory analysis the task to justify all this.

This simple approach to the Wick rotation technique is enough for our purposes in most of this book. One can consider however Euclidean field theories in their own right, as an autonomous object of investigation. This is, of course, a vast subject, and it is impossible to exhaust it in a satisfactory way in a short section. However we think a short introduction is necessary, in part to justify some of the previous statements which might seem otherwise too peremptory, in part to set the ground for some aspects of the global anomaly discussion in Chap. 13.

1.5 Euclidean Fermion Field Theories

This is a short introduction to Euclidean field theory for fermions, with the warning that the problem of the transition from a Minkowski fermion theory to a Euclidean counterpart does not admit a unique solution. Below we outline a simple approach.

Let us start from the theory of a massive Dirac fermion and its spinor causal Green's function, or Schwinger function.[4] We can safely perform the natural replacements

$$ip_0 = p_4 \quad ix_0 = x_4 \tag{1.126}$$

so as to obtain a positive definite denominator in the Fourier transform, that is

$$S^F_{\alpha\beta}(-ix_4, \mathbf{x}) = i\,(2\pi)^{-4} \int \mathrm{d}\mathbf{p} \int_{-\infty}^{\infty} \mathrm{d}p_4\, \frac{\exp\{ip_4x_4 + ip_jx_j\}}{p_4^2 + \mathbf{p}^2 + M^2}\, (\gamma^0 p_4 - i\,\gamma^k p^k + iM\,)_{\alpha\beta}$$

Next, we define the *Euclidean Dirac Matrices*

$$\overline{\gamma}_4 \equiv \gamma^0, \quad \overline{\gamma}_k \equiv i\,\gamma_k = -i\,\gamma^k \quad (k = 1, 2, 3) \tag{1.127}$$

$$\overline{\gamma}_\mu = \left(\overline{\gamma}_k, \overline{\gamma}_4\right) = \overline{\gamma}_\mu^\dagger, \quad \left\{\overline{\gamma}_\mu, \overline{\gamma}_\nu\right\} = 2\,\delta_{\mu\nu} \tag{1.128}$$

Then we can write

$$\begin{aligned} S^F_{\alpha\beta}(-ix_4, \mathbf{x}) &= \frac{i}{(2\pi)^4} \int \mathrm{d}^4p_E\, \exp\{ip_E \cdot x_E\} \left[\frac{\overline{\gamma}_\mu p_{E\mu} + iM}{p_E^2 + M^2}\right]_{\alpha\beta} \\ &= \frac{i}{(2\pi)^4} \int \mathrm{d}^4p_E\, \exp\{ip_E \cdot x_E\} \left[\frac{1}{\not{p}_E - iM}\right]_{\alpha\beta} \\ &\equiv -\,S^E_{\alpha\beta}(x_E) \end{aligned} \tag{1.129}$$

[4] The n−point Green's functions in a Euclidean spacetime are usually named n−point Schwinger functions.

where we understand

$$\mathbf{p} = (p^1, p^2, p^3) = (p_x, p_y, p_z) = \vec{p}_E = (\bar{p}_1, \bar{p}_2, \bar{p}_3)$$

together with

$$\begin{aligned} p_4\overline{\gamma}_4 + \overline{\gamma}_k \bar{p}_k &= \overline{\gamma}_\mu p_{E\mu} \equiv \not{p}_E \\ \not{\partial}_E \equiv \overline{\gamma}_\mu \partial_{E\mu} &= \overline{\gamma}_k \partial_k + \overline{\gamma}_4 \partial_4 \end{aligned}$$

The above move, however, is not cost-free. The Schwinger function (1.129) conflicts with hermiticity. The way out suggested by Osterwalder and Schrader, which ensures reflection positivity, is to assume that the proper classical variables for a Euclidean formulation of the Dirac spinor field theory, should be two distinct Euclidean bispinors, which we may denote again ψ_E and $\bar{\psi}_E$, but obey

$$\{\psi_E(x),\, \psi_E(y)\} = \{\bar{\psi}_E(x),\, \bar{\psi}_E(y)\} = \{\psi_E(x),\, \bar{\psi}_E(y)\} = 0 \tag{1.130}$$

for all points x and y of the four-dimensional Euclidean space $\mathbb{R}^4$, while there is a non-trivial equal time canonical bracket between ψ_E and the Hermitean conjugate of $\overline{\psi}_E$. The last of the (1.130) relations is crucial, for it implies that $\bar{\psi}_E$ does not coincide with the Hermitean conjugate of ψ_E times some matrix γ_4. Thus, if we want to set up a meaningful Euclidean formulation for the Dirac spinor field theory, then we can treat ψ_E and $\bar{\psi}_E$ as *totally independent classical Grassmann valued fields* . This independence is the main novelty of the Euclidean fermion field theory, the rest of the construction being straightforward.

For instance, we use the definition of the Euclidean matrices $\overline{\gamma}_\mu$ to derive the $O(4)$ transformation law for ψ_E in the usual way. As a matter of fact, we have the six Hermitean generators

$$\Sigma_{\mu\nu} = \frac{i}{4}\left[\overline{\gamma}_\mu, \overline{\gamma}_\nu\right] \quad \Sigma_\imath \equiv \varepsilon_{\imath\jmath k}\Sigma_{\jmath k} \quad (\imath, \jmath, k = 1, 2, 3) \tag{1.131}$$

$$\Sigma_{23} \equiv \Sigma_1 = \tfrac{1}{2}\begin{pmatrix} \sigma_1 & 0 \\ 0 & \sigma_1 \end{pmatrix} \quad \Sigma_{31} \equiv \Sigma_2 = \tfrac{1}{2}\begin{pmatrix} \sigma_2 & 0 \\ 0 & \sigma_2 \end{pmatrix} \tag{1.132}$$

$$\Sigma_{12} \equiv \Sigma_3 = \tfrac{1}{2}\begin{pmatrix} \sigma_3 & 0 \\ 0 & \sigma_3 \end{pmatrix} \quad \Sigma_{k4} \equiv \overline{\Sigma}_k = \tfrac{1}{2}\begin{pmatrix} \sigma_k & 0 \\ 0 & -\sigma_k \end{pmatrix} \tag{1.133}$$

$so(4)$−`Lie algebra`. The six $O(4, \mathbb{R})$ generators satisfy the following Lie algebra commutation relations, *viz.*,

$$[\,\Sigma_\jmath,\, \Sigma_k\,] = i\varepsilon_{\jmath k\ell}\,\Sigma_\ell = [\,\overline{\Sigma}_\jmath,\, \overline{\Sigma}_k\,] \qquad [\,\Sigma_\jmath,\, \overline{\Sigma}_k\,] = i\varepsilon_{\jmath k\ell}\,\overline{\Sigma}_\ell$$

which can be suitably recast into the four-vector notations to yield

$$[\,\Sigma_{\mu\nu},\, \Sigma_{\kappa\lambda}\,] = i\delta_{\mu\kappa}\,\Sigma_{\nu\lambda} - i\delta_{\lambda\mu}\,\Sigma_{\nu\kappa} + i\delta_{\nu\lambda}\,\Sigma_{\mu\kappa} - i\delta_{\kappa\nu}\,\Sigma_{\mu\lambda} \tag{1.134}$$

It follows that we have the following orthogonal transformation rules for the two-component spinors and four-component bispinors of the Euclidean formulation,

$$\psi'_{\uparrow}(x'_E) = \exp\{(i/2)\,\vec{\sigma}\cdot(\vec{\alpha}+\vec{\eta})\}\,\psi_{\uparrow}(x_E) \tag{1.135}$$
$$\psi'_{\downarrow}(x'_E) = \exp\{(i/2)\,\vec{\sigma}\cdot(\vec{\alpha}-\vec{\eta})\}\,\psi_{\downarrow}(x_E) \tag{1.136}$$

with $x'_E = R\,x_E,\ R \in O(4)$[5] while $0 \le |\,\vec{\alpha}\,| < 2\pi$ and $0 \le |\,\vec{\eta}\,| < 2\pi$, that yields

$$\psi_E(x_E) = \begin{Bmatrix} \psi_{\uparrow}(x_E) \\ \psi_{\downarrow}(x_E) \end{Bmatrix}$$

$$\psi'(x'_E) = \exp\left\{(i/2)\Sigma_{\mu\nu}\theta_{\mu\nu}\right\}\psi_E(x_E) \tag{1.137}$$
$$\theta_{k4} = -\,\eta_k \quad \theta_{\imath\jmath} = -\,\varepsilon_{\imath\jmath k}\,\alpha_k \tag{1.138}$$

It is crucial to gather that the upper and lower spinor components of a bispinor transform according to equivalent fundamental representations $D_{\frac{1}{2}}$ of SU(2), at variance with left and right Weyl spinors on the Minkowski space, so that the Euclidean bispinor $\psi_E(x_E)$ belongs to the 4d reducible representation $D_{\frac{1}{2}} \oplus D_{\frac{1}{2}}$. Thus, the very concept of non-equivalent Weyl spinors of different chirality is lost in this Euclidean formulation of the spinor field theory. Moreover, $\overline{\psi}_E$ *transforms just like the Hermitean conjugated of* ψ_E, *by definition.* As a matter of fact, from the unitary property

$$U(\theta) = \exp\{(i/2)\,\overline{\Sigma}_{\mu\nu}\,\theta_{\mu\nu}\} \qquad U^{\dagger}(\theta) = U^{-1}(\theta) = U(-\,\theta) \tag{1.139}$$

we have

$$\left(\overline{\psi}'_E,\psi'_E\right) = \int \mathrm{d}^4x'_E\,\overline{\psi}'_E(x'_E)\,\psi'_E(x'_E) = \int \mathrm{d}^4x_E\,\overline{\psi}_E(x_E)\,U^{\dagger}(\theta)\,U(\theta)\,\psi_E(x_E) = (\,\overline{\psi}_E,\psi_E\,)$$

with $x'_E = R\,x_E,\ R^{\top} = R^{-1}$. In this regard, let us focus on the conversion of the Minkowski space chiral matrix

$$\gamma_5 = i\gamma_0\gamma^1\gamma^2\gamma^3 = \gamma_5^{\dagger} \quad \{\gamma_5,\gamma^{\mu}\} = 0$$

The standard definition in the Euclidean $D-$dimensional Clifford algebra is

$$\overline{\gamma}_5 \equiv \overline{\gamma}_1\overline{\gamma}_2\overline{\gamma}_3\overline{\gamma}_4 = -\,\gamma_5 \quad \{\overline{\gamma}_5,\overline{\gamma}_{\mu}\} = 0 \tag{1.140}$$

in such a manner that the projector on the upper and lower components of a Euclidean bispinor become

$$\mathbb{P}_{\uparrow} = \tfrac{1}{2}(1+\overline{\gamma}_5) \quad \mathbb{P}_{\downarrow} = \tfrac{1}{2}(1-\overline{\gamma}_5)$$

---

[5] The orthogonal group $O(4)$ of the rotations in the Euclidean space $\mathbb{R}^4$ is a semi-simple Lie group $O(4)$ isomorphic to $O(3)\times O(3)$.

the upper and lower spinor components belonging to equivalent fundamental representations of SU(2).

The Euclidean action integral for the bispinor field is given by

$$S_E[\,\psi_E, \overline{\psi}_E\,] \;=\; \int \mathrm{d}^4x_E\; \overline{\psi}_E(x_E)\,(\not{\partial}_E + M)\;\psi_E(x_E) \tag{1.141}$$

where ψ_E is a column bispinor, whereas $\overline{\psi}_E$ is an independent transposed or row bispinor. Here the overall sign as well as any overall factor are actually irrelevant and arbitrary, so that we could always absorb them into e.g. $\overline{\psi}_E$—remember that we are allowed to change $\overline{\psi}_E$ and not ψ_E or *vice-versa*. Conversely, the lack of the factor i in front of the derivative term is not at all conventional: it is there just to ensure that the Euclidean fermion propagator, a.k.a. the 2-point *Schwinger's function*, is proportional to $(i\not{p}_E - M)/(p_E^2 + M^2)$; if it were not for this i, then we would have tachyon poles after transition back to the momentum space.

It is worth to notice that the above spinor Euclidean action integral can be obtained from the corresponding one in the Minkowski space, after the customary standard replacements (1.126) and (1.127)

$$x_4 = i\,x_0, \quad \overline{\gamma}_k \;\equiv\; -\,i\gamma^k \quad (k = 1, 2, 3\,), \quad \overline{\gamma}_4 \;\equiv\; \gamma_0$$

$$\psi(x) \;\mapsto\; \psi_E\,(x_E)\,, \quad \overline{\psi}(x) \;\mapsto\; \mathrm{e}^{i\theta}\,\overline{\psi}_E(x_E) \quad (\,0 \le \theta < 2\pi\,)$$

in such a manner that we can always set

$$iS[\,\psi, \overline{\psi}\,] \mapsto S_E[\,\psi_E, \overline{\psi}_E\,] = \int \mathrm{d}^4x_E\; \overline{\psi}_E(x_E)\;(\,\not{\partial}_E + M\,)\;\psi_E(x_E) \tag{1.142}$$

Furthermore, the Euclidean Dirac operator $(\,\not{\partial}_E + M\,)$ is precisely that one which gives, according to the definition (1.129), the 2-point Schwinger function inversion formula

$$(\,\not{\partial}_E + M\,)_{\alpha\beta}\; S^E_{\beta\eta}(x_E) \;=\; \delta\,(x_E)\;\delta_{\alpha\eta}$$

Finally we summarize the basic formulas

$$S_E[\,\overline{\psi},\, \psi_E\,] = \int \mathrm{d}^4x_E\; \overline{\psi}(x_E)\,(\,\not{\partial}_E + M\,)\,\psi_E(x_E) \tag{1.143}$$

$$S^F_{\alpha\beta}(\,-ix_4, \mathbf{x}) \;\to\; -\,S^E_{\alpha\beta}(x_E)$$

$$S^E_{\alpha\beta}(x_E) \;=\; \int \frac{\mathrm{d}^4p_E}{(2\pi)^4}\; \exp\{ip_E\cdot x_E\}\; \left[\frac{i}{-\not{p}_E + iM}\right]_{\alpha\beta} \tag{1.144}$$

$$(\,\not{\partial}_E + M\,)_{\alpha\beta}\; S^E_{\beta\eta}(x_E) \;=\; \delta(x_E)\,\delta_{\alpha\eta}$$

It is important to highlight that the Euclidean action integral for the bispinor field is by no means a real quantity.

What we have just presented is a possible solution for a Euclidean field theory obtained via a Wick rotation from a Minkowski field theory. The conclusion is that as long as we consider the prescription introduced here, we end up with the doubling of the fermionic fields: the $\bar{\psi}$ in the action has to be replaced with a bispinor independent of ψ; moreover, we have to abandon the expectation of a real action. In addition to this presentation, there is no room for a Euclidean field theory of Weyl fermions. There are however more refined prescriptions, [11–15], which, for instance, accompany the Wick rotation with a rotation of the spinor fields. In these new approaches, one can impose hermiticity on the action for a Euclidean Dirac spinor. It is also possible to define a Euclidean action for Weyl spinors, but it remains impossible to comply with hermiticity: in the latter case, the action is not Hermitean and, as a consequence, the integrand in the path integral is complex. This point will become relevant in Sect. 13.6.

References

1. E. Merzbacher, *Quantum Mechanics* (Wiley, 1961)
2. M.E. Peskin, D.V. Schroeder, *An Introduction to Quantum Field Theory* (Westview Press, 1995)
3. N.N. Bogoliubov, D.V. Shirkov, *Introduction to the Theory of Quantized Fields* (Interscience Publisher, New York, 1959)
4. H.B. Lawson Jr., M.-L. Michelsohn, *Spin Geometry* (Princeton University Press, Princeton, 1989)
5. J. Vaz Jr., R. da Rocha Jr., *An Introduction to Clifford Algebras and Spinors* (Oxford University Press, Oxford, 2016)
6. J. Schwinger, Euclidean quantum electrodynamics. Phys. Rev. **115**, 721, and 117 (1960) 1407; On the Euclidean structure of relativistic field theory. Proc. Nat. Acad. Sci. **44** (1958). 956 (1959)
7. K. Osterwalder, R. Schrader, Feynman-Kac formula for Euclidean Fermi and Bose fields. Phys. Rev. Lett. **29**, 1423 (1972); Euclidean Fermi fields and a Feynman-Kac formula for boson-fermion models. Helv. Phys. Acta **46**, 277 (1973); Axioms for Euclidean Green's functions. Comm. Math. Phys. **31**, 83 (1973) and Axioms for Euclidean Green's functions. 2. Comm. Math. Phys. **42**, 281 (1975). J. Fröhlich, K. Osterwalder, Is there a Euclidean field theory for fermions. Helv. Phys. Acta **47**, 781 (1974)
8. B. Zumino, Euclidean supersymmetry and the many-instanton problem. Phys. Lett. **69B**, 369 (1977)
9. H. Nicolai, A possible constructive approach to (SUPER ϕ**3) in four-dimensions. 1. Euclidean formulation of the model. Nucl. Phys. **B 140**, 294–300 (1978)
10. J. Lukierski, A. Nowicki, On superfield formulation of Euclidean supersymmetry. J. Math. Phys. **25**, 2545 (1984)
11. M.R. Mehta, Euclidean continuation of the dirac fermion. Phys. Rev. Lett. **65**, 1983 (1990)
12. J. Kupsch, W.D. Thacker, Euclidean Majorana and Weyl Spinors. Fortsch. Phys. **38**, 35 (1990). Z. Haba, J. Kupsch, Supersymmetry in Euclidean quantum field theory. Fortsch. Phys. **43**, 41–66 (1995)
13. P. van Nieuwenhuizen, A. Waldron, On Euclidean spinors and wick rotations. Phys. Lett. **B389**, 29 (1996). e-print: 9608174 [hep-th]
14. C. Wetterich, Spinors in Euclidean field theory, complex structures and discrete symmetries. Nucl. Phys. **B 852**, 174–234 (2011). e-Print:1002.3556 [hep-th]
15. R.S. Streator, A.S. Wightman, *PCT, Spin, Statistics, and all that* (Princeton University Press, Princeton, 2000)

Chapter 2
Classical and BRST Symmetries

In this chapter, we introduce the classical symmetries whose anomalies we wish to analyze in the next chapters: classical gauge symmetries, diffeomorphisms and local Lorentz transformations. Each symmetry has a corresponding conserved current. After quantization, they change nature and become BRST symmetries. For each of them, we write down the (anticommuting) transformations and show that they are nilpotent. They consequently define nilpotent functional operators, which later on will be identified with coboundary operators.

From now on, we use the natural units $\hbar = 1, c = 1$, unless otherwise specified. References and further readings for this chapter are [1–8].

2.1 Local Gauge Symmetry and Local Gauge Theories

Since the classical local gauge field theories are a well-known subjects, illustrated in a number of textbooks, we will limit ourselves to a short introduction, meant essentially to fix the notation.

2.1.1 Gauge Symmetry and Gauge Theories

A standard (pure) gauge theory is defined by a gauge potential $V_\mu = V_\mu^a T^a$, where T^a are the generators of the Lie algebra $\mathfrak{g}$ of a Lie group G characterizing the theory. We choose the generators to be anti-Hermitean, with commutation relations

$$[T^a, T^b] = f^{abc}T^c, \quad f^{abc} \in \mathbb{R} \tag{2.1}$$

The corresponding field strenght is

$$F_{\mu\nu} = \partial_\mu V_\nu - \partial_\nu V_\mu + [V_\mu, V_\nu] \tag{2.2}$$

L. Bonora, *Fermions and Anomalies in Quantum Field Theories*, Theoretical and Mathematical Physics, https://doi.org/10.1007/978-3-031-21928-3_2

It is often expedient to employ the compact form notation $V = V_\mu \mathrm{d}x^\mu$. It is well-known that V can be viewed as a connection form of a principal fiber bundle with structure group G. This fact will be extensively used later on, but, for the time being, V is simply a useful compact notation. The field strength becomes a two-form $F = \mathrm{d}\,V + V \wedge V = \frac{1}{2} F_{\mu\nu} \mathrm{d}x^\mu \wedge \mathrm{d}x^\nu$, where $\mathrm{d} = \frac{\partial}{\partial x^\mu}\mathrm{d}x^\mu$ is the exterior differential and $\wedge$ denotes the wedge product $\mathrm{d}x^\mu \wedge \mathrm{d}x^\nu = \frac{1}{2}\,(\mathrm{d}x^\mu \otimes \mathrm{d}x^\nu - \mathrm{d}x^\nu \otimes \mathrm{d}x^\mu)$.

The potential V_μ is endowed with an infinitesimal gauge transformation

$$\delta_\lambda V_\mu = \partial_\mu \lambda + [V_\mu, \lambda] \equiv D_\mu \lambda, \quad \lambda = \lambda^a(x) T^a \tag{2.3}$$

so that

$$\delta_\lambda F_{\mu\nu} = [F_{\mu\nu}, \lambda]. \tag{2.4}$$

$\lambda^a(x)$ are infinitesimal arbitrary smooth functions. The corresponding finite gauge transformation takes the form

$$V_\mu \to \mathrm{e}^{-\lambda}(\partial_\mu + V_\mu)\mathrm{e}^{\lambda}, \quad F_{\mu\nu} \to \mathrm{e}^{-\lambda} F_{\mu\nu} \mathrm{e}^{\lambda} \tag{2.5}$$

The Yang-Mills action

$$S[V] = \frac{1}{4g_{YM}^2} \int \mathrm{d}^{\mathsf{d}}x \,\mathrm{tr}\left(F_{\mu\nu} F^{\mu\nu}\right) \tag{2.6}$$

is invariant both under infinitesimal and finite gauge transformations. g_{YM} is the gauge coupling, which is dimensionless only in $\mathsf{d} = 4$.

To quantize this action, we have to fix the gauge and apply the Faddeev-Popov recipe. For instance, if we impose the Lorenz gauge with parameter α and apply the standard approach, the gauge-fixed action becomes

$$S[V, c, B] = \frac{1}{g_{YM}^2} \int \mathrm{d}^{\mathsf{d}}x \left(\frac{1}{4} F_{\mu\nu} F^{\mu\nu} - V^\mu \partial_\mu B + \partial^\mu \bar{c} D_\mu c - \frac{\alpha}{2} B B\right) \tag{2.7}$$

where c, $\bar{c}$ and B are the (FP) ghost, antighost and Nakanishi-Lautrup fields, respectively. c, $\bar{c}$ are anticommuting, while B is commuting.

The action (2.7) is symmetric under the BRST transformations

$$\mathfrak{s} V_\mu = D_\mu c, \quad \mathfrak{s} c = -\frac{1}{2}[c, c], \quad \mathfrak{s}\bar{c} = B, \quad \mathfrak{s} B = 0 \tag{2.8}$$

The second transformation means that $\mathfrak{s} c^a = -\frac{1}{2} f^{abc} c^b c^c$. These transformations are nilpotent, i.e. $\mathfrak{s}^2 = 0$. In particular

$$\mathfrak{s}(D_\mu c) = 0, \quad \mathfrak{s}[c, c] = 0$$

One should not be misled by the previous historical presentation into thinking that the BRST transformations and symmetry are determined by the particular gauge fixing used. The BRST transformations are independent of the gauge fixing; they are rather determined by the classical gauge transformations and are related only to the gauge group structure, as will become evident in Chap. 11.

In the sequel, we are mostly interested in models where spinor matter fields interact with the gauge field V_μ. The effect of an infinitesimal gauge transformation on a spinor field ψ is

$$\delta_\lambda \psi = -\lambda \psi \tag{2.9}$$

So the gauge invariant matter action, to be added to (2.6), is

$$S_m[V, \psi] = \int \mathrm{d}^\mathrm{d} x \, \overline{\psi} \not{D} \psi \tag{2.10}$$

where $\not{D} = i\gamma^\mu \left(\partial_\mu + V_\mu\right)$. This invariance, according to the Noether procedure, brings about the covariant conservation of the current $j^a_\mu = i\overline{\psi}\gamma_\mu T^a \psi$

$$\left(D^\mu j_\mu\right)^a = \partial^\mu j^a_\mu + f^{abc} V^{b\mu} j^c_\mu = 0 \tag{2.11}$$

The Lorenz gauge fixing does not drive any change in the matter action.

2.1.2 *Gravitational Symmetries: Diffeomorphisms and Local Lorentz Transformations*

Another important family of local gauge theories are the theories describing gravitation. The basic field is the metric $g_{\mu\nu}(x)$ and the basic symmetry is the symmetry under diffeomorphisms. The latter correspond to the general coordinate transformations $x^\mu \to x^\mu + \xi^\mu(x)$, where $\xi^\mu(x)$ are generic infinitesimal smooth functions of the coordinates. They act on the metric as follows

$$\delta_\xi g_{\mu\nu} = D_\mu \xi_\nu + D_\nu \xi_\mu, \quad \xi_\mu = g_{\mu\nu} \xi^\nu \tag{2.12}$$

D_μ is the covariant derivative with respect to the metric: $D_\mu \xi_\nu = \partial_\mu \xi_\nu - \Gamma^\lambda_{\mu\nu} \xi_\lambda$, and $\Gamma^\lambda_{\mu\nu}$ are the Christoffel symbols, whose transformation properties are

$$\delta_\xi \Gamma^\lambda_{\mu\nu} = \xi^\sigma \partial_\sigma \Gamma^\lambda_{\mu\nu} + \partial_\mu \xi^\sigma \Gamma^\lambda_{\sigma\nu} + \partial_\nu \xi^\sigma \Gamma^\lambda_{\mu\sigma} - \partial_\sigma \xi^\lambda \Gamma^\sigma_{\mu\nu} + \partial_\mu \partial_\nu \xi^\lambda \tag{2.13}$$

The basic covariant geometrical object in this context is the Riemann tensor

$$R^\rho{}_{\lambda\mu\nu} = \partial_\mu \Gamma^\rho_{\nu\lambda} - \partial_\nu \Gamma^\rho_{\mu\lambda} + \Gamma^\rho_{\mu\sigma} \Gamma^\sigma_{\nu\lambda} - \Gamma^\rho_{\nu\sigma} \Gamma^\sigma_{\mu\lambda} \tag{2.14}$$

whose transformation property is

$$\delta_\xi R^\rho{}_{\lambda\mu\nu} = \xi^\sigma \partial_\sigma R^\rho{}_{\lambda\mu\nu} - \partial_\sigma \xi^\rho R^\sigma{}_{\lambda\mu\nu} + \partial_\lambda \xi^\sigma R^\rho{}_{\sigma\mu\nu} + \partial_\mu \xi^\sigma R^\rho{}_{\lambda\sigma\nu} + \partial_\nu \xi^\sigma R^\rho{}_{\lambda\mu\sigma} \tag{2.15}$$

Also in this case, it is convenient to introduce a compact matrix-form notation for the Christoffel symbols

$$\Gamma \equiv \{\Gamma_\nu{}^\lambda\}, \quad \Gamma_\nu{}^\lambda = \mathrm{d}x^\mu \, \Gamma_{\mu\nu}{}^\lambda \tag{2.16}$$

and for the Riemann curvature tensor

$$\mathcal{R} = \mathrm{d}\Gamma + \Gamma \wedge \Gamma, \quad \mathcal{R} = \{\mathcal{R}^\rho{}_\lambda\}, \quad \mathcal{R}^\rho{}_\lambda = \frac{1}{2} \mathrm{d}x^\mu \wedge \mathrm{d}x^\nu \, R_{\mu\nu}{}^\rho{}_\lambda \tag{2.17}$$

The product between adjacent entries is the matrix product: $(XY)^\lambda{}_\rho = X^\lambda{}_\nu Y^\nu{}_\rho$.

The BRST transformations for the gravitational case are obtained by promoting the parameters ξ^μ to anticommuting fields. In order to avoid a proliferation of symbols, we denote them with the same symbol ξ^μ and, henceforth, unless otherwise specified, it is understood that this symbol denotes anticommuting fields. The BRST transformations of $g_{\mu\nu}$ are the same as above, (2.12), but now ξ^μ are anticommuting fields, which transform themselves as

$$\delta_\xi \xi^\mu = \xi^\lambda \partial_\lambda \xi^\mu \tag{2.18}$$

It is not hard to show that, due to this transformation, δ_ξ becomes a nilpotent operation:

$$\delta_\xi^2 = 0 \tag{2.19}$$

In this book, we are interested mostly in fermions ψ interacting with gravity. The relevant action is

$$S[g, \psi] = \int \mathrm{d}^\mathrm{d}x \, \sqrt{g} \, i\overline{\psi}\gamma^\mu \left(D_\mu + \frac{1}{2}\omega_\mu \right) \psi \tag{2.20}$$

where $\gamma^\mu = \mathrm{e}_a^\mu \gamma^a$, and e_a^μ are the vielbein ($\mu, \nu, \ldots$ are world indices, $a, b, \ldots$ are flat indices). D is the covariant derivative with respect to the world indices and ω_μ is the spin connection:

$$\omega_\mu = \omega_\mu^{ab} \Sigma_{ab}$$

where $\Sigma_{ab} = \frac{1}{4}[\gamma_a, \gamma_b]$ are the Lorentz generators. In Eq.(2.20) ψ is a generic fermion (Dirac, Weyl or Majorana). We will often consider the case of Weyl fermions. A right-handed Weyl fermion will be more explicitly denoted $\psi_R = \frac{1+\gamma_*}{2}\psi$, where ψ

is a Dirac field. Here, γ_* denotes the chirality matrix (in 4d it is γ_5). Classically the energy-momentum tensor

$$T_{\mu\nu} = \frac{i}{4}\overline{\psi}\gamma_\mu \overleftrightarrow{\nabla}_\nu \psi + \{\mu \leftrightarrow \nu\}, \quad \nabla_\mu = D_\mu + \frac{1}{2}\omega_\mu, \tag{2.21}$$

is both conserved on shell and traceless, i.e.

$$D^\mu T_{\mu\nu}(x) = 0, \quad T^\mu_\mu(x) = 0 \tag{2.22}$$

A fermion field transforms, under diffeomorphisms, as

$$\delta_\xi \psi = \xi^\mu \partial_\mu \psi, \tag{2.23}$$

and it is easy to prove that (2.20) is invariant under diffeomorphisms. But the introduction of fermions in a gravitational theory brings about the appearance of a new local symmetry, the local Lorentz symmetry. It is defined by transformations

$$\delta_\Lambda \psi = -\frac{1}{2}\Lambda\psi, \quad \Lambda = \Sigma_{ab}\Lambda^{ab} \tag{2.24}$$

where $\Lambda^{ab}(x)$ are arbitrary infinitesimal parameters, antisymmetric in a, b. This implies

$$\delta_\Lambda \omega_\mu = \partial_\mu \Lambda + \frac{1}{2}[\omega_\mu, \Lambda] \tag{2.25}$$

On the vielbein e^a_μ the transformation is

$$\delta_\Lambda e^a_\mu = \Lambda^a{}_b e^b_\mu \tag{2.26}$$

It is straightforward to prove that (2.20) is invariant under these transformations. As usual, this corresponds to a current conservation. The current is:

$$j^{ab}_\mu = i\overline{\psi}\gamma_\mu \Sigma^{ab}\psi \tag{2.27}$$

and the covariant current conservation is

$$\nabla^\mu j_\mu = 0, \quad j_\mu = j^{ab}_\mu \Sigma_{ab}, \quad \nabla_\mu = D_\mu + \frac{1}{2}[\omega_\mu, \ ] \tag{2.28}$$

As before, we will promote also $\Lambda_{ab}(x)$ to an anticommuting field, denoted by the same symbol, and endowed with the transformation property

$$\delta_\Lambda \Lambda_{ab} = -\Lambda_a{}^c \Lambda_{cb} \tag{2.29}$$

With this property δ_Λ becomes a nilpotent operation:

$$\delta_\Lambda^2 = 0 \tag{2.30}$$

Consistency requires that

$$\delta_\xi \Lambda_{ab} = \xi^\mu \partial_\mu \Lambda_{ab}, \quad \delta_\Lambda \xi^\mu = 0 \tag{2.31}$$

which implies that

$$(\delta_\xi + \delta_\Lambda)^2 = 0 \tag{2.32}$$

as well.

Also for the spin connection, we introduce the Lie algebra-valued one-form compact notation

$$\boldsymbol{\omega} = \omega_\mu \mathrm{d}x^\mu, \quad \omega_\mu = \omega_\mu^{ab} \Sigma_{ab} \tag{2.33}$$

whose transformation law is

$$\delta_\Lambda \boldsymbol{\omega} = \mathrm{d}\boldsymbol{\Lambda} + \frac{1}{2}[\boldsymbol{\omega}, \boldsymbol{\Lambda}], \quad \boldsymbol{\Lambda} = \Lambda^{ab} \Sigma_{ab} \tag{2.34}$$

The curvature of $\boldsymbol{\omega}$ is the Riemann two-form

$$\mathbf{R} = \mathbf{d}\boldsymbol{\omega} + \frac{1}{4}[\boldsymbol{\omega}, \boldsymbol{\omega}], \quad \mathbf{R} = \mathbf{R}^{ab} \Sigma_{ab} \tag{2.35}$$

which transforms as

$$\delta_\Lambda \mathbf{R} = \frac{1}{2}[\mathbf{R}, \boldsymbol{\Lambda}] \tag{2.36}$$

Notice that if we refer, instead, to the matrix one-form $\omega = \{\omega^{ab}\}$, $\omega^{ab} = \omega_\mu^{ab} \mathrm{d}x^\mu$ and the matrix $\Lambda = \{\Lambda^{ab}\}$, the previous formulas take the form

$$\delta_\Lambda \omega = \mathrm{d}\Lambda + [\omega, \Lambda], \tag{2.37}$$

$$R = \mathrm{d}\omega + \frac{1}{2}[\omega, \omega], \quad R = \{R^{ab}\} \tag{2.38}$$

and

$$\delta_\Lambda R = [R, \Lambda]. \tag{2.39}$$

Remark. The relation between R and $\mathcal{R}$ is

$$\mathcal{R}^{\mu}{}_{\nu} = e^{\mu}_{a}\, R^{ab}\, \mathrm{e}_{b\nu} \tag{2.40}$$

References

1. C. Becchi, A. Rouet, R. Stora, The Abelian Higgs Kibble model, unitarity of the S-operator. Phys. Lett. B **52**, 344 (1974)
2. C. Becchi, A. Rouet, R. Stora, Renormalization of the Abelian Higgs-Kibble model. Commun. Math. Phys. **42**, 127 (1975)
3. C. Becchi, A. Rouet, R. Stora, Renormalization of gauge theories. Ann. Phys. **98**, 287 (1976)
4. C. Itzykson, B. Zuber, *Quantum Field Theory* (Dover Publications, 2006)
5. M.E. Peskin, D.V. Schroeder, *An Introduction to Quantum Field Theory* (Westview Press, 1995)
6. R. Stora, in *Continuum Gauge Theories*, ed. by M. Levy. New Developments in Quantum Field Theory and Statistical Mechanics, (Cargèse 1976), Proceedings NATO ASI, Ser. B—Mathematical and Physical Sciences, vol. 26 (Plenum Press, 1977)
7. A. Trautman, On gauge transformations and symmetries. Bull. Acad. Pol. Sci. **XXVII**(1), 7 (1979)
8. I.V. Tyutin, Gauge invariance in field theory and statistical physics in operator formalism, Lebedev Institute Preprint, Report No.: FIAN-39 (1975) e-Print: 0812.0580 [hep-th]

Chapter 3
Conformal Symmetry

The trace or conformal anomalies are the second important family of anomalies. They affect the trace of the energy-momentum tensor of a theory and represent the violation of a classical property: the on shell vanishing of the trace of the energy-momentum tensor when the (classical) theory is conformal invariant. Since conformal invariance is less frequented than gauge invariance in the textbooks, we will devote a more detailed introduction to conformal symmetry and conformal field theories.

3.1 Introduction

There are systems in nature that are characterized by *scale invariance*, that is they appear to have the same features if we blow up or down distances between points (or elements) of the system. This is the case for statistical systems at critical points. In such systems, singular thermodynamical quantities are related to the correlation length by a set of critical exponents (universality principle). At critical points, the correlation length becomes infinite, and the systems become scale invariant, since there is no scale left to measure distances.

The situation just outlined occurs for systems such as the vapor-liquid one at the critical point, characterized by a critical pressure p_c and temperature T_c. Below T_c density is not well-defined, while it is well-defined at and above T_c. A similar picture holds as well for a ferromagnetic material with magnetization m. In ferromagnetic materials, the source of magnetization is the spin of the electrons in incomplete atomic shells, each electron carrying one unit of magnetic moment. Such spins can be imagined to be attached to lattice points and to interact with the neighboring ones in such a way that the state of lowest energy corresponds to all the spins being parallel. At $T = 0$, all the spins are aligned. When the temperature increases, the thermal motion destroys this order, but not completely if the temperature is low enough; there remains patches where the spins are all aligned, with the result that a finite magnetization survives. As T reaches T_c and goes beyond it, order is completely destroyed and magnetization vanishes (disordered or paramagnetic phase).

L. Bonora, *Fermions and Anomalies in Quantum Field Theories*, Theoretical and Mathematical Physics, https://doi.org/10.1007/978-3-031-21928-3_3

In nature, there are many other systems with analogous characteristics. The universality and simplicity of the critical phenomena just illustrated prompt us to suppose that such systems can be described by scale invariant Euclidean field theories (time is not involved). Such theories contain fields $A_i(x)$ that, under a coordinate dilatation

$$x^a \longrightarrow \lambda x^a, \qquad \lambda > 0, \qquad a = 1, \dots d,$$

transform as

$$A_i(x) \longrightarrow \lambda^{\Delta_i} A_i(\lambda x)$$

The number Δ_i is called the *scaling weight* of $A_i(x)$. The example of ferromagnetism, discussed above, suggests that what is at work at the critical point is not simply scale invariance, but a larger symmetry, *confomal invariance*. For the correlation length is interpreted as the average size of the polarized spin patches; the fact that at the critical point, this becomes infinite, which means that we have patches of any size. Therefore, the physical picture does not change, not only when we rescale the system rigidly, but also when we rescale it with a scale varying from point to point. This is precisely conformal symmetry (which locally changes the size without changing the shape).

Beside, the successes of CFT in 2d, conformal symmetry and conformal field theories are at the forefront also in higher dimensions. The truth is that conformal symmetry is intrinsic to quantum field theory via the renormalization group (RG) and the Wilsonian philosophy. Suppose we start from a microscopic system represented by a distribution of spins. We are interested in seeing how the system evolves when our resolution decreases (i.e. the typical energy, or scale, decreases). To this end, we can replace the system with another one constructed as follows: we subdivide the system into small patches and replace the spins in each patch by their average over the patch. If we next rescale the distances and the resulting spins, we obtain a system of the same type as before. We can repeat the process over and over and accordingly decrease our resolution. In many cases, this process will converge to a new theory. Although such a process is more effectively carried out in momentum space, the spacetime description may help intuition of what the renormalization group approach is. The RG in quantum field theory is mathematically described by the Callan-Symanzik equation (written down here for a massless theory), valid for any correlator among (in general, composite) fields $A_i(x)$:

$$\begin{aligned}&\left(\frac{\partial}{\partial \log \Lambda} + \beta^I \frac{\partial}{\partial g^I}\right) \langle A_{i_1}(x_1) \dots A_{i_n}(x_n)\rangle \\ &= \gamma_{i_1}^{j_1} \langle A_{j_1}(x_1) \dots A_{i_n}(x_n)\rangle + \dots + \gamma_{i_n}^{j_n} \langle A_{i_1}(x_1) \dots A_{j_n}(x_n)\rangle \qquad (3.1)\end{aligned}$$

where Λ is the scale of energy, g^I are the couplings in the theory, $\beta^I = \frac{\partial g^I}{\partial \log \Lambda}$ the beta functions and γ_i^j the matrix of anomalous dimensions. This equation says that when the energy scale changes, the correlators remain formally the same provided the couplings and the anomalous dimensions of the fields suitably change. The β^Is are the crucial objects of our concern. Suppose for simplicity there is only one coupling. When the corresponding β-function vanishes, it means that the coupling does not evolve with the scale of energy. Therefore, a fixed point of the beta function means that the theory has evolved to scale invariance. Since we expect scale invariance to imply conformal invariance, fixed points of the RG provide plenty of examples of (not necessarily Lagrangian) conformal field theories. The fixed points can be UV or IR, and we have no guarantee that a theory flowing from the UV to the IR along an RG trajectory will end up in a fixed point. However, this turns out to be the case in many physically interesting examples.

References and further readings for this chapter are [1–4].

3.2 Conformal Algebra

3.2.1 $d \geq 3$

Conformal transformations are coordinate transformations that 'change the size without changing the shape', that is change the distances between points without changing the angles. The conformal group in $d \geq 3$ dimensions is finite dimensional. It encompasses Poincaré transformations, dilatations and special conformal transformations (sct's). The latter are

$$x^\mu \to x'^\mu = \frac{x^\mu - b^\mu x^2}{1 - 2b\cdot x + b^2 x^2} \tag{3.2}$$

where b^μ are constant parameters. This of course can be viewed as a particular general coordinate transformation. For infinitesimal b^μ

$$\frac{x^\mu - b^\mu x^2}{1 - 2b\cdot x + b^2 x^2} \approx x^\mu - b^\mu x^2 + 2b\cdot x\, x^\mu \tag{3.3}$$

and the general coordinate transformation takes the form: $x^\mu \to x^\mu + \xi^\mu$ where $\xi^\mu = -b^\mu x^2 + 2b\cdot x\, x^\mu$.

Looking at how the square line element $\mathrm{d}s^2 = \mathrm{d}x^\mu \eta_{\mu\nu} \mathrm{d}x^\nu$ transforms,

$$\mathrm{d}s^2 \longrightarrow (1 + 4b\cdot x)\mathrm{d}s^2, \tag{3.4}$$

Equation (3.2) can be understood as a special local rescaling of the metric. Conformal transformations can in fact be defined as those coordinate transformations that correspond to a (local) rescaling of the metric.

The Lie algebra generators of the conformal group (CG) are

$$\begin{aligned} iP_\mu &= \partial_\mu \\ iD &= x^\mu \partial_\mu \\ iL_{\mu\nu} &= (x_\mu\partial_\nu - x_\nu\partial_\mu) \\ iK_\mu &= (2x_\mu x^\nu\partial_\nu - x^2\partial_\mu) \end{aligned}$$

They generate translations, dilatations, Lorentz and SCT's, respectively. The Lie brackets among them are

$$\begin{aligned} [L_{\mu\nu}, L_{\lambda\rho}] &= i\left(\eta_{\mu\lambda}L_{\nu\rho} - \eta_{\mu\rho}L_{\nu\lambda} - \eta_{\nu\lambda}L_{\mu\rho} + \eta_{\nu\rho}L_{\mu\lambda}\right) \\ [P^\mu, P^\nu] &= 0 \\ [L_{\mu\nu}, P_\lambda] &= i\left(\eta_{\mu\lambda}P_\nu - \eta_{\nu\lambda}P_\mu\right) \\ [P^\mu, D] &= -iP^\mu \\ [K^\mu, D] &= iK^\mu \\ [P^\mu, K^\nu] &= -2i\eta^{\mu\nu}D + 2iL^{\mu\nu} \\ [K^\mu, K^\nu] &= 0 \\ [L^{\mu\nu}, D] &= 0 \\ [L^{\mu\nu}, K^\lambda] &= i\eta^{\lambda\nu}K^\mu - i\eta^{\lambda\mu}K^\nu \end{aligned} \tag{3.5}$$

They form the conformal algebra (CA), which is the maximally extended bosonic algebra of the Poincaré algebra. So far, we have not specified what background metric we understand in the previous formulas. It can be either Lorentzian or Euclidean. In the first case, the conformal algebra turns out to be isomorphic to $so(d, 2)$; in the latter case to $so(d+1, 1)$, the Lie algebra of the orthogonal group $SO(d, 2)$ and $SO(d+1, 1)$, respectively.

3.2.2 d = 2

In two dimensions, the conformal group is infinite dimensional for the diffeomorphisms that correspond to local rescalings of the metric which are represented by arbitrary functions (see Appendix 3A). Here, we introduce the notation when the background metric is Euclidean. We will adopt a complex variable notation by defining $z = \frac{1}{\sqrt{2}}(x^1 + ix^2)$. Infinitesimal conformal transformations are defined by

$$z \to z + \epsilon(z), \qquad \bar{z} \to \bar{z} + \bar{\epsilon}(\bar{z}) \tag{3.6}$$

where $\epsilon(z)(\bar{\epsilon}(\bar{z}))$ are arbitrary infinitesimal holomorphic (antiholomorphic) functions in the entire complex plane with the possible exclusion of the origin and the point at infinity. The group of infinitesimal conformal transformations is thus a direct product of its holomorphic and antiholomorphic subgroups. This induces the characteristic chirality splitting of conformal field theories in $\mathsf{d} = 2$, with the consequence that the holomorphic and antiholomorphic sectors of these theories can be studied separately: for instance, the results for the full theory correlators are obtained by putting together the results of the two sectors and imposing reality.

Let us see a few properties of the holomorphic conformal transformations. From now on, we will limit our attention to the $\epsilon(z)$ parameters that admit a Laurent expansion

$$\epsilon(z) = \sum_{n\in\mathbb{Z}} \epsilon_{-n} z^{n+1} \tag{3.7}$$

The generator corresponding to the nth mode of this expansion is the vector field

$$\ell_n = -z^{n+1} \frac{d}{dz} \tag{3.8}$$

as can be seen by noting that, for any holomorphic function $f(z)$,

$$\delta_\epsilon f(z) \approx f(z+\epsilon) - f(z) \approx \mathcal{L}_{\sum_n \epsilon_{-n}\ell_n} f(z)$$

where $\mathcal{L}_X$ denotes the Lie derivative with respect to the vector field X. The Lie bracket between the ℓ_n's define the *Witt algebra* of the holomorphic vector fields on the unit circle

$$[\ell_n, \ell_m] = (n-m)\ell_{n+m} \tag{3.9}$$

In a quantum field theory, this algebra appears deformed by a central extension

$$[L_n, L_m] = (n-m)L_{n+m} + \frac{c}{12}(n^3 - n)\delta_{m+n,0}, \tag{3.10}$$

which is known as the *Virasoro algebra*. c is the central charge of the given model (representation). It counts the number of (free) degrees of freedom.

A finite subalgebra of both algebras is spanned by $\ell_{-1}, \ell_0, \ell_1$ (or L_{-1}, L_0, L_1), which generate $s\ell(2,\mathbb{C})$, the Lie algebra of the group $SL(2,\mathbb{C})$. The latter is the group of fractional transformations of the complex plane with unit determinant:

$$z \to \frac{az+b}{cz+d}, \qquad ad - bc = 1 \tag{3.11}$$

If we set $a = 1+\alpha,\ b = \beta,\ c = \gamma,\ d = 1+\delta$, with $\alpha, \beta, \gamma, \delta$ infinitesimal parameters, the corresponding fractional transformation becomes, to first order,

$$z \to z + \beta + (\alpha - \delta)z - \gamma z^2$$

This is to be compared with the transformation $z \to z + \epsilon_1 + \epsilon_0 z + \epsilon_{-1} z^2$ generated by $\ell_1, \ell_0, \ell_{-1}$. One can easily recognize the parameters of the 2d conformal group: the translation parameters are given by the two components of ϵ_1; $\epsilon_0 = \rho e^{i\theta}$ gives the dilatation and rotation parameters; finally the sct parameters b_1 and b_2 correspond to $-\Re(\epsilon_{-1})$ and to $\Im(\epsilon_{-1})$, respectively.

3.2.3 Conformal Algebra Representations

The representations of the CA we are interested in are those in terms of local fields. Fields that transform according to an irreducible representation of the CA are called *primary*. In particular, a primary field has a definite weight Δ. For a generic scalar or fermion primary field $\mathcal{O}(x)$ of weight Δ (we understand the tensor or spinor labels), we have

$$
\begin{aligned}
&[P_\mu, \mathcal{O}(x)] = i\partial_\mu O(x)\\
&[L_{\mu\nu}, \mathcal{O}(x)] = i(x_\mu\partial_\nu - x_\nu\partial_\mu)\mathcal{O}(x) + \Sigma_{\mu\nu}\mathcal{O}(x)\\
&[D, \mathcal{O}(x)] = i(\Delta + x^\mu\partial_\mu)\mathcal{O}(x)\\
&[K_\mu, \mathcal{O}(x)] = i(2\Delta x_\mu + 2x_\mu x^\lambda\partial_\lambda - x^2\partial_\mu - 2ix^\lambda\Sigma_{\lambda\mu})\mathcal{O}(x)
\end{aligned}
\tag{3.12}
$$

where $\Sigma_{\mu\nu}$ are the spinor part of the Lorentz generators. In 2d primary fields correspond to irreducible representations of (3.10). Since the relevant formulas will not be explicitly used in the sequel, we dispense with writing them down.

For a current and the e.m. tensor, in particular, the explicit form of the commutators is

$$
\begin{aligned}
&[D, J_\mu(x)] = i(\mathrm{d} - 1 + x^\lambda\partial_\lambda)J_\mu(x)\\
&[K_\lambda, J_\mu] = i\left(2(\mathrm{d}-1)x_\lambda + 2x_\lambda\, x\cdot\partial - x^2\partial_\lambda\right) J_\mu\\
&\qquad + 2i\left(x^\alpha J_\alpha\eta_{\lambda\mu} - x_\mu J_\lambda\right)
\end{aligned}
\tag{3.13}
$$

$$
\begin{aligned}
&[D, T_{\mu\nu}(x)] = i(\mathrm{d} + x^\lambda\partial_\lambda)T_{\mu\nu}(x)\\
&[K_\lambda, T_{\mu\nu}] = i\left(2\mathrm{d}x_\lambda + 2x_\lambda\, x\cdot\partial - x^2\partial_\lambda\right) T_{\mu\nu}\\
&\qquad + 2i\left(x^\alpha T_{\alpha\nu}\eta_{\lambda\mu} + x^\alpha T_{\mu\alpha}\eta_{\lambda\nu} - x_\mu T_{\lambda\nu} - x_\nu T_{\mu\lambda}\right)
\end{aligned}
\tag{3.14}
$$

The third line and the sixth line will be often referred to as the Lorentz part of the commutators. Let us remark that if we disregard the spin part, the last equation in (3.12) for infinitesimal b_μ, can be viewed as a local rescaling, for it is equivalent to

$$
\tilde{\mathcal{O}}(x^\mu) \approx (1 + 4b\cdot x)^{\frac{\Delta}{2}}\mathcal{O}(x^\mu - b^\mu x^2 + 2b\cdot x x^\mu)
\tag{3.15}
$$

For this reason, it is often said in the literature that a sct is made of a local rescaling plus a Lorentz transformation.

Note. The commutators (3.12, 3.13, 3.14) can be obtained as follows. For any tensorial primary field $\Phi_{\mu_1\ldots\mu_n}$ of weight Δ, one can write the invariance relation

$$\tilde{\Phi}_{\mu_1\ldots\mu_n}(x)\mathrm{d}x^{\mu_1}\ldots\mathrm{d}x^{\mu_n}(\mathrm{d}s)^{\Delta-n} = \Phi_{\mu_1\ldots\mu_n}(x')dx'^{\mu_1}\ldots dx'^{\mu_n}(\mathrm{d}s')^{\Delta-n} \tag{3.16}$$

for any transformation $x^\mu \to x'^\mu$. If the transformation is a sct generated by K_μ, for instance, we can then write

$$\tilde{\Phi}_{\mu_1\ldots\mu_n}(x) \equiv e^{ibK}\Phi_{\mu_1\ldots\mu_n}(x)e^{-ibK} = \Phi_{\nu_1\ldots\nu_n}(x')\frac{\partial x'^{\nu_1}}{\partial x^{\mu_1}}\cdots\frac{\partial x'^{\nu_n}}{\partial x^{\mu_n}}\left(\frac{\mathrm{d}s'}{\mathrm{d}s}\right)^{\Delta-n}, \tag{3.17}$$

from which the commutators follow using (3.3, 3.4). Similarly one can proceed for the other transformations.

3.3 The Energy-Momentum Tensor and Its Properties

Among the transformations of the conformal group, there are dilatations and special conformal tranformations. Both induce a rescaling of the metric, the first a constant and the second a local rescaling. Moreover, a sct is a particular type of non-trivial local general coordinate transformation or diffeomorphism. In this way, the metric is inevitably called into question, even if we start from a theory defined in a flat background. Now, in any theory, the metric sources the energy-momentum (e.m.) tensor, and the trace of the metric sources the trace of the energy-momentum tensor. Therefore, invariance under the conformal group brings in both conservation of the e.m. and vanishing of its trace.

The previous qualitative statements can be made precise as follows.

In a local field theory, Poincaré invariance implies the existence of a symmetric and conserved energy-momentum (e.m.) tensor

$$T_{\mu\nu} - T_{\nu\mu} = 0, \qquad \partial^\mu T_{\mu\nu} = 0 \tag{3.18}$$

This implies in particular that the Lorentz current $\mathcal{M}_\lambda^{\mu\nu} = x^{[\mu}T_\lambda^{\nu]}$ is conserved.

Scale invariance requires the existence of a current J^μ (the *virial* current) such that

$$T_\mu^\mu - \partial_\mu J^\mu = 0 \tag{3.19}$$

This implies that the current $D_\mu = x^\lambda T_{\lambda\mu} - J_\mu$ is conserved.

Invariance under special conformal transformations requires

$$T_\mu^\mu = 0 \tag{3.20}$$

so that the current $K_\mu^{(\rho)} = [\rho_\nu x^2 - 2x_\nu(\rho\cdot x)]T_\mu^\nu$ is conserved for any constant vector ρ_μ.

For a conformal field theory, these three properties (3.18, 3.19, 3.20) are true classically. At a quantum level, they appear in a weaker form, as constraints for the correlators where, for instance, T^μ_μ is inserted.

In a theory with a Lagrangian density $\mathcal{L}$, a canonical e.m. tensor is obtained via the functional formula

$$T^{\mu\nu} = \sum_i \frac{\partial \mathcal{L}}{\partial \partial_\mu \varphi_i} \partial^\nu \varphi^i - \eta^{\mu\nu} \mathcal{L} \tag{3.21}$$

where φ^i denote the generic fields in the theory. This e.m. tensor is, in general, not unique. There is some freedom to redefine it without breaking the conservation law $\partial_\mu T^{\mu\nu} = 0$. For instance, it may happen that $\partial_\mu T^{\mu\nu} = 0$ and $T^\mu_\mu = \partial_\mu \partial_\nu L^{\mu\nu}$. Then one can define

$$\begin{aligned}\Theta^{\mu\nu} &= T^{\mu\nu} + \frac{1}{d-2}\left(\partial^\mu \partial_\lambda L^{\lambda\nu} + \partial^\nu \partial_\lambda L^{\lambda\mu} - \eta^{\mu\nu} \partial_\lambda \partial_\rho L^{\lambda\rho} - \Box L^{\mu\nu}\right) \\ &+ \frac{1}{(d-2)(d-1)}\left(\eta^{\mu\nu}\Box - \partial^\mu \partial^\nu\right) L_\lambda{}^\lambda\end{aligned} \tag{3.22}$$

which satisfies both $\partial_\mu \Theta^{\mu\nu} = 0$ and $\Theta^\mu_\mu = 0$.

A practical way to derive the e.m. tensor of a (matter) theory defined by a local action S is to couple it to an external metric $g_{\mu\nu} = \eta_{\mu\nu} + h_{\mu\nu}$ in a covariant way and define the e.m. tensor as

$$T_{\mu\nu} = \frac{2}{\sqrt{g}} \frac{S[g]}{\delta g^{\mu\nu}}\bigg|_{g^{\mu\nu}=\eta^{\mu\nu}} \qquad T^{\mu\nu} = -\frac{2}{\sqrt{g}} \frac{S[g]}{\delta g_{\mu\nu}}\bigg|_{g_{\mu\nu}=\eta_{\mu\nu}} \tag{3.23}$$

where $S[\eta] = S$. For instance, in the case of a Yang-Mills theory

$$\begin{aligned}S &= \frac{1}{4g_{YM}} \int d^{\mathrm{d}}x \,\mathrm{Tr}\left(F_{\mu\nu} F^{\mu\nu}\right) \\ &\longrightarrow S[g] = \frac{1}{4g_{YM}} \int d^{\mathrm{d}}x \sqrt{g} g^{\mu\mu'} g^{\nu\nu'} \mathrm{Tr}\left(F_{\mu\nu} F_{\mu'\nu'}\right)\end{aligned} \tag{3.24}$$

and in the case of a free massless fermion theory

$$S = \int d^{\mathrm{d}}x \, i\bar{\psi}\gamma^\mu \partial_\mu \psi \to S[g] = \int d^{\mathrm{d}}x \,\sqrt{g}\, i\bar{\psi}\gamma^\mu (D_\mu + \frac{1}{2}\omega_\mu)\psi \tag{3.25}$$

where $\gamma^\mu = e^\mu_a \gamma^a$, D_μ is the covariant derivative and ω_μ is the spin connection:

$$\omega_\mu = \omega_\mu^{ab} \Sigma_{ab}$$

where $\Sigma_{ab} = \frac{1}{4}[\gamma_a, \gamma_b]$ are the Lorentz generators.

In this type of theories, Poincaré invariance extends to diffeomorphism invariance, i.e. invariance under $x^\mu \to x^\mu + \xi^\mu(x)$, and conformal invariance to Weyl invariance, i.e. invariance under local metric rescalings. In the first case $\delta_\xi g_{\mu\nu} = D_\mu \xi_\nu + D_\nu \xi_\mu$, with $\xi_\mu = g_{\mu\nu}\xi^\nu$ and

$$\begin{aligned} 0 &= \delta_\xi S[g] = \int d^{\mathrm{d}}x \, \frac{\delta S[g]}{\delta g_{\mu\nu}(x)} \delta_\xi g_{\mu\nu}(x) \\ &= -\int d^{\mathrm{d}}x \left(\frac{\sqrt{g}}{2} T^{\mu\nu} \left(D_\mu \xi_\nu + D_\nu \xi_\mu \right) \right) \\ &= \int d^{\mathrm{d}}x \sqrt{g}\, \xi_\nu D_\mu T^{\mu\nu} \to D_\mu T^{\mu\nu} = 0 \end{aligned} \tag{3.26}$$

after integration by parts, because ξ^μ is arbitrary. In flat spacetime, the RHS becomes $\partial_\mu T^{\mu\nu} = 0$. In the second step of (3.26), we have skipped the variation with respect to the matter fields in the theory. These variations are proportional to the corresponding equations of motion. Having disregarded them means that the result holds only on shell.

In the case of Weyl invariance, i.e. invariance under local metric rescalings $g_{\mu\nu}(x) \to 2\omega(x) g_{\mu\nu}(x)$, we have analogously

$$\begin{aligned} 0 &= \delta_\omega S[g] = \int d^{\mathrm{d}}x \, \frac{\delta S[g]}{\delta g_{\mu\nu}(x)} \delta_\omega g_{\mu\nu}(x) \\ &= -\int d^{\mathrm{d}}x \, \omega \, T^\mu_\mu \quad \to \quad T^\mu_\mu = 0 \end{aligned} \tag{3.27}$$

This also holds on shell.

3.3.1 A Few Examples

Here are some examples of classically conformal invariant field theories.

Scalar field theory. Let us consider the scalar field ϕ and the action

$$S[g] = \frac{1}{2} \int d^{\mathrm{d}}x \, \sqrt{g} g^{\mu\nu} \partial_\mu \phi \partial_\nu \phi \tag{3.28}$$

Proceeding as above we find $T_{\mu\nu} = \partial_\mu \phi \partial_\nu \phi - \frac{\eta_{\mu\nu}}{2} \partial_\lambda \phi \partial^\lambda \phi$. So,

$$T^\mu_\mu = \frac{2-\mathrm{d}}{4} \Box \phi^2 = \partial^\mu \partial^\nu L_{\mu\nu}, \qquad L_{\mu\nu} = \frac{2-\mathrm{d}}{4} \eta_{\mu\nu} \phi^2 \tag{3.29}$$

Now we can use formula (3.22) to define a traceless e.m. tensor or, better, remark that (3.28) can be made conformal invariant by adding a term involving the Ricci scalar R:

$$S'[g] = \frac{1}{2}\int d^{\mathsf{d}}x\,\sqrt{g}\left(g^{\mu\nu}\partial_\mu\phi\partial_\nu\phi + \frac{\mathsf{d}-2}{4(\mathsf{d}-1)}R\phi^2\right) \tag{3.30}$$

Now the model is Weyl invariant under the metric rescaling $\delta_\omega R = -2\omega R - 2(\mathsf{d}-1)\Box\omega$ and $\delta_\omega\phi = -\frac{\mathsf{d}-2}{2}\omega\phi$; consequently $T'^{\mu}_{\ \mu} = 0$.

Free massless Dirac fermion. This is the theory introduced above, (3.25). It is conformal invariant under the Weyl transformation (with $\psi \to e^{-\frac{\mathsf{d}-1}{2}\omega}\psi$). The flat e.m. tensor is

$$T_{\mu\nu} = \frac{i}{4}\left(\bar\psi\gamma_\mu\overset{\leftrightarrow}{\partial}_\nu\psi + \bar\psi\gamma_\nu\overset{\leftrightarrow}{\partial}_\mu\psi\right) \tag{3.31}$$

It is traceless on shell in any dimension (see Sect. 7.1 for a discussion on this e.m. tensor).

3.3.2 *Correlators in Free Field Theories*

In CFT, the relevant objects to be calculated are the correlators with various field insertions. In free field theories, this can be done by means of the Wick theorem. For this, we need the propagators. For instance, in 4d, in configuration space representation, we have:

- for a free scalar field ϕ: $\langle 0|\mathcal{T}\phi(x)\phi(y)|0\rangle = \frac{1}{(x-y)^2}$
- for a free Dirac fermion ψ: $\langle 0|\mathcal{T}\psi(x)\bar\psi(y)|0\rangle = \frac{\not x-\not y}{(x-y)^4} = \gamma_\mu\frac{(x^\mu-y^\mu)}{(x-y)^4}$

since $\Box\frac{1}{x^2} = -4\pi^2\delta^{(4)}(x)$, etc.

Let us consider two simple examples, the two-point correlators of two normal ordered composite operators $\Phi(x) =: \phi^2 : (x)$ and $J_\mu(x) =: \bar\psi\gamma_\mu\psi : (x)$, of conformal weight 2 and 3, respectively. Using the Wick theorem, we can straightforwardly compute

$$\langle 0|\mathcal{T}\Phi(x)\Phi(y)|0\rangle = \frac{2}{(x-y)^4}, \tag{3.32}$$

$$\langle 0|\mathcal{T}J_\mu(x)J_\nu(y)|0\rangle = 4\frac{2(x_\mu-y_\mu)(x_\nu-y_\nu)-\eta_{\mu\nu}(x-y)^2}{(x-y)^8} \tag{3.33}$$

3.4 Conformal WI's and Correlators

The purpose of CFT is to compute conformal correlators. Our purpose here is to focus on conformal correlators of currents and e.m. tensors. Poincaré and scale covariance is rather easy to comply with. The most difficult part is to implement covariance with respect to special conformal transformations (sct's). To this, end correlators must satisfy the appropriate Ward identities (WI's). In this section, we show how to derive the relevant Ward identities (a.k.a. Ward-Takahashi identities) for the e.m. tensor.

The special conformal transformation (sct) of the e.m. tensor $T_{\mu\nu}$ in coordinate representation is given by

$$
\begin{aligned}
-i[K_\lambda, T_{\mu\nu}] = &\left(2\mathrm{d}x_\lambda + 2x_\lambda\, x\cdot\partial - x^2\partial_\lambda\right) T_{\mu\nu} \\
&+2\left(x^\alpha T_{\alpha\nu}\eta_{\lambda\mu} + x^\alpha T_{\mu\alpha}\eta_{\lambda\nu} - x_\mu T_{\lambda\nu} - x_\nu T_{\mu\lambda}\right) \qquad (3.34)
\end{aligned}
$$

The simplest way to derive the corresponding WI is to couple $T_{\mu\nu}$ to an external source $h_{\mu\nu}$, a 0-weight symmetric tensor field, and to write down the corresponding generating function of connected Green's functions, which in a Minkowski spacetime, is

$$
\begin{aligned}
W[h_{\mu\nu}] = &\, W[0] \qquad (3.35)\\
&+\sum_{n=1}^{\infty}\frac{i^{n-1}}{2^n n!}\int\prod_{i=1}^{n}\mathrm{d}x_i\; h^{\mu_i\nu_i}(x_i)\,\langle 0|\mathcal{T}\{T_{\mu_1\nu_1}(x_1)\ldots T_{\mu_n\nu_n}(x_n)\}|0\rangle_c .
\end{aligned}
$$

In order for W to be invariant under sct's the external source $h_{\mu\nu}$ must transform as $\delta_b h_{\mu\nu} = -i[b^\lambda K_\lambda(x), h_{\mu\nu}(x)] \equiv -i[b\cdot K(x), h_{\mu\nu}(x)]$, where

$$
\begin{aligned}
i[K_\lambda(x), h_{\mu\nu}(x)] = &\left(2x_\lambda\, x\cdot\partial - x^2\partial_\lambda\right) h_{\mu\nu} \qquad (3.36)\\
&+2\left(x^\alpha h_{\alpha\nu}\eta_{\lambda\mu} + x^\alpha h_{\mu\alpha}\eta_{\lambda\nu} - x_\mu h_{\lambda\nu} - x_\nu h_{\mu\lambda}\right)
\end{aligned}
$$

Invariance of $W[h]$ leads to

$$
\begin{aligned}
0 = \delta_b W &= \int d^{\mathrm{d}}x\frac{\delta W}{\delta h^{\mu\nu}}\delta h^{\mu\nu} = -i\int d^{\mathrm{d}}x\,[b\cdot K(x), h^{\mu\nu}(x)]\langle\!\langle T_{\mu\nu}(x)\rangle\!\rangle \\
&= i\int d^{\mathrm{d}}x\, h^{\mu\nu}(x)[b\cdot K(x), \langle\!\langle T_{\mu\nu}(x)\rangle\!\rangle] = 0 \qquad (3.37)
\end{aligned}
$$

where

$$
\begin{aligned}
\langle\!\langle T_{\mu\nu}(x)\rangle\!\rangle = 2\frac{\delta W[h]}{\delta h^{\mu\nu}(x)} = &\sum_{n=1}^{\infty}\frac{i^n}{2^n n!}\int \mathrm{d}x_1\ldots\int \mathrm{d}x_n h^{\mu_1\nu_1}(x_1)\ldots h^{\mu_n\nu_n}(x_n)\\
&\times\langle 0|\mathcal{T}T_{\mu\nu}(x)T_{\mu_1\nu_1}(x_1)\ldots T_{\mu_n\nu_n}(x_n)|0\rangle_c \qquad (3.38)
\end{aligned}
$$

Differentiating (3.37) with respect to $h_{\mu\nu}(x)$ and setting $h_{\mu\nu} = 0$ we get $0 = 0$, because $\langle 0|T_{\mu\nu}(x)|0\rangle = \langle 0|T_{\mu\nu}(0)|0\rangle$. Differentiating twice (3.37) and integrating by parts we get

$$
(b\cdot K(x) + b\cdot K(y))\langle 0|\mathcal{T}T_{\mu\nu}(x)T_{\lambda\rho}(y)\rangle|0\rangle = 0 \qquad (3.39)
$$

Differentiating three times (3.37)

$$(b\cdot K(x)+b\cdot K(y)+b\cdot K(z))\langle 0|\mathcal{T}T_{\mu\nu}(x)T_{\lambda\rho}(y)T_{\alpha\beta}(z)\rangle|0\rangle=0 \tag{3.40}$$

In both equations, it is understood that the Lorentz part of $b\cdot K(x)$ acts on the indices μ,ν only, $b\cdot K(y)$ on the indices λ, ρ and $b\cdot K(z)$ on α, β alone.

Due to translational invariance, we can set $y=0$ in (3.39) and $z=0$ in (3.40). These equations become

$$b\cdot K(x)\langle 0|\mathcal{T}T_{\mu\nu}(x)T_{\lambda\rho}(0)\rangle|0\rangle=0 \tag{3.41}$$

and

$$(b\cdot K(x)+b\cdot K(y))\langle 0|\mathcal{T}T_{\mu\nu}(x)T_{\lambda\rho}(y)T_{\alpha\beta}(0)\rangle|0\rangle=0 \tag{3.42}$$

with the previous prescription for the Lorentz part.

3.4.1 General WI's for Sct's

Given a Poincaré covariant correlator for primary fields $\mathcal{O}_i$ of conformal weight Δ_i, $i=1,\dots,n$, for it to be conformally covariant, it must satisfy the WI for scaling transformation

$$\sum_{i=1}^{n}\left(\Delta_i+x_i^{\mu}\frac{\partial}{\partial x_i^{\mu}}\right)\langle\mathcal{O}_1(x_1)\dots\mathcal{O}_n(x_n)\rangle=0, \tag{3.43}$$

and the WI for sct's

$$\sum_{i=1}^{n}\left(b\cdot\tilde{K}(x_i)+b\cdot L(x_i)\right)\langle\mathcal{O}_1(x_1)\dots\mathcal{O}_n(x_n)\rangle=0 \tag{3.44}$$

where $b\cdot K(x_i)$ is split into two parts: $b\cdot\tilde{K}(x_i)$ is a differential operator, independent of the tensor structure of the correlator, and $b\cdot L(x_i)$ a multiplication operator, both linear in b. For scalar operators, the latter is absent, while the former coincides with $b\cdot K(x_i)$:

$$b\cdot K(x_i)=2\,\Delta\,b\cdot x_i+2\,b\cdot x_i\,x_i\cdot\frac{\partial}{\partial x_i}-x_i^2\,b\cdot\frac{\partial}{\partial x_i} \tag{3.45}$$

Some examples. For a scalar field of weight Δ, the two-point function is $\langle\Phi(x)\Phi(y)\rangle\sim\frac{1}{(x-y)^{2\Delta}}$. The Ward identity reads

$$(b\cdot K(x) + b\cdot K(y))\frac{1}{(x-y)^{2\Delta}} = \frac{1}{(x-y)^{2\Delta}}\, b\cdot(x-y)(2\Delta - 2\Delta) = 0 \qquad (3.46)$$

Less simple examples are the two-point functions of a conserved current $j_\mu(x)$, which we have computed above in the free field case, eq.(3.33),

$$\langle j_\mu(x) j_\nu(y)\rangle = \frac{c'}{(x-y)^{2\Delta}} I_{\mu\nu}(x-y), \qquad I_{\mu\nu}(x) = \delta_{\mu\nu} - 2\frac{x_\mu x_\nu}{x^2} \qquad (3.47)$$

and of an e.m. tensor $T_{\mu\nu}(x)$

$$\begin{aligned} &\langle T_{\mu\nu}(x)\, T_{\rho\sigma}(y)\rangle \qquad (3.48)\\ &= \frac{c/2}{(x-y)^{2d}} \left(I_{\mu\rho}(x-y)\, I_{\nu\sigma}(x-y) + I_{\nu\rho}(x-y)\, I_{\mu\sigma}(x-y) - \frac{2}{\mathsf{d}} \eta_{\mu\nu}\eta_{\rho\sigma} \right) \end{aligned}$$

Proving that they satisfy the conformal WI's is not as straightforward as (3.46), but it can be done. In fact, there are more efficient indirect methods, like the null-cone method, which is however beyond the scope of this short introduction.

Appendix 3A. Conformal Transformations in d = 2

In a Euclidean 2d space, diffeomorphisms are transformations $x^a \to x^a + \xi^a(x)$, with $a = 1, 2$, where ξ^a are infinitesimal generic functions. The metric transforms according to

$$\delta_\xi g_{ab} = \nabla_a \xi_b + \nabla_b \xi_a \qquad (3.49)$$

where $\xi_a = g_{ab}\xi^b$. Under a Weyl transformation it transforms according to

$$\delta_\omega g_{ab} = 2\omega\, g_{ab} \qquad (3.50)$$

where $\omega(x)$ is a generic positive function. A conformal transformation is a diffeomorphism which coincides with a Weyl transformation. So if we start from a flat metric η_{ab}, it is specified by

$$\partial_a \xi_b + \partial_b \xi_a = 2\omega\eta_{ab} \qquad (3.51)$$

from which it follows that

$$\partial_1\xi_2 + \partial_2\xi_1 = 0, \qquad \partial_1\xi_1 - \partial_2\xi_2 = 0 \qquad (3.52)$$

where $\partial_i = \partial_{x^i}$. Let us introduce the complex coordinates $z = \frac{1}{\sqrt{2}}(x^1 + ix^2)$, $\bar{z} = \frac{1}{\sqrt{2}}(x^1 - ix^2)$ and the combinations $\epsilon(z, \bar{z}) = \frac{1}{\sqrt{2}}(\xi_1 + i\xi_2)$ and $\bar{\epsilon}(z, \bar{z}) = \frac{1}{\sqrt{2}}(\xi_1 - i\xi_2)$. One can verify that

$$\partial_{\bar{z}}\epsilon(z, \bar{z}) = \frac{1}{2}(\partial_1 + i\partial_2)(\xi_1 + i\xi_2) = 0, \qquad \partial_z\bar{\epsilon}(z, \bar{z}) = 0 \tag{3.53}$$

due to (3.52). So ϵ is a function only of z, and $\bar{\epsilon}$ only of $\bar{z}$. Thus, in a 2d Euclidean framework, conformal transformations take the form (3.6).

References

1. G. Mack, A. Salam, Finite component field representations of the conformal group. Ann. Phys. **53**, 174–202 (1969)
2. A.A. Belavin, A.M. Polyakov, A.B. Zamolodchikov, Infinite conformal symmetry in two-dimensional quantum field theory. Nucl. Phys. B **241**, 333 (1984)
3. N. Yu, *Scale Invariance vs Conformal Invariance* [arXiv:1302.0884]
4. S. Rychkov, *EPFL Lectures on Conformal Field Theories in $D \geq 3$ Dimensions* (Lausanne, École Polytechnique, 2012)

Part II
Anomalies and Cohomology

Cohomology is a mathematical structure underlying consistent anomalies. For any classically conserved local or rigid symmetry, we can define a Ward identity for the corresponding quantum theory. We show in this Part II that any such identity can be given the form of a nilpotent, or coboundary, operator acting on the effective action. Any such operator defines a cohomology. The cohomology we are interested in in a local quantum field theory is the BRST, or local, cohomology, the coboundary operator being precisely the nilpotent generator of the BRST transformations introduced in a previous chapter. Any (local) violation of the Ward identity written in this form must consequently satisfy a consistency condition, the Wess-Zumino consistency condition, and identifies a potential anomaly. We show that such consistency conditions have celebrated solutions based on a Chern formula, [1], which gives rise to the descent equations. We construct solutions for gauge, diffeomorphisms and local Lorentz symmetry. Then we turn to conformal transformations and corresponding Weyl symmetry. Even in this case, we can define a corresponding cohomology problem. There is nothing similar to the Chern formula, but we can, nevertheless, define an algorithmic procedure to find trivial and non-trivial cocycles, and, thus, identify the potential Weyl anomalies.

Reference

1. S.S. Chern, *Complex Manifolds without Potential Theory* (Springer, Berlin, 1969)

Chapter 4
Effective Actions and Anomalies

In this chapter, we define effective actions and Ward identities corresponding to gauge, diffeormorphism, local Lorentz invariance and conformal invariance. Anomalies are violation of the latter. Nevertheless, they are not arbitrary violations since they must satisfy consistency conditions, the celebrated Wess-Zumino (WZ) consistency conditions, [1] (see also [2, 3]). In this chapter, we formulate the WZ consistency conditions in the equivalent mathematical language of cohomology.

4.1 Effective Action for Gauge Theories and Consistent Anomalies

In the path, integral approach to quantization the effective action is the (mostly non-local) expression of the potentials we obtain by integrating out the matter fields in quadratic actions such as (2.10) and (2.20).

In perturbative quantization, the definition of effective action[1] for the action (2.10) is

$$W[V] = W[0] + \sum_{n=1}^{\infty} \frac{i^{n-1}}{n!} \int \prod_{i=1}^{n} \mathrm{d}^{\mathrm{d}} x_i \; V_{\mu_i}(x_i) \langle 0|\mathcal{T} j_{\mu_1}(x_1) \ldots j_{\mu_n}(x_n)|0\rangle_c \quad (4.1)$$

in terms of the connected time-ordered Green's functions of the currents $j_\mu(x)$. Later on, we will need the effective (full one-loop) vector current

[1] In general an effective action is defined to be the generating function of the 1-particle irreducible connected Green functions. However 1-particle reducible amplitudes do not show up in this book. Therefore connectedness is a sufficient qualification.

L. Bonora, *Fermions and Anomalies in Quantum Field Theories*, Theoretical and Mathematical Physics, https://doi.org/10.1007/978-3-031-21928-3_4

$$\langle\langle j_\mu(x)\rangle\rangle = \frac{\delta W[V]}{\delta V_\mu(x)} = \sum_{n=0}^{\infty} \frac{i^n}{n!} \int \prod_{i=1}^{n} \mathrm{d}^d x_i$$
$$V_{\mu_i}(x_i)\langle 0|\mathcal{T} j_\mu(x) j_{\mu_1}(x_1)\dots j_{\mu_n}(x_n)|0\rangle_c \tag{4.2}$$

Gauge invariance of the effective action can be expressed via the functional operator $X^a(x)$ defined by

$$X^a(x) = \partial_\mu \frac{\delta}{\delta V_\mu^a(x)} + f^{abc} V_\mu^b(x) \frac{\delta}{\delta V_\mu^c(x)}, \tag{4.3}$$

as follows

$$X^a(x) W[V] = 0 \tag{4.4}$$

This is the full one-loop Ward identity (WI) for gauge symmetry. This equation can be obtained by expressing the invariance of the effective action as follows

$$0 = \delta_\lambda W[V] = \int \mathrm{d}^d x\, \delta_\lambda V_\mu^a(x) \frac{\delta}{\delta V_\mu^a(x)} W[V] \tag{4.5}$$

integrating by parts and noting that $\lambda^a(x)$ are arbitrary functions. Equation (4.4) is equivalent to the one-loop covariant conservation law

$$[D^\mu \langle\langle j_\mu(x)\rangle\rangle]^a = 0 \tag{4.6}$$

In a number of cases, the WI (4.4) is violated

$$X^a(x) W[V] = \kappa \Delta^a(x) \tag{4.7}$$

Here, κ is a small dimensionless expansion parameter characterizing the one-loop calculation, which, from now on, for simplicity, we shall set =1.

Applying $X^b(y)$ to both sides of (4.7) and then inverting the two operations, we find a remarkable relation of group-theoretical nature

$$X^a(x)\Delta^b(y) - X^b(y)\Delta^a(x) + f^{abc}\Delta^c(x)\delta(x-y) = 0, \tag{4.8}$$

that $\Delta^a(x)$ must satisfy. These are the Wess-Zumino (WZ) consistency conditions.

Now, it may happen that

$$\Delta^a(x) = X^a(x)\,\mathcal{C} \tag{4.9}$$

where $\mathcal{C}$ is the spacetime integral of a local expression of $V^a(x)$, briefly, a local counterterm. In this case, we can redefine the effective action as $W'[V] = W[V] - \mathcal{C}$, so that

$$X^a(x)\, W'[V] = 0 \tag{4.10}$$

and invariance is recovered. If (4.9) is not possible for any local counterterm, then $\Delta^a(x)$ is a true anomaly. Saturating $\Delta^a(x)$ with $\lambda^a(x)$ and integrating over spacetime, we obtain the integrated anomaly

$$\mathcal{A}_\lambda = \int \mathrm{d}^\mathrm{d}x \, \lambda^a(x)\Delta^a(x) \tag{4.11}$$

4.2 Anomalies and BRST Cohomology

The WZ consistency condition (4.8) is a beautiful but rather complicated relation from a practical point of view. It can, however, be cast into a mathematically more appealing and practically easier-to-deal-with form by using the BRST formalism. To this end, let us recall the nilpotent BRST transformations (2.8), and let us denote by $\mathfrak{s}$ the functional operator that generates them (here and in the sequel we disregard the fields $\bar{c}$ and B)

$$\mathfrak{s} = \int \mathrm{d}^\mathrm{d}x \left(\mathfrak{s}V_\mu^a(x)\frac{\partial}{\partial V_\mu^a(x)} + \mathfrak{s}c^a(x)\frac{\partial}{\partial c^a(x)} \right). \tag{4.12}$$

It is a nilpotent operator

$$\mathfrak{s}^2 = 0 \tag{4.13}$$

Using this, we can rewrite the WZ consistency condition as follows

$$\mathfrak{s}\mathcal{A}_c = 0 \tag{4.14}$$

where we have replaced the gauge parameter λ with the anticommuting field c. In order to recover (4.8), we should simply differentiate (4.14) with respect to $c^a(x)$ and $c^b(y)$. In this language, the anomalous WI (4.7) can be rewritten

$$\mathfrak{s}W[V] = \mathcal{A}_c, \tag{4.15}$$

from which (4.14) is obtained by simply applying $\mathfrak{s}$ to both sides. Similarly, we can rewrite the triviality condition (4.9) as

$$\mathcal{A}_c = \mathfrak{s}\,\mathcal{C} \tag{4.16}$$

What we have just achieved is the definition of a cohomology problem. $\mathfrak{s}$ is the coboundary operator of this problem. The cochain space is the space of local polynomials in V_μ^a and c^a and their derivatives with the right canonical dimension. A cochain satisfying (4.15) is a cocycle, and a cocycle satisfying (4.16) is a coboundary. Sometimes, it is also said that a cocycle is a closed cochain and a coboundary is an exact cochain. In this way, we have transformed the anomaly problem into a cohomology problem. Anomalies are non-trivial (non-exact) cocycles of the operator $\mathfrak{s}$.

4.3 Gravitational Effective Actions and Diffeomorphism Anomalies

The perturbative effective action corresponding to the classical action (2.20) is expressed in terms of the metric fluctuation $h_{\mu\nu}$, where $g_{\mu\nu} = \eta_{\mu\nu} + h_{\mu\nu}$:

$$W[g] = W[0] + \sum_{n=1}^{\infty} \frac{i^{n-1}}{2^n n!} \int \prod_{i=1}^{n} \mathrm{d}^{\mathrm{d}} x_i \sqrt{g(x_i)} h^{\mu_i \nu_i}(x_i) \langle 0 | \mathcal{T} T_{\mu_1\nu_1}(x_1) \dots T_{\mu_n\nu_n}(x_n) | 0 \rangle \tag{4.17}$$

As already remarked, the coefficients of this expansion must be consistent with the definition of graviton emission vertices. Remember that $\frac{\delta S}{\delta g_{\mu\nu}}|_{g=\eta} = \frac{1}{2} T_{\mu\nu}$. This explain the factor 2^n in the denominator of (4.17). From (4.17), we can extract the one-loop one-point function of the energy-momentum tensor.[2]

$$\langle\!\langle T_{\mu\nu}(x) \rangle\!\rangle = \sum_{n=0}^{\infty} \frac{i^n}{2^n n!} \int \prod_{i=1}^{n} \mathrm{d}^{\mathrm{d}} x_i \sqrt{g(x_i)} h^{\mu_i \nu_i}(x_i) \langle 0 | \mathcal{T} T_{\mu\nu}(x) T_{\mu_1\nu_1}(x_1) \dots T_{\mu_n\nu_n}(x_n) | 0 \rangle \tag{4.18}$$

The one-loop Ward identity for diffeomorphisms is

$$D^{\mu} \langle\!\langle T_{\mu\nu}(x) \rangle\!\rangle = 0 \tag{4.19}$$

which is the one-loop quantum version of the first equation in (2.22).

Now, there is no point in repeating the same derivation as in the gauge case, by introducing the analogous of the functional operator $X^a(x)$, etc. We simply use the nilpotent transformation δ_ξ introduced above, see Eqs. (2.12,2.18,2.19), and denote with the same symbol the corresponding functional operator. The Ward identity takes the form

$$\delta_\xi \, W[g] = 0 \tag{4.20}$$

If, on the other hand, this relation is violated at one-loop, we will have

$$\delta_\xi \, W[g] = \mathcal{A}_\xi \tag{4.21}$$

The term in the RHS is linear in ξ and, due to (2.19), will satisfy the consistency relation

---

[2] The presence or absence of the factors $\sqrt{g(x_i)}$ in (4.18) will be discussed further on, in Chap. 7.

$$\delta_\xi \mathcal{A}_\xi = 0 \tag{4.22}$$

As before, we may have

$$\mathcal{A}_\xi = \delta_\xi \mathcal{C}[g] \tag{4.23}$$

where $\mathcal{C}[g]$ is a local expression of the metric, its inverse and their derivatives. In this case, the cocycle is a coboundary and the WI can be restored. Or, there is no such local $C[g]$ and $\mathcal{A}_\xi$ is a true anomaly of the diffeomorphisms. In this way, we have reduced the identification of anomalies to a cohomology problem.

4.4 Effective Actions and Local Lorentz Anomalies

As pointed out above, when fermions are present in a theory, another symmetry, beside diffeomorphisms, becomes relevant: the local Lorentz symmetry. The perturbative effective action can be represented as follows

$$\begin{aligned} W[\omega] &= W[0] \\ &+ \sum_{n=1}^{\infty} \frac{i^{n-1}}{n!} \int \prod_{i=1}^{n} \mathrm{d}^{\mathrm{d}} x_i \sqrt{g(x_i)}\, \overset{\circ}{\omega}{}_{\mu_i}^{a_i b_i}(x_i) \langle 0 | \mathcal{T} j^{\mu_1 a_1 b_1}(x_1) \ldots j^{\mu_n a_n b_n}(x_n) | 0 \rangle \end{aligned} \tag{4.24}$$

where $\overset{\circ}{\omega}$ is the lowest order approximant of the spin connection.

Without further ado, we recall the nilpotent transformations defined above by Eqs. (2.26, 2.29, 2.30) and denote with the same symbol δ_Λ the corresponding functional operator. The WI for local Lorentz transformations is

$$\delta_\Lambda W[\omega] = 0 \tag{4.25}$$

However, at one-loop, we may have

$$\delta_\Lambda W[g] = \mathcal{A}_\Lambda \tag{4.26}$$

where $\mathcal{A}_\Lambda$ is a local expression linear in Λ^{ab}. Due to (2.30), it will satisfy the consistency relation

$$\delta_\Lambda \mathcal{A}_\Lambda = 0 \tag{4.27}$$

As before we may have

$$\mathcal{A}_\Lambda = \delta_\Lambda \mathcal{C}[\omega] \tag{4.28}$$

where $\mathcal{C}[\omega]$ is a local expression of ω_μ and its derivatives. In this case, the cocycle is a coboundary and the WI can be restored. Or, there is no such local $C[\omega]$, and $\mathcal{A}_\omega$ is a true local Lorentz anomaly.

4.5 Special Conformal Transformations and Cohomology

In Chap. 3, we have defined the conformal algebra in flat spacetime. Apart from the Poincaré transformations, it contains dilatations and special conformal transformations (sct's). We have also defined the corresponding WI's, (3.43) or (3.44). Now, violations of these WI's define conformal anomalies, more precisely dilatation and sct anomalies. However, a sleek method for deriving them from their explicit WI's, like in the gauge or diffeomorphisms case (see next chapter), does not exist. We can certainly define the BRST transformations for the conformal Lie algebra in the standard way. For instance, for the special conformal transformations we proceed as follows. We take the infinitesimal transformation

$$\delta_b x^\mu = -b^\mu x^2 + 2b\cdot x x^\mu \tag{4.29}$$

and define the corresponding BRST transformation by promoting the (constant) parameter b^μ to an anticommuting parameter, which we denote with the same symbol. Simultaneously, for any local operator $\mathcal{O}$ we define

$$\delta_b \mathcal{O}(x) = -i[b\cdot K, \mathcal{O}(x)] \tag{4.30}$$

where the commutator is defined in (3.12). It is easy to prove that, for instance $\delta_b b_\mu = 0$ and, in general,

$$\delta_b^2 = 0 \tag{4.31}$$

We denote with the same symbol the corresponding functional operator, which is also nilpotent. Therefore, we obtain a coboundary operator, and we can define a cohomology problem for it. For instance, we can easily identify (local) invariants: take a scalar local expression $\mathcal{O}(x)$ of weight Δ and integrate it over the spacetime, then

$$\delta_b \int \mathrm{d}^4x\, \mathcal{O}(x) = 2\int \mathrm{d}^4x\, (\Delta - 4)\, b\cdot x \mathcal{O}(x) \tag{4.32}$$

This vanishes only for $\Delta = 4$. On the other hand, if we consider a local vector expression $\mathcal{J}_\mu$ of weight Δ and construct the 1-cochain $\Delta_b[\mathcal{J}] = \int d^4x\, b^\mu \mathcal{J}_\mu(x)$ we have

$$\delta_b \int \mathrm{d}^4x\, b\cdot\mathcal{J} x = -2\int \mathrm{d}^4x\, (\Delta - 3) b\cdot x\, b\cdot \mathcal{J} \tag{4.33}$$

which vanishes for $\Delta = 3$. Therefore, $\Delta_b[\mathcal{J}]$ is a potential cocycle for $\Delta = 3$. However, this is not the type of cocycles relevant for local field theory. Let us recall that invariance under sct's requires the divergence of $K_\mu^{(\rho)} = [\rho_\nu x^2 - 2x_\nu(\rho\cdot x)]T_\mu^\nu$ to vanish for any constant ρ_μ. An anomaly, if any, is obtained by inserting an e.m. tensor in any correlator in the combination given by $K_\mu^{(\rho)}$; therefore, it is bound to be linear in b_μ, which we may identify with ρ_μ, and quadratic in x^μ. It follows that a cocycle like $\Delta_b[\mathcal{J}]$ is of purely academical interest. In order to find relevant anomalies, we should look at the cohomology based on the parameter $\xi^\mu \equiv -b^\mu x^2 + 2b\cdot x\, x^\mu = \delta_b x^\mu$, but this is tantamount to analyzing diffeomorphisms. Similarly, we might analyze the transformation of the flat metric $\delta_b \eta_{\mu\nu} = 4b\cdot x\, \eta_{\mu\nu}$, but this means studying Weyl transformations.

Here, we choose to simplify the problem. Since the conformal group contains particular cases of nontrivial diffeomorphisms and local rescalings of the metric (Weyl transformations), it is much more efficient to study the problem in general, that is, much as we have done in order to derive the e.m. tensor from an action, to couple minimally the theory to a generic metric and study diffeomorphisms and Weyl transformations and their local cohomology (their anomalies) in this augmented theory. At the end, one can return to the flat background. From now on, therefore, we will consider this more general problem. That is we consider beside (2.12) and

$$\delta_\xi \psi = \xi^\lambda \partial_\lambda \psi, \quad \delta_\xi \phi = \xi^\lambda \partial_\lambda \phi, \tag{4.34}$$

the (infinitesimal) Weyl transformations

$$\delta_\omega g_{\mu\nu} = 2\omega g_{\mu\nu}, \quad \delta_\omega \psi = -\frac{\mathsf{d}-1}{2}\omega\psi, \quad \delta_\omega \phi = -\frac{\mathsf{d}-2}{2}\phi \tag{4.35}$$

This transformation is Abelian; therefore, passing to the corresponding BRST transformations does not require any change, and, promoting ω to an anticommuting field, we have

$$\delta_\omega^2 = 0 \tag{4.36}$$

On the other hand, considering Weyl transformations together with diffeomorphisms requires

$$\delta_\omega \xi^\mu = 0, \quad \delta_\xi \omega = \xi^\lambda \partial_\lambda \omega \tag{4.37}$$

Then, one can prove that

$$(\delta_\xi + \delta_\omega)^2 = 0 \tag{4.38}$$

4.6 Gravitational Effective Actions and Trace Anomalies

It remains, therefore, for us to study conformal transformations and relevant anomalies. The perturbative effective action is the same as in Eq. (4.17), and the one-point one-loop e.m. tensor is as in Eq. (4.18). The invariance under Weyl transformations, or conformal invariance, is expressed by

$$0 = \delta_\omega W[g] = \int \mathrm{d}^{\mathsf{d}}x\, \frac{\delta W[g]}{\delta g^{\mu\nu}(x)} \delta_\omega g^{\mu\nu}(x) = -\int \mathrm{d}^{\mathsf{d}}x\, \omega(x) \langle\!\langle T_{\mu\nu}(x) \rangle\!\rangle g^{\mu\nu}(x) \tag{4.39}$$

Since ω is a generic infinitesimal function, this implies

$$\langle\langle T_{\mu\nu}(x)\rangle\rangle g^{\mu\nu}(x) = 0 \tag{4.40}$$

But we may have violations of this classical invariance

$$\delta_\omega W[g] = \mathcal{A}_\omega \tag{4.41}$$

Due to the nilpotence of δ_ω, the consistency condition

$$\delta_\omega \mathcal{A}_\omega = 0 \tag{4.42}$$

follows. As before, we can have

$$\mathcal{A}_\omega = \delta_\omega \mathcal{C}[g] \tag{4.43}$$

where $\mathcal{C}[g]$ is a local counterterm. In this case, the cocycle $\mathcal{A}_\omega$ is a coboundary and the WI can be recovered. If no such counterterm exists $\mathcal{A}_\omega$ is a trace (or conformal, or Weyl) anomaly.

As we shall see in Chap. 7, the previous definition of trace anomaly is not devoid of ambiguity, and a more refined one will be needed.

References

1. J. Wess, B. Zumino, Consequences of anomalous Ward identities. Phys. Lett. B **37**, 95–97 (1971)
2. R. Stora, Algebraic structure and topological origin of anomalies, in *Progress in Gauge Field Theory* ed. by G. 't Hooft, A. Jaffe, G. Lehmann, P.K. Mitter, I.M. Singer, NATO ASI, Ser. B, vol. 115 (Plenum Press, 1984)
3. J. Mañes, R. Stora, B. Zumino, Algebraic study of chiral anomalies. Comm. Math. Phys. **102**, 157 (1985)

Chapter 5
Cohomological Analysis of Anomalies

In quantum field theory anomalies are met when regularizing the logarithmic divergent parts of the relevant WI's, an operation that gives rise to local expressions. Very often the approach is perturbative, most of the times by means of Feynman diagrams. Perturbative calculations are usually very handy at the lowest order of approximation, but they become soon unpractical if one wants to proceed to the next orders. It is therefore of tremendous help to know in advance the form anomalies can take to all orders, so that knowing the first order can uniquely identify them. This is where the cohomological analysis becomes precious. The first advantage of the cohomological analysis is that, given the symmetries of a theory, it provides a classification of its possible anomalies.

In this chapter we study the cohomology of various symmetries and construct their non-trivial cocycles.

5.1 Cohomology of Joint Symmetries

In quantum field theory we have normally to do with several symmetries of the action simultaneously. In this section we analyze the effects of joint symmetries on the relevant cohomologies. For the sake of simplicity let us consider two symmetries, which we shall refer to as R and S (for instance vector and axial gauge transformations, or diffeomorphisms and local Lorentz or Weyl transformations). Let δ_R and δ_S be the corresponding coboundary operators. They are functional differential operators linear in the relevant ghost fields, and satisfy

$$(\delta_R + \delta_S)^2 = 0 \tag{5.1}$$

(see for instance (2.32)), which splits into

$$\delta_R^2 = 0 \qquad \delta_S^2 = 0, \qquad \delta_R\delta_S + \delta_S\delta_R = 0 \tag{5.2}$$

L. Bonora, *Fermions and Anomalies in Quantum Field Theories*, Theoretical and Mathematical Physics, https://doi.org/10.1007/978-3-031-21928-3_5

A cocycle of $\delta_R + \delta_S$ is the sum of two integrated local expressions Δ_R and Δ_S

$$(\delta_R + \delta_S)\, W = \Delta_R + \Delta_S \tag{5.3}$$

Therefore

$$(\delta_R + \delta_S)\, (\Delta_R + \Delta_S) = 0 \tag{5.4}$$

This equation split into three distinct ones

$$\delta_R \Delta_R = 0, \qquad \delta_S \Delta_R + \delta_R \Delta_S = 0, \qquad \delta_S \Delta_S = 0 \tag{5.5}$$

Focusing on Δ_R, we are faced with two possibilities. Either it is a non-trivial R-cocycle or it is an R-coboundary. In the latter case $\Delta_R = \delta_R \mathcal{C}$, where $\mathcal{C}$ is a local expression of the fields excluding ghosts. Then we can redefine $W \to W' = W - \mathcal{C}$, so that

$$\delta_R W' = \Delta'_R = 0 \tag{5.6}$$
$$\delta_S W' = \Delta_S - \delta_S \mathcal{C} \equiv \Delta'_S \tag{5.7}$$

We have, of course,

$$\delta_S \Delta'_S = 0 \tag{5.8}$$

and, using the middle equation in Eq. (5.5)

$$0 = \delta_S \Delta_R + \delta_R \Delta_S = \delta_R \Delta_S + \delta_S \delta_R \mathcal{C} = \delta_R\, (\Delta_S - \delta_S \mathcal{C}|) = \delta_R \Delta'_S \tag{5.9}$$

This means that the cocycle Δ'_S is R-invariant. It might happen instead that Δ_S is an S-coboundary, i.e., $\Delta_S = \delta_S \mathcal{C}'$. We could repeat the previous steps and come to the conclusion that $\Delta'_S = 0$ and Δ'_R is S-invariant. One further possibility is that, simultaneously Δ_R is an R-coboundary and Δ_S is an S-coboundary. We get $\Delta'_R = 0$ and $\Delta'_S = 0$ provided the counterterms satisfy the condition

$$\delta_S \delta_R \mathcal{C} + \delta_R \delta_S \mathcal{C}' = 0 \tag{5.10}$$

which is verified, for instance, if $\mathcal{C} = \mathcal{C}'$ or if $\mathcal{C}$ is S-invariant and $\mathcal{C}'$ is R- invariant, but not in general. I.e., the fact that Δ_R and Δ_S are separately trivial, does not necessarily imply that the sum $\Delta_R + \Delta_S$ is a coboundary of the cohomology of $\delta_R + \delta_S$. One last possibility is that neither Δ_R is an R-coboundary nor Δ_S is an S-coboundary.

Remark. In the previous argument we have not specified what kind of symmetry R and S are. In the sequel we will use it with R and S being local gauge symmetries. But in fact it is more general, it holds also for *rigid* symmetries i.e., for symmetries whose transformation parameters are spacetime constants.[1] For instance, if the action is invariant under a constant Lie algebra valued transformation parameter $\lambda = \lambda^a T^a$ (for instance $\delta_\lambda \psi = -\lambda \psi$) we can promote it to a constant ghost $c = c^a T^a$ and endow it with the transformation property $\mathfrak{s} c = -\frac{1}{2}[c, c]$. This generates a nilpotent functional operator, which is a coboundary operator and, in turn, defines a cohomology problem. Another example is the group of dilatations, while sct's are not rigid transformations, although the parameter b_μ is constant, see the discussion in the previous chapter. The cohomological analysis proceeds in the same way as for the local cohomology, but it is much less 'constrained' than the local one: the nontrivial rigid cocycles are generically much more numerous than the local ones. Rigid cohomology will not be treated in this book.

5.1.1 Anomaly Problem Setup

Let us summarize the anomaly formalism introduced above. We can have

- gauge anomalies,

$$(D_\mu \langle\langle J_\mu(x) \rangle\rangle)^a \equiv X^a(x) W[A] = \mathcal{A}^a[A](x), \tag{5.11}$$

  (among which we include also the local Lorentz anomalies) where

$$X^a(x) = \partial_\mu \frac{\delta}{\delta A^a_\mu(x)} + f^{abc} A^b_\mu(x) \frac{\delta}{\delta A^c_\mu(x)}; \tag{5.12}$$

- diffeomorphism anomalies,

$$D^\mu \langle\langle T_{\mu\nu} \rangle\rangle \equiv D^\mu \frac{\delta}{\delta g^{\mu\nu}(x)} W[g] = \mathcal{A}_\nu[g](x), \tag{5.13}$$

- trace anomalies,

$$\langle\langle T^\mu_\mu \rangle\rangle \equiv g^{\mu\nu}(x) \frac{\delta}{\delta g^{\mu\nu}(x)} W[g] = T[g](x). \tag{5.14}$$

For simplicity from now on, as we have done above, we will denote in the same way the variation of a field and the corresponding functional operator (δ_λ is mostly denoted with the traditional symbol $\mathfrak{s}$)

[1] In the literature such transformations are often called global, but such a qualifier in this book would interfere with the locution 'global anomalies', where it has a completely different meaning.

$$\delta_\lambda A^a = \mathrm{d}\lambda^a + f^{abc} A^b \lambda^c, \qquad \delta_\lambda = \int \mathrm{d}x\, \delta_\lambda A^a(x) \frac{\delta}{\delta A^a(x)} = -\int \mathrm{d}x\, \lambda^a(x) X^a(x)$$
$$\delta_\xi g_{\mu\nu} = D_\mu \xi_\nu + D_\nu \xi_\mu, \qquad \delta_\xi = \int \mathrm{d}x\, \delta_\xi g_{\mu\nu} \frac{\delta}{\delta g_{\mu\nu}(x)} \tag{5.15}$$
$$\delta_\omega g_{\mu\nu} = 2\omega g_{\mu\nu}, \qquad \delta_\omega = \int \mathrm{d}x\, \delta_\omega g_{\mu\nu} \frac{\delta}{\delta g_{\mu\nu}(x)}$$

Using this and integrating over spacetime both sides of (5.11 and 5.13) and (5.14) multiplied by their respective gauge parameters $\lambda^a(x)$, $\xi^\mu(x)$ and $\omega(x)$, we obtain the integrated anomalous WI's:

$$\delta_\lambda W[A] = -\mathcal{A}_\lambda, \qquad \mathcal{A}_\lambda = \int d^{\mathrm{d}}x\, \sqrt{g}\, \lambda^a(x) \mathcal{A}^a[A](x) \tag{5.16}$$

$$\delta_\xi W[g] = \mathcal{A}_\xi, \qquad \mathcal{A}_\xi = \int d^{\mathrm{d}}x\, \sqrt{g}\, \xi^\mu(x) \mathcal{A}_\mu[g](x) \tag{5.17}$$

$$\delta_\omega W[g] = -\mathcal{A}_\omega, \qquad \mathcal{A}_\omega = \int d^{\mathrm{d}}x\, \sqrt{g}\, \omega(x) T[g](x) \tag{5.18}$$

Anomalies must satisfy consistency conditions which come from the group theoretical nature of the symmetry they violate. Stated more precisely, we can operate another transformation (say λ', ξ', ω') on both sides of (5.16, 5.17 and 5.18), and we can repeat the transformations in reverse order: then the way the RHS's of these equations are related under such reversing is dictated by the formal rules of the LHS's. For instance for gauge anomalies

$$\delta_{\lambda_1} \mathcal{A}_{\lambda_2} - \delta_{\lambda_2} \mathcal{A}_{\lambda_2} = \mathcal{A}_{[\lambda_1, \lambda_2]}, \tag{5.19}$$

and likewise for the other cocycles. If the group is Abelian the RHS vanishes. This is the case for Weyl transformations (which are Abelian). There is an elegant way to express these otherwise very complicated relations by promoting the gauge parameters to *anticommuting fields* (or *ghost*) and endowing them with their own transformation property. If we assume

$$\delta_\xi \xi^\mu = \xi^\nu \partial_\nu \xi^\mu, \qquad \delta_\xi \lambda^a = \xi^\nu \partial_\nu \lambda^a, \qquad \delta_\xi \omega = \xi^\nu \partial_\nu \omega, \tag{5.20}$$
$$\delta_\lambda \xi^\mu = 0, \qquad \delta_\lambda \lambda^a = -f^{abc} \lambda^b \lambda^c, \qquad \delta_\lambda \omega = 0, \tag{5.21}$$
$$\delta_\omega \xi^\mu = 0, \qquad \delta_\omega \lambda^a = 0, \qquad \delta_\omega \omega = 0, \tag{5.22}$$

one can easily verify that

$$(\delta_\xi + \delta_\lambda + \delta_\omega)^2 = 0 \tag{5.23}$$

by applying the LHS to any field in the game. This means $\delta_\xi + \delta_\lambda + \delta_\omega$ is a coboundary operator and defines a cohomology problem. It follows that

$$(\delta_\lambda + \delta_\xi + \delta_\omega)(\mathcal{A}_\lambda + \mathcal{A}_\xi + \mathcal{A}_\omega) = 0, \tag{5.24}$$

i.e., the anomaly is a cocycle of $\delta_\xi + \delta_\lambda + \delta_\omega$. Equation (5.24) are the (Wess-Zumino) *consistency conditions*. The LHS is bilinear in the various ghost fields. Thus (5.24) splits into separate homogeneous consistency conditions:

$$\delta_\lambda \mathcal{A}_\lambda = 0, \qquad \delta_\lambda \mathcal{A}_\xi + \delta_\xi \mathcal{A}_\lambda = 0, \tag{5.25}$$

$$\delta_\xi \mathcal{A}_\xi = 0, \qquad \delta_\omega \mathcal{A}_\xi + \delta_\xi \mathcal{A}_\omega = 0, \tag{5.26}$$

$$\delta_\omega \mathcal{A}_\omega = 0, \qquad \delta_\lambda \mathcal{A}_\omega + \delta_\omega \mathcal{A}_\lambda = 0. \tag{5.27}$$

The cocycle $\mathcal{A}_\lambda + \mathcal{A}_\xi + \mathcal{A}_\omega$ actually defines a class of cohomology, because we can add to it an arbitrary coboundary $(\delta_\lambda + \delta_\xi + \delta_\omega)\mathcal{C}$, where $\mathcal{C}$ is a local counterterm, and it still satisfies the consistency conditions (5.24).

We shall say that a cocycle $\mathcal{A}_\lambda + \mathcal{A}_\xi + \mathcal{A}_\omega$ has a *minimal form* if, possibly by subtracting suitable counterterms, it can be reduced to a cocycle of the form $\mathcal{A}_\lambda$ or $\mathcal{A}_\xi$ or $\mathcal{A}_\omega$ alone. All the known anomalies analyzed till now can be cast into a minimal form.

5.2 Construction of Non-trivial Cocycles

As already pointed out it is important to know in advance explicit expressions of non-trivial cocycles, i.e., of local expressions that satisfy the relevant consistency conditions and cannot be absorbed into a redefinition of the effective action. Solving in general the consistency conditions is not an easy task, and has been done so far only in few particular cases. But it is perhaps not such an important problem. From a physical point of view there is an obvious hierarchy: if a gauge theory has (non-trivial) chiral gauge anomalies satisfying the Wess-Zumino consistency conditions, [1], it is a doomed theory, and it does not make much sense to search for possible anomalies of other (if any) classical symmetries of the theory; the same can be said for diffeomorphism anomalies in theories coupled to gravity: if diffeomorphisms are irremediably anomalous the theory is inconsistent and it does not make sense to look for, say, Weyl anomalies. Therefore the foremost cases to be considered are the chiral gauge anomalies and the diffeomorphism anomalies. The method to produce the corresponding cocycles is, in principle, very simple: one writes down all the possible 1-cochains (with the appropriate canonical dimensions), i.e., local expressions of the fields involved together with their derivatives, linear in the appropriate ghost field and of the right canonical dimensions; then one applies to them the coboundary operator and determines the combinations that satisfy the consistency conditions (cocycles); finally one writes down all the 0-cochains (with the appropriate canonical dimensions), i.e., local expressions without ghosts, applies to them the coboundary operator and identifies the cocycles that are coboundaries. The trouble is that this method is far from efficient, especially when we deal with diffeomorphisms or with higher dimensions. Fortunately a most welcome shortcut in the case of gauge and diffeomorphisms anomalies is provided by a general formula due to Chern [2]. This is the first and, probably, most important case in which we have compact formulas for non-trivial cocycles, see [3–17]. Another important instance is when one can use (gauge or diffeomorphism) covariance in order to construct cocycles for another symmetry, for instance Weyl cocycles. We are going to illustrate these two cases in turn.

5.2.1 The Chern Formula and Descent Equations

Let us consider a generic gauge theory, with connection $A^a_\mu T^a$, valued in a Lie algebra $\mathfrak{g}$ with anti-hermitean generators T^a, such that $[T^a, T^b] = f^{abc} T^c$. In the following it is convenient to use the compact form notation and represent the connection as a one-form $A = A^a_\mu T^a \mathrm{d}x^\mu$, so that the gauge transformation becomes

$$\delta_\lambda A = d\lambda + [A, \lambda] \tag{5.28}$$

with $\lambda(x) = \lambda^a(x) T^a$ and $d = \mathrm{d}x^\mu \frac{\partial}{\partial x^\mu}$. As explained above the mathematical problem is better formulated if we promote the gauge parameter λ to an anticommuting ghost field $c = c^a T$ and define the BRST transform as

$$\mathfrak{s}A = dc + [A, c], \qquad \mathfrak{s}c = -\frac{1}{2}[c, c] \tag{5.29}$$

The operation $\mathfrak{s}$ is nilpotent. As usual, we represent with the same symbol $\mathfrak{s}$ the corresponding functional operator, i.e.,

$$\mathfrak{s} = \int d^{\mathrm{d}}x \left(\mathfrak{s}A^a(x) \frac{\partial}{\partial A^a(x)} + \mathfrak{s}c^a(x) \frac{\partial}{\partial c^a(x)} \right) \tag{5.30}$$

To construct the descent equations we start from a symmetric polynomial in the Lie algebra of order n, $P_n(T^{a_1}, \ldots, T^{a_n})$, invariant under the adjoint transformations (see Appendix 5A):

$$P_n([X, T^{a_1}], \ldots, T^{a_n}) + \ldots + P_n(T^{a_1}, \ldots, [X, T^{a_n}]) = 0 \tag{5.31}$$

for any element X of $\mathfrak{g}$. In many cases these polynomials are symmetric traces of the generators in the corresponding representation

$$P_n(T^{a_1}, \ldots, T^{a_n}) = \mathrm{Str}(T^{a_1} \ldots T^{a_n}) \tag{5.32}$$

(Str stands for symmetric trace). Then one can construct the $2n$-form

$$\Delta_{2n}(A) = P_n(F, F, \dots F) \tag{5.33}$$

where $F = d\,A + \frac{1}{2}[A, A]$. In this expression the product of forms is understood to be the exterior product. It is easy to prove that

$$P_n(F, F, \dots F) = d\left(n\int_0^1 dt\, P_n(A, F_t, \dots, F_t)\right) = d\Delta^{(0)}_{2n-1}(A) \tag{5.34}$$

where we have introduced the symbols $A_t = tA$ and its curvature $F_t = dA_t + \frac{1}{2}[A_t, A_t]$, where $0 \leq t \leq 1$. Equation (5.34) is easily proven by noting that $\frac{d}{dt}F_t = d_{A_t}A \equiv dA + [A_t, A]$ and $d_{A_t}F_t = 0$, and exploiting in an essential way the symmetry of P_n and the graded commutativity of the exterior product of forms.

Equation (5.34) is the first of a sequence of equations that can be proven

$$\Delta_{2n}(A) - d\Delta^{(0)}_{2n-1}(A) = 0 \tag{5.35}$$

$$\mathfrak{s}\Delta^{(0)}_{2n-1}(A) - d\Delta^{(1)}_{2n-2}(A, c) = 0 \tag{5.36}$$

$$\mathfrak{s}\Delta^{(1)}_{2n-2}(A, c) - d\Delta^{(2)}_{2n-3}(A, c) = 0 \tag{5.37}$$

$$\dots\dots$$

$$\mathfrak{s}\Delta^{(2n-1)}_0(c) = 0 \tag{5.38}$$

All the expressions $\Delta^{(p)}_k(A, c)$ are polynomials of c, $\mathrm{d}c$, A, $\mathrm{d}A$ and their commutators. Their explicit form can be found in Appendix 5B. The lower index k is the form degree and the upper one p is the ghost number, i.e., the number of c factors. The last polynomial $\Delta^{(2n-1)}_0(c)$ is a 0-form and clearly a function only of c. All these polynomials have explicit compact form. For instance, the next interesting case after Eq. (5.35) is

$$\mathfrak{s}\,\Delta^{(0)}_{2n-1}(A) = d\left(n(n-1)\int_0^1 dt(1-t)P_n(dc, A, F_t, \dots F_t)\right) \tag{5.39}$$

This means in particular that integrating $\Delta^{(0)}_{2n-1}(A)$ over spacetime in $\mathsf{d} = 2n - 1$ dimensions we obtain an invariant local expression. This provides the gauge CS action in any odd dimension. But what matters here is that the RHS contains the general expression of the consistent gauge anomaly in $\mathsf{d} = 2n - 2$ dimension, for, integrating (5.37) over spacetime, one gets

$$\mathfrak{s}\,\mathcal{A}[c, A] = 0 \tag{5.40}$$
$$\mathcal{A}[c, A] = \int d^{\mathsf{d}}x\, \Delta^1_{\mathsf{d}}(c, A), \qquad \text{where}$$
$$\Delta^{(1)}_{\mathsf{d}}(c, A) = n(n-1)\int_0^1 dt(1-t)P_n(dc, A, F_t, \dots F_t)$$

$\mathcal{A}[c, A]$ identifies the anomaly up to an overall numerical coefficient.

Thus the existence of chiral gauge anomalies relies on the existence of the adjoint-invariant polynomials P_n. The properties of the latter are reviewed in Appendix 5A, below. One may wonder if the so-obtained cocycles are non-trivial. We can show that they are with a *reductio ad absurdum* argument. Let us suppose that (5.40) is trivial. Then we can write

$$\Delta^{(1)}_{\mathsf{d}}(c, A) = \mathfrak{s}\, C^{(0)}_{2n-2}(A, c) + dC^{(1)}_{2n-3}(A, c) \tag{5.41}$$

where here and below $C^{(p)}_k(A, c)$ denotes a polynomial k-form of ghost number p. Applying $\mathfrak{s}$ to (5.41) and using (5.37) we get

$$d\,\mathfrak{s}\, C^{(1)}_{2n-3}(A, c) - d\,\Delta^{(2)}_{2n-3}(A, c) = 0 \tag{5.42}$$

Applying the local Poincaré theorem one gets

$$\Delta^{(2)}_{2n-3}(A, c) = \mathfrak{s}\, C^{(1)}_{2n-3}(A, c) + d\, C^{(2)}_{2n-4}(A, c) \tag{5.43}$$

Repeating the procedure down to the 0-form degree (this is done explicitly in Appendix 5B) we will eventually find that there must exist a 0-form $C^{(2n-2)}_0(c)$ such that

$$\Delta^{(2n-1)}_0(c) = \mathfrak{s}\, C^{(2n-2)}_0(c) \tag{5.44}$$

However this is impossible, for the expression for $\Delta^{(2n-1)}_0(c)$ is

$$\Delta^{(2n-1)}_0(c) \sim P_n(c, [c, c]_+, \dots, [c, c]_+) \tag{5.45}$$

and the only possibility for $C^{(2n-2)}_0(c)$ to satisfy (5.44) is to have the form

$$C^{(2n-2)}_0(c) \sim P_n(c, c, [c, c]_+, \dots, [c, c]_+) \tag{5.46}$$

which, however, vanishes due to the symmetry of P_n and the anticommutativity of c.

It has been proven that the expressions (5.40) are the only consistent non-trivial gauge cocycles.

Remark. If a nontrivial background metric $g_{\mu\nu}$ is present we have to insert a $\sqrt{g} \equiv \sqrt{\det(g_{\mu\nu})}$ factor in the integrand in order to guarantee diffeomorphism invariance

$$\mathcal{A}[c, A] = \int d^{\mathrm{d}}x \sqrt{g}\, \Delta^1_{\mathrm{d}}(c, A) \tag{5.47}$$

The expression $\Delta^1_{\mathrm{d}}(c, A)$ is a d-form, because A is a 1-form, F_t is a 2-form, while c is geometrically a scalar (i.e., $\delta_\xi c = \xi \cdot \partial c$), so that dc is a 1-form too. Under a diffeomorphism, which is the action of a vector field $\xi = \xi^\mu \frac{\partial}{\partial x^\mu}$, any form Φ transforms as $\delta_\xi \Phi = (i_\xi d + \partial\, i_\xi)\Phi$, where i_ξ is the inner product by the vector $\xi^\mu \frac{\partial}{\partial x^\mu}$. Therefore, using (2.12)

$$\delta_\xi \mathcal{A}[c, A] = 0$$

5.2.2 Diffeomorphism and Lorentz Cocycles

In theories including gravity symmetry under diffeomorphisms is a fundamental invariance. If the theory involves fermions also the local Lorentz symmetry enters the game. They are both fundamental symmetries which are not allowed to be broken by anomalies, the price being the inconsistency of the theory. The study of the corresponding anomalies is thus of utmost importance.

Anomalies of the local Lorentz symmetry are not formally different from the local gauge anomalies analysed so far. The role of gauge connection is played by the spin connection, which is denoted below in the matrix notation

$$\omega = \{\omega^{ab}\} \qquad \omega^{ab} = \omega^{ab}_\mu \mathrm{d}x^\mu \tag{5.48}$$

whose transformation law is

$$\delta_\Lambda \omega = d\Lambda + [\omega, \Lambda], \qquad \Lambda = \{\Lambda^{ab}\} \tag{5.49}$$

The curvature of ω is the Riemann two-form (see above)

$$R = d\omega + \frac{1}{2}[\omega, \omega], \qquad R = \{R^{ab}\} \tag{5.50}$$

Thus the general cocycle in dimension $\mathrm{d} = 2n - 2$ is obtained from formula (5.40) by simply making the replacements $c \to \Lambda$, $A \to \omega$ and $F \to R$:

$$\Delta^1_{\mathrm{d}}(\Lambda, \omega) = n(n-1) \int_0^1 dt (1-t) P_n(d\Lambda, \omega, R_t, \dots R_t) \tag{5.51}$$

In this case the anomaly is (up to an overall coefficient)

$$\mathcal{A}_L[\Lambda, \omega] = \int d^{\mathrm{d}}x \sqrt{g}\, \Delta^1_{\mathrm{d}}(\Lambda, \omega), \qquad \delta_\Lambda \mathcal{A}_L[\Lambda, \omega] = 0 \tag{5.52}$$

The $\sqrt{g}$ is introduced in order to guarantee diffeomorphism invariance, as explained in the above remark:

$$\delta_\xi \boldsymbol{A}_L[\Lambda, \omega] = 0 \tag{5.53}$$

Let us come next to the cocycles of diffeomorphisms. In order to exploit the parallelism with the gauge case we introduce, as before, for the Christoffel symbols the matrix-form notation

$$\Gamma \equiv \{\Gamma_\nu{}^\lambda\}, \qquad \Gamma_\nu{}^\lambda = dx^\mu\, \Gamma_{\mu\nu}{}^\lambda \tag{5.54}$$

and for the Riemannian curvature

$$\mathcal{R} = d\,\Gamma + \Gamma^2, \qquad \Gamma_t, \qquad \mathcal{R}_t = d\,\Gamma_t + \Gamma_t^2 = t\mathcal{R} + (t^2 - t)\Gamma^2 \tag{5.55}$$

The product between adjacent entries is the matrix product: $(XY)^\rho{}_\lambda = X^\rho{}_\nu Y^\nu{}_\lambda$. More explicitly,

$$\mathcal{R} = \{\mathcal{R}_\lambda{}^\rho\}, \qquad \mathcal{R}_\lambda{}^\rho = \frac{1}{2} dx^\mu \wedge dx^\nu R_{\mu\nu\lambda}{}^\rho$$
$$R^\rho{}_{\lambda\mu\nu} = \partial_\mu \Gamma^\rho_{\nu\lambda} - \partial_\nu \Gamma^\rho_{\mu\lambda} + \Gamma^\rho_{\mu\sigma}\Gamma^\sigma_{\nu\lambda} - \Gamma^\rho_{\nu\sigma}\Gamma^\sigma_{\mu\lambda} \tag{5.56}$$

Now we can adapt the previous cocycle formulas to this case

$$\Delta^1_{\mathrm{d}}(\xi, \Gamma) = n(n-1) \int_0^1 dt (1-t) \mathrm{Str}(d\,\Xi\, \Gamma\, \mathcal{R}_t \ldots \mathcal{R}_t) \tag{5.57}$$

where $\Xi_\lambda{}^\rho = \partial_\lambda \xi^\rho$, and Str denotes the symmetric trace of the matrix entries. This formula is justified because, using (2.12) and (2.18), one gets

$$\delta_\xi \Gamma^\rho{}_\lambda = (i_\xi d + d i_\xi)\Gamma^\rho{}_\lambda + d\,\Xi^\rho{}_\lambda + [\Gamma, \Xi]^\rho{}_\lambda \tag{5.58}$$
$$\delta_\xi \mathcal{R}^\rho{}_\lambda = (i_\xi d + d i_\xi)\mathcal{R}^\rho{}_\lambda + [\mathcal{R}, \Xi]^\rho{}_\lambda \tag{5.59}$$
$$\delta_\xi d\,\Xi^\rho{}_\lambda = (i_\xi d + d \imath_\xi) d\,\Xi^\rho{}_\lambda{}^\rho + [d\,\Xi, \Xi]^\rho{}_\lambda \tag{5.60}$$

The corresponding anomaly (up to an overall coefficient) is:

$$\boldsymbol{A}_{\mathrm{d}}[\xi, \Gamma] = \int d^{\mathrm{d}}x\, \sqrt{g}\, \Delta^1_{\mathrm{d}}(\xi, \Gamma) \tag{5.61}$$

which is obviously local Lorentz invariant.

The relation between R and $\mathcal{R}$ is given by Eq. (2.40).

Another Family of Diffeomorphism Cocycles

From the previous formulas (5.58, 5.59 and 5.60) it is evident that any diffeomorphism, acting on a field, splits into two parts: one looks like an ordinary gauge transformation, the other corresponds to the Lie derivative $\mathcal{L}_\xi = i_\xi d + d i_\xi$. The first transformation gives rise to the anomaly (5.61). The second can also give rise to cocycles of the form

$$\mathcal{A}_{\mathsf{d}}[\xi, B] = \int d^{\mathsf{d}}x\, \sqrt{g}\, \partial\cdot\xi\; B \tag{5.62}$$

where B is a local expression of the fields that transforms like a scalar: $\delta_\xi B = \xi \cdot \partial B$. This cocycle is consistent and, usually, has a Weyl partner (see next section), and form with it a non-trivial cocycle of the coupled cohomology.

5.3 Weyl Cocycles

In the following section it will become clear that for trace anomalies, even less than for chiral anomalies, it is not advisable to proceed blindly with perturbative calculations. It is of tremendous help to know in advance the form trace anomalies can take. This is provided by the cohomological analysis (for references see [18–22]).

The approach we will follow has been outlined in the previous section. In the trace anomaly case we are not so lucky as to have a general formula like the Chern one for gauge anomalies, which is valid in any dimension, incorporates the relevant descent equations, and contain, in particular, the consistency conditions. Therefore we have to proceed in a more down-to-earth way. The context where trace anomalies become important is in theories that preserve diffeomorphisms. So we will set out to determine diffeomorphism invariant cocycles of the nilpotent Weyl coboundary operator δ_ω, defined by (4.35) with the associated transformation rules (5.20) and (5.22). In a spacetime of dimension d the procedure consists in

1. listing all the diffeomorphism invariant 1-cochains (i.e., cochains linear in ω) and all the diffeomorphisms invariant 0-cochains (i.e., cochains independent of it) having d mass dimension;
2. upon acting with δ_ω on them, determining the combinations of 1-cochains that are annihilated by it (cocycles);
3. separating among the latter those that can be obtained by acting with δ_ω on 0-cochains (coboundaries) and those that cannot (non-trivial cocycles).

To carry out this program we will need the following transformation formulas

$$\delta_\omega \Gamma^\lambda_{\mu\nu} = \partial_\mu\omega\, \delta^\lambda_\nu + \partial_\nu\omega\, \delta^\lambda_\mu - g^{\lambda\rho}\partial_\rho\omega\, g_{\mu\nu} \tag{5.63}$$

$$\delta_\omega R_{\mu\nu\lambda}{}^{\rho} = -D_\mu\partial_\lambda\omega\, \delta^\rho_\nu + D_\nu\partial_\lambda\omega\, \delta^\rho_\mu + g^{\rho\sigma} D_\mu\partial_\sigma\omega\, g_{\nu\lambda} - g^{\rho\sigma} D_\nu\partial_\sigma\omega\, g_{\mu\lambda} \tag{5.64}$$

$$\delta_\omega R_{\mu\nu} = (2-\mathsf{d}) D_\mu D_\nu\omega - \Box\omega\, g_{\mu\nu} \tag{5.65}$$

$$\delta_\omega R = 2(1-\mathsf{d})\,\square\omega\, R - 2\omega\, R \tag{5.66}$$

$$\delta_\omega \sqrt{g} = \mathsf{d}\,\omega\,\sqrt{g} \tag{5.67}$$

where $R_{\mu\nu}$ and R are the Ricci tensor and scalar.

Weyl Cocycles in d = 2

In 2d the only diffeomorphism invariant 1-cochain is $\Delta^{(1)}[\omega, g] = \int d^2x\,\sqrt{g}\,\omega\, R$, and the only diffeomorphism-invariant 0-cochain is $\Delta^{(0)}[g] = \int d^2x\,\sqrt{g}\,R$. From (5.66) it is easy to prove that $\delta_\omega\Delta^{(1)}[\omega, g] = 0$ and $\delta_\omega\Delta^{(0)}[g] = 0$. Therefore $\Delta^{(1)}[\omega, g]$ is a non-trivial cocycle.

In 2d there is no possibility to construct a diff-invariant odd parity 1- or 0-cochain of dimension 2.

Weyl Cocycles in d = 4

In 4d the relevant 1-cochains are listed below together with their Weyl transforms

i	$\Delta_i^{(1)}[\omega, g]$	$\delta_\omega\Delta_i^{(1)}[\omega, g]$
1	$\int d^4x\,\sqrt{g}\,\omega\, R_{\mu\nu\lambda\rho}R^{\mu\nu\lambda\rho}$	$4\int d^4x\sqrt{g}\;R\,\omega\square\omega$
2	$\int d^4x\,\sqrt{g}\,\omega\, R_{\mu\nu}R^{\mu\nu}$	$4\int d^4x\,\sqrt{g}\,R\,\omega\square\omega$
3	$\int d^4x\,\sqrt{g}\,\omega\, R^2$	$12\int d^4x\,\sqrt{g}\,R\,\omega\square\omega$
4	$\int d^4x\,\sqrt{g}\,\omega\,\square R$	0
5	$\int d^4x\,\sqrt{g}\,\omega\,\varepsilon^{\mu\nu\mu'\nu'}R_{\mu\nu\lambda\rho}R_{\mu'\nu'}{}^{\lambda\rho}$	0

It is clear that $\Delta_4^{(1)}$ and $\Delta_5^{(1)}$ are cocycles, as is the combination $\sum_{i=1}^3 a_i\,\Delta_i^{(1)}$ provided $a_1 + a_2 + 3a_3 = 0$.

In order to see whether they are trivial or not let us consider the list of 0-cocycles with respective Weyl transformations

i	$C_i[g]$	$\delta_\omega C_i[g]$	
1	$\int d^4x\,\sqrt{g}\,R_{\mu\nu\lambda\rho}R^{\mu\nu\lambda\rho}$	$4\int d^4x\sqrt{g}\,\omega\,\square R$	(5.68)
2	$\int d^4x\,\sqrt{g}\,R_{\mu\nu}R^{\mu\nu}$	$4\int d^4x\,\sqrt{g}\,\omega\,\square R$	
3	$\int d^4x\,\sqrt{g}\,\omega\,R^2$	$12\int d^4x\,\sqrt{g}\omega\,\square R$	

Notice that we have not included $\int d^4x\,\sqrt{g}\,\square R$ and $\int d^4x\,\sqrt{g}\,\varepsilon^{\mu\nu\mu'\nu'}R_{\mu\nu\lambda\rho}R_{\mu'\nu'}{}^{\lambda\rho}$ among the 0-cocycles, because on a spacetime manifold without boundary they vanish identically.

From this it follows that $\Delta_4^{(1)}$ is a coboundary, while $\Delta_5^{(1)}$ as well as the combination $\sum_{i=1}^{3} a_i \, \Delta_i^{(1)}$ with $a_1 + a_2 + 3a_3 = 0$, are non-trivial cocycles. The latter therefore contains two independent linear combinations, which are chosen to be defined by the quadratic Weyl density

$$\mathcal{W}^2 = R_{\mu\nu\lambda\rho} R^{\mu\nu\lambda\rho} - 2R_{\mu\nu} R^{\mu\nu} + \frac{1}{3} R^2 \tag{5.69}$$

and the Gauss-Bonnet (or Euler) density,

$$E = R_{\mu\nu\lambda\rho} R^{\mu\nu\lambda\rho} - 4R_{\mu\nu} R^{\mu\nu} + R^2, \tag{5.70}$$

while $\Delta_5^{(1)}$ is characterized by the Pontryagin density

$$P = \frac{1}{2} \left(\varepsilon^{\mu\nu\mu'\nu'} R_{\mu\nu\lambda\rho} R_{\mu'\nu'}{}^{\lambda\rho} \right) \tag{5.71}$$

So that the trace of the energy-momentum tensor in 4d is expected to take the form

$$T_\mu{}^\mu = aE + c\mathcal{W}^2 + eP \tag{5.72}$$

The coefficients a, c and e are model-dependent.

Weyl Cocycles in d = 6

In d $= 6$ the most general 1-cochain is the superposition of 17 different elements

$$\Delta[\omega, g] = \sum_{i=1}^{17} a_i \Delta^{(i)}[\omega, g], \qquad \Delta^{(i)}[\omega, g] = \int d^6x \sqrt{g}\, \omega\, K_i[g] \tag{5.73}$$

where $K_i[g]$ are the following covariant expressions numbered from 1 to 17:

$$\begin{aligned}
&R^3, \quad R R_{\mu\nu} R^{\mu\nu}, \quad R R_{\mu\nu\lambda\rho} R^{\mu\nu\lambda\rho}, \quad R_{\mu\nu} R^{\nu\lambda} R_\lambda{}^\mu, \quad R_{\mu\nu} R^{\mu\lambda\rho\nu} R_{\lambda\rho} \\
&R_{\mu\nu} R^{\mu\lambda\rho\sigma} R^\nu{}_{\lambda\rho\sigma}, \quad R_{\mu\nu\lambda\rho} R^{\lambda\rho\sigma\tau} R^{\mu\nu}{}_{\sigma\tau}, \quad R_{\mu\nu\lambda\rho} R^{\tau\nu\lambda\sigma} R^\mu{}_{\tau\sigma}{}^\rho, \quad R \Box R \\
&R_{\mu\nu} \Box R^{\mu\nu}, \quad R^{\mu\lambda\rho\sigma} \Box R_{\mu\lambda\rho\sigma}, \quad R_{\mu\nu} D^\mu D^\nu R, \quad D^\rho R^{\mu\nu} D_\rho R_{\mu\nu} \\
&D^\rho R^{\mu\nu} D_\mu R_{\nu\rho}, \quad D^\mu R^{\nu\lambda\rho\sigma} D_\mu R_{\nu\lambda\rho\sigma}, \quad \Box R^2, \quad \Box^2 R
\end{aligned} \tag{5.74}$$

When applying δ_ω to any $\Delta^{(i)}[\omega, g]$ we get a sum of terms quadratic in ω. We will express all of them in terms of a basis of nine independent 2-cochains $\Omega^{(k)}[\omega, g] = \int d^6x \sqrt{g} H^{(k)}[\omega, g]$, with $k = 1, \ldots, 9$. The $H^{(k)}[\omega, g]$ are ordered from 1 to 9 in the following list

$$R^2\omega\Box\omega, \quad R_{\mu\nu}R^{\mu\nu}\omega\Box\omega, \quad RR_{\mu\nu}\,\omega D^\mu D^\nu\omega, \quad R^{\mu\lambda}R_\lambda{}^\nu\omega D_\mu D_\nu\omega \tag{5.75}$$
$$R^{\tau\nu\lambda\sigma}R_{\tau\nu\lambda\sigma}\omega\Box\omega, \quad \Box R\,\omega\Box\omega, \quad R\,\omega\Box^2\omega, \quad R^{\mu\nu}\omega D_\mu D_\nu\Box\omega, \quad \omega D_\mu D_\nu\omega R^{\mu\lambda\rho\nu}R_{\lambda\rho}$$

The consistency conditions can be written as follows

$$0 = \delta_\omega \Delta^{(i)}[\omega, g] = \sum_{k=1}^{9} f_k(a_i)\Omega^{(k)}[\omega, g] \tag{5.76}$$

where $f_k(a_i)$ are numerical linear combinations of the coefficients a_i. Equations (5.76) are a linear system of 9 equations in 17 unknowns a_i with characteristic 7. Thus the number of cocycles is 10. Now we have to determine which are the coboundaries. To this end we write down the most general diffeomorphism invariant

$$C[g] = \sum_{i=0}^{11} b_i C^{(i)}[g], \qquad C^{(i)}[g] = \int d^6x\,\sqrt{g}K^{(i)}[g] \tag{5.77}$$

These $K^{(i)}[g]$ are the first eleven expressions in the list (5.74) (the remaining 6 are easily seen to reduce upon partial integration to the first 10 or a combination thereof). The invariance condition is written

$$0 = \delta_\omega C[g] = \sum_{k=1}^{17} f'_k(b_i)\Delta^{(k)}[\omega, g] \tag{5.78}$$

where $f'_k(b_i)$ are numerical linear combinations of the unknowns b_i. This system has characteristic 6, which means that we have five invariants. Therefore we are left with 6 coboundaries. In total we have therefore $10 - 6 = 4$ non-trivial cocycles, which can be chosen to be

$$\mathcal{A}_i[\omega, g] = \int d^6x\,\sqrt{g}\,\omega\, M_i[g], \qquad i = 1, \ldots, 4 \tag{5.79}$$

where

$$\begin{aligned}
M_1[g] &= W_{\mu\nu\lambda\rho}W^{\tau\nu\lambda\sigma}W^\mu{}_{\tau\sigma}{}^\rho = \frac{19}{800}K_1 - \frac{57}{160}K_2 + \frac{3}{40}K_3 + \frac{7}{16}K_4 \\
&\quad -\frac{9}{8}K_5 - \frac{3}{4}K_6 + K_8 \\
M_2[g] &= W_{\mu\nu\lambda\rho}W^{\lambda\rho\sigma\tau}W_{\sigma\tau}{}^{\mu\nu} = \frac{9}{200}K_1 - \frac{27}{40}K_2 + \frac{3}{10}K_3 + \frac{5}{4}K_4 - \frac{3}{2}K_5 - 3K_6 + K_7 \\
M_3[g] &= K_1 - 12K_2 + 3K_3 + 16K_4 - 24K_5 - 24K_6 + 4K_7 + 8K_8 \\
M_4[g] &= -\frac{1}{3}K_1 - 8K_2 - 2K_3 + 10K_4 - 10K_5 + \frac{1}{2}K_9 - 5K_{10} + 5K_{11}
\end{aligned} \tag{5.80}$$

The tensor $W_{\mu\nu\lambda\rho}$ is the Weyl tensor in six dimensions and $M_3[g]$ corresponds to the Euler density. In 6d there cannot exist odd parity cocycles.

The expressions

$$\mathcal{I}_i[g] = \int d^6x \sqrt{g} M_i[g], \qquad i = 1, 2, 3 \tag{5.81}$$

are invariants. The other two invariants contain derivatives of the Riemann tensor or the Ricci tensor and scalar.

The most general solutions of (5.76) and (5.78) are recorded in Appendix 5C below. The linear systems in this section have been analyzed with Mathematica.

5.4 Perturbative Anomalies and Perturbative Cohomology

The cocycles of the previous section for different cohomologies have two characteristics that are generally not present in the perturbative approach: they are in a minimal form, that is they are full cocycles of a unique coboundary operator (for instance, of the gauge or diffeomorphism coboundary operator), and are invariant under the remaining symmetries of the theory. In general this is not the way anomalies appear in perturbative calculations. In fact (1) in perturbative calculations, except in a few fortunate cases, we do not obtain full cocycles but, rather, approximate expressions, usually at the lowest significant order of approximation; (2) in perturbative calculations we often obtain cocycles that satisfy, in an approximate way, the consistency conditions of the symmetry in question, but are not invariant under the other symmetries (not even in an approximate way). It is clear that we have to be more precise about the qualifier 'approximate': what do we mean by the sentence 'the consistency conditions are satisfied in an approximate way'? This question is answered by introducing *perturbative cohomology*. Perturbative cohomology is tailor-made to dovetail with the series expansion of perturbative field theory.

Let us start from the gauge transformations

$$\delta A = d\lambda + [A, \lambda], \qquad \delta\lambda = -\frac{1}{2}[\lambda, \lambda]_+, \qquad \delta^2 = 0, \qquad \lambda = \lambda^a(x)T^a \tag{5.82}$$

To adapt to the perturbative expansion it is useful to split δ into $\delta^{(0)} + \delta^{(1)}$ and consider A and λ infinitesimal (the latter anticommuting), defining

$$\begin{aligned}
&\delta^{(0)} A = d\lambda, \qquad \delta^{(0)}\lambda = 0, \qquad (\delta^{(0)})^2 = 0\\
&\delta^{(1)} A = [A, \lambda], \qquad \delta^{(1)}\lambda = -\frac{1}{2}[\lambda, \lambda]_+\\
&\delta^{(0)}\delta^{(1)} + \delta^{(1)}\delta^{(0)} = 0, \qquad (\delta^{(1)})^2 = 0
\end{aligned} \tag{5.83}$$

A transformation δ arranged as a sum of transformations, for instance $\delta = \delta^{(0)} + \delta^{(1)}$, is also called a filtration.

The full coboundary operator for diffeomorphisms is given by the transformations

$$\delta_\xi g_{\mu\nu} = D_\mu \xi_\nu + D_\nu \xi_\nu, \qquad \delta_\xi \xi^\mu = \xi^\lambda \partial_\lambda \xi^\mu \tag{5.84}$$

with $\xi_\mu = g_{\mu\nu}\xi^\nu$. We can introduce a perturbative cohomology, or graded cohomology, using as grading the order of infinitesimal, as follows

$$g_{\mu\nu} = \eta_{\mu\nu} + h_{\mu\nu}, \qquad g^{\mu\nu} = \eta^{\mu\nu} - h^{\mu\nu} + h^\mu_\lambda h^{\lambda\nu} + \cdots \tag{5.85}$$

Inserting the above expansions in (5.84) we see that we have a grading in the transformations, given by the order of infinitesimals. So we can define an (infinite) sequence of transformations (filtration)

$$\delta_\xi = \delta_\xi^{(0)} + \delta_\xi^{(1)} + \delta_\xi^{(2)} + \cdots$$

At the lowest level we find immediately

$$\delta_\xi^{(0)} h_{\mu\nu} = \partial_\mu \xi_\nu + \partial_\nu \xi_\mu, \qquad \delta_\xi^{(0)} \xi_\mu = 0 \tag{5.86}$$

and $\xi_\mu = \xi^\mu$. Since $(\delta_\xi^{(0)})^2 = 0$ this defines a cohomology problem in its own right.
At the next level we get

$$\delta_\xi^{(1)} h_{\mu\nu} = \xi^\lambda \partial_\lambda h_{\mu\nu} + \partial_\mu \xi^\lambda h_{\lambda\nu} + \partial_\nu \xi^\lambda h_{\mu\lambda}, \qquad \delta_\xi^{(1)} \xi^\mu = \xi^\lambda \partial_\lambda \xi^\mu \tag{5.87}$$

One can verify that

$$(\delta_\xi^{(0)})^2 = 0 \qquad \delta_\xi^{(0)} \delta_\xi^{(1)} + \delta_\xi^{(1)} \delta_\xi^{(0)} = 0, \qquad (\delta^{(1)})^2 = 0 \tag{5.88}$$

Proceeding in the same way we can define an analogous sequence of transformations for the Weyl transformations. From $g_{\mu\nu} = \eta_{\mu\nu} + h_{\mu\nu}$ and $\delta_\omega h_{\mu\nu} = 2\omega g_{\mu\nu}$ we find

$$\delta_\omega^{(0)} h_{\mu\nu} = 2\omega \eta_{\mu\nu}, \qquad \delta_\omega^{(1)} h_{\mu\nu} = 2\omega h_{\mu\nu}, \qquad \delta_\omega^{(2)} h_{\mu\nu} = 0, \ldots \tag{5.89}$$

as well as $\delta_\omega^{(0)} \omega = \delta_\omega^{(1)} \omega = 0, \ldots$.

Notice that we have $\delta_\xi^{(0)} \omega = 0$, $\delta_\xi^{(1)} \omega = \xi^\lambda \partial_\lambda \omega$. As a consequence we can extend (5.88) to

$$(\delta_\xi^{(0)} + \delta_\omega^{(0)})(\delta_\xi^{(1)} + \delta_\omega^{(1)}) + (\delta_\xi^{(1)} + \delta_\omega^{(1)})(\delta_\xi^{(0)} + \delta_\omega^{(0)}) = 0 \tag{5.90}$$

and $\delta_\xi^{(1)} \delta_\omega^{(1)} + \delta_\omega^{(1)} \delta_\xi^{(1)} = 0$, which, together with the previous relations, make

$$(\delta_\xi^{(0)} + \delta_\omega^{(0)} + \delta_\xi^{(1)} + \delta_\omega^{(1)})^2 = 0 \tag{5.91}$$

We consider next cases involving frame fields and local Lorentz transformations. The perturbative expansion for the vielbein is

$$e^a_\mu = \delta^a_\mu + \chi^a_\mu + \frac{1}{2}\psi^a_\mu + \cdots,$$

The fluctuation fields χ^a_μ contains a symmetric and an antisymmetric part, $\chi^a_\mu = s^a_\mu + a^a_\mu$. Since $e^a_\mu \eta_{ab} e^b_\nu = g_{\mu\nu}$, we must have

$$s_{\mu\nu} = \frac{1}{2}h_{\mu\nu}, \quad \psi_{\mu\nu} + \psi_{\nu\mu} = -\left(s^a_\mu s^a_\nu + a^a_\mu s^a_\nu + s^a_\mu a^a_\nu + a^a_\mu a_\nu\right), \ldots \tag{5.92}$$

while the antisymmetric parts are unconstrained. Since the local Lorentz transformation is $\delta_\Lambda e^a_\mu = e^c_\mu \Lambda_c{}^a$, where Λ is a local antisymmetric matrix, its perturbative expansion gives, in particular,

$$\delta^{(0)}_\Lambda s^a_\mu = 0, \qquad \delta^{(0)}_\Lambda a^a_\mu = \Lambda^a_\mu \tag{5.93}$$

In calculations where local Lorentz invariance is assumed to hold, i.e., local Lorentz transformations are a spectator, we usually set the antisymmetric part of the vielbein to 0, i.e., we fix the local Lorentz gauge.

Having defined a perturbative series for each BRST transformation, let us come now to the definition of perturbative cohomology. Let δ represent any of the above coboundary operator and $\delta\mathcal{A} = 0$ any of the consistency conditions considered above. We expand both terms in the LHS according to the just illustrated criteria (A, $h_{\mu\nu}$ and λ, ξ^μ, ω are infinitesimal of the first order), then we get

$$(\delta^{(0)} + \delta^{(1)} + \delta^{(2)} \cdots)(\mathcal{A}^{(0)} + \mathcal{A}^{(1)} + \mathcal{A}^{(2)} \cdots) = 0 \tag{5.94}$$

This identity decomposes into homogeneous ones

$$\delta^{(0)}\mathcal{A}^{(0)} = 0 \tag{5.95}$$

$$\delta^{(0)}\mathcal{A}^{(1)} + \delta^{(1)}\mathcal{A}^{(0)} = 0 \tag{5.96}$$

$$\cdots$$

Usually (but not always), as we shall see, the lowest order consistency condition (5.95) is enough to identify an anomaly.

Appendix 5A. Invariant Polynomials

This Appendix is devoted to a review of the adjoint-invariant symmetric polynomials used to define chiral anomalies. Let us consider a finite dimensional compact Lie group G with Lie algebra $\mathfrak{g}$ and let us introduce real-valued symmetric multi-linear mappings with n entries

$$P_n : \underbrace{\mathfrak{g} \times \cdots \times \mathfrak{g}}_{n \text{ factors}} \to \mathbb{R} \tag{5.97}$$

which are invariant under the adjoint action of G:

$$P_n(ad_g X_1, \ldots, ad_g X_n) = P_n(X_1, \ldots, X_n) \tag{5.98}$$

for any $g \in \mathsf{G}$. The infinitesimal version of (5.98) is

$$\sum_{i=1}^{n} P_n(X_1, \ldots, [Y, X_i], \ldots, X_n) = 0 \tag{5.99}$$

for any $Y \in \mathfrak{g}$. Let T^a, $(a = 1, \ldots, \dim\mathfrak{g})$, be a basis of $\mathfrak{g}$. If we set

$$h^{a_1 \ldots a_n} = P_n(T^{a_1}, \ldots, T^{a_n}) \tag{5.100}$$

and $X_i = X_i^a T^a$, we get $P_n(X_1, \ldots, X_n) = h^{a_1 \ldots a_n} X_1^{a_1} \ldots X_n^{a_n}$ (summation over repeated indices is understood).

Now take n $\mathfrak{g}$-valued forms $\omega_1, \ldots, \omega_n$ of order $q_1, \ldots, q_n$ respectively. Then we can write the $q_1 + \cdots + q_n$-form

$$P_n(\omega_1, \ldots, \omega_n) = h^{a_1 \ldots a_n} \omega_1^{a_1} \wedge \ldots \wedge \omega_n^{a_n} \tag{5.101}$$

A well-known realization of completely symmetric tensors is given by

$$d^{a_1 \ldots a_n} = \mathrm{Str}(T^{a_1} \ldots T^{a_n}) \tag{5.102}$$

whenever the generators are given in matrix form and Str means symmetrized trace. For instance

$$\mathrm{Tr}(T^a T^b) = c_2(R)\delta^{ab} \tag{5.103}$$

$$\frac{1}{2}\mathrm{Tr}(T^a \{T^b, T^c\}) = c_3(R) d^{abc} \tag{5.104}$$

where $c_2(R)$, $c_3(R)$ are representation-dependent numerical coefficients. As these two examples show, the tensors $d^{a_1 \ldots a_n}$ are universal, they are characteristic of the Lie algebra; changing the representation only modifies the numerical coefficients in front of them.

In the field theory literature these ad-invariant tensors are also called the *Casimirs* of the corresponding representation.

The case of non-compact version of the group G is treated in an analogous way. For instance the passage from SO(d) to SO(1, d − 1), is obtained by multiplying by i some of the generators, therefore the corresponding tensors (5.102) will be proportional to the compact ones by suitable coefficients.

The space of symmetric ad-invariant polynomials of the simple group G is usually denoted by $\mathbb{I}(\mathsf{G})$. It is a commutative algebra. If G has rank r, then $\mathbb{I}(\mathsf{G})$ has r algebraically independent generators of order $m_1, \ldots, m_r$. In other words we have r algebraically independent polynomials P_n , or symmetric tensors $h^{a_1 \ldots a_m}$, of order $m_1, \ldots, m_r$. The values $m_1, \ldots, m_r$ for the simple Lie algebras are as follows

Lie algebra	$m_1, \ldots, m_r$
A_r	$2, 3, \ldots r+1$
B_r	$2, 4, \ldots, 2r$
C_r	$2, 4, \ldots, 2r$
D_r	$2, 4, \ldots 2r-2, r$
G_2	$2, 6$
F_4	$2, 6, 8, 12$
E_6	$2, 5, 6, 8, 9, 12$
E_7	$2, 6, 8, 10, 12, 14, 18$
E_8	$2, 8, 12, 14, 18, 20, 24, 30$

This table tells a lot about anomalies. If a symmetric tensor is absent in this table, the corresponding cocycle, and thus the corresponding anomaly does not exist. This is the case, for instance, for SU(2) in 4d. The corresponding Lie algebra is A_1, which has only the second order symmetric tensor, while in order to construct a consistent chiral anomaly in 4d one needs the third order tensor. Analogously, the third order symmetric tensor does not exist for D_2, which is the Lie algebra of SO(4), the compact version of the Lorentz group in 4d. In fact the local Lorentz symmetry is not anomalous in 4d.

Appendix 5B. Descent Equations

In this Appendix we record the explict expressions of all the elements of the descent equations (5.35–5.38).

$$\Delta^{(0)}_{2n-1}(A) = n \int_0^1 dt\, P_n(A, F_t, \ldots, F_t) \tag{5.105}$$

$$\Delta^{(1)}_{2n-2}(A, c) = n(n-1) \int_0^1 dt\, (1-t) P_n(dc, A, F_t, \ldots, F_t) \tag{5.106}$$

$$\Delta^{(2)}_{2n-3}(A, c) = \frac{n(n-1)(n-2)}{2} \int_0^1 dt\, (1-t)^2 P_n(dc, dc, A, F_t, \ldots, F_t) \tag{5.107}$$

$$\ldots = \ldots$$

$$\Delta^{(k)}_{2n-k-1}(A, c) = \frac{n(n-1)\ldots(n-k)}{k!} \int_0^1 dt\, (1-t)^k P_n(\underbrace{dc, \ldots, dc}_{k \text{ entries}}, A, F_t, \ldots, F_t) \tag{5.108}$$

$$\ldots = \ldots$$

$$\Delta^{(n)}_{n-1}(c) = n \int_0^1 dt\, (t-1)^{n-1} P_n(c, dc, \ldots, dc) \tag{5.109}$$

$$\Delta^{(n+1)}_{n-2}(c) = -\frac{1}{2}\frac{n(n-1)}{n+1} \int_0^1 dt\, (t-1)^{n-1} P_n(c, [c, c], dc, \ldots, dc) \tag{5.110}$$

$$\ldots = \ldots$$

$$\Delta^{(n+k)}_{n-k-1}(c) = \left(-\frac{1}{2}\right)^k \frac{n(n-1)\ldots(n-k)}{(n+1)\ldots(n+k)} \int_0^1 dt\, (t-1)^{n-1} P_n(c, \underbrace{[c, c], \ldots, [c, c]}_{k \text{ entries}}, dc, \ldots, dc) \tag{5.111}$$

$$\ldots = \ldots$$

$$\Delta^{(2n-1)}_0(c) = \left(-\frac{1}{2}\right)^{n-1} \frac{n!}{(n+1)\ldots(2n-1)} \int_0^1 dt\, (t-1)^{n-1} P_n(c, [c, c], \ldots, [c, c]) \tag{5.112}$$

In Sect. 5.2 above we have argued that the so-obtained cocycles are non-trivial with a *reductio ad absurdum* argument.

Appendix 5C. Cocycles in 6d

In this Appendix we record the general invariants and cocycles for the cohomology in $\mathsf{d} = 6$, i.e., the most general solutions of (5.76) and (5.78). These linear systems have been solved with Mathematica. The most general invariant solutions can be represented by the following values of the coefficients b_i:

$$\begin{aligned}
b_6 &= -\frac{375}{2}b_1 - \frac{119}{4}b_2 - \frac{27}{2}b_3 - \frac{33}{8}b_4 + \frac{29}{8}b_5 \\
b_7 &= \frac{575}{4}b_1 + \frac{175}{8}b_2 + \frac{35}{4}b_3 + \frac{41}{16}b_4 - \frac{37}{16}b_5 \\
b_8 &= -175b_1 - \frac{47}{2}b_2 - 11b_3 - \frac{9}{4}b_4 + \frac{5}{4}b_5 \\
b_9 &= -15b_1 - 2b_2 - b_3 + \frac{1}{4}b_5 \\
b_{10} &= \frac{75}{2}b_1 + \frac{19}{4}b_2 + \frac{7}{2}b_3 - \frac{3}{8}b_4 - \frac{5}{8}b_5
\end{aligned}$$

$$b_{11} = \frac{75}{4}b_1 + \frac{23}{8}b_2 - \frac{1}{4}b_3 + \frac{9}{16}b_4 - \frac{5}{16}b_5 \tag{5.113}$$

while $b_1, \ldots, b_5$ are arbitrary. The invariants (5.80) are obtained by imposing that $b_9 = b_{10} = b_{11} = 0$. Therefore the other two independent invariants must contain terms with explicit derivatives.

The most general solutions of (5.76) in terms of the coefficients a_i are

$$\begin{aligned}
a_8 &= 8a_2 + 8a_3 + 4a_4 - 4a_5 + 4a_6 + 4a_7, \\
a_9 &= 3a_1 + \frac{87a_5}{100} - \frac{37a_2}{25} - \frac{107a_3}{25} - \frac{51a_4}{50} - \frac{93a_7}{50} - \frac{153a_6}{100}, \\
a_{11} &= -15a_1 + \frac{39a_2}{10} + \frac{77a_3}{5} + \frac{57a_4}{20} + \frac{103a_6}{20} + \frac{63a_7}{10} - a_{10} - \frac{57a_5}{20}, \\
a_{12} &= -6a_1 + \frac{79a_2}{25} + \frac{194a_3}{25} + \frac{117a_4}{50} + \frac{69a_6}{25} + \frac{78a_7}{25} - \frac{77a_5}{50}, \\
a_{13} &= -\frac{13a_2}{2} - 14a_3 + \frac{15a_5}{4} + a_{10} - \frac{15a_4}{4} - \frac{33a_7}{4} - \frac{49a_6}{8}, \\
a_{14} &= 7a_2 + 12a_3 + \frac{9a_4}{2} + \frac{23a_6}{4} + \frac{15a_7}{2} - \frac{7a_5}{2}, \\
a_{15} &= -15a_1 + \frac{27a_2}{5} + \frac{97a_3}{5} + \frac{18a_4}{5} + \frac{32a_6}{5} + \frac{39a_7}{5} - a_{10} - \frac{18a_5}{5}
\end{aligned} \tag{5.114}$$

while $a_1, \ldots, a_8$ and a_{10} are arbitrary. The three solutions (5.79) are obtained by imposing $a_9 = a_{10} = \cdots = a_{17} = 0$. The values (5.114) define 8 solutions. It is easily seen that $\Delta^{(16)}[\omega, g]$ and $\Delta^{(17)}[\omega, g]$ are also solutions of (5.79), but they are coboundaries. So, in fact, we have altogether 10 solutions of (5.79) and 6 coboundaries. The four nontrivial cocycles are recorded in (5.79). We remark that $M_4[g]$ contains explicit derivatives of the Riemann tensor and the Ricci tensor and scalar. It is easy to prove that the four cochains $\mathcal{A}_i[\omega, g]$, $i = 1, 2, 3, 4$ are indeed cocycles: they cannot be reproduced by acting with δ_ω on $C[g]$ for any value of the coefficients b_i.

References

1. J. Wess, B. Zumino, Consequences of anomalous Ward identities. Phys. Lett. B **37**, 95–97 (1971)
2. S.S. Chern, *Complex Manifolds without Potential Theory* (Springer, Berlin, 1969)
3. R. Grimm, S. Marculescu, The structure of anomalies for arbitrary dimension of the space-time. Nucl. Phys. B **68**, 203–213 (1974)
4. R. Stora, Continuum gauge theories, in *New Developments in Quantum Field Theory and Statistical Mechanics, (Cargèse 1976)*, ed. by M. Lévy Proceedings NATO ASI, Ser. B—Mathematical and Physical Sciences, vol. 26 (Plenum Press, 1977)
5. L. Bonora, P. Cotta-Ramusino, ABJ anomalies and superfield formalism in gauge theories. Phys. Lett. B **107**, 87 (1981)

6. L. Bonora, P. Cotta-Ramusino, Some remarks on BRS transformations, anomalies and the cohomology of the Lie algebra of the group of gauge transformations. Commun. Math. Phys. **87**, 589 (1982)
7. P.H. Frampton, T.W. Kephart, The analysis of anomalies in higher space-time dimensions. Phys. Rev. D **28**, 1010 (1983)
8. F. Langouche, T. Schücker, R. Stora, Gravitational anomalies of the Adler-Bardeen type. Phys. Lett. B **145**, 342 (1984)
9. L. Baulieu, Algebraic structure of quantum gravity and the classification of the gravitational anomalies. Phys. Lett. B **145**, 53–60 (1984)
10. R. Stora, Algebraic structure and topological origin of anomalies, in *Progress in Gauge Field Theory*, NATO ASI, Ser. B, vol. 115, ed. by G. 't Hooft, A. Jaffe, G. Lehmann, P.K. Mitter, I.M. Singer (Plenum Press, 1984)
11. W.A. Bardeen, B. Zumino, Consistent and covariant anomalies in gauge and gravitational theories. Nucl. Phys. B **244**, 421 (1984)
12. B. Zumino, W. Yong-Shi, A. Zee, Chiral anomalies, higher dimensions, and differential geometry. Nucl. Phys. B **239**, 477–507 (1984)
13. J. Mañes, R. Stora, B. Zumino, Algebraic study of chiral anomalies. Comm. Math. Phys. **102**, 157 (1985)
14. L. Bonora, P. Pasti, M. Tonin, The anomaly structure of theories with external gravity. J. Math. Phys. **27**, 2259 (1986)
15. M. DuBois-Violette, M. Talon, C.M. Viallet, B.R.S. algebras. Analysis of the consistency equations in gauge theories. Comm. Math. Phys. **102**, 105 (1985)
16. M. DuBois-Violette, M. Talon, C.M. Viallet, Anomalous terms in gauge theory: relevance of the structure group. Ann. Inst. H. Poincare Phys. Theor. **44**, 103–114 (1986)
17. R. Stora, The Wess Zumino consistency condition: a paradigm in renormalized perturbation theory. Fortsch. Phys. **54**, 175–182 (2006)
18. L. Bonora, P. Cotta-Ramusino, C. Reina, Conformal anomaly and cohomology. Phys. Lett. **126B**, 305 (1983)
19. L. Bonora, M. Bregola, P. Pasti, Weyl cocycles class. Quantum Grav. **3**, 635 (1986)
20. H. Osborn, Weyl consistency conditions and a local renormalization group equation for general renormalizable field theories. Nucl. Phys. B **363**, 486 (1991)
21. S. Deser, A. Schwimmer, Geometric classification of conformal anomalies in arbitrary dimensions. Phys. Lett. B **309**, 279 (1993). [hep-th/9302047]
22. F.M. Ferreira, I.L. Shapiro, Integration of trace anomaly in 6D. Phys. Lett. **B 772**, 174–178 (2017). e-Print: 1702.06892 [hep-th]

Part III
Perturbative Methods for Anomalies

This part of the book introduces and illustrates with many examples some perturbative methods to compute anomalies. Certainly, the most well-known, and historically the first, approach is the one based on Feynman diagrams. It led to the discovery of the ABJ anomaly and opened the way to all the subsequent developments. While calculating a triangle diagram in QED with fermions propagating in the internal lines, it was found that the relevant integral is logarithmically divergent. To assign a meaning to it, a regularization was introduced, which led to a finite result. This result, the anomaly, is unambiguous since it does not depend on the type of regulator used. In other words, it is an intrinsic property of the theory. The second discovery was that in theories of Weyl fermions, the anomaly is a signal that the theory is ill-defined. This of course spurred further investigations and the search for other methods of calculation that could confirm these discoveries. In this third part of the book, we illustrate some of these perturbative methods, mostly, but not exclusively, based on Feynman diagrams.

The first subject we deal with is the definition of Dirac-Weyl operator and its inverse. It is in fact in a theory of a Weyl fermion coupled to a vector gauge potential that we find the first example of consistent gauge anomaly. It is clear from the very beginning that one such theory is problematic. The problem of inverting the Dirac-Weyl operator in a perturbative approach can be formally solved by adding a free Weyl fermion of opposite chirality. This move does not jeopardize the chirality of the model. But it cannot avoid the chiral consistent anomaly, which shows up at one-loop. Such an anomaly puts out of contention the usual arguments that ensure unitarity and renormalizability. In the case the gauge potential is non-Abelian, there are more chances that it may happen that the anomaly vanishes identically if the group or the representation is suitably chosen. Another possibility is that in the theory, there are different species of fermions so that cancelations among different representations can occur. In this case, therefore, the inverse of the Dirac-Weyl operator, i.e. the corresponding propagator, exists and makes sense.

In Chap. 6 we compute this anomaly (the consistent chiral anomaly). In addition, we compute it in the case of a Dirac fermion coupled to a vector and an axial potential. This is also a consistent anomaly, from which, in particular, the previous consistent anomaly for Weyl fermions can be easily derived as a particular case. From it, we can also derive the ABJ anomaly in the limit in which the axial potential vanishes.

In Chap. 7 we continue the analysis with other examples of perturbative anomalies. We start with the 2d case, intended as a playground for (even and odd parity) trace and diffeomorphism anomaly calculations. We analyze them with two perturbative methods: a differential regularization of conformal correlators and a Feynman diagram approach. This allows us to appreciate the interplay between Weyl and diffeomorphism cohomology. Next we move on to 4d and study the trace and diffeomorphism anomalies generated by coupling fermions to vector and axial potentials. While diffeomorphisms are always preserved, we meet non-vanishing even and odd parity gauge-induced trace anomalies. Finally, we consider the case of Weyl fermions in a non-at metric background. We reproduce the well-known result that diffeomorphism anomalies are absent, while, again, one meets non-trivial trace anomalies. We actually compute only the odd-parity anomalies, the even parity ones being more accessible within non-perturbative approaches.

Chapter 6
Feynman Diagrams and Regularizations

The method based on Feynman diagrams is the first historically used and the one that led to the discovery of anomalies (the ABJ anomalies). Feynman diagrams are introduced and illustrated with plenty of examples in a number of textbooks. The history of the discovery of the ABJ anomaly in 4d is also well known. Therefore we will dispense with a detailed introduction to Feynman diagrams. As for the ABJ anomaly, it will be re-derived as a byproduct of a more general calculation. Here, instead, we focus directly on the derivation of chiral anomalies in gauge theories in 4d. The basic ingredient in this context is the study of triangle graphs, their regularization and the inherent problems. As will become more clear in later parts of the book the origin of anomalies is related to the non-existence of the fermion propagator. They will in fact appear as an obstruction in the context of the index theorem. But the existence of Weyl fermion propagators shows up as a serious threat already in the perturbative approach. The first part of this chapter is devoted to this problem and to how it can be circumvented. Then we deal with the explicit calculation of the consistent chiral anomaly in gauge theories in 4d with different regularizations. Further on we consider the theory of Dirac fermions coupled to a vector and an axial-vector potential, compute its anomalies and show how to recover from them both the ABJ anomaly and the consistent anomaly for Weyl fermions.

The bibliography for this chapter is rather vast. Here we limit ourselvs to an essential selection: [1–7].

6.1 Weyl Fermion Catastrophe and Rescue

The purpose of this section is to justify the use of the Dirac propagator for Weyl fermions in the internal lines of Feynman diagrams. To this end it is useful to take the general vantage point offered by the path integral.

L. Bonora, *Fermions and Anomalies in Quantum Field Theories*, Theoretical and Mathematical Physics, https://doi.org/10.1007/978-3-031-21928-3_6

Let us denote by $\not{D}$ the standard Dirac operator

$$\not{D} = i\left(\not{\partial} + \not{V}\right) = i\gamma^{\mu}\left(\partial_{\mu} + V_{\mu}\right) \tag{6.1}$$

where V_{μ} is any anti-Hermitean vector potential, including a spin connection when in the presence of a non-trivial background metric. We use here the four component formalism for fermions. The functional integral, i.e., the effective action for a quantum Dirac spinor in the presence of a classical background potential

$$\mathcal{Z}[V] = \int \mathcal{D}\psi\mathcal{D}\bar{\psi}\; e^{i\int d^4x\,\sqrt{g}\,\bar{\psi}\not{D}\psi} \tag{6.2}$$

is formally understood as the determinant of $\not{D}$: $\det(\not{D}) = \prod_i^{\infty} \lambda_i$, where λ_i are its eigenvalues.[1] From a concrete point of view, the latter can be operationally defined in two alternative ways: either in perturbation theory, i.e., as the sum of an infinite number of one-loop Feynman diagrams, some of which contain UV divergences by naive power counting, or by a non-perturbative approach, i.e., as the suitably regularized infinite product of the eigenvalues of $\not{D}$ by analytic continuation. The perturbative approach requires some UV regulator in order to assign a meaning to a finite number of UV one-loop diagrams which are divergent by naive power counting.

In many practical calculations one has to take variations of (6.2) with respect to V. In turn, any such variation requires the existence of an inverse of the kinetic operator, as follows from the abstract formula for the determinant of an operator A

$$\delta \det A = \det A \operatorname{tr}\left(A^{-1}\delta A\right). \tag{6.3}$$

It turns out that an inverse of $\not{D}$ does exist and, if full causality is required in forwards and backwards time evolution on e.g. Minkowski space, it is the Feynman propagator or Schwinger distribution $\$$. The latter is unique and characterized by the well-known Feynman prescription, in such a manner that

$$\not{D}_x \$(x-y) = \delta(x-y), \qquad \not{D}\$ = 1. \tag{6.4}$$

The second equation is a shortcut operator notation, which we are going to use again in the sequel.[2]

For instance, the scheme to extract the trace of the stress-energy tensor from the functional integral is well-known. It is its response under a Weyl transformation $\delta_{\omega} g_{\mu\nu} = 2\omega g_{\mu\nu}$:

$$\delta_{\omega} \log \mathcal{Z} = -i \int d^4x\, \omega(x)\, g^{\mu\nu}(x) \,\langle\!\langle\, T_{\mu\nu}(x) \rangle\!\rangle \tag{6.5}$$

---

[1] Of course this makes proper sense only in a Euclidean background.

[2] For simplicity we understand factors of $\sqrt{g}$, which should be there in the case of a non-trivial metric, but are inessential in this discussion.

where $g_{\mu\nu}(x)\langle\langle T^{\mu\nu}(x)\rangle\rangle$ is the quantum trace of the energy-momentum tensor. Analogously, the divergence of the vector current $j_\mu = i\bar{\psi}\gamma_\mu\psi$ is the response of $\log \mathcal{Z}$ under the Abelian gauge transformation $\delta_\lambda V_\mu = \partial_\mu \lambda$:

$$\delta_\lambda \log \mathcal{Z} = -i \int d^4x\, \lambda(x)\partial_\mu \langle\langle j^\mu(x)\rangle\rangle \tag{6.6}$$

and so on. These quantities can be calculated in various ways with perturbative or non-perturbative methods. As pointed out before, the most frequently used ones are the Feynman diagram technique and the so-called analytic functional method, respectively. The latter denomination actually includes a collection of approaches, based on the heat kernel equation, and includes the Schwinger's proper-time method and the Seeley-DeWitt method (with zeta-function regularization), as well as the Fujikawa method frequently used in field theory. The central tool in these approaches is the (full) kinetic operator of the fermion action (or the square thereof), and its inverse, the full fermion propagator. All these methods yield well-known results with no disagreement among them.

On the contrary, when we have to deal with Weyl fermions things drastically change. The classical action on the 4d Minkowski space for a left-handed Weyl fermion reads

$$S_L = \int \mathrm{d}^4x\, \bar{\psi}_L \not{D} \psi_L. \tag{6.7}$$

The Dirac operator, acting on left-handed spinors maps them to right-handed ones. Hence, the Sturm-Liouville or eigenvalue problem itself is not well posed, so that the Weyl determinant cannot even be defined. This is reflected in the fact that the inverse of $\not{D}_L = \not{D} P_L = P_R \not{D}$ does not exist, since it is the product of an invertible operator times a projector. As a consequence the full propagator of a Weyl fermion does not exist in this naive form.

The remedy for the Weyl fermion disaster consists in using as kinetic operator

$$i\gamma^\mu \left(\partial_\mu + P_L V_\mu\right), \tag{6.8}$$

which is invertible. In the Feynman diagram approach this means using for Weyl fermions the ordinary free Feynman propagator for Dirac fermions. This corresponds to the intuition that the free right-handed fermions added to the left-handed theory in this way do not interfere with the conservation of chirality and do not alter the left-handed nature of the theory. The reason is that the information about chirality is preserved by the fermion-boson-fermion vertex, which contains the P_L projector. On the contrary, the full (non-perturbative) propagator must contain the full chiral information, including the information contained in the vertex, i.e., the potential. Needless to say, in this problem there is no simple shortcut such as pretending to replace the full Weyl propagator with the full Dirac propagator multiplied by a chiral projector, because this would counterfeit the information concerning chirality. One has to use (6.8) and its inverse, as will be further motivated in the sequel.

Returning to the perturbative approach, it is important to explicitly check that the free right-handed fermions added to the left-handed theory do not interfere with the conservation of chirality and do not alter the left-handed nature of the theory. The rest of this section is devoted to a close inspection of this problem and its solution via a relatively easy example, [8].

6.1.1 Regularisations for Weyl Spinors

To show that replacing the (non-existing) Weyl fermion propagator with a Dirac propagator in Feynman diagrams does not jeopardize the conservation of chirality, let us consider in this section the model of a left-handed Weyl fermion $\chi_L = \begin{pmatrix} \chi(x) \\ 0 \end{pmatrix}$ minimally coupled to an Abelian gauge potential $A^\mu(x)$. Using the formula of Sect. 1.1 of Chap. 1, the classical action is

$$S = \int \mathrm{d}^4x \, \left(\chi_L^\dagger \, \alpha^\nu i \partial_\nu \chi_L + g A^\nu \, \chi_L^\dagger \, \alpha_\nu \chi_L - \tfrac{1}{4} \, F^{\mu\nu} \, F_{\mu\nu} \right) \tag{6.9}$$

where $F_{\mu\nu} = \partial_\mu A_\nu - \partial_\nu A_\mu$. This action is invariant under the gauge transformation

$$\chi_L'(x) = e^{ig\theta(x)} \chi_L(x) \qquad A_\nu'(x) = A_\nu(x) + \partial_\nu \theta(x)$$

The question naturally arises as to whether the symmetries of (6.9) hold true after the transition to the quantum theory and, in particular, if they are protected against loop radiative corrections within the perturbative approach. Now in order to develop perturbation theory, one faces the problem of the lack of an inverse for both the Dirac-Weyl operator, as anticipated above, but also of the gauge field kinetic operator, owing, respectively, to chirality and gauge invariance. In order to solve it, it is expedient to add to the Lagrangian non-interacting terms, which are fully decoupled from any physical quantity. They break chirality and gauge invariance, albeit in a harmless way, just to allow us to define a Feynman propagator, or causal Green's function, for both the Weyl and gauge quantum fields. The simplest choice, which preserves Poincaré and internal U(1) phase change symmetries, is provided by

$$\Delta \mathcal{L} = \varphi_R^\dagger \, \alpha^\nu i \partial_\nu \varphi_R - \tfrac{1}{2} (\partial \cdot A)^2 \tag{6.10}$$

where

$$\varphi_R(x) = \begin{bmatrix} 0 \\ \varphi(x) \end{bmatrix}$$

is a left-chirality breaking right-handed Weyl spinor field. Notice incidentally that the modified Lagrangian $\mathcal{L} + \Delta \mathcal{L}$ exhibits a further U(1) internal symmetry under the so called chiral phase transformations

$$\psi'(x) = (\cos\theta + i\sin\theta\,\gamma_5)\psi(x) \qquad \psi(x) = \begin{Bmatrix} \chi(x) \\ \varphi(x) \end{Bmatrix}$$

so that the modified theory involves another conserved charge at the classical level. From the modified Lagrangian density we get the Feynman propagators for the massless Dirac field $\psi(x)$, as well as for the massless vector field in the so called Feynman gauge: namely,

$$S(p) = \frac{i\not{p}}{p^2 + i\varepsilon} \qquad D_{\mu\nu}(k) = \frac{-i\eta_{\mu\nu}}{k^2 + i\epsilon} \tag{6.11}$$

and the vertex $ig\gamma^\nu P_L$, with $k + p - q = 0$, which involves a vector particle of momentum k and a Weyl pair of particle and anti-particle of momenta p and q respectively and opposite helicity.

Our purpose hereafter is to show that, notwithstanding the use of the non-chiral propagators (6.11), a mass in the Weyl kinetic term cannot arise as a consequence of quantum corrections. The lowest order one-loop correction to the kinetic term $\not{k} P_L$ is provided, by the Feynman rules in Minkowski space, in the following form

$$\Sigma_2(\not{k}) = -ig^2 \int \frac{\mathrm{d}^4 p}{(2\pi)^4}\, \gamma^\mu\, D_{\mu\nu}(k-p)\, S(p)\, \gamma^\nu P_L, \tag{6.12}$$

which comes from a loop diagram formed by a fermion propagator S and a gauge propagator $D_{\mu\nu}$. A mass term in this context should be proportional to the identity matrix (in the spinor space).

By naïve power counting the above one-loop integral turns out to be UV divergent. Hence a regularisation procedure is mandatory in order to give a meaning and evaluate the radiative correction $\Sigma_2(\not{k})$ to the Weyl kinetic operator. Hereafter we shall examine in detail the dimensional, Pauli-Villars and UV cut-off regularizations.

6.1.2 Dimensional, PV and Cutoff Regularizations

In a $2\omega-$dimensional spacetime the dimensionally regularized radiative correction to the Weyl kinetic term takes the form

$$\mathtt{reg}\,\Sigma_2(\not{k}) = -ig^2\mu^{2\epsilon} \int \frac{\mathrm{d}^{2\omega} p}{(2\pi)^{2\omega}}\, D_{\mu\nu}(p)\,\gamma^\mu\, S(p+k)\,\gamma^\nu P_L \tag{6.13}$$

where $\epsilon = 2 - \omega < 0$ is the shift with respect to the physical space time dimensions. μ is a mass scale parameter introduced in order to guarantee dimensionlessness for quantites to be analytically continued. Since the above expression is traceless and has the canonical engineering dimension of a mass in natural units, it is quite apparent that the latter cannot generate any mass term, which, as anticipated above, would be proportional to the unit matrix. Hence, mass is forbidden and it remains for us to evaluate

$$\mathtt{reg}\,\Sigma_2(\not{k}) \equiv f(k^2)\,\not{k} P_L \qquad \mathrm{tr}\,[\not{k}\mathtt{reg}\,\Sigma_2(\not{k})] = \tfrac{1}{2}\,2^{\omega} k^2 f(k^2) \tag{6.14}$$

$$\mathrm{tr}\,[\not{k}\;\mathtt{reg}\,\Sigma_2(\not{k})] = -ig^2\mu^{2\epsilon}\int \frac{\mathrm{d}^{2\omega}p}{(2\pi)^{2\omega}} \frac{\mathrm{tr}\left(\not{k}\gamma^{\lambda}\not{p}/\gamma_{\lambda}P_L\right)}{\left[\,(p-k)^2+i\epsilon\,\right]\left(p^2+i\epsilon\right)}. \tag{6.15}$$

(6.14) follows from the Breitenlohner-Maison rules, [9], for $2^{\omega}\times 2^{\omega}$ γ−matrices in a 2ω−dimensional spacetime with a Minkowski signature. They are

$$\gamma^{\mu} = \begin{cases} \bar{\gamma}^{\mu} & \mu = 0,1,2,3 \\ \hat{\gamma}^{\mu} & \mu = 4,\ldots,2\omega-4 \end{cases}$$
$$\{\bar{\gamma}^{\mu},\bar{\gamma}^{\nu}\} = 2\bar{\eta}^{\mu\nu}\,\mathbb{I} \qquad \{\hat{\gamma}^{\mu},\hat{\gamma}^{\nu}\} = 2\hat{\eta}^{\mu\nu}\,\mathbb{I} \qquad \{\bar{\gamma}^{\mu},\hat{\gamma}^{\nu}\} = 0$$
$$\|\,\bar{\eta}\,\| = \mathrm{diag}\,(+,-,-,-) \qquad \|\,\hat{\eta}\,\| = -\,\hat{\mathbb{I}}$$
$$\gamma_5 \equiv i\bar{\gamma}^0\bar{\gamma}^1\bar{\gamma}^2\bar{\gamma}^3 \qquad \gamma_5^2 = \mathbb{I} \qquad \{\bar{\gamma}^{\mu},\gamma_5\} = 0 = \left[\,\hat{\gamma}^{\mu},\gamma_5\,\right]$$

from which follows

$$\begin{aligned} \mathrm{tr}\,(\gamma^{\mu}\gamma^{\nu}) &= \eta^{\mu\nu}\,\mathrm{tr}\,\mathbb{I} \;=\; 2^{\omega}\,\eta^{\mu\nu} \\ 2^{-\omega}\mathrm{tr}\left(\gamma^{\kappa}\gamma^{\lambda}\gamma^{\mu}\gamma^{\nu}\right) &= \eta^{\kappa\lambda}\,\eta^{\mu\nu} \;-\; \eta^{\kappa\mu}\,\eta^{\lambda\nu} \;+\; \eta^{\kappa\nu}\,\eta^{\lambda\mu} \\ \mathrm{tr}\left(\bar{\gamma}^{\kappa}\bar{\gamma}^{\lambda}\hat{\gamma}^{\mu}\hat{\gamma}^{\nu}\right) &= 2^{\omega}\,\bar{\eta}^{\kappa\lambda}\,\hat{\eta}^{\mu\nu} \\ \mathrm{tr}\left(\gamma_5\bar{\gamma}^{\mu}\bar{\gamma}^{\lambda}\bar{\gamma}^{\rho}\bar{\gamma}^{\nu}\right) &= -\,i\,2^{\omega}\varepsilon^{\mu\lambda\rho\nu} \end{aligned}$$

Tracing the gamma matrices in the RHS of (6.15) according to these definitions, the γ_5 piece does not contribute and the rest gives

$$k^2\,f(k^2) = ig^2\mu^{2\epsilon}\int \frac{\mathrm{d}^{2\omega}p}{(2\pi)^{2\omega}} \frac{2p\cdot k}{\left[\,(p-k)^2+i\epsilon\,\right]\left(p^2+i\epsilon\right)}. \tag{6.16}$$

Turning to the Feynman parametric representation we obtain

$$k^2\,f(k^2) = ig^2\mu^{2\epsilon}\int_0^1 \mathrm{d}x \int \frac{\mathrm{d}^{2\omega}p}{(2\pi)^{2\omega}} \frac{2p\cdot k}{\left[\,p^2-2x\,k\cdot p+xk^2+i\epsilon\,\right]^2}. \tag{6.17}$$

Completing the square in the denominator and after shifting to the momentum $p' \equiv p - xk$, dropping the linear term in p' in the numerator owing to symmetric integration, we have

$$f(k^2) = 2ig^2\mu^{2\epsilon}\int_0^1 \mathrm{d}x\,x \int \frac{\mathrm{d}^{2\omega}p}{(2\pi)^{2\omega}} \frac{1}{\left[\,p^2+x(1-x)k^2+i\epsilon\,\right]^2}. \tag{6.18}$$

Now we perform a Wick rotation $p^0 \to ip^0, k^0 \to ik^0$ and use the integrals in Appendix 7A (with $\omega = 2 + \frac{\delta}{2}$). Then we expand around $\epsilon = 0$ and get

$$f(k^2) = 2\left(\frac{g}{4\pi}\right)^2 \int_0^1 \mathrm{d}x\, x \left(\frac{1}{\epsilon} - \gamma - \log\left(\frac{x(1-x)k^2}{4\pi\mu^2}\right)\right) \tag{6.19}$$

where γ denotes the Euler-Mascheroni constant. Integrating in x finally we get

$$f(k^2) = \left(\frac{g}{4\pi}\right)^2 \left[\frac{1}{\epsilon} + 2 - \gamma + \ln\left(\frac{4\pi\mu^2}{k^2}\right)\right] + \mathcal{O}(\epsilon) \tag{6.20}$$

Similar results are obtained with the Pauli-Villars and cut-off regularizations. In the PV case the latter is simply implemented by the following replacement of the massless Dirac propagator

$$\texttt{reg}\,\Sigma_2(\not{k}) = -ig^2 \int \frac{\mathrm{d}^4 l}{(2\pi)^4}\, \gamma^\mu D_{\mu\nu}(k-l) \sum_{s=0}^{S} C_s\, S(l,\, M_s)\, \gamma^\nu P_L, \tag{6.21}$$

where $M_0 = 0,\ C_0 = 1$ while $\{\, M_s \equiv \lambda_s M \mid \lambda_s \gg 1\ (s = 1, 2, \ldots, S)\,\}$ is a collection of very large auxiliary masses. The constants C_s are required to satisfy:

$$\sum_{s=1}^{S} C_s = -1 \qquad \sum_{s=1}^{S} C_s \lambda_s = 0$$

and the following identification with the divergent parameter is made

$$\frac{1}{\epsilon} = \sum_{s=1}^{S} C_s \ln \lambda_s .$$

The result for $f(k^2)$ is

$$f(k^2) = \left(\frac{g}{4\pi}\right)^2 \left[\sum_{s=1}^{S} C_s \ln \lambda_s + \frac{1}{4} + \frac{1}{2}\ln\left(\frac{M^2}{k^2}\right)\right] + \text{evanescent}. \tag{6.22}$$

The same calculation can be repeated with an UV cutoff K. To sum up, we have verified that the one-loop correction to the (left) Weyl spinor self-energy has the general form, which is universal, i.e., regularisation independent: namely, after an inverse Wick rotation,

$$\begin{aligned}
\mathtt{reg}\,\Sigma_2(\not k) &\equiv f(k^2)\,\not k P_L \\
f(k^2) &:= \left(\frac{g}{4\pi}\right)^2\left[\frac{1}{\epsilon}+2-\gamma+\ln\left(-\frac{4\pi\mu^2}{k^2}\right)\right] \qquad (\mathrm{DR}) \\
&:= \left(\frac{g}{4\pi}\right)^2\left[\sum_{s=1}^{S} C_s\,\ln\lambda_s+\frac{1}{4}+\frac{1}{2}\ln\left(-\frac{M^2}{k^2}\right)\right] \qquad (\mathrm{PV}) \\
&:= \left(\frac{g}{4\pi}\right)^2\ln\left[-\frac{(\Lambda K)^2}{k^2}\right] \qquad (\mathrm{CUT-OFF})
\end{aligned}$$

Comments

1. In the present model of a left Weyl spinor minimally coupled to a gauge vector potential, *no mass term can be generated by the radiative corrections* in any regularization scheme. The left-handed part of the classical kinetic term does renormalize, while its right-handed part does not undergo any radiative correction and keeps on being free. The latter has to be necessarily introduced in order to define a Feynman propagator for the massless spinor field, much like the gauge fixing term is introduced in order to invert the kinetic term of the gauge potential.
2. The interaction definitely preserves left chirality and scale invariance of the counterterms in the transition from the classical to the (perturbative) quantum theory: no mass coupling between the left-handed (interacting) Weyl spinor χ_L and right-handed (free) Weyl spinor φ_R can be generated by radiative loop corrections.
3. Of course this is not the end of the story. The previous result guarantees that chirality is preserved by quantization, at least at the lowest perturbative order. But it does not guarantee that unitarity is preserved. As we shall see next an anomaly may arise, and does arise, that put in jeopardy the whole theory. The problem of inverting the Dirac-Weyl operator strikes back in another form.
4. While the cut-off and dimensional regularized theories admit a local formulation in $\mathsf{d}=4$ or $\mathsf{d}=2\omega$ spacetime dimensions, there is no such local formulation for the Pauli-Villars regularization. The reason is that the PV spinor propagator

$$\sum_{s=0}^{S} C_s\,S(\ell,\,M_s)$$

where $M_0=0,\ C_0=1$ while $\{M_s\equiv\lambda_s\,M\mid\lambda_s\gg 1\ (s=1,2,\ldots,S)\}$, cannot be the inverse of any local differential operator of the Calderon-Zygmund type. Hence, there is no local action involving a bilinear spinor term that can produce, after a suitable inversion, the Pauli-Villars regularized spinor propagator. So, in view of the application of the Seeley-Schwinger-DeWitt method in later chapters, we can anticipate that the Pauli-Villars regularization cannot be applied to the construction of a regularised full kinetic operator for such non-perturbative method, nor, of course, to its inverse.

6.2 Consistent Chiral Gauge Anomalies for Weyl Fermions

In the present and forthcoming sections we aim to calculate all the chiral anomalies of Weyl, Dirac and Majorana fermions coupled to gauge potentials, in flat background, with the basic method of Feynman diagrams, [8]. Having clarified the issue of the Weyl fermion propagator we start with the calculation of the consistent chiral anomaly for Weyl fermions, which was historically the first and of outstanding importance for its dramatic consequences.

Let us consider the classical action integral for a right-handed Weyl fermion coupled to an external gauge field $V_\mu = V_\mu^a T^a$, T^a being Hermitean generators, $[T^a, T^b] = i f^{abc} T^c$ (in the Abelian case $T = 1$, $f = 0$) in the fundamental representation of a compact group G:

$$S_R[V] = \int d^4x\, i\, \overline{\psi_R}\left(\partial\!\!\!/ - i V\!\!\!/\right)\psi_R. \tag{6.23}$$

This action is invariant under the gauge transformation $\delta V_\mu = D_\mu \lambda \equiv \partial_\mu \lambda - i[V_\mu, \lambda]$, which implies the conservation of the non-Abelian current $J^a_{R\mu} = \bar{\psi}_R \gamma_\mu T^a \psi_R$, i.e.,

$$(D\cdot J_R)^a \equiv (\partial^\mu \delta^{ac} + i f^{abc} V^{b\mu}) J^c_{R\mu} = 0. \tag{6.24}$$

The quantum effective action for this theory is given by the generating functional of the connected Green's functions of such currents in the presence of the source $V^{a\mu}$

$$W[V] = W[0] + \sum_{n=1}^{\infty} \frac{i^{n-1}}{n!} \int \prod_{i=1}^{n} d^4x_i\, V^{a_i\mu_i}(x_i)\, \langle 0|\mathcal{T} J^{a_1}_{R\mu_1}(x_1)\ldots J^{a_n}_{R\mu_n}(x_n)|0\rangle_c \tag{6.25}$$

and the full one-loop 1-point function of $J^a_{R\mu}$ is

$$\langle\!\langle J^a_{R\mu}(x)\rangle\!\rangle = \frac{\delta W[V]}{\delta V^{a\mu}(x)} = \sum_{n=0}^{\infty} \frac{i^n}{n!} \int \prod_{i=1}^{n} d^4x_i\, V^{a_i\mu_i}(x_i) \langle 0|\mathcal{T} J^a_{R\mu}(x) J^{a_1}_{R\mu_1}(x_1)\ldots J^{a_n}_{R\mu_n}(x_n)|0\rangle_c \tag{6.26}$$

Our purpose here is to calculate the odd parity anomaly of the divergence $D\cdot\langle\!\langle J^a_R\rangle\!\rangle$ (the even parity divergence vanishes identically, see Appendix 6B). The RHS of (6.26) contains an infinite series of current amplitudes. The one-point amplitude vanishes as a consequence of translational invariance. The two-point correlator is non-vanishing but, for algebraic reasons, it cannot contain an odd parity part (and its divergence anyhow vanishes, see Appendix 6B). Therefore the first nontrivial contribution to the anomaly comes from the divergence of the three-point function in the RHS of (6.26). For simplicity we will denote it $\langle \partial\cdot J_R\, J_R\, J_R\rangle$. Below we evaluate it in some detail as a sample for other similar subsequent calculations.

6.2.1 The Calculation

Let us start with dimensional regularization. The fermion propagator is $\frac{i}{\not p}$ (here and below in general the $i\epsilon$ prescription is understood) and the vertex $i\gamma_\mu P_R T^a$. The Fourier transform of the three currents amplitude $\langle J_R\, J_R\, J_R\rangle$ is given by

$$\begin{aligned}\widetilde{F}^{(R)abc}_{\mu\lambda\rho}(k_1,k_2) &= \int \frac{d^4p}{(2\pi)^4}\mathrm{Tr}\left\{\frac{1}{\not p}\gamma_\lambda\frac{1+\gamma_5}{2}T^b\frac{1}{\not p-\not k_1}\gamma_\rho\frac{1+\gamma_5}{2}T^c\frac{1}{\not p-\not q}\gamma_\mu\frac{1+\gamma_5}{2}T^a\right\}\\ &\equiv \mathrm{Tr}(T^aT^bT^c)\widetilde{F}^{(R)}_{\mu\lambda\rho}(k_1,k_2) \end{aligned} \tag{6.27}$$

where $q = k_1 + k_2$. To it we must add the cross term (for $\lambda \leftrightarrow \rho$ and $k_1 \leftrightarrow k_2$).

From now on we focus on the Abelian part $\widetilde{F}^{(R)}_{\mu\lambda\rho}(k_1,k_2)$. We dimensionally regularize it (in a slightly more practical way than before) by introducing δ additional dimensions and corresponding momentum $\ell_{\hat\mu}$, $\hat\mu = 4,\dots,3+\delta$ to be added to the original running momentum p_μ, with the properties

$$\begin{aligned}&\not\ell\not p+\not p\not\ell=0,\qquad [\not\ell,\gamma_5]=0,\qquad \not p^2=p^2,\qquad \not\ell^2=-\ell^2\\ &\mathrm{tr}(\gamma_\mu\gamma_\nu\gamma_\lambda\gamma_\rho\gamma_5)=-2^{2+\frac{\delta}{2}}\,i\,\varepsilon_{\mu\nu\lambda\rho},\qquad \mathrm{tr}(\gamma_\mu\gamma_\nu)=2^{2+\frac{\delta}{2}}\eta_{\mu\nu},\qquad \dots\end{aligned}$$

so the relevant expression to be calculated is

$$\begin{aligned} q^\mu\widetilde{F}^{(R)}_{\mu\lambda\rho}(k_1,k_2) &= \int\frac{d^4p\,d^\delta\ell}{(2\pi)^{4+\delta}}\mathrm{tr}\left\{\frac{1}{\not p+\not\ell}\gamma_\lambda\frac{1+\gamma_5}{2}\frac{1}{\not p+\not\ell-\not k_1}\gamma_\rho\frac{1+\gamma_5}{2}\frac{1}{\not p+\not\ell-\not q}\not q\frac{1+\gamma_5}{2}\right\}\\ &= \int\frac{d^4p\,d^\delta\ell}{(2\pi)^{4+\delta}}\mathrm{tr}\left\{\frac{\not p}{p^2-\ell^2}\gamma_\lambda\frac{\not p-\not k_1}{(p-k_1)^2-\ell^2}\gamma_\rho\frac{\not p-\not q}{(p-q)^2-\ell^2}\frac{1-\gamma_5}{2}\not q\right\}\\ &\equiv \widetilde{F}^{(R)}_{\lambda\rho}(k_1,k_2,\delta). \end{aligned} \tag{6.28}$$

The relevant Feynman diagram is shown in Fig. 6.1.

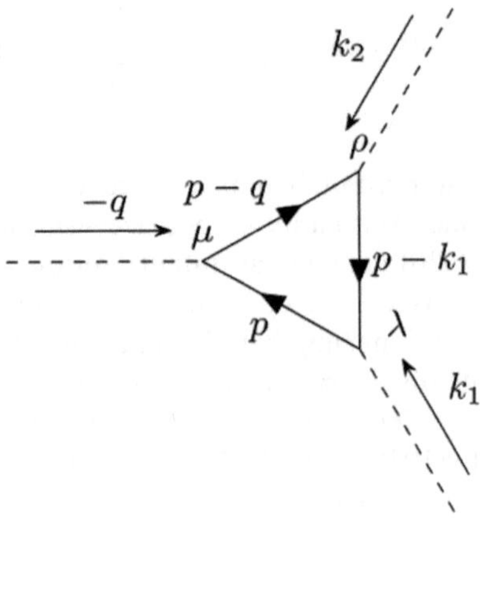

Fig. 6.1 The Feynman diagram corresponding to $\widetilde{F}^{(R)}_{\mu\lambda\rho}(k_1,k_2)$

Note. Dimensional regularization is based on analytic continuation in δ, which makes sense only for dimensionless quantities. To make the quantities one wishes to evaluate dimensionless it is customary to introduce a mass scale parameter μ, like we have done in the previous section. In general, in the sequel, whenever μ does not play a significant role, we understand it, or, alternatively, we set $\mu = 1$. Whenever opportune we reinsert it in the final results, by suitably rescaling the momenta.

Now we focus on the odd part (the even part vanishes, see Appendix 6B) and work out the gamma traces:

$$\widetilde{F}^{(R,odd)}_{\lambda\rho}(k_1, k_2, \delta) = -2^{1+\frac{\delta}{2}}\, i\varepsilon_{\mu\nu\lambda\rho} \int \frac{d^4p d^\delta \ell}{(2\pi)^{4+\delta}} \frac{\left(p^2 q^\mu + (q^2 - 2p\cdot q)p^\mu\right)\left(p^\nu - k_1^\nu\right)}{(p^2 - \ell^2)((p-k_1)^2 - \ell^2)((p-q)^2 - \ell^2)} \tag{6.29}$$

Let us write the numerator on the RHS as follows:

$$p^2 q^\mu + (q^2 - 2p\cdot q)p^\mu = -(p^2 - \ell^2)(p-q)^\mu + ((p-q)^2 - \ell^2)p^\mu + \ell^2 q^\mu. \tag{6.30}$$

Then (6.29) can be rewritten as

$$\begin{aligned}
\widetilde{F}^{(R,odd)}_{\lambda\rho}(k_1, k_2, \delta) &= -2^{1+\frac{\delta}{2}}\, i\varepsilon_{\mu\nu\lambda\rho} \int \frac{d^4p d^\delta \ell}{(2\pi)^{4+\delta}} \Bigg\{ \ell^2 \frac{q^\mu \left(p^\nu - k_1^\nu\right)}{(p^2 - \ell^2)((p-k_1)^2 - \ell^2)((p-q)^2 - \ell^2)} \\
&\quad + \frac{(q-p)^\mu (p-k_1)^\nu}{((p-k_1)^2 - \ell^2)((p-q)^2 - \ell^2)} + \frac{p^\mu (p-k_1)^\nu}{(p^2 - \ell^2)((p-k_1)^2 - \ell^2)} \Bigg\} \\
&= -2^{1+\frac{\delta}{2}}\, i\varepsilon_{\mu\nu\lambda\rho} \int \frac{d^4p d^\delta \ell}{(2\pi)^{4+\delta}} \Bigg\{ \ell^2 \frac{q^\mu \left(p^\nu - k_1^\nu\right)}{(p^2 - \ell^2)((p-k_1)^2 - \ell^2)((p-q)^2 - \ell^2)} \\
&\quad - \frac{(p-k_2)^\mu p^\nu}{(p^2 - \ell^2)((p-k_2)^2 - \ell^2)} + \frac{p^\mu (p-k_1)^\nu}{(p^2 - \ell^2)((p-k_1)^2 - \ell^2)} \Bigg\}.
\end{aligned} \tag{6.31}$$

The last two terms do not contribute because of the antisymmetric ε tensor, as one can easily see by introducing a Feynman parameter x and making the shift $p \to p + xk_2$ and $p \to p + xk_1$, respectively. The first term can be evaluated by introducing two Feynman parameters x and y, and making the shift $p \to p + (x+y)k_1 + yk_2$,

$$\begin{aligned}
&\widetilde{F}^{(R,odd)}_{\lambda\rho}(k_1, k_2, \delta) \\
&= -2^{2+\frac{\delta}{2}}\, i\varepsilon_{\mu\nu\lambda\rho} \int_0^1 \mathrm{d}x \int_0^{1-x} \mathrm{d}y \int \frac{d^4p d^\delta \ell}{(2\pi)^{4+\delta}} \ell^2 \frac{q^\mu (p^\nu + (x+y-1)k_1 + yk_2)^\nu)}{(p^2 - \ell^2 + \Delta(x,y))^3}
\end{aligned} \tag{6.32}$$

where $\Delta(x, y) = (x+y)(1-x-y)k_1^2 + y(1-y)k_2^2 + 2y(1-x-y)k_1\cdot k_2$. Now we make a Wick rotation on the integration momentum, $p^0 \to ip^0$, and the same on k_1, k_2 (although, for simplicity, we avoid introducing new symbols and stick to the same symbols).

Then, using (see Appendix 7A)

$$\int \frac{d^4p}{(2\pi)^4} \int \frac{d^\delta \ell}{(2\pi)^\delta} \frac{\ell^2}{(p^2 + \ell^2 + \Delta)^3} = -\frac{1}{2(4\pi)^2}, \tag{6.33}$$

and taking the limit $\delta \to 0$, we find

$$\widetilde{F}^{(R,odd)}_{\lambda\rho}(k_1,k_2) = \frac{2}{(4\pi)^2}\varepsilon_{\mu\nu\lambda\rho}k_1^\mu k_2^\nu \int_0^1 \mathrm{d}x \int_0^{1-x} \mathrm{d}y\,(1-x) = \frac{1}{24\pi^2}\varepsilon_{\mu\nu\lambda\rho}k_1^\mu k_2^\nu. \tag{6.34}$$

We must add the cross term (for $\lambda \leftrightarrow \rho$ and $k_1 \leftrightarrow k_2$), so that the total result is

$$\widetilde{F}^{(R,odd)}_{\lambda\rho}(k_1,k_2) + \widetilde{F}^{(R,odd)}_{\rho\lambda}(k_2,k_1) = \frac{1}{12\pi^2}\varepsilon_{\mu\nu\lambda\rho}k_1^\mu k_2^\nu. \tag{6.35}$$

In order to return to configuration space we have to insert this result into (6.26). We consider here, for simplicity, the Abelian case. Since the even part vanishes (see Appendix 6B), we have

$$\begin{aligned}\partial^\mu \langle\!\langle J_{R\mu}(x)\rangle\!\rangle &= \int \frac{d^4q}{(2\pi)^4}e^{-iqx}\,(-iq^\mu)\langle\!\langle \tilde{J}_{R\mu}(q)\rangle\!\rangle = \frac{i}{2}\int \frac{d^4q}{(2\pi)^4}\int \frac{d^4k_1}{(2\pi)^4}\int \frac{d^4k_2}{(2\pi)^4}\int d^4y d^4z \\ &\times q^\mu\, e^{i(k_1y+k_2z-qx)}\left(\widetilde{F}^{(R,odd)}_{\lambda\rho}(k_1,k_2) + \widetilde{F}^{(R,odd)}_{\rho\lambda}(k_2,k_1)\right)V^\lambda(y)V^\rho(z).\end{aligned} \tag{6.36}$$

After an inverse Wick rotation we can replace (6.35) inside the integrals

$$\begin{aligned}&\partial^\mu \langle\!\langle J_{R\mu}(x)\rangle\!\rangle \\ &= -\frac{1}{24\pi^2}\int \frac{d^4q\, d^4k_1\, d^4k_2}{(2\pi)^{12}}\int d^4y d^4z\, e^{i(qx-k_1y-k_2z)}\delta(q-k_1-k_2)\varepsilon_{\mu\nu\lambda\rho}k_1^\mu k_2^\nu\, V^\lambda(y)V^\rho(z) \\ &= -\frac{1}{24\pi^2}\int \frac{d^4k_1}{(2\pi)^4}\int \frac{d^4k_2}{(2\pi)^4}\int d^4yd^4z\; e^{ik_1(x-y)}e^{-ik_2(x-z)}\varepsilon_{\mu\nu\lambda\rho}\,\partial^\mu V^\lambda(y)\partial^\nu V^\rho(z) \\ &= \frac{1}{24\pi^2}\varepsilon_{\mu\nu\lambda\rho}\,\partial^\mu V^\nu(x)\partial^\lambda V^\rho(x).\end{aligned} \tag{6.37}$$

The same result can be obtained with the Pauli-Villars regularization, see Appendix 6A.

6.2.2 Comments

Equation(6.37) is the consistent gauge anomaly of a right-handed Weyl fermion coupled to an Abelian vector field $V_\mu(x)$. It is well known that the consistent anomaly (6.37) destroys consistency of the Abelian gauge theory. One way to see it is by recalling that the Lorentz invariant quantum theory of a gauge vector field unavoidably involves a Fock space of states with indefinite norm. Now, in order to select a physical Hilbert subspace of the Fock space, a subsidiary condition is necessary. In the Abelian case, when the fermion current satisfies the continuity equation, the equations of motion lead to $\Box(\partial\cdot V) = 0$, so that a subspace of states of non-negative norm can be selected through the auxiliary condition

$$\partial\cdot V^{(-)}(x)\,|\,\mathrm{phys}\,\rangle = 0$$

$V^{(-)}(x)$ being the annihilation operator, the positive frequency part of a d'Alembert quantum field. On the contrary, in the present chiral model we find

$$\Box(\partial\cdot V) = -\frac{1}{3}\left(\frac{1}{4\pi}\right)^2 F_*^{\mu\nu}F_{\mu\nu} \neq 0$$

in such a manner that it is impossible to select a physical subspace of states with non-negative norm, where a unitary restriction of the collision operator S could be defined.

Another way of viewing the problem created by the consistent anomaly is to remark that, for instance, $J_{R\mu}$ couples minimally to V^μ at the fermion-fermion-gluon vertex. Unitarity and renormalizability rely on the Ward identity that guarantees current conservation at any such vertex. But this is impossible in the presence of a consistent anomaly.

The consistent anomaly in the non-Abelian case would require the calculation of the four-current correlators, but it can be obtained in a simpler way from the Abelian case using the Wess-Zumino consistency conditions and their solutions already found in Chap. 5. In the non-Abelian case the three-point correlators are multiplied by

$$\mathrm{Tr}(T^aT^bT^c) = \frac{1}{2}\mathrm{Tr}(T^a[T^b,T^c]) + \frac{1}{2}\mathrm{Tr}(T^a\{T^b,T^c\}) = f^{abc} + d^{abc} \tag{6.38}$$

where the normalization used is $\mathrm{Tr}(T^aT^b) = \delta^{ab}$. Since the three-point function is the sum of two equal pieces with $\lambda \leftrightarrow \rho$, $k_1 \leftrightarrow k_2$, the first term in the RHS of (6.38) drops out and only the second remains. For the right-handed current $J^a_{R\mu}$ we have

$$(D\cdot\langle\langle J_R\rangle\rangle)^a = \frac{1}{24\pi^2}\varepsilon_{\mu\nu\lambda\rho}\mathrm{Tr}\left[T^a\partial^\mu\left(V^\nu\partial^\lambda V^\rho + \frac{i}{2}V^\nu V^\lambda V^\rho\right)\right]. \tag{6.39}$$

This is a particular case of (5.40).

The previous results are well-known. Equation (6.39) coincides with the gauge cocycle introduced in the previous chapter, identified by the ad-invariant polynomial P_3 (the cubic term in V_μ is uniquely determined by the consistency conditions once the quadratic term is known). However this does not tell the whole story about gauge anomalies in a theory of Weyl fermions. To delve into this we have to enlarge the parameter space by coupling the fermions to an additional potential, namely to an axial vector field.

6.3 $V-A$ Anomalies

The action of a Dirac fermion coupled to a vector V_μ and an axial potential A_μ (for simplicity, to start with, we consider only the Abelian case) is

$$S[V, A] = \int d^4x\, i\,\overline{\psi}\,(\not\partial - i\not V - i\not A\gamma_5)\,\psi. \tag{6.40}$$

It is invariant under the following vector and axial-vector transformations

$$\begin{aligned}\delta_\alpha V_\mu &= \partial_\mu \alpha, \qquad \delta_\alpha A_\mu = 0\\ \delta_\beta V_\mu &= 0, \qquad \delta_\beta A_\mu = \partial_\mu \beta\end{aligned} \tag{6.41}$$

The generating functional of the connected Green's functions is

$$W[V, A] = W[0, 0] + \sum_{n,m=1}^{\infty} \frac{i^{n+m-1}}{n!m!} \int \prod_{i=1}^{n} d^4x_i\, V^{\mu_i}(x_i) \prod_{j=1}^{m} d^4y_j A^{\nu_j}(y_j)$$
$$\times \langle 0|\mathcal{T} J_{\mu_1}(x_1)\ldots J_{\mu_n}(x_n) J_{5\nu_1}(y_1)\ldots J_{5\nu_m}(x_m)|0\rangle_c. \tag{6.42}$$

We can extract the full one-loop one-point function for two currents: the vector current $J_\mu = \bar{\psi}\gamma_\mu\psi$

$$\langle\!\langle J_\mu(x)\rangle\!\rangle = \frac{\delta W[V, A]}{\delta V^\mu(x)} = \sum_{n,m=0}^{\infty} \frac{i^{n+m}}{n!m!} \int \prod_{i=1}^{n} d^4x_i\, V^{\mu_i}(x_i) \prod_{j=1}^{m} d^4y_j A^{\nu_j}(y_j)$$
$$\times \langle 0|\mathcal{T} J_\mu(x) J_{\mu_1}(x_1)\ldots J_{\mu_n}(x_n) J_{5\nu_1}(y_1)\ldots J_{5\nu_m}(x_m)|0\rangle_c \tag{6.43}$$

and the axial current $J_\mu = \bar{\psi}\gamma_\mu\gamma_5\psi$

$$\langle\!\langle J_{5\mu}(x)\rangle\!\rangle = \frac{\delta W[V, A]}{\delta A^\mu(x)} = \sum_{n,m=0}^{\infty} \frac{i^{n+m}}{n!m!} \int \prod_{i=1}^{n} d^4x_i\, V_{\mu_i}(x_i) \prod_{j=1}^{m} d^4y_j A^{\nu_j}(y_j)$$
$$\times \langle 0|\mathcal{T} J_{5\mu}(x) J_{\mu_1}(x_1)\ldots J_{\mu_n}(x_n) J_{5\nu_1}(y_1)\ldots J_{5\nu_m}(x_m)|0\rangle_c. \tag{6.44}$$

These currents are conserved except for possible anomaly contributions. The aim of this section is to study the continuity equations for these currents, that is to compute the 4-divergences of the correlators on the RHS of (6.43) and (6.44). The even parity part of these divergences vanish, see Appendix 6B.

As for the odd parity part, for the same reason explained above, we focus on the three current correlators: they are all we need in the Abelian case (and the starting point to compute the full anomaly expression by means of the Wess-Zumino consistency conditions in the non-Abelian case). So for $\partial \cdot J(x)$ the first relevant contributions are

$$\begin{aligned}\partial^\mu \langle\!\langle J_\mu(x)\rangle\!\rangle = -\Big(\frac{1}{2}\int & d^4x_1 d^4x_2 V^{\mu_1}(x_1) V^{\mu_2}(x_2) \partial^\mu \langle 0|\mathcal{T} J_\mu(x) J_{\mu_1}(x_1) J_{\mu_2}(x_2)|0\rangle\\ &+ \int d^4x_1 d^4y_1 V^{\mu_1}(x_1) A^{\nu_1}(y_1) \partial^\mu \langle 0|\mathcal{T} J_\mu(x) J_{\mu_1}(x_1) J_{5\nu_1}(y_1)|0\rangle\\ &+ \frac{1}{2}\int d^4y_1 d^4y_2 A^{\nu_1}(y_1) A^{\nu_2}(y_2) \partial^\mu \langle 0|\mathcal{T} J_\mu(x) J_{5\nu_1}(y_1) J_{5\nu_2}(y_2)|0\rangle\Big)\end{aligned} \tag{6.45}$$

and for $\partial^\mu J_{5\mu}(x)$

$$\partial^\mu\langle\langle J_{5\mu}(x)\rangle\rangle = -\Big(\frac{1}{2}\int d^4x_1 d^4x_2 V^{\mu_1}(x_1)V^{\mu_2}(x_2)\partial^\mu\langle 0|\mathcal{T} J_{5\mu}(x)J_{\mu_1}(x_1)J_{\mu_2}(x_2)|0\rangle$$
$$+\int d^4x_1 d^4y_1 V^{\mu_1}(x_1)A^{\nu_1}(y_1)\partial^\mu\langle 0|\mathcal{T} J_{5\mu}(x)J_{\mu_1}(x_1)J_{5\nu_1}(y_1)|0\rangle$$
$$+\frac{1}{2}\int d^4y_1 d^4y_2 A^{\nu_1}(y_1)A^{\nu_2}(y_2)\partial^\mu\langle 0|\mathcal{T} J_{5\mu}(x)J_{5\nu_1}(y_1)J_{5\nu_2}(y_2)|0\rangle\Big) \tag{6.46}$$

Since we are focusing on odd parity anomalies, the only possible contribution to (6.45) may come from the term in the second line, which we denote concisely $\langle\partial\cdot J\,J\,J_5\rangle$. As for (6.46) there are two possible contributions from the first and third lines, i.e., $\langle\partial\cdot J_5\,J\,J\rangle$ and $\langle\partial\cdot J_5\,J_5\,J_5\rangle$. Below we report the results for the corresponding amplitudes, obtained with dimensional regularization.

The amplitude for $\langle\partial\cdot J_5\,J\,J\rangle^{(odd)}$ is

$$q^\mu \widetilde{F}^{(5,odd)}_{\mu\lambda\rho}(k_1,k_2) = \int \frac{d^4p d^\delta\ell}{(2\pi)^{4+\delta}}\mathrm{tr}\left\{\frac{1}{\not p+\not\ell}\gamma_\lambda\frac{1}{\not p+\not\ell-\not k_1}\gamma_\rho\frac{1}{\not p+\not\ell-\not q}\not q\gamma_5\right\}. \tag{6.47}$$

The relevant Feynman diagram is the same as in Fig. 6.1, with a γ_5 inserted in the leftmost vertex.

Adding the cross contribution one gets

$$q^\mu\left(\widetilde{F}^{(5,odd)}_{\mu\lambda\rho}(k_1,k_2)+\widetilde{F}^{5,odd}_{\mu\rho\lambda}(k_2,k_1)\right) = \frac{1}{2\pi^2}\varepsilon_{\mu\nu\lambda\rho}k_1^\mu k_2^\nu. \tag{6.48}$$

The amplitude for $\langle\partial\cdot J_5\,J_5\,J_5\rangle^{(odd)}$ is given by

$$\begin{aligned} q^\mu \widetilde{F}^{(555,odd)}_{\mu\lambda\rho}(k_1,k_2) &= \int \frac{d^4p d^\delta\ell}{(2\pi)^{4+\delta}}\mathrm{tr}\left\{\frac{1}{\not p+\not\ell}\gamma_\lambda\gamma_5\frac{1}{\not p+\not\ell-\not k_1}\gamma_\rho\gamma_5\frac{1}{\not p+\not\ell-\not q}\not q\gamma_5\right\}\\ &= -2^{2+\frac{\delta}{2}}i\varepsilon_{\mu\nu\lambda\rho}\int \frac{d^4p d^\delta\ell}{(2\pi)^{4+\delta}}\ell^2\frac{q^\mu(3p^\nu-k_1^\nu)}{(p^2-\ell^2)((p-k_1)^2-\ell^2)((p-q)^2-\ell^2)}\\ &\quad -\int \frac{d^4p d^\delta\ell}{(2\pi)^{4+\delta}}\frac{\mathrm{tr}(\not q\not p\gamma_\lambda(\not p-\not k_1)\gamma_\rho(\not p-\not q)\gamma_5)}{(p^2-\ell^2)((p-k_1)^2-\ell^2)((p-q)^2-\ell^2)} \end{aligned} \tag{6.49}$$

The first line in the last expression, after introducing the Feynman parameters x and y and shifting p as usual, yields a factor $\int_0^1 \mathrm{d}x\int_0^{1-x}\mathrm{d}y\,(1-3x)=0$, so it vanishes. The last line is $2\times\widetilde{F}^{(R,odd)}_{\lambda\rho}(k_1,k_2,\delta)$, cf. (6.28,6.29). Therefore, using (6.35), we get

$$q^\mu\left(\widetilde{F}^{(555,odd)}_{\mu\lambda\rho}(k_1,k_2)+\widetilde{F}^{(555,odd)}_{\mu\rho\lambda}(k_2,k_1)\right) = \frac{1}{6\pi^2}\varepsilon_{\mu\nu\lambda\rho}k_1^\mu k_2^\nu. \tag{6.50}$$

Finally the amplitude for $\langle \partial \cdot J\, J\, J_5 \rangle^{(odd)}$ is

$$q^\mu \widetilde{F}^{(5',odd)}_{\mu\lambda\rho}(k_1,k_2) = \int \frac{d^4p d^\delta \ell}{(2\pi)^{4+\delta}} \mathrm{tr}\left\{ \frac{1}{\not p + \not \ell} \gamma_\lambda \frac{1}{\not p + \not \ell - \not k_1} \gamma_\rho \gamma_5 \frac{1}{\not p + \not \ell - \not q} \not q \right\} = 0. \tag{6.51}$$

All the above results have been obtained also with PV regularization.

Plugging in these results in (6.43) and (6.44), and taking account that the even part vanishes, we find

$$\partial^\mu \langle\!\langle J_\mu(x) \rangle\!\rangle = 0 \tag{6.52}$$

and

$$\partial^\mu \langle\!\langle J_{5\mu}(x) \rangle\!\rangle = \frac{1}{4\pi^2} \varepsilon_{\mu\nu\lambda\rho} \left(\partial^\mu V^\nu(x) \partial^\lambda V^\rho(x) + \frac{1}{3} \partial^\mu A^\nu(x) \partial^\lambda A^\rho(x) \right), \tag{6.53}$$

which is Bardeen's result in the Abelian case, [3]. From (6.53) we can derive the covariant chiral anomaly by setting $A_\mu = 0$, then

$$\partial^\mu \langle\!\langle J_{5\mu}(x) \rangle\!\rangle = \frac{1}{4\pi^2} \varepsilon_{\mu\nu\lambda\rho}\, \partial^\mu V^\nu(x) \partial^\lambda V^\rho(x). \tag{6.54}$$

Of course this is nothing but (6.48). For the $J_{5\mu}(x)$ current is obtained by differentiating the action with respect to $A_\mu(x)$ and its divergence leads to the covariant anomaly.

From the Abelian case, as we have already noted, using the consistency conditions one can derive the corresponding non-Abelian anomaly. In the $V-A$ case this derivation, although possible, is not so straightforward. Later on we will present a non-perturbative calculation of the non-Abelian anomaly. For the time being we limit ourselves to writing down the result. The vector current is conserved

$$[D^\mu_V j_\mu]^a + [A^\mu, j_{5\mu}]^a = 0 \tag{6.55}$$

while the axial conservation is anomalous:

$$\begin{aligned} [D^\mu_V j_{5\mu}]^a + [A^\mu, j_\mu]^a = \frac{1}{4\pi^2} \varepsilon_{\mu\nu\lambda\rho} \mathrm{tr}\Big[T^a \Big(&\frac{1}{4} F^{\mu\nu}_V F^{\lambda\rho}_V + \frac{1}{12} F^{\mu\nu}_A F^{\lambda\rho}_A - \frac{i}{6} F^{\mu\nu}_V A^\lambda A^\rho \\ &- \frac{i}{6} A^\mu A^\nu F^{\lambda\rho}_V - i\frac{2}{3} A^\mu F^{\nu\lambda}_V A^\rho + \frac{1}{3} A^\mu A^\nu A^\lambda A^\rho \Big) \Big] \end{aligned} \tag{6.56}$$

where $F^{\mu\nu}_V = \partial^\mu V^\nu - \partial^\nu V^\mu + i[V^\mu, V^\nu]$, and $F^{\mu\nu}_A = \partial^\mu A^\nu - \partial^\nu A^\mu + i[V^\mu, A^\nu] + i[A^\mu, V^\nu]$.

6.3.1 Some Conclusions

Let us recall that in the chiral limit $V \to V/2$, $A \to V/2$ in the action (6.40) we recover the theory of a right-handed Weyl fermion (with the addition of a free left-handed part, as explained at length above). Now $J_\mu(x) = J_{R\mu}(x) + J_{L\mu}(x)$ and $J_{5\mu}(x) = J_{R\mu}(x) - J_{L\mu}(x)$. In the chiral limit we find

$$\partial^\mu \langle\langle J^{(cs)}_{R\mu}(x)\rangle\rangle = \frac{1}{24\pi^2}\varepsilon_{\mu\nu\lambda\rho}\, \partial^\mu V^\nu(x)\partial^\lambda V^\rho(x). \tag{6.57}$$

Similarly

$$\partial^\mu \langle\langle J^{(cs)}_{L\mu}(x)\rangle\rangle = -\frac{1}{24\pi^2}\varepsilon_{\mu\nu\lambda\rho}\, \partial^\mu V^\nu(x)\partial^\lambda V^\rho(x). \tag{6.58}$$

These are the consistent right and left gauge anomalies - the label $^{(cs)}$ stands for consistent, to be distinguished from the covariant anomaly. As a matter of fact, application of the same chiral current splitting to the covariant anomaly of Eq. (6.54) yields instead

$$\partial^\mu \langle\langle J^{(cv)}_{R\mu}(x)\rangle\rangle = \frac{1}{8\pi^2}\varepsilon_{\mu\nu\lambda\rho}\, \partial^\mu V^\nu(x)\partial^\lambda V^\rho(x) \tag{6.59}$$

and

$$\partial^\mu \langle\langle J^{(cv)}_{L\mu}(x)\rangle\rangle = -\frac{1}{8\pi^2}\varepsilon_{\mu\nu\lambda\rho}\, \partial^\mu V^\nu(x)\partial^\lambda V^\rho(x). \tag{6.60}$$

The label $^{(cv)}$ stands for covariant, and it is in order to tell apart these anomalies from the previous consistent ones. The two cases should not be confused: the consistent anomalies appears in the divergence of a current minimally coupled in the action to the vector potential V_μ. They represent the response of the effective action under a gauge transformation of V_μ, which is supposed to propagate in the internal lines of the corresponding gauge theory. The covariant anomalies represent the response of the effective action under a gauge transformation of the external axial current A_μ.

It goes without saying that, both for right and left currents in the chiral limit, in the non-Abelian case the consistent anomaly takes the form (6.39) with opposite sign for the left, while the covariant one, obtained from (9.27) in the limit $A_\mu \to 0$, reads[3]

$$(D \cdot \langle\langle J_R(x)\rangle\rangle)^a = \frac{1}{32\pi^2}\varepsilon_{\mu\nu\lambda\rho}\,\mathrm{Tr}\left(T^a F^{\mu\nu}(x)F^{\lambda\rho}(x)\right) \tag{6.61}$$

where $F_{\mu\nu}(x) = F^a_{\mu\nu}(x)T^a$ denotes the usual non-Abelian field strength. This is also called the *multiplet* covariant anomaly.

[3] This anomaly is called covariant because under a gauge transformation it transforms in the adjoint way (in the Abelian case it is invariant). For a comment on covariant anomalies, see Appendix 11A.

6.4 The Case of Majorana Fermions

As we have seen above, Majorana fermions are defined by the condition

$$\Psi = \widehat{\Psi}, \qquad \text{where} \qquad \widehat{\Psi} = \gamma_0 C \Psi^*. \tag{6.62}$$

Let $\psi_R = P_R \psi$ be a generic Weyl fermion. We have

$$P_R \psi_R = \psi_R \qquad P_L \widehat{\psi_R} = \widehat{\psi_R}$$

i.e., $\widehat{\psi_R}$ is left-handed. We have already remarked that $\psi_M = \psi_R + \widehat{\psi_R}$ is a Majorana fermion and any Majorana fermion can be represented in this way. Using this correspondence one can transfer the results for Weyl fermions to Majorana fermions[4]. The vector current is defined by $J^\mu_M = \bar{\psi}_M \gamma^\mu \psi_M$ and the axial current by $J^\mu_{5M} = \bar{\psi}_M \gamma^\mu \gamma_5 \psi_M$. We can write

$$J^\mu_M(x) = \overline{\psi_R}(x)\gamma^\mu \psi_R(x) + \overline{\widehat{\psi_R}}(x)\gamma^\mu \widehat{\psi_R}(x) \equiv J^\mu_R(x) + J^\mu_L(x) \tag{6.63}$$

and

$$J^\mu_{5M}(x) = \overline{\psi_R}(x)\gamma^\mu \psi_R(x) - \overline{\widehat{\psi_R}}(x)\gamma^\mu \widehat{\psi_R}(x) \equiv J^\mu_R(x) - J^\mu_L(x). \tag{6.64}$$

Using (6.57) and (6.58) one concludes that, as far as the consistent anomaly is concerned,

$$\partial_\mu \langle\!\langle J^\mu_M(x) \rangle\!\rangle = 0 \tag{6.65}$$

This shows the consistency of our procedure, for one can show that, in general, $J_L(x) = -J_R(x)$, and $J_M(x) = 0$, as it should be for a Majorana fermion. On the other hand for the axial current we have

$$\partial_\mu \langle\!\langle J^\mu_{5M}(x) \rangle\!\rangle = \frac{1}{8\pi^2} \varepsilon_{\mu\nu\lambda\rho}\, \partial^\mu V^\nu(x) \partial^\lambda V^\rho(x) \tag{6.66}$$

where the naive sum has been divided by 2, because the two contributions come from the same degrees of freedom (which are half those of a Dirac fermion). From these results we see that, apart from the coefficient, the anomalies of a massless Majorana fermion are the same as those of a massless Dirac fermion. These obviously descend from the fact that both Dirac and Majorana fermions contain two opposite chiralities, at variance with a Weyl fermion, which is characterized by one single chirality.

[4] The converse is not true. It is impossible to reconstruct Weyl fermion anomalies from those of a massless Majorana fermion.

Appendix 6A. Consistent Gauge Anomaly with PV Regularization

To implement a PV regularization we replace $\widetilde{F}^{(R)}_{\mu\nu\lambda}(k_1, k_2)$ with

$$\widetilde{F}^{(R)}_{\mu\nu\lambda}(k_1, k_2) = \int \frac{d^4 p}{(2\pi)^4} \mathrm{tr} \left\{ \frac{1}{\not p + m} \frac{1-\gamma_5}{2} \gamma_\nu \frac{1}{\not p - \not k_1 + m} \frac{1-\gamma_5}{2} \gamma_\lambda \frac{1}{\not p - \not q + m} \frac{1-\gamma_5}{2} \gamma_\mu \right.$$
$$\left. - \frac{1}{\not p + M} \frac{1-\gamma_5}{2} \gamma_\nu \frac{1}{\not p - \not k_1 + M} \frac{1-\gamma_5}{2} \gamma_\lambda \frac{1}{\not p - \not q + M} \frac{1-\gamma_5}{2} \gamma_\mu \right\} \quad (6.67)$$

m and M are IR and UV regulators, respectively, and tr is the trace of gamma matrices.

Contracting with q^μ and working out the traces one gets

$$q^\mu \widetilde{F}^{(R)}_{\mu\nu\lambda}(k_1, k_2)$$
$$= -2i\varepsilon_{\mu\nu\rho\lambda} \int \frac{d^4 p}{(2\pi)^4} \left(2p{\cdot}q\, p^\mu - p^2 q^\mu - q^2 p^\mu\right) (p-k_1)^\rho \left(\frac{1}{\Delta_{m^2}} - \frac{1}{\Delta_{M^2}} \right) \quad (6.68)$$

where

$$\Delta_{m^2} = (p^2 - m^2)((p-k_1)^2 - m^2)((p-q)^2 - m^2),$$
$$\Delta_{M^2} = (p^2 - M^2)((p-k_1)^2 - M^2)((p-q)^2 - M^2)$$

For later use we introduce also

$$\Omega_{m^2} = ((p-k_1)^2 - m^2)((p-q)^2 - m^2), \quad \Lambda_{m^2} = (p^2 - m^2)((p-k_1)^2 - m^2),$$
$$\Omega_{M^2} = ((p-k_1)^2 - M^2)((p-q)^2 - M^2), \quad \Lambda_{M^2} = (p^2 - M^2)((p-k_1)^2 - M^2). \quad (6.69)$$

Now all the integrals are convergent because the divergent terms have been subtracted away. Let us proceed

$$q^\mu \widetilde{F}^{(R)}_{\mu\nu\lambda}(k_1, k_2) = -2i\varepsilon_{\mu\nu\rho\lambda} \int \frac{d^4 p}{(2\pi)^4} \left\{ \left[\frac{-k_2^\mu (p-k_1)^\rho}{\Omega_{m^2}\Delta_{M^2}} + \frac{p^\mu k_1^\rho}{\Lambda_{m^2}\Delta_{M^2}} \right] \right. \quad (6.70)$$
$$\times \Big(m^6 - M^6 + \left(M^4 - m^4\right)\left(p^2 + (p-k_1)^2 + (p-q)^2\right)$$
$$+ \left(m^2 - M^2\right)\left((p-k_1)^2(p-q)^2 + (p-k_1)^2 p^2 + p^2(p-q)^2\right) \Big)$$
$$\left. + m^2 \left(p^\mu k_1^\rho - q^\mu (p-k_1)^\rho\right) \left(\frac{1}{\Delta_{m^2}} - \frac{1}{\Delta_{M^2}} \right) \right\}$$

The last line does not contribute, for the integrals converge (separately) and give a finite result, but since they are multiplied by m^2 they vanish in the limit $m \to 0$. So the last line can be dropped.

Now the strategy consists in simplifying separately each monomials in the numerator with a corresponding term in the denominator. For instance, if in a term proportional to M^* there is the ratio $p^2/(p^2 - m^2)$, write p^2 as $p^2 - m^2 + m^2$. The $p^2 - m^2$ can be simplified with a corresponding term in the denominator. If $p^2 - m^2$ in the denominator is missing, there will be $p^2 - M^2$. So we write p^2 as $p^2 - M^2 + M^2$, and $p^2 - M^2$ can be simplified, while the term proportional to M^2 remains and contributes to the term of order M^{*+2}. Proceed in the same way also with $(p-q)^2$ and $(p-k_1)^2$. Many terms (such as those of order M^6) cancel out. What remains is

$$\begin{aligned}
q^\mu \widetilde{F}^{(R)}_{\mu\nu\lambda}(k_1,k_2) &= -2i\varepsilon_{\mu\nu\rho\lambda}\int \frac{d^4p}{(2\pi)^4} \\
&\left\{ p^\mu k_1^\rho \left[\frac{\left(M^2-m^2\right)^2}{\Lambda_{m^2}\Lambda_{M^2}} - \frac{\left(M^2-m^2\right)}{\Lambda_{M^2}}\left(\frac{1}{p^2-m^2}+\frac{1}{(p-k_1)^2-m^2}\right) \right.\right. \\
&\left. - \frac{M^2-m^2}{\Delta_{M^2}} \right] - k_2^\mu (p-k_1)^\rho \left[\frac{\left(M^2-m^2\right)^2}{\Omega_{m^2}\Omega_{M^2}} \right. \\
&\left.\left. - \frac{\left(M^2-m^2\right)}{\Omega_{M^2}}\left(\frac{1}{(p-k_1)^2-m^2}+\frac{1}{(p-q)^2-m^2}\right) - \frac{M^2-m^2}{\Delta_{M^2}} \right]\right\}
\end{aligned} \tag{6.71}$$

It is easy to verify that, after introducing the relevant Feynman parameters, most of the terms vanish either because the numerator is linear in p or because of the anti-symmetric tensor ε. Only the last term inside each square bracket remains, so that:

$$q^\mu \widetilde{F}^{(R)}_{\mu\nu\lambda}(k_1,k_2) = 2i\, M^2 \varepsilon_{\mu\nu\rho\lambda}\int \frac{d^4p}{(2\pi)^4}\frac{p^\mu k_1^\rho - k_2^\mu (p-k_1)^\rho}{\Delta_{M^2}} \tag{6.72}$$

Next we introduce two Feynman parameters x and y, shift p like in Sect. 6.2.1 and make a Wick rotation on the momenta. Then (6.71) becomes

$$\begin{aligned}
q^\mu \widetilde{F}^{(R)}_{\mu\nu\lambda}(k_1,k_2) &= -4\, M^2 \varepsilon_{\mu\nu\rho\lambda}\int_0^1 \mathrm{d}x \int_0^{1-x} \mathrm{d}y \int \frac{d^4p}{(2\pi)^4}\frac{(1-x)k_1^\mu k_2^\rho}{(p^2+M^2+\Delta(x,y))^3} \\
&= -\frac{1}{8\pi^2}\, M^2 \varepsilon_{\mu\nu\rho\lambda}\int_0^1 \mathrm{d}x \int_0^{1-x} \mathrm{d}y \frac{(1-x)k_1^\mu k_2^\rho}{M^2+\Delta(x,y)} \\
&= -\frac{1}{24\pi^2}\varepsilon_{\mu\nu\rho\lambda}k_1^\mu k_2^\rho
\end{aligned} \tag{6.73}$$

Adding the cross term we get

$$q^\mu \widetilde{T}^{(R)}_{\mu\nu\lambda}(k_1,k_2) = -\frac{1}{12\pi^2}\varepsilon_{\mu\nu\rho\lambda}k_1^\mu k_2^\rho \tag{6.74}$$

which is the same result as in Sect. 6.2.1.

Appendix 6B. Even Gauge Current Divergences

In this Appendix we show that in a theory of Dirac or Weyl fermions, the even part of the divergence of the relevant gauge currents (i.e., $\langle\langle J_{R\mu}\rangle\rangle$ in the right-handed case and $\langle\langle J_\mu\rangle\rangle$ and $\langle\langle J_{5\mu}\rangle\rangle$ in the $V-A$ case) vanish.

A Preliminary Calculation

As a starting calculation we want to prove that the even triangle diagram contribution

$$q^\mu \widetilde{T}^{(\mathrm{even1})}_{\mu\lambda\rho}(k_1,k_2) = \int \frac{d^4p\, d^\delta \ell}{(2\pi)^{4+\delta}} \mathrm{tr}\left\{\frac{\not p}{p^2-\ell^2}\gamma_\lambda \frac{\not p - \not k_1}{(p-k_1)^2-\ell^2}\gamma_\rho \frac{\not p - \not q}{(p-q)^2-\ell^2}\not q\right\} \tag{6.75}$$

vanishes. Using $(\not p - \not q)\not q \not p = (2p\cdot q - q^2)\not p - p^2 \not q$, the integrand becomes

$$q^\mu \widetilde{T}^{(\mathrm{even1})}_{\mu\lambda\rho}(k_1,k_2) = \int \frac{d^4p\, d^\delta \ell}{(2\pi)^{4+\delta}} \frac{(2p\cdot q - q^2)\mathrm{tr}\left(\not p \gamma_\lambda(\not p - \not k_1)\gamma_\rho\right) - p^2 \mathrm{tr}\left(\not q \gamma_\lambda(\not p - \not k_1)\gamma_\rho\right)}{(p^2-\ell^2)((p-k_1)^2-\ell^2)((p-q)^2-\ell^2)} \tag{6.76}$$

Evaluating traces:

$$\begin{aligned}
&= 2^{2+\frac{\delta}{2}} \int \frac{d^4p\, d^\delta \ell}{(2\pi)^{4+\delta}} \Big[-(p-q)^2\big(p_\lambda(p-k_1)_\rho - \eta_{\lambda\rho}\, p\cdot(p-k_1) + p_\rho(p-k_1)_\lambda\big) \\
&+ p^2\big((p-q)_\lambda(p-k_1)_\rho - \eta_{\lambda\rho}(p-q)\cdot(p-k_1) + (p-q)_\rho(p-k_1)_\lambda\big)\Big] \\
&\times \frac{1}{(p^2-\ell^2)((p-k_1)^2-\ell^2)((p-q)^2-\ell^2)}
\end{aligned} \tag{6.77}$$

To this we have to add the cross contribution $k_1 \leftrightarrow k_2$, $\lambda \leftrightarrow \rho$

$$\begin{aligned}
&q^\mu \widetilde{T}^{(\mathrm{even2})}_{\mu\rho\lambda}(k_2,k_1) \\
&= 2^{2+\frac{\delta}{2}} \int \frac{d^4p\, d^\delta \ell}{(2\pi)^{4+\delta}} \Big[-(p-q)^2\big(p_\lambda(p-k_2)_\rho - \eta_{\lambda\rho}\, p\cdot(p-k_2) + p_\rho(p-k_2)_\lambda\big) \\
&+ p^2\big((p-q)_\lambda(p-k_2)_\rho - \eta_{\lambda\rho}(p-q)\cdot(p-k_2) + (p-q)_\rho(p-k_2)_\lambda\big)\Big] \\
&\times \frac{1}{(p^2-\ell^2)((p-k_2)^2-\ell^2)((p-q)^2-\ell^2)}
\end{aligned} \tag{6.78}$$

Now shift $p \to p+q$ and change $p \to -p$, then (6.78) becomes

$$q^\mu \widetilde{T}^{(\text{even2})}_{\mu\rho\lambda}(k_2,k_1) = 2^{2+\frac{\delta}{2}} \int \frac{d^4p d^\delta \ell}{(2\pi)^{4+\delta}} \Big[-p^2\big((p-q)_\lambda (p-k_1)_\rho - \eta_{\lambda\rho}(p-q)\cdot(p-k_1)$$
$$+(p-q)_\rho(p-k_1)_\lambda\big) + (p-q)^2\big(p_\lambda(p-k_1)_\rho - \eta_{\lambda\rho} p\cdot(p-k_1) + p_\rho(p-k_1)_\lambda\big)\Big]$$
$$\times \frac{1}{(p^2-\ell^2)((p-k_1)^2-\ell^2)((p-q)^2-\ell^2)} \tag{6.79}$$

which is the opposite of (6.77). Therefore

$$q^\mu \widetilde{T}^{(\text{even1})}_{\mu\lambda\rho}(k_1,k_2) + q^\mu \widetilde{T}^{(\text{even2})}_{\mu\rho\lambda}(k_2,k_1) = 0 \tag{6.80}$$

Now we can look at the various cases

Even Part of $\langle \partial \cdot j_R\, j_R\, j_R \rangle$

The three point function $\langle \partial \cdot j_R\, j_R\, j_R \rangle$ is

$$q^\mu \widetilde{T}^{(R)}_{\mu\lambda\rho}(k_1,k_2) = \int \frac{d^4p d^\delta \ell}{(2\pi)^{4+\delta}} \mathrm{tr}\left\{ \frac{1}{\not p + \not \ell} \frac{1-\gamma_5}{2} \gamma_\lambda \frac{1}{\not p + \not \ell - \not k_1} \frac{1-\gamma_5}{2} \gamma_\rho \frac{1}{\not p + \not \ell - \not q} \frac{1-\gamma_5}{2} \not q \right\}$$
$$= \int \frac{d^4p d^\delta \ell}{(2\pi)^{4+\delta}} \mathrm{tr}\left\{ \frac{\not p}{p^2-\ell^2} \gamma_\lambda \frac{\not p - \not k_1}{(p-k_1)^2-\ell^2} \gamma_\rho \frac{\not p - \not q}{(p-q)^2-\ell^2} \frac{1-\gamma_5}{2} \not q \right\}$$
$$\equiv \widetilde{F}^{(R)}_{\lambda\rho}(k_1,k_2,\delta) \tag{6.81}$$

The even part thereof is 1/2 of (6.75), therefore it vanishes.

In view of an application to the $V-A$ system, let us consider other even 3pt correlator.

Even Part of $\langle \partial \cdot j\, j\, j \rangle$

Let us consider the correlator

$$q^\mu \widetilde{T}^{(VVV)}_{\mu\lambda\rho}(k_1,k_2) = \int \frac{d^4p}{(2\pi)^4} \mathrm{tr}\left\{ \frac{1}{\not p} \gamma_\lambda \frac{1}{\not p - \not k_1} \gamma_\rho \frac{1}{\not p - \not q} \not q \right\}$$
$$= \int \frac{d^4p d^\delta \ell}{(2\pi)^{4+\delta}} \mathrm{tr}\left\{ \frac{\not p + \not \ell}{p^2-\ell^2} \gamma_\lambda \frac{\not p - \not k_1 + \not \ell}{(p-k_1)^2-\ell^2} \gamma_\rho \frac{\not p - \not q + \not \ell}{(p-q)^2-\ell^2} \not q \right\} \tag{6.82}$$

Using (6.80), this reduces to

$$q^\mu \widetilde{T}^{(VVV)}_{\mu\lambda\rho}(k_1,k_2) = \int \frac{d^4p d^\delta \ell}{(2\pi)^{4+\delta}} \ell^2 \frac{\mathrm{tr}\left[\gamma_\lambda \gamma_\rho (\not p - \not q)\not q + \gamma_\lambda (\not p - \not k_1)\gamma_\rho \not q + \not p \gamma_\lambda \gamma_\rho \not q\right]}{(p^2-\ell^2)((p-k_1)^2-\ell^2)((p-q)^2-\ell^2)} \tag{6.83}$$
$$= 2^{2+\frac{\delta}{2}} \int \frac{d^4p d^\delta \ell}{(2\pi)^{4+\delta}} \ell^2 \frac{\eta_{\lambda\rho}(p-k_2)\cdot q + q_\lambda (p-2k_1-k_2)_\rho + q_\rho (p+k_2)_\lambda}{(p^2-\ell^2)((p-k_1)^2-\ell^2)((p-q)^2-\ell^2)}$$

Introducing x, y Feynman parameters and shifting $p \to p + (x+y)k_1 + yk_2$, integrating over p and ℓ and taking $\delta \to 0$, one gets

$$q^\mu \widetilde{T}^{(VVV)}_{\mu\lambda\rho}(k_1,k_2) = -\frac{8i}{(4\pi)^2}\int_0^1 dx \int_0^{1-x} dy\Big[\eta_{\lambda\rho}((x+y)k_1+(y-1)k_2)\cdot q$$
$$+q_\lambda((x+y-2)k_1+(y-1)k_2)_\rho + q_\rho((x+y)k_1+(y+1)k_2)_\lambda\Big]$$
$$= \frac{i}{12\pi^2}\Big[\eta_{\lambda\rho}(k_1-k_2)\cdot q - q_\lambda(2k_1+k_2)_\rho + q_\rho(k_1+2k_2)_\lambda\Big] \tag{6.84}$$

Adding the cross term ($k_1 \leftrightarrow k_2$, $\lambda \leftrightarrow \rho$) one gets 0.

Even Part of $\langle \partial \cdot j\ j_5\ j_5 \rangle$

Let us write down the triangle contribution to $\langle \partial \cdot j\ j_5\ j_5 \rangle$. It is

$$q^\mu \widetilde{T}^{(VAA)}_{\mu\lambda\rho}(k_1,k_2) = \int \frac{d^4p}{(2\pi)^4}\mathrm{tr}\left\{\frac{1}{\not p}\gamma_\lambda\gamma_5\frac{1}{\not p-\not k_1}\gamma_\rho\gamma_5\frac{1}{\not p-\not q}\not q\right\} \tag{6.85}$$
$$= \int \frac{d^4p d^\delta\ell}{(2\pi)^{4+\delta}}\mathrm{tr}\left\{\frac{\not p+\not \ell}{p^2-\ell^2}\gamma_\lambda\gamma_5\frac{\not p-\not k_1+\not\ell}{(p-k_1)^2-\ell^2}\gamma_\rho\gamma_5\frac{\not p-\not q+\not\ell}{(p-q)^2-\ell^2}\not q\right\}$$

Using (6.80), this reduces to

$$q^\mu \widetilde{T}^{(VAA)}_{\mu\lambda\rho}(k_1,k_2) = \int \frac{d^4p d^\delta\ell}{(2\pi)^{4+\delta}}\ell^2\frac{\mathrm{tr}\left[-\gamma_\lambda\gamma_\rho(\not p-\not q)\not q + \gamma_\lambda(\not q-\not k_1\gamma_\rho\not q - \not p\gamma_\lambda\gamma_\rho\not q\right]}{(p^2-\ell^2)((p-k_1)^2-\ell^2)((p-q)^2-\ell^2)}$$
$$= 2^{2+\frac{\delta}{2}}\int \frac{d^4p d^\delta\ell}{(2\pi)^{4+\delta}}\ell^2\frac{-\eta_{\lambda\rho}(3p-2k_1-k_2)\cdot q + q_\lambda(p+k_2)_\rho + q_\rho(p-2k_1-k_2)_\lambda}{(p^2-\ell^2)((p-k_1)^2-\ell^2)((p-q)^2-\ell^2)} \tag{6.86}$$

Introducing x, y Feynman parameters and shifting $p \to p + (x+y)k_1 + yk_2$, integrating over p and ℓ and taking $\delta \to 0$, one gets

$$q^\mu \widetilde{T}^{(VAA)}_{\mu\lambda\rho}(k_1,k_2) = -\frac{8i}{(4\pi)^2}\int_0^1 dx \int_0^{1-x} dy\Big[\eta_{\lambda\rho}((3x+3y-2)k_1+(3y-1)k_2)\cdot q$$
$$+q_\lambda((x+y)k_1+(y+1)k_2)_\rho + q_\rho((x+y-2)k_1+(y-1)k_2)_\lambda\Big]$$
$$= \frac{i}{12\pi^2}\Big[-k_{1\lambda}k_{1\rho} + k_{2\lambda}k_{2\rho}\Big] \tag{6.87}$$

Adding the cross term ($k_1 \leftrightarrow k_2$, $\lambda \leftrightarrow \rho$) one gets 0.

Even Part of $\langle \partial \cdot j_5\; j\; j_5 \rangle$ *and* $\langle \partial \cdot j_5\; j_5\; j \rangle$

Let us consider next the triangle contribution to $\langle \partial\cdot j_5\; j\; j_5\rangle$. It is

$$\begin{aligned} q^\mu \widetilde{T}^{(AVA)}_{\mu\lambda\rho}(k_1,k_2) &= \int \frac{d^4p}{(2\pi)^4} \mathrm{tr}\left\{ \frac{1}{\not p}\gamma_\lambda\gamma_5 \frac{1}{\not p - \not k_1}\gamma_\rho \frac{1}{\not p - \not q}\gamma_5 \not q \right\} \\ &= \int \frac{d^4p d^\delta \ell}{(2\pi)^{4+\delta}} \mathrm{tr}\left\{ \frac{\not p + \not \ell}{p^2-\ell^2}\gamma_\lambda\gamma_5 \frac{\not p - \not k_1 + \not \ell}{(p-k_1)^2-\ell^2}\gamma_\rho \frac{\not p - \not q + \not \ell}{(p-q)^2-\ell^2}\gamma_5 \not q \right\} \end{aligned} \tag{6.88}$$

Using (6.80), this reduces to

$$\begin{aligned} q^\mu \widetilde{T}^{(AVA)}_{\mu\lambda\rho}(k_1,k_2) &= \int \frac{d^4p d^\delta \ell}{(2\pi)^{4+\delta}} \ell^2 \frac{\mathrm{tr}\left[-\gamma_\lambda\gamma_\rho(\not p - \not q)\not q - \gamma_\lambda(\not q - \not k_1\gamma_\rho \not q + \not p \gamma_\lambda\gamma_\rho \not q\right]}{(p^2-\ell^2)((p-k_1)^2-\ell^2)((p-q)^2-\ell^2)} \\ &= 2^{2+\frac{\delta}{2}} \int \frac{d^4p d^\delta \ell}{(2\pi)^{4+\delta}} \ell^2 \frac{\eta_{\lambda\rho}(p+k_2)\cdot q - q_\lambda(3p-2k_1-k_2)_\rho + q_\rho(p-k_2)_\lambda}{(p^2-\ell^2)((p-k_1)^2-\ell^2)((p-q)^2-\ell^2)} \end{aligned} \tag{6.89}$$

Introducing x, y Feynman parameters and shifting $p \to p + (x+y)k_1 + yk_2$, integrating over p and ℓ and taking $\delta \to 0$, one gets

$$\begin{aligned} q^\mu \widetilde{T}^{(AVA)}_{\mu\lambda\rho}(k_1,k_2) &= -\frac{4i}{(4\pi)^2} \int_0^1 \mathrm{d}x \int_0^{1-x} \mathrm{d}y \Big[-\eta_{\lambda\rho}((x+y)k_1 + (y+1)k_2)\cdot q \\ &\quad + q_\lambda((3x+3y-2)k_1 + (3y-1)k_2)_\rho + q_\rho((x+y)k_1 + (y-1)k_2)_\lambda \Big] \\ &= \frac{i}{4\pi^2}\Big[-\frac{1}{3}\eta_{\lambda\rho}(k_1+2k_2)\cdot q + \frac{1}{3} q_\rho (k_1-k_2)_\lambda \Big] \end{aligned} \tag{6.90}$$

Adding the cross term one obtains

$$q^\mu \widetilde{T}^{(AVA)}_{\mu\lambda\rho}(k_1,k_2) + q^\mu \widetilde{T}^{(AVA)}_{\mu\rho\lambda}(k_2,k_1) = \frac{i}{4\pi^2}\Big[-\eta_{\lambda\rho} q^2 + \frac{2}{3}(k_{1\lambda}k_{2\rho} - k_{2\lambda}k_{1\rho}) \Big] \tag{6.91}$$

The even part of $\langle \partial\cdot j_5\; j\; j_5\rangle$ and $\langle \partial\cdot j_5\; j_5\; j\rangle$ are non-vanishing, but opposite, for repeating the calculation of

$$\begin{aligned} q^\mu \widetilde{T}^{(AAV)}_{\mu\lambda\rho}(k_1,k_2) &= \int \frac{d^4p}{(2\pi)^4} \mathrm{tr}\left\{ \frac{1}{\not p}\gamma_\lambda \frac{1}{\not p - \not k_1}\gamma_\rho\gamma_5 \frac{1}{\not p - \not q}\gamma_5 \not q \right\} \\ &= \int \frac{d^4p d^\delta \ell}{(2\pi)^{4+\delta}} \mathrm{tr}\left\{ \frac{\not p + \not \ell}{p^2-\ell^2}\gamma_\lambda \frac{\not p - \not k_1 + \not \ell}{(p-k_1)^2-\ell^2}\gamma_\rho\gamma_5 \frac{\not p - \not q + \not \ell}{(p-q)^2-\ell^2}\gamma_5 \not q \right\} \end{aligned} \tag{6.92}$$

we find

$$q^\mu \widetilde{T}^{(AAV)}_{\mu\lambda\rho}(k_1,k_2) + q^\mu \widetilde{T}^{(AAV)}_{\mu\rho\lambda}(k_2,k_1) = \frac{i}{4\pi^2}\Big[\eta_{\lambda\rho} q^2 - \frac{2}{3}(k_{1\lambda}k_{2\rho} - k_{2\lambda}k_{1\rho}) \Big] \tag{6.93}$$

Thus

$$\frac{1}{2}\left(\langle \partial\cdot j_5\; j\; j_5\rangle + \langle \partial\cdot j_5\; j_5\; j\rangle\right) = 0 \tag{6.94}$$

Let us come to the conclusions. Consider first the $V-A$ case, with the formulas (6.43) and (6.44). We have already remarked that the one-point current amplitudes vanish. The even parity two-point current correlators do not vanish. However in a Minkowski background they are proportional to $k^2\eta_{\mu\nu} - k_\mu k_\nu$, where k_μ is the (unique) relevant momentum. Contracting them with k^μ we obtain a vanishing divergence. It remains for us to consider the three-point functions and focus on (6.45) and (6.46). As we have just shown the even parity amplitudes vanish. Therefore we can conclude

$$\partial^\mu \langle\!\langle J_\mu(x) \rangle\!\rangle^{(even)} = 0 \tag{6.95}$$

and

$$\partial^\mu \langle\!\langle J_{5\mu}(x) \rangle\!\rangle^{(even)} = 0 \tag{6.96}$$

These are the gauge Ward identities up to order two in the potentials. A similar argument leads to

$$\partial^\mu \langle\!\langle J_{R\mu}(x) \rangle\!\rangle^{(even)} = 0 \tag{6.97}$$

As expected there are no even parity anomalies.

A side remark here is that there is no ambiguity in passing from the regularization of $\langle 0|j_\mu(x) j_\lambda(y) j_\rho(z)|0\rangle$ (where j is any of the currents considered above) to the regularization of $\langle 0|\partial^\mu j_\mu(x) j_\lambda(y) j_\rho(z)|0\rangle$, in other words $\partial^\mu \langle 0|j_\mu(x) j_\lambda(y) j_\rho(z)|0\rangle$ is the same as $\langle 0|\partial^\mu j_\mu(x) j_\lambda(y) j_\rho(z)|0\rangle$. As we shall see the situation is different when we have to compute the trace of the e.m. tensor instead of its divergence.

References

1. S.L. Adler, Axial-vector vertex in spinor electrodynamics. Phys. Rev. **177**, 2426 (1969)
2. S.L. Adler, W.A. Bardeen, Absence of higher order corrections in the anomalous axial vector divergence equation. Phys. Rev. **182**, 1517–1536 (1969)
3. W.A. Bardeen, Anomalous ward identities in spinor field theories. Phys. Rev. **184**, 1848 (1969)
4. J.S. Bell, R. Jackiw, A PCAC puzzle: $\pi^0 \to \gamma\gamma$ in the σ model. Nuovo Cim. A **60**, 47–61 (1969)
5. R. Jackiw, K. Johnson, Anomalies of the axial vector current. Phys. Rev. **182**, 1459–1469 (1969)
6. R.A. Bertlmann, *Anomalies in Quantum Field Theory* (Oxford Science Publications, 1996)
7. S.B. Treiman, E. Witten, R. Jackiw, B. Zumino, *Current Algebra and Anomalies* (1986)
8. L. Bonora, R. Soldati, S. Zalel, *Dirac, Majorana, Weyl in 4d*. Universe 6 **8**, 111 (2020). arXiv:2006.04546 [hep-th]
9. P. Breitenlohner, D. Maison, Dimensional renormalization and the action principle. Comm. Math. Phys. **52**, 11–38 (1977); Dimensionally renormalized Green's functions for theories with massless particles. I. Comm. Math. Phys. **52**, 39–54 (1977); Dimensionally renormalized Green's functions for theories with massless particles. II. Comm. Math. Phys. **52**, 55–75 (1977)

Chapter 7
Perturbative Diffeomorphism and Trace Anomalies

When the spacetime is not the flat Minkowski spacetime, but we are in the presence of a non-trivial background metric, the anomaly calculation becomes more complicated for various reasons. First, a new character enters the game: the e.m. tensor. This means that in addition to the current correlators we have to consider correlators containing at least one insertion of the e.m. tensor. It has one canonical dimension more than a current. This implies that Feynman diagrams with one or more e.m. tensor insertions are more divergent than the corresponding diagrams containing only currents. Incidentally, this fact definitely favors the dimensional regularization as compared to others. Another considerable complication is that when the background metric is non-trivial, diffeomorphism invariance becomes a fundamental symmetry and we have to face the problem of diffeomorphism anomalies. Whatever other anomaly we are looking for, we must take care that diffeomorphism invariance is preserved. In addition, a new difficulty arises from the very definition of the e.m. tensor. As we have seen in Chap. 3, the e.m. tensor of a theory is not uniquely defined. It is clear that, before starting any calculation, this ambiguity must be resolved. This is the first issue we cope with in this chapter. Once this is clarified we deem it useful to consider the examples of anomalies in two-dimensional theories. These examples are important in connection with string theory, but they are also very useful as a playground in view of higher dimensional cases. After the 2d parenthesis, we return to four dimensions and start with the study of possible diffeomorphism and trace anomalies in a theory of fermions defined on a nontrivial metric background, which originate from their couplings to vector and axial gauge potentials. This means studying (vector and axial) current correlators with one e.m. tensor insertion. Finally, we cope with the problem of computing the anomalies due to diffeomorphisms and the trace anomalies of fermion theories generated by a non-trivial metric background, which requires the study of three e.m. tensor correlators. While calculating odd parity trace anomalies, we find that they have a particular characteristic: contrary to what happens for all the anomalies considered so far, it is not possible to compute them only on the basis of the lowest perturbative order approximation: higher order or nonperturbative calculations are needed.

L. Bonora, *Fermions and Anomalies in Quantum Field Theories*, Theoretical and Mathematical Physics, https://doi.org/10.1007/978-3-031-21928-3_7

7.1 Definition of e.m. Tensors and Their Traces

As we have seen before, chiral anomalies appear in the divergence of chiral currents as a result of regularizing, in the perturbative approach, fermion loops. No essential ambiguities arise in the process of defining the relevant objects to be computed. The case of trace anomalies is somewhat more involved. As we shall see in various examples, *a priori* different definitions of the trace anomaly are possible in a perturbative approach. Such different definitions lead in general to different results. There are in fact several sources of ambiguity. To eliminate these ambiguities, our definition of trace anomaly *in the perturbative case* will be the following: if $T_{\mu\nu}$ is the stress-energy tensor of a theory, the trace anomaly is given by the difference, [1–3],

$$g^{\mu\nu}\langle\!\langle T_{\mu\nu}(x)\rangle\!\rangle - \langle\!\langle g^{\mu\nu}T_{\mu\nu}(x)\rangle\!\rangle \tag{7.1}$$

This formula needs suitable elucidations, which will be given in due time below. But, just to make an initial example, when a theory is conformal invariant, the field operator $T^{\mu}_{\mu}(x)$ vanishes on shell, while in the case a theory contains a conformal soft breaking term (a mass term, for instance) $T^{\mu}_{\mu}(x) \neq 0$ even on shell. The second term of (7.1) is certainly present in such a case and the subtraction in (7.1) is needed in order to exclude this unwanted term from the anomaly. As a matter of fact, as we shall see, this term is non-vanishing in many other instances and in subtler ways. The purpose of this section is to discuss formula (7.1) in its own right, abstracting for the time being from explicit calculations, which will be considered in the next sections.

To illustrate the problem with its peculiarities we shall refer to a concrete example, the generalization to other instances being straightforward. We consider the action of a Dirac fermion coupled to a metric and an Abelian vector field

$$S = \int \mathrm{d}^4x\, \sqrt{g}\, i\overline{\psi}\gamma^{\mu}\left(\partial_{\mu} + \frac{1}{2}\omega_{\mu} + V_{\mu}\right)\psi \tag{7.2}$$

where $\gamma^{\mu} = e^{\mu}_{a}\,\gamma^{a}$, and ω_{μ} is the spin connection, $\omega_{\mu} = \omega^{ab}_{\mu}\Sigma_{ab}$ and $\Sigma_{ab} = \frac{1}{4}[\gamma_a, \gamma_b]$ are the Lorentz generators. The vector current is $j_{\mu} = i\overline{\psi}\gamma_{\mu}\psi$ and the stress-energy tensor is

$$T_{\mu\nu} = \frac{i}{4}\overline{\psi}\gamma_{\mu}\overset{\leftrightarrow}{\nabla}_{\nu}\psi + \{\mu \leftrightarrow \nu\} \tag{7.3}$$

They are both conserved on shell. $T_{\mu\nu}$ is also traceless on shell. At this juncture, however, the first ambiguity appears. There is in fact another definition of the e.m. tensor, which corresponds to the general formula

$$T_{\mu\nu}(x) = \frac{2}{\sqrt{g}}\frac{\delta S}{\delta g^{\mu\nu}} \tag{7.4}$$

and reads

$$\widehat{T}_{\mu\nu} = \frac{i}{4}\overline{\psi}\gamma_\mu \overleftrightarrow{\nabla}_\nu \psi + (\mu \leftrightarrow \nu) - g_{\mu\nu}\frac{i}{2}\overline{\psi}\gamma^\lambda \overleftrightarrow{\nabla}_\lambda \psi = T_{\mu\nu} - g_{\mu\nu} T_\lambda{}^\lambda, \qquad (7.5)$$

which is also conserved and traceless on shell. This ambiguity in the definition of the e.m. tensor gives rise to an ambiguity in the definition of the trace anomaly (as well as of the diffeomorphism anomaly). Such an uncertainty is in fact resolved by the definition (7.1): thanks to it, the second term of $\widehat{T}_{\mu\nu}$ drops out. For this reason, in the rest of this section, we proceed with (7.3).

This also resolves a notation problem. So far we have generically used the notation $T_{\mu\nu}$ for the e.m. tensor. From now on we have to be more precise. For instance, as we shall see in detail in the following sections, the difference between the two definitions of the e.m. tensor can be reproduced in the definition of the corresponding effective action by inserting or not the factor $\sqrt{g(x_i)}$ in the integral $\int \mathrm{d}^\mathrm{d} x_i$ of such formulas as (3.35) (4.17) and similar ones. Precisely, when this factor is inserted one should use $\widehat{T}_{\mu\nu}$, while, if it is not inserted, the definition is valid for $T_{\mu\nu}$.

The gauge and diffeomorphism anomalies are violations of the classical conservation laws $\partial^\mu j_\mu(x) = 0$ and $\nabla^\mu T_{\mu\nu}(x) = 0$, respectively. The trace anomaly is a violation of the classical tracelessness condition $T_\mu^\mu(x) = 0$. The basic point here is that these equations are all valid on shell, while off-shell they do not hold in general (except possibly in dimension 2 for the latter). Another important point to be kept in mind is that, in terms of representations of the Lorentz group, $T_{\mu\nu}(x)$ is a reducible tensor of which the trace $T_\mu^\mu(x)$ is an irreducible component. In the expression of the effective action, the latter is coupled to the field $h(x) = h_\mu^\mu(x)$. Likewise, the amplitude $\langle 0|\mathcal{T} T_\mu^\mu(x)\Phi(y)\Psi(z)|0\rangle$, where Φ and Ψ are generic fields, is an irreducible component of $\langle 0|\mathcal{T} T_{\mu\nu}(x)\Phi(y)\Psi(z)|0\rangle$.

We shall see several examples where, when calculated with Feynman diagrams, the amplitudes $\langle 0|\mathcal{T} T_\mu^\mu(x)\Phi(y)\Psi(z)|0\rangle$ and $\eta^{\mu\nu}\langle 0|\mathcal{T}_{\mu\nu}\Phi(y)\Psi(z)|0\rangle$ are generally different. One possible attitude is to declare that the true value of the amplitude is given by the latter and ignore the former, considering their difference to be an oddity of the regularization. However, there is a difficulty on the way of such a cavalier solution. On the one hand, we shall see that the definition (7.1) for the trace anomaly yields results coincident with non-perturbative approaches. On the other hand, we will see in several examples, that the amplitudes of the type $\langle 0|\mathcal{T} T_\mu^\mu(x) j_\lambda(y) j_\rho(z)|0\rangle$ are related to the Adler-Bell-Jackiw anomalies, and, moreover, it is possible to prove that the amplitude $\langle 0|\mathcal{T} T_\mu^\mu(x) T_{\lambda\rho}(y) T_{\sigma\tau}(z)|0\rangle$ are rigidly related to the Kimura-Delbourgo-Salam anomaly. Therefore, if we decide to ignore $\langle 0|\mathcal{T} T_\mu^\mu(x) j_\lambda(y) j_\rho(z)|0\rangle$ and $\langle 0|\mathcal{T} T_\mu^\mu(x) T_{\lambda\rho}(y) T_{\sigma\tau}(z)|0\rangle$, the calculation of KDS and ABJ anomalies is called into question. In both cases, the connection is visible at the level of the corresponding lowest order Feynman diagrams. But also in the non-perturbative approach of Chap. 9, this link is clear.[1] That simplification is clearly unacceptable.

[1] This link does not exist for even trace anomalies.

Therefore, we have to live with the trace anomaly (perturbatively) defined by the difference

$$\eta^{\mu\nu}\langle 0|\mathcal{T}T_{\mu\nu}(x)\Phi(y)\Psi(z)|0\rangle - \langle 0|\mathcal{T}T^{\mu}_{\mu}(x)\Phi(y)\Psi(z)|0\rangle \tag{7.6}$$

This difference means in particular that the (regularized) effective action has discontinuities: differentiating it with respect to $h_{\mu\nu}$ and then saturating the result with $\eta_{\mu\nu}$ may not be the same as differentiating it with respect to $h(x) = h^{\mu}_{\mu}(x)$, which is the conjugate source of $T^{\mu}_{\mu}(x)$. This is something we have to accept. Like in many other situations in quantum theories, we are not allowed to make an arbitrary choice such as ignoring the second term in (7.6). We have to let the theory speak, keep all the information provided by it and, eventually, interpret it. Therefore, let us try to interpret formula (7.1).

A model of the situation we are facing, which can help our intuition, is provided by the formula

$$\Box\frac{1}{x^2} \sim \delta^{(4)}(x)\,. \tag{7.7}$$

valid in distribution theory in a Euclidean 4d space. The derivatives ∂_μ of $\frac{1}{x^2}$ or $\frac{x^\nu}{x^4}$ are well-defined for $x^\mu \neq 0$ where $\Box\frac{1}{x^2} = 0$, but they are ill-defined at $x^\mu = 0$. On the other hand, the differentiation with respect to $r = \sqrt{x^2}$ makes sense even at $r = 0$ and gives rise to the formula (7.7). In our case, the analog of x^μ is $h_{\mu\nu}$, the analog of r is h^μ_μ and the analog of $x = 0$ is the classical on shell condition $T^\mu_\mu = 0$. We are therefore forced to take into account this discontinuity of the effective action. In sum, in the perturbative approach, we are obliged to refine the naive definition of the trace anomaly as follows:

$$g^{\mu\nu}(x)\langle\!\langle T_{\mu\nu}(x)\rangle\!\rangle - \langle\!\langle g^{\mu\nu}(x)T_{\mu\nu}(x)\rangle\!\rangle = T[g](x). \tag{7.8}$$

Besides verifying that this definition works properly, one may ask what its physical meaning is. It is clear that the reason for taking the difference in the LHS of (7.8) is that two correlators may in general contain extra terms which have nothing to do with the anomaly. These terms are

- possible soft terms that classically violate conformal invariance;
- the term $i\eta^{\mu\nu}\bar\psi\overleftrightarrow{\partial\!\!\!/}\psi$ in the modified definition of the e.m. tensor;
- the semi-local terms in the conformal WI (see below);
- possible off-shell contributions to the anomaly: contrary to the example above where $\Box$ applied to the argument yields 0 for $x \neq 0$, the derivative with respect to $h_{\mu\nu}$ contracted with $\eta^{\mu\nu}$, or the derivative with respect to h^μ_μ, do not automatically vanish off-shell. In fact, the operator T^μ_μ identically vanishes on shell, and therefore, its contribution can only be off-shell. This means that in formula (7.8) the off-shell contributions to the anomaly are subtracted away. In other words, the trace anomaly (7.8) receives only on shell contributions.

All these terms cancel out in (7.8).

Let us expand a bit on the last point. First of all, let us notice that

$$T_{\mu}^{\mu}(x) \sim \bar{\psi} \overleftrightarrow{\nabla} \psi \tag{7.9}$$

Therefore, $\langle 0|\mathcal{T} T_{\mu}^{\mu}(x)\Phi(y)\Phi(z)|0\rangle$ is linear in the LHS of the equation of motion. So it represents a contribution to the quantum object $\langle\langle\bar{\psi} \overleftrightarrow{\nabla} \psi\rangle\rangle$ off-shell (because on shell it vanishes).

On the other hand, $\eta^{\mu\nu}\langle 0|\mathcal{T} T_{\mu\nu}(x)\Phi(y)\Phi(z)|0\rangle$ represents the contribution coming from the differentiation of the effective action with respect to $h^{\mu\nu}(x)$, which, after contraction of the indices, may not coincide with the differentiation with respect to $h(x)$ (there is a discontinuity). Therefore, we interpret it as the on shell plus off-shell contribution to $\langle\langle\bar{\psi} \overleftrightarrow{\nabla} \psi\rangle\rangle$. The difference (7.6) measures the one-loop violation to the equation of motion contracted with $\bar{\psi}$. It represents, so to speak, the quantization of 0, a genuine quantum effect. We call this violation *the trace anomaly*.

This has to be compared with the other anomalies (gauge and diffeomorphism). For instance, in the case of gauge anomalies we have, similar to the trace case,

$$\partial_{\mu} j^{\mu}(x) \sim \partial_{\mu}\left(\bar{\psi}\gamma^{\mu}\psi\right) \tag{7.10}$$

which is proportional to the LHS of the Dirac eom. However, in this case the amplitude $\langle 0|\mathcal{T}\partial^{\mu} j_{\mu}(x) j_{\nu}(y) j_{\lambda}(z)|0\rangle$ gives the same result as $\partial_x^{\mu}\langle 0|\mathcal{T} j_{\mu}(x) j_{\nu}(y) j_{\lambda}(z)|0\rangle$, which is not surprising because $\langle 0|\mathcal{T}\partial^{\mu} j_{\mu}(x) j_{\nu}(y) j_{\lambda}(z)|0\rangle$ is not an amplitude independent of $\langle 0|\mathcal{T} j_{\mu}(x) j_{\nu}(y) j_{\lambda}(z)|0\rangle$. There is no way to disentangle the on shell from the off-shell part, if any. Therefore, we simply set

$$\partial_{\mu}\langle\langle j^{\mu}(x)\rangle\rangle = \langle\langle\partial_{\mu} j^{\mu}(x)\rangle\rangle = \text{anomaly} \tag{7.11}$$

Finally, we have insisted throughout our discussion that the definition (7.1) has to be applied in a perturbative framework. In the non-perturbative approaches of the heat kernel type, the trace anomaly is simply the response of the effective action to a Weyl rescaling of the metric. The connection of the two definitions is not simple, and in the non-perturbative cases, it is hard if not impossible to separately evaluate the two terms of this definition.[2] In all cases considered in this book, the two definitions lead to the same (final) results. We may nevertheless ask: what is the relation of the two terms in the LHS of (7.1) with the result of a corresponding non-perturbative calculation? As we have noted before the perturbative approach starts from the lowest order of the perturbative cohomology, because higher order calculations are more difficult and often inaccessible. Now the lowest order perturbative cohomology is a much looser mathematical structure than the full BRST cohomology. The former has plenty of nontrivial cocycles, while in the latter non-trivial cocycles are very limited in

[2] For instance, in the perturbative approach on shell configurations refer to solutions of the eom in the space of asymptotic plane waves, while in the non-perturbative cases they refer to the zero modes. The two concepts should not be confused.

number (this issue will be further discussed in Sect. 7.4.6 below). The definition (7.1) is designed to channel the lowest order perturbative results in the right track so as to coincide with the non-perturbative approaches. In a more formal language, one could say that each term of (7.1) is separately unstable in the perturbative cohomology, while their difference is stable.

7.2 A 2d Playground

In this section, we deal with examples in 2d, specifically with two-point correlators of the e.m. tensor, [4]. Due to the relative simplicity, we can use regularizations which may become forbidding in higher dimensions. Below we regularize this two-point function using the techniques of differential regularization and we derive the 2d diffeomorphism and trace anomalies. We also discuss the ambiguities implicit in the regularization procedure which allow us to exhibit the interplay between diffeomorphism and trace anomalies. In the spirit of 2d CFT, all correlators considered in this part are Euclidean. Next we redo the same derivation by means of Feynman diagrams in a $1+1$ Minkowski spacetime. In both cases, the method is perturbative; i.e. we represent the metric as $g_{\mu\nu} = \eta_{\mu\nu} + h_{\mu\nu}$ and consider series expansions in $h_{\mu\nu}$. In this regard, a clarification is in order. In (3.35), the generating functional is written as a function of $h_{\mu\nu}$, $W[h]$, while elsewhere we have denoted it $W[g]$. There is no disagreement between the two notations. The notation $W[h]$ emphasizes the fact that the RHS of (3.35) is a series in $h_{\mu\nu}$. We will use the symbol h whenever we want to stress the perturbative aspect, and the symbol g whenever we want to indicate the complete result. It should be remarked that anomalies, in this chapter, are obtained via perturbative calculations, and therefore, we must refer to perturbative cohomology; see Sect. 5.4.

7.2.1 *Differential Regularization*

Throughout this and the following subsection, the background metric is supposed to be Euclidean. The e.m. tensor conformal 2-point function in 2d is very well-known and is given by

$$\langle T_{\mu\nu}(x)\, T_{\rho\sigma}(0)\rangle = \frac{c/2}{x^4}\left(I_{\mu\rho}(x)\, I_{\nu\sigma}(x) + I_{\nu\rho}(x)\, I_{\mu\sigma}(x) - \eta_{\mu\nu}\eta_{\rho\sigma}\right) \tag{7.12}$$

where c is the central charge of the model. For $x \neq 0$ it satisfies the Ward identities

$$\partial^{\mu}\langle T_{\mu\nu}(x)\, T_{\rho\sigma}(0)\rangle = 0, \tag{7.13}$$

$$\langle T^{\mu}_{\mu}(x)\, T_{\rho\sigma}(0)\rangle = 0. \tag{7.14}$$

We shall refer to (7.12) as the *bare* correlator.

This 2-point function is UV singular for $x \to 0$, and such a divergence has to be dealt with for the correlator to be well-defined everywhere. In this context, the most convenient way to regularize it is with the technique of *differential regularization.* The recipe of differential regularization is: given a function $f(x)$ that needs to be regularized at $x = 0$, find the most general function $F(x)$ such that $\mathcal{D}F(x) = f(x)$ for $x \neq 0$, where $\mathcal{D}$ is some differential operator, and such that the Fourier transform of $\mathcal{D}F(x)$ is well-defined (alternatively, $\mathcal{D}F(x)$ has integrable singularities).

In our case, we have two guiding principles: Ward identities and dimensional analysis. Differential regularization tells us that our 2-point function should be some differential operator applied to a function, i.e.

$$\langle T_{\mu\nu}(x)\, T_{\rho\sigma}(0)\rangle = \mathcal{D}_{\mu\nu\rho\sigma}(f(x)), \tag{7.15}$$

while conservation requires that the differential operator $\mathcal{D}_{\mu\nu\rho\sigma}$ be transverse, i.e.

$$\partial^{\mu}\mathcal{D}_{\mu\nu\rho\sigma} = \cdots = \partial^{\sigma}\mathcal{D}_{\mu\nu\rho\sigma} = 0.$$

The most general transverse operator with four derivatives, symmetric in μ,ν and in ρ,σ and in the exchange of the couple μ, ν with ρ, σ, one can write, is

$$\mathcal{D}_{\mu\nu\rho\sigma} = \alpha\mathcal{D}^{(1)}_{\mu\nu\rho\sigma} + \beta\mathcal{D}^{(2)}_{\mu\nu\rho\sigma}, \tag{7.16}$$

where

$$\mathcal{D}^{(1)}_{\mu\nu\rho\sigma} = \partial_\mu\partial_\nu\partial_\rho\partial_\sigma - \left(\eta_{\mu\nu}\partial_\rho\partial_\sigma + \eta_{\rho\sigma}\partial_\mu\partial_\nu\right)\Box + \eta_{\mu\nu}\eta_{\rho\sigma}\Box\Box, \tag{7.17}$$

$$\begin{aligned}\mathcal{D}^{(2)}_{\mu\nu\rho\sigma} &= \partial_\mu\partial_\nu\partial_\rho\partial_\sigma - \frac{1}{2}\left(\eta_{\mu\rho}\partial_\nu\partial_\sigma + \eta_{\nu\rho}\partial_\mu\partial_\sigma + \eta_{\mu\sigma}\partial_\nu\partial_\rho + \eta_{\nu\sigma}\partial_\mu\partial_\rho\right)\Box \\ &\quad + \frac{1}{2}\left(\eta_{\mu\rho}\eta_{\nu\sigma} + \eta_{\nu\rho}\eta_{\mu\sigma}\right)\Box\Box. \end{aligned} \tag{7.18}$$

and α, β are numerical parameters.

Dimensional analysis tells us that the function $f(x)$ in (7.15) can be at most a function of $\log\mu^2x^2$ since the lhs of (7.15) scales like $1/x^4$ and this scaling is already saturated by the differential operator with four derivatives. We have introduced an arbitrary mass scale μ to make the argument of log dimensionless. Let us write the most general ansatz for (7.15):

$$\begin{aligned}\langle T_{\mu\nu}(x)\, T_{\rho\sigma}(0)\rangle &= +\mathcal{D}^{(1)}_{\mu\nu\rho\sigma}\left[\alpha_1\log\mu^2x^2 + \alpha_2\left(\log\mu^2x^2\right)^2 + \cdots\right] \\ &\quad + \mathcal{D}^{(2)}_{\mu\nu\rho\sigma}\left[\beta_1\log\mu^2x^2 + \beta_2\left(\log\mu^2x^2\right)^2 + \cdots\right]. \end{aligned} \tag{7.19}$$

Now our task is to fix the coefficients α_i and β_j for (7.19) to match (7.12) for $x \neq 0$. As it turns out, we only need terms up to $\log^2$ (otherwise one cannot avoid unwanted logarithmic terms for $x \neq 0$). The matching gives us

$$\alpha_1 + \beta_1 = -\frac{c}{24}, \quad \alpha_2 = -\beta_2 = -\frac{c}{96},$$

Only the sum $\alpha_1 + \beta_1$ is fixed. So, to simplify, we can set, for instance $\beta_1 = 0$. Finally

$$\langle T_{\mu\nu}(x) T_{\rho\sigma}(0) \rangle = -\frac{c}{24} \mathcal{D}^{(1)}_{\mu\nu\rho\sigma} \left(\log \mu^2 x^2\right) - \frac{c}{96} \left(\mathcal{D}^{(1)}_{\mu\nu\rho\sigma} - \mathcal{D}^{(2)}_{\mu\nu\rho\sigma}\right) \left(\log \mu^2 x^2\right)^2. \tag{7.20}$$

Equation (7.20) gives back (7.12) for $x \neq 0$. But if we take the trace, we find that

$$\langle T^{\mu}_{\mu}(x) T_{\rho\sigma}(0) \rangle = -\frac{c}{24} \eta^{\mu\nu} \mathcal{D}^{(1)}_{\mu\nu\rho\sigma} \left(\log \mu^2 x^2\right) = -\frac{c}{24} \left(\partial_\rho \partial_\sigma - \eta_{\rho\sigma} \Box\right) \Box \log \mu^2 x^2.$$

These terms have support only at $x = 0$, for in 2d

$$\Box \log \mu^2 x^2 = -4\pi \delta^2(x). \tag{7.21}$$

Therefore, we find the anomalous Ward identity

$$\langle T^{\mu}_{\mu}(x) T_{\rho\sigma}(y) \rangle = c\frac{\pi}{6} \left(\partial_\rho \partial_\sigma - \eta_{\rho\sigma} \Box\right) \delta^2(x - y), \tag{7.22}$$

Now, if we insert this result in (3.35), i.e. saturate with $h^{\rho\sigma}(y)$ and integrate over y, we get

$$\langle\langle T^{\mu}_{\mu} \rangle\rangle = c\frac{\pi}{12} (\partial_\mu \partial_\nu - \eta_{\mu\nu} \Box) h^{\mu\nu}, \tag{7.23}$$

which coincides with the lowest contribution of the expansion in h of the Ricci scalar, i.e.

$$R = (\partial_\mu \partial_\nu - \eta_{\mu\nu} \Box) h^{\mu\nu} + \mathcal{O}(h^2). \tag{7.24}$$

Covariance requires the higher order corrections in h to the trace of the e.m. tensor, in the presence of a background metric g, to be such that we recover the covariant expression

$$\langle\langle T^{\mu}_{\mu} \rangle\rangle = c\frac{\pi}{12} R. \tag{7.25}$$

For instance, for a free Dirac fermion $c = \frac{1}{8\pi^2}$. We are authorized to use the covariant expression (7.25) because the energy-momentum tensor is conserved (there are no diffeomorphism anomalies). We can also say that requiring the regularized correlator to be conserved at $x = 0$ implies the appearance of a trace anomaly. However, this is not the end of the story, since there are ambiguities in the regularization process we have so far disregarded.

But, before that, let us pause to consider the cohomological aspects. It is easy to see that

$$\delta_\omega^{(0)} \int d^2x\, \omega(\partial_\mu\partial_\nu - \eta_{\mu\nu}\Box)h^{\mu\nu} = 0 \tag{7.26}$$

and, using $\delta_\omega R = -2\omega R - 2\Box\omega$,

$$\delta_\omega \int d^2x\, \sqrt{g}\, \omega\, R = 0 \tag{7.27}$$

as already pointed out in Sect. 5.3. This is the first simple example of how perturbative cohomology and full cohomology (with the help of covariance) operate. In this case, the diffeomorphisms do not enter the game because we have preserved e.m. tensor conservation all through the construction.

Ambiguities

The ambiguity arises from the fact that we can add to (7.20) terms that have support only at $x = 0$. The most general modification of the parity even part that would affect only its expression for $x = 0$ is given by

$$\begin{aligned} A^{(e)}_{\mu\nu\rho\sigma} &= A\left(\eta_{\mu\nu}\partial_\rho\partial_\sigma + \eta_{\rho\sigma}\partial_\mu\partial_\nu\right)\Box \log\mu^2x^2 \\ &+ B\left(\eta_{\mu\rho}\partial_\nu\partial_\sigma + \eta_{\nu\rho}\partial_\mu\partial_\sigma + \eta_{\mu\sigma}\partial_\nu\partial_\rho + \eta_{\nu\sigma}\partial_\mu\partial_\rho\right)\Box\log\mu^2x^2 \\ &+ C\left(\eta_{\mu\rho}\eta_{\nu\sigma} + \eta_{\nu\rho}\eta_{\mu\sigma}\right)\Box\Box\log\mu^2x^2 \\ &+ D\eta_{\mu\nu}\eta_{\rho\sigma}\Box\Box\log\mu^2x^2. \end{aligned} \tag{7.28}$$

where A, B, C, D are arbitrary parameters. We remark that this term is in general neither conserved nor traceless

$$\begin{aligned} \partial^\mu A^{(e)}_{\mu\nu\rho\sigma}(x) &= 4\pi\left((A+2B)\partial_\nu\partial_\rho\partial_\sigma + (A+D)\eta_{\rho\sigma}\partial_\nu\Box \right.\\ &\left.\quad + (B+C)\left(\eta_{\rho\nu}\partial_\sigma\Box + \eta_{\sigma\nu}\partial_\rho\Box\right)\right)\delta^{(2)}(x) \end{aligned} \tag{7.29}$$

$$A^{(e)\mu}{}_{\mu\rho\sigma}(x) = 4\pi\left((2A+4B)\partial_\rho\partial_\sigma + (A+2C+2D)\eta_{\rho\sigma}\Box\right)\delta^{(2)}(x) \tag{7.30}$$

Thus, these terms are potential anomalies. For them to be true anomalies, they must satisfy the consistency conditions.

We notice that by imposing (7.29) to vanish implies that also (7.30) will vanish. We may wonder whether using this ambiguity we can cancel the trace anomaly. Using Eqs. (5.17) and (5.18), let write down these potential anomalies in integrated form.

$$\mathcal{A}_\xi^{(e,p)} = -\int d^2x d^2y\, \xi^\nu \partial^\mu A_{\mu\nu\rho\sigma}(x)\delta(x-y)h^{\rho\sigma}(y) \tag{7.31}$$

$$\mathcal{A}_\omega^{(e,p)} = \int \mathrm{d}^2x\mathrm{d}^2y \; \omega A^\mu_{\mu\rho\sigma}(x)\delta(x-y)h^{\rho\sigma}(y) \tag{7.32}$$

Now, imposing $\delta_\xi^{(0)} \mathcal{A}_\xi^{(e,p)} = 0$ we find the condition $2A + 3B + C + D = 0$. This means that $\mathcal{A}_\xi^{(e,p)}$ can be rearranged as follows

$$\mathcal{A}_\xi^{(e,p)} = -4\pi \int \mathrm{d}^2x \, \xi^\nu \Big(E \, \partial_\nu \big(\partial_\rho \partial_\sigma h^{\rho\sigma} - \Box h\big) + F \, \partial_\sigma \big(\partial_\nu \partial_\rho h^{\rho\sigma} - 2\Box h^\sigma_\nu\big)\Big) \tag{7.33}$$

where $E = A + 3B + C$ and $F = 2A + 2B + D$ are unconstrained parameters.

Next, if we impose $\delta_\omega^{(0)} \mathcal{A}_\omega^{(e,p)} = 0$, we find the condition $3A + 4B + 2C + 2D = 0$ and

$$\mathcal{A}_\omega^{(e,p)} = 8\pi(A + 2B) \int \mathrm{d}^2x \, \omega\big(\partial_\rho \partial_\sigma h^{\rho\sigma} - \Box h\big) \tag{7.34}$$

with unconstrained $A + 2B$. Finally, we have to impose $\delta_\xi^{(0)} \mathcal{A}_\omega^{(e,p)} + \delta_\omega^{(0)} \mathcal{A}_\xi^{e,(p)} = 0$. It is easy to verify that $\delta_\xi^{(0)} \mathcal{A}_\omega^{(e,p)} = 0$ without any constraint, while $\delta_\omega^{(0)} \mathcal{A}_\xi^{(e,p)} = 0$ requires $F = 0$, which in turn implies $E = A + 2B$.

In conclusion, the ambiguities (7.28) are allowed provided $C = -B$ and $D = -2A - 2B$, where A and B are free parameters. They give rise to the trace anomaly (7.34) and to the diffeomorphism anomaly

$$\mathcal{A}_\xi^{(e,p)} = -4\pi(A + 2B) \int \mathrm{d}^2x \, \xi^\nu \, \partial_\nu \big(\partial_\rho \partial_\sigma h^{\rho\sigma} - \Box h\big) \tag{7.35}$$

The cocycles (7.35) and (7.34) satisfy separately the overall consistency conditions

$$\left(\delta_\xi^{(0)} + \delta_\omega^{(0)}\right) \left(\mathcal{A}_\xi^{(e,p)}, \mathcal{A}_\omega^{(e,p)}\right) = 0.$$

Returning now to the initial question of whether we can cancel the trace anomaly (7.23), or (7.25), by means of these ambiguities, the answer is yes. We can do it by subtracting a combination of ambiguities that satisfy $C = -B$ and $D = -2A - 2B$, and adjusting the overall coefficient. But this operation triggers the diffeomorphism anomaly (7.35). In other words, the anomaly (7.25) is a *minimal* non-trivial cocycle of the overall symmetry under diffeomorphisms plus Weyl transformations. It may take different forms, either as a pure diffeomorphism anomaly or a pure trace anomaly. In general, both components may be non-vanishing. It is obvious that it is preferable to preserve diffeomorphism invariance, so that the cocycle takes the form (7.25).

7.2.2 Parity Odd Terms in 2d

In this section, we compute all possible 'bare' parity odd terms in the 2-point function of the energy-momentum tensor in 2d. To this end, we write the most general expression $\mathcal{T}^{\rm odd}_{\mu\nu\rho\sigma}(x)$ linear in the antisymmetric tensor $\varepsilon_{\alpha\beta}$ with the right dimensions, which is symmetric and traceless in μ, ν and ρ, σ separately, is symmetric in the exchange $(\mu, \nu) \leftrightarrow (\rho, \sigma)$, and is conserved. The calculation is tedious but straightforward. The result is as follows. Let us define

$$T_{\mu\nu\rho\sigma} = \frac{1}{x^4}\left(I_{\mu\rho}(x)I_{\nu\sigma}(x) + I_{\mu\sigma}(x)I_{\nu\rho}(x) - \eta_{\mu\nu}\eta_{\rho\sigma}\right), \tag{7.36}$$

which is proportional to the parity even 2-point function, and

$$\mathcal{T}^{\rm odd}_{\mu\nu\rho\sigma}(x) = \frac{e}{2}\left(\varepsilon_{\alpha\mu}T^{\alpha}{}_{\nu\rho\sigma}(x) + \varepsilon_{\alpha\nu}T_{\mu}{}^{\alpha}{}_{\rho\sigma}(x) + \varepsilon_{\alpha\rho}T_{\mu\nu}{}^{\alpha}{}_{\sigma}(x) + \varepsilon_{\alpha\rho}T_{\mu\nu\rho}{}^{\alpha}(x)\right), \tag{7.37}$$

where e is an undetermined constant. We assume (7.37) to represent the two-point correlator $\langle T_{\mu\nu}(x)T_{\rho\sigma}(0)\rangle_{\rm odd}$. It satisfies all the desired properties (it is traceless and conserved). In order to make sure that it is conformal covariant, we can check that it is chirally split. To this end, we can, for instance, introduce the light-cone coordinates $x_\pm = x^0 \pm x^1$. It is not hard to verify that

$$\langle T_{++}(x)T_{--}(0)\rangle_{\rm odd} = 0. \tag{7.38}$$

The task of regularizing the parity odd terms is very much simplified if we write them in terms of the parity even part. We can therefore use the same regularization as above. Let us start by the regularization that preserves diffeomorphisms for the parity even part, Eq. (7.20):

$$T_{\mu\nu\rho\sigma}(x) = -\frac{1}{12}\mathcal{D}^{(1)}_{\mu\nu\rho\sigma}\left(\log\mu^2x^2\right) - \frac{1}{48}\left(\mathcal{D}^{(1)}_{\mu\nu\rho\sigma} - \mathcal{D}^{(2)}_{\mu\nu\rho\sigma}\right)\left(\log\mu^2x^2\right)^2. \tag{7.39}$$

Regularizing (7.37) leads to both a trace anomaly

$$\left\langle T^{\mu}_{\mu}(x)\,T_{\rho\sigma}(0)\right\rangle_{\rm odd} = \frac{\pi e}{24}\left(\varepsilon_{\rho\alpha}\partial^\alpha\partial_\sigma + \varepsilon_{\sigma\alpha}\partial^\alpha\partial_\rho\right)\delta^2(x), \tag{7.40}$$

and a diffeomorphism anomaly

$$\partial^\mu\left\langle T_{\mu\nu}(x)\,T_{\rho\sigma}(0)\right\rangle_{\rm odd} = \frac{\pi e}{24}\varepsilon_{\nu\alpha}\partial^\alpha\left(\eta_{\rho\sigma}\Box - \partial_\rho\partial_\sigma\right)\delta^2(x). \tag{7.41}$$

Inserting these results in (3.35) gives rise to the following lowest order anomalies

$$\mathcal{A}_\xi^{(0)} = -\frac{\pi e}{48}\int \mathrm{d}^2x\, \xi^\nu \varepsilon_{\nu\alpha}\partial^\alpha \left(\eta_{\rho\sigma}\Box - \partial_\rho\partial_\sigma\right) h^{\sigma\rho} \tag{7.42}$$

and

$$\mathcal{T}_\omega^{(0)} = -\frac{\pi e}{48}\int \mathrm{d}^2x\, \omega\, \varepsilon_{\lambda\alpha}\partial^\alpha\partial_\rho h^{\lambda\rho} \tag{7.43}$$

It is a useful exercise to verify that

$$\delta_\xi^{(0)}\mathcal{A}_\xi^{(0)} = 0, \quad \delta_\xi^{(0)}\mathcal{T}_\omega^{(0)} = 0 = \delta_\omega^{(0)}\mathcal{A}_\xi^{(0)}, \quad \delta_\omega^{(0)}\mathcal{T}_\omega^{(0)} = 0. \tag{7.44}$$

We see that the diffeomorphism anomaly is accompanied by a trace partner.

It is worth pointing out that (7.41) is the lowest order approximation of the following 'full one-loop' relation

$$\langle\langle \nabla_\mu T^{\mu\nu}(x)\rangle\rangle = \frac{\pi e}{48}\varepsilon^{\nu\alpha}\partial_\alpha R, \tag{7.45}$$

Equation (7.45) is known as the covariant form of the diffeomorphism anomaly. At the lowest order, however, also the consistent anomaly

$$\langle\langle D^\mu T_{\mu\nu}(x)\rangle\rangle = \frac{\pi e}{24}\frac{\varepsilon^{\mu\rho}}{\sqrt{g}}\partial_\mu\partial_\alpha\Gamma^\alpha_{\rho\nu} = \frac{\pi e}{48}\varepsilon_{\alpha\rho}\partial^\alpha\left(\Box h_\nu^\rho - \partial_\nu\partial_\sigma h^{\rho\sigma}\right) + \mathcal{O}\left(h^2\right) \tag{7.46}$$

takes the same form due to the 2d identity

$$2\varepsilon_{\mu\nu}\partial^\mu\left(\partial_\alpha\partial_\beta - \eta_{\alpha\beta}\Box\right) = \varepsilon_{\mu\alpha}\left(\partial^\mu\partial_\nu\partial_\beta - \eta_{\nu\beta}\partial^\mu\Box + (\alpha\leftrightarrow\beta)\right) \tag{7.47}$$

We recall that saturating (7.46) with the ghost ξ^ν, and integrating it over the 2d spacetime, it can be written in the compact form

$$\sim \int \mathrm{d}^2x\,\sqrt{g}\,\mathrm{Tr}(\mathrm{d}\Xi\,\Gamma) \tag{7.48}$$

where Ξ is the matrix with components $\Xi_\rho{}^\sigma = \partial_\rho\xi^\sigma$ and Γ is the matrix one-form $\Gamma^\rho_{\mu\sigma}\mathrm{d}x^\mu$. The trace is over the matrix indices. In the form (7.48), proving consistency is a very simple exercise, as was shown in Chap. 5. The anomaly (7.48) coincides with the diffeomorphism cocycle introduced there; see Eq. (5.57) with $\mathrm{d} = 2$.

The cocycle (7.46) is not invariant under Weyl transformations. Therefore, it must be accompanied by a trace anomaly partner, whose lowest term is given by (7.43). It can also be cast in minimal form (i.e. with vanishing conformal partner), see just below. For the 'covariant' anomaly (7.45) this possibility is irrelevant since it does not satisfy the diffeomorphism consistency condition. In fact, it is an anomaly analogous to the ABJ one in gauge theories.

Ambiguities in the Parity Odd Part

We know that the regularization used above is not the ultimate one, because there are ambiguities. They entail a modification of the parity odd part given by

$$A^{(o)}_{\mu\nu\rho\sigma} = \varepsilon_{\alpha\mu} A^{\alpha}{}_{\nu\rho\sigma} + \varepsilon_{\alpha\nu} A_{\mu}{}^{\alpha}{}_{\rho\sigma} + \varepsilon_{\alpha\rho} A_{\mu\nu}{}^{\alpha}{}_{\sigma} + \varepsilon_{\alpha\rho} A_{\mu\nu\rho}{}^{\alpha}, \tag{7.49}$$

where the RHS is written in terms of (7.28), which explicitly is

$$\begin{aligned} A^{(o)}_{\mu\nu\rho\sigma} &= A\left[\eta_{\mu\nu}\left(\varepsilon_{\rho\alpha}\partial^{\alpha}\partial_{\sigma} + \varepsilon_{\sigma\alpha}\partial^{\alpha}\partial_{\rho}\right) + \eta_{\rho\sigma}\left(\varepsilon_{\mu\alpha}\partial^{\alpha}\partial_{\nu} + \varepsilon_{\nu\alpha}\partial^{\alpha}\partial_{\mu}\right)\right]\Box \log \mu^2 x^2 \\ &+ B\left[\varepsilon_{\mu\alpha}\left(\eta_{\nu\rho}\partial^{\alpha}\partial_{\sigma} + \eta_{\nu\sigma}\partial^{\alpha}\partial_{\rho}\right) + \varepsilon_{\nu\alpha}\left(\eta_{\mu\rho}\partial^{\alpha}\partial_{\sigma} + \eta_{\mu\sigma}\partial^{\alpha}\partial_{\rho}\right)\right. \\ &\left.+\varepsilon_{\rho\alpha}\left(\eta_{\sigma\mu}\partial^{\alpha}\partial_{\nu} + \eta_{\sigma\nu}\partial^{\alpha}\partial_{\mu}\right) + \varepsilon_{\sigma\alpha}\left(\eta_{\rho\mu}\partial^{\alpha}\partial_{\nu} + \eta_{\rho\nu}\partial^{\alpha}\partial_{\mu}\right)\right]\Box \log \mu^2 x^2. \end{aligned} \tag{7.50}$$

The trace and the divergence of (7.50) are given by:

$$\eta^{\mu\nu} A^{(o)}_{\mu\nu\rho\sigma} = 8\pi\,(A+2B)\left(\varepsilon_{\rho\alpha}\partial^{\alpha}\partial_{\sigma} + \varepsilon_{\sigma\alpha}\partial^{\alpha}\partial_{\rho}\right)\delta^2(x), \tag{7.51}$$

$$\begin{aligned} \partial^{\mu} A^{(o)}_{\mu\nu\rho\sigma} &= 4\pi\left(B\eta_{\nu\rho}\Box + (A+B)\,\partial_{\nu}\partial_{\rho}\right)\varepsilon_{\sigma\alpha}\partial^{\alpha}\delta^2(x) \\ &+ 4\pi\left(B\eta_{\nu\sigma}\Box + (A+B)\,\partial_{\nu}\partial_{\sigma}\right)\varepsilon_{\rho\alpha}\partial^{\alpha}\delta^2(x) \\ &+ 4\pi\left(A\eta_{\rho\sigma}\Box + 2B\partial_{\rho}\partial_{\sigma}\right)\varepsilon_{\nu\alpha}\partial^{\alpha}\delta^2(x). \end{aligned} \tag{7.52}$$

As before let us define

$$\mathcal{A}^{(o,p)}_{\xi} = \int \mathrm{d}^2x\mathrm{d}^2y\,\xi^{\nu}\partial^{\mu} A^{(o)}_{\mu\nu\rho\sigma}(x)\delta(x-y)h^{\rho\sigma}(y) \tag{7.53}$$

$$\mathcal{A}^{(o,p)}_{\omega} = -\int \mathrm{d}^2x\mathrm{d}^2y\,\omega A^{(o)\mu\rho\sigma}_{\mu}(x)\delta(x-y)h^{\rho\sigma}(y) \tag{7.54}$$

It is not hard to see that

$$\begin{aligned} \delta^{(0)}_{\xi}\mathcal{A}^{(o,p)}_{\xi} &= 0 = \delta^{(0)}_{\omega}\mathcal{A}^{(o,p)}_{\omega} \\ \delta^{(0)}_{\omega}\mathcal{A}^{(o,p)}_{\xi} = -\delta^{(0)}_{\xi}\mathcal{A}^{(o,p)}_{\omega} &= -16\pi(A+2B)\int \mathrm{d}^2x\,\xi^{\nu}\varepsilon_{\nu\alpha}\partial^{\alpha}\Box\omega \end{aligned} \tag{7.55}$$

(7.53) and (7.54), coming from the local ambiguity (7.50), satisfy the diffeomorphism + Weyl consistency conditions, for any A and B.

Using these ambiguities, we can modify the previously computed trace anomaly (7.40) and diffeomorphism anomaly (7.41). In particular, it is interesting to see whether we can cancel the latter. To achieve this, we have to adjust the coefficient

A and B so that (7.53) takes the form (7.41). This is possible only if $A + 2B = 0$, but this entails the vanishing of both (7.53) and (7.54), as one can verify by means of the identity (7.47). Thus, it is impossible to cancel in this way the diffeomorphism anomaly. By suitably adjusting the coefficient, we can instead cancel the trace anomaly (7.40) at the expenses of a diffeomorphism anomaly.

Remark. In the previous section, we have introduced the definition (7.1) of the trace anomaly. How does that definition relate to the calculations of this section? What we have done here is to compute the first term of (7.1), because we have started from the (unregulated) conformal two-point function $\langle T_{\mu\nu}(x)T_{\rho\sigma}(0)\rangle$ and then taken the trace of the first entry. In conformal field theory, when computing amplitudes like (7.12) it is assumed that the field operator $T_\mu^\mu(x)$ is identically vanishing, so that the equivalent of $\langle T_\mu^\mu(x)T_{\rho\sigma}(0)\rangle$ is 0. Therefore, the trace anomaly calculated above fits completely the definition (7.1). In conclusion, the minimal form of the even trace anomaly is given by (7.25) with a vanishing diffeomorphism partner. The odd parity anomaly is defined by the diffeomorphism cocycle (7.48) and by its Weyl partner (for the explicit expression of the latter; see (7.85)).

7.2.3 *The Feynman Diagrams Method in* 2d

We wish now to derive the previous results using Feynman diagrams. The action for a massless Dirac fermion in 2d Minkowski spacetime is

$$S = \int \mathrm{d}^2x\, \sqrt{g}\, i\overline{\psi}\gamma^\mu \left(\partial_\mu + \frac{1}{2}\omega_\mu\right)\psi \tag{7.56}$$

where $g = \det(g_{\mu\nu})$, $\gamma^\mu = \mathrm{e}_a^\mu \gamma^a$, ($\mu, \nu, \ldots$ are world indices, $a, b, \ldots$ are flat indices) and ω_μ is the spin connection:

$$\omega_\mu = \omega_\mu^{ab}\Sigma_{ab}$$

where $\Sigma_{ab} = \frac{1}{4}[\gamma_a, \gamma_b]$ are the Lorentz generators.

Let us recall a few basic definitions concerning 2d gamma matrices:

$$\left\{\gamma^a, \gamma^b\right\} = 2\eta^{ab} \Rightarrow \left(\gamma^0\right)^2 = 1, \quad \left(\gamma^1\right)^2 = -1. \tag{7.57}$$

Clearly, $\gamma^0 = \gamma_0$ and $\gamma^1 = -\gamma_1$. The chirality matrix γ_* is defined by $\gamma_* = -\gamma_0\gamma_1$.

It is straightforward to check that the following relations are true:

$$\gamma_a = \varepsilon_{ab}\gamma^b\gamma_*, \quad \varepsilon_{ab}\gamma^b = \gamma_a\gamma_*, \tag{7.58}$$

where we are using the convention $\varepsilon_{01} = 1$. It follows

$$\mathrm{tr}(\gamma_a \gamma_b \gamma_*) = -2\varepsilon_{ab}. \tag{7.59}$$

Two-Point e.m. Correlator for Chiral Fermions

In two dimensions, due to the anticommutativity of spinors and the proportionality between the unique Lorentz generator and the chiral matrix γ_*, the spin connection drops out of the action (7.56). Let us consider first the case of a right-handed spinor: $\psi_R = P_R \psi$, where $P_R = \frac{1+\gamma_*}{2}$. For ψ_R, the action becomes

$$S_R = i \int \mathrm{d}^2x \, \sqrt{g}\, \overline{\psi}_R \gamma^\mu \partial_\mu \psi_R \tag{7.60}$$

We can further simplify it by absorbing the $\sqrt{g}$ into a redefinition of ψ: $\psi \to \widetilde{\psi} = g^{\frac{1}{4}} \psi$. In the path integral spirit, this is allowed because the Jacobian of this transformation factors out and can be disregarded. We can therefore replace S_R with

$$\widetilde{S}_R = i \int \mathrm{d}^2x \, \overline{\widetilde{\psi}}_R \gamma^\mu \partial_\mu \widetilde{\psi}_R = i \int \mathrm{d}^2x \, \overline{\widetilde{\psi}}_R \gamma^a e_a^\mu \partial_\mu \widetilde{\psi}_R \tag{7.61}$$

Now let us write $e_a^\mu \approx \delta_a^\mu - \chi_a^\mu$ and make the identification $2\chi_a^\mu = h_a^\mu$,[3] where $h_{\mu\nu}$ is the gravitational fluctuation field: $g_{\mu\nu} = \eta_{\mu\nu} + h_{\mu\nu}$. The fermion propagator is

$$\frac{i}{\not{p} + i\epsilon} \tag{7.62}$$

and there is only one graviton-fermion-fermion vertex (see Fig. 7.1) given by

$$\frac{i}{8}\left[(p+p')_\mu \gamma_\nu + (p+p')_\nu \gamma_\mu\right] \frac{1+\gamma_*}{2}. \tag{7.63}$$

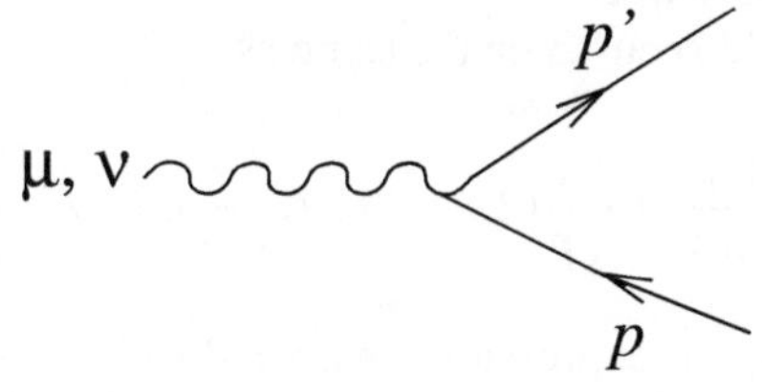

Fig. 7.1 V_{gff} vertex

[3] This implies a choice of local Lorentz gauge because in this way χ_a^μ is symmetric.

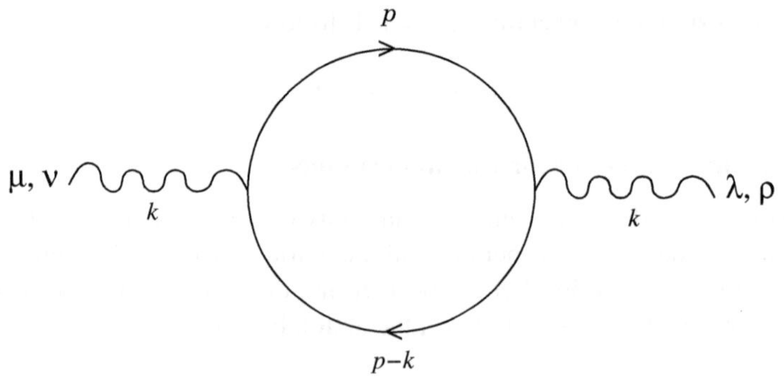

Fig. 7.2 Relevant Feynman diagram for the computation

There is only one non-trivial contribution that comes from the bubble diagram with one incoming and one outgoing line with momentum k and an internal momentum p (see Fig. 7.2).

The relevant 2-point function is

$$\langle T_{\mu\nu}(x) T_{\lambda\rho}(y)\rangle = 4 \int \frac{d^2 k}{(2\pi)^2} e^{-ik(x-y)} \widetilde{\mathcal{T}}_{\mu\nu\lambda\rho}(k) \tag{7.64}$$

with

$$\widetilde{\mathcal{T}}_{\mu\nu\lambda\rho}(k) = -\frac{1}{64} \int \frac{d^2 p}{(2\pi)^2} \mathrm{tr}\left(\frac{1}{\not p}(2p-k)_\mu \gamma_\nu \frac{1+\gamma_*}{2} \frac{1}{\not p - \not k}(2p-k)_\lambda \gamma_\rho \frac{1+\gamma_*}{2}\right) + \left\{\begin{matrix} \mu \leftrightarrow \nu \\ \lambda \leftrightarrow \rho \end{matrix}\right\}. \tag{7.65}$$

The last line means that we have to add three more terms like in the first line so as to realize a symmetry under the exchanges $\mu \leftrightarrow \nu$, $\lambda \leftrightarrow \rho$. Moreover, we have to symmetrize with respect to the exchange $(\mu, \nu) \leftrightarrow (\lambda, \rho)$ (bosonic symmetry). This will be understood in the sequel.

Equation (7.65) would seem to be the same as

$$\widetilde{\mathcal{T}}'_{\mu\nu\lambda\rho}(k) = -\frac{1}{64} \int \frac{d^2 p}{(2\pi)^2} \mathrm{tr}\left(\frac{1}{\not p}(2p-k)_\mu \gamma_\nu \frac{1}{\not p - \not k}(2p-k)_\lambda \gamma_\rho \frac{1+\gamma_*}{2}\right). \tag{7.66}$$

But they are both divergent expressions that need to be regularized. It is hardly a surprise that, once this is done, they lead to different results. The difference consists, however, as we shall see, of local terms.

Let us proceed to regularize these expressions, starting from (7.65), with the method of dimensional regularization. To this end, as usual, we introduce extra space components of the momentum running around the loop, $p_\mu \to p_\mu + \ell_{\bar{\mu}}$ ($\ell_{\bar{\mu}} = \ell_2, \ldots, \ell_{\delta+1}$) . So (7.65) becomes

$$\widetilde{\mathcal{T}}^{(\mathrm{reg})}_{\mu\nu\lambda\rho}(k) = -\frac{1}{64}\int \frac{\mathrm{d}^2p\,\mathrm{d}^\delta\ell}{(2\pi)^{2+\delta}}\mathrm{tr}\Big(\frac{1}{\not{p}+\not{\ell}}(2p-k)_\mu\gamma_\nu\frac{1+\gamma_*}{2}\frac{1}{\not{p}+\not{\ell}-\not{k}} \times (2p-k)_\lambda\gamma_\rho\frac{1+\gamma_*}{2}\Big). \tag{7.67}$$

Now let us recall that $\not{p}^2 = p^2, \not{\ell}^2 = -\ell^2$, $\not{p}\not{\ell} + \not{\ell}\not{p} = 0$ and $[\gamma_*, \not{\ell}] = 0$, while $\{\gamma_*, \not{p}\} = 0$. Moreover, Eq. (7.59) is replaced by

$$\mathrm{tr}\big(\gamma_\mu\gamma_\nu\gamma_*\big) = -2^{1+\frac{\delta}{2}}\varepsilon_{\mu\nu}. \tag{7.68}$$

Thus, (7.67) can be rewritten (for simplicity we drop from now on the $^{(reg)}$ tag)

$$\widetilde{\mathcal{T}}_{\mu\nu\lambda\rho}(k) = -\frac{1}{64}\int \frac{\mathrm{d}^2p\,\mathrm{d}^\delta\ell}{(2\pi)^{2+\delta}}\mathrm{tr}\Big(\frac{\not{p}+\not{\ell}}{p^2-\ell^2}(2p-k)_\mu\gamma_\nu\frac{1+\gamma_*}{2}\frac{\not{p}+\not{\ell}-\not{k}}{(p-k)^2-\ell^2} \times (2p-k)_\lambda\gamma_\rho\frac{1+\gamma_*}{2}\Big). \tag{7.69}$$

$$= -\frac{1}{64}\int \frac{\mathrm{d}^2p\,\mathrm{d}^\delta\ell}{(2\pi)^{2+\delta}}\mathrm{tr}\Big(\frac{\not{p}}{p^2-\ell^2}(2p-k)_\mu\gamma_\nu\frac{1+\gamma_*}{2}\frac{\not{p}-\not{k}}{(p-k)^2-\ell^2} \times (2p-k)_\lambda\gamma_\rho\frac{1+\gamma_*}{2}\Big). \tag{7.70}$$

$$= -\frac{1}{64}\int \frac{\mathrm{d}^2p\,\mathrm{d}^\delta\ell}{(2\pi)^{2+\delta}}\mathrm{tr}\Big(\frac{\not{p}}{p^2-\ell^2}(2p-k)_\mu\gamma_\nu\frac{\not{p}-\not{k}}{(p-k)^2-\ell^2} \times (2p-k)_\lambda\gamma_\rho\frac{1+\gamma_*}{2}\Big). \tag{7.71}$$

Next we evaluate the gamma matrix traces and introduce a Feynman parameter u, $0 \le u \le 1$, in order to evaluate the p integral. $\widetilde{\mathcal{T}}_{\mu\nu\lambda\rho}(k)$ becomes

$$
\begin{aligned}
\widetilde{\mathcal{T}}_{\mu\nu\lambda\rho}(k) = & -\frac{2^{\frac{\delta}{2}}}{64}\int \frac{\mathrm{d}^2 p\, \mathrm{d}^\delta \ell}{(2\pi)^{2+\delta}} \int_0^1 \mathrm{d}u \Big((p+uk)_\nu (p+(u-1)k)_\rho \\
& + (p+uk)_\rho (p+(u-1)k)_\nu - (p+uk)\cdot(p+(u-1)k)\eta_{\nu\rho} \\
& + \varepsilon_{\rho\sigma}\big((p+uk)_\nu (p+(u-1)k)^\sigma + (p+uk)^\sigma (p+(u-1)k)_\nu\big)\Big) \\
& \times \frac{(2p+(2u-1)k)_\mu (2p+(2u-1)k)_\lambda}{[p^2+u(1-u)k^2-\ell^2]^2}.
\end{aligned} \tag{7.72}
$$

The integral can be evaluated after a Wick rotation of the momenta $p_0 \to i p_0^E$, $k_0 \to i k_0^E$, using the results of Appendix 7A. The complete result is recorded in Appendix 7B.

Trace and Diffeomorphism Cocycles

From the previous result, we can compute the trace and the divergence (k_μ denotes the Euclidean momentum, in particular $\mathsf{k}^2 = -k^2$):

$$
\widetilde{\mathcal{T}}^{E\mu}_{\mu\lambda\rho}(\mathsf{k}) = \frac{i}{192\pi}\Big[\mathsf{k}_\lambda \mathsf{k}_\rho + \mathsf{k}^2 \eta_{\lambda\rho} + \frac{1}{2}\big(\mathsf{k}_\lambda \varepsilon_{\rho\sigma}\mathsf{k}^\sigma + \mathsf{k}_\rho \varepsilon_{\lambda\sigma}\mathsf{k}^\sigma\big)\Big] \tag{7.73}
$$

and

$$
\begin{aligned}
\mathsf{k}^\mu \widetilde{\mathcal{T}}^E_{\mu\nu\lambda\rho}(\mathsf{k}) = & -\frac{i}{384\pi}\Big[\mathsf{k}_\nu \mathsf{k}_\lambda \mathsf{k}_\rho + \frac{1}{2}\mathsf{k}^2\big(\eta_{\nu\lambda}\mathsf{k}_\rho + \eta_{\nu\rho}\mathsf{k}_\lambda\big) \\
& + \frac{1}{2}\mathsf{k}_\nu\big(\mathsf{k}_\lambda \varepsilon_{\rho\sigma} + \mathsf{k}_\rho \varepsilon_{\lambda\sigma}\big)\mathsf{k}^\sigma + \frac{1}{2}\mathsf{k}^2\big(\eta_{\nu\lambda}\varepsilon_{\rho\sigma} + \eta_{\nu\rho}\varepsilon_{\lambda\sigma}\big)\mathsf{k}^\sigma\Big]
\end{aligned} \tag{7.74}
$$

Thus trace and divergence are non-vanishing both for the even and odd part. Using (4.18) and (5.14), we can derive the trace cocycle in coordinates space

$$
\begin{aligned}
\Delta_\omega &= \frac{1}{2}\int \mathrm{d}^2 x\, \omega(x) \int \mathrm{d}^2 y\, h^{\lambda\rho}(y) \langle 0|\mathcal{T} T^\mu_\mu(x) T_{\lambda\rho}(y)|0\rangle_c \\
&= 2\int \mathrm{d}^2 x\, \omega(x) \int \mathrm{d}^2 y\, h^{\lambda\rho}(y) \int \frac{\mathrm{d}^2 k}{(2\pi)^2}\, \mathrm{e}^{-ik\cdot(x-y)} \widetilde{\mathcal{T}}^\mu_{\mu\lambda\rho}(k) \\
&= \frac{1}{96\pi}\int \mathrm{d}^2 x\, \omega(x)\big[\partial_\lambda \partial_\rho h^{\lambda\rho}(x) - \Box h^\lambda_\lambda(x) - \varepsilon_{\lambda\sigma}\partial^\sigma \partial_\rho h^{\lambda\rho}\big]
\end{aligned} \tag{7.75}
$$

where, in the last step, an inverse Wick rotation has been performed after inserting (7.73).

In the same way, the diffeomorphism cocycle can be obtained

$$
\begin{aligned}
\Delta_\xi &= -\frac{1}{2}\int d^2x\,\xi^\nu(x)\int d^2y\,h^{\lambda\rho}(y)\partial_x^\mu\langle 0|TT_{\mu\nu}(x)T_{\lambda\rho}(y)|0\rangle_c \qquad (7.76)\\
&= -\frac{1}{2}\int d^2x\,\xi^\nu(x)\int d^2y\,h^{\lambda\rho}(y)(-ik^\mu)e^{-ik\cdot(x-y)}\widetilde{\mathcal{T}}_{\mu\nu\lambda\rho}(k)\\
&= \frac{1}{192\pi}\int d^2x\,\xi^\nu(x)\Big[\partial_\nu\partial_\lambda\partial_\rho h^{\lambda\rho}(x) - \partial_\lambda\Box h_\nu^\lambda(x)\\
&\qquad\qquad +\varepsilon_{\rho\sigma}\partial^\sigma\left(\partial_\nu\partial_\lambda - \eta_{\nu\lambda}\Box\right)h^{\lambda\rho}(x)\Big]
\end{aligned}
$$

It is easy to prove that

$$
\delta_\omega^{(0)}\Delta_\omega = 0,\quad \delta_\xi^{(0)}\Delta_\xi = 0,\quad \delta_\omega^{(0)}\Delta_\xi + \delta_\xi^{(0)}\Delta_\omega = 0. \qquad (7.77)
$$

Now let us focus on the even parity part. Both trace and diffeomorphism cocycles are non-vanishing. However, the diffeomorphism one is trivial. For let us consider the local counterterm

$$
\mathcal{C}^{(even)} = \frac{1}{384\pi}\int d^2x\,\left(h_\rho^\nu(x)\partial_\lambda\partial_\nu h^{\lambda\rho}(x) - h_{\lambda\rho}(x)\Box h^{\lambda\rho}(x)\right) \qquad (7.78)
$$

It is easy to prove that

$$
\Delta_\xi^{'(even)} \equiv \Delta_\xi^{(even)} + \delta_\xi^{(0)}\mathcal{C}^{(even)} = 0 \qquad (7.79)
$$

On the other hand, the overall even trace cocycle becomes

$$
\Delta_\omega^{'(even)} \equiv \Delta_\omega^{(even)} + \delta_\omega^{(0)}\mathcal{C}^{(even)} = \frac{1}{48\pi}\int d^2x\,\omega\left[\partial_\lambda\partial_\rho h^{\lambda\rho} - \Box h_\lambda^\lambda\right] \qquad (7.80)
$$

Contrary to the even case, the cocycle $\Delta_\xi^{(odd)}$ is non-trivial. Consequently, we cannot impose diffeomorphism covariance by subtracting some local counterterm. With a suitable counterterm, we can cancel instead $\Delta_\omega^{(odd)}$, but it is a mere academical exercise.

The Second Term of (7.1)

Instead of calculating the two-point correlator of the energy-momentum tensor and subsequently taking its trace, we proceed directly to the computation of a two-point correlator containing one insertion of the trace of the e.m. tensor. The integral to be regulated is (the symmetrization $\lambda \leftrightarrow \rho$ is understood)

$$\widetilde{\widetilde{\mathcal{T}}}^{\mu}{}_{\mu\lambda\rho}(k) \tag{7.81}$$
$$= -\frac{1}{32}\int \frac{d^2 p}{(2\pi)^2}\,\mathrm{tr}\left(\frac{\not p}{p^2}\,(2\not p-\not q)\,\frac{1+\gamma_*}{2}\,\frac{\not p-\not k}{(p-k)^2}(2p-k)_\lambda\gamma_\rho\frac{1+\gamma_*}{2}\right)$$

We regularize it as follows

$$\widetilde{\widetilde{\mathcal{T}}}^{\mu}{}_{\mu\lambda\rho}(k) = -\frac{1}{32}\int \frac{d^2 p\, d^\delta \ell}{(2\pi)^{2+\delta}} \tag{7.82}$$
$$\times \mathrm{tr}\left(\frac{\not p+\not\ell}{p^2-\ell^2}\,(2\not p+2\not\ell-\not k)\,\frac{1+\gamma_*}{2}\,\frac{\not p+\not\ell-\not k}{(p-k)^2-\ell^2}(2p-k)_\lambda\gamma_\rho\frac{1+\gamma_*}{2}\right)$$
$$= -\frac{1}{64}\int \frac{d^2 p\, d^\delta \ell}{(2\pi)^{2+\delta}}$$
$$\times \mathrm{tr}\left(\frac{\not p+\not\ell}{p^2-\ell^2}\,(2\not p+2\not\ell-\not k)\,\frac{\not p-\not k}{(p-k)^2-\ell^2}(2p-k)_\lambda\gamma_\rho\frac{1+\gamma_*}{2}\right)$$

A direct calculation yields

$$\widetilde{\widetilde{\mathcal{T}}}^{E\mu}{}_{\mu\lambda\rho}(k) = 0. \tag{7.83}$$

which corresponds to a vanishing cocycle $\widehat{\Delta}_\omega$, both for the even and odd part.

Therefore, according to the definition (7.1) the even trace anomaly is

$$\mathcal{A}^{(\mathrm{even})}_\omega = \Delta'^{(even)}_\omega - \widehat{\Delta}^{(\mathrm{even})}_\omega = \frac{1}{48\pi}\int d^2x\,\omega\left[\partial_\lambda\partial_\rho h^{\lambda\rho} - \Box h^\lambda_\lambda\right] \tag{7.84}$$

As for the odd trace anomaly, from (7.75), we have

$$\mathcal{A}^{(odd)}_\omega = \Delta^{(odd)}_\omega - \widehat{\Delta}^{(odd)}_\omega = -\frac{1}{96\pi}\int d^2x\,\omega(x)\varepsilon_{\lambda\sigma}\partial^\sigma\partial_\rho h^{\lambda\rho}, \tag{7.85}$$

unless we wish to eliminate it by adding a counterterm $\frac{1}{768\pi}\int d^2x h\varepsilon_{\lambda\sigma}\partial^\sigma\partial^\rho h^{\lambda\rho}$.

Regularization Ambiguities and Rightmost γ_* Prescription

We have mentioned above the possibility to regulate the expression (7.66) instead of (7.65). They look formally identical, but they are unregulated expressions. If we start from (7.66) we get

$$\widetilde{\mathcal{T}}^{(\mathrm{reg})'}_{\mu\nu\lambda\rho}(k) = -\frac{1}{32}\int \frac{d^2 p\, d^\delta \ell}{(2\pi)^{2+\delta}}\mathrm{tr}\left(\frac{1}{\not p+\not\ell}(2p-k)_\mu\gamma_\nu\frac{1}{\not p+\not\ell-\not k}(2p-k)_\lambda\gamma_\rho\frac{1+\gamma_*}{2}\right) \tag{7.86}$$

We recall that here we understand for the moment the symmetrization with respect to the exchanges $\mu \leftrightarrow \nu$, $\lambda \leftrightarrow \rho$ and $(\mu, \nu) \leftrightarrow (\lambda, \rho)$ (bosonic symmetry).

The difference between (7.86) and (7.67) is

$$\begin{aligned}\Delta\widetilde{\mathfrak{T}}^{(\mathrm{reg})}_{\mu\nu\lambda\rho}(k) &= \widetilde{\mathfrak{T}}^{(\mathrm{reg})'}_{\mu\nu\lambda\rho}(k) - \widetilde{\mathfrak{T}}^{(\mathrm{reg})}_{\mu\nu\lambda\rho}(k) \\ &= -\frac{2^{\frac{\delta}{2}}}{64}(\eta_{\nu\rho} - \varepsilon_{\nu\rho}) \int \frac{d^2p\, d^\delta\ell}{(2\pi)^{2+\delta}}\, \ell^2 \, \frac{(2p-k)_\mu (2p-k)_\lambda}{(p^2-\ell^2)((p-k)^2-\ell^2)}\end{aligned} \tag{7.87}$$

After a Wick rotation, one can carry out the integrals and symmetrize. The result is (dropping the $^{(\mathrm{reg})}$ label)

$$\begin{aligned}\Delta\widetilde{\mathfrak{T}}^{E}_{\mu\nu\lambda\rho}(k) = \frac{i}{768\pi}\Big[&2\mathsf{k}^2(\eta_{\mu\lambda}\eta_{\nu\rho} + \eta_{\nu\lambda}\eta_{\mu\rho}) + \eta_{\nu\rho}\mathsf{k}_\mu\mathsf{k}_\lambda \\ &+\eta_{\mu\rho}\mathsf{k}_\nu\mathsf{k}_\lambda + \eta_{\nu\lambda}\mathsf{k}_\mu\mathsf{k}_\rho + \eta_{\mu\rho}\mathsf{k}_\nu\mathsf{k}_\lambda\Big]\end{aligned} \tag{7.88}$$

Inserted in (4.18), this gives rise to local terms in the effective action. From this result, we learn that different regularization procedures may lead to discrepancies in the effective action. These differences however are expressed by local terms. The contribution to the trace from (7.88) is

$$\Delta\widetilde{\mathfrak{T}}^{E\mu}_{\mu\lambda\rho}(\mathsf{k}) = \frac{i}{192\pi}\Big[\mathsf{k}_\lambda\mathsf{k}_\rho + \mathsf{k}^2\eta_{\lambda\rho}\Big] \tag{7.89}$$

and the contribution to the divergence is

$$\mathsf{k}^\mu \Delta\widetilde{\mathfrak{T}}^{E}_{\mu\nu\lambda\rho}(\mathsf{k}) = \frac{i}{768\pi}\Big[\mathsf{k}^2\big(\eta_{\nu\lambda}\mathsf{k}_\rho + \eta_{\nu\rho}\mathsf{k}_\lambda\big) + 2\mathsf{k}_\nu\mathsf{k}_\lambda\mathsf{k}_\rho\Big] \tag{7.90}$$

These additions modify (only) the even parts of both (7.73) and (7.74), in fact they double the even part of (7.73) and exactly cancel the even part of (7.74). This implies diffeomorphism covariance for even part of the trace anomaly. As for the odd parity part, it does not change (see the end of previous subsection).

We shall refer to the regularization prescription (7.86) as the rightmost γ_* prescription. It is clearly the most convenient one because it yields the conservation of the e.m. one-point tensor without the need of any subtraction.

So far our calculations with the rightmost γ_* prescription refer to the first term of (7.1). Now we have to consider the second term. But in fact, the result does not change if we start from

$$
\begin{aligned}
&\widetilde{\widetilde{\mathcal{T}}}'^{\mu}{}_{\mu\lambda\rho}(k) && (7.91)\\
&= -\frac{1}{32}\int \frac{d^2p}{(2\pi)^2}\, \mathrm{tr}\left(\frac{\not p}{p^2}\,(2\not p-\not q)\,\frac{1+\gamma_*}{2}\,\frac{\not p-\not k}{(p-k)^2}(2p-k)_\lambda\gamma_\rho\frac{1+\gamma_*}{2}\right)\\
&= -\frac{1}{32}\int \frac{d^2p}{(2\pi)^2}\, \mathrm{tr}\left(\frac{\not p}{p^2}\,(2\not p-\not q)\,\frac{\not p-\not k}{(p-k)^2}(2p-k)_\lambda\gamma_\rho\frac{1+\gamma_*}{2}\right),
\end{aligned}
$$

which leads to regularization prescription

$$
\begin{aligned}
\widetilde{\widetilde{\mathcal{T}}}'^{\mu}{}_{\mu\lambda\rho}(k) &\stackrel{\text{reg}}{=} -\frac{1}{32}\int \frac{d^2p\, d^\delta\ell}{(2\pi)^{2+\delta}} && (7.92)\\
&\times \mathrm{tr}\left(\frac{\not p+\not \ell}{p^2-\ell^2}\,(2\not p+2\not\ell-\not k)\,\frac{\not p+\not\ell-\not k}{(p-k)^2-\ell^2}(2p-k)_\lambda\gamma_\rho\frac{1+\gamma_*}{2}\right),
\end{aligned}
$$

because the calculation gives 0, again. Therefore using one or another prescription does not affect the cohomological result, but is only a question of opportunity.

Conclusion

In conclusion, the even trace anomaly is determined by

$$
g^{\mu\nu}\langle\langle T_{\mu\nu}(x)\rangle\rangle^{(\text{even})} - \langle\langle g^{\mu\nu}T_{\mu\nu}(x)\rangle\rangle^{(\text{even})} = g^{\mu\nu}\langle\langle T_{\mu\nu}(x)\rangle\rangle^{(\text{even})} \tag{7.93}
$$

and the RHS has been calculated above. As already noticed, (7.84) is the first-order approximation of

$$
\mathcal{A}_\omega^{(\text{even})} = \frac{1}{48\pi}\int d^2x\,\sqrt{g}\,\omega\,R \tag{7.94}
$$

Comparing with (7.25), we can determine the central charge for a Weyl spinor

$$
c = c_W = \frac{1}{4\pi^2} \tag{7.95}
$$

In view of the fact that the theory of a single (complex) Weyl spinor has a diffeomorphism anomaly, the result (7.95) is only formal. However, a (complex) free Dirac spinor can be seen as a couple of opposite chirality free Weyl spinors. The chirality affects only the odd parity results, which have opposite sign for opposite chirality. As a consequence the odd parity anomalies cancel out, while the even parity ones, having the same sign, will add up. Therefore, for a Dirac spinor the diffeomorphism anomaly vanishes and the even trace anomaly is twice (7.84). So the central charge for a free (real) Dirac spinor is

$$
c = c_D = \frac{1}{2\pi^2}, \tag{7.96}
$$

Eqs. (7.95) and (7.96) represent unnormalized central charges. In order to normalize them, one considers the fermion propagator in coordinate space

$$\left\langle \psi(x)\overline{\psi}(y)\right\rangle = \frac{1}{2\pi}\frac{\gamma\cdot(x-y)}{(x-y)^2}, \tag{7.97}$$

This implies that in Eq. (7.12) the parameter c contains the factor $\frac{1}{4\pi^2}$. Stripping off this factor, the normalized central charge $\mathsf{c} = 4\pi^2 c$ for a (complex) Weyl fermion is

$$\mathsf{c}_W = 1 \tag{7.98}$$

and twice as much for a Dirac fermion.[4]

As for the odd parity part of the anomalies, one can repeat the same analysis as in 7.2.2, since the cocycles are the same but with precise coefficients and conclude that for diffeomorphisms the result is (proportional to) the non-trivial diffeomorphism anomaly (7.45) or (7.48). One can get rid of the trivial Weyl cocycle proportional to (7.85) by shifting the anomaly to the diffeomorphism sector. More about these two anomalies in Sect. 17.1.

Remark on the local Lorentz anomaly

The action (7.60) is invariant under local Lorentz transformations $\delta_\Lambda \mathrm{e}_\mu^a = \mathrm{e}_\mu^b \Lambda_b{}^a$, $\delta_\Lambda \psi = -\frac{\Lambda}{2}\psi$. The reason is the same for which one can drop the spin connection: the current $j_\mu^{01} = \overline{\psi}_R \Sigma^{01}\gamma_\mu \psi_R$, where $\Sigma^{01} = -2\gamma_*$ is the unique Lorentz generator, vanishes identically. Since there is no current, there is no local Lorentz anomaly. Putting together this result with the anomaly (7.48), we get the simplest example of a system which has a diffeomorphism anomaly but is local Lorentz anomaly free. In general, as we shall see in Sect. 16.3, we can trade a diffeomorphism anomaly with a local Lorentz one by means of a Wess-Zumino term. But it is clear that in this case such an operation does not make sense because the Lorentz current is identically 0. On the other hand, we shall see in Sect. 17.1.6 that the anomaly (7.48) is a pure gauge construct, and in the conformal gauge, it vanishes. In the same gauge, also the corresponding local Lorentz anomaly vanishes.

7.3 Trace Anomalies Due to Gauge Fields

In this section, we return to 4d. The simplest case study in which diffeomorphism and conformal anomalies are involved is in a theory of fermions coupled to a vector gauge potential in a non-trivial metric background. In this context, we analyze even and odd parity amplitudes so as to encompass both even and odd parity diffeomorphism

---

[4] We recall that in 2d reality can be imposed on a Weyl spinor, giving rise to a Majorana-Weyl spinor, whose normalized central charge is therefore $\mathsf{c}_{MW} = \frac{1}{2}$.

and trace anomalies. To be more explicit the candidate anomalies we wish to search is the odd parity anomaly with density $\varepsilon^{\mu\nu\lambda\rho} F_{\mu\nu} F_{\lambda\rho}$ but also the even parity anomaly with density $F_{\mu\nu} F^{\mu\nu}$. These densities can show up as trace anomalies in the form $\int d^4x \sqrt{g}\, \omega(x)\, F_{\mu\nu}(x) F^{\mu\nu}(x)$ and $\int d^4x \sqrt{g}\, \omega(x)\, \varepsilon^{\mu\nu\lambda\rho} F_{\mu\nu}(x) F_{\lambda\rho}(x)$, since they are consistent under the conformal transformations $\delta_\omega g_{\mu\nu} = 2\omega g_{\mu\nu}$ and $\delta_\omega V_\mu = 0$. We will consider two examples where these anomalies do appear: the even trace anomaly in the theory of a Dirac fermion, and the odd one in the theory of a Weyl fermion, both coupled to a vector potential V_μ, [5].

7.3.1 Even Parity Trace Anomalies Due to Vector Gauge Field

We consider the action (7.2), i.e. the action of a Dirac fermion coupled to a metric and an Abelian vector potential. The extension to the non-Abelian potential case is straightforward: $V_\mu = V_\mu^a T^a$ and ψ transforms according to the representation of the gauge group whose Lie algebra is spanned by the T^a's. For a uniform treatment with the non-Abelian case where the generators are anti-Hermitean, we use an imaginary vector field V_μ. Eventually one can make the replacement $V_\mu \to -i V_\mu$. We recall that the vector current is $j_\mu = i\overline{\psi}\gamma_\mu\psi$ and the stress-energy tensor is

$$T_{\mu\nu} = \frac{i}{4}\overline{\psi}\gamma_\mu \overset{\leftrightarrow}{\nabla}_\nu \psi + \{\mu \leftrightarrow \nu\} \tag{7.99}$$

They are both conserved on shell. $T_{\mu\nu}$ is also traceless on shell. In this section, we use perturbative methods with dimensional regularization and focus on the possibility for $F_{\mu\nu}F^{\mu\nu}$ to appear as a trace anomaly density. As expected such a density cannot appear in the divergence of a current. In any case, we have shown it for completeness in Appendix 6B.

Let us return also to the definition (7.99) of the e.m. tensor. As already pointed out, another definition is possible, which corresponds to the general formula

$$\widehat{T}_{\mu\nu}(x) = \frac{2}{\sqrt{g}} \frac{\delta S}{\delta g^{\mu\nu}} \tag{7.100}$$

and reads

$$\widehat{T}_{\mu\nu} = \frac{i}{4}\overline{\psi}\gamma_\mu \overset{\leftrightarrow}{\nabla}_\nu \psi + (\mu \leftrightarrow \nu) - g_{\mu\nu}\frac{i}{2}\overline{\psi}\gamma^\lambda \overset{\leftrightarrow}{\nabla}_\lambda \psi = T_{\mu\nu} - g_{\mu\nu} T_\lambda{}^\lambda \tag{7.101}$$

This ambiguity in defining the e.m. tensor gives rise to an ambiguity in the definition of the trace anomaly. As already pointed out, such an uncertainty is actually resolved by the definition (7.1): thanks to the latter, the second term of $\widehat{T}_{\mu\nu}$ drops out. However, the two definitions lead to two different WI's and we cannot ignore this fact. For the time being, we have to keep track of both. Luckily the relation between the two WIs is simple and, as will become clear, eventually we can refer to $T_{\mu\nu}$ alone.

The anomaly we are after may appear at one-loop in the trace of the e.m. tensor. Therefore, we have to compute the correlators that contain one insertion of the latter plus insertions of the vector currents. With reference to the definition (7.1), at the lowest order we have two possibilities: the correlator $\eta^{\mu\nu}\langle 0|\mathcal{T}T_{\mu\nu}(x)j_\lambda(y)j_\rho(z)|0\rangle$ and the correlator $\langle 0|\mathcal{T}T^\mu_\mu(x)j_\lambda(y)j_\rho(z)|0\rangle$. The first means that we regularize and compute the correlator $\langle 0|\mathcal{T}T_{\mu\nu}(x)j_\lambda(y)j_\rho(z)|0\rangle$ and afterwards we contract the two indices μ and ν. The second means that we consider the correlator with T^μ_μ inserted at x and two currents at y and z, regularize and compute it. Generally speaking, the two procedures lead to different results. Notice that $T^\mu_\mu(x)$ vanishes on shell in conformal invariant theories, but in general does not vanish off-shell. An important remark is that the trace $T^\mu_\mu(x)$ is an irreducible component of $T_{\mu\nu}(x)$. So, the two above-mentioned amplitudes, when regulated, are generally different. This situation is to be contrasted with the case of the three-point current amplitude $\langle 0|\mathcal{T}j_\mu(x)j_\lambda(y)j_\rho(z)|0\rangle$, whose divergence (see Appendix 6B) is the same as $\langle 0|\mathcal{T}\partial^\mu j_\mu(x)j_\lambda(y)j_\rho(z)|0\rangle$. In this case, there is no regularization ambiguity.

On the basis of the definition (7.1), to the lowest order[5] in the perturbative approach, the trace anomaly will be given by:

$$\langle\langle T^\mu_\mu(x)\rangle\rangle \approx \eta^{\mu\nu}\langle 0|\mathcal{T}T_{\mu\nu}(x)j_\lambda(y)j_\rho(z)|0\rangle - \langle 0|\mathcal{T}T^\mu_\mu(x)j_\lambda(y)j_\rho(z)|0\rangle \quad (7.102)$$

In the sequel, we compute these two amplitudes by means of Feynman diagrams. The Feynman rule for the vector-fermion-fermion vertex V_{vff} is $i\gamma_\mu$ and the graviton-fermion-fermion vertex V_{hff} is

$$-\frac{i}{8}\left[(p-p')_\mu\gamma_\nu + (p-p')_\nu\gamma_\mu\right] \quad (7.103)$$

There are also other vertices, but they are not relevant to the present calculation. The fermion propagator is $\frac{i}{\not p}$. In order to regularize the Feynman integrals (and only to that purpose), we will add, to the ordinary 3+1, δ additional dimensions. The conventions for the gamma matrix traces (in Minkowski background) are

$$\mathrm{tr}(\gamma_\mu\gamma_\nu) = 2^{2+\frac{\delta}{2}}\eta_{\mu\nu}, \quad \mathrm{tr}(\gamma_\mu\gamma_\nu\gamma_\lambda\gamma_\rho\gamma_5) = -i\, 2^{2+\frac{\delta}{2}}\varepsilon_{\mu\nu\lambda\rho} \quad (7.104)$$

[5] One should consider also the two-point amplitudes, but it is easy to show that

$$\langle 0|\mathcal{T}T^\mu_\mu(x)j_\lambda(y)|0\rangle = \eta^{\mu\nu}\langle 0|\mathcal{T}T_{\mu\nu}(x)j_\lambda(y)|0\rangle = 0$$

.

The First Correlator

We start by evaluating the second term in the RHS of (7.102), i.e. the amplitude schematically denoted $\langle t\, j\, j\rangle$, where t represents the trace of the e.m. tensor,

$$\widetilde{F}^{(tjj)}_{\lambda\rho}(k_1,k_2) = \frac{1}{2}\int \frac{\mathrm{d}^4 p}{(2\pi)^4}\mathrm{tr}\left\{\frac{1}{\not p}\gamma_\lambda \frac{1}{\not p-\not k_1}\gamma_\rho\frac{1}{\not p-\not q}(2\not p-\not q)\right\}$$
$$\stackrel{\mathrm{reg}}{=} \frac{1}{2}\int \frac{\mathrm{d}^4 p \mathrm{d}^\delta \ell}{(2\pi)^{4+\delta}}\mathrm{tr}\left\{\frac{\not p+\not\ell}{p^2-\ell^2}\gamma_\lambda\frac{\not p-\not k_1+\not\ell}{(p-k_1)^2-\ell^2}\gamma_\rho\right.$$
$$\left. +\gamma_\lambda\frac{\not p-\not k_1+\not\ell}{(p-k_1)^2-\ell^2}\gamma_\rho\frac{\not p-\not q+\not\ell}{(p-q)^2-\ell^2}\right\} \tag{7.105}$$

Here and in the sequel in this section p is the internal momentum, k_1, k_2 are the outgoing gluon momenta and $q = k_1 + k_2$. In the second and third lines, we have regularized the integral by introducing, as usual, an additional momentum $\ell_{\bar\mu}$, with $\bar\mu = 4, \ldots, 3+\delta$. Then we have rewritten $2\not p - \not q$ as $\not p + \not\ell + \not p - \not q + \not\ell$ and simplified.

We recall that in the non-Abelian gauge case the amplitude (7.105) is multiplied by $\mathrm{tr}(T^a T^b)$, a factor to be saturated with the sources V^a and V^b.

Adding the cross term $k_1 \leftrightarrow k_2$, $\lambda \leftrightarrow \rho$, we get

$$\widetilde{F}^{(tjj)}_{\lambda\rho}(k_1,k_2) + \widetilde{F}^{(tjj)}_{\rho\lambda}(k_2,k_1)$$
$$= 4\times 2^{\frac{\delta}{2}}\int \frac{\mathrm{d}^4 p \mathrm{d}^\delta \ell}{(2\pi)^{4+\delta}}\left\{\frac{\ell^2\eta_{\lambda\rho} + p_\lambda(p-k_1)_\rho - \eta_{\lambda\rho}\, p\cdot(p-k_1) + (p-k_1)_\lambda p_\rho}{(p^2-\ell^2)((p-k_1)^2-\ell^2)}\right.$$
$$\left. + \frac{\ell^2\eta_{\lambda\rho} + p_\lambda(p-k_2)_\rho - \eta_{\lambda\rho}\, p\cdot(p-k_2) + (p-k_2)_\lambda p_\rho}{(p^2-\ell^2)((p-k_2)^2-\ell^2)}\right\} \tag{7.106}$$

Let us deal first with the first line of the last integral. Introducing a Feynman parameter x and shifting $p \to p + k_1$ and Wick rotating ($p^0 \to ip^0$), one gets[6]

$$= 4i\times 2^{\frac{\delta}{2}}\int\frac{\mathrm{d}^4 p}{(2\pi)^4}\int_0^1 \mathrm{d}x \int \frac{\mathrm{d}^4 p \mathrm{d}^\delta \ell}{(2\pi)^{4+\delta}}\frac{\left(\ell^2+\frac{p^2}{2}\right)\eta_{\lambda\rho} + x(x-1)\left(2k_{1\lambda}k_{1\rho}+\eta_{\lambda\rho}k_1^2\right)}{\left(p^2+\ell^2+x(1-x)k_1^2\right)^2}$$
$$= \frac{i}{6\pi^2}\left(k_{1\lambda}k_{1\rho}+\eta_{\lambda\rho}k_1^2\right)\left(\frac{2}{\delta}+\gamma-\frac{5}{3}+\log\frac{k_1^2}{2\pi\mu^2}\right)$$

[6] In order to avoid clogging our formulas with the Euclidean label such as p^E, ..., we keep the same symbols for the Wick-rotated momenta p, k_1, k_2 and q, with the only change represented by the momentum square acquiring the opposite sign.

Adding also the second line, we get

$$\widetilde{F}^{(tjj)}_{\lambda\rho}(k_1,k_2)+\widetilde{F}^{(tjj)}_{\rho\lambda}(k_2,k_1)=\frac{i}{12\pi^2}\left(k_{1\lambda}k_{1\rho}+\eta_{\lambda\rho}k_1^2\right)\left(\frac{2}{\delta}+\gamma-\frac{5}{3}+\log\frac{k_1^2}{2\pi\mu^2}\right)$$
$$+\frac{i}{12\pi^2}\left(k_{2\lambda}k_{2\rho}+\eta_{\lambda\rho}k_2^2\right)\left(\frac{2}{\delta}+\gamma-\frac{5}{3}+\log\frac{k_2^2}{2\pi\mu^2}\right) \quad (7.107)$$

This contains a non-local part. It is a (so-called) semi-local term which is necessary in order to satisfy the conformal Ward identity.

The Conformal Ward Identity

To find the WI's for our case, we have to start from the effective action that include both the vector current and the energy-momentum tensor, which is born out of the action (7.2),

$$W[h,V]=W[0]+\sum_{n,r=1}^{\infty}\frac{i^{n+r-1}}{2^n n!r!}\int\prod_{i=1}^{n}\mathrm{d}x_i\sqrt{g(x_i)}h^{\mu_i\nu_i}(x_i)\prod_{l=1}^{r}\mathrm{d}y_l\sqrt{g(y_l)}e^{\lambda_l}_{a_l}(y_l)V_{\lambda_l}(y_l)$$
$$\cdot\langle 0|\mathcal{T}\widehat{T}_{\mu_1\nu_1}(x_1)\ldots\widehat{T}_{\mu_n\nu_n}(x_n)j^{a_1}(y_1)\ldots j^{a_r}(y_r)|0\rangle \quad (7.108)$$

The introduction of the vierbein is dictated by the coupling between the current and the gauge potential in the presence of a nontrivial background metric: $V_\lambda(x)\bar{\psi}(x)e^\lambda_a(x)\gamma^a\psi(x)$. Notice that in this definition we have used $\widehat{T}_{\mu\nu}$; this is because, in the subsequent manipulations, we will refer to (7.5). Let us recall a previous remark about the coefficients of this expansion. They must be consistent with the definition of graviton emission vertices. Remember that $\frac{\delta S}{\delta g_{\mu\nu}}|_{g=\eta}=\frac{1}{2}\widehat{T}_{\mu\nu}$. This explains the factor 2^n in the denominator of (7.108).

Differentiating with respect to $h_{\mu\nu}(x)$ and setting $h=0$, we obtain $\frac{1}{2}\langle\langle\widehat{T}_{\mu\nu}(x)\rangle\rangle$. Therefore,

$$2\frac{\delta W}{\delta h_{\mu\nu}(x)}\bigg|_{h=0}=i\int \mathrm{d}^4y_2\,\eta_{\mu\nu}\,V_{\lambda_1}(x)V_{\lambda_2}(y_2)\delta^{\lambda_1}_{a_1}\delta^{\lambda_2}_{a_2}\langle 0|\mathcal{T}j^{a_1}(x)j^{a_2}(y_2)|0\rangle$$
$$-\frac{i}{2}\int \mathrm{d}^4y_2\,V_{\lambda_1}(x)V_{\lambda_2}(y_2)\left(\delta^{\lambda_1}_{\mu}\eta_{a_1\nu}+\delta^{\lambda_1}_{\nu}\eta_{a_1\mu}\right)\langle 0|\mathcal{T}j^{a_1}(x)j^{a_2}(y_2)|0\rangle$$
$$-\frac{1}{2}\int \mathrm{d}^4y_1\mathrm{d}^4y_2\,V^{\lambda_1}(y_1)V^{\lambda_2}(y_2)\langle 0|\mathcal{T}\widehat{T}_{\mu\nu}(x)j_{\lambda_1}(y_1)j_{\lambda_2}(y_2)|0\rangle$$
$$(7.109)$$

Next differentiating with respect to $V^\lambda(y)$ and $V^\rho(z)$ and setting $V_\lambda = 0$, we get

$$\begin{aligned}\left.\frac{\delta^2 \langle\!\langle \widehat{T}_{\mu\nu}(x)\rangle\!\rangle}{\delta V^\lambda(y)\delta V^\rho(z)}\right|_{V=0} &= i\eta_{\mu\nu}\big(\delta(x-y)\langle 0|\mathcal{T} j_\lambda(x) j_\rho(z)|0\rangle + \delta(x-z)\langle 0|\mathcal{T} j_\rho(x) j_\lambda(y)|0\rangle\big)\\ &-\frac{i}{2}\eta_{\mu\lambda}\,\delta(x-y)\langle 0|\mathcal{T} j_\nu(x) j_\rho(z)|0\rangle - \frac{i}{2}\eta_{\mu\rho}\,\delta(x-z)\langle 0|\mathcal{T} j_\nu(x) j_\lambda(y)|0\rangle\\ &-\frac{i}{2}\eta_{\nu\lambda}\,\delta(x-y)\langle 0|\mathcal{T} j_\mu(x) j_\rho(z)|0\rangle - \frac{i}{2}\eta_{\nu\rho}\,\delta(x-z)\langle 0|\mathcal{T} j_\mu(x) j_\lambda(y)|0\rangle\\ &-\langle 0|\mathcal{T}\widehat{T}_{\mu\nu}(x) j_\lambda(y) j_\rho(z)|0\rangle \end{aligned} \tag{7.110}$$

Now we saturate this relation with $\eta^{\mu\nu}$ and notice that $\widehat{T}_\mu{}^\mu = -3T_\mu{}^\mu$. It is easy to see that the numerical factor 3 factors out and can be dropped from the equation obtained by equating the RHS of (7.110) to zero. Thus, we obtain

$$\begin{aligned}&\langle 0|\mathcal{T} T^\mu_\mu(x) j_\lambda(y) j_\rho(z)|0\rangle \qquad (7.111)\\ &\quad + i\Big(\delta(x-y)\langle 0|\mathcal{T} j_\lambda(x) j_\rho(z)|0\rangle + \delta(x-z)\langle 0|\mathcal{T} j_\lambda(y) j_\rho(x)|0\rangle\Big) = 0\end{aligned}$$

This is the appropriate conformal WI for a fermionic system coupled to an external gauge potential V_μ, at level 2, in the absence of anomalies. Fourier transforming and Wick rotating it (under a Wick rotation the two-point function changes sign, the three-point function gets multiplied by $-i$) we get

$$-i\left(\widetilde{F}_{\lambda\rho}(k_1,k_2) + \widetilde{F}_{\rho\lambda}(k_2,k_1) + \langle 0|\mathcal{T}\tilde{j}_\lambda(k_1)\tilde{j}_\rho(-k_1)|0\rangle + \langle 0|\mathcal{T}\tilde{j}_\lambda(k_2)\tilde{j}_\rho(-k_2)|0\rangle\right) = 0 \tag{7.112}$$

multiplied by $\delta(q-k_1-k_2)$. This is the WI if no anomaly is present.

The 2-Point Current Correlator

In order to verify (7.111) (or (7.112)), we need the two-point current correlators. The 2-current correlator is given by a bubble diagram with a fermion propagating in the internal lines and two gluons, one incoming and the other outgoing, with the same momentum k_μ:

$$\begin{aligned}\langle 0|\mathcal{T}\tilde{j}_\lambda(k)\tilde{j}_\rho(-k)|0\rangle &= \int \frac{d^4p}{(2\pi)^4}\mathrm{tr}\left\{\frac{1}{\not p}\gamma_\lambda\frac{1}{\not p-\not k}\gamma_\rho\right\}\\ &\overset{\text{reg}}{=} \int \frac{d^4p\,d^\delta\ell}{(2\pi)^{4+\delta}}\mathrm{tr}\left\{\frac{1}{\not p+\not \ell}\gamma_\lambda\frac{1}{\not p-\not k+\not \ell}\gamma_\rho\right\}\end{aligned} \tag{7.113}$$

This equals the first term in the third line of (7.105), apart from the coefficient $\frac{1}{2}$. Therefore,

$$\langle 0|\mathcal{T}\tilde{j}_\lambda(k)\tilde{j}_\rho(-k)|0\rangle = -\frac{i}{12\pi^2}\left(k_\lambda k_\rho + \eta_{\lambda\rho}k^2\right)\left(\frac{2}{\delta}+\gamma-\frac{5}{3}+\log\frac{k^2}{2\pi\mu^2}\right) \tag{7.114}$$

in Euclidean background. From this, one can see that the WI (7.112) is satisfied.

The Second Correlator

Replacing (7.107) and (7.114) in (7.112) we see that the WI is satisfied. One could conclude therefore that in this case there is no anomaly. However, (7.107) corresponds to $\langle\langle g^{\mu\nu}T_{\mu\nu}(x)\rangle\rangle$, and, comparing it with $g^{\mu\nu}\langle\langle T_{\mu\nu}(x)\rangle\rangle$, whose relevant amplitude is

$$\stackrel{\approx}{F}{}^{(tjj)}_{\lambda\rho}(k_1,k_2) = \frac{1}{2}\int \frac{\mathrm{d}^4 p \mathrm{d}^\delta \ell}{(2\pi)^{4+\delta}} \mathrm{tr}\left\{\frac{1}{\not p+\not\ell}\gamma_\lambda \frac{1}{\not p-\not k_1+\not\ell}\gamma_\rho \frac{1}{\not p-\not q+\not\ell}(2\not p-\not q)\right\}, \tag{7.115}$$

one finds a difference

$$\Delta\widetilde{F}^{(tjj)}_{\lambda\rho}(k_1,k_2) = \frac{1}{2}\int \frac{\mathrm{d}^4 p \mathrm{d}^\delta \ell}{(2\pi)^{4+\delta}} \mathrm{tr}\left\{\frac{1}{\not p+\not\ell}\gamma_\lambda \frac{1}{\not p-\not k_1+\not\ell}\gamma_\rho \frac{1}{\not p-\not q+\not\ell}2\not\ell\right\}, \tag{7.116}$$

which is a local term. Adding the cross term, after a Wick rotation, one finds

$$\Delta\widetilde{F}^{(tjj)}_{\lambda\rho}(k_1,k_2) + \Delta\widetilde{F}^{(tjj)}_{\rho\lambda}(k_2,k_1) = -\frac{i}{6\pi^2}\left[\eta_{\lambda\rho}k_1\cdot k_2 + k_{2\lambda}k_{1\rho}\right] \tag{7.117}$$

Beware, here the metric is Euclidean!

In accordance with (7.1), the integrated anomaly is defined as follows

$$\mathcal{A}_\omega = \int \mathrm{d}^4x\sqrt{g}\,\omega\,\left(g^{\mu\nu}\langle\langle T_{\mu\nu}\rangle\rangle - \langle\langle T_\mu^\mu\rangle\rangle\right) \tag{7.118}$$

We have already remarked that if we use the definition (7.5) instead of (7.99), the second piece of the former drops out in Eq. (7.118).

In order to obtain the form of the anomaly in coordinate representation, we proceed as follows. We have in general

$$-\frac{1}{2}\int \mathrm{d}^4y\mathrm{d}^4z\, V^\lambda(y)V^\rho(z)\langle 0|T^\mu_\mu(x)j_\lambda(y)j_\rho(z)|0\rangle = -\frac{1}{2}\int \mathrm{d}^4y\mathrm{d}^4z\, V^\lambda(y)V^\rho(z)$$
$$\cdot\int \frac{\mathrm{d}^4q}{(2\pi)^4}\frac{\mathrm{d}^4k_1}{(2\pi)^4}\frac{\mathrm{d}^4k_2}{(2\pi)^4}\mathrm{e}^{-iq\cdot x+ik_1\cdot y+ik_2\cdot z}\delta^{(4)}(q-k_1-k_2)\langle 0|\tilde{T}^\mu_\mu(q)\tilde{j}_\lambda(-k_1)\tilde{j}_\rho(k_2)|0\rangle$$

Inserting (7.117) in the RHS gives the contribution corresponding to the anomaly in coordinates:

$$\begin{aligned}
&-\frac{1}{2}\int \mathrm{d}^4y\mathrm{d}^4z\, V^\lambda(y)V^\rho(z)\int \frac{\mathrm{d}^4q}{(2\pi)^4}\frac{\mathrm{d}^4k_1}{(2\pi)^4}\frac{\mathrm{d}^4k_2}{(2\pi)^4}\mathrm{e}^{-iq\cdot x+ik_1\cdot y+ik_2\cdot z}\delta^{(4)}(q-k_1-k_2)\\
&\cdot\left(-\frac{i}{6\pi^2}\left[\eta_{\lambda\rho}k_1\cdot k_2+k_{2\lambda}k_{1\rho}\right]\right)\\
&=\frac{i}{12\pi^2}\int \mathrm{d}^4y\mathrm{d}^4z\, V^\lambda(y)V^\rho(z)\left(-\eta_{\lambda\rho}\partial_x\cdot\partial_z-\partial^y_\rho\partial^z_\lambda\right)\delta^{(4)}(x-y)\delta^{(4)}(x-z)\\
&=-\frac{i}{12\pi^2}\left(\partial_\lambda V_\rho\partial^\lambda V^\rho+\partial_\lambda V_\rho\partial^\rho V^\lambda\right)
\end{aligned} \tag{7.119}$$

Therefore, after migrating back to Minkowski, the trace anomaly at the (non-trivial) lowest level of approximation is

$$\mathcal{A}_\omega = -\frac{1}{12\pi^2}\int d^4x\,\omega\left(-\partial_\nu V_\lambda \partial^\nu V^\lambda + \partial_\nu V_\lambda \partial^\lambda V^\nu\right) = \frac{1}{24\pi^2}\int d^4x\,\omega\, F_{\nu\lambda}F^{\nu\lambda} \quad (7.120)$$

This is a (trivially) consistent conformal anomaly because in perturbative cohomology (see Sect. 5.4) at the lowest order, we have

$$\delta_\omega^{(0)}\omega = 0, \quad \delta_\omega^{(0)} V_\mu = 0. \quad (7.121)$$

So

$$\delta_\omega^{(0)}\mathcal{A}_\omega = 0 \quad (7.122)$$

It is clear that the all-order expression of this anomaly is

$$\mathcal{A}_\omega = \frac{1}{24\pi^2}\int d^4x\,\sqrt{g}\,\omega\, F_{\nu\lambda}F^{\nu\lambda}, \quad (7.123)$$

which is invariant both under diffeomorphisms and under gauge transformations. The non-Abelian extension of (7.123) is

$$\mathcal{A}_\omega = \frac{1}{24\pi^2}\int d^4x\,\sqrt{g}\,\omega\,\mathrm{tr}\left(F_{\nu\lambda}F^{\nu\lambda}\right). \quad (7.124)$$

When diffeomorphisms are involved, however, this is not the end of the story. It is usually assumed that diffeomorphisms are conserved in 4d. This is due to the fact that consistent chiral diffeomorphism anomalies are uniquely linked to the existence of the third-order symmetric adjoint invariant tensor of the relevant Lie algebra, which, in 4d, is the Lie algebra of $GL(3,1)$ and, thus, the relevant tensor vanishes. But here we are considering the possibility that diffeomorphisms are violated by the coupling of a Dirac fermion to a vector potential, and an anomaly proportional to $\int d^4x\,\partial\cdot\xi\,F_{\mu\nu}F^{\mu\nu}$ (where ξ^μ is the general coordinate transformation parameter) is consistent and algebraically not excluded. To cancel, it a counterterm $\int d^4x\,h\,F_{\mu\nu}F^{\mu\nu}$ would be necessary. Since $h = h_\mu^\mu$ transforms as $\delta_\omega h = 8\omega$ under Weyl transformations, such a counterterm would generate an anomaly of the same kind as (7.120) and, thus, modify its coefficient. It is therefore important to verify that such a diffeomorphism anomaly generated by the coupling to a gauge field V_μ is absent in our theory. This is what we intend to show next.

7.3.2 Diffeomorphisms Are Conserved

We have to show that the WI for diffeomorphisms is respected when coupling our Dirac fermion to V_μ. In order to derive the relevant WI, we return to Subsect. 7.3 and, precisely, to Eq. (7.110). We differentiate the RHS of the latter with respect to x^μ and equate it to zero. The WI we obtain is formulated in terms of $\widehat{T}_{\mu\nu}$. The identity simplifies considerably if we express it in terms of $T_{\mu\nu}$, for, using (7.111), the first line in the RHS of (7.110) drops out. Therefore, we are left with

$$\begin{aligned}0 =& -\frac{i}{2}\Big(\partial_\lambda^x\left(\delta(x-y)\langle 0|\mathcal{T} j_\nu(x) j_\rho(z)|0\rangle\right) + \partial_\rho^x\left(\delta(x-z)\langle 0|\mathcal{T} j_\nu(x) j_\lambda(y)|0\rangle\right)\Big)\\ &-\frac{i}{2}\Big(\eta_{\nu\lambda}\,\partial_x^\mu\left(\delta(x-y)\langle 0|\mathcal{T} j_\mu(x) j_\rho(z)|0\rangle\right) + \eta_{\nu\rho}\,\partial_x^\mu\left(\delta(x-z)\langle 0|\mathcal{T} j_\mu(x) j_\lambda(y)|0\rangle\right)\Big)\\ &-\partial_x^\mu\langle 0|\mathcal{T} T_{\mu\nu}(x) j_\lambda(y) j_\rho(z)|0\rangle \end{aligned} \tag{7.125}$$

If we denote by $\widetilde{T}_{\mu\nu\lambda\rho}(k_1,k_2)$ the Fourier transform of $\langle 0|\mathcal{T} T_{\mu\nu}(x) j_\lambda(y) j_\rho(z)|0\rangle$, the Fourier transform of Eq. (7.125) is

$$\begin{aligned}&-iq^\mu\left(\widetilde{T}_{\mu\nu\lambda\rho}(k_1,k_2) + \widetilde{T}_{\mu\nu\rho\lambda}(k_2,k_1)\right)\\ &\quad= \frac{1}{2}\Big(q_\rho\langle 0|\mathcal{T}\tilde{j}_\nu(k_1)\tilde{j}_\lambda(-k_1)|0\rangle + q_\lambda\langle 0|\mathcal{T}\tilde{j}_\nu(k_2)\tilde{j}_\rho(-k_2)|0\rangle\\ &\qquad+\eta_{\nu\rho}q^\mu\langle 0|\mathcal{T}\tilde{j}_\mu(k_1)\tilde{j}_\lambda(-k_1)|0\rangle + \eta_{\nu\lambda}q^\mu\langle 0|\mathcal{T}\tilde{j}_\mu(k_2)\tilde{j}_\rho(-k_2)|0\rangle\Big)\end{aligned} \tag{7.126}$$

where $q = k_1 + k_2$. Thus, we have to compute $q^\mu\widetilde{T}_{\mu\nu\lambda\rho}(k_1,k_2)$:

$$\begin{aligned}q^\mu\widetilde{T}_{\mu\nu\lambda\rho}(k_1,k_2) &= \frac{1}{2}\int\frac{d^4p}{(2\pi)^4}\mathrm{tr}\left[\frac{1}{\not p}\gamma_\lambda\frac{1}{\not p-\not k_1}\gamma_\rho\frac{1}{\not p-\not q}\left(q\cdot(2p-q)\gamma_\nu + (2p-q)_\nu\not q\right)\right]\\ &\overset{\text{reg}}{=}\frac{1}{2}\int\frac{d^4p\,d^\delta\ell}{(2\pi)^{4+\delta}}\mathrm{tr}\left[\frac{1}{\not p+\not\ell}\gamma_\lambda\frac{1}{\not p-\not k_1+\not\ell}\gamma_\rho\frac{1}{\not p-\not q+\not\ell}\left(q\cdot(2p-q)\gamma_\nu + (2p-q)_\nu\not q\right)\right]\end{aligned} \tag{7.127}$$

Using now $\not q = \not p+\not\ell-(\not p-\not q+\not\ell)$, and $q\cdot(2p-q) = p^2-\ell^2-((p-q)^2-\ell^2)$, we obtain an easier-to-deal-with expression of the integrand

$$\begin{aligned}q^\mu\widetilde{T}_{\mu\nu\lambda\rho}(k_1,k_2) &= \frac{1}{2}\int\frac{d^4p\,d^\delta\ell}{(2\pi)^{4+\delta}}\Big[\mathrm{tr}\Big((\not p+\not\ell)\gamma_\lambda(\not p-\not k_1+\not\ell)\gamma_\rho(\not p-\not q+\not\ell)\gamma_\nu\Big)\\ &\cdot\left(\frac{1}{((p-k_1)^2-\ell^2)((p-q)^2-\ell^2)} - \frac{1}{(p^2-\ell^2)((p-k_1)^2-\ell^2)}\right)\\ &-\mathrm{tr}\left(\gamma_\lambda\frac{1}{\not p-\not k_1+\not\ell}\gamma_\rho\frac{1}{\not p-\not q+\not\ell} - \frac{1}{\not p+\not\ell}\gamma_\lambda\frac{1}{\not p-\not k_1+\not\ell}\gamma_\rho\right)(2p-q)_\nu\Big]\end{aligned} \tag{7.128}$$

The − sign in front of the last line (opposite to the previous equation, (7.127)) needs an explanation. It is due to our minimalistic prescription for a Wick rotation which consists simply in changing the signs of the squared momenta: this leads to a global + sign in the first term and a global − sign in the second of (7.127)), but the subsequent simplifications between numerators and denominators imply a global + sign in both terms. Therefore, the rule has to be corrected by changing 'by hand' the sign of the second term.

The expression (7.128) contains four integrations, but two of them are copies of the other two. To see it, one can proceed by changing variables as follows: $p \to p + q$, followed by $p \to -p$, then $k_1 \leftrightarrow k_2$, $\lambda \leftrightarrow \rho$; finally using the invariance of the trace under transposition.

Next, selecting two independent terms, one works out the gamma matrix algebra, introduces a Feynman parameter and performs a Wick rotation, after which the integrals can be easily calculated. After some algebra, the final result is

$$\begin{aligned} & q^\mu \widetilde{T}_{\mu\nu\lambda\rho}(k_1, k_2) \\ &= \frac{i}{24\pi^2}\left(k_{1\lambda}k_{1\nu}q_\rho - \eta_{\lambda\rho}k_1^2 q_\rho + \eta_{\nu\rho}(k_1^2 k_{2\lambda} - k_1\cdot k_2\, k_{1\lambda})\right)\left(\frac{2}{\delta} + \gamma - \frac{5}{3} + \log\frac{k_1^2}{2\pi\mu^2}\right) \end{aligned} \tag{7.129}$$

to which we have to add the cross term with $k_1 \leftrightarrow k_2$, $\lambda \leftrightarrow \rho$. As a consequence, we are left with two semi-local terms, which are exactly what is needed in order to satisfy the WI (7.126).

In conclusion, the WI for diffeomorphisms is satisfied and there is no anomaly. The trace anomaly (7.123) is therefore confirmed.

7.3.3 *The Case of a Right-Handed Weyl Fermion. Odd Parity*

The second example we wish to consider is the case of a Weyl fermion coupled to a vector potential and compute its odd parity trace anomaly on the basis of the definition (7.1). That is we intend now to compute the relation between the (odd parity) trace $\eta^{\mu\nu}\langle 0|\mathcal{T}T_{R\mu\nu}(x)j_{R\lambda}(y)j_{R\rho}(z)|0\rangle$ and the (odd parity) correlator $\langle 0|\mathcal{T}T^{\mu}_{R\mu}(x)j_{R\lambda}(y)j_{R\rho}(z)|0\rangle$.

Let us start from the following remark. For a right-handed fermion, the triangle contribution to the gauge trace anomaly is

$$\begin{aligned} \widetilde{T}^{(R)\mu}_{\mu\lambda\rho}(k_1, k_2) &= \frac{1}{2}\int \frac{d^4 p}{(2\pi)^4}\mathrm{tr}\left[\frac{1}{\not p}\gamma_\lambda P_R \frac{1}{\not p - \not k_1}\gamma_\rho P_R \frac{1}{\not p - \not k_1 - \not k_2}(2\not p - \not q)P_R\right] \\ &\overset{\mathrm{reg}}{=} \frac{1}{2}\int \frac{d^4 p d^\delta \ell}{(2\pi)^{4+\delta}}\mathrm{tr}\left[\frac{1}{\not p + \not \ell}\gamma_\lambda P_R \frac{1}{\not p - \not k_1 + \not \ell}\gamma_\rho P_R \frac{1}{\not p - \not q + \not \ell}(2\not p + 2\not \ell - \not q)P_R\right] \\ &= \frac{1}{2}\int \frac{d^4 p d^\delta \ell}{(2\pi)^{4+\delta}}\mathrm{tr}\left\{\frac{\not p}{p^2 - \ell^2}\gamma_\lambda \frac{\not p - \not k_1}{(p - k_1)^2 - \ell^2}\gamma_\rho \frac{\not p - \not q + \not \ell}{(p - q)^2 - \ell^2}(2\not p + 2\not \ell - \not q)P_R\right\} \end{aligned} \tag{7.130}$$

to which the cross term must be added.

In Section 6.2.1, we have already calculated the amplitude $\langle \partial \cdot j_R\, j_R\, j_R \rangle$

$$\widetilde{F}^{(R,odd)}_{\lambda\rho}(k_1, k_2, \delta) = \int \frac{d^4 p d^\delta \ell}{(2\pi)^{4+\delta}} \mathrm{tr} \left\{ \frac{\not p}{p^2 - \ell^2} \gamma_\lambda \frac{\not p - \not k_1}{(p-k_1)^2 - \ell^2} \gamma_\rho \frac{\not p - \not q}{(p-q)^2 - \ell^2} \not q P_R \right\} \Bigg|_{\text{odd}}$$
$$= \frac{1}{24\pi^2} \varepsilon_{\mu\nu\lambda\rho} k_1^\mu k_2^\nu \qquad (7.131)$$

which, multiplied by 2, gives $\langle 0|\mathcal{T} \partial^\mu j_{R\mu}(x) j_{R\lambda}(y) j_{R\rho}(z)|0\rangle$.

The difference with (7.130) apart from the factor $\frac{1}{2}$, is the $2\not p - \not q$ factor in the RHS, instead of the $\not q$ one. On the other hand, let us remark that the odd part of the expression

$$\Delta \widetilde{T}^{(R)\mu}_{\mu\lambda\rho}(k_1, k_2)$$
$$= \frac{1}{2} \int \frac{d^4 p d^\delta \ell}{(2\pi)^{4+\delta}} \mathrm{tr} \left[\frac{1}{\not p + \not \ell} \gamma_\lambda P_R \frac{1}{\not p - \not k_1 + \not \ell} \gamma_\rho P_R \frac{1}{\not p - \not q + \not \ell} (2\not p + 2\not \ell - 2\not q) P_R \right]$$
$$= \frac{1}{2} \int \frac{d^4 p d^\delta \ell}{(2\pi)^{4+\delta}} \frac{\mathrm{tr}\left[\gamma_\lambda \not p \gamma_\rho (\not p - \not k_1) P_R\right]}{(p^2 - \ell^2)((p-k_1)^2 - \ell^2)} \qquad (7.132)$$

vanishes by symmetry. Now, since $2\not p + 2\not \ell - \not q = 2\not p + 2\not \ell - 2\not q + \not q$, it follows that

$$\widetilde{T}^{(R)\mu}_{\mu\lambda\rho}(k_1, k_2) + \widetilde{T}^{(R)\mu}_{\mu\rho\lambda}(k_2, k_1) \Big|_{\text{odd}} = \frac{1}{24\pi^2} \varepsilon_{\mu\nu\lambda\rho} k_1^\mu k_2^\nu, \qquad (7.133)$$

which is the result for $\langle 0|\mathcal{T} T^\mu_{R\mu}(x) j_{R\lambda}(y) j_{R\rho}(z)|0\rangle$. Now, it is easy to compute

$$\frac{1}{2} \int \frac{d^4 p d^\delta \ell}{(2\pi)^{4+\delta}} \mathrm{tr} \left\{ \frac{\not p}{p^2 - \ell^2} \gamma_\lambda \frac{\not p - \not k_1}{(p-k_1)^2 - \ell^2} \gamma_\rho \frac{\not p - \not q + \not \ell}{(p-q)^2 - \ell^2} 2\not \ell P_R \right\} \Bigg|_{\text{odd}}$$
$$= \frac{1}{48\pi^2} \varepsilon_{\mu\nu\lambda\rho} k_1^\mu k_2^\nu \qquad (7.134)$$

Subtracting (7.134) from (7.133), we get

$$\frac{1}{2} \int \frac{d^4 p d^\delta \ell}{(2\pi)^{4+\delta}} \mathrm{tr} \left\{ \frac{\not p}{p^2 - \ell^2} \gamma_\lambda \frac{\not p - \not k_1}{(p-k_1)^2 - \ell^2} \gamma_\rho \frac{\not p - \not q}{(p-q)^2 - \ell^2} (2\not p - \not q) P_R \right\} \Bigg|_{\text{odd}} + \mathit{cross}$$
$$= \frac{1}{48\pi^2} \varepsilon_{\mu\nu\lambda\rho} k_1^\mu k_2^\nu \qquad (7.135)$$

which gives the odd parity part of $\eta^{\mu\nu} \langle 0|\mathcal{T} T_{R\mu\nu}(x) j_{R\lambda}(y) j_{R\rho}(z)|0\rangle$. Therefore, we can claim that (after returning to a Lorentzian background)

$$g^{\mu\nu} \langle\langle T_{R\mu\nu}(x) \rangle\rangle \Big|_{\text{odd}} - \langle\langle g^{\mu\nu} T_{R\mu\nu}(x) \rangle\rangle \Big|_{\text{odd}} = -\frac{i}{96\pi^2} \varepsilon_{\mu\nu\lambda\rho} \partial^\mu V^\nu(x) \partial^\lambda V^\rho(x) \quad (7.136)$$

The non-Abelian version of this result is

$$\mathcal{A}_{\omega}^{(odd,R)} = -\frac{i}{384\pi^2}\varepsilon_{\mu\nu\lambda\rho}\text{tr}(F^{\mu\nu}F^{\lambda\rho}) \tag{7.137}$$

This is the gauge-induced odd parity trace anomaly for a right-handed Weyl fermion. For a left-handed Weyl fermion, the only difference is the sign in front of γ_5 in the initial formula (7.130). Therefore, the only change for the odd parity part is an overall sign:

$$\mathcal{A}_{\omega}^{(odd,L)} = \frac{i}{384\pi^2}\varepsilon_{\mu\nu\lambda\rho}\text{tr}(F^{\mu\nu}F^{\lambda\rho}) \tag{7.138}$$

This perturbative derivation will be further commented upon in Sects. 7.4.5 and 7.4.6. In fact, we have not justified the cubic and quartic terms in (7.137) and (7.138), which would require the calculation of quadrangle and pentagon amplitudes. We prefer to derive these results with non-perturbative methods: the same results (7.137) and (7.138) will be obtained with such methods in Sect. 10.4.9.

Remark. On the basis of a not uncommon prejudice in the literature, the previous result sounds unexpected. It is often stated that the theory of gravity is chirally blind, meaning that the relevant charge, the mass, is positive, and is thus different from the typical case of a U(1) interaction. This is certainly a basic peculiarity of gravity with several important consequences. However, one should reflect on the fact that the coupling between gravity and matter is given by the juxtaposition of the metric and the energy-momentum tensor, and the energy-momentum tensors of fermions with opposite chiralities are different. One can suspect therefore that at some stage differences might emerge between fermions with opposite chiralities in their interaction with gravity. A privileged place where such differences may show up is the anomalies, and in this case, the candidate is the trace anomaly, because it involves precisely the coupling between the metric and the energy-momentum tensor. Another privileged calculation could be the consistent diffeomorphism anomaly, which, however, vanishes in 4d for group theoretical reasons, as pointed out elsewhere. The result of this section confirms that the theory of gravity is not chirally blind. This will be confirmed later on by the calculation of metric-induced odd parity trace anomaly.

7.3.4 Gauge-Induced Trace Anomalies and Diffeomorphisms

As already noted, in order to validate the previous result, it is necessary to verify the invariance under diffeomorphisms in the presence of a gauge field V, for we cannot exclude a priori an anomaly of the type $\int d^4x\sqrt{g}\,\partial\cdot\xi\,\varepsilon^{\mu\nu\lambda\rho}\partial_\mu V_\nu \partial_\lambda V_\rho$. This means that we need to analyze amplitudes $\langle \partial\cdot T\, J\, J\rangle$, where T and J are the appropriate e.m. tensor and currents. The proof of the absence of odd parity anomalies can be given in a general form, applicable also to a $V-A$ background. For this reason, we consider amplitudes that involve $T_{\mu\nu}$ and j_μ, but also $T_{5\mu\nu}$ and $j_{5\mu}$, where the last two are obtained by suitably inserting γ_5 in the former.

The Amplitude $\langle \partial \mathrm{T}_5 \mathrm{j}\,\mathrm{j}\rangle$

Let us start from the amplitude $\langle \partial \cdot T_5\, j\, j\rangle$, given by

$$\begin{aligned} & q^\mu \widetilde{T}^{(5VV)}_{\mu\nu\lambda\rho}(k_1,k_2) \\ &= \frac{1}{4}\int \frac{\mathrm{d}^4 p}{(2\pi)^4}\mathrm{tr}\left[\frac{1}{\not p}\gamma_\lambda \frac{1}{\not p-\not k_1}\gamma_\rho \frac{1}{\not p-\not q}\left(q\cdot(2p-q)\gamma_\nu+(2p-q)_\nu \not q\right)\gamma_5\right] \\ &= \frac{1}{4}\int \frac{\mathrm{d}^4 p d^\delta \ell}{(2\pi)^{4+\delta}}\mathrm{tr}\left[\frac{1}{\not p+\not\ell}\gamma_\lambda \frac{1}{\not p-\not k_1+\not\ell}\gamma_\rho \frac{1}{\not p-\not q+\not\ell}\left(q\cdot(2p-q)\gamma_\nu+(2p-q)_\nu \not q\right)\gamma_5\right] \end{aligned} \tag{7.139}$$

We call $\widetilde{A}_{\nu\lambda\rho}(k_1,k_2)$ the piece proportional to $q\cdot(2p-q)\gamma_\nu$, and $\widetilde{B}_{\nu\lambda\rho}(k_1,k_2)$ the rest, and write $q\cdot(2p-q) = p^2-\ell^2-((p-q)^2-\ell^2)$. Then

$$\begin{aligned} & \widetilde{A}_{\nu\lambda\rho}(k_1,k_2) \\ &= \frac{1}{4}\int \frac{\mathrm{d}^4 p d^\delta \ell}{(2\pi)^{4+\delta}}\left\{\frac{-2^{2+\frac{\delta}{2}} i\varepsilon_{\mu\nu\lambda\rho}\ell^2(p+k_1-k_2)^\mu + \mathrm{tr}\left((\not p+\not k_1)\gamma_\lambda \not p\gamma_\rho(\not p-\not k_2)\gamma_\nu\gamma_5\right)}{(p^2-\ell^2)((p-k_2)^2-\ell^2)}\right. \\ & \left. + \frac{2^{2+\frac{\delta}{2}} i\varepsilon_{\mu\nu\lambda\rho}\ell^2(p-k_2)^\mu - \mathrm{tr}\left(\not p\gamma_\lambda(\not p-\not k_1)\gamma_\rho(\not p-\not q)\gamma_\nu\gamma_5\right)}{(p^2-\ell^2)((p-k_1)^2-\ell^2)}\right\} \end{aligned} \tag{7.140}$$

where the first line has been obtained with a shift $p\to p+k_1$. To this, we have to add the cross contribution $\lambda \leftrightarrow k_1 \leftrightarrow k_2$. Now, since the trace of a matrix equals the trace of its transpose we get

$$\mathrm{tr}\left(\not p\gamma_\lambda(\not p-\not k_1)\gamma_\rho(\not p-\not q)\gamma_\nu\gamma_5\right) = \mathrm{tr}\left(\gamma_5\gamma_\nu(\not p-\not q)\gamma_\rho(\not p-\not k_1)\gamma_\lambda\not p\right) \tag{7.141}$$

Next, one can prove that, when integrated over p,

$$\frac{\mathrm{tr}\left(\gamma_5\gamma_\nu(\not p-\not q)\gamma_\rho(\not p-\not k_1)\gamma_\lambda\not p\right)}{(p^2-\ell^2)((p-k_1)^2-\ell^2)} = \frac{\mathrm{tr}\left((\not p+\not k_1)\gamma_\lambda\not p\gamma_\rho(\not p-\not q)\gamma_\nu\gamma_5\right)}{(p^2-\ell^2)((p-k_2)^2-\ell^2)}$$

This is obtained, once again, with the exchanges $\lambda \leftrightarrow \rho, k_1 \leftrightarrow k_2$, followed by the shift $p\to p+k_2$ and $p\to -p$. Therefore, since eventually we have to add the cross contribution, we see that the second term in the second line of (7.140) cancels the second in the first line. In a similar fashion, one can prove that, upon integration,

$$\frac{(p-k_2)^\mu}{(p^2-\ell^2)((p-k_1)^2-\ell^2)} = \frac{(p+k_1-k_2)^\mu}{(p^2-\ell^2)((p-k_2)^2-\ell^2)}$$

by shifting $p\to p+k_1$ and changing $p\to -p$, followed by the cross exchange $\lambda \leftrightarrow \rho, k_1 \leftrightarrow k_2$. It follows that the first term in the first line of (7.140) cancels the first term of the second line. Therefore,

$$\widetilde{A}_{\nu\lambda\rho}(k_1,k_2) + \widetilde{A}_{\nu\rho\lambda}(k_2,k_1) = 0 \tag{7.142}$$

Now let us consider $\widetilde{B}$:

$$\widetilde{B}_{\nu\lambda\rho}(k_1,k_2) = \frac{1}{4}\int \frac{d^4p d^\delta \ell}{(2\pi)^{4+\delta}} \mathrm{tr}\left[\frac{1}{\not p+\not\ell}\gamma_\lambda \frac{1}{\not p-\not k_1+\not\ell}\gamma_\rho \frac{1}{\not p-\not q+\not\ell}(2p-q)_\nu \not q\gamma_5\right]$$
$$= \frac{1}{4}\int \frac{d^4p d^\delta \ell}{(2\pi)^{4+\delta}} \frac{-2^{2+\frac{\delta}{2}} i\varepsilon_{\mu\sigma\lambda\rho}\ell^2(p+k_1)^\mu q^\sigma + \mathrm{tr}\left((\not p+\not q)\gamma_\lambda(\not p+\not k_2)\gamma_\rho \not p\not q\gamma_5\right)}{(p^2-\ell^2)((p+k_2)^2-\ell^2)((p+q)^2-\ell^2)}(2p+q)_\nu \tag{7.143}$$

after a shift $p \to p+q$. Now using transposition inside the trace

$$\mathrm{tr}\left((\not p+\not q)\gamma_\lambda(\not p+\not k_2)\gamma_\rho \not p\not q\gamma_5\right) = -\mathrm{tr}\left(\not p\gamma_\rho(\not p+\not k_2)\gamma_\lambda(\not p+\not q)\not q\gamma_5\right)$$

the cross contribution $\lambda \leftrightarrow \rho,\ k_1 \leftrightarrow k_2$ is

$$\widetilde{B}_{\nu\rho\lambda}(k_2,k_1) \tag{7.144}$$
$$= \frac{1}{8}\int \frac{d^4p d^\delta \ell}{(2\pi)^{4+\delta}} \frac{2^{2+\frac{\delta}{2}} i\varepsilon_{\mu\sigma\lambda\rho}\ell^2(p+k_2)^\mu q^\sigma - \mathrm{tr}\left(\not p\gamma_\lambda(\not p+\not k_1)\gamma_\rho(\not p+\not q)\not q\gamma_5\right)}{(p^2-\ell^2)((p+k_1)^2-\ell^2)((p+q)^2-\ell^2)}(2p+q)_\nu$$
$$= \frac{1}{4}\int \frac{d^4p d^\delta \ell}{(2\pi)^{4+\delta}} \frac{2^{2+\frac{\delta}{2}} i\varepsilon_{\mu\sigma\lambda\rho}\ell^2(p-k_2)^\mu q^\sigma - \mathrm{tr}\left(\not p\gamma_\lambda(\not p-\not k_1)\gamma_\rho(\not p+\not q)\not q\gamma_5\right)}{(p^2-\ell^2)((p-k_1)^2-\ell^2)((p-q)^2-\ell^2)}(2p-q)_\nu$$

Thus

$$\widetilde{B}_{\nu\lambda\rho}(k_1,k_2) + \widetilde{B}_{\nu\rho\lambda}(k_2,k_1) = 0 \tag{7.145}$$

Finally

$$\langle \partial\cdot T_5\, j\, j\rangle = 0 \tag{7.146}$$

Remark. For later use, it is important to remark that the amplitude (7.139) vanishes separately for the ℓ^2-dependent and the ℓ^2-independent parts.

The Amplitude $\langle \partial \mathbf{T_R}\, \mathbf{j_R}\, \mathbf{j_R}\rangle$

The triangle contribution is

$$q^\mu \widetilde{T}^{(RRR)}_{\mu\nu\lambda\rho}(k_1,k_2) \tag{7.147}$$
$$= \frac{1}{4}\int \frac{d^4p}{(2\pi)^4} \mathrm{tr}\left[\frac{1}{\not p}\gamma_\lambda P_R \frac{1}{\not p-\not k_1}\gamma_\rho P_R \frac{1}{\not p-\not q}\left(q\cdot(2p-q)\gamma_\nu + (2p-q)_\nu \not q\right) P_R\right]$$
$$\stackrel{\mathrm{reg}}{=} \frac{1}{4}\int \frac{d^4p d^\delta \ell}{(2\pi)^{4+\delta}} \mathrm{tr}\left[\frac{\not p}{p^2-\ell^2}\gamma_\lambda \frac{\not p-\not k_1}{(p-k_1)^2-\ell^2}\gamma_\rho \frac{\not p-\not q}{(p-q)^2-\ell^2}\left(q\cdot(2p-q)\gamma_\nu + (2p-q)_\nu \not q\right) P_R\right]$$

It is evident that the odd part is half the ℓ^2 independent part of (7.127), and thus, it vanishes.

The Amplitude $\langle \partial \mathbf{T}_5\, \mathbf{j}_5\, \mathbf{j}_5 \rangle$

The amplitude is

$$\begin{aligned} & q^\mu \widetilde{T}^{(555)}_{\mu\nu\lambda\rho}(k_1, k_2) \\ &= \frac{1}{4}\int \frac{d^4p}{(2\pi)^4} \mathrm{tr}\left[\frac{1}{\not p}\gamma_\lambda\gamma_5 \frac{1}{\not p - \not k_1}\gamma_\rho\gamma_5 \frac{1}{\not p - \not q}\left(q\cdot(2p-q)\gamma_\nu + (2p-q)_\nu \not q\right)\gamma_5\right] \\ &\overset{\mathrm{reg}}{=} \frac{1}{4}\int \frac{d^4p\, d^\delta\ell}{(2\pi)^{4+\delta}} \mathrm{tr}\left[\frac{1}{\not p + \not \ell}\gamma_\lambda\gamma_5 \frac{1}{\not p - \not k_1 + \not \ell}\gamma_\rho\gamma_5 \frac{1}{\not p - \not q + \not \ell}\left(q\cdot(2p-q)\gamma_\nu + (2p-q)_\nu \not q\right)\gamma_5\right] \end{aligned} \tag{7.148}$$

The ℓ^2-independent part is the same as before and vanishes. The ℓ^2-dependent part is

$$\begin{aligned} & q^\mu \widetilde{T}^{(555)}_{\mu\nu\lambda\rho}(k_1, k_2) \\ &= 2^{\frac{\delta}{2}-2} i \int \frac{d^4p\, d^\delta\ell}{(2\pi)^{4+\delta}} \ell^2 \frac{\varepsilon_{\mu\nu\lambda\rho}(3p-2k_1-k_2)^\mu q\cdot(2p-q) + \varepsilon_{\mu\sigma\lambda\rho}(3p-2k_1-k_2)^\mu q^\sigma(2p-q)_\nu)}{(p^2-\ell^2)((p-k_1)^2-\ell^2)((p-q)^2-\ell^2)} \end{aligned} \tag{7.149}$$

If we shift $p \to p+q$, change $p \to -p$ and make the exchange $\lambda \leftrightarrow \rho, k_1 \leftrightarrow k_2$, we obtain the same expression with opposite sign. Therefore, the amplitude $\langle \partial\cdot T_5\; j_5\; j_5 \rangle$ vanishes too.

The Amplitudes $\langle \partial \mathbf{T}\, \mathbf{j}\, \mathbf{j}_5 \rangle$ and $\langle \partial\cdot T\; j_5\; j \rangle$

The amplitudes $\langle \partial\cdot T\; j\; j_5 \rangle$ and $\langle \partial\cdot T\; j_5\; j \rangle$ are non-vanishing, but in fact, due to the bosonic symmetry we have to compute the sum $\langle \partial\cdot T\; j\; j_5 \rangle + \langle \partial\cdot T\; j_5\; j \rangle$, that is

$$\begin{aligned} & q^\mu \left(\widetilde{T}^{(VV5)}_{\mu\nu\lambda\rho}(k_1, k_2) + \widetilde{T}^{(V5V)}_{\mu\nu\lambda\rho}(k_1, k_2)\right) \\ &= \frac{1}{4}\int \frac{d^4p\, d^\delta\ell}{(2\pi)^{4+\delta}} \mathrm{tr}\Big[\frac{1}{\not p + \not \ell}\gamma_\lambda \frac{1}{\not p - \not k_1 + \not \ell}\gamma_\rho\gamma_5 \frac{1}{\not p - \not q + \not \ell}\left(q\cdot(2p-q)\gamma_\nu + (2p-q)_\nu \not q\right) \\ &\quad + \frac{1}{\not p + \not \ell}\gamma_\lambda\gamma_5 \frac{1}{\not p - \not k_1 + \not \ell}\gamma_\rho \frac{1}{\not p - \not q + \not \ell}\left(q\cdot(2p-q)\gamma_\nu + (2p-q)_\nu \not q\right)\Big] \\ &= -2^{\frac{\delta}{2}} i \int \frac{d^4p\, d^\delta\ell}{(2\pi)^{4+\delta}} \ell^2 \frac{\varepsilon_{\mu\nu\lambda\rho}(p-k_1)^\mu\, q\cdot(2p-q) + \varepsilon_{\mu\sigma\lambda\rho}(p-k_1)^\mu q^\sigma(2p-q)_\nu}{(p^2-\ell^2)((p-k_1)^2-\ell^2)((p-q)^2-\ell^2)} \end{aligned} \tag{7.150}$$

Now if we shift $p \to p+q$, change $p \to -p$ and make the exchange $\lambda \leftrightarrow \rho, k_1 \leftrightarrow k_2$ we obtain the same expression with opposite sign. Therefore, (7.150) vanishes.

This is enough to conclude that the WI's for diffeomorphisms concerning odd amplitudes are conserved. The odd parity 2pt correlators vanish, and therefore, the odd parity WI's reduce to the vanishing of the divergence of 3pt functions containing one e.m. tensor insertion. These 3pt correlator divergences must vanish, up to anomalies. And this is what we have just proven to be true even in a gauge background including an axial potential. In conclusion, diffeomorphisms are conserved as far as the gauge fields are concerned: no odd parity anomalies appear.

Remark. A well-known result, [6], has it that any odd parity conformal amplitude of the type $\langle T\, J\, J\rangle$ vanishes identically for algebraic reasons (this fact will be further commented upon in Sect. 7.4). Therefore, the results of the present subsection are not surprising. They are nevertheless important because they show that, as pointed out above, regularizing divergences of correlators is not subject to ambiguities.

7.3.5 Link Between Chiral Gauge and Odd Trace Anomalies

This is a good point to highlight a result already implicit in the previous calculations, but worth stressing: the rigid connection between chiral and odd trace anomalies in a gauge background.

Let us start from the case of the right-handed fermion. The correlator is, symbolically, $\langle t_R\, J_R\, J_R\rangle$, i.e. $\langle 0|\mathcal{T}\, T^{\mu}_{R\mu}(x)\, J_{R\lambda(y)} J_{R\rho(z)}|0\rangle$, its Fourier transform being given by

$$\widetilde{F}^{(R)\mu}_{\mu\lambda\rho}(k_1,k_2) = \frac{1}{4}\int \frac{d^4p}{(2\pi)^4}\mathrm{tr}\left[\frac{1}{\not p}\frac{1-\gamma_5}{2}\gamma_\lambda\frac{1}{\not p-\not k_1}\frac{1-\gamma_5}{2}\gamma_\rho\frac{1}{\not p-\not k_1-\not k_2}\frac{1-\gamma_5}{2}(2\not p-\not q)\right]. \tag{7.151}$$

The difference with respect to the Fourier transform of $\langle\partial\cdot J_R\, J_R\, J_R\rangle$—see Eq. (6.81)—apart from the factor $\frac{1}{4}$, is the $(2\not p-\not q)$ factor in the RHS, instead of $\not q$. The relevant difference is therefore twice

$$\begin{aligned}\Delta\widetilde{F}^{(R)\mu}_{\mu\lambda\rho}(k_1,k_2) &= \frac{1}{4}\int \frac{d^4p}{(2\pi)^4}\mathrm{tr}\left[\frac{1}{\not p}\frac{1-\gamma_5}{2}\gamma_\lambda\frac{1}{\not p-\not k_1}\frac{1-\gamma_5}{2}\gamma_\rho\frac{1}{\not p-\not k_1-\not k_2}\frac{1-\gamma_5}{2}\not p\right]\\ &= \frac{1}{4}\int \frac{d^4p\,d^\delta\ell}{(2\pi)^{4+\delta}}\frac{\mathrm{tr}\left[\gamma_\lambda(\not p-\not k_1)\gamma_\rho(\not p-\not q)\frac{1-\gamma_5}{2}\right]}{((p-k_1)^2-\ell^2)((p-q)^2-\ell^2)}.\end{aligned} \tag{7.152}$$

We can now replace $p\to p+k_1$

$$\Delta\widetilde{F}^{(R)\mu}_{\mu\lambda\rho}(k_1,k_2) = \frac{i}{4}\int \frac{d^4p\,d^\delta\ell}{(2\pi)^{4+\delta}}\frac{\mathrm{tr}\left[\gamma_\lambda\not p\gamma_\rho(\not p-\not k_2)\frac{1-\gamma_5}{2}\right]}{(p^2-\ell^2)((p-k_2)^2-\ell^2)}. \tag{7.153}$$

The odd part vanishes by symmetry.

If we consider instead the amplitude for $\langle\partial\cdot J_5\, J\, J\rangle$, (6.47), the result does not change. In that case for the odd part, we get

$$\begin{aligned}\Delta\widetilde{F}^{(5)\mu}_{\mu\lambda\rho}(k_1,k_2) &\sim \int \frac{d^4p\,d^\delta\ell}{(2\pi)^{4+\delta}}\mathrm{tr}\left(\gamma_\lambda\frac{\not p+\not\ell-\not k_1}{(p-k_1)^2-\ell^2}\gamma_\rho\frac{\not p+\not\ell-\not q}{(p-q)^2-\ell^2}\gamma_5\right)\\ &= \int \frac{d^4p\,d^\delta\ell}{(2\pi)^{4+\delta}}\frac{\left(-\ell^2\mathrm{tr}(\gamma_\lambda\gamma_\rho\gamma_5)+\mathrm{tr}\left(\gamma_\lambda(\not p-\not k_1)\gamma_\rho(\not p-\not q)\gamma_5\right)\right)}{((p-k_1)^2-\ell^2)((p-q)^2-\ell^2)}\end{aligned} \tag{7.154}$$

The first term in the numerator vanishes. The rest can be rewritten as

$$\Delta \widetilde{F}^{(5)\mu}_{\mu\lambda\rho}(k_1, k_2) \sim \int \frac{d^4p d^\delta \ell}{(2\pi)^{4+\delta}} \frac{\mathrm{tr}\left(\gamma_\lambda \not{p} \gamma_\rho (\not{p} - \not{k}_2)\gamma_5\right)}{(p^2 - \ell^2)((p-k_2)^2 - \ell^2)} = 0 \qquad (7.155)$$

for the same reason as above. In the same way, one can easily prove that

$$\Delta \widetilde{F}^{(5')\mu}_{\mu\lambda\rho}(k_1, k_2) = \Delta \widetilde{F}^{(5'')\mu}_{\mu\lambda\rho}(k_1, k_2) = 0. \qquad (7.156)$$

In conclusion, the amplitude for the chiral anomalies and those for the odd trace anomalies due to couplings with gauge fields are rigidly related, the corresponding coefficients exhibiting a fixed ratio, i.e. the former are minus four times the latter.

7.4 Diffeomorphisms and Trace Anomalies in 4d

Above we have studied possible diffeomorphism anomalies induced by the coupling of fermions to gauge potentials in the presence of a non-trivial metric in 4d. We have seen that such anomalies do not exist. Now we have to consider the possibility that consistent anomalies are generated by the non-trivial metric itself. It is a well-known fact that consistent chiral, diffeomorphism or local Lorentz, anomalies such as those of Sect. 5.2.2 vanish identically in 4d. However, there could be anomalies of the second type; see (5.62). In this section, we would like to exclude this possibility. To start with let us lay down the problem in more detail.

Let us consider the case of a right-handed Weyl fermion coupled to a metric. The action is

$$S = \int d^4x \sqrt{g}\, i \overline{\psi_R} \gamma^\mu (D_\mu + \frac{1}{2}\omega_\mu)\psi_R \qquad (7.157)$$

where $\gamma^\mu = e^\mu_a \gamma^a$, D_μ ($\mu, \nu...$ are world indices, $a, b, ...$ are flat indices) is the covariant derivative with respect to the world indices and ω_μ is the spin connection:

$$\omega_\mu = \omega^{ab}_\mu \Sigma_{ab}$$

where $\Sigma_{ab} = \frac{1}{4}[\gamma_a, \gamma_b]$ are the Lorentz generators. Finally, $\psi_R = \frac{1+\gamma_5}{2}\psi$. Classically the energy-momentum tensor

$$T^{(R)}_{\mu\nu} = \frac{i}{4}\overline{\psi_R}\gamma_\mu \overleftrightarrow{\nabla}_\nu \psi_R + \{\mu \leftrightarrow \nu\} \qquad (7.158)$$

is both conserved and traceless on shell. To start with from (7.157), we have to extract the Feynman rules. More explicitly the action (7.157) can be written as

$$S = \int \mathrm{d}^4x \sqrt{|g|} \left[\frac{i}{2} \overline{\psi_R} \gamma^\mu \overleftrightarrow{\partial}_\mu \psi_R - \frac{1}{4} \varepsilon^{\mu abc} \omega_{\mu ab} \overline{\psi_R} \gamma_c \gamma_5 \psi_R \right] \tag{7.159}$$

where it is understood that the derivative applies to ψ_R and $\overline{\psi_R}$ only. We have used the relation $\{\gamma^a, \Sigma^{bc}\} = i\, \varepsilon^{abcd} \gamma_d \gamma_5$.

Now we expand

$$e_\mu^a = \delta_\mu^a + \chi_\mu^a + ..., \quad e_a^\mu = \delta_a^\mu + \hat{\chi}_a^\mu + ..., \quad \text{and} \quad g_{\mu\nu} = \eta_{\mu\nu} + h_{\mu\nu} + ... \tag{7.160}$$

and make a local Lorentz gauge choice by dropping the antisymmetric part of the vierbein. Inserting these expansions in the defining relations $\mathrm{e}_\mu^a \mathrm{e}_b^\mu = \delta_b^a$, $g_{\mu\nu} = e_\mu^a e_\nu^b \eta_{ab}$, we find

$$\hat{\chi}_\nu^\mu = -\chi_\nu^\mu \quad \text{and} \quad h_{\mu\nu} = 2\, \chi_{\mu\nu}. \tag{7.161}$$

Let us expand accordingly the spin connection. Using

$$\omega_{\mu ab} = e_{\nu a}(\partial_\mu e_b^\nu + \mathrm{e}^\sigma{}_b \Gamma_\sigma{}^\nu{}_\mu), \quad \Gamma_\sigma{}^\nu{}_\mu = \frac{1}{2}\eta^{\nu\lambda}(\partial_\sigma h_{\lambda\mu} + \partial_\mu h_{\lambda\sigma} - \partial_\lambda h_{\sigma\mu}) + ...$$

after some algebra we get

$$\omega_{\mu ab}\, \varepsilon^{\mu abc} = -\varepsilon^{\mu abc}\, \partial_\mu \chi_{a\lambda}\, \chi_b^\lambda + ... \tag{7.162}$$

For later use let us quote the following approximation for the Pontryagin density

$$\varepsilon^{\mu\nu\lambda\rho} R_{\mu\nu}{}^{\sigma\tau} R_{\lambda\rho\sigma\tau} = 8\varepsilon^{\mu\nu\lambda\rho} \left(\partial_\mu \partial_\sigma \chi_\nu^a\, \partial_\lambda \partial_a \chi_\rho^\sigma - \partial_\mu \partial_\sigma \chi_\nu^a\, \partial_\lambda \partial^\sigma \chi_{a\rho} \right) + ... \tag{7.163}$$

Therefore, up to second order the action can be written (by incorporating $\sqrt{|g|}$ in a redefinition of the ψ field[7])

$$S \approx \int \mathrm{d}^4x \left[\frac{i}{2} \left(\delta_a^\mu - \chi_a^\mu + \frac{3}{2}(\chi^2)_a^\mu \right) \overline{\psi_L} \gamma^a \overleftrightarrow{\partial}_\mu \psi_L + \frac{1}{4} \varepsilon^{\mu abc}\, \partial_\mu \chi_{a\lambda}\, \chi_b^\lambda\, \overline{\psi}_L \gamma_c \gamma_5 \psi_L \right]$$

As a consequence, the Feynman rules are as follows (momenta are incoming and the external gravitational field is assumed to be $h_{\mu\nu}$). The fermion propagator is

$$\frac{i}{\not{p} + i\epsilon} \tag{7.164}$$

[7] This is the simplest way to deal with $\sqrt{|g|}$. Alternatively, one can keep it explicitly in the action and approximate it as $1 + \frac{1}{2}h_\mu^\mu$; this would produce two additional vertices, which however do not contribute to our final result. The effect of $\sqrt{|g|}$ has been studied more thoroughly in [7].

The two-fermion-one-graviton vertex (V_{ffh}) is

$$-\frac{i}{8}\left[(p-p')_\mu\gamma_\nu+(p-p')_\nu\gamma_\mu\right]\frac{1+\gamma_5}{2} \tag{7.165}$$

The momenta p, p' are as in Fig. 7.1. There are two two-fermion-two-graviton vertices: V_{ffhh}

$$\frac{3i}{64}\Big[\left((p+p')_\mu\gamma_{\mu'}\eta_{\nu\nu'}+(p+p')_\mu\gamma_{\nu'}\eta_{\nu\mu'}+\{\mu\leftrightarrow\nu\}\right) \tag{7.166}$$
$$+\left((p+p')_{\mu'}\gamma_\mu\eta_{\nu\nu'}+(p+p')_{\mu'}\gamma_\nu\eta_{\mu\nu'}+\{\mu'\leftrightarrow\nu'\}\right)\Big]\frac{1+\gamma_5}{2}$$

and V^{ε}_{ffhh}

$$\frac{1}{64}t_{\mu\nu\mu'\nu'\kappa\lambda}(k-k')^\lambda\gamma^\kappa\frac{1+\gamma_5}{2} \tag{7.167}$$

where

$$t_{\mu\nu\mu'\nu'\kappa\lambda}=\eta_{\mu\mu'}\varepsilon_{\nu\nu'\kappa\lambda}+\eta_{\nu\nu'}\varepsilon_{\mu\mu'\kappa\lambda}+\eta_{\mu\nu'}\varepsilon_{\nu\mu'\kappa\lambda}+\eta_{\nu\mu'}\varepsilon_{\mu\nu'\kappa\lambda} \tag{7.168}$$

and the graviton momenta k, k' are incoming.

For later use let us introduce also the tensors

$$t^{(ij)}_{\mu\nu\mu'\nu'\kappa\lambda}=k_{i\mu}k_{j\mu'}\varepsilon_{\nu\nu'\kappa\lambda}+k_{i\nu}k_{j\nu'}\varepsilon_{\mu\mu'\kappa\lambda}+k_{i\mu}k_{j\nu'}\varepsilon_{\nu\mu'\kappa\lambda}+k_{i\nu}k_{j\mu'}\varepsilon_{\mu\nu'\kappa\lambda} \tag{7.169}$$

7.4.1 No Diffeomorphism Anomalies

In the sequel, we will consider the possibility of odd diffeomorphism anomalies, assuming that even diffeomorphism anomalies are not possible. In this sense, we have already seen a few examples before. To prove it, we should do the analog of Sects. 7.3.2 and 7.3.4 (with two currents replaced by two e.m. tensors). In the analog of 7.3.2, the (even) WI involves semi-local terms, due to the presence of even two-point correlator of the e.m. tensor (7.158). For simplicity, we will assume that the even WI is satisfied. This hypothesis will be confirmed further on by nonperturbative calculations and by the index theorem approach. In the odd WI, instead there are no semi-local terms in the game because the odd two-point e.m. tensor correlator vanishes. Therefore, we will limit ourselves to the analog of 7.3.4. The effective action is

$$W^{(R)}[g]=W^{(R)}[0]+\sum_{n=1}^{\infty}\frac{i^{n-1}}{2^n n!}\int\prod_{i=1}^{n}d^{\mathrm{d}}x_i h^{\mu_i\nu_i}(x_i)\langle 0|\mathcal{T}T^{(R)}_{\mu_1\nu_1}(x_1)\dots T^{(R)}_{\mu_n\nu_n}(x_n)|0\rangle \tag{7.170}$$

Notice that the factors $\sqrt{g(x_i)}$ are absent in the integral because we are going to use the e.m. (7.158), not the analog of (7.5). As explained around Eq. (7.111) this simplifies the WI's (and corresponds to absorbing the factor $g^{\frac{1}{4}}$ in a redefinition of the fermion field ψ, as we have done above, which corresponds to starting from $T_{\mu\nu}(x) = 2\frac{\delta S}{\delta g^{\mu\nu}(x)}\Big|_{g=\eta}$).

After these premises, the WI to be verified, to the lowest significant order, is

$$\langle\!\langle D^\mu T^{(R)}_{\mu\nu}(x)\rangle\!\rangle^{(odd)} \tag{7.171}$$
$$\approx -\frac{1}{2}\int dx_1 dx_2 h^{\mu_1\nu_1}(x_1)h^{\mu_2\nu_2}(x_2)\partial^\mu\langle 0|\mathcal{T}T^{(R)}_{\mu\nu}(x)T^{(R)}_{\mu_1\nu_1}(x_1)T^{(R)}_{\mu_2\nu_2}(x_2)|0\rangle^{(odd)} = 0$$

On the basis of a general theorem, the odd parity conformal amplitude of three e.m. tensors vanishes identically, [6]. Since the theory (7.159) is conformal invariant and since we do not expect any significant intrusion of the regularization process on the end result, we expect this theorem to apply also in this case and, as a consequence, the Ward identity (7.171) to hold at one-loop order. However, it is instructive and interesting to verify it directly.

As a second remark, in the following calculation we will adopt the rightmost γ_5 prescription. We have seen in 2d that this is the simplest prescription when associated with the dimensional regularization. On the other hand, since the three-point function vanishes identically we do not expect any change in the final result due to a modification of the regularization process. We will see that this is indeed the case.

Let us start from the triangle diagram. The diagram with three e.m. insertions is constructed by joining three vertices V_{ffh} with three fermion lines. The external momenta are q (incoming) associated with the labels μ and ν, and k_1, k_2 (outgoing), with labels λ, ρ and α, β, respectively. Of course $q = k_1 + k_2$. The internal momenta are $p, p-k_1$ and $p-k_1-k_2$, respectively. Following the rightmost γ_5 prescription, we move P_R to the rightmost position and drop the $\frac{1}{2}$ addend in it. The odd part of the relevant triangle contribution for a right-handed fermion is given by

$$q^\mu\widetilde{\mathfrak{T}}^{(1odd)}_{\mu\nu\lambda\rho\alpha\beta}(k_1,k_2) = -\frac{1}{512}\int\frac{d^4p}{(2\pi)^4}\,\mathrm{tr}\left[\left(\frac{1}{\not p}((2p-k_1)_\lambda\gamma_\rho + (\lambda\leftrightarrow\rho))\,\frac{1}{\not p-\not k_1}\right.\right.$$
$$\left.\left.\times((2p-2k_1-k_2)_\alpha\gamma_\beta + (\alpha\leftrightarrow\beta))\,\frac{1}{\not p-\not q}((2p-q)\cdot q\,\gamma_\nu + (2p-q)_\nu\not q)\right)\gamma_5\right] \tag{7.172}$$

to which the cross contribution $q^\mu\mathfrak{T}^{(2odd)}_{\mu\nu\alpha\beta\lambda\rho}(k_2,k_1)$ has to be added. We regularize (7.172), as usual, as follows (we leave aside for the moment the symmetrization $\alpha\leftrightarrow\beta$ and $\lambda\leftrightarrow\rho$)

$$q^\mu\widetilde{\mathfrak{T}}^{(1odd)}_{\mu\nu\lambda\rho\alpha\beta}(k_1,k_2) = -\frac{1}{512}\int\frac{d^4p\,d^\delta\ell}{(2\pi)^{4+\delta}}\,\mathrm{tr}\left[\left(\frac{\not p+\not\ell}{p^2-\ell^2}(2p-k_1)_\lambda\gamma_\rho\,\frac{\not p-\not k_1+\not\ell}{(p-k_1)^2-\ell^2}\right.\right.$$
$$\left.\left.\times\,(2p-2k_1-k_2)_\alpha\gamma_\beta\,\frac{\not p-\not q+\not\ell}{(p-q)^2-\ell^2}((2p-q)\cdot q\,\gamma_\nu - (2p-q)_\nu\not q)\right)\gamma_5\right]$$
$$\equiv \widetilde{\mathcal{A}}_{\nu\lambda\rho\alpha\beta}(k_1,k_2) + \widetilde{\mathfrak{B}}_{\nu\lambda\rho\alpha\beta}(k_1,k_2) \tag{7.173}$$

where

$$\widetilde{\mathcal{A}}_{\nu\lambda\rho\alpha\beta}(k_1,k_2) = -\frac{1}{512}\int \frac{d^4p d^\delta\ell}{(2\pi)^{4+\delta}} \operatorname{tr}\left[\left(\frac{\not p}{p^2-\ell^2}(2p-k_1)_\lambda\gamma_\rho \frac{\not p - \not k_1}{(p-k_1)^2-\ell^2}\right.\right.$$
$$\left.\left.\times(2p-2k_1-k_2)_\alpha\gamma_\beta \frac{\not p - \not q}{(p-q)^2-\ell^2}((2p-q)\cdot q\,\gamma_\nu - (2p-q)_\nu \not q)\right)\gamma_5\right] \quad (7.174)$$

and

$$\widetilde{\mathcal{B}}_{\nu\lambda\rho\alpha\beta}(k_1,k_2) = -\frac{1}{512}\int \frac{d^4p d^\delta\ell}{(2\pi)^{4+\delta}}\,\ell^2\left[\operatorname{tr}\left(\gamma_\rho\gamma_\beta\gamma_\nu(\not p - \not k_2)\gamma_5\right) q\cdot(2p-q)\right. \quad (7.175)$$
$$\left. - \operatorname{tr}\left(\gamma_\rho\gamma_\beta \not q(\not p - \not k_2)\gamma_5\right)(2p-q)_\nu\right]\frac{(2p-k_1)_\lambda(2p-2k_1-k_2)_\alpha}{(p^2-\ell^2)((p-k_1)^2-\ell^2)((p-q)^2-\ell^2)}$$

The rationale for the change of sign in (7.173, 7.174) and (7.175) has been explained after Eq. (7.128), but in the present case, they are irrelevant since the corresponding addends will vanish separately.

Let us consider first $\widetilde{\mathcal{A}}$. Since the integral is regulated we can shift $p \to p + k_1$ and get

$$\widetilde{\mathcal{A}}_{\nu\lambda\rho\alpha\beta}(k_1,k_2) = -\frac{1}{512}\int \frac{d^4p d^\delta\ell}{(2\pi)^{4+\delta}} \operatorname{tr}\left[\left(\frac{\not p + \not k_1}{(p+k_1)^2-\ell^2}(2p+k_1)_\lambda\gamma_\rho \frac{\not p}{p^2-\ell^2}\right.\right.$$
$$\left.\left.\times(2p-k_2)_\alpha\gamma_\beta \frac{\not p - \not k_2}{(p-k_2)^2-\ell^2}((2p+k_1-k_2)\cdot q\,\gamma_\nu + (2p+k_1-k_2)_\nu \not q)\right)\gamma_5\right]$$
$$= -\frac{1}{512}\int \frac{d^4p d^\delta\ell}{(2\pi)^{4+\delta}}\left[\operatorname{tr}\left((\not p + \not k_1)\gamma_\rho \not p \gamma_\beta(\not p - \not k_2)\gamma_\nu\gamma_5\right)(2p+k_1-k_2)\cdot q\right.$$
$$\left. + \operatorname{tr}\left((\not p + \not k_1)\gamma_\rho \not p\gamma_\beta(\not p - \not k_2)\not q\gamma_5\right)(2p+k_1-k_2)_\nu\right]$$
$$\times\frac{(2p+k_1)_\lambda(2p-k_2)_\alpha}{((p+k_1)^2-\ell^2)(p^2-\ell^2)((p-k_2)^2-\ell^2)} \quad (7.176)$$

Now, as we have done several times before, we add the cross contribution ($k_1 \leftrightarrow k_2$, $(\lambda,\rho) \leftrightarrow (\alpha,\beta)$), which leads to

$$\widetilde{\mathcal{A}}_{\nu\alpha\beta\lambda\rho}(k_2,k_1) = -\frac{1}{512}\int \frac{d^4p d^\delta\ell}{(2\pi)^{4+\delta}}\left[\operatorname{tr}\left((\not p - \not k_2)\gamma_\beta \not p\gamma_\rho(\not p + \not k_1)\gamma_\nu\gamma_5\right)(2p+k_1-k_2)\cdot q\right.$$
$$\left. + \operatorname{tr}\left((\not p - \not k_2)\gamma_\beta \not p\gamma_\rho(\not p + \not k_1)\not q\gamma_5\right)(2p+k_1-k_2)_\nu\right]$$
$$\times\frac{(2p-k_2)_\alpha(2p+k_1)_\lambda}{((p-k_2)^2-\ell^2)(p^2-\ell^2)((p+k_1)^2-\ell^2)} = -\widetilde{\mathcal{A}}_{\nu\lambda\rho\alpha\beta}(k_1,k_2) \quad (7.177)$$

To prove the last equality one has to change $p \to -p$ and to use the property that any trace of gamma matrices in 4d does not change if we reverse the order of the latter. In the same way, one can prove that the analogous contributions coming from the symmetrization $\alpha \leftrightarrow \beta$ and $\lambda \leftrightarrow \rho$ vanish. In conclusion, the $\widetilde{\mathcal{A}}$ contribution to the triangle diagram vanishes.

Let us consider next the $\widetilde{\mathcal{B}}$ part which can be rewritten as

$$\widetilde{\mathcal{B}}_{\nu\lambda\rho\alpha\beta}(k_1,k_2) = i\frac{2^{\frac{\delta}{2}}}{128}\int\frac{\mathrm{d}^4p d^\delta\ell}{(2\pi)^{4+\delta}}\,\ell^2\Big[\varepsilon_{\rho\beta\nu\sigma}(p-k_2)^\sigma(2p-q)\cdot q \tag{7.178}$$
$$+\varepsilon_{\rho\beta\sigma\tau}q^\sigma(p-k_2)^\tau(2p-q)_\nu\Big]\frac{(2p-k_1)_\lambda(2p-2k_1-k_2)_\alpha}{(p^2-\ell^2)((p-k_1)^2-\ell^2)((p-q)^2-\ell^2)}$$
$$= i\frac{2^{\frac{\delta}{2}}}{128}\int\frac{\mathrm{d}^4p d^\delta\ell}{(2\pi)^{4+\delta}}\,\ell^2\Big[\varepsilon_{\rho\beta\nu\sigma}(p+k_1-k_2)^\sigma(2p+k_1-k_2)\cdot q$$
$$+\varepsilon_{\rho\beta\sigma\tau}q^\sigma(p+k_1-k_2)^\tau(2p+k_1-k_2)_\nu\Big]$$
$$\frac{(2p+k_1)_\lambda(2p-k_2)_\alpha}{((p+k_1)^2-\ell^2)(p^2-\ell^2)((p-k_2)^2-\ell^2)}$$

The cross contribution ($k_1 \leftrightarrow k_2$, $(\lambda,\rho)\leftrightarrow(\alpha,\beta)$), via the change $p\to -p$, yields

$$\widetilde{\mathcal{B}}_{\nu\alpha\beta\lambda\rho}(k_2,k_1) = -\widetilde{\mathcal{B}}_{\nu\lambda\rho\alpha\beta}(k_1,k_2) \tag{7.179}$$

and similarly for the contributions coming from the symmetrization $\alpha \leftrightarrow \beta$ and $\lambda \leftrightarrow \rho$.

Therefore, neither $\widetilde{\mathcal{B}}$ contributes to the triangle diagram. As far as the triangle diagram is concerned, we can conclude

$$q^\mu\widetilde{\mathcal{T}}^{(1odd)}_{\mu\nu\lambda\rho\alpha\beta}(k_1,k_2) + q^\mu\widetilde{\mathcal{T}}^{(2odd)}_{\mu\nu\lambda\rho\alpha\beta}(k_1,k_2) = 0 \tag{7.180}$$

A remark: the integrand of the amplitude (7.173) contains a factor $(2p-q)\cdot q\,\gamma_\nu - (2p-q)_\nu\not{q}$. As anticipated above, the vanishing of (7.173) holds separately for $(2p-q)\cdot q\,\gamma_\nu$ and $(2p-q)_\nu\not{q}$.

There are other graphs, besides the triangle one, that are relevant in this case and, for completeness, must be evaluated. After a general introduction to these other diagrams in Appendix 7C and a process of trimming the irrelevant ones, in Appendix 7E we focus on the odd bubble diagrams and show that their contributions vanish too. Therefore, we can conclude that the WI (7.171) is satisfied.

Finally, let us return to the question of the rightmost γ_5 prescription. Had we used the usual approach to regularization, like in Eq. (7.71) for instance, the calculation would be actually simplified because the ℓ^2 dependent term $\widetilde{\mathcal{B}}_{\nu\lambda\rho\alpha\beta}(k_1,k_2)$ would be absent. The same is true for the bubble diagrams. Therefore, our conclusion on the vanishing of the e.m. tensor divergence holds *a fortiori*.

7.4.2 Odd Parity Trace Anomaly

Even trace anomalies for any kind of tensor fields are well-known since the 70s of the last century. But except in two dimensions, they are exceedingly complicated to be computed by means of perturbative methods. The algebra of momenta in the relevant Feynman diagrams is daunting and the polynomial of external momenta to be computed, which in 2d is made of two terms, in 4d contains a large number of monomials. Therefore, we postpone the derivation of some even parity trace anomalies to the next part of the book, after the introduction of non-perturbative methods. On the contrary, the odd parity trace anomaly in 4d is relatively easily accessible with perturbative methods. In the sequel, we show how to derive it (references: [7–9])

The Triangle Diagram

The contributions to the odd parity trace anomaly may come from different diagrams. In this section, we compute the odd parity contribution of the triangle diagram and postpone to Appendix 7E the treatment of the others, which yields vanishing contributions. At first sight, this calculation does not seem to make sense, because, as already mentioned, a well-known result of CFT states that a conformal odd parity three-point function $\langle 0|\mathcal{T} T_{\mu\nu}(x) T_{\mu'\nu'}(y) T_{\alpha\beta}(z)|0\rangle^{(odd)}$ vanishes identically for algebraic reasons. Therefore, at the lowest perturbative order, we can write

$$\eta^{\mu\nu}\langle\langle T_{\mu\nu}(x)\rangle\rangle^{(odd)} = 0, \tag{7.181}$$

as can be proven also by a direct calculation. However, according to the definition (7.1) we must compute also the second term with one insertion of a trace of the e.m. tensor. For the Lorentz group representation theory, the trace of the e.m. tensor is a scalar, and there is no vanishing theorem for an amplitude of a scalar and two e.m. tensors. Therefore, to settle this question, it remains for us to evaluate the triangle diagram in which one of the insertion is $T_\mu^\mu(x)$.

The triangle diagram with three e.m. insertions is constructed by joining three vertices V_{ffh} with three fermion lines. The external momenta are q (incoming) with labels σ and τ, and k_1, k_2 (outgoing), with labels μ, ν and μ', ν', respectively. Of course $q = k_1 + k_2$. The internal momenta are p, $p - k_1$ and $p - k_1 - k_2$, respectively. In this section, we will use the rightmost γ_5 prescription and, eventually, we will put the external legs on shell, $k_1^2 = k_2^2 = 0$. This simplifies the calculations without reducing the generality of the result. Justifying it will lead us however to an unexpected turn.

Employing the above Feynman rules of the free chiral fermion coupled to an external gravitational field, we can write down the Fourier transform of the three-point e.m. tensor amplitude $\widetilde{\mathcal{T}}^{(odd)}_{\mu\nu\mu'\nu'\alpha\beta}(k_1, k_2)$. Contracting α and β the correlator, we are looking for is

$$
\begin{aligned}
\widetilde{\mathsf{T}}_{\mu\nu\mu'\nu'}(k_1,k_2) &\equiv \widetilde{\mathfrak{T}}^{(odd)\alpha}_{\mu\nu\mu'\nu'\alpha}(k_1,k_2) \\
&= \int \frac{\mathrm{d}^4 p}{(2\pi)^4} \mathrm{tr}\left\{ \frac{i}{8}\left[(2p-k_1)_\mu \gamma_\nu + (\mu \leftrightarrow \nu)\right]\left(\frac{1+\gamma_5}{2}\right)\frac{i}{(\not p - \not k_1)} \right. \\
&\times \frac{i}{8}\left[(2p-2k_1-k_2)_{\mu'}\gamma_{\nu'} + (\mu' \leftrightarrow \nu')\right]\left(\frac{1+\gamma_5}{2}\right)\frac{i}{(\not p - \not k_1 - \not k_2)} \\
&\times \left. \frac{i}{4}(2\not p - \not k_1 - \not k_2)\left(\frac{1+\gamma_5}{2}\right)\frac{i}{\not p}\right\}.
\end{aligned}
\tag{7.182}
$$

Moving P_R to the rightmost position, one obtains

$$
\begin{aligned}
\widetilde{\mathsf{T}}_{\mu\nu\mu'\nu'}(k_1,k_2) &= -\frac{1}{256}\int \frac{\mathrm{d}^4 p}{(2\pi)^4}\mathrm{Tr}\left\{\left[\frac{\not p}{p^2}(2p-k_1)_\mu\gamma_\nu + (\mu \leftrightarrow \nu)\right]\frac{(\not p - \not k_1)}{(p-k_1)^2}\right. \\
&\times \left[(2p-2k_1-k_2)_{\mu'}\gamma_{\nu'}(\mu' \leftrightarrow \nu')\right]\frac{(\not p - \not k_1 - \not k_2)}{(p-k_1-k_2)^2}(2\not p - \not k_1 - \not k_2)\left.\left(\frac{1+\gamma_5}{2}\right)\right\}.
\end{aligned}
\tag{7.183}
$$

Clearly, such an integral is ultraviolet divergent. In order to proceed with the computation, we employ dimensional regularization; that is, as usual, additional components are added to the momentum, namely $p \to p + \ell$, where $\ell = (\ell_4, \ldots, \ell_{\delta+3})$.

Hence, Eq. (7.183) is replaced by

$$
\begin{aligned}
\widetilde{\mathsf{T}}_{\mu\nu\mu'\nu'}(k_1,k_2) &= -\frac{1}{256}\int \frac{\mathrm{d}^4 p}{(2\pi)^4}\int \frac{\mathrm{d}^\delta \ell}{(2\pi)^\delta}\mathrm{Tr}\left\{\left[\frac{\not p + \not \ell}{p^2-\ell^2}(2p-k_1)_\mu\gamma_\nu + (\mu \leftrightarrow \nu)\right]\right. \\
&\times \frac{(\not p + \not \ell - \not k_1)}{(p-k_1)^2-\ell^2}\left[(2p-2k_1-k_2)_{\mu'}\gamma_{\nu'} + (\mu' \leftrightarrow \nu')\right] \\
&\times \left.\underbrace{\frac{(\not p + \not \ell - \not k_1 - \not k_2)}{(p-k_1-k_2)^2-\ell^2}(2\not p + 2\not \ell - \not k_1 - \not k_2)}_{(*)}\left(\frac{1+\gamma_5}{2}\right)\right\}.
\end{aligned}
\tag{7.184}
$$

Expression (7.184) is now regularized and we can continue with the computation of the diagram. Now we ignore the identity in the projector $(1+\gamma_5)/2$ since we are concerned with the odd parity contribution of the diagram, which is contained in the γ_5 sector. Also, for the time being, we omit the symmetrization in $(\mu \leftrightarrow \nu)$, and in $(\mu' \leftrightarrow \nu')$, we will reintroduce it later on.

Let us take the term $(*)$ and set $q = k_1 + k_2$. It is simple to check that

$$
(*) = \frac{(\not p + \not \ell - \not q)}{(p-q)^2-\ell^2}(2\not p + 2\not \ell - \not q) = 1 + \frac{\not p - \not \ell}{\not p + \not \ell - \not q} + \frac{2\not \ell}{\not p + \not \ell - \not q},
\tag{7.185}
$$

and plugging it into Eq. (7.184), one ends up with

$$\widetilde{\mathsf{T}}_{\mu\nu\mu'\nu'}(k_1,k_2) = \mathsf{T}^{(1)}_{\mu\nu\mu'\nu'}(k_1,k_2) + \mathsf{T}^{(2)}_{\mu\nu\mu'\nu'}(k_1,k_2) + \widehat{\mathsf{T}}_{\mu\nu\mu'\nu'}(k_1,k_2), \qquad (7.186)$$

with

$$\begin{aligned}
\mathsf{T}^{(1)}_{\mu\nu\mu'\nu'}(k_1,k_2) &= -\frac{1}{256}\int\frac{\mathrm{d}^4p}{(2\pi)^4}\int\frac{\mathrm{d}^\delta\ell}{(2\pi)^\delta}\\
&\times\mathrm{Tr}\left[\frac{\not p+\not\ell}{p^2-\ell^2}(2p-k_1)_\mu\gamma_\nu\frac{(\not p+\not\ell-\not k_1)}{(p-k_1)^2-\ell^2}(2p-2k_1-k_2)_{\mu'}\gamma_{\nu'}\frac{\gamma_5}{2}\right],\\
\mathsf{T}^{(2)}_{\mu\nu\mu'\nu'}(k_1,k_2) &= -\frac{1}{256}\int\frac{\mathrm{d}^4p}{(2\pi)^4}\int\frac{\mathrm{d}^\delta\ell}{(2\pi)^\delta}\mathrm{Tr}\left[\frac{\not p+\not\ell}{p^2-\ell^2}(2p-k_1)_\mu\gamma_\nu\right.\\
&\left.\times\frac{(\not p+\not\ell-\not k_1)}{(p-k_1)^2-\ell^2}(2p-2k_1-k_2)_{\mu'}\gamma_{\nu'}\frac{(\not p-\not\ell)}{(p-q)^2-\ell^2}\frac{\gamma_5}{2}\right],\\
\widehat{\mathsf{T}}_{\mu\nu\mu'\nu'}(k_1,k_2) &= -\frac{1}{256}\int\frac{\mathrm{d}^4p}{(2\pi)^4}\int\frac{\mathrm{d}^\delta\ell}{(2\pi)^\delta}\mathrm{Tr}\left[\frac{\not p+\not\ell}{p^2-\ell^2}(2p-k_1)_\mu\gamma_\nu\right.\\
&\left.\times\frac{(\not p+\not\ell-\not k_1)}{(p-k_1)^2-\ell^2}(2p-2k_1-k_2)_{\mu'}\gamma_{\nu'}\frac{\not\ell}{(p-q)^2-\ell^2}\gamma_5\right]. \qquad (7.187)
\end{aligned}$$

We postpone the computation of $T^{(1)}$, $T^{(2)}$ to Appendix 7E. Their contribution vanishes. Here we continue with the explicit calculation of the most important piece, $\widehat{\mathsf{T}}$.

$$\begin{aligned}
\widehat{\mathsf{T}}_{\mu\nu\mu'\nu'}(k_1,k_2) &= -\frac{1}{256}\int\frac{\mathrm{d}^4p}{(2\pi)^4}\int\frac{\mathrm{d}^\delta\ell}{(2\pi)^\delta}\\
&\times\frac{(2p-k_1)_\mu(2p-2k_1-k_2)_{\mu'}}{(p^2-\ell^2)\left[(p-k_1)^2-\ell^2\right]\left[(p-q)^2-\ell^2\right]}\\
&\times\underbrace{\mathrm{Tr}\left[(\not p+\not\ell)\gamma_\nu(\not p+\not\ell-\not k_1)\gamma_{\nu'}(\not p+\not\ell-\not q)\not\ell\gamma_5\right]}_{2^{2+\frac{\delta}{2}}ik_1^\alpha k_2^\beta\varepsilon_{\nu\alpha\nu'\beta}\,\ell^2}. \qquad (7.188)
\end{aligned}$$

The Feynman parametrization leads to

$$\begin{aligned}
\widehat{\mathsf{T}}_{\mu\nu\mu'\nu'}(k_1,k_2) &= -i\frac{2^{\frac{\delta}{2}}}{32}k_1^\alpha k_2^\beta\varepsilon_{\nu\alpha\nu'\beta}\int\frac{\mathrm{d}^4p}{(2\pi)^4}\int\frac{\mathrm{d}^\delta\ell}{(2\pi)^\delta}\int_0^1\mathrm{d}x \qquad (7.189)\\
\times\int_0^{1-x}\mathrm{d}y&\frac{(2p-k_1)_\mu(2p-2k_1-k_2)_{\mu'}}{\left\{\left[(p-k_1)^2-\ell^2\right]x+\left[(p-q)^2-\ell^2\right]y+(p^2-\ell^2)(1-x-y)\right\}^3}\ell^2.
\end{aligned}$$

Making the shift $p \to p + xk_1 + yq$ and few algebraic manipulations, Eq. (7.189) becomes

$$\widehat{\mathsf{T}}_{\mu\nu\mu'\nu'}(k_1, k_2) = -i\,\frac{2^{\frac{\delta}{2}}}{32}\,k_1^\alpha k_2^\beta \varepsilon_{\nu\alpha\nu'\beta} \int_0^1 \mathrm{d}x \int_0^{1-x} \mathrm{d}y \int \frac{\mathrm{d}^4 p}{(2\pi)^4}$$
$$\times \int \frac{\mathrm{d}^\delta \ell}{(2\pi)^\delta} \frac{(2p + 2xk_1 + 2yq - k_1)_\mu (2p + 2xk_1 + 2yq - 2k_1 - k_2)_{\mu'}}{\left[p^2 - \ell^2 + A(x, y)\right]^3} \ell^2. \tag{7.190}$$

where $A(x, y) = (x + y)(1 - x - y)k_1^2 + y(1 - y)k_2^2 + 2y(1 - x - y)k_1 \cdot k_2$. Expanding the numerator of (7.190), using $p_\mu p_{\mu'} = \eta_{\mu\nu'} \frac{p^2}{4}$, and keeping only the non-vanishing terms, Eq. (7.190) becomes

$$\widehat{\mathsf{T}}_{\mu\nu\mu'\nu'}(k_1, k_2) = -i\,\frac{2^{\frac{\delta}{2}}}{32}\,k_1^\alpha k_2^\beta \varepsilon_{\nu\alpha\nu'\beta} \int_0^1 \mathrm{d}x \int_0^{1-x} \mathrm{d}y \int \frac{\mathrm{d}^4 p}{(2\pi)^4}$$
$$\times \int \frac{\mathrm{d}^\delta \ell}{(2\pi)^\delta} \frac{p^2 \eta_{\mu\mu'} + 4y(x + y - 1)k_{1\mu'} k_{2\mu}}{\left[p^2 - \ell^2 + A(x, y)\right]^3} \ell^2. \tag{7.191}$$

To make sense of the integrals present in (7.191), we make as usual a Wick rotation $k^0 \to i k_E^0$ for any momentum k^μ, as well as $p^0 \to i p_E^0$: so, for instance, in the previous integral $p^2 \to -p_E^2$, etc., and, as usual, we dispense from explicitly indicating the Euclidean label $_E$. So (7.191) is replaced by

$$\widehat{\mathsf{T}}_{\mu\nu\mu'\nu'}(k_1, k_2) = \frac{2^{\frac{\delta}{2}}}{32}\,k_1^\alpha k_2^\beta \varepsilon_{\nu\alpha\nu'\beta} \int_0^1 \mathrm{d}x \int_0^{1-x} \mathrm{d}y \int \frac{\mathrm{d}^4 p}{(2\pi)^4}$$
$$\times \int \frac{\mathrm{d}^\delta \ell}{(2\pi)^\delta} \frac{p^2 \eta_{\mu\mu'} - 4y(x + y - 1)k_{1\mu'} k_{2\mu}}{\left[p^2 + \ell^2 + A(x, y)\right]^3} \ell^2. \tag{7.192}$$

Now the integrals are well-defined, and we can use the Euclidean integrals collected in Appendix 7A. Using them and performing the integration over the Feynman parameters (x, y) and returning to the Lorentzian metric, one obtains

$$\widehat{\mathsf{T}}_{\mu\nu\mu'\nu'}(k_1, k_2) = -\frac{i}{6144\pi^2} k_1^\alpha k_2^\beta \varepsilon_{\nu\nu'\alpha\beta} \left(\eta_{\mu\mu'} \left(k_1^2 + k_2^2 + k_1 \cdot k_2\right) - k_{1\mu'} k_{2\mu}\right). \tag{7.193}$$

Of course, as previously mentioned, one should symmetrize expression (7.193) with respect to $(\mu \leftrightarrow \nu)$ and $(\mu' \leftrightarrow \nu')$. Then, (7.193) becomes

$$\widehat{\mathsf{T}}_{\mu\nu\mu'\nu'}(k_1, k_2) = -\frac{i}{6144\pi^2} k_1^\alpha k_2^\beta \left(\left(k_1^2 + k_2^2 + k_1 \cdot k_2\right) t_{\mu\nu\mu'\nu'\alpha\beta} - t^{(21)}_{\mu\nu\mu'\nu'\alpha\beta}\right), \tag{7.194}$$

The tensors t and $t^{(21)}$ have been defined in (7.168) and (7.169).

Finally, we have to add the cross graph contribution, obtained by $k_1, \mu, \nu \leftrightarrow k_2, \mu', \nu'$. Under this exchange, the t tensors transform as follows:

$$t \leftrightarrow -t, \quad t^{(21)} \leftrightarrow -t^{(21)}, \tag{7.195}$$

Hence, the sum of (7.194) with the cross diagram contribution gives rise to

$$\widehat{\mathsf{T}}^{(\text{tot})}_{\mu\nu\mu'\nu'}(k_1, k_2) = -\frac{i}{3072\pi^2} k_1^\alpha k_2^\beta \left(\left(k_1^2 + k_2^2 + k_1 \cdot k_2\right) t_{\mu\nu\mu'\nu'\alpha\beta} - t^{(21)}_{\mu\nu\mu'\nu'\alpha\beta} \right). \tag{7.196}$$

To simplify the derivation, we set the external lines on shell, $k_1^2 = k_2^2 = 0$. This requires a comment.

On Shell Conditions

Putting the external lines on shell means that the corresponding fields have to satisfy the eom of Einstein-Hilbert gravity $R_{\mu\nu} = 0$. In the linearized form, this means

$$\Box \chi_{\mu\nu} = \partial_\mu \partial_\lambda \chi_\nu^\lambda + \partial_\nu \partial_\lambda \chi_\mu^\lambda - \partial_\mu \partial_\nu \chi_\lambda^\lambda = 0 \tag{7.197}$$

We also choose the De Donder gauge: $\Gamma^\lambda_{\mu\nu} g^{\mu\nu} = 0$, which at the linearized level becomes $2\partial_\mu \chi_\lambda^\mu - \partial_\lambda \chi_\mu^\mu = 0$. In this gauge, (7.197) becomes

$$\Box \chi_{\mu\nu} = 0 \tag{7.198}$$

In momentum space, this implies that $k_1^2 = k_2^2 = 0$. We remark that this is not an *ad hoc* trick to arrive at the result. We are not simply disrupting the cohomology, but we are in fact defining a restricted cohomology of the diffeomorphisms and the Weyl transformations: a cohomology defined up to terms $\Box h_{\mu\nu}$ and $\Box \xi^\mu$. This is a well-defined cohomology, under which we have, in particular,

$$\delta_\xi \left(2\partial_\mu \chi_\lambda^\mu - \partial_\lambda \chi_\mu^\mu\right) = 2 \, \Box \xi_\lambda \approx 0 \tag{7.199}$$

i.e. in this restricted cohomology the De Donder gauge fixing is irrelevant. Similarly, the term corresponding in momentum space to $k_1^\alpha k_2^\beta \, (k_1^2 + k_2^2) t_{\mu\nu\mu'\nu'\alpha\beta}$ remains null after a restricted diffeomorphism transformation. The restricted cohomology has the same odd class (the Pontryagin one) as the unrestricted one; i.e. it completely determines it (this is not true for the even classes). Since we know that the final result must be covariant and that there is no covariant extension to all order of the term $k_1^\alpha k_2^\beta \, (k_1^2 + k_2^2) t_{\mu\nu\mu'\nu'\alpha\beta}$, the simplification of considering it null does not jeopardize it. This means that this term must be superfluous in some way. We will comment on this below.

Overall Contribution

The overall one-loop contribution to the trace anomaly in momentum space, *as far as the parity violating part is concerned*, is given by (7.196). After returning to the Minkowski metric and Fourier-antitransforming it, we can extract the local expression of the trace anomaly, by replacing the results found so far in (4.18). The result, to lowest order, is

$$\langle\!\langle T^{\mu}_{\mu}(x)\rangle\!\rangle^{(odd)} \approx -\frac{i}{768\pi^2}\varepsilon^{\mu\nu\lambda\rho}\left(\partial_\mu\partial_\sigma h^{\tau}_{\nu}\,\partial_\lambda\partial_\tau h^{\sigma}_{\rho} - \partial_\mu\partial_\sigma h^{\tau}_{\nu}\,\partial_\lambda\partial^{\sigma} h_{\tau\rho}\right) \quad (7.200)$$

The factor in front is due to a factor of 8 that comes from the coefficient $\frac{1}{2^n n!}$ with $n = 2$ in the denominator and a factor of 8 in the numerator because the Fourier transform of the three-point function of the e.m. tensor is 8 times $\widetilde{\mathcal{T}}_{\mu\nu\mu'\nu'\alpha\beta}(k_1, k_2)$; another factor of 4 in the numerator is due to the symmetry of the tensors t and $t^{(21)}$ in (7.196), which yields four times the same term. Comparing with the expansion

$$\varepsilon^{\mu\nu\lambda\rho}R_{\mu\nu}{}^{\sigma\tau}R_{\lambda\rho\sigma\tau} = \varepsilon^{\mu\nu\lambda\rho}\left(\partial_\mu\partial_\sigma\chi^{a}_{\nu}\,\partial_\lambda\partial_a\chi^{\sigma}_{\rho} - \partial_\mu\partial_\sigma\chi^{a}_{\nu}\,\partial_\lambda\partial^{\sigma}\chi_{a\rho}\right) + \ldots \quad (7.201)$$

we obtain

$$\langle\!\langle T^{\mu}_{\mu}(x)\rangle\!\rangle^{(odd)} = -\frac{i}{768\pi^2}\,\frac{1}{2}\,\varepsilon^{\mu\nu\lambda\rho}R_{\mu\nu}{}^{\sigma\tau}R_{\lambda\rho\sigma\tau} \quad (7.202)$$

Now applying the definition (7.1) and recalling (7.181), we obtain the covariant expression of the parity violating part of the trace anomaly

$$T[g](x) = \frac{i}{768\pi^2}\,\frac{1}{2}\,\varepsilon^{\mu\nu\lambda\rho}R_{\mu\nu}{}^{\sigma\tau}R_{\lambda\rho\sigma\tau}. \quad (7.203)$$

7.4.3 The KDS Anomaly

Consider a Dirac fermion ψ coupled to a metric and vector potential V_μ. The action is

$$S = \int d^4x\,\sqrt{g}\,i\overline{\psi}\gamma^{\mu}(\partial_\mu - iV_\mu + \frac{1}{2}\omega_\mu)\psi \quad (7.204)$$

where the remaining symbols are the same as in (7.157). It is invariant under the Abelian local transformation $\psi \to e^{i\gamma_5\eta}\psi$ and $V_\mu \to V_\mu + \gamma_5\partial_\mu\eta$. The corresponding axial current $j_{5\mu} = \overline{\psi}\gamma_\mu\gamma_5\psi$ is classically conserved. Is it conserved in the quantum theory? Or is it violated due to the gravitational interaction? To answer this question, one has to consider the amplitude $\langle\partial\cdot j_5\, T\, T\rangle$. The amplitude in question is the same as (7.183) with the factor $(2\not{p} - \not{q})$ replaced by $\not{q}$ and P_R replaced by γ_5, and an overall coefficient $\frac{1}{64}$ instead of $\frac{1}{256}$. If we rewrite $\not{q} = 2\not{p} - (2\not{p} - \not{q})$, we see that the second term reproduces the calculation of the previous subsection multiplied by a suitable coefficient, while the term corresponding to $2\not{p}$, once regularized, is easily seen to vanish (an explicit proof can be found in Appendix 7F). Therefore, we can conclude that

$$\partial^\mu j_{5\mu} = \frac{1}{768\pi^2}\varepsilon^{\mu\nu\lambda\rho} R_{\mu\nu}{}^{\sigma\tau} R_{\lambda\rho\sigma\tau}. \tag{7.205}$$

This is the Kimura-Delbourgo-Salam anomaly, [10–12]. It is an anomaly of the ABJ type, which means that the problem could be better formulated in terms of an axial potential A_μ, in the combination $V_\mu + \gamma_5 A_\mu$, endowed with the transformation property $A_\mu \to A_\mu + \partial_\mu \eta$, and eventually setting $A_\mu = 0$. We recall that by the same procedure one can derive the original ABJ (gauge) anomaly, the only difference being that the ABJ anomaly is expressed in terms of V_μ instead of the metric.

The previous calculation of the KDS anomaly is another example of the rigid link that connects chiral ABJ type anomalies to odd parity trace anomalies, as already pointed out at the beginning of this chapter and explicitly shown in Sect. 7.3.5.

7.4.4 Consistent Mixed Anomaly

The previous remark gives us the opportunity to compute a consistent mixed gauge-gravity anomaly. Consider the action

$$S = \int d^4x\, \sqrt{g}\, i\overline{\psi}_R \gamma^\mu (\partial_\mu - iV_\mu + \frac{1}{2}\omega_\mu)\psi_R \tag{7.206}$$

and its invariance under $V_\mu \to V_\mu + \partial_\mu \lambda$. The relevant current $j_\mu = \bar{\psi}_R \gamma_\mu \psi_R$ is conserved and to find the corresponding anomaly, if any, we have to evaluate the amplitude $\langle \partial \cdot j_R\, T_R\, T_R \rangle$. The latter is the same as (7.183) with the factor $(2\not{p} - \not{q})$ replaced by $\not{q}$. Therefore, the final result can be easily extracted from the previous calculation

$$\partial^\mu j_{R\mu} = \frac{1}{1536\pi^2}\varepsilon^{\mu\nu\lambda\rho} R_{\mu\nu}{}^{\sigma\tau} R_{\lambda\rho\sigma\tau}. \tag{7.207}$$

Its integrated form is $\sim \Delta_G(\lambda, g) = \int d^4x \lambda\, \varepsilon^{\mu\nu\lambda\rho} R_{\mu\nu}{}^{\sigma\tau} R_{\lambda\rho\sigma\tau}$, which is a diffeomorphism invariant (trivially) consistent Abelian gauge cocycle. This cocycle can take different forms, as explained in Sect. 5.1. Anticipating the conclusions of Sect. 11.6.3, $\Delta_G(\lambda, g)$ is equivalent to

$$\Delta_d(\xi, g, V) = \int d^4x \varepsilon^{\mu\nu\lambda\rho} \mathrm{tr}\left(\partial_\mu \Xi\, \Gamma_\nu\right) F_{\lambda\rho} \tag{7.208}$$

where $F_{\mu\nu} = \partial_\mu V_\nu - \partial_\nu V_\mu$, Ξ represents the matrix $\Xi_\tau{}^\sigma = \partial_\tau \xi^\sigma$ and Γ_μ represents the matrix $\Gamma^\tau_{\mu\sigma}$. They can be obtained from each other by subtracting a suitable conterterm (see Sect. 11.6.3)

This anomaly can also take the form of a (gauge+diffeomorphism invariant) Lorentz anomaly. As will be explained in Chap. 11 the reason is that all these cocycles descend from the same 6-form $\mathrm{tr}(RR)F$ of the classifying space of the group $SO(4) \times U(1)$.

7.4.5 An Unexpected Obstacle

To end this section, we would like to comment about the reduced cohomology we have used above in order to derive the previous results, that is the choice of putting on shell the external graviton legs [introduced after Eq. (7.196)]. The involved term of the triangle diagram is

$$\sim k_2^\beta \left(k_1^2 + k_2^2\right) t_{\mu\nu\mu'\nu'\alpha\beta} \tag{7.209}$$

Inserted in the formula for the effective action this gives rise to the cocycle Δ_ω

$$\sim \int \mathrm{d}^4x\, \omega\, \varepsilon^{\mu\nu\lambda\rho} \partial_\mu \Box h_\nu^\alpha \partial_\lambda h_{\rho\alpha} \tag{7.210}$$

This is a consistent Weyl cocycle (at the lowest significant order), but it is not invariant under a diffeomorphism transformation. On the other hand, the careful scrutiny carried out in Appendix 7C and 7D shows that the triangle result (7.196) is not corrected by other contributions. One possibility would be that the term (7.209) is canceled by a contribution from the first term in (7.1). But this is not possible because we know on a general ground that this term vanishes. On the basis of the lesson learnt in Sect. 5.1, we would expect therefore a companion diffeomorphism cocycle Δ_ξ such that $\delta_\xi \Delta_\omega + \delta_\omega \Delta_\xi = 0$. But in Sect. 7.4.1 and Appendix 7D, we have seen that, at the lowest order, diffeomorphisms are conserved. We can consider the problem also from another viewpoint. We can cancel the cocycle (7.210) by subtracting from the effective action a counterterm

$$\sim \int \mathrm{d}^4x\, h\, \varepsilon^{\mu\nu\lambda\rho} \partial_\mu \Box h_\nu^\alpha \partial_\lambda h_{\rho\alpha} \tag{7.211}$$

where $h = h_\mu^\mu$. But this counterterm is not invariant under diffeomorphisms, therefore it would generate a diffeomorphism anomaly. This means that at the lowest perturbative order the cross consistency conditions for diffeomorphisms and Weyl transformations are not satisfied.

In sum, we are facing a puzzle. What we have just seen is that the consistency conditions are not satisfied at the lowest perturbative order. What is the solution? Well, for the first time in this book we have to surrender to the evidence that this problem is *undecidable* at the lowest perturbative order. Let us see why.

So far we have been using a precise regularization scheme, the rightmost γ_5 scheme. One may wonder whether changing scheme the problem might disappear. However, a moment's reflection leads us to the conclusion that this is not the root of the puzzle. To this end, let us remark in the following: the vanishing of the odd e.m. divergence, as shown in Sect. 7.4.1 and Appendix 7D, is an obvious result, which we have shown explicitly only for the sake of pedagogy, but could have assumed as true without any calculation. In fact, as we have already pointed out, the conformal odd three-point function of the e.m. tensor vanishes for algebraic reasons. As we have already stressed the regularization of the divergence, very differently from the regularization of the trace, does not lead to any surprise: in particular, if the correlator vanishes, its divergence vanishes too, no matter what regularization scheme we use. So we may change regularization scheme but the problem remains: we do not report explicit calculations, but with a different regularization the form of trace anomaly in general changes (for instance, we may get a nonzero output from bubble diagrams), but the divergence of the three-point e.m. tensor correlator remains zero.

The origin of the puzzle is not the regularization scheme, but the accidental vanishing of the odd three-point function of the e.m. tensor, [13]. We have already seen several examples (and more we will see in the sequel) where, as a result of an explicit calculation, nonzero trace anomalies and nonzero diffeomorphism anomalies appear in couples, and it often happens that by subtracting a counterterm we can recover diffeomorphism invariance and modify the trace anomaly to a minimal form. In other words, diffeomorphisms play a 'repairing' or 'stabilizing' role in cohomology; i.e. the diffeomorphism cohomology does accompany the regularization scheme in such a way that the latter preserves the cohomology class. However, this role can be effective only if the relevant amplitude is not identically vanishing, which is not what happens in our puzzling case.

The next question is: Does that mean that the perturbative calculation of the trace anomaly is impossible? The answer is: no, it is only more difficult. In our particular case, the problem arises from the vanishing of the odd three-point function of the e.m. tensor. However the three-point function corresponds to the lowest possible order yielding a significant contribution to the calculation of the trace anomaly. But of course one should consider also the four-point function, the five-point function, and so on. There is no such accidental vanishing for the odd four-point function and, in general, for the higher order functions. Therefore, we should calculate the odd four-point function of the energy-momentum tensor and compute both the trace and the divergence in the same way we have done for the three-point function. In this way, the stabilizing effect of diffeomorphisms (together with the possible contribution of other graphs, such as the bubble ones) would unfold undisturbed. It is clear that in order to concretely solve the puzzle and master this singular case we have to find an accessible method that, on the one hand, guarantees diffeomorphism invariance, on the other hand, puts us in the condition to effectively evaluate a sufficient number of approximants, so that the cohomologically 'repairing' effect of diffeomorphisms can operate. The trouble here is the technical complexity. The difficulties in this direction advise us to take another path and to turn to a non-perturbative method. Such a non-perturbative method exists, it is the Seeley-Schwinger-DeWitt method: the diffeomorphism invariance is inbuilt in it and, being non-perturbative, it encompasses all the relevant higher order amplitudes. In Chap. 10, it will be applied precisely to the problem of the odd trace anomaly in 4d, and it will confirm the results obtained above. Until then the result (7.203) remains *sub judice*.

Having come to this conclusion we should reconsider also the gauge induced odd parity trace anomaly (7.137), computed in Sect. 7.3.3. We remarked there that any odd parity correlator of the type $\langle TJJ \rangle$ vanishes in conformal field theory for an algebraic impossibility to be constructed. It follows that the divergence $\langle \partial \cdot T\, J\, J \rangle$ vanishes, no matter what regularization scheme we use. Therefore, also in that case the repairing mechanism of diffeomorphisms cannot be at work. Although in that case there is no need to restrict the cohomology (no residual coboundary to be subtracted), it remains that the result (7.137) must be confirmed by an independent calculation. This is what we will do in Sect. 10.4.8.

Remark It is worth pointing out that a missing contribution from the perturbative calculation of an anomaly, such as the one we have come across above, accompanied by a repairing role of a companion symmetry, is not unique. Let us consider, for instance, the (multiplet) non-Abelian covariant anomaly $\sim \varepsilon^{\mu\nu\lambda\rho}\mathrm{tr}(T^a F_{\mu\nu} F_{\lambda\rho})$ we encountered in Sect. 6.3. This anomaly contains a quartic term in the potential $V_\mu = V_\mu^a T^a$, which can come only from a pentagon diagram. This diagram however is UV convergent. Therefore, the quartic term cannot be produced through a perturbative calculation. It is nevertheless required by the conservation of the vector current in order to guarantee the invariance of the vector gauge symmetry (which plays a role analogous to the diffeomorphisms in solving the above puzzle).

7.4.6 Regularization Scheme, Cohomology and Exceptional Cases

To end this chapter, it is opportune to dwell on the conclusions of the previous subsection and highlight what is a piece of wisdom gained in this book. To be honest, we have been lucky obtaining at the first strike the result (7.203) for the trace anomaly. This is the right result, as will be confirmed by the nonperturbative calculation of Chap. 10. But, had we adopted another regularization scheme instead of the rightmost γ_5, for instance the one extensively used in the previous chapter, at the lowest order we would have likely obtained a result different from (7.196). Does that mean that one regularization scheme is 'correct' and the other is 'incorrect'? No, *there is no incorrect regularization scheme as long as the result we obtain satisfies the consistency conditions*. Let us expand on this. In general, we have to consider two or more intertwined symmetries (see Chap. 5), for instance, Weyl, diffeomorphism and gauge, or, vector and axial gauge symmetries plus diffeomorphism. If we compute trace anomalies within a given regularization scheme the first crucial test consists in verifying that the result satisfies the lowest order Weyl consistency conditions. This must be so if the calculations are correctly carried out. The result however may violate the other symmetry, in our case the diffeomorphisms (assuming that gauge symmetry is absent or satisfied). If this is the case we have to compute the divergence of the energy-momentum tensor and find the associated anomaly. If the calculations are correctly carried out, the Weyl and diffeomorphism cocycles will satisfy the (lowest order) coupled consistency conditions. In general, a suitable counterterm in the effective action will cancel one of the two anomalies by modifying the other, as we have seen and will see in many examples. As for the trace anomalies in 4d considered above, we expect the diffeomorphisms to be anomaly free and the trace anomaly to take the form of the Pontryagin density. However, we repeat, at the lowest order we cannot expect, in general, to find the lowest order of the Pontryagin density (with the right coefficient) together with conserved diffeomorphisms. As we have pointed out before, this is due to the looser structure of the lowest order cohomology with respect to the full cohomology: at the lowest order there are more cocycles than in the full cohomology. The lowest order perturbative result, if the calculations are correct, will coincide with one cocycle of the lowest order cohomology. In what cocycle, will depend on the regularization scheme one uses. But, whatever the regularization scheme is, using cohomology (i.e. counterterms) it must always be possible to reach the final covariant result. In this sense, we have used above the expression 'stabilizing effect of the diffeomorphisms' on the trace anomaly.

The one just outlined is the most favorable (and generic) situation, where a slogan frequently met in the literature makes sense: if a theory is anomaly free in a particular regularization (or regularization scheme), it is anomaly free for all of them (but one had better specify: if all the gauge and diffeomorphism symmetries of the theory are simultaneously anomaly free). However, we have seen that this regular pattern may not persist in some specific cases, let us call them *exceptional*. If, like in our example above, a crucial amplitude vanishes for some accidental reason extraneous to cohomology, the compensating mechanism between Weyl and diffeomorphisms cocycles breaks down, at least at the lowest order of approximation. In this case, as we have seen, there are two ways out: the first is to go to the next order of approximation, which may be technically impervious, the second is to use a nonperturbative method that guarantees invariance under all symmetries except one (in our case invariance under diffeomorphisms and, possibly, gauge transformations, but not under Weyl transformations). Without a confirmation from one of these two methods, the anomalous results obtained at the lowest order of approximation are not reliable.

There is a lesson to be learned from all this about the reliability of perturbative calculations of anomalies. It is something already implicit in what precedes, but it is time to highlight it. We will start with an example. The lowest order results in the calculation of the trace anomaly are often such that we can cancel them by adding local counterterms to the effective action. Therefore, one might conclude that the trace anomaly vanishes. These terms however usually break other symmetries (typically, the diffeomorphisms), so that the addition is not permitted. This is an example of the factors that may affect the perturbative results. That is, our perturbative results to lowest order may be affected by the following factors: choice of the regularization scheme and possibility to add local counterterms permitted by cohomology, appropriate or inappropriate definition of the trace anomaly (with reference to formula

(7.1)), exceptional cases. All together they form a tangle of problems not easy and in some cases (as we have seen) even impossible to resolve at lowest order: the fact that (7.203) turns out to be the right result at the lowest order of approximation should be better attributed to serendipity. So a provoking counterpoint to the above slogan could be: the absence of anomalies for a particular symmetry and a particular regularization scheme, at the lowest perturbative order, does not mean anything.[8] Fortunately, the situation is not so grim and in most cases perturbative calculations (with all the guarantees for all the remaining symmetries) are reliable. But the problem, in particular for the trace anomalies, remains and is a basic one. It forms a structural part of the anomaly calculus. The piece of wisdom alluded to above is: lowest order perturbative calculations of the trace anomalies must not be trusted at face value, but must be interpreted with great care.

Appendix 7A: Regularization formulas in 2d and 4d

In this appendix, we collect the regularized integrals that are needed to evaluate the Feynman diagrams in the text both in 2d and 4d. The integrals below are *Euclidean integrals*. They are intermediate results. Since the starting points and the final results are Lorentzian, it is understood that one has to do the appropriate Wick rotations in order to be able to use them. In the formulas below, we have reinserted the scale parameter μ.

In 2d, after introducing δ extra dimensions in the internal momentum and a Feynman parameter u ($0 \leq u \leq 1$), in the limit $\delta \to 0$, we have

$$\begin{aligned}\int \frac{d^2p}{(2\pi)^2}\int \frac{d^\delta \ell}{(2\pi)^\delta}\frac{\ell^2}{(p^2+\ell^2+\Delta)^2} &= -\frac{1}{4\pi}\\ \int \frac{d^2p}{(2\pi)^2}\int \frac{d^\delta \ell}{(2\pi)^\delta}\frac{\ell^2 p^2}{(p^2+\ell^2+\Delta)^2} &= \frac{1}{4\pi}\Delta\end{aligned} \tag{7.212}$$

and

$$\begin{aligned}\int \frac{d^2p}{(2\pi)^2}\int \frac{d^\delta \ell}{(2\pi)^\delta}\frac{1}{(p^2+\ell^2+\Delta)^2} &= \frac{1}{4\pi}\frac{1}{\Delta}\\ \int \frac{d^2p}{(2\pi)^2}\int \frac{d^\delta \ell}{(2\pi)^\delta}\frac{p^2}{(p^2+\ell^2+\Delta)^2} &= \frac{1}{4\pi}\left(-\frac{2}{\delta}-\gamma-\log\frac{\Delta}{4\pi\mu^2}\right)\\ \int \frac{d^2p}{(2\pi)^2}\int \frac{d^\delta \ell}{(2\pi)^\delta}\frac{p^4}{(p^2+\ell^2+\Delta)^2} &= \frac{1}{2\pi}\Delta\left(\frac{2}{\delta}-1+\gamma+\log\frac{\Delta}{4\pi\mu^2}\right)\end{aligned} \tag{7.213}$$

where $\Delta = u(1-u)k^2$.

[8] For instance, multiplying (7.200) by $h = h_\mu^\mu$ and integrating over spacetime gives rise to a counterterm that kills the lowest order anomaly itself. But this is a higher order term that violates diffeomorphisms.

Proceeding in the same way in 4d, with two Feynman parameters u and v, in the limit $\delta \to 0$, we find

$$\int \frac{\mathrm{d}^4 p}{(2\pi)^4} \int \frac{\mathrm{d}^\delta \ell}{(2\pi)^\delta} \frac{1}{(p^2+\ell^2+\Delta)^2} = \frac{1}{(4\pi)^2}\left(-\frac{2}{\delta} - \gamma - \log\frac{\Delta}{4\pi\mu^2}\right)$$
$$\int \frac{\mathrm{d}^4 p}{(2\pi)^4} \int \frac{\mathrm{d}^\delta \ell}{(2\pi)^\delta} \frac{p^2}{(p^2+\ell^2+\Delta)^2} = \frac{\Delta}{8\pi^2}\left(\frac{2}{\delta} + \gamma - 1 + \log\frac{\Delta}{4\pi\mu^2}\right) \quad (7.214)$$

and

$$\int \frac{\mathrm{d}^4 p}{(2\pi)^4} \int \frac{\mathrm{d}^\delta \ell}{(2\pi)^\delta} \frac{\ell^2}{(p^2+\ell^2+\Delta)^2} = \frac{1}{(4\pi)^2}\Delta.$$

Next

$$\int \frac{\mathrm{d}^4 p}{(2\pi)^4} \int \frac{\mathrm{d}^\delta \ell}{(2\pi)^\delta} \frac{1}{(p^2+\ell^2+\Delta)^3} = \frac{1}{2(4\pi)^2}\frac{1}{\Delta}$$
$$\int \frac{\mathrm{d}^4 p}{(2\pi)^4} \int \frac{\mathrm{d}^\delta \ell}{(2\pi)^\delta} \frac{p^2}{(p^2+\ell^2+\Delta)^3} = \frac{1}{(4\pi)^2}\left(-\frac{2}{\delta} - \gamma - \log\frac{\Delta}{4\pi\mu^2}\right)$$
$$\int \frac{\mathrm{d}^4 p}{(2\pi)^4} \int \frac{\mathrm{d}^\delta \ell}{(2\pi)^\delta} \frac{p^4}{(p^2+\ell^2+\Delta)^3} = \frac{\Delta}{2(4\pi)^2}\left(-\frac{2}{\delta} - \gamma + 4 - \log\frac{\Delta}{4\pi\mu^2}\right)$$
$$(7.215)$$

and

$$\int \frac{\mathrm{d}^4 p}{(2\pi)^4} \int \frac{\mathrm{d}^\delta \ell}{(2\pi)^\delta} \frac{\ell^2}{(p^2+\ell^2+\Delta)^3} = -\frac{1}{2(4\pi)^2}$$
$$\int \frac{\mathrm{d}^4 p}{(2\pi)^4} \int \frac{\mathrm{d}^\delta \ell}{(2\pi)^\delta} \frac{\ell^2 p^2}{(p^2+\ell^2+\Delta)^3} = \frac{1}{(4\pi)^2}\Delta$$
$$\int \frac{\mathrm{d}^4 p}{(2\pi)^4} \int \frac{\mathrm{d}^\delta \ell}{(2\pi)^\delta} \frac{\ell^4}{(p^2+\ell^2+\Delta)^3} = \frac{1}{2(4\pi)^2}\Delta \quad (7.216)$$

where $\Delta = u(1-u)k_1^2 + v(1-v)k_2^2 + 2uv\, k_1 k_2$, with $0 \le u \le 1$ and $0 \le v \le 1-u$.

Appendix 7B: The $\widetilde{\mathfrak{T}}_{\mu\nu\lambda\rho}(\mathsf{k})$ Correlator

Here we record the full expression $\widetilde{\mathfrak{T}}_{\mu\nu\lambda\rho}(\mathsf{k})$ in terms of the Euclidean momentum (in 2d). Let us set $a(\delta, \mathsf{k}) = \frac{2}{\delta} + \gamma + \log\left(\frac{\mathsf{k}^2}{2\pi\mu^2}\right)$ and separate even and odd parts,

$$\widetilde{\mathfrak{T}}_{\mu\nu\lambda\rho}(\mathsf{k}) = \widetilde{\mathfrak{T}}^{(\mathrm{even})}_{\mu\nu\lambda\rho}(\mathsf{k}) + \widetilde{\mathfrak{T}}^{(odd)}_{\mu\nu\lambda\rho}(\mathsf{k}).$$

Then

$$768\pi\,\widetilde{\mathcal{T}}^{(\text{even})}_{\mu\nu\lambda\rho}(\mathsf{k})=\left(a(\delta,\mathsf{k})-\frac{8}{3}\right)\Big[-\left(\mathsf{k}^2\eta_{\mu\nu}\eta_{\lambda\rho}+\eta_{\mu\nu}\mathsf{k}_\lambda\mathsf{k}_\rho+\eta_{\lambda\rho}\mathsf{k}_\mu\mathsf{k}_\nu\right) \\ +\frac{3}{4}\left(\eta_{\mu\lambda}\mathsf{k}_\nu\mathsf{k}_\rho+\eta_{\nu\lambda}\mathsf{k}_\mu\mathsf{k}_\rho+\eta_{\mu\rho}\mathsf{k}_\nu\mathsf{k}_\lambda+\eta_{\nu\rho}\mathsf{k}_\mu\mathsf{k}_\lambda\right)\Big] \\ -\frac{1}{4}\left(a(\delta,\mathsf{k})-\frac{14}{3}\right)\left(\eta_{\mu\lambda}\mathsf{k}_\nu\mathsf{k}_\rho+\eta_{\nu\lambda}\mathsf{k}_\mu\mathsf{k}_\rho+\eta_{\mu\rho}\mathsf{k}_\nu\mathsf{k}_\lambda+\eta_{\nu\rho}\mathsf{k}_\mu\mathsf{k}_\lambda\right) \\ +\frac{1}{2}\left(a(\delta,\mathsf{k})-\frac{5}{3}\right)\mathsf{k}^2\left(\eta_{\mu\lambda}\eta_{\nu\rho}+\eta_{\nu\lambda}\eta_{\mu\rho}\right) \\ +\frac{2}{\mathsf{k}^2}\mathsf{k}_\mu\mathsf{k}_\nu\mathsf{k}_\lambda\mathsf{k}_\rho+\frac{1}{4}\left(\eta_{\mu\lambda}\mathsf{k}_\nu\mathsf{k}_\rho+\eta_{\nu\lambda}\mathsf{k}_\mu\mathsf{k}_\rho+\eta_{\mu\rho}\mathsf{k}_\nu\mathsf{k}_\lambda+\eta_{\nu\rho}\mathsf{k}_\mu\mathsf{k}_\lambda\right) \tag{7.217}$$

and

$$768\pi\,\widetilde{\mathcal{T}}^{(odd)}_{\mu\nu\lambda\rho}(\mathsf{k})=-\frac{1}{4}\left(a(\delta,\mathsf{k})-\frac{8}{3}\right)\Big[\eta_{\mu\nu}\mathsf{k}_\lambda\varepsilon_{\rho\sigma}\mathsf{k}^\sigma+\eta_{\mu\nu}\mathsf{k}_\rho\varepsilon_{\lambda\sigma}\mathsf{k}^\sigma+\eta_{\lambda\rho}\mathsf{k}_\mu\varepsilon_{\nu\sigma}\mathsf{k}^\sigma \\ +\eta_{\lambda\rho}\mathsf{k}_\nu\varepsilon_{\mu\sigma}\mathsf{k}^\sigma+\frac{1}{2}\big(\eta_{\mu\lambda}\mathsf{k}_\nu\varepsilon_{\rho\sigma}\mathsf{k}^\sigma+\eta_{\mu\rho}\mathsf{k}_\nu\varepsilon_{\lambda\sigma}\mathsf{k}^\sigma+\eta_{\nu\lambda}\mathsf{k}_\mu\varepsilon_{\rho\sigma}\mathsf{k}^\sigma+\eta_{\nu\rho}\mathsf{k}_\mu\varepsilon_{\lambda\sigma}\mathsf{k}^\sigma \\ +\eta_{\mu\lambda}\mathsf{k}_\rho\varepsilon_{\nu\sigma}\mathsf{k}^\sigma+\eta_{\lambda\nu}\mathsf{k}_\rho\varepsilon_{\mu\sigma}\mathsf{k}^\sigma+\eta_{\rho\nu}\mathsf{k}_\lambda\varepsilon_{\nu\sigma}\mathsf{k}^\sigma+\eta_{\nu\rho}\mathsf{k}_\lambda\varepsilon_{\mu\sigma}\mathsf{k}^\sigma\big)\Big] \\ +\frac{1}{4}\left(a(\delta,\mathsf{k})-\frac{5}{3}\right)\Big[\eta_{\mu\lambda}\mathsf{k}_\nu\varepsilon_{\rho\sigma}\mathsf{k}^\sigma+\eta_{\mu\rho}\mathsf{k}_\nu\varepsilon_{\lambda\sigma}\mathsf{k}^\sigma+\eta_{\nu\lambda}\mathsf{k}_\mu\varepsilon_{\rho\sigma}\mathsf{k}^\sigma+\eta_{\nu\rho}\mathsf{k}_\mu\varepsilon_{\lambda\sigma}\mathsf{k}^\sigma \\ +\eta_{\mu\lambda}\mathsf{k}_\rho\varepsilon_{\nu\sigma}\mathsf{k}^\sigma+\eta_{\lambda\nu}\mathsf{k}_\rho\varepsilon_{\mu\sigma}\mathsf{k}^\sigma+\eta_{\rho\nu}\mathsf{k}_\lambda\varepsilon_{\nu\sigma}\mathsf{k}^\sigma+\eta_{\nu\rho}\mathsf{k}_\lambda\varepsilon_{\mu\sigma}\mathsf{k}^\sigma\Big] \\ +\frac{1}{2\mathsf{k}^2}\Big[\mathsf{k}_\mu\mathsf{k}_\lambda\mathsf{k}_\rho\varepsilon_{\nu\sigma}\mathsf{k}^\sigma+\mathsf{k}_\nu\mathsf{k}_\lambda\mathsf{k}_\rho\varepsilon_{\mu\sigma}\mathsf{k}^\sigma+\mathsf{k}_\mu\mathsf{k}_\lambda\mathsf{k}_\nu\varepsilon_{\rho\sigma}\mathsf{k}^\sigma+\mathsf{k}_\nu\mathsf{k}_\mu\mathsf{k}_\rho\varepsilon_{\lambda\sigma}\mathsf{k}^\sigma\Big] \tag{7.218}$$

The trace of these two expressions is

$$\eta^{\mu\nu}\widetilde{\mathcal{T}}^{(\text{even})}_{\mu\nu\lambda\rho}(\mathsf{k})=\frac{i}{768\pi}\left(\mathsf{k}_\lambda\mathsf{k}_\rho+\mathsf{k}^2\eta_{\lambda\rho}\right) \tag{7.219}$$

$$\eta^{\mu\nu}\mathcal{T}^{(odd)}_{\mu\nu\lambda\rho}(\mathsf{k})=\frac{i}{1536\pi}\left(\mathsf{k}_\lambda\varepsilon_{\rho\sigma}\mathsf{k}^\sigma+\mathsf{k}_\rho\varepsilon_{\lambda\sigma}\mathsf{k}^\sigma\right) \tag{7.220}$$

and the divergence

$$\mathsf{k}^\mu\widetilde{\mathcal{T}}^{(\text{even})}_{\mu\nu\lambda\rho}(\mathsf{k})=-\frac{i}{1536\pi}\Big[\mathsf{k}_\nu\mathsf{k}_\lambda\mathsf{k}_\rho+\frac{1}{2}\mathsf{k}^2\big(\eta_{\nu\lambda}\mathsf{k}_\rho+\eta_{\nu\rho}\mathsf{k}_\lambda\big)\Big] \tag{7.221}$$

$$\mathsf{k}^\mu\mathcal{T}^{(odd)}_{\mu\nu\lambda\rho}(\mathsf{k})=-\frac{i}{3072\pi}\Big[\mathsf{k}_\nu\big(\mathsf{k}_\lambda\varepsilon_{\rho\sigma}+\mathsf{k}_\rho\varepsilon_{\lambda\sigma}\big)+\mathsf{k}^2\big(\eta_{\nu\lambda}\varepsilon_{\rho\sigma}+\eta_{\nu\rho}\varepsilon_{\lambda\sigma}\big)\mathsf{k}^\sigma\Big]. \tag{7.222}$$

Appendix 7C: Derivation of Feynman Rules with Background Gravity

Consider a free fermion theory coupled to ordinary gravity. We assume that the action has the expansion

$$
\begin{aligned}
S = \sum_{n=0}^{\infty} S_n \equiv S_0 + \sum_{n=1}^{\infty} \int \prod_{i=1}^{n} \mathrm{d}x_i \, \frac{1}{n!} \, \frac{\delta^n S}{\delta h_{\mu_1\nu_1}(x_1)\ldots\delta h_{\mu_n\nu_n}(x_n)}\bigg|_{h=0} h_{\mu_1\nu_1}(x_1)\ldots h_{\mu_n\nu_n}(x_n) \\
= S_0 + \int \mathrm{d}x \, \frac{\delta S}{\delta h_{\mu\nu}(x)}\bigg|_{h=0} h_{\mu\nu}(x) \qquad (7.223) \\
+\frac{1}{2}\int \mathrm{d}x_1\mathrm{d}x_2 \, \frac{\delta^2 S}{\delta h_{\mu_1\nu_1}(x_1)\delta h_{\mu_2\nu_2}(x_2)}\bigg|_{h=0} h_{\mu_1\nu_1}(x_1)h_{\mu_2\nu_2}(x_2) + \ldots
\end{aligned}
$$

In the sequel, we have in mind the free fermion theory in 4d defined by (7.56), and set $g_{\mu\nu} \approx \eta_{\mu\nu} + h_{\mu\nu}$. For simplicity, as we have done several times, we will absorb the density $\sqrt{g}$ in the fermion fields.[9] In this case, the e.m. tensor is defined as

$$
T_{\mu\nu}(x) = 2\frac{\delta S}{\delta g^{\mu\nu}(x)} \qquad (7.224)
$$

Now, using this definition, one can write

$$
\begin{aligned}
T_{\mu\nu}(x) = 2\Bigg(\frac{\delta S}{\delta h^{\mu\nu}(x)}\bigg|_{h=0} + \int \mathrm{d}x_1 \, \frac{\delta^2 S}{\delta h^{\mu\nu}(x)\delta h^{\mu_1\nu_1}(x_2)}\bigg|_{h=0} h^{\mu_1\nu_1}(x_1) \\
+ \frac{1}{2}\int \mathrm{d}x_1\mathrm{d}x_2 \, \frac{\delta^3 S}{\delta h^{\mu\nu}(x)\delta h^{\mu_1\nu_1}(x_1)\delta h^{\mu_2\nu_2}(x_2)}\bigg|_{h=0} h^{\mu_1\nu_1}(x_1)h^{\mu_2\nu_2}(x_2) + \ldots\Bigg) \\
\equiv T^{(0)}_{\mu\nu}(x) + T^{(1)}_{\mu\nu}(x) + \ldots \qquad (7.225)
\end{aligned}
$$

We define

$$
T^{(n)}_{\mu\nu}(x) = \frac{2}{n!}\int \prod_{i=1}^{n} \mathrm{d}x_i \, \frac{\delta^{m+1} S}{\delta h^{\mu\nu}(x)\delta h^{\mu_1\nu_1}(x_1)\ldots\delta h^{\mu_m\nu_m}(x_m)}\bigg|_{h=0} h^{\mu_1\nu_1}(x_1)\ldots h^{\mu_m\nu_m}(x_m)
$$

So we can rewrite

$$
S_n = \frac{1}{2n}\int \mathrm{d}x \, T^{(n-1)}_{\mu\nu}(x) h^{\mu\nu}(x) \qquad (7.226)
$$

[9] In [7], this simplification has been avoided, obtaining nevertheless the same results.

One-Loop One-Point Function

Representing by ϕ the matter fields in the model, the one-loop one-point function of $T_{\mu\nu}$ in the presence of a metric $g_{\mu\nu} \approx \eta_{\mu\nu} + h_{\mu\nu}$ is

$$\begin{aligned} \langle\langle T_{\mu\nu}(x)\rangle\rangle &= \int \mathcal{D}\phi\, T_{\mu\nu}(x)\, e^{iS[\phi,h]} \\ &= \int \mathcal{D}\phi\, \left(T^{(0)}_{\mu\nu}(x) + T^{(1)}_{\mu\nu}(x) + T^{(2)}_{\mu\nu}(x) + \ldots\right) e^{i(S_0+S_1+S_2+\ldots)} \\ &= \int \mathcal{D}\phi\, \left[\left(T^{(0)}_{\mu\nu}(x) + T^{(1)}_{\mu\nu}(x) + T^{(2)}_{\mu\nu}(x) + \ldots\right) e^{i(S_1+S_2+\ldots)}\right] e^{iS_0} \end{aligned} \tag{7.227}$$

e^{iS_0} has been singled out as the free part of the integration measure. The rest of S (the interaction) is treated perturbatively.

Rearranging (7.227) order by order in h:

$$\begin{aligned} \langle\langle T_{\mu\nu}(x)\rangle\rangle &= \int \mathcal{D}\phi\, T^{(0)}_{\mu\nu}(x)\, e^{iS_0} + \int \mathcal{D}\phi\, \left(iS_1\, T^{(0)}_{\mu\nu}(x) + T^{(1)}_{\mu\nu}(x)\right) e^{iS_0} \\ &+ \int \mathcal{D}\phi \left((iS_2 - \frac{1}{2}S_1^2)\, T^{(0)}_{\mu\nu}(x) + iS_1\, T^{(1)}_{\mu\nu}(x) + T^{(2)}_{\mu\nu}(x)\right) e^{iS_0} \\ &+ \int \mathcal{D}\phi \left((iS_3 - S_1S_2 - \frac{i}{3!}S_1^3)\, T^{(0)}_{\mu\nu}(x) + (iS_2 - \frac{1}{2}S_1^2)\, T^{(1)}_{\mu\nu}(x) + iS_1 T^{(2)}_{\mu\nu}(x) + T^{(3)}_{\mu\nu}(x)\right) e^{iS_0} \\ &+ \ldots \end{aligned} \tag{7.228}$$

Next, the general procedure is to introduce auxiliary external currents and couple them to the free fields in S_0. For instance if the free fields are ψ, $\bar{\psi}$, we introduce j, $\bar{j}$ and add a term

$$\langle\langle T_{\mu\nu}(x)\rangle\rangle[j,\bar{j}] = \int \mathcal{D}\phi\, (\ldots\ldots\ldots)\exp[iS_0 + i\int(\bar{j}\psi + \overline{\psi}j)]$$

and set at the end $j = \bar{j} = 0$. At this point in $\left(\ldots\ldots\ldots\right)$ one can replace ψ by $\frac{\delta}{\delta\bar{j}}$ and $\overline{\psi}$ by $-\frac{\delta}{\delta j}$, so that the only remaining dependence on ψ and $\overline{\psi}$ is in the factor $\exp[iS_0 + \int(\bar{j}\psi + \overline{\psi}j)]$. Since the exponent is a quadratic expression, one can formally integrate over ψ and $\overline{\psi}$ by completing the square. This leads to an irrelevant infinite constant times

$$\exp[-i\int \bar{j}\, P\, j] \tag{7.229}$$

where P is the inverse of the kinetic differential operator in S_0, i.e. the propagator in configuration space. Finally

$$
\begin{aligned}
\langle\!\langle T_{\mu\nu}(x)\rangle\!\rangle = &\Big[T^{(0)}_{\mu\nu}(x)\exp[-i\int \bar{j}\,P\,j] + \left(iS_1\,T^{(0)}_{\mu\nu}(x)+T^{(1)}_{\mu\nu}(x)\right)\exp[-i\int \bar{j}\,P\,j] \qquad (7.230)\\
&+\left((iS_2-\frac{1}{2}S_1^2)\,T^{(0)}_{\mu\nu}(x)+iS_1\,T^{(1)}_{\mu\nu}(x)+T^{(2)}_{\mu\nu}(x)\right)\exp[-i\int \bar{j}\,P\,j]\\
&+\left((iS_3-S_1S_2-\frac{i}{3!}S_1^3)\,T^{(0)}_{\mu\nu}(x)+(iS_2-\frac{1}{2}S_1^2)\,T^{(1)}_{\mu\nu}(x)+iS_1T^{(2)}_{\mu\nu}(x)+T^{(3)}_{\mu\nu}(x)\right)\\
&\cdot\exp[-i\int \bar{j}\,P\,j]\Big]\Big|_{j=\bar{j}=0}+\dots
\end{aligned}
$$

where all the ψ, $\overline{\psi}$ fields in $T_{(n)}$, S_n are understood to be replaced by $\frac{\delta}{\delta\bar{j}}$ and $-\frac{\delta}{\delta j}$, respectively. This is the final expression of the one-point one-loop correlator from which the Feynman rules are extracted. Equation (7.230) is thus rewritten as

$$
\begin{aligned}
\langle\!\langle T_{\mu\nu}(x)\rangle\!\rangle &= \langle 0|T^{(0)}_{\mu\nu}(x)|0\rangle \qquad (7.231)\\
&+\langle 0|\mathcal{T}\left(iS_1\,T^{(0)}_{\mu\nu}(x)+T^{(1)}_{\mu\nu}(x)\right)|0\rangle\\
&+\langle 0|\mathcal{T}\left((iS_2-\frac{1}{2}S_1^2)\,T^{(0)}_{\mu\nu}(x)+iS_1\,T^{(1)}_{\mu\nu}(x)+T^{(2)}_{\mu\nu}(x)\right)|0\rangle,\\
&+\langle 0|\mathcal{T}\left((iS_3-S_1S_2-\frac{i}{3!}S_1^3)\,T^{(0)}_{\mu\nu}(x)+(iS_2-\frac{1}{2}S_1^2)\,T^{(1)}_{\mu\nu}(x)+iS_1T^{(2)}_{\mu\nu}(x)+T^{(3)}_{\mu\nu}(x)\right)|0\rangle\\
&+\dots
\end{aligned}
$$

and the time-ordered amplitudes are computed by means of Feynman diagrams.

Expansion in $h_{\mu\nu}$

We now expand the action (7.56) in series of $h_{\mu\nu}$. To this end, we use[10]

$$
\begin{aligned}
g_{\mu\nu} &= \eta_{\mu\nu}+h_{\mu\nu} \qquad (7.232)\\
g^{\mu\nu} &= \eta^{\mu\nu}-h^{\mu\nu}+(h^2)^{\mu\nu}+\dots\\
e^{\mu}_{a} &= \delta^{\mu}_{a}-\frac{1}{2}h^{\mu}_{a}+\frac{3}{8}(h^2)^{\mu}_{a}-\frac{5}{16}(h^3)^{\mu}_{a}+\dots\\
\mathrm{e}^{a}_{\mu} &= \delta^{a}_{\mu}+\frac{1}{2}h^{a}_{\mu}-\frac{1}{8}(h^2)^{a}_{\mu}+\frac{1}{16}(h^3)^{a}_{\mu}+\dots,
\end{aligned}
$$

$$
\Gamma^{\lambda}_{\mu\nu} = \frac{1}{2}\left(\partial_\mu h^\lambda_\nu+\partial_\nu h^\lambda_\mu-\partial^\lambda h_{\mu\nu}\right)-\frac{1}{2}(h-h^2)^{\lambda\rho}\left(\partial_\mu h_{\rho\nu}+\partial_\nu h_{\rho\mu}-\partial_\rho h_{\mu\nu}\right)+\dots \qquad (7.233)
$$

and

$$
\begin{aligned}
\omega^{ab}_{\mu} = &\frac{1}{2}\left(\partial^b h^a_\mu-\partial^a h^b_\mu\right)+\frac{1}{4}\left(h^{\sigma a}\partial_\sigma h^b_\mu-h^{\sigma b}\partial_\sigma h^a_\mu+h^{b\sigma}\partial^a h_{\sigma\mu}-h^{a\sigma}\partial^b h_{\sigma\mu}\right)\\
&-\frac{1}{8}\left(h^{a\sigma}\partial_\mu h^b_\sigma-h^{b\sigma}\partial_\mu h^a_\sigma\right)+\dots \qquad (7.234)
\end{aligned}
$$

[10] Like in other occasions, we ignore the antisymmetric part of the vierbein.

Clearly, Latin and Greek indices in the RHS of these equations are to be treated on the same footing. Up to third order in $h_{\mu\nu}$, the action (7.60) becomes

$$S = \int d^4x \left[\frac{i}{2}\overline{\psi_R}\gamma^\mu \overleftrightarrow{\partial}_\mu \psi_R - \frac{i}{4}\overline{\psi_R} h^\mu_a \gamma^a \overleftrightarrow{\partial}_\mu \psi_R + \frac{3i}{16}\overline{\psi_R}(h^2)^\mu_a \gamma^a \overleftrightarrow{\partial}_\mu \psi_R - \frac{5i}{32}\overline{\psi_R}(h^3)^\mu_a \gamma^a \overleftrightarrow{\partial}_\mu \psi_R \right.$$
$$\left. -\frac{1}{16}\varepsilon^{\mu abc}\overline{\psi_R}\gamma_c\gamma_5\psi_R \left(h^\sigma_\mu \partial_a h_{b\sigma} + (h^2)^\sigma_\mu \partial_b h_{a\sigma} - h^\rho_\mu h^\sigma_a \partial_\sigma h_{\rho b} - \frac{1}{2} h^\rho_\mu \partial_a h_{\rho\sigma} h^\sigma_c\right)\right] \quad (7.235)$$

From the first term in the RHS we extract the propagator, from the second term the vertex V_{ffh}, from the third term the vertex V_{ffhh} and from the first term in the second line the vertex V^ε_{ffhh}. The fermion propagator, the vertices V_{ffh}, V_{ffhh} and V^ε_{ffhh} have already been written down above. There are also other two-fermion-three-graviton vertices , but they will not be written down explicitly because they are irrelevant for the odd parity trace anomaly.

The Weyl Ward Identity

The Ward identity for Weyl invariance, in the absence of anomalies, is:

$$\mathcal{T}(x) \equiv g^{\mu\nu}(x)\langle\langle T_{\mu\nu}(x)\rangle\rangle = \langle\langle T^\mu_\mu(x)\rangle\rangle + h^{\mu\nu}(x)\langle\langle T_{\mu\nu}(x)\rangle\rangle = 0 \quad (7.236)$$

Writing

$$\langle\langle T_{\mu\nu}(x)\rangle\rangle = \langle 0\, T^{(0)}_{\mu\nu}(x)|0\rangle \quad (7.237)$$
$$+\sum_{n=1}^{\infty}\frac{1}{2^n n!}\int\prod_{i=0}^{n} dx_i\, h^{\mu_1\nu_1}(x_1)\dots h^{\mu_n\nu_n}(x_n)\mathcal{T}_{\mu\nu\mu_1\nu_1\dots\mu_n\nu_n}(x, x_1, \dots, x_n),$$

order by order in h, Eq. (7.236) breaks down to

$$\langle\langle \mathcal{T}(x)\rangle\rangle = \langle 0|\mathcal{T}^{(0)}(x)|0\rangle$$
$$+\sum_{n=1}^{\infty}\frac{1}{2^n n!}\int\prod_{i=0}^{n} dx_i\, h^{\mu_1\nu_1}(x_1)\dots h^{\mu_n\nu_n}(x_n)\mathcal{T}^{(n)}_{\mu_1\nu_1\dots\mu_n\nu_n}(x, x_1, \dots, x_n),$$
$$(7.238)$$

where

$$\mathcal{T}^{(0)}(x) \equiv \langle 0\, T^{(0)\mu}{}_\mu(x)|0\rangle = 0 \quad (7.239)$$
$$\mathcal{T}^{(1)}_{\mu_1\nu_1}(x, x_1) \equiv \mathcal{T}^\mu_{\mu\mu_1\nu_1}(x, x_1) + 2\delta(x - x_1)\langle 0\, T^{(0)}_{\mu_1\nu_1}(x)|0\rangle = 0 \quad (7.240)$$
$$\mathcal{T}^{(2)}_{\mu_1\nu_1,\mu_2,\nu_2}(x, x_1, x_2) \equiv \mathcal{T}^\mu_{\mu\mu_1\nu_1\mu_2\nu_2}(x, x_1, x_2) + 2\delta(x - x_1)\mathcal{T}_{\mu_1\nu_1\mu_2\nu_2}(x, x_2)$$
$$+2\delta(x - x_2)\mathcal{T}_{\mu_2\nu_2\mu_1\nu_1}(x, x_1) = 0 \quad (7.241)$$

Here

$$T^{(0)}_{\mu\nu} = 2\frac{\delta S}{\delta h^{\mu\nu}(x)}\bigg|_{h=0} = -\frac{i}{4}\left(\overline{\psi_R}\gamma_\mu \overleftrightarrow{\partial}_\nu \psi_R + \mu \leftrightarrow \nu\right), \tag{7.242}$$

$$\begin{aligned}\mathcal{T}_{\mu\nu\mu_1\nu_1}(x, x_1) &= i\langle 0|\mathcal{T}T^{(0)}_{\mu\nu}(x)T^{(0)}_{\mu_1\nu_1}(x_1)|0\rangle - \eta_{\mu_1\nu_1}\delta(x-x_1)\langle 0\, T^{(0)}_{\mu\nu}(x)|0\rangle \\ &+4\langle 0|\frac{\delta^2 S}{\delta h^{\mu\nu}(x)\delta h^{\mu_1\nu_1}(x_1)}|0\rangle\end{aligned} \tag{7.243}$$

and

$$\begin{aligned}&\mathcal{T}_{\mu\nu\mu_1\nu_1\mu_2\nu_2}(x, x_1, x_2) \\ &= -\langle 0|\mathcal{T}T^{(0)}_{\mu\nu}(x)T^{(0)}_{\mu_1\nu_1}(x_1)T^{(0)}_{\mu_2\nu_2}(x_2)|0\rangle + 4i\langle 0|\mathcal{T}T^{(0)}_{\mu\nu}(x)\frac{\delta^2 S}{\delta h^{\mu_1\nu_1}(x_1)\delta h^{\mu_2\nu_2}(x_2)}|0\rangle \\ &- i\eta_{\mu_1\nu_1}\delta(x-x_1)\langle 0|\mathcal{T}T^{(0)}_{\mu\nu}(x)T^{(0)}_{\mu_2\nu_2}(x_2)|0\rangle - i\eta_{\mu_2\nu_2}\delta(x-x_2)\langle 0|\mathcal{T}T^{(0)}_{\mu\nu}(x)T^{(0)}_{\mu_1\nu_1}(x_1)|0\rangle \\ &+4i\langle 0|\mathcal{T}T^{(0)}_{\mu_1\nu_1}(x_1)\frac{\delta^2 S}{\delta h^{\mu\nu}(x)\delta h^{\mu_2\nu_2}(x_2)}|0\rangle + 4i\langle 0|\mathcal{T}T^{(0)}_{\mu_2\nu_2}(x_2)\frac{\delta^2 S}{\delta h^{\mu_1\nu_1}(x_1)\delta h^{\mu\nu}(x)}|0\rangle \\ &+ \left(\eta_{\mu_1\nu_1}\eta_{\mu_2\nu_2} + \eta_{\mu_1\nu_2}\eta_{\mu_2\nu_1} + \eta_{\mu_1\mu_2}\eta_{\nu_1\nu_2}\right)\delta(x-x_1)\delta(x-x_2)\langle 0\, T^{(0)}_{\mu\nu}(x)|0\rangle \\ &-4\eta_{\mu_1\nu_1}\delta(x-x_1)\langle 0|\frac{\delta^2 S}{\delta h^{\mu\nu}(x)\delta h^{\mu_2\nu_2}(x_2)}|0\rangle - 4\eta_{\mu_2\nu_2}\delta(x-x_2)\langle 0|\frac{\delta^2 S}{\delta h^{\mu\nu}(x)\delta h^{\mu_1\nu_1}(x_1)}|0\rangle \\ &+8\langle 0|\frac{\delta^3 S}{\delta h^{\mu\nu}(x)\delta h^{\mu_1\nu_1}(x_1)h^{\mu_2\nu_2}(x_2)}|0\rangle\end{aligned} \tag{7.244}$$

The functional derivatives of S with respect to h are understood to be evaluated at $h = 0$.

The Weyl Ward identity is obtained by saturating the previous amplitudes with $\eta^{\mu\nu}$ recalling that there are two modalities of doing so; see Sect. 7.1. The diffeomorphism Ward identity is obtained by applying ∂^μ_x to the same expressions.

Before starting with explicit evaluations let us remark that the above expansions (7.242, 7.243) and (7.244) carry into the game a number of new Feynman diagrams. We have not only the familiar triangle diagram and bubble diagrams (with a loop formed by two fermion propagators), but also tadpole and seagull diagrams (with one fermion propagator). The latter are produced by vertices with two fermion legs and two or more graviton legs, by contracting the fermion legs with a propagator, thus forming a fermion loop.

So far we have been completely general. From now on, we consider only odd parity correlators, that is only correlators linear in $\varepsilon_{\mu\nu\lambda\rho}$. To start with, to $\langle 0|T^{(0)\mu}{}_\mu(x)|0\rangle$, a constant, only a tadpole can contribute, but its odd part vanishes because there is no scalar one can construct with ε and η. For the same reason, also $\langle 0\, T^{(0)}_{\mu\nu}(x)|0\rangle$ vanishes. The odd part of the two-point function $\langle 0|\mathcal{T}T^{(0)}_{\mu\nu}(x)T^{(0)}_{\mu_1\nu_1}(x_1)|0\rangle$ also vanishes, because in momentum space it must be a 4-tensor linear in ε and formed with η and the momentum k: there is no such tensor, symmetric in $\mu \leftrightarrow \nu$, $\mu_1 \leftrightarrow \nu_1$ and $(\mu, \nu) \leftrightarrow (\mu_1, \nu_1)$. As for the terms $\langle 0|\frac{\delta^2 S}{\delta h^{\mu\nu}(x)\delta h^{\mu_1\nu_1}(x_1)}|0\rangle$, they might also produce odd non-vanishing contribution from tadpole diagrams, but like in the previous case it is impossible to satisfy the combinatorics. In conclusion, (7.239) and (7.240) are identically satisfied, while (7.241) becomes

$$\begin{aligned}
\mathcal{T}^{(2)}(x) &= T^{\mu}_{\mu\mu_1\nu_1\mu_2\nu_2}(x, x_1, x_2)\\
&= \eta^{\mu\nu}\Big(- \langle 0|\mathcal{T}T^{(0)}_{\mu\nu}(x)T^{(0)}_{\mu_1\nu_1}(x_1)T^{(0)}_{\mu_2\nu_2}(x_2)|0\rangle + 4i\,\langle 0|\mathcal{T}T^{(0)}_{\mu\nu}(x)\frac{\delta^2 S}{\delta h^{\mu_1\nu_1}(x_1)\delta h^{\mu_2\nu_2}(x_2)}|0\rangle\\
&+4i\,\langle 0|\mathcal{T}T^{(0)}_{\mu_1\nu_1}(x_1)\frac{\delta^2 S}{\delta h^{\mu\nu}(x)\delta h^{\mu_2\nu_2}(x_2)}|0\rangle + 4i\,\langle 0|\mathcal{T}T^{(0)}_{\mu_2\nu_2}(x_2)\frac{\delta^2 S}{\delta h^{\mu_1\nu_1}(x_1)\delta h^{\mu\nu}(x)}|0\rangle\\
&+8\langle 0|\frac{\delta^3 S}{\delta h^{\mu\nu}(x)\delta h^{\mu_1\nu_1}(x_1)h^{\mu_2\nu_2}(x_2)}|0\rangle\Big)
\end{aligned} \tag{7.245}$$

To proceed further let us focus on the last term. This corresponds to seagull terms, with three external graviton lines attached to the same point of a fermion loop. They are generated by the fourth term in first line of (7.235) and by the last three terms in the second line. It is clear that the former cannot produce an ε because of the lack of enough γ matrices. The latter three instead already contain an ε tensor, and therefore, they might contribute an odd parity term to the anomaly. The relevant vertex has two fermion legs, with the usual momenta p and p', and three graviton legs, with incoming momenta k_1, k_2, k_3 and labels $\mu_1, \nu_1, \mu_2, \nu_2$ and μ_3, ν_3, respectively. For instance, a typical contribution of the second term in the second line of (7.235) is proportional to

$$\sim \varepsilon_{\mu_2\mu_3\lambda\rho}(k_2 - k_3)^\lambda \gamma^\rho \eta_{\mu_1\nu_3}\eta_{\nu_1\nu_2} \tag{7.246}$$

symmetrized in $\mu_1 \leftrightarrow \nu_1$, $\mu_2 \leftrightarrow \nu_2$, $\mu_3 \leftrightarrow \nu_3$, and with respect to the exchange of any two couples (μ_i, ν_i). The seagull term is therefore proportional to

$$\int \mathrm{d}^4 p\, \frac{p^\rho}{p^2}$$

which vanishes. All the other similar terms vanish likewise.

To proceed further, we focus now on the terms containing the second derivative of S in (7.245). Looking at (7.235) we see that they can be generated by the third term in the first line and by the first term in the second line. Once again the former, involving the vertex V_{ffhh}, cannot contribute for lack of enough γ matrices, while the second in principle could because the corresponding vertex, V^{ε}_{ffhh}, contains the t tensor. But we remark that contractions produced by $\eta^{\mu\nu}$ in the third and fourth terms in the RHS contracts precisely couples of indices of the t tensor for which it is traceless.

In summary, the odd trace anomaly may receive non-trivial contributions only from

$$\mathcal{T}^{(2)}(x) = T_\mu^{\mu\mu_1\nu_1\mu_2\nu_2}(x, x_1, x_2)$$
$$= \eta_{\mu\nu}\Big(- \langle 0|\mathcal{T}T_{(0)}^{\mu\nu}(x)T_{(0)}^{\mu_1\nu_1}(x_1)T_{(0)}^{\mu_2\nu_2}(x_2)|0\rangle + 4i\,\langle 0|\mathcal{T}T_{(0)}^{\mu\nu}(x)\frac{\delta^2 S}{\delta h_{\mu_1\nu_1}(x_1)\delta h_{\mu_2\nu_2}(x_2)}|0\rangle\Big) \tag{7.247}$$

These amplitudes have been computed either in the main text or in the following Appendix 7E.

Appendix 7D: Bubble Diagram Contributions to the Odd Parity Divergence

In this Appendix, we evaluate the odd parity contribution of the bubble diagrams to the divergence of the e.m. tensor $T_{\mu\nu}^{(R)}$. The potential contributions to the divergence of the energy-momentum tensor come from the diagrams: $V_{ffh} \leftrightarrows V^{\varepsilon}_{ffhh}$ and $V_{ffh} \leftrightarrows V_{ffhh}$, where the arrows represent propagators and the vertices have been defined in (7.103) and (7.167). The first takes the form (dropping the addend $\frac{1}{2}$ in P_R)

$$q^\sigma D^{(R,\varepsilon,odd)\tau}_{\mu\nu\mu'\nu'\sigma}(k_1,k_2) = \frac{i}{512}\int\frac{d^4p}{(2\pi)^4}\mathrm{tr}\Big[\frac{1}{\not p}\, t_{\mu\nu\mu'\nu'\lambda\rho}\,(k_2-k_1)^\lambda$$
$$\times\,\gamma^\rho\frac{1}{\not p-\not q}\left((2p-q)\cdot q\gamma^\tau + (2p-q)^\tau\not q\right)\Big] \tag{7.248}$$

$$= \frac{i}{512}\int\frac{d^4p d^\delta\ell}{(2\pi)^{4+\delta}}\mathrm{tr}\Big[\frac{\not p+\not\ell}{p^2-\ell^2}\, t_{\mu\nu\mu'\nu'\lambda\rho}\,(k_2-k_1)^\lambda$$
$$\times\,\gamma^\rho\frac{\not p-\not q+\not\ell}{(p-q)^2-\ell^2}\left((2p-q)\cdot q\gamma^\tau + (2p-q)^\tau\not q\right)\Big] \tag{7.249}$$

which is

$$q^\sigma D^{(R,\varepsilon,odd)\tau}_{\mu\nu\mu'\nu'\sigma}(k_1,k_2) = i\frac{2^{\frac{\delta}{2}}}{512}\int\frac{d^4p d^\delta\ell}{(2\pi)^{4+\delta}}t_{\mu\nu\mu'\nu'\lambda\rho}\,(k_2-k_1)^\lambda$$
$$\times\Big\{\ell^2\left[\mathrm{tr}\left(\gamma^\rho\gamma^\tau\right)(2p-q)\cdot q + \mathrm{tr}\left(\gamma^\rho\not q\right)(2p-q)^\tau\right]$$
$$+\,\mathrm{tr}\left(\not p\gamma^\rho(\not p-\not q)\gamma^\tau\right)(2p-q)\cdot q + \mathrm{tr}\left(\not p\gamma^\rho(\not p-\not q)\not q\right)(2p-q)^\tau\Big\} \tag{7.250}$$
$$\frac{1}{(p^2-\ell^2)((p-q)^2-\ell^2)}$$

Now shift $p \to p+q$

$$
\begin{aligned}
q^\sigma D^{(R,\varepsilon,odd)\tau}_{\mu\nu\mu'\nu'\sigma}(k_1,k_2) &= i\frac{2^{\frac{\delta}{2}}}{512}\int \frac{d^4p d^\delta \ell}{(2\pi)^{4+\delta}} t_{\mu\nu\mu'\nu'\lambda\rho}\,(k_2-k_1)^\lambda \\
&\times \Big\{\ell^2\left[\mathrm{tr}\,(\gamma^\rho\gamma^\tau)\,(2p+q)\cdot q + \mathrm{tr}\,(\gamma^\rho \not q)\,(2p+q)^\tau\right] \\
&+\mathrm{tr}\left((\not p+\not q)\gamma^\rho \not p\gamma^\tau\right)(2p+q)\cdot q + \mathrm{tr}\left((\not p+\not q)\gamma^\rho \not p \not q\right)(2p+q)^\tau\Big\} \quad (7.251) \\
&\frac{1}{((p+q)^2-\ell^2)(p^2-\ell^2)}
\end{aligned}
$$

Upon changing integration variable $p \to -p$, one gets

$$
q^\sigma D^{(R,\varepsilon,odd)}_{\mu\nu\mu'\nu'\sigma\tau}(k_1,k_2) = -q^\sigma D^{(R,\varepsilon,odd)}_{\mu\nu\mu'\nu'\sigma\tau}(k_1,k_2) = 0 \quad (7.252)
$$

The other bubble contribution to the e.m. tensor divergence is

$$
\begin{aligned}
q^\sigma D^{(R,odd)\tau}_{\mu\nu\mu'\nu'\sigma}(k_1,k_2) &= \frac{3i}{1024}\int \frac{d^4p d^\delta \ell}{(2\pi)^{4+\delta}} \mathrm{tr}\Big(\frac{1}{\not p+\not\ell}(2p-q)_\mu \gamma_{\mu'}\eta_{\nu\nu'}\frac{1}{\not p-\not q+\not \ell} \\
&\times \left((2p\cdot q-q^2)\gamma^\tau + (2p-q)^\tau \not q\right)\gamma_5\Big) \quad (7.253)
\end{aligned}
$$

to which we have to add the terms with $\mu \leftrightarrow \nu$, $\mu' \leftrightarrow \nu'$ and $\{\mu,\nu\} \leftrightarrow \{\mu',\nu'\}$. These additional terms can be treated in the same way as the first. Now let us write $\not q = \not p + \not\ell - (\not p - \not q + \not\ell)$. It is evident that the term linear in $\not q$ gives a vanishing contribution because by simplifying the denominators and noting that the ℓ^2 terms cannot contribute we are left with a trace of γ_5 with only two gamma matrices. Next we write $2p\cdot q - q^2 = -(p-q)^2 + p^2$ and simplify with the denominators, so that

$$
\begin{aligned}
q^\sigma D^{(R,odd)\tau}_{\mu\nu\mu'\nu'\sigma}(k_1,k_2) &= \frac{3i}{1024}\int \frac{d^4p d^\delta \ell}{(2\pi)^{4+\delta}} \mathrm{tr}\Big(\not p(2p-q)_\mu\gamma_{\mu'}\eta_{\nu\nu'}\frac{\not p-\not q}{(p-q)^2+\ell^2}\gamma^\tau\gamma_5 \\
&-\frac{\not p}{p^2-\ell^2}(2p-q)_\mu\gamma_{\mu'}\eta_{\nu\nu'}(\not p-\not q)\gamma^\tau\gamma_5\Big) \quad (7.254)
\end{aligned}
$$

Due to the ε tensor the cubic powers of p in the numerator reduce to quadratic powers. Using this we shift $p \to p+q$, followed by $p \to -p$, in the second line. It reduces to the first with opposite sign.

Appendix 7E: The Bubble Diagrams and Other Vanishing Contributions to the Trace Anomaly

In this appendix, we give a few details of the calculations concerning the odd trace anomaly in 4d.

Let us consider first the bubble diagrams. They are obtained by joining two vertices, V_{ffh} (on the left) and V^{ε}_{ffhh} or V_{ffhh} (on the right) with two fermion propagators. The incoming graviton in V_{ffh} has momentum q and Lorentz labels σ, τ and the two outgoing gravitons in V^{ε}_{ffhh} or V_{ffhh} are specified by k_1, μ, ν and k_2, μ', ν', respectively, with $q = k_1 + k_2$. The two fermion propagators form a loop. The running momentum is clockwise oriented. We denote the momentum in the upper branch of the loop by p and in the lower branch by $p - k_1 - k_2$. The first bubble diagram gives

$$2\frac{i}{512}\int \frac{d^4 p}{(2\pi)^4} \operatorname{tr}\left[\frac{1}{\not p} t_{\mu\nu\mu'\nu'\lambda\rho}\,(k_2 - k_1)^{\lambda}\gamma^{\rho}\frac{1}{\not p - \not q}\left((2p^{\sigma} - q^{\sigma})\,\gamma^{\tau} + (\sigma \leftrightarrow \tau)\right)\frac{1+\gamma_5}{2}\right] \tag{7.255}$$

The factor of two in front comes from the combinatorics of diagrams: this one must contribute twice. Its possible contribution to the trace anomaly comes from contracting the indices σ and τ with a Kronecker delta (in principle we should consider contracting also the other couple of indices μ, ν and μ', ν', but this gives 0 due to the symmetry properties of the t tensor). The integral is divergent and needs to be regularized. As usual, we introduce additional components of the momentum running on the loop: $p \to p + \ell, \ell = (\ell_4, \ldots, \ell_{\delta+3})$. The relevant integral (dropping the even part) becomes

$$\begin{aligned}\mathrm{D}^{(R,\varepsilon,odd)}_{\mu\nu\mu'\nu'}(k_1, k_2) &= \frac{i}{256}\int \frac{d^4 p}{(2\pi)^4}\int \frac{d^{\delta}\ell}{(2\pi)^{\delta}}\, t_{\mu\nu\mu'\nu'\lambda\rho}(k_2 - k_1)^{\lambda}\\ &\cdot \operatorname{tr}\left(\frac{\not p + \not l}{p^2 - \ell^2}\gamma^{\rho}\frac{\not p - \not q + \not \ell}{(p-q)^2 - \ell^2}(2\not p + 2\not \ell - \not q)\right)\end{aligned} \tag{7.256}$$

After some algebra and introducing a parametric representation for the denominators, one finally gets

$$\begin{aligned}\mathrm{D}^{(R,\varepsilon,odd)}_{\mu\nu\mu'\nu'}(k_1, k_2) &= -i\frac{2^{\frac{\delta}{2}}}{64}\, t_{\mu\nu\mu'\nu'\lambda\rho}(k_2 - k_1)^{\lambda}\int_0^1 dx \int \frac{d^4 p}{(2\pi)^4}\int \frac{d^{\delta}\ell}{(2\pi)^{\delta}} \\ &\cdot\left[\left(\frac{3}{2}(2x-1)p^2 + x(x-1)(2x-1)q^2 - (2x-1)\ell^2\right)q^{\rho}\right]\frac{1}{(p^2 + x(1-x)q^2 - \ell^2)^2}\end{aligned} \tag{7.257}$$

This vanishes because of the x integration.

The second bubble diagram contributes to the odd e.m. trace with

$$
\begin{aligned}
& D^{(R,odd)}_{\mu\nu\mu'\nu'}(k_1,k_2) \qquad (7.258)\\
&= \frac{3}{256}\int \frac{d^4p d^\delta \ell}{(2\pi)^{4+\delta}} \mathrm{tr}\left(\frac{\not p+\not\ell}{p^2-\ell^2}\left[\left((2p-q)_\mu \gamma_{\mu'}\eta_{\nu\nu'} + (2p-q)_\mu \gamma_{\nu'}\eta_{\nu\mu'} + \{\mu \leftrightarrow \nu\}\right)\right.\right.\\
&+ \left.\left.\left((2p-q)_{\mu'}\gamma_\mu \eta_{\nu\nu'} + (2p-q)_{\mu'}\gamma_\nu \eta_{\mu\nu'} + \{\mu' \leftrightarrow \nu'\}\right)\right] \frac{\not p - \not q + \not\ell}{(p-q)^2-\ell^2}(2\not p + 2\not\ell - \not q)\gamma_5\right)
\end{aligned}
$$

The $\not\ell$-dependent terms in the numerator do not contribute. Now take for instance the first term in the square bracket and work out the traces. One gets

$$
= \frac{3\,2^{\frac{\delta}{2}}}{256}\int \frac{d^4p d^\delta \ell}{(2\pi)^{4+\delta}} \varepsilon_{\lambda\rho\mu'\tau}(2p-q)^\lambda p^\rho (p-q)^\tau \eta_{\nu\nu'} \frac{(2p-q)_\mu}{(p^2-\ell^2)((p-q)^2-\ell^2)} = 0 \qquad (7.259)
$$

because of the epsilon tensor.

We complete this Appendix with the calculation of $\mathsf{T}^{(1)}$ and $\mathsf{T}^{(2)}$.

7F1. $\mathsf{T}^{(1)}_{\mu\nu\mu'\nu'}(k_1,k_2)$

The contribution $\mathsf{T}^{(1)}$ can be expressed as

$$
\begin{aligned}
\mathsf{T}^{(1)}_{\mu\nu\mu'\nu'}(k_1,k_2) &= -\frac{1}{256}\int \frac{d^4p}{(2\pi)^4}\int \frac{d^\delta \ell}{(2\pi)^\delta} \frac{(2p-k_1)_\mu (2p-2k_1-k_2)_{\mu'}}{2(p^2-\ell^2)\left[(p-k_1)^2-\ell^2\right]}\\
&\times \underbrace{\mathrm{Tr}\left[(\not p+\not\ell)\gamma_\nu(\not p+\not\ell-\not k_1)\gamma_{\nu'}\gamma_5\right]}_{2^{2+\frac{\delta}{2}}\, i\, p^\alpha k_1^\beta \varepsilon_{\alpha\nu\beta\nu'}}. \qquad (7.260)
\end{aligned}
$$

Employing the Feynman parametrization, expression (7.260) is written as

$$
\begin{aligned}
\mathsf{T}^{(1)}_{\mu\nu\mu'\nu'}(k_1,k_2) &= -i\frac{2^{\frac{\delta}{2}}}{128}\int \frac{d^4p}{(2\pi)^4}\int \frac{d^\delta \ell}{(2\pi)^\delta}\int_0^1 dx \qquad (7.261)\\
&\times \frac{(2p-k_1)_\mu (2p-2k_1-k_2)_{\mu'}}{\left\{\left[(p-k_1)^2-\ell^2\right]x + (1-x)(p^2-\ell^2)\right\}^2} p^\alpha k_1^\beta \varepsilon_{\alpha\nu\beta\nu'}.
\end{aligned}
$$

Performing the shift $p \to p + xk_1$ and taking into account that just even powers of p in the numerator will result in non-vanishing contributions to $\mathsf{T}^{(1)}$, one obtains

$$
\begin{aligned}
\mathsf{T}^{(1)}_{\mu\nu\mu'\nu'}(k_1,k_2) &= i\frac{2^{\frac{\delta}{2}}}{128}\int_0^1 dx \int \frac{d^\delta \ell}{(2\pi)^\delta}\int \frac{d^4p}{(2\pi)^4} \qquad (7.262)\\
&\times \frac{2p_{\mu'}(1-2x)k_{1\mu} + 2p_\mu\left[2(1-x)k_1+k_2\right]_\mu}{\left[p^2+x(1-x)k_1^2-\ell^2\right]^2} p^\alpha k_1^\beta \varepsilon_{\alpha\nu\beta\nu'}.
\end{aligned}
$$

Making use of Lorentz symmetry, one can make the replacement $p^\mu p^\nu \longrightarrow \frac{1}{4}\eta^{\mu\nu}p^2$, which gives rise to

$$\mathsf{T}^{(1)}_{\mu\nu\mu'\nu'}(k_1,k_2) = i\frac{2^{\frac{\delta}{2}}}{256}\int_0^1 \mathrm{d}x \int \frac{\mathrm{d}^\delta \ell}{(2\pi)^\delta}\int \frac{\mathrm{d}^4 p}{(2\pi)^4} \times \frac{\delta^\alpha_{\mu'}(1-2x)k_{1\mu} + \delta^\alpha_\mu\,[2(1-x)k_1 + k_2]_{\mu'}}{\left[p^2 + x(1-x)k_1^2 - \ell^2\right]^2} p^2 k_1^\beta \varepsilon_{\alpha\nu\beta\nu'}. \tag{7.263}$$

After taking into account the contraction of the Kronecker deltas with the ε-tensor and imposing the symmetrization of $(\mu \leftrightarrow \nu)$ and $(\mu' \leftrightarrow \nu')$, one immediately sees that the contribution from $\mathsf{T}^{(1)}$ vanishes.

7F2. $\mathsf{T}^{(2)}_{\mu\nu\mu'\nu'}(k_1,k_2)$

$$\mathsf{T}^{(2)}_{\mu\nu\mu'\nu'}(k_1,k_2) = \frac{1}{256}\int \frac{\mathrm{d}^4 p}{(2\pi)^4}\int \frac{\mathrm{d}^\delta \ell}{(2\pi)^\delta} \times \frac{(2p+k_1)_\mu(2p-k_2)_{\mu'}}{2(p^2-\ell^2)\left[(p-k_2)^2-\ell^2\right]} \underbrace{\mathrm{Tr}\left[\gamma_\nu(\not p + \not\ell)\gamma_{\nu'}(\not p + \not\ell - \not k_2)\gamma_5\right]}_{2^{2+\frac{\delta}{2}} i p^\alpha k_2^\beta \varepsilon_{\nu\alpha\nu'\beta}}. \tag{7.264}$$

As before, one employs the Feynman parametrization and in very strict analogy performs the shift $p \to p + xk_2$. This yields

$$T^{(2)}_{\mu\nu\mu'\nu'}(k_1,k_2) = i\frac{2^{\frac{\delta}{2}}}{128}\int_0^1 \mathrm{d}x \int \frac{\mathrm{d}^\delta \ell}{(2\pi)^\delta}\int \frac{\mathrm{d}^4 p}{(2\pi)^4} \times \frac{(2p+k_1+2xk_2)_\mu(2p-(1-x)k_2)_{\mu'}}{\left[p^2-\ell^2-x(x-1)k_2^2\right]^2} p^\alpha k_2^\beta \varepsilon_{\nu\alpha\nu'\beta}. \tag{7.265}$$

Collecting just the even power of p in the numerator of (7.265) and using Lorentz covariance as before, one immediately obtains

$$\mathsf{T}^{(2)}_{\mu\nu\mu'\nu'}(k_1,k_2) = i\frac{2^{\frac{\delta}{2}}}{256}\int_0^1 \mathrm{d}x \int \frac{\mathrm{d}^\delta \ell}{(2\pi)^\delta}\int \frac{\mathrm{d}^4 p}{(2\pi)^4} \times \frac{\delta^\alpha_\mu(x-1)k_{2\mu'} + \delta^\alpha_{\mu'}(k_1+2xk_2)_\mu}{\left[p^2-\ell^2-x(x-1)k_2^2\right]^2} k_2^\beta \varepsilon_{\nu\alpha\nu'\beta}. \tag{7.266}$$

For the same reasons described previously, after symmetrizations, the contribution from $\mathsf{T}^{(2)}$ vanishes.

Appendix 7F: The KDS Anomaly: Explicit Calculation

The triangle contribution to the Kimura-Delbourgo-Salam anomaly, symbolically represented by $\langle \partial \cdot j_5\, T\, T\rangle$, is

$$\widetilde{T}^{(j5)}_{\mu\nu\mu'\nu'}(k_1,k_2) = \frac{1}{64}\int \frac{d^4p}{(2\pi)^4}\mathrm{tr}\left[\frac{1}{\not p}(2p-k_1)_\mu\gamma_\nu\frac{1}{\not p-\not k_1}(2p-2k_1-k_2)_{\mu'}\gamma_{\nu'}\right.$$
$$\left.\times\,\frac{1}{\not p-\not q}\not q\gamma_5\frac{1}{\not p}+\mathrm{sym}\begin{Bmatrix}\mu\leftrightarrow\nu\\ \mu'\leftrightarrow\nu'\end{Bmatrix}\right] \tag{7.267}$$

We wish to compare it with the $\langle t_5\, T\, T\rangle$ diagram

$$\widetilde{T}^{(5)}_{\mu\nu\mu'\nu'}(k_1,k_2) = -\frac{1}{128}\int \frac{d^4p}{(2\pi)^4}\mathrm{Tr}\left\{\left[\frac{\not p}{p^2}(2p-k_1)_\mu\gamma_\nu+(\mu\leftrightarrow\nu)\right]\frac{(\not p-\not k_1)}{(p-k_1)^2}\right.$$
$$\left.\left[(2p-2k_1-k_2)_{\mu'}\gamma_{\nu'}+(\mu'\leftrightarrow\nu')\right]\frac{(\not p-\not q)}{(p-q)^2}(2\not p-\not q)\gamma_5\right\}. \tag{7.268}$$

The difference between the two is determined by the term

$$\Delta\widetilde{T}^{(5)}_{\mu\nu\mu'\nu'}(k_1,k_2) = -\frac{1}{64}\int \frac{d^4p}{(2\pi)^4}\mathrm{Tr}\left\{\left[\frac{\not p}{p^2}(2p-k_1)_\mu\gamma_\nu+(\mu\leftrightarrow\nu)\right]\frac{(\not p-\not k_1)}{(p-k_1)^2}\right.$$
$$\left.\left[(2p-2k_1-k_2)_{\mu'}\gamma_{\nu'}+(\mu'\leftrightarrow\nu')\right]\frac{(\not p-\not q)}{(p-q)^2}\not p\gamma_5\right\}. \tag{7.269}$$

After regularizing and simplifying, this becomes

$$\Delta\widetilde{T}^{(5)}_{\mu\nu\mu'\nu'}(k_1,k_2) = -\frac{1}{64}\int \frac{d^4p\,d^\delta\ell}{(2\pi)^{4+\delta}}\mathrm{Tr}\left(\gamma_\nu(\not p-\not k_1+\not\ell)\gamma_{\nu'}(\not p-\not q+\not\ell)\gamma_5\right) \tag{7.270}$$
$$\times\frac{(2p-k_1)_\mu(2p-q-k_1)_{\mu'}}{((p-k_1)^2-\ell^2)((p-q)^2-\ell^2)}$$

where the symmetrization $\mu\leftrightarrow\nu$, $\mu'\leftrightarrow\nu'$ has been understood. The ℓ^2 term in the numerator does not contribute. What remains is

$$\Delta\widetilde{T}^{(5)}_{\mu\nu\mu'\nu'}(k_1,k_2) = i\frac{2^{\frac{\delta}{2}}}{16}\int \frac{d^4p\,d^\delta\ell}{(2\pi)^{4+\delta}}\varepsilon_{\nu\nu'\lambda\rho}p^\lambda k_2^\rho\frac{(2p+k_1)_\mu(2p-k_2)_{\mu'}}{(p^2-\ell^2)((p-k_2)^2-\ell^2)} \tag{7.271}$$

after a shift $p \to p + k_1$. Introducing now a Feynman parameter x and a shift $p \to p + xk_2$

$$\Delta\widetilde{T}^{(5)}_{\mu\nu\mu'\nu'}(k_1,k_2) = i\frac{2^{\frac{\delta}{2}}}{16}\int_0^1 \mathrm{d}x \int \frac{\mathrm{d}^4p\mathrm{d}^\delta\ell}{(2\pi)^{4+\delta}}\varepsilon_{\nu\nu'\lambda\rho}p^\lambda k_2^\rho \frac{(2p+k_1+2xk_2)_\mu(2p+(x-1)k_2)_{\mu'}}{(p^2-\ell^2+x(1-x)k_2^2)^2}$$
$$= -\frac{2^{\frac{\delta}{2}}}{32}\int_0^1 \mathrm{d}x \int \frac{\mathrm{d}^4p\mathrm{d}^\delta\ell}{(2\pi)^{4+\delta}}\varepsilon_{\nu\nu'\lambda\rho}k_2^\rho\, p^2 \frac{\left(\delta^\lambda_{\mu'}(k_1+2xk_2)_\mu - \delta^\lambda_\mu(1-x)k_{2\mu'}\right)}{(p^2+\ell^2+x(1-x)k_2^2)^2} \tag{7.272}$$

which yields the factor either $\varepsilon_{\nu\nu'\mu\rho}$ or $\varepsilon_{\nu\nu'\mu'\rho}$. Upon imposing the symmetrization $\mu \leftrightarrow \nu$, $\mu' \leftrightarrow \nu'$, one gets 0.

Let us see now consider the bubble diagrams. With reference to (7.256) and (7.258), we have

$$\Delta D^{(\varepsilon,odd)}_{\mu\nu\mu'\nu'}(k_1,k_2) = \frac{i}{256}\int \frac{\mathrm{d}^4p\mathrm{d}^\delta\ell}{(2\pi)^{4+\delta}} t_{\mu\nu\mu'\nu'\lambda\rho}(k_2-k_1)^\lambda \,\mathrm{tr}\left(\gamma^\rho\gamma_5\frac{\not p - \not q + \not\ell}{(p-q)^2-\ell^2}\gamma_5\right)$$
$$= -i\frac{2^{\frac{\delta}{2}}}{128}\int \frac{\mathrm{d}^4p\mathrm{d}^\delta\ell}{(2\pi)^{4+\delta}} t_{\mu\nu\mu'\nu'\lambda\rho}(k_2-k_1)^\lambda \frac{p^\rho}{p^2-\ell^2} = 0 \tag{7.273}$$

after a shift $p \to p + q$, and

$$\Delta D^{(odd)}_{\mu\nu\mu'\nu'}(k_1,k_2) = \frac{3}{128}\int \frac{\mathrm{d}^4p\mathrm{d}^\delta\ell}{(2\pi)^{4+\delta}}$$
$$\times\mathrm{tr}\left(\left[(2p-q)_\mu\gamma_{\mu'}\eta_{\nu\nu'} + (2p-q)_\mu\gamma_{\nu'}\eta_{\nu\mu'} + \dots\right]\frac{\not p - \not q + \not\ell}{(p-q)^2-\ell^2}\gamma_5\right) = 0, \tag{7.274}$$

respectively.

Conclusion. The metric trace anomaly and the KDS anomaly are the same up to a multiplicative factor.

References

1. M.J. Duff, Observations on conformal anomalies. Nucl. Phys. B **125**, 334 (1977)
2. M.J. Duff, Twenty years of the Weyl anomaly. Class. Quant. Grav. **11**, 1387 (1994). ([hep-th,9308075])
3. M.J. Duff, Weyl, Pontryagin, Euler, Eguchi and Freund. J. Phys. A: Math. Theor. **53**, 301001 (2020). arXiv:2006.03574
4. L. Bonora, A.D. Pereira, B.L. de Souza, Regularization of energy-momentum tensor correlators and parity-odd terms. JHEP **1506**, 24 (2015). arXiv:1503.03326 [hep-th]
5. L. Bonora, Perturbative and non-perturbative trace anomalies. Symmetry **137**, 1292 (2021). e-Print: arXiv:2107.07918 [hep-th]
6. A. Zhiboedov, *A note on three-point functions of conserved currents*. arXiv:1206.6370 [hep-th]. S. Jain, R.R. John, A. Mehta, A.A. Nizami, A. Suresh, Momentum space parity-odd CFT 3-point functions. JHEP **889** (2021). e-Print arXiv:2101.11635 [hep-th]

7. L. Bonora, M. Cvitan, P. Dominis Prester, A. Duarte Pereira, S. Giaccari, T. Štemberga, Axial gravity, massless fermions and trace anomalies. Eur. Phys. J. C **77**, 511 (2017). arXiv:1703.10473 [hep-th]
8. L. Bonora, S. Giaccari, B. Lima de Souza, Trace anomalies in chiral theories revisited. JHEP **1407**, 117 (2014). arXiv:1403.2606 [hep-th]
9. C.-Y. Liu, Investigation of Pontryagin trace anomaly using Pauli-Villars regularization. Nucl. Phys. **B 980**, 115840 (2022). e-Print arXiv:2202.13893 [hep-th]
10. T. Kimura, Divergence of axial-vector current in the gravitational field. Prog. Theor. Phys. **42**, 1191 (1969)
11. R. Delbourgo, A. Salam, PCAC anomalies and gravitation, preprint IC/72/86
12. R. Delbourgo, A. Salam, The gravitational correctioin to PCAC. Phys. Lett. **B 40**, 381 (1972)
13. L. Bonora, Elusive anomalies. EPL **139**, 44001 (2022). e-Print arXiv:2207.03279 [hep-th]

Part IV
Nonperturbative Methods. (A) Heat Kernel

Part IV of the book is devoted to non-perturbative methods for deriving the same anomalies we have calculated before in perturbative ways. In this first sector (A), we consider the heat kernel-like methods: the Schwinger-DeWitt proper time method, the analytical method à la Seeley and the zeta function methods. All these methods are characterized by the fact that they start from the definition of the kinetic operator of a theory, and they square it if necessary, like in the case of the Dirac operator, in order to obtain a quadratic elliptic operator after a Wick rotation. Then they heuristically adopt some mathematical formula in order to define the determinant of this quadratic operator. The logarithm of the determinant is identified with the effective action corresponding to the initial classical action. Taking its variation with respect to a symmetry parameter of the classical action, one can verify whether it vanishes or not. In the second case, we have to do with an anomaly. The difference with respect to the perturbative derivations, where one obtains an approximate expression (usually the lowest order approximation), is that in many important cases, one is able to derive the complete expression of the anomaly. One can therefore compare the results obtained with the two methods, at least in the simplest cases, and check that they coincide. Then one can *bona fide* extend the reach of non-perturbative methods to cases inaccessible to perturbative ones. The coincidence between perturbative and non-perturbative results is a guarantee that the inevitable heuristic steps that characterize both methods are reasonable and acceptable. We shall refer to the methods considered so far, perturbative and non-perturbative, as algorithmic methods.

Chapter 8
Functional Non-perturbative Methods

By analytic functional methods, we mean a collection of approaches, including the Schwinger's proper-time method, the heat kernel method, the Seeley-DeWitt and the zeta function regularization (a similar one is the Fujikawa method, see [1, 2]). As we have explained before, the central tool in these approaches is the (full) kinetic operator of the action (or the square thereof in the case of Dirac-like operators). It is crucial for these methods to work that the operator in question, after passing from a Minkowski to a Euclidean background metric, be a quadratic elliptic operator. In this chapter, we first construct various quadratic elliptic operators starting from the Dirac operators of various models. In the second part of the chapter, we explain the terminology and introduce the methods referred to above: heat kernel, Schwinger proper time, Seeley-DeWitt and zeta function, [3–6]. We end this chapter with some general considerations about the anomaly calculus in this non-perturbative context.

8.1 Square Dirac Operators

Below we introduce the Dirac operator $\not{D}$ in the presence of various backgrounds. We analyze in particular its properties under Hermitean conjugation and gauge transformations, in Lorentzian and Euclidean metric background. We next consider the candidates for the square: $\not{D}^2$, $\not{D}^\dagger \not{D}$ and $\overline{\not{D}}\not{D}$.

8.1.1 $i\not{\partial}$

We use a metric $\eta_{\mu\nu}$ with mostly $(-)$ signature. The gamma matrices satisfy $\{\gamma^\mu, \gamma^\nu\} = 2\eta^{\mu\nu}$ and

$$\gamma_\mu^\dagger = \gamma_0 \gamma_\mu \gamma_0.$$

L. Bonora, *Fermions and Anomalies in Quantum Field Theories*, Theoretical and Mathematical Physics, https://doi.org/10.1007/978-3-031-21928-3_8

The generators of the Lorentz group are $\Sigma_{\mu\nu} = \frac{1}{4}[\gamma_\mu, \gamma_\nu]$.

The chirality matrix $\gamma_5 = i\gamma^0\gamma^1\gamma^2\gamma^3$ has the properties

$$\gamma_5^\dagger = \gamma_5, \quad (\gamma_5)^2 = 1, \quad C^{-1}\gamma_5 C = \gamma_5^T.$$

and

$$\mathrm{tr}(\gamma_5\gamma_\mu\gamma_\nu\gamma_\lambda\gamma_\rho) = -4i\varepsilon_{\mu\nu\lambda\rho} \tag{8.1}$$

A Wick rotation means: $x^0 \to \tilde{x}^0 = ix^0$, $k^0 \to \tilde{k}^0 = ik^0$ and $\gamma^0 \to \tilde{\gamma}^0 = i\gamma^0$, while x^i, k_i, γ^i remain unchanged. From now on, a tilde represents a Euclidean object. We have in particular

$$\{\tilde{\gamma}^\mu, \tilde{\gamma}^\nu\} = 2\tilde{\eta}^{\mu\nu} = -2\delta^{\mu\nu} \tag{8.2}$$

and

$$\tilde{\gamma}_\mu^\dagger = -\tilde{\gamma}_\mu \tag{8.3}$$

As we have explained in Sect. 1.5, the chiral matrix γ_5 changes only sign under a Wick rotation.

Let us consider next the operator $\not{D} = i\gamma^\mu\partial_\mu$. Since $(i\partial_\mu)^\dagger = i\partial_\mu$, we have

$$\not{D}^\dagger = \gamma_0\not{D}\gamma_0 \tag{8.4}$$

So $\not{D}$ is not self-adjoint in Lorentzian metric.

Let us consider next $\tilde{\not{D}}$. Using (8.3), we have

$$\tilde{\not{D}}^\dagger = -\tilde{\not{D}} \tag{8.5}$$

The operator $\tilde{\not{D}}$ is anti-self-adjoint in Euclidean metric.

Let us consider next the square operators. We have

$$\left(\not{D}^2\right)^\dagger = \gamma_0\not{D}^2\gamma_0 = \not{D}^2 = -\partial_0^2 + \sum_{i=1}^{3}\partial_i^2 \tag{8.6}$$

while

$$\left(\tilde{\not{D}}^2\right)^\dagger = \tilde{\not{D}}^2 = \tilde{\partial}_0^2 + \sum_{i=1}^{3}\tilde{\partial}_i^2 \tag{8.7}$$

Therefore, the square operator is self-adjoint in both backgrounds, but it is an elliptic operator only in Euclidean metric.

8.1.2 Coupling to a Vector Potential

Let us add to $i\not\partial$ a vector potential $\not V = \gamma^\mu V_\mu$, where $V_\mu = V_\mu^a T^a$ and T^a are anti-Hermitean generators of a Lie algebra. The operator $\not D$ becomes $\not\mathcal{D} = i(\not\partial + \not V)$. In Lorentzian metric, we have

$$(i\not V)^\dagger = \gamma_0 i\not V\gamma_0, \quad \text{so} \quad \not\mathcal{D}^\dagger = \gamma_0\not\mathcal{D}\gamma_0 \tag{8.8}$$

In Euclidean metric, for the same reason as for $\tilde{\not D}$,

$$(i\tilde{\not V})^\dagger = -i\tilde{\not V}, \quad \text{so} \quad \tilde{\not\mathcal{D}}^\dagger = -\tilde{\not\mathcal{D}} \tag{8.9}$$

Therefore, for the squares we have again

$$\left(\not\mathcal{D}^2\right)^\dagger = \gamma_0\not\mathcal{D}^2\gamma_0 \tag{8.10}$$

Here, $\not\mathcal{D}^2 = -\eta^{\mu\nu}(\partial_\mu + V_\mu)(\partial_\nu + V_\nu) - \Sigma^{\mu\nu}V_{\mu\nu}$, where $V_{\mu\nu} = \partial_\mu V_\nu - \partial_\nu V_\mu + [V_\mu, V_\nu]$. In the Euclidean

$$\left(\tilde{\not\mathcal{D}}^2\right)^\dagger = \tilde{\not\mathcal{D}}^2 \tag{8.11}$$

where $\tilde{\not\mathcal{D}}^2 = \delta^{\mu\nu}(\tilde\partial_\mu + \tilde V_\mu)(\tilde\partial_\nu + \tilde V_\nu) - \tilde\Sigma^{\mu\nu}\tilde V_{\mu\nu}$. The $\tilde\Sigma^{\mu\nu}$ are generators of the Lie algebra of SO(4).

Therefore, the square operator is self-adjoint in Euclidean metric, while it is not in Lorentzian metric.

8.1.3 Adding an Axial-Vector Potential

Let us couple to $\not\mathcal{D}$ an axial-vector field $\not A = \gamma^\mu A_\mu$ with $A_\mu = A_\mu^a T^a$. We have $\not\mathscr{D} = \not\mathcal{D} + i\not A\gamma_5$. Now,

$$(i\not A)^\dagger = \gamma_0(i\not A)\gamma_0, \quad (i\not A\gamma_5)^\dagger = \gamma_0(i\not A\gamma_5)\gamma_0, \quad \text{so} \quad \not\mathscr{D}^\dagger = \gamma_0\not\mathscr{D}\gamma_0 \tag{8.12}$$

On the other hand, in a Euclidean background, we assume[1]

$$(i\tilde{\not{A}})^\dagger = i\tilde{\not{A}}, \quad (i\tilde{\not{A}}\gamma_5)^\dagger = -i\tilde{\not{A}}\gamma_5, \quad \text{so} \quad \tilde{\not{\mathscr{D}}}^\dagger = -\tilde{\not{\mathscr{D}}} \tag{8.13}$$

For the squares, we have again

$$\left(\not{\mathscr{D}}^2\right)^\dagger = \gamma_0 \not{\mathscr{D}}^2 \gamma_0 \tag{8.14}$$

where

$$\begin{aligned}\not{\mathscr{D}}^2 &= -\eta^{\mu\nu}\overline{\mathscr{D}}_\mu \mathscr{D}_\nu \\ &- \Sigma^{\mu\nu}\big(V_{\mu\nu} - [A_\mu, A_\nu] \\ &+ \big(\partial_\mu A_\nu - \partial_\nu A_\mu + \{V_\mu, A_\nu\} - \{A_\mu, V_\nu\} + 2(A_\nu\partial_\mu - A_\mu\partial_\nu)\big)\gamma_5\big)\end{aligned} \tag{8.15}$$

and $\mathscr{D}_\mu = \partial_\mu + V_\mu + A_\mu\gamma_5$, $\overline{\mathscr{D}}_\mu = \partial_\mu + V_\mu - A_\mu\gamma_5$.

In a Euclidean background,

$$\left(\tilde{\not{\mathscr{D}}}^2\right)^\dagger = \tilde{\not{\mathscr{D}}}^2 \tag{8.16}$$

where

$$\begin{aligned}\tilde{\not{\mathscr{D}}}^2 &= \delta^{\mu\nu}\overline{\tilde{\mathscr{D}}}_\mu \tilde{\mathscr{D}}_\nu \\ &- \tilde{\Sigma}^{\mu\nu}\big(\tilde{V}_{\mu\nu} - [\tilde{A}_\mu, \tilde{A}_\nu] + \big(\tilde{\partial}_\mu \tilde{A}_\nu \\ &- \tilde{\partial}_\nu \tilde{A}_\mu + \{\tilde{V}_\mu, \tilde{A}_\nu\} - \{\tilde{A}_\mu, \tilde{V}_\nu\} + 2(\tilde{A}_\nu\tilde{\partial}_\mu - \tilde{A}_\mu\tilde{\partial}_\nu)\big)\gamma_5\big)\end{aligned} \tag{8.17}$$

and $\tilde{\mathscr{D}}_\mu = \tilde{\partial}_\mu + \tilde{V}_\mu + \tilde{A}_\mu\gamma_5$, $\overline{\tilde{\mathscr{D}}}_\mu = \tilde{\partial}_\mu + \tilde{V}_\mu - \tilde{A}_\mu\gamma_5$.

Therefore, the square operator is self-adjoint in Euclidean metric, while it is not in Lorentzian metric.

We can consider also the quadratic operator $\overline{\not{\mathscr{D}}}\not{\mathscr{D}}$, where $\overline{\not{\mathscr{D}}} = i\left(\not{\partial} + \not{V} - \not{A}\gamma_5\right)$, which has a complicated Hermitean conjugate:

$$(\overline{\not{\mathscr{D}}}\not{\mathscr{D}})^\dagger = \gamma_0(\not{\mathscr{D}}\overline{\not{\mathscr{D}}})\gamma_0$$

but

$$\overline{\not{\mathscr{D}}}\not{\mathscr{D}} = -\eta^{\mu\nu}\mathscr{D}_\mu\mathscr{D}_\nu - \Sigma^{\mu\nu}\left(V_{\mu\nu} + [A_\mu, A_\nu] + \left(\partial_\mu A_\nu - \partial_\nu A_\mu + [V_\mu, A_\nu] + [A_\mu, V_\nu]\right)\gamma_5\right) \tag{8.18}$$

[1] This means that $\tilde{A}_0 = A_0$ and $\tilde{A}_k = -iA_k$ (while $\tilde{V}_0 = iV_0$ and $\tilde{V}_k = V_k$), which guarantees self-adjointness of $\tilde{\not{\mathscr{D}}}^2$. This 'complexification' of A^a_μ is not the unique possible choice, but considerably simplifies the formalism.

In the Euclidean, we have

$$(\overline{\tilde{\not{\mathscr{D}}}}\tilde{\not{\mathscr{D}}})^{\dagger} = \tilde{\not{\mathscr{D}}}\overline{\tilde{\not{\mathscr{D}}}} \tag{8.19}$$

and

$$\overline{\tilde{\not{\mathscr{D}}}}\tilde{\not{\mathscr{D}}} = \delta^{\mu\nu}\tilde{\mathscr{D}}_{\mu}\tilde{\mathscr{D}}_{\nu} - \tilde{\Sigma}^{\mu\nu}\left(\tilde{V}_{\mu\nu} + [\tilde{A}_{\mu}, \tilde{A}_{\nu}] + \left(\tilde{\partial}_{\mu}\tilde{A}_{\nu} - \tilde{\partial}_{\nu}\tilde{A}_{\mu} + [\tilde{V}_{\mu}, \tilde{A}_{\nu}] + [\tilde{A}_{\mu}, \tilde{V}_{\nu}]\right)\gamma_5\right) \tag{8.20}$$

8.2 Gauge Transformations

The ordinary gauge transformation for a vector field V_{μ} is

$$\delta_{\lambda}V_{\mu} = \partial_{\mu}\lambda + [V_{\mu}, \lambda] \tag{8.21}$$

where $\lambda = \lambda^{a}(x)T^{a}$. Then

$$\delta_{\lambda}\not{D} = [\not{D}, \lambda], \quad \delta_{\lambda}\not{D}^{2} = [\not{D}^{2}, \lambda] \tag{8.22}$$

The operator $\not{D}^{\dagger}\not{D}$ is not covariant.

In Euclidean metric, the transformations are a transcription of those in Lorentz metric.

$$\delta_{\lambda}\tilde{V}_{\mu} = \tilde{\partial}_{\mu}\lambda + [\tilde{V}_{\mu}, \lambda] \tag{8.23}$$

and

$$\delta_{\lambda}\tilde{\not{D}} = [\tilde{\not{D}}, \lambda], \quad \delta_{\lambda}\tilde{\not{D}}^{2} = [\tilde{\not{D}}^{2}, \lambda] \tag{8.24}$$

8.2.1 *The Axial-Vector Case*

Let us introduce $\mathscr{V}_{\mu} = V_{\mu} + A_{\mu}\gamma_5$. The axial-vector gauge transformations are expressed in terms of the parameter $\Upsilon = \lambda + \rho\gamma_5$, where $\rho = \rho^{a}(x)T^{a}$,

$$\delta_{\Upsilon}\mathscr{V}_{\mu} = \partial_{\mu}\Upsilon + [\mathscr{V}_{\mu}, \Upsilon] \tag{8.25}$$

which means

$$\begin{aligned}\delta_\Upsilon V_\mu &= \partial_\mu\lambda + [V_\mu, \lambda] + [A_\mu, \rho]\\ \delta_\Upsilon A_\mu &= \partial_\mu\rho + [V_\mu, \rho] + [A_\mu, \lambda]\end{aligned} \tag{8.26}$$

Let a Dirac spinor transform like $\delta_\Upsilon\Psi = -\Upsilon\psi$. Then,

$$\delta_\Upsilon \not{\mathcal{D}}\psi = -\overline{\Upsilon}\not{\mathcal{D}}\psi \tag{8.27}$$

where $\overline{\Upsilon} = \lambda - \rho\gamma_5$. This implies

$$\delta_\Upsilon \not{\mathcal{D}} = [\not{\mathcal{D}}, \Upsilon] + 2\gamma_5\rho\not{\mathcal{D}} \tag{8.28}$$

It follows that $\not{\mathcal{D}}^2$ is not adjoint-covariant, while, on the contrary, we have

$$\delta_\Upsilon\left(\overline{\not{\mathcal{D}}}\not{\mathcal{D}}\psi\right) = -\Upsilon\left(\overline{\not{\mathcal{D}}}\not{\mathcal{D}}\psi\right), \quad \text{i.e.} \quad \delta_\Upsilon\left(\overline{\not{\mathcal{D}}}\not{\mathcal{D}}\right) = [\overline{\not{\mathcal{D}}}\not{\mathcal{D}}, \Upsilon] \tag{8.29}$$

Since in proving (8.27) only the property $\{\gamma_5, \gamma_\mu\} = 0$ is used, and $\{\gamma_5, \tilde{\gamma}_\mu\} = 0$ is still valid in Euclidean metric, we have immediately

$$\delta_\Upsilon\widetilde{\not{\mathcal{D}}}\psi = -\overline{\Upsilon}\widetilde{\not{\mathcal{D}}}\psi, \quad \delta_\Upsilon\widetilde{\not{\mathcal{D}}} = [\widetilde{\not{\mathcal{D}}}, \Upsilon] + 2\gamma_5\rho\widetilde{\not{\mathcal{D}}} \tag{8.30}$$

and

$$\delta_\Upsilon\left(\overline{\widetilde{\not{\mathcal{D}}}}\widetilde{\not{\mathcal{D}}}\right) = [\overline{\widetilde{\not{\mathcal{D}}}}\widetilde{\not{\mathcal{D}}}, \Upsilon] \tag{8.31}$$

8.2.2 Non-trivial Background Metric

In, the case of a non-trivial metric background the relevant operator is $\not{\nabla} = \not{\mathfrak{D}} + \frac{1}{2}\not{\Omega}$, where $\not{\mathfrak{D}} = i\gamma^a e_a^\mu \mathfrak{D}_\mu$ and $\not{\Omega} = i\gamma^a e_a^\mu \Omega_\mu^{ab}\Sigma_{ab}$. e_a^μ is the inverse vierbein and $\Sigma_{ab} = \frac{1}{4}[\gamma_a, \gamma_b]$ are again the anti-Hermitean Lorentz generators.

$$\left(\not{\nabla}^2\right)^\dagger = \gamma_0\not{\nabla}^2\gamma_0, \tag{8.32}$$

and

$$\left(\tilde{\not{\nabla}}^2\right)^\dagger = \tilde{\not{\nabla}}^2 \tag{8.33}$$

when the metric is Euclidean. Concerning the gauge transformations nothing changes with respect to the operators $\not{\mathfrak{D}}^2$ and $\not{\mathcal{D}}^2$. We have

$$\delta_\lambda \tilde{\nabla}\!\!\!/^2 = [\tilde{\nabla}\!\!\!/^2, \lambda] \tag{8.34}$$

In this case, on the basis of a subsequent discussion (see below), no gauge anomaly is expected from the effective action generated by $\tilde{\nabla}\!\!\!/^2$.

Since the metric background is non-trivial, we have to take account of diffeomorphisms. Under a general coordinate transformation $x^\mu \to x^\mu + \xi^\mu(x)$, we have

$$\delta_\xi \tilde{\nabla}\!\!\!/ = [\xi \cdot \partial, \tilde{\nabla}\!\!\!/], \quad \delta_\xi \tilde{\nabla}\!\!\!/^2 = [\xi \cdot \partial, \tilde{\nabla}\!\!\!/^2] \tag{8.35}$$

So also in this case, no diffeomorphism anomalies are expected, but there can be a trace anomaly because the corresponding transformation is not of the adjoint type:

$$\delta_\omega \tilde{\nabla}\!\!\!/ = -\frac{1}{2} \{\tilde{\nabla}\!\!\!/, \omega\} \tag{8.36}$$

Next, we can add an axial potential A to the vector one. The relevant operator is $\boldsymbol{\nabla}\!\!\!/ = \mathcal{D}\!\!\!/ + \frac{1}{2}\Omega\!\!\!/ = \tilde{\nabla}\!\!\!/ + i A\!\!\!/\gamma_5$, where $A\!\!\!/ = \gamma^a e_a^\mu A_\mu$. We have

$$\left(\boldsymbol{\nabla}\!\!\!/^2\right)^\dagger = \gamma_0 \boldsymbol{\nabla}\!\!\!/^2 \gamma_0, \tag{8.37}$$

and

$$\left(\tilde{\boldsymbol{\nabla}\!\!\!/}^2\right)^\dagger = \tilde{\boldsymbol{\nabla}\!\!\!/}^2 \tag{8.38}$$

when the metric is Euclidean. Moreover,

$$(\overline{\boldsymbol{\nabla}\!\!\!/}\boldsymbol{\nabla}\!\!\!/)^\dagger = \gamma_0 (\boldsymbol{\nabla}\!\!\!/\overline{\boldsymbol{\nabla}\!\!\!/})\gamma_0 \tag{8.39}$$

and

$$(\overline{\tilde{\boldsymbol{\nabla}\!\!\!/}}\tilde{\boldsymbol{\nabla}\!\!\!/})^\dagger = (\tilde{\boldsymbol{\nabla}\!\!\!/}\overline{\tilde{\boldsymbol{\nabla}\!\!\!/}}) \tag{8.40}$$

when the metric is Euclidean. A bar denotes, as usual, changing sign of A. As for the axial-complex gauge transformations $\Upsilon = \lambda + \gamma_5 \rho$, we have

$$\delta_\Upsilon \boldsymbol{\nabla}\!\!\!/ \psi = -\overline{\Upsilon} \boldsymbol{\nabla}\!\!\!/ \psi \tag{8.41}$$

It follows that $\boldsymbol{\nabla}\!\!\!/^2$ is not adjoint covariant, which may generate anomalies. On the other hand,

$$\delta_\Upsilon (\overline{\boldsymbol{\nabla}\!\!\!/}\boldsymbol{\nabla}\!\!\!/) = [\overline{\boldsymbol{\nabla}\!\!\!/}\boldsymbol{\nabla}\!\!\!/, \Upsilon], \quad \text{and} \quad \delta_\Upsilon (\overline{\tilde{\boldsymbol{\nabla}\!\!\!/}}\tilde{\boldsymbol{\nabla}\!\!\!/}) = [\overline{\tilde{\boldsymbol{\nabla}\!\!\!/}}\tilde{\boldsymbol{\nabla}\!\!\!/}, \Upsilon]. \tag{8.42}$$

which leads to vector and axial current conservation (no anomalies expected).

There are no diffeomorphism anomalies because

$$\delta_\xi \not{\nabla} = [\xi \cdot \partial, \not{\nabla}], \quad \delta_\xi \overline{\not{\nabla}} = [\xi \cdot \partial, \overline{\not{\nabla}}] \tag{8.43}$$

but there can be trace anomalies because

$$\delta_\omega \not{\nabla} = -\frac{1}{2}\left\{\not{\nabla}, \omega\right\} \tag{8.44}$$

8.3 Heat Kernel, Schwinger Proper Time and Seeley-DeWitt Method

Let us call $\mathcal{F}$ any previously defined quadratic Dirac operator

We identify the effective action for Dirac fermions with

$$W = -\frac{i}{2}\mathrm{Tr}\,(\ln\,\mathcal{F}) \tag{8.45}$$

Tr includes all the traces plus the spacetime integration. Any variation of (8.45) is given by

$$\delta W = \frac{i}{2}\mathrm{Tr}\,(\mathcal{G}\,\delta\,\mathcal{F}) \tag{8.46}$$

where, formally,

$$\mathcal{F}\,\mathcal{G} = -1 \tag{8.47}$$

So, we can write formally

$$\delta W = \delta\left(-\frac{1}{2}\int_0^\infty \frac{\mathrm{d}s}{is}\mathrm{Tr}\left(\mathrm{e}^{i\mathcal{F}s}\right)\right) = -\frac{1}{2}\mathrm{Tr}\left(\int_0^\infty \mathrm{d}s\,\mathrm{e}^{i\mathcal{F}s}\delta\mathcal{F}\right). \tag{8.48}$$

It follows that W can be represented as

$$W = -\frac{1}{2}\int_0^\infty \frac{\mathrm{d}s}{is}\mathrm{Tr}\left(\mathrm{e}^{i\mathcal{F}s}\right) + \mathrm{const} \equiv \mathsf{L} + \mathrm{const}, \tag{8.49}$$

where L is the relevant effective action

$$\mathsf{L} = \int \mathrm{d}^\mathrm{d}x\, L(x) \equiv \int \mathrm{d}^\mathrm{d}x\, \langle x|\mathsf{L}|x\rangle, \tag{8.50}$$

which is better written as

$$L(x) = -\frac{1}{4} \lim_{x' \to x} \operatorname{tr} \int_0^\infty \frac{ds}{is} K(x, x', s), \tag{8.51}$$

where the kernel K is defined by

$$K(x, x', s) = \langle x | \mathrm{e}^{i\mathcal{F}s} | x' \rangle \tag{8.52}$$

Inserted in δW, under the symbol Tr, it means integrating over x after taking the limit $x' \to x$.

Remark. In all the previous, as well as subsequent, s integrals we understand the $i\epsilon$ prescription, i.e. the substitution $\mathcal{F} \to \mathcal{F} + i\epsilon$ with infinitesimal $\epsilon > 0$. Equation (8.45) is suggested by the formula

$$\frac{i}{A + i\epsilon} = \int_0^\infty \mathrm{d}s\, \mathrm{e}^{is(A+i\epsilon)} \tag{8.53}$$

which holds for a real number A and extended to an operator by representing it by its eigenvalues. The factor $\frac{1}{2}$ is, of course, because our formulas are applied to a squared operator.

With such an understanding, the previous *Schwinger proper-time representation* holds if the quadratic operator is elliptic, therefore in a Euclidean background. For definiteness from now on we understand that $\mathcal{F} = -\mathfrak{D}^2$, which, in a Euclidean background, becomes the elliptic operator $-\tilde{\mathfrak{D}}^2$.

Schwinger-DeWitt Method

Let us define the amplitude

$$\langle x, s | x', 0 \rangle = \langle x | \mathrm{e}^{i\mathcal{F}s} | x' \rangle, \tag{8.54}$$

which satisfies the (heat kernel) differential equation

$$i \frac{\partial}{\partial s} \langle x, s | x', 0 \rangle = -\mathcal{F}_x \langle x, s | x', 0 \rangle, \tag{8.55}$$

where $\mathfrak{F}_x$ is, for instance, the differential operator

$$\mathfrak{F}_x = \nabla_\mu g^{\mu\nu}\nabla_\nu - \frac{1}{4}R + \mathcal{V}, \tag{8.56}$$

and $\mathcal{V} = \Sigma^{ab}e^\mu_a e^\nu_b \left(\partial_\mu V_\nu - \partial_\nu V_\mu + [V_\mu, V_\nu]\right)$. Here, we have introduced also the dependence on a non-trivial metric $g_{\mu\nu}$ and $\nabla_\mu = \mathfrak{D}_\mu + \frac{1}{2}\Omega_\mu$, where $\Omega_\mu = \Omega^{ab}_\mu \Sigma_{ab}$ is the spin connection. The insertion of a non-trivial background metric calls for a covariant limit $x' \to x$. Then, we make the ansatz

$$\langle x, s|x', 0\rangle = -\lim_{m\to 0} \frac{i}{(4\pi s)^{\frac{\mathsf{d}}{2}}} \sqrt{\mathfrak{D}(x, x')} e^{i\left(\frac{\sigma(x,x')}{2s} - m^2 s\right)} \Phi(x, x', s), \tag{8.57}$$

where $\mathfrak{D}(x, x')$ is the Van Vleck-Morette (VVM) determinant and $\sigma(x, x')$ is the world function (a detailed presentation of these objects in a more general setting will be given further on in Chap. 10). $\Phi(x, x', s)$ is a function to be determined. It is useful to introduce also the mass parameter m, to guarantee convergence for the s integration at infinity.[2] In the limit $s \to 0$ the RHS of (8.57) becomes the definition of a delta function multiplied by Φ. More precisely, since it must be $\langle x, 0|x', 0\rangle = \delta(x, x')$, and

$$\lim_{s\to 0} \frac{i}{(4\pi s)^{\frac{\mathsf{d}}{2}}} \sqrt{\mathfrak{D}(x, x')}\, e^{i\left(\frac{\sigma(x,x')}{2s} - m^2 s\right)} = \sqrt{|g(x)|}\, \delta^{(\mathsf{d})}(x, x'), \tag{8.58}$$

we must have

$$\lim_{s\to 0} \Phi(x, x', s) = \mathbf{1}. \tag{8.59}$$

The use of (8.57) and (8.58) guarantees that covariance with respect to diffeomorphisms is preserved. Now, the next step consists in determining $\Phi(x, x', s)$.

Looking at (8.57), in dimension d, we can make the identification

$$K(x, x, s) = \frac{i}{(4\pi i s)^{\frac{\mathsf{d}}{2}}} \sqrt{g}\, e^{-im^2 s}[\Phi](x, x, s). \tag{8.60}$$

From now on the symbol $[H](x, \ldots)$ will denote the coincidence limit $\lim_{x'\to x} H(x, x', \ldots)$.

Let us focus on $\mathsf{d} = 4$. Equation (8.55) becomes an equation for $\Phi(x, x', s)$. Using

$$\frac{1}{2}\sigma_{;\mu}\sigma_{;}{}^{\mu} = \sigma \tag{8.61}$$

[2] This requires a suitable $i\epsilon$ prescription also for m^2.

and

$$(\mathfrak{D}(x,x')\sigma^{;\mu})_{;\mu} = 4\,\mathfrak{D}(x,x'), \tag{8.62}$$

after some algebra one gets

$$i\frac{\partial \Phi}{\partial s} + \frac{i}{s}\nabla^\mu \Phi \nabla_\mu \sigma + \frac{1}{\sqrt{\mathfrak{D}}}\nabla^\mu \nabla_\mu \left(\sqrt{\mathfrak{D}}\Phi\right) - \left(\frac{1}{4}R - \mathcal{V} - m^2\right)\Phi = 0 \tag{8.63}$$

Now, we expand

$$\Phi(x,x',s) = \sum_{n=0}^{\infty} a_n(x,x')(is)^n \tag{8.64}$$

with the boundary condition $[a_0] = 1$. The a_n must satisfy the recursive relations:

$$\begin{aligned}(n+1)a_{n+1} + \nabla^\mu a_{n+1}\nabla_\mu \sigma - \frac{1}{\sqrt{\mathfrak{D}}}\nabla^\mu \nabla_\mu \left(\sqrt{\mathfrak{D}}a_n\right) \\ + \left(\frac{1}{4}R - \mathcal{V} - m^2\right)a_n = 0\end{aligned} \tag{8.65}$$

Using these relations and the coincidence results (see below), it is possible to compute each coefficient a_n at the coincidence limit $x' \to x$. The latter are denoted by $[a_n](x)$. Once we know the coefficients, we can reconstruct the effective action by partially integrating over s and using d as a regulator. For the purpose of the analytic continuation in d, we multiply L by $\mu^{-\mathsf{d}}$, where μ is a mass parameter, so as to render L dimensionless. For a Dirac fermion the properly normalized formula is

$$\frac{L(x)}{\mu^{\mathsf{d}}} = -\frac{i}{2}(4\pi\mu^2)\mathrm{tr}\int_0^\infty ds\,(4\pi i\mu^2 s)^{-\frac{\mathsf{d}}{2}-1}\sqrt{g}e^{-im^2 s}[\Phi(x,x,s)] \tag{8.66}$$

For instance, near $\mathsf{d} = 2$, this can be rewritten as

$$\begin{aligned}L(x) = &\frac{1}{4\pi}\left(\frac{1}{\mathsf{d}-2} - \frac{1}{2}\right)\mathrm{tr}\left(([a_1] - m^2)\sqrt{g}\right) \\ &-\frac{i}{8\pi}\mathrm{tr}\int_0^\infty ds\,\ln(4\pi i\mu^2 s)\sqrt{g}\frac{\partial^2}{\partial(is)^2}\left(e^{-im^2 s}[\Phi(x,x,s)]\right)\end{aligned} \tag{8.67}$$

where tr denotes the trace over gamma matrices and possible Lie algebra generators. The anomaly is produced by taking the variation with respect to the symmetry parameter. If the variation turns out to be proportional to $\mathsf{d} - 2$, it will give a finite result for $\mathsf{d} \to 2$ (see the calculations of trace anomalies, for instance). The second line involves terms with higher canonical dimensions, compensated for by negative powers of μ^2, which cannot contribute to the $\mathsf{d} = 2$ anomaly. In 4d, we have a pole $\frac{1}{\mathsf{d}-4}$, and so on.

We refer to the previous method as the *Schwinger-DeWitt method* . This method is fit for trace anomalies. Another application of the SDW method will be illustrated in the next chapter, while deriving chiral anomalies.

ζ-Function Regularization. Seeley-DeWitt Method

Given a differential operator A in analogy with the Riemann ζ function, the expression A^{-z}, for complex z, is called ζ function regularization of A:

$$\zeta(z, A) = A^{-z} = \frac{1}{\Gamma(z)} \int_0^\infty \mathrm{d}t\, t^{z-1}\, \mathrm{e}^{-tA}. \tag{8.68}$$

We will apply this representation to the operator $\mathcal{F}(x)$ defined by :

$$(\mathcal{F}(x))^{-z} = \frac{1}{\Gamma(z)} \int_0^\infty \mathrm{d}t\, t^{z-1}\, \langle x|\mathrm{e}^{-t\mathcal{F}}|x\rangle, \tag{8.69}$$

where $\langle x|\mathrm{e}^{-t\mathcal{F}}|x\rangle$ means the coincidence limit of $\langle x|\mathrm{e}^{-t\mathcal{F}}|x'\rangle$. Equation (8.69) is not quite accurate because only dimensionless quantities can be raised to an arbitrary power. Moreover, the object of interest will be $\mathcal{G}$, rather than $\mathcal{F}$. Thus, we introduce a mass parameter μ and shift from t to $is\mu$.

$$\zeta(x, z) \equiv (\mu^2\mathcal{G}(x, x))^z = \frac{1}{\Gamma(z)} \int_0^\infty (i\mu^2)\mathrm{d}s\, (is\mu^2)^{z-1}\, \langle x|\mathrm{e}^{is\mathcal{F}}x\rangle. \tag{8.70}$$

Finally, we replace $\langle x|\mathrm{e}^{is\mathcal{F}}|x\rangle$ with $K(x, x, s) = \lim_{x'\to x} K(x, x', s)$, see (8.52). The result is

$$\zeta(x, z) = (\mu^2\mathcal{G}(x, x))^z = \frac{i}{\Gamma(z)} \frac{\mu^d}{(4\pi)^{\frac{d}{2}}} \sqrt{g} \int_0^\infty (i\mu^2)\mathrm{d}s\, (is\mu^2)^{z-1-\frac{d}{2}}\, \mathrm{e}^{-im^2 s}[\Phi(x, x, s)], \tag{8.71}$$

which can be rewritten as

$$\zeta(x, z) = (\mu^2\mathcal{G}(x, x))^z = -\frac{i}{\Gamma(z)} \frac{\mu^{\mathsf{d}-4}}{(4\pi)^{\frac{\mathsf{d}}{2}}} \frac{\sqrt{g}}{(z - \frac{\mathsf{d}}{2})(z - \frac{\mathsf{d}}{2} + 1)(z - \frac{\mathsf{d}}{2} + 2)}$$
$$\times \int_0^\infty d(is)\, (is\mu^2)^{z-\frac{\mathsf{d}}{2}+2} \frac{\partial^3}{\partial(is)^3} \left(\mathrm{e}^{-im^2 s}[\Phi(x, x, s)]\right). \tag{8.72}$$

This is well defined for $\mathsf{d} = 4$ at $z = 0$.

$$\zeta(x, 0) = \frac{i\sqrt{g}}{2(4\pi)^2} \left[\frac{\partial^2}{\partial(is)^2} \left(\mathrm{e}^{-im^2 s} [\Phi(x, x, s)] \right) \right]_{s=0} \tag{8.73}$$

Now, differentiating (8.68) with respect to z and evaluating at $z = 0$, we get formally

$$\frac{\mathrm{d}}{\mathrm{d}z} \zeta(z, A)|_{z=0} = -\mathrm{Tr} \ln A \tag{8.74}$$

This suggest the procedure to regularize W (which is the trace of a log). More precisely

$$W \to W_\zeta = \frac{i}{2} \zeta'(0), \quad \text{where} \quad \zeta(z) = \int \mathrm{tr}\, \zeta(x, z) d^{\mathsf{d}} x \tag{8.75}$$

Now, suppose that the operator A, under a symmetry transformation with parameter ϵ, transforms as

$$\delta_\epsilon A = \{A, \epsilon\}. \tag{8.76}$$

Then,

$$\delta_\epsilon \mathrm{Tr} A^{-z} = -2z \mathrm{Tr} \left(A^{-z} \epsilon \right) = -2z \mathrm{Tr} \left(\zeta(z, A) \epsilon \right) \tag{8.77}$$

Since the relevant result is obtained by differentiating with respect to z and setting $z = 0$, once the functional is regularized, the anomalous part of the effective action is very easy to derive:

$$L_{\mathcal{A}} = -2\mathrm{Tr} \left(\zeta(0, A) \epsilon \right) \tag{8.78}$$

8.3.1 General Considerations Concerning Anomalies

It is evident from (8.48) that if $\delta\mathfrak{F}$ is a variation of the adjoint type, like (8.22) and (8.24), the variation δW will vanish due to the trace. In that case, therefore, we do not expect any anomaly. This is true also for the axial-vector case, provided $\mathfrak{F}$ is represented by $\overline{\mathfrak{D}}\mathfrak{D}$ or $\widetilde{\overline{\mathfrak{D}}}\widetilde{\mathfrak{D}}$, see (8.29) or (8.31). In all such cases, no anomalies can arise.

Anomalies may arise when the variation of $\delta\mathfrak{F}$ is not of the adjoint type. For instance in the case of Weyl transformations, $\delta g_{\mu\nu} = 2\omega g_{\mu\nu}$, we have

$$\delta_\omega \left(\not{\partial} + \frac{1}{2} \not{\Omega} \right) = -\frac{1}{2} \left\{ \not{\partial} + \frac{1}{2} \not{\Omega}, \omega \right\} \tag{8.79}$$

Or in the case of a vector-axial-vector gauge background, if we use for $\mathfrak{F}$ the square operator $\not{\mathcal{D}}^2$ or $\widetilde{\not{\mathcal{D}}}^2$, as is evident from the expressions (8.28) compared to (8.29). In such cases, anomalies are expected, and do in fact appear.

This raises a problem. Both $\widetilde{\not{\mathcal{D}}}^2$ and $\overline{\widetilde{\not{\mathcal{D}}}}\widetilde{\not{\mathcal{D}}}$ are elliptic operators, but while $\widetilde{\not{\mathcal{D}}}^2$ is self-adjoint, $\overline{\widetilde{\not{\mathcal{D}}}}\widetilde{\not{\mathcal{D}}}$ is not. The question is:

- *do we have to use* $\widetilde{\not{\mathcal{D}}}^2$ *or* $\overline{\widetilde{\not{\mathcal{D}}}}\widetilde{\not{\mathcal{D}}}$ *in order to define the fermion determinant*?

As already pointed out, the operator $\widetilde{\not{\mathcal{D}}}^2$ is not covariant with respect to the variation δ_Υ, so it may (and in fact it does, see below) produce an anomaly. The operator $\overline{\widetilde{\not{\mathcal{D}}}}\widetilde{\not{\mathcal{D}}}$, instead, is adjoint-covariant [(see (8.31)], and it does not produce any anomaly. They are both elliptic, their other important difference is that $\widetilde{\not{\mathcal{D}}}^2$ is self-adjoint, while the other is not. But, as one can see from (8.19), the obstruction to self-adjointness of $\overline{\widetilde{\not{\mathcal{D}}}}\widetilde{\not{\mathcal{D}}}$ is due to the A-dependent part. If $A = 0$, $\overline{\widetilde{\not{\mathcal{D}}}}\widetilde{\not{\mathcal{D}}} = \widetilde{\not{\mathcal{D}}}^2$ becomes self-adjoint. If $A \neq 0$, Hermitean conjugation changes, the sign of A. If A is small the A-dependent part of $\overline{\widetilde{\not{\mathcal{D}}}}\widetilde{\not{\mathcal{D}}}$ is small, as a consequence the structure of eigenvalues will not change significantly with respect to $\widetilde{\not{\mathcal{D}}}^2$. It would seem, therefore, that we could use both quadratic operators. However, the fact that $\overline{\widetilde{\not{\mathcal{D}}}}\widetilde{\not{\mathcal{D}}}$ is not self-adjoint means that a part of the eigenvalues may be outside the real axis. In view of the remark after (8.52) on the ϵ-prescription, one sees immediately that the presence of such non-real eigenvalues puts in danger the convergence of such integrals as (8.49). This crucial observation compels us to use $\widetilde{\not{\mathcal{D}}}^2$.

Similar considerations will prevail in analogous cases.

A final important remark concerns diffeomorphisms. It is impossible to derive diffeomorphism anomalies with the SDW method presented above. The reason is that, as already observed, this method is taylor-made to preserve diffeomorphism invariance: One uses the point-splitting method based on a covariant square line element (the world function), as a consequence all the derivations are covariant and the results cannot be but covariant. Any attempt to break this covariance in such a framework is completely 'unnatural'. We cannot exclude that the heat kernel equation might be utilized to derive diffeomorphism anomalies, but, no doubt, this would require a different approach. This marks an important difference between the SDW and all the other methods, because diffeomorphism anomalies can be derived both with perturbative methods and with the family's index theorem (see below). Our present prescription while using the SDW method is that the choice of the square Dirac operator must be made in such a way as to conserve diffeomorphisms (also axial diffeomorphisms, as in Chap. 10 below). It trivially follows that the SDW method is applicable only when we know that diffeomorphisms are conserved.

Similar conclusions hold also for local Lorentz transformations and symmetry, see Chap. 10.

References

1. K. Fujikawa, On the evaluation of chiral anomaly in gauge theories with γ_5 couplings. Phys. Rev. D **29**, 285 (1984)
2. K. Fujikawa, H. Suzuki, *Path Integrals and Quantum Anomalies* (Oxford Science Publications, 2004)
3. R.T. Seeley, Complex powers of an elliptic operator, in *Proceedings of Symposia in Pure Mathematics*, vol. 10 (American Mathematical Society, 1967), pp. 288–307
4. B.S. DeWitt, *Global Approach to Quantum Field Theory*, vols. I and II (Oxford University Press, Oxford, 2003)
5. K. Kirsten, Heat kernel asymptotics: more special case calculations. Nucl. Phys. B (Proc. Suppl.) **104**, 119–126 (2002)
6. D.V. Vassilevich, Heat Kernel expansion: user's manual. Phys. Rep. **388**, 279–360 (2003)

Chapter 9
Explicit Non-perturbative Derivations

In this chapter, we apply the non-perturbative approaches outlined above to explicit anomaly calculations. As a first example, we compute the anomaly of a Dirac fermion coupled to a vector and an axial potential in a flat spacetime, and obtain Bardeen's result. We use two procedures: first a somewhat simplified heat kernel approach, although in the same mood as the Schwinger-DeWitt one; then, we employ the full arsenal of the Schwinger-DeWitt method. Eventually, anyhow one realizes that the first method is a sort of 'diagonal' version of the SDW (Appendix 9A). Next, we apply the Schwinger-DeWitt method to extract directly the consistent chiral anomaly of a Weyl fermion coupled to a vector potential (without passing through the $V-A$ system).

9.1 Bardeen's Anomaly

In, this section we derive in a non-perturbative way the chiral anomaly of a Dirac fermion coupled to a vector and an axial-vector potential, [1–5]. Let us first analyze the definition of the functional integral. In a Minkowski spacetime, the Dirac operator $\mathcal{D} = i(\partial\!\!\!/ + V\!\!\!/)$ is not self-adjoint, but $\mathcal{D}^\dagger = \gamma_0 \mathcal{D} \gamma_0$. What is needed in defining the determinant of the Dirac operator, however, is, in fact, the corresponding Euclidean operator. In a Euclidean background, we have a natural candidate for the eigenvalue problem, the elliptic self-adjoint operator $\tilde{\mathcal{D}}^2$. If the Euclidean spacetime is compact the eigenvalues are discrete (for simplicity we are using here the alternative Wick rotation—see Sect. 1.5—for which $\tilde{\mathcal{D}}$ is Hermitean):

$$\tilde{\mathcal{D}}^2 \tilde{\chi}_i = \tilde{\mu}_i^2 \tilde{\chi}_i \tag{9.1}$$

With the usual heuristic attitude of quantum field theory, we can therefore define the corresponding functional integral as follows

L. Bonora, *Fermions and Anomalies in Quantum Field Theories*, Theoretical and Mathematical Physics, https://doi.org/10.1007/978-3-031-21928-3_9

$$Z[\widetilde{\not{D}}] = \sqrt{\det(\widetilde{\not{D}}^2)} = \sqrt{\prod_i \tilde{\mu}_i^2} \tag{9.2}$$

Thanks to Eq. (8.22) the eigenvalues $\tilde{\mu}_i$ are gauge invariant. No gauge anomaly is expected in this context.

Let us consider next the case of a vector-axial-vector background. Proceeding as before, the obvious candidate to build a Dirac determinant in a Euclidean spacetime is $\widetilde{\mathcal{D}}^2$, which is self-adjoint. Therefore,

$$\widetilde{\mathcal{D}}^2 \psi_i = \lambda_i^2 \psi_i \tag{9.3}$$

Then we can set

$$Z[\widetilde{\mathcal{D}}] = \sqrt{\det(\widetilde{\mathcal{D}}^2)} = \sqrt{\prod_i \lambda_i^2} \tag{9.4}$$

where λ_i are real numbers. The difference with the previous case is that $\mathcal{D}^2$ (and $\widetilde{\mathcal{D}}^2$) does not behave covariantly under axial gauge transformations, see Eq. (8.28). This gives rise to anomalies, as we shall see below.

Now, in order to regularize the infinite product (9.4), the idea is to put a cutoff on the number of eigenvalues. We refer to this regularization as *mode-cutoff*. Its relation to the heat kernel method is explained in Appendix 9A.

As long as the Euclidean spacetime is compact the eigenvalues are discrete. The compactness hypothesis is not strictly necessary but it simplifies the presentation. We order the λ_i^2 in increasing order and define the projector

$$\widetilde{P}_\Lambda = \sum_{\lambda_i^2 \le \Lambda^2} |\psi_i\rangle\langle\psi_i| \tag{9.5}$$

It obviously commute with $\widetilde{\mathcal{D}}$: $[\widetilde{\mathcal{D}}, \widetilde{P}_\Lambda] = 0$, and it can be analytically represented as follows

$$P_\Lambda = \int_{-\infty}^{+\infty} \frac{d\alpha}{2\pi i\alpha} e^{i\alpha\left(1-\frac{\widetilde{\mathcal{D}}^2}{\Lambda^2}\right)} = \theta\left(1 - \frac{\widetilde{\mathcal{D}}^2}{\Lambda^2}\right) \tag{9.6}$$

The contour integral passes just below the origin and closes with the upper or lower infinite circle in the α-plane according to whether the expression in round brackets is negative or positive. In fact, applying $\widetilde{P}_\Lambda$ to a completeness $\sum_i |\psi_i\rangle\langle\psi_i|$ we obtain exactly (9.5). This allows us to regularize the fermion determinant (9.4)

$$Z_\Lambda[\widetilde{\mathcal{D}}] = \sqrt{\det\left(1 - \widetilde{P}_\Lambda + \widetilde{P}_\Lambda \widetilde{\mathcal{D}}^2 P_\Lambda\right)} \tag{9.7}$$

Given the property $\widetilde{P}^2_\Lambda = \widetilde{P}_\Lambda$, it is easy to prove that, for any variation of the fields we have

$$\widetilde{P}_\Lambda \, \delta\widetilde{P}_\Lambda \, \widetilde{P}_\Lambda = 0 \tag{9.8}$$

Moreover,

$$\left(1 - \widetilde{P}_\Lambda + \widetilde{P}_\Lambda \widetilde{\mathcal{D}}^2 \widetilde{P}_\Lambda\right)^{-1} = 1 - \widetilde{P}_\Lambda + \frac{\widetilde{P}_\Lambda}{\widetilde{\mathcal{D}}^2} \tag{9.9}$$

From the above properties, we can derive the general expression

$$\delta \ln Z_\Lambda = \mathsf{Tr}\left(\widetilde{P}_\Lambda \delta \widetilde{\mathcal{D}} \frac{1}{\widetilde{\mathcal{D}}}\right) \tag{9.10}$$

where Tr includes also spacetime integration. By differentiating with respect to V^a_μ and A^a_μ, one gets the (one-loop) quantum currents

$$\langle\!\langle j^{a\mu}(x)\rangle\!\rangle = \mathsf{Tr}\left(\widetilde{P}_\Lambda \frac{\delta \widetilde{\mathcal{D}}}{\delta V^a_\mu(x)} \frac{1}{\widetilde{\mathcal{D}}}\right), \quad \langle\!\langle j^{a\mu}_5(x)\rangle\!\rangle = \mathsf{Tr}\left(\widetilde{P}_\Lambda \frac{\delta \widetilde{\mathcal{D}}}{\delta A^a_\mu(x)} \frac{1}{\widetilde{\mathcal{D}}}\right) \tag{9.11}$$

Let us consider next the gauge transformations (8.25) and (8.26). First notice that

$$\delta_\lambda \widetilde{\mathcal{D}} = [\widetilde{\mathcal{D}}, \lambda] \tag{9.12}$$

while

$$\delta_\rho \widetilde{\mathcal{D}} = \{\widetilde{\mathcal{D}}, \gamma_5 \rho\} \tag{9.13}$$

Therefore,

$$\delta_\lambda \ln Z_\Lambda = \mathsf{Tr}\left(\widetilde{P}_\Lambda [\widetilde{\mathcal{D}}, \lambda] \frac{1}{\widetilde{\mathcal{D}}}\right) = 0 \tag{9.14}$$

while

$$\delta_\rho \ln Z_\Lambda = \mathsf{Tr}\left(\widetilde{P}_\Lambda \{\widetilde{\mathcal{D}}, \gamma_5 \rho\} \frac{1}{\widetilde{\mathcal{D}}}\right) = 2\mathsf{Tr}\left(\widetilde{P}_\Lambda \gamma_5 \rho\right) = 2 \int d^4x \, Tr\left(\rho(x)\gamma_5 \langle x|\widetilde{P}_\Lambda|x\rangle\right) \tag{9.15}$$

where Tr denotes the trace over gamma matrices and Lie algebra generators. Equation(9.14) says that the vector current is conserved, while (9.15) tells us that the axial current may be anomalous. More precisely

$$\left(D_V^{\mu}\langle\langle j_{\mu}(x)\rangle\rangle + [A^{\mu}(x), \langle\langle j_{5\mu}(x)\rangle\rangle]\right)^a = 0 \tag{9.16}$$

$$\left(D_V^{\mu}\langle\langle j_{5\mu}(x)\rangle\rangle + [A^{\mu}(x), \langle\langle j_{\mu}(x)\rangle\rangle]\right)^a = -2\, Tr\left(T^a \gamma_5 \langle x|\widetilde{P}_{\Lambda}|x\rangle\right) \tag{9.17}$$

where D_V is the covariant derivative with respect to the gauge vector feld V_{μ} and $j_{\mu} = j_{\mu}^a T^a, j_{5\mu} = j_{5\mu}^a T^a$.

From now on, in order to simplify the notation, we will drop the tilde symbol and use only the corresponding Minkowski notation. Actually, we use the parallelism between the two formalisms, replacing Euclidean symbols with Minkowski symbols in formulas derived and valid in a Euclidean background. As already pointed out in Sect. 1.4, this is the safest way to recover the correct results in Minkowski form.

In order to evaluate the anomaly, we insert (9.6) and a momentum completeness $\int \frac{d^4k}{(2\pi)^4}|k\rangle\langle k|$ into the RHS of (9.17). This means a continuous distribution of momenta; i.e. we are abandoning the idea of a compact spacetime and understanding an infinite volume limit. This procedure is somewhat heuristic, but justified, because local anomalies are a local phenomenon and do not depend on the global spacetime topology (see Chap. 13); so, they do not depend on the infinite volume limit. Proceeding therefore this way, we can write

$$\langle x|P_{\Lambda}|x\rangle = \int_{-\infty}^{+\infty} \frac{d\alpha}{2\pi i\alpha} \int \frac{d^4k}{(2\pi)^4} \langle x|e^{i\alpha\left(1-\frac{\not{\mathcal{D}}^2}{\Lambda^2}\right)}|k\rangle\langle k|x\rangle \tag{9.18}$$

Using the representation $\langle k|x\rangle = e^{ik\cdot x}$, we get

$$\langle x|P_{\Lambda}|x\rangle = \int_{-\infty}^{+\infty} \frac{d\alpha}{2\pi i\alpha} \int \frac{d^4k}{(2\pi)^4} e^{i\alpha\left(1-\frac{(\not{k}+\not{\mathcal{D}}_x)^2}{\Lambda^2}\right)} = \Lambda^4 \int_{-\infty}^{+\infty} \frac{d\alpha}{2\pi i\alpha} \int \frac{d^4k}{(2\pi)^4} e^{i\alpha\left(1-\left(\not{k}+\frac{\not{\mathcal{D}}_x}{\Lambda}\right)^2\right)} \tag{9.19}$$

Inserting this into (9.17), factoring out $e^{i\alpha(1-k^2)}$ from the exponential and expanding the rest in power series, then collecting the terms with the same power of $1/\Lambda$, one obtains a series in $1/\Lambda^2$ whose coefficients are polynomial in k^2. The relevant terms in 4d are

$$\langle x|P_{\Lambda}|x\rangle = \Lambda^4 \int_{-\infty}^{+\infty} \frac{d\alpha}{2\pi i\alpha} \int \frac{d^4k}{(2\pi)^4} e^{i\alpha(1-k^2)} \left(\ldots - \frac{\alpha^2}{2}\left(\frac{\not{\mathcal{D}}_x^2}{\Lambda^2}\right)^2 + \frac{\alpha^4}{4!}\left(\frac{\not{k}}{\not{\mathcal{D}}_x} + \not{\mathcal{D}}_x\not{k}\Lambda\right)^4 \right.$$
$$-i\frac{\alpha^3}{3!}\left(\left(\frac{\not{k}}{\not{\mathcal{D}}_x} + \not{\mathcal{D}}_x\not{k}\Lambda\right)^2 \frac{\not{\mathcal{D}}_x^2}{\Lambda^2} + \frac{\not{k}}{\not{\mathcal{D}}_x} + \not{\mathcal{D}}_x\not{k}\Lambda\,\frac{\not{\mathcal{D}}_x^2}{f}\Lambda^2\,\frac{\not{k}}{\not{\mathcal{D}}_x} + \not{\mathcal{D}}_x\not{k}\Lambda + \frac{\not{\mathcal{D}}_x^2}{\Lambda^2}\left(\frac{\not{k}}{\not{\mathcal{D}}_x} + \not{\mathcal{D}}_x\not{k}\Lambda\right)^2\right)$$
$$\left. + \ldots \right) \tag{9.20}$$

Integrating over k first, one uses

$$\int \frac{d^4k}{(2\pi)^4} e^{i\alpha k^2} = -\frac{1}{16\pi^2\alpha^2}, \quad \int \frac{d^4k}{(2\pi)^4} k^2 e^{i\alpha k^2} = -\frac{1}{8\pi^2}\frac{i}{\alpha^3}, \quad \int \frac{d^4k}{(2\pi)^4} k^4 e^{i\alpha k^2} = \frac{1}{16\pi^2}\frac{6}{\alpha^4}.$$

The integration over α reduces to

$$\int_{-\infty}^{+\infty} \frac{d\alpha}{2\pi i\alpha} e^{i\alpha} = 1$$

The final result for the RHS of (9.17), after an inverse Wick rotation, is

$$\mathcal{A}^a_\Lambda(x) = -2\, Tr\left(T^a\gamma_5\langle x|P_\Lambda|x\rangle\right) = \Lambda^2\mathcal{B}^a_0(x) + \mathcal{B}^a_1(x) + \mathcal{A}^a(x) + O\left(\frac{1}{\Lambda^2}\right) \tag{9.21}$$

where

$$\mathcal{B}^a_0(x) = \frac{4}{\pi^2}\mathrm{tr}\left(T^a D^\mu_V A_\mu\right) \tag{9.22}$$

$$\mathcal{B}^a_1(x) = \frac{1}{4\pi^2}\mathrm{tr}\Big(T^a\Big(\frac{2}{3}[D^\mu_V V_{\mu\lambda}, A^\lambda] - \frac{1}{3}[D^\mu_V A^\lambda, V_{\mu\lambda}] + \frac{1}{3}D^\mu_V D_{V\mu} D^\lambda_V A_\lambda$$
$$+\frac{1}{3}\{D^\mu_V A_\mu, A^\lambda A_\lambda\} + \frac{2}{3}\{\{D^\mu_V A^\lambda, A_\lambda\}, A_\mu\} + \frac{2}{3}A^\mu(D^\lambda_V A_\lambda)A_\mu + \frac{1}{6}[[A^\mu, A^\lambda], A_{\mu\lambda}]\Big)\Big) \tag{9.23}$$

$$\mathcal{A}^a(x) = \frac{1}{4\pi^2}\varepsilon^{\mu\nu\lambda\rho}\mathrm{tr}\Big(T^a\Big(\frac{1}{4}V_{\mu\lambda}V_{\mu\rho} + \frac{1}{12}A_{\mu\nu}A_{\lambda\rho} - \frac{1}{6}V_{\mu\nu}A_\lambda A_\rho - \frac{1}{6}A_\mu A_\nu V_{\lambda\rho}$$
$$-\frac{2}{3}A_\mu V_{\nu\lambda}A_\rho - \frac{1}{3}A_\mu A_\nu A_\lambda A_\rho\Big)\Big) \tag{9.24}$$

where $V_{\mu\nu} = \partial_\mu V_\nu + \partial_\nu V_\mu + [V_\mu, V_\nu]$ and $A_{\mu\nu} = \partial_\mu A_\nu + \partial_\nu A_\mu + [V_\mu, A_\nu] - [V_\nu, A_\mu]$, and tr represents the trace with respect to the Lie algebra generators.

The terms $\mathcal{B}^a_0$ and $\mathcal{B}^a_1$ can be subtracted away by adding the following counterterms to the effective action $\ln Z_\Lambda$:

$$-2\frac{\Lambda^2}{\pi^2}\int d^4x\,\mathrm{tr}\left(A^\mu A_\mu\right) - \frac{1}{12\pi^2}\int d^4x\Big(-\frac{1}{4}V_{\mu\lambda}V^{\mu\lambda} + A_\mu A_\lambda V^{\mu\lambda}$$
$$+\frac{1}{2}(D^\mu_V A_\mu)(D^\lambda_V A_\lambda) + \frac{1}{2}A^\mu A^\lambda A_\mu A_\lambda - \frac{3}{2}(A_\mu A^\mu)^2\Big) \tag{9.25}$$

Now, one can safely take the limit $\Lambda \to \infty$ and finally get

$$\left(D^\mu_V\langle\langle j_\mu(x)\rangle\rangle + [A^\mu(x), \langle\langle j_{5\mu}(x)\rangle\rangle]\right)^a = 0 \tag{9.26}$$

while the axial conservation equation is anomalous:

$$\left(D_V^\mu \langle\langle j_{5\mu}(x)\rangle\rangle + [A^\mu(x), \langle\langle j_\mu(x)\rangle\rangle]\right)^a = \frac{1}{4\pi^2}\varepsilon_{\mu\nu\lambda\rho}\mathrm{tr}\left[T^a\left(\frac{1}{4}F_V^{\mu\nu}F_V^{\lambda\rho} + \frac{1}{12}F_A^{\mu\nu}F_A^{\lambda\rho}\right.\right.$$
$$\left.\left. -\frac{1}{6}F_V^{\mu\nu}A^\lambda A^\rho - \frac{1}{6}A^\mu A^\nu F_V^{\lambda\rho} - \frac{2}{3}A^\mu F_V^{\nu\lambda}A^\rho - \frac{1}{3}A^\mu A^\nu A^\lambda A^\rho\right)\right] \tag{9.27}$$

where $F_V^{\mu\nu} = \partial^\mu V^\nu - \partial^\nu V^\mu + [V^\mu, V^\nu]$, and $F_A^{\mu\nu} = \partial^\mu A^\nu - \partial^\nu A^\mu + [V^\mu, A^\nu] + [A^\mu, V^\nu]$. This anomaly coincides with the result of Bardeen.

Remark 1 The anomaly (9.27) is consistent. To prove it, first one constructs the functional operator δ_Υ that generate the transformations (8.26). Then, promoting λ and ρ to anticommuting fields and endowing them with the transformations properties

$$\delta_\Upsilon\lambda = -\frac{1}{2}[\lambda, \lambda]_+ - \frac{1}{2}[\rho, \rho]_+, \quad \delta_\Upsilon\rho = -[\rho, \lambda]_+ \tag{9.28}$$

one can easily prove that $\delta_\Upsilon^2 = 0$. Writing the integrated anomaly $\mathcal{A} = \int d^4x\lambda^a(x)\, \mathcal{A}^a(x)$, the relevant WZ consistency condition is

$$\delta_\Upsilon\mathcal{A} = 0 \tag{9.29}$$

Although it is not a trivial exercise, one can prove that these conditions are satisfied. This has been explicitly done, for instance, in [6]).

Remark 2 From this expression, we can derive two results in particular. Setting $A_\mu = 0$ we get the covariant anomaly

$$[D_V^\mu j_{5\mu}]^a = \frac{1}{16\pi^2}\varepsilon_{\mu\nu\lambda\rho}\mathrm{tr}\left(T^a V^{\mu\nu}V^{\lambda\rho}\right) \tag{9.30}$$

Taking the chiral limit $V \to \frac{V}{2}, A \to \frac{V}{2}$, and adding (9.26) to (9.27) we get

$$[D_{V\mu} j_R^\mu]^a = \frac{1}{24\pi^2}\varepsilon_{\mu\nu\lambda\rho}\mathrm{tr}\left[T^a\partial^\mu\left(V^\nu\partial^\lambda V^\rho + \frac{1}{2}V^\nu V^\lambda V^\rho\right)\right] \tag{9.31}$$

where $j_{R\mu}^a = i\overline{\psi}_R\gamma_\mu T^a\psi_R$ with $\psi_R = \frac{1+\gamma_5}{2}\psi$, which is the consistent non-Abelian gauge anomaly. It is easy to verify that the chiral limit in the original action maps it to the action for a right-handed Weyl fermion coupled to a vector potential plus the action of a free left-handed Weyl fermion. This not only confirms the results of Chap. 6, in particular Eq. (6.39) and Sect. 6.2. But it also adds confidence to the idea that adding to a Weyl fermion a free one with opposite chirality is the correct thing to do.

Remark 3 Bardeen's anomaly (9.27), the covariant anomaly (9.30) and the consistent anomaly (9.31) are all proportional to the third-order ad-invariant symmetric tensor $d^{abc} = \frac{1}{2}\mathrm{tr}\left(T^a\{T^b, T^c\}\right)$. Therefore, they vanish simultaneously when the latter vanishes.

9.1.1 d = *6*

We can repeat the same derivation in 6d (where γ_7 is the chirality matrix). The result is (see [6])

$$\left(D_V^\mu \langle\!\langle j_\mu(x)\rangle\!\rangle + [A^\mu(x), \langle\!\langle j_{7\mu}(x)\rangle\!\rangle]\right)^a = 0 \tag{9.32}$$

and

$$\begin{aligned}\left(D_V^\mu \langle\!\langle j_{7\mu}(x)\rangle\!\rangle + [A^\mu(x), \langle\!\langle j_\mu(x)\rangle\!\rangle]\right)^a &= \frac{1}{24\pi^2}\,\varepsilon\,\mathrm{tr}\left[T^a\left(\frac{1}{8}F_VF_VF_V - \frac{1}{4}AF_VF_VA\right.\right.\\
&-\frac{1}{4}F_VAF_VA - \frac{1}{4}AF_VAF_V + \frac{1}{16}F_AF_AF_V + \frac{1}{16}F_VF_AF_A - \frac{1}{5}AAAAF_V\\
&-\frac{1}{5}AAAF_VA + \frac{3}{10}AAF_VAA - \frac{1}{5}AF_VAAA - \frac{1}{5}F_VAAAA\\
&-\frac{1}{40}F_AF_AAA + \frac{1}{20}F_AAF_AA + \frac{1}{20}F_AAAF_A - \frac{3}{20}AF_AF_AA\\
&\left.\left.+\frac{1}{20}AF_AAF_A - \frac{1}{40}AAF_AF_A - \frac{1}{5}AAAAAA\right)\right]\end{aligned} \tag{9.33}$$

where the six world indices attached to ε, F_V, F_A and A have been understood.

9.2 Dirac Fermions in V-A Background. The Calculation with SDW

In this section, we re-derive Bardeen's anomaly by means of the standard SDW method. The relevant Dirac operator is $\mathcal{D} = i(\partial\!\!\!/ + V\!\!\!/ + A\!\!\!/\gamma_5)$ and its Euclidean version $\tilde{\mathcal{D}}$. However, as anticipated above, we will avoid from now on weighing down the formulas with Euclidean symbols (the tilde). Instead, as previously noted, we can exploit the parallelism between the two formalisms to use all the time Minkowski symbols. This does not endanger the final results, which come out in the expected Minkowski form.

We use the formulas of Sects. 8.1.3 and 8.1.1. Then, the relevant quadratic operator is (8.15), which we rewrite in the more compact form

$$\mathcal{D}^2 = -\eta^{\mu\nu}\overline{\mathscr{D}}_\mu\mathscr{D}_\nu - \Sigma^{\mu\nu}\left(\mathcal{B}_{\mu\nu} + \left(\mathcal{C}_{\mu\nu} + 2(A_\nu\partial_\mu - A_\mu\partial_\nu)\right)\gamma_5\right) \tag{9.34}$$

where $\Sigma_{\mu\nu} = \frac{1}{4}[\gamma_\mu, \gamma_\nu]$ and

$$\mathcal{B}_{\mu\nu} = F_{V\mu\nu} - [A_\mu, A_\nu], \quad \mathcal{C}_{\mu\nu} = \partial_\mu A_\nu - \partial_\nu A_\mu + \{V_\mu, A_\nu\} - \{A_\mu, V_\nu\} \tag{9.35}$$

We use also the notation

$$\mathcal{E} = \mathcal{B} + \gamma_5 \mathcal{C}, \quad \text{where} \quad \mathcal{B} = \Sigma^{\mu\nu}\mathcal{B}_{\mu\nu}, \quad \mathcal{C} = \Sigma^{\mu\nu}\mathcal{C}_{\mu\nu} \tag{9.36}$$

Taking the variation with respect to the axial transformation ρ of the path integral Z, we get, formally,

$$\delta_\rho \ln Z = 2i\mathsf{Tr}\,(\gamma_5 \rho) \equiv \mathcal{A}^{(unreg)} \tag{9.37}$$

where Tr includes all traces plus spacetime integration and $\rho = \rho^a(x)T^a$ is the axial gauge parameter. $\mathcal{A}^{(unreg)}$ is the unregulated anomaly. We regularize this expression with the Schwinger proper time approach. This means using the procedure outlined in the previous chapter with $\mathcal{F} = -\not{\mathcal{D}}^2$ with regard in particular Eq. (8.48).

$$\begin{aligned}\mathcal{A} \equiv \delta_\rho W &= \frac{1}{2}\mathsf{Tr}\left(\int_0^\infty ds\, \mathrm{e}^{-i\not{\mathcal{D}}^2 s}\delta\not{\mathcal{D}}^2\right) = i\mathsf{Tr}\left(\rho\gamma_5 \int_0^\infty ds\, \mathrm{e}^{-i\not{\mathcal{D}}^2 s}\,\not{\mathcal{D}}^2\right)\\ &= -\mathsf{Tr}\left(\rho\gamma_5 \int_0^\infty ds\, \frac{\partial}{\partial s}\mathrm{e}^{-i\not{\mathcal{D}}^2 s}\right) = \mathsf{Tr}\left(\rho\gamma_5\, \mathrm{e}^{-i\not{\mathcal{D}}^2 s}\right)\Big|_{s=0}\end{aligned} \tag{9.38}$$

In the last integration convergence at infinity is guaranteed by the factor $\mathrm{e}^{im^2 s}$. A remark is in order here: looking at Eq. (8.30) these manipulations are possible because the operator $\not{\mathcal{D}}$ in it commute with $\mathrm{e}^{i\not{\mathcal{D}}^2 s}$. This is to be contrasted with a situation in the next chapter where a similar operation will not be allowed.

Now, we write (9.38) in coordinate representation

$$\mathcal{A} = -\lim_{x' \to x} \mathsf{Tr}\left(\rho\gamma_5\, \langle x|\mathrm{e}^{-i\not{\mathcal{D}}^2 s}|x'\rangle\right)\Big|_{s=0} \tag{9.39}$$

and insert

$$\langle x, s|x', 0\rangle = \lim_{m\to 0} \frac{i}{(4\pi s)^2}\sqrt{\mathfrak{D}(x,x')}\mathrm{e}^{i\left(\frac{\sigma(x,x')}{2s} - m^2 s\right)}\Phi(x,x',s), \tag{9.40}$$

where $\mathfrak{D}(x,x')$ is the Van Vleck-Morette (VVM) determinant and $\sigma(x,x')$ is the world function. Since for our problem we can choose a flat metric, we have $\sqrt{\mathcal{D}} = 1$ and $\sigma(x,x') = \frac{(x-x')^2}{2}$. Next we expand $\Phi(x,x',s)$ in powers of is, as in (8.64), and integrate over s. Setting $s = 0$ kills all the higher terms in the expansion, except the 0th-, first- and second-order term. The 0th-order term is annihilated by the trace over γ_5. The first-order term, involving $[a_1]$, diverges for $s \to 0$. $[a_1]$ is calculated below, see Eq. (9.52). It is evident that the odd parity part vanishes due to the gamma matrix trace, while the even parity part survives the tracing, but is trivial and can be subtracted with a counterterm.

Therefore, the finite part of (9.39) becomes

$$\mathcal{A} = -\frac{i}{16\pi^2}\int d^4x\mathrm{Tr}\left(\gamma_5\,\rho\,[a_2](x)\right), \quad [a_2(x)] = \lim_{x'\to x} a_2(x,x') \tag{9.41}$$

In order to compute the coefficient a_2 in a flat background, we have at our disposal the heat kernel equation

$$i\frac{\partial}{\partial s}\langle x, s|x', 0\rangle = \mathcal{D}\!\!\!/^2\langle x, s|x', 0\rangle \tag{9.42}$$

with (in this application we set $m = 0$)

$$\langle x, s|x', 0\rangle = \frac{i}{(4\pi s)^2} e^{i\frac{(x-x')^2}{4s}}\,\Phi(x, x'; s) \tag{9.43}$$

Replacing (9.43) into (9.42), we get the equation

$$i\frac{\partial\Phi}{\partial s} + \Box\Phi + \frac{i}{s}(x-x')^\mu \mathfrak{D}_\mu\Phi + \overline{\mathcal{D}}^\mu\mathcal{D}_\mu\Phi + \mathcal{E}\Phi + 4\gamma_5\Sigma^{\mu\nu}A_\nu\left(\frac{i}{2s}(x-x')_\mu\Phi + \partial_\mu\Phi\right) = 0 \tag{9.44}$$

where $\mathfrak{D}_\mu = \partial_\mu + V_\mu$. Using the expansion

$$\Phi(x, x', s) = \sum_{n=0}^{\infty} a_n(x, x')(is)^n \tag{9.45}$$

we obtain the recursion relation

$$\begin{aligned}&(n+1)a_{n+1} + (x-x')^\mu\left(\mathfrak{D}_\mu a_{n+1} + 2\gamma_5\Sigma^{\mu\nu}A_\nu a_{n+1}\right) - \overline{\mathcal{D}}^\mu\mathcal{D}_\mu a_n\\ &-4\gamma_5\Sigma^{\mu\nu}A_\nu\partial_\nu a_n - \mathcal{E}\,a_n = 0\end{aligned} \tag{9.46}$$

Setting $n = -1$, we obtain

$$(x-x')^\mu\left(\mathfrak{D}_\mu a_0 + 2\gamma_5\Sigma^{\mu\nu}A_\nu a_0\right) = 0 \tag{9.47}$$

Differentiating with respect to x^μ, we get

$$\begin{aligned}&\partial_\mu a_0 + V_\mu a_0 + 2\gamma_5\Sigma_\mu{}^\nu A_\nu a_0 + (x-x')^\lambda\big(\partial_\mu\partial_\lambda a_0\\ &+\partial_\mu V_\lambda a_0 + V_\lambda\partial_\mu a_0 + 2\gamma_5\Sigma_\lambda{}^\nu\partial_\mu A_\nu a_0 + 2\gamma_5\Sigma_\lambda{}^\nu A_\nu\partial_\mu a_0\big) = 0\end{aligned} \tag{9.48}$$

Starting from $[a_0](x) = 1$ we compute the coincidence limit

$$[\partial_\mu a_0] = -V_\mu - 2\gamma_5 \Sigma_\mu{}^\nu A_\nu \tag{9.49}$$

Differentiating (9.48) with respect to x^λ, we obtain

$$\begin{aligned}[\partial_\mu \partial_\lambda a_0] = &-\frac{1}{2}\left(\partial_\lambda V_\mu + \partial_\mu V_\lambda\right) + \frac{1}{2}\left(V_\mu V_\lambda + V_\lambda V_\mu\right) + 2\left(\Sigma_\lambda{}^\nu \Sigma_\mu{}^\rho + \Sigma_\lambda{}^\rho \Sigma_\mu{}^\nu\right) A_\nu A_\rho \\ &-\gamma_5\big(\Sigma_\lambda{}^\nu \left(\partial_\mu A_\nu - \{V_\mu, A_\nu\}\right) + \Sigma_\mu{}^\nu \left(\partial_\lambda A_\nu - \{V_\lambda, A_\nu\}\right)\big)\end{aligned} \tag{9.50}$$

Contracting λ with μ and taking the coincidence limit gives

$$[\Box a_0] = -\partial\cdot V + V^2 - 3A^2 - 4\Sigma^{\mu\nu} A_\mu A_\nu - 2\gamma_5 \Sigma^{\mu\nu}\left(\partial_\mu A_\nu - \{V_\mu, A_\nu\}\right) \tag{9.51}$$

where $V^2 = V^\mu V_\mu$, etc. and we have used $\Sigma^{\mu\nu}\Sigma_\mu{}^\lambda = -\frac{3}{4}\eta^{\nu\lambda} - \Sigma^{\nu\lambda}$.

Selecting now $n = 0$ and proceeding the same, way we get

$$\begin{aligned}[a_1] = &\, 2A^2 + 4\Sigma^{\mu\nu} A_\mu A_\nu + \Sigma^{\mu\nu}\mathcal{B}_{\mu\nu} \\ &+\gamma_5\left(\partial\cdot A + [V^\mu, A_\mu] - 2\Sigma^{\mu\nu}\left(\partial_\mu A_\nu + \{V_\mu, A_\nu\}\right) + \Sigma^{\mu\nu}\mathcal{C}_{\mu\nu}\right)\end{aligned} \tag{9.52}$$

and, with $n = 1$,

$$\begin{aligned}[a_2] = &\,\frac{1}{2}[\Box a_1] + V^\lambda[\partial_\lambda a_1] + \frac{1}{2}\left(V^2 - A^2 + \gamma_5[V^\mu, A_\mu]\right)[a_1] \\ &+\frac{1}{2}\left(\partial\cdot V + \gamma_5 \partial\cdot A\right)[a_1] + 2\gamma_5\Sigma^{\lambda\nu} A_\nu [\partial_\lambda a_1] + \frac{1}{2}\mathcal{E}\,[a_1]\end{aligned} \tag{9.53}$$

Moreover,

$$\begin{aligned}[\partial_\lambda a_1] = &-\frac{1}{2}V_\lambda[a_1] - \gamma_5\Sigma_\lambda{}^\nu A_\nu[a_1] + \frac{1}{2}[\partial_\lambda \Box a_0] + \partial_\lambda V^\nu[\partial_\nu a_0] + V^\nu[\partial_\lambda\partial_\nu a_0] \\ &+\frac{1}{2}\partial_\lambda\left(V^2 - A^2 + \gamma_5[V^\mu, A_\mu]\right) + \frac{1}{2}\left(V^2 - A^2 + \gamma_5[V^\mu, A_\mu]\right)[\partial_\lambda a_0] \\ &+\frac{1}{2}\partial_\lambda\left(\partial\cdot V + \gamma_5\partial\cdot A\right) + \frac{1}{2}\left(\partial\cdot V + \gamma_5\partial\cdot A\right)[\partial_\lambda a_0] \\ &+2\gamma_5\Sigma^{\mu\nu}\partial_\lambda A_\nu[\partial_\mu a_0] + 2\gamma_5\Sigma^{\mu\nu}A_\nu[\partial_\lambda\partial_\mu a_0] + \frac{1}{2}\mathcal{E}[\partial_\lambda a_0] + \frac{1}{2}\partial_\lambda\mathcal{E}\end{aligned} \tag{9.54}$$

and

$$\begin{aligned}[\Box a_1] = &-\frac{2}{3}\partial\cdot V[a_1] - \frac{2}{3}V^\lambda[\partial_\lambda a_1] - \frac{4}{3}\gamma_5\Sigma^{\lambda\nu}\partial_\lambda A_\nu[a_1] - \frac{4}{3}\gamma_5\Sigma^{\lambda\nu}A_\nu[\partial_\lambda a_1] \\ &+\frac{1}{3}[\Box\overline{\mathscr{D}}^\mu \mathscr{D}_\mu a_0] + \frac{2}{3}\partial_\lambda\mathcal{E}[\Box a_0] + \frac{1}{3}\Box\mathcal{E} \\ &+\frac{4}{3}\gamma_5\Sigma^{\mu\nu}\big(\Box A_\nu[\partial_\mu a_0] + 2\partial^\lambda A_\nu[\partial_\lambda\partial_\mu a_0] + A_\nu[\Box\partial_\mu a_0]\big)\end{aligned} \tag{9.55}$$

where

$$
\begin{aligned}
[\square\overline{\mathscr{D}}^{\mu}\mathscr{D}_{\mu}a_0] = & \, 2\square\partial\cdot V + \square V^{\lambda}V_{\lambda} + V^{\lambda}\square V_{\lambda} + 2\partial^{\lambda}V^{\mu}\partial_{\lambda}V_{\mu} - \square A^{\lambda}A_{\lambda} - A^{\lambda}\square A_{\lambda} \\
& -2\partial^{\lambda}A^{\mu}\partial_{\lambda}A_{\mu} + [\square^2 a_0] + 2\partial\cdot V[\square a_0] + 4\partial^{\lambda}\partial\cdot V[\partial_{\lambda}a_0] \\
& +\gamma_5\Big(\square(\partial\cdot A + [V^{\mu}, A_{\mu}]) + [V^{\mu}, A_{\mu}][\square a_0] + 2\partial^{\lambda}([V^{\mu}, A_{\mu}])[\partial_{\lambda}a_0] \\
& +\partial\cdot A[\square a_0] + 2\partial^{\lambda}\partial\cdot A[\partial_{\lambda}a_0]\Big) \qquad (9.56)
\end{aligned}
$$

Next

$$
\begin{aligned}
[\square^2 a_0] = & -\square\partial\cdot V - 2\partial^{\lambda}\partial\cdot V[\partial_{\lambda}a_0] - 2\square V^{\lambda}[\partial_{\lambda}a_0] \\
& -2\partial^{\lambda}V^{\rho}[\partial_{\lambda}\partial_{\rho}a_0] - \partial\cdot V[\square a_0] - V^{\lambda}[\partial_{\lambda}\square a_0] \\
& +\gamma_5\Sigma^{\mu\nu}\Big(\frac{1}{2}\square\partial_{\mu}A_{\nu} + \partial^{\lambda}\partial_{\mu}A_{\nu}[\partial_{\lambda}a_0] + \frac{1}{2}\square A_{\nu}[\partial_{\mu}a_0] \\
& +\frac{1}{2}\partial_{\mu}A_{\nu}[\square a_0] + \partial^{\lambda}A_{\nu}[\partial_{\lambda}\partial_{\mu}a_0] + \frac{1}{2}a_{\nu}[\partial_{\mu}\square A_0]\Big) \qquad (9.57)
\end{aligned}
$$

and

$$
\begin{aligned}
[\square\partial_{\sigma}a_0] = & -\frac{1}{3}\Big[\square V_{\sigma} + 2\partial_{\sigma}\partial\cdot V - 2\partial\cdot VV_{\sigma} - 2\partial_{\sigma}V^{\lambda}V_{\lambda} - 2V^{\lambda}\partial_{\sigma}V_{\lambda} \\
& -8\Sigma^{\lambda\nu}\Sigma_{\sigma}{}^{\mu}\partial_{\lambda}A_{\nu}A_{\mu} - 8\Sigma^{\lambda\nu}\Sigma_{\lambda}{}^{\mu}\partial_{\sigma}A_{\nu}A_{\mu} - 8\Sigma_{\sigma}{}^{\nu}\Sigma_{\lambda}{}^{\mu}\partial^{\lambda}A_{\nu}A_{\mu} \\
& -4\gamma_5\Big(\Sigma_{\sigma}{}^{\nu}\partial\cdot VA_{\nu} + \Sigma^{\lambda\nu}\partial_{\sigma}V_{\lambda}A_{\nu} + \Sigma^{\lambda\nu}\partial_{\lambda}V_{\sigma}A_{\nu} - \Sigma^{\lambda\nu}\partial_{\sigma}\partial_{\lambda}A_{\nu} - \Sigma^{\lambda\nu}\partial_{\lambda}A_{\nu}V_{\sigma} \\
& -\Sigma^{\lambda\nu}\partial_{\sigma}A_{\nu}V_{\lambda} + 2\Sigma_{\sigma}{}^{\nu}\square A_{\nu} - \Sigma_{\sigma}{}^{\nu}\partial^{\lambda}A_{\nu}V_{\lambda}\Big) + (V_{\sigma} + 2\gamma_5\Sigma_{\sigma}{}^{\nu}A_{\nu})[\square a_0] \\
& +2(V^{\lambda} + 2\gamma_5\Sigma^{\lambda\nu}A_{\nu})[\partial_{\lambda}\partial_{\sigma}a_0]\Big] \qquad (9.58)
\end{aligned}
$$

After repeated substitutions, we can write down an explicit expression for $[a_2]$ and replace it into (9.41). The resulting formula is huge, but here we focus on the part of it which is relevant to the odd parity anomaly. The monomials of $[a_2]$ are of two types: the axial type ones with an odd number of γ_5 so that one γ_5 can be factored out, and the vector ones, which do not contain γ_5. The chiral anomaly arises from the latter: the really relevant ones are those that contain four distinct γ's. In fact all products of gamma matrices in these monomials may be reduced to products of at most four γ's by means of the following relations

$$
\begin{aligned}
\Sigma^{\mu\nu}\Sigma_{\mu}{}^{\lambda} &= -\frac{3}{4}\eta^{\nu\lambda} - \Sigma^{\nu\lambda}, \\
\Sigma^{\lambda\nu}\Sigma^{\sigma\mu}\Sigma_{\lambda}{}^{\rho} &= \frac{3}{4}(\eta^{\nu\mu}\eta^{\sigma\rho} - \eta^{\mu\rho}\eta^{\nu\sigma}) - \frac{3}{4}\eta^{\nu\rho}\Sigma^{\sigma\mu} - \eta^{\nu\sigma}\Sigma^{\mu\rho} + \eta^{\nu\mu}\Sigma^{\sigma\rho} \\
& \quad -\Sigma^{\sigma\mu}\Sigma^{\nu\rho} - \Sigma^{\nu\mu}\Sigma^{\sigma\rho} + \Sigma^{\nu\sigma}\Sigma^{\mu\rho} \qquad (9.59)
\end{aligned}
$$

In sum, the terms that may contribute to the chiral anomaly are

$$
\begin{aligned}
[a_2] = & -\frac{2}{3}\partial_\lambda \mathcal{C}\,\Sigma^{\lambda\rho}A_\rho - \frac{2}{3}\mathcal{B}\,\Sigma^{\lambda\rho}A_\lambda A_\rho - \frac{1}{6}(\Sigma^{\mu\nu}\partial_\mu A_\nu + 2\mathcal{C})\Sigma^{\lambda\rho}(\partial_\lambda A_\rho - \{V_\lambda, A_\rho\}) \\
& + \frac{2}{3}\partial^\lambda V^\mu\left(\Sigma_\lambda{}^\sigma\Sigma_\mu{}^\tau + \Sigma_\mu{}^\sigma\Sigma_\lambda{}^\tau\right)A_\sigma A_\tau - \frac{3}{2}\Sigma^{\mu\nu}\partial_\lambda A_\nu \Sigma^{\lambda\sigma}(\partial_\mu A_\sigma - \{V_\mu, A_\sigma\}) \\
& + \frac{1}{2}\mathcal{B}^2 + 2\mathcal{B}\,\Sigma^{\mu\nu}A_\mu A_\nu - \frac{2}{3}\Sigma^{\mu\nu}\partial_\mu A_\nu(\mathcal{C} - 2\Sigma^{\mu\nu}(\partial_\mu A_\nu + \{V_\mu, A_\nu\}) \\
& + \frac{3}{4}\Sigma^{\sigma\nu}A_\nu\Big(\frac{4}{3}\Sigma^{\lambda\rho}\partial_\lambda V_\sigma A_\rho + \frac{1}{3}\Sigma^{\lambda\rho}A_\rho(\partial_\sigma V_\lambda + \partial_\lambda V_\sigma) + \frac{2}{3}\Sigma^{\lambda\rho}\partial_\sigma V_\lambda A_\rho \\
& + \frac{2}{3}\Sigma^{\lambda\rho}(\partial_\lambda A_\rho V_\sigma + \partial_\sigma A_\rho V_\lambda) + \frac{1}{3}\Sigma^{\lambda\rho}V_\sigma(\partial_\lambda A_\rho - \{V_\lambda, A_\rho\}) + \frac{1}{3}\Sigma^{\lambda\rho}V^\lambda(\partial_\sigma A_\rho - \{V_\sigma, A_\rho\}) \\
& - \frac{2}{3}\Sigma^{\lambda\rho}A_\rho(V_\lambda V_\sigma + V_\sigma V_\lambda) + \frac{8}{3}\Sigma_\sigma{}^\mu\Sigma^{\lambda\rho}A_\mu A_\lambda A_\rho + \frac{8}{3}\Sigma_\sigma{}^\mu\Sigma^{\lambda\rho}A_\lambda A_\mu A_\rho + \frac{8}{3}\Sigma^{\mu\rho}A_\sigma A_\mu A_\rho \\
& + \frac{16}{3}\Sigma^{\lambda\mu}\Sigma_\sigma{}^\rho A_\lambda A_\mu A_\rho + \frac{8}{3}\Sigma_\sigma{}^\lambda\Sigma^{\mu\rho}A_\lambda A_\mu A_\rho\Big) \\
& + \frac{4}{3}\Sigma^{\lambda\nu}A_\nu\Big(\frac{4}{3}\Sigma^{\mu\nu}V_\lambda\partial_\mu A_\nu - \frac{2}{3}\Sigma^{\mu\nu}A_\nu\partial_\lambda V_\mu - \frac{2}{3}\Sigma^{\mu\nu}A_\nu\partial_\mu V_\lambda + \frac{2}{3}\Sigma^{\mu\nu}\partial_\mu A_\nu V_\lambda \\
& - \frac{4}{3}\Sigma^{\mu\nu}\partial_\lambda A_\nu V_\mu - \frac{2}{3}\Sigma^{\mu\nu}V_\mu\partial_\lambda A_\nu - \frac{4}{3}\Sigma^{\mu\nu}\partial_\lambda V_\mu A_\nu + \frac{2}{3}\Sigma^{\mu\nu}\partial_\mu V_\lambda A_\nu \\
& + \frac{2}{3}\Sigma^{\mu\nu}V_\lambda\{V_\mu, A_\nu\} + \frac{2}{3}\Sigma^{\mu\nu}V_\mu\{A_\nu, V_\lambda\} + \frac{2}{3}\Sigma^{\mu\nu}A_\nu\{V_\mu, V_\lambda\} - \frac{8}{3}\Sigma^{\mu\nu}A_\lambda A_\mu A_\nu \\
& - \frac{16}{3}\Sigma_\lambda{}^\rho\Sigma^{\mu\nu}A_\rho A_\mu A_\nu - \frac{8}{3}\Sigma_\lambda{}^\mu\Sigma^{\nu\rho}A_\nu A_\mu A_\rho - \frac{16}{3}\Sigma^{\nu\mu}\Sigma_\lambda{}^\rho A_\nu A_\mu A_\rho \\
& + \frac{1}{2}(\partial_\lambda\mathcal{C} - \{V_\lambda, \mathcal{C}\}) - \{\Sigma_\lambda{}^\rho A_\rho, \mathcal{B}\}\Big) + \ldots
\end{aligned}
\tag{9.60}
$$

where ellipses denote terms not relevant to the odd parity anomaly. Inserting this into (9.41) and using as usual $\mathrm{tr}(\gamma_5\gamma_\mu\gamma_\nu\gamma_\lambda\gamma_\rho) = -4i\varepsilon_{\mu\nu\lambda\rho}$ we obtain 19 monomials of $V_\mu, A_\mu, V_{\mu\nu} = \partial_\mu V_\nu - \partial_\nu V_\mu$ and $A_{\mu\nu} = \partial_\mu A_\nu - \partial_\nu A_\mu$. More precisely we get

$$
\begin{aligned}
\mathcal{A} = & -\frac{1}{4\pi^2}\varepsilon_{\mu\nu\lambda\rho}\mathrm{tr}\Big[\rho\Big(\frac{1}{4}V^{\mu\nu}V^{\lambda\rho} + \frac{1}{2}V^{\mu\nu}V^\lambda V^\rho + \frac{1}{2}V^\mu V^\nu V^{\lambda\rho} - \frac{1}{6}V^{\mu\nu}A^\lambda A^\rho - \frac{1}{6}A^\mu A^\nu V^{\lambda\rho} \\
& + \frac{1}{12}A^{\mu\nu}A^{\lambda\rho} + \frac{1}{6}A^{\mu\nu}V^\lambda A^\rho + \frac{1}{6}A^{\mu\nu}A^\lambda V^\rho + \frac{1}{6}V^\mu A^\nu A^{\lambda\rho} + \frac{1}{6}A^\mu V^\nu A^{\lambda\rho} \\
& - \frac{2}{3}A^\mu V^{\nu\lambda}A^\rho + V^\mu V^\nu V^\lambda V^\rho - \frac{1}{3}V^\mu V^\nu A^\lambda A^\rho - \frac{1}{3}A^\mu A^\nu V^\lambda V^\rho + \frac{1}{3}V^\mu A^\nu V^\lambda A^\rho \\
& + \frac{1}{3}A^\mu V^\nu A^\lambda V^\rho + \frac{1}{3}V^\mu A^\nu A^\lambda V^\rho - A^\mu V^\nu V^\lambda A^\rho - \frac{1}{3}A^\mu A^\nu A^\lambda A^\rho\Big)\Big]
\end{aligned}
\tag{9.61}
$$

For completeness: the coefficients of the monomials $\varepsilon_{\mu\nu\lambda\rho}V^\mu A^{\nu\lambda}A^\rho$ and $\varepsilon_{\mu\nu\lambda\rho}A^\mu A^{\nu\lambda}V^\rho$ (which are possible candidate terms being quadratic in A) vanish.

As an example, consider the term $\varepsilon_{\mu\nu\lambda\rho}A^{\mu\nu}A^\lambda V^\rho$. It comes from

$$\sim 2\text{Tr}\Big[\gamma_5\rho\Big(\frac{5}{6}\Sigma^{\mu\nu}\partial_\mu A_\nu\Sigma^{\lambda\rho}A_\rho V_\lambda + \frac{3}{2}\Sigma^{\mu\nu}\partial_\lambda\Sigma^{\lambda\sigma}A_\sigma V_\mu$$
$$+\frac{4}{3}\Sigma^{\mu\nu}\partial_\mu A_\nu\Sigma^{\lambda\rho}A_\rho V_\lambda + \frac{4}{3}\Sigma^{\mu\nu}\partial_\mu A_\nu\Sigma^{\lambda\rho}A_\lambda V_\rho\Big)\Big] \tag{9.62}$$

The last two terms cancel. The first two are opposite to each other but have different coefficients ($-\frac{5}{6}+\frac{3}{2}=\frac{2}{3}$). Therefore, (9.62) becomes

$$\sim \frac{4}{3}\text{Tr}(\gamma_5\rho\Sigma^{\mu\nu}\partial_\mu A_\nu\Sigma^{\lambda\rho}A_\lambda V_\rho) = -\frac{i}{3}\varepsilon_{\mu\nu\lambda\rho}\partial^\mu A^\nu A^\lambda V^\rho = -\frac{i}{6}\varepsilon_{\mu\nu\lambda\rho}A^{\mu\nu}A^\lambda V^\rho$$

Equation (9.61) corresponds precisely to Bardeen's anomaly (9.27).

Contrary to the previous section, we do not explicitly compute the even part of the anomaly, which is obtained from the monomials in $[a_2]$ linear in γ_5. We know that there exists no even parity anomaly, therefore there certainly are counterterms that cancel it.

9.3 Weyl Fermion Consistent Chiral Anomalies

It is important and, as we shall see, useful to derive the consistent chiral anomaly for a right-handed Weyl fermion coupled to a vector field, directly with the Schwinger-DeWitt method, [7]. This is what we will do in this section. We are again in 4d, and like in the previous section, we are going to use throughout Minkowski symbols.

As we have learned above, to put the problem in the right setting we consider a Dirac fermion in a (V, A) background and take the chiral limit $V \to V/2, A \to V/2$. Then, the relevant kinetic operator becomes

$$\mathcal{D}\!\!\!/ = i\gamma^\mu\mathcal{D}_\mu, \quad \mathcal{D}_\mu = \partial_\mu + P_+V_\mu, \quad \overline{\mathcal{D}}_\mu = \partial_\mu + P_-V_\mu, \quad P_\pm = \frac{1\pm\gamma_5}{2} \tag{9.63}$$

This implies that the gauge transformation parameters in the chiral limit become: $\lambda \to \rho/2, \rho \to \rho/2$, and $\Upsilon \to P_+\rho$. Then,

$$\mathcal{D}\!\!\!/^2 = -\eta^{\mu\nu}\overline{\mathcal{D}}_\mu\mathcal{D}_\nu - \Sigma^{\mu\nu}\left(\overline{\mathcal{D}}_\mu\mathcal{D}_\nu - \overline{\mathcal{D}}_\nu\mathcal{D}_\mu\right)$$
$$= -\left(\Box + P_+\partial\cdot V + V\cdot\partial\right) - \Sigma^{\mu\nu}\left(\partial_\mu V_\nu - \partial_\nu V_\mu + \gamma_5\left(V_\nu\partial_\mu - V_\mu\partial_\nu\right)\right) \tag{9.64}$$

Next, we use the definition (9.4). Taking the variation with respect to the limiting transformation ρ and using (8.30), we get, formally,

$$\delta_\rho \ln Z = 2i\text{Tr}\left(\gamma_5\rho\right) \equiv \mathcal{A}^{(unreg)} \tag{9.65}$$

where Tr includes all traces plus spacetime integration and $\rho = \rho^a(x)T^a$ is the axial gauge parameter. $\mathcal{A}^{(unreg)}$ is the unregulated anomaly. We regularize this expression using the Schwinger proper time formulas. This means repeating the previous argument starting from (8.49) with $\mathfrak{F} = -\not{\mathcal{D}}^2$. Proceeding exactly as before we find

$$\mathcal{A} = -\frac{i}{16\pi^2}\int \mathrm{d}^4x \mathrm{Tr}\,(\gamma_5\,\rho\,[a_2](x))\,,\quad [a_2(x)] = \lim_{x'\to x} a_2(x,x') \tag{9.66}$$

In order to compute the coefficient a_2 in a flat background, we use again the heat kernel equation

$$i\frac{\partial}{\partial s}\langle x,s|x',0\rangle = \not{\mathcal{D}}^2\langle x,s|x',0\rangle \tag{9.67}$$

with (we set $m = 0$)

$$\langle x,s|x',0\rangle = \frac{i}{(4\pi s)^2}\mathrm{e}^{i\frac{(x-x')^2}{4s}}\,\Phi(x,x';s) \tag{9.68}$$

Replacing (9.68) into (9.67) and using (9.64), we get the equation

$$\begin{aligned} i\frac{\partial\Phi}{\partial s} &+ \Box\Phi + V^\mu\partial_\mu\Phi + \frac{i}{s}(x-x')^\mu\partial_\mu\Phi + \frac{i}{2s}(x-x')^\mu V_\mu\Phi + \frac{i}{s}\gamma_5\Sigma^{\mu\nu}(x-x')_\mu V_\nu\Phi \\ &-2\gamma_5\Sigma^{\mu\nu}V_\mu\partial_\nu\Phi + P_+\big(\partial\cdot V + \Sigma^{\mu\nu}(\partial_\mu V_\nu - \partial_\nu V_\mu)\big)\Phi = 0 \end{aligned} \tag{9.69}$$

Using the expansion

$$\Phi(x,x',s) = \sum_{n=0}^{\infty} a_n(x,x')(is)^n \tag{9.70}$$

we obtain the recursion relation

$$\begin{aligned} &(n+1)a_{n+1} - \Box a_n - V^\mu\partial_\mu a_n + (x-x')^\mu\partial_\mu a_{n+1} + \frac{1}{2}(x-x')^\mu V_\mu a_{n+1} \\ &+\gamma_5\Sigma^{\mu\nu}(x-x')_\mu V_\nu a_{n+1} - 2\gamma_5\Sigma^{\mu\nu}V_\mu\partial_\nu a_n - P_+\big(\partial\cdot V + \Sigma^{\mu\nu}(\partial_\mu V_\nu - \partial_\nu V_\mu)\big)a_n = 0 \end{aligned} \tag{9.71}$$

This equation does not depend on the spacetime dimension except for the replacement of the γ_5 with the appropriate chiral matrix. Setting $n = -1$, we obtain

$$(x-x')^\mu\partial_\mu a_0 + \frac{1}{2}(x-x')^\mu V_\mu a_0 + \gamma_5\Sigma^{\mu\nu}(x-x')_\mu V_\nu a_0 = 0 \tag{9.72}$$

Differentiating with respect to x^μ, we get

$$\partial_\mu a_0 + \frac{1}{2}V_\mu a_0 + \gamma_5 \Sigma_\mu{}^\nu V_\nu a_0 + (x-x')^\lambda \partial_\mu \partial_\lambda a_0 + \frac{1}{2}(x-x')^\lambda \partial_\mu V_\lambda a_0$$
$$+ \frac{1}{2}(x-x')^\lambda V_\lambda \partial_\mu a_0 + \gamma_5 \Sigma^{\lambda\nu}(x-x')_\lambda \partial_\mu V_\nu a_0 + \gamma_5 \Sigma^{\lambda\nu}(x-x')_\lambda V_\nu \partial_\mu a_0 = 0 \quad (9.73)$$

Starting from $[a_0](x) = 1$ we compute the coincidence limit

$$[\partial_\mu a_0] = -\frac{1}{2}V_\mu - \gamma_5 \Sigma_\mu{}^\nu V_\nu \quad (9.74)$$

Differentiating (9.73) with respect to x^λ, contracting λ with μ and taking the coincidence limit gives

$$[\Box a_0] = -\frac{1}{2}\partial\cdot V - \frac{1}{2}V^2 - \Sigma^{\mu\nu}V_\mu V_\nu - \frac{1}{2}\gamma_5 \Sigma^{\mu\nu}\left(\partial_\mu V_\nu - \partial_\nu V_\mu\right) \quad (9.75)$$

where we have used $\Sigma^{\mu\nu}\Sigma_\mu{}^\lambda = -\frac{3}{4}\eta^{\nu\lambda} - \Sigma^{\nu\lambda}$.

Selecting now $n = 0$ and proceeding the same way, we get

$$\begin{aligned}[a_1] &= -\frac{1}{2}\partial\cdot V + \frac{1}{2}V^2 + \Sigma^{\mu\nu}V_\mu V_\nu - \frac{1}{2}\gamma_5 \Sigma^{\mu\nu}\left(\partial_\mu V_\nu - \partial_\nu V_\mu\right) \\ &\quad + P_+\left(\partial\cdot V + \gamma_5 \Sigma^{\mu\nu}\left(\partial_\mu V_\nu - \partial_\nu V_\mu\right)\right) \\ &= \frac{1}{2}V^2 + \Sigma^{\mu\nu}V_\mu V_\nu + \frac{1}{2}\left(\gamma_5 \partial\cdot V + \Sigma^{\mu\nu}\left(\partial_\mu V_\nu - \partial_\nu V_\mu\right)\right) \end{aligned} \quad (9.76)$$

and

$$[a_2] = \frac{1}{2}[\Box a_1] + \frac{1}{2}V^\lambda[\partial_\lambda a_1] + \gamma_5 \Sigma^{\lambda\nu}V_\nu[\partial_\lambda a_1] + \frac{1}{2}P_+\left(\partial\cdot V + \Sigma^{\mu\nu}C_{\mu\nu}\right)[a_1] \quad (9.77)$$

where $C_{\mu\nu} = \partial_\mu V_\nu - \partial_\nu V_\mu$. Moreover

$$\begin{aligned}[\partial_\lambda a_1] &= -\frac{1}{4}V_\lambda[a_1] - \frac{1}{2}\gamma_5 \Sigma^{\lambda\nu}V_\nu[a_1] + \frac{1}{2}[\partial_\lambda \Box a_0] + \frac{1}{2}\partial_\lambda V^\nu[\partial_\nu a_0] + \frac{1}{2}V^\nu[\partial_\lambda \partial_\nu a_0] \\ &\quad + \gamma_5 \Sigma^{\mu\nu}\partial_\lambda V_\nu[\partial_\mu a_0] + \gamma_5 \Sigma^{\mu\nu}V_\nu[\partial_\lambda \partial_\mu a_0] + \frac{1}{2}P_+\left(\partial_\lambda \partial\cdot V + \Sigma^{\mu\nu}\partial_\lambda C_{\mu\nu}\right)[a_0] \\ &\quad + \frac{1}{2}P_+\left(\partial\cdot V + \Sigma^{\mu\nu}C_{\mu\nu}\right)[\partial_\lambda a_0] \end{aligned} \quad (9.78)$$

and

$$
\begin{aligned}
[\Box a_1] = & -\frac{1}{3}\partial\cdot V[a_1] - \frac{1}{3}V^\lambda[\partial_\lambda a_1] - \frac{2}{3}\gamma_5\Sigma^{\lambda\nu}\partial_\lambda V_\nu[a_1] - \frac{2}{3}\gamma_5\Sigma^{\lambda\nu}V_\nu[\partial_\lambda a_1] + \frac{1}{3}[\Box\Box a_0] \\
& + \frac{1}{3}\Box V^\lambda[\partial_\lambda a_0] + \frac{1}{3}V^\lambda[\partial_\lambda\Box a_0] + \frac{2}{3}\partial^\lambda V^\mu[\partial_\lambda\partial_\mu a_0] + \frac{2}{3}\gamma_5\Sigma^{\mu\nu}\Box V_\nu[\partial_\mu a_0] \\
& + \frac{4}{3}\Sigma^{\mu\nu}\partial_\lambda V_\nu[\partial^\lambda\partial_\mu a_0] + \frac{2}{3}\gamma_5\Sigma^{\mu\nu}V_\nu[\partial_\mu\Box a_0] + \frac{1}{3}P_+(\Box\partial\cdot V + \Sigma^{\mu\nu}\Box C_{\mu\nu})[a_0] \\
& + \frac{2}{3}P_+(\partial^\lambda\partial\cdot V + \Sigma^{\mu\nu}\partial^\lambda C_{\mu\nu})[\partial_\lambda a_0] + \frac{1}{3}P_+(\partial\cdot V + \Sigma^{\mu\nu}C_{\mu\nu})[\Box a_0] \qquad (9.79)
\end{aligned}
$$

The coincident limits $[\Box\Box a_0]$, $[\partial_\mu\Box a_0]$ and $[\partial_\lambda\partial_\nu a_0]$ can be found in Appendix 8B. Using these formulas and replacing them into (9.77), we find an enormous expression. However, only few terms can contribute to the odd parity part: they must be proportional to four gamma's, so it is not hard to figure out the candidates. For $[a_2]$, they are the following ones

$$
\begin{aligned}
[a_2] = & \frac{1}{6}V^\lambda V^\mu[\partial_\lambda\partial_\mu a_0] + \frac{1}{3}\Sigma^{\nu\rho}V_\nu V_\rho[a_1] \\
& + \frac{2}{3}\Sigma^{\lambda\nu}\Sigma^{\mu\rho}V_\nu V_\rho[\partial_\lambda\partial_\mu a_0] + \frac{1}{3}\gamma_5\Sigma^{\mu\nu}\{V^\lambda, V_\nu\}[\partial_\lambda\partial_\mu a_0] \\
& + \left(\frac{1}{4} + \frac{1}{12}\gamma_5\right)\Sigma^{\mu\nu}C_{\mu\nu}[a_1] + \frac{1}{6}\partial^\lambda V^\mu[\partial_\lambda\partial_\mu a_0] + \frac{1}{3}\gamma_5\Sigma^{\mu\nu}\partial_\lambda V_\nu[\partial_\lambda\partial_\mu a_0] \\
& + \frac{1}{4}V^\lambda[\partial_\lambda\Box a_0] + \frac{1}{2}\gamma_5\Sigma^{\mu\nu}V_\nu[\partial_\lambda\Box a_0] + \frac{1}{12}\Sigma^{\mu\nu}C_{\mu\nu}[\Box a_0] + \ldots \qquad (9.80)
\end{aligned}
$$

where ellipses denote terms that cannot contribute to the odd parity part.

Using the formulas above and in Appendix 9B to evaluate (9.66) and then using $\mathrm{tr}(\gamma_5\gamma_\mu\gamma_\nu\gamma_\lambda\gamma_\rho) = -4i\varepsilon_{\mu\nu\lambda\rho}$ we find that the term with 4 V's vanish. The coefficients of the terms ∂VVV, $V\partial VV$ and $VV\partial V$ are all equal and one half the coefficient of the term $\partial V\partial V$. In conclusion, we obtain

$$
\begin{aligned}
\mathcal{A} &= -\frac{i}{16\pi^2}\int \mathrm{d}^4x \mathrm{Tr}\,(\gamma_5\,\rho\,[a_2(x)]) \\
&= -\frac{1}{24\pi^2}\int \mathrm{d}^4x\,\varepsilon^{\mu\nu\lambda\rho}\,\mathrm{tr}\left(\rho\left(\partial_\mu V_\nu\partial_\lambda V_\rho + \frac{1}{2}\left(\partial_\mu V_\nu V_\lambda V_\rho + V_\mu\partial_\nu V_\lambda V_\rho + V_\mu V_\nu\partial_\lambda V_\rho\right)\right)\right) \\
&= \frac{1}{24\pi^2}\int \mathrm{d}^4x\,\varepsilon^{\mu\nu\lambda\rho}\,\mathrm{tr}\left(\partial_\mu\rho\left(V_\nu\partial_\lambda V_\rho + \frac{1}{2}V_\nu V_\lambda V_\rho\right)\right) \\
&= \frac{1}{4\pi^2}\int \mathrm{d}^4x\int_0^1 \mathrm{d}t(1-t)P_3\left(d\rho, V, tdV + \frac{t^2}{2}[V, V]\right) \qquad (9.81)
\end{aligned}
$$

where $\rho = \rho^a(x)T^a$ and V is the form $V = V_\mu \mathrm{d}x^\mu$.

9.4 Remarks on Local Lorentz Anomalies

So far, we have not presented any explicit derivation of local Lorentz anomalies. With reference to a theory (2.20), which we rewrite here for convenience,

$$S[g,\psi] = \int d^dx \sqrt{|g|}\, i\overline{\psi}\gamma^\mu (D_\mu + \frac{1}{2}\omega_\mu)\psi \tag{9.82}$$

a local Lorentz anomaly appears as a non-conservation of the current.

$$j_\mu^{ab} = i\overline{\psi}\gamma_\mu \Sigma^{ab}\psi \tag{9.83}$$

where $\Sigma^{ab} = \frac{1}{4}[\gamma^a, \gamma^b]$ are the Lorentz Lie algebra generators, see Eq. (2.28). The perturbative approach is quite similar to the gauge case, with a basic difference for we remark that in this case the invariance of (9.82) under a local Lorentz transformation $\delta_\Lambda = -\frac{1}{2}\Lambda\psi$, with $\Lambda = \Lambda^{ab}\Sigma_{ab}$, needs an additional step with respect to ordinary gauge theories. This is due to the presence of the vierbein in $\gamma^\mu = \gamma^a e_a^\mu$, which transforms under a local Lorentz transformation[1]. Such an additional term is compensated for by the commutator $[\gamma^\mu, \Lambda]$, a property due to the soldering in the geometry of the frame bundle. All this is absent in ordinary gauge theories, where the Lie algebra generators commute with γ_μ, while it induces a complication in the perturbative approach to local Lorentz anomalies.

It is nevertheless true that the expected lowest order contribution to the covariant divergence of (9.83) comes from the triangle diagram, which is analogous to Fig. 6.1, only the Lie algebra generators being the specific ones for this case. More precisely, it takes the form

$$\begin{aligned} & q^\mu \widetilde{F}^{(R)a_1a_2b_1b_2c_1c_2}_{\mu\lambda\rho}(k_1,k_2) \\ &= \frac{1}{8}\int \frac{d^4p}{(2\pi)^4}\mathrm{tr}\left\{\frac{1}{\not p}\gamma_\lambda \frac{1+\gamma_5}{2}\Sigma^{b_1b_2}\frac{1}{\not p-\not k_1}\gamma_\rho\frac{1+\gamma_5}{2}\Sigma^{c_1c_2}\frac{1}{\not p-\not q}\not q\frac{1+\gamma_5}{2}\Sigma^{a_1a_2}\right\} \end{aligned} \tag{9.84}$$

where $q = k_1 + k_2$, and tr means the trace over γ matrices.

It is immediately clear that we cannot simply factorize the Lie algebra generators as in Eq. (6.27), since the Lorentz generators do not commute with the γ's carrying world indices. Equation (6.27) is replaced by

$$q^\mu \widetilde{F}^{(R)a_1a_2b_1b_2c_1c_2}_{\mu\lambda\rho}(k_1,k_2) = \frac{1}{8}\mathrm{tr}(\Sigma^{a_1a_2}\Sigma^{b_1b_2}\Sigma^{c_1c_2})\, q^\mu \widetilde{F}^{(R)}_{\mu\lambda\rho}(k_1,k_2) + \ldots \tag{9.85}$$

[1] By the way, in this framework we cannot assume that the antisymmetric part of the vierbein vanishes, as we have done many times, because, as is clear from Sect. 5.4, the local Lorentz transformation of the vierbein is generated exactly by its antisymmetric piece.

where $q^\mu \widetilde{F}^{(R)}_{\mu\lambda\rho}(k_1,k_2)$ is given by Eq. (6.28) and the ellipses represent the additional terms generated by commuting the Σ's to the right, plus other terms that come from the ellipses in the formula $\mathrm{tr}(AB) = \mathrm{tr}A\,\mathrm{tr}B + \ldots$ for any two matrices A and B.

The first term in the RHS of (9.85) reproduces (6.34) multiplied by the trace of Σ's. Once this result is replaced into (4.24) and the cross-terms is added, we get, to the lowest order,

$$\partial^\mu \langle\!\langle j^{a_1 a_2}_{R\mu}(x)\rangle\!\rangle = \frac{1}{192\pi^2}\varepsilon^{\mu\nu\lambda\rho}\mathrm{tr}\left(\Sigma^{a_1a_2}\,\partial_\mu\omega_\nu(x)\partial_\lambda\omega_\rho(x)\right) + \ldots \tag{9.86}$$

The ellipses here correspond to the additional terms signaled in and after (9.85). Let us consider the first term in the RHS. It is proportional to $\mathrm{tr}\left(\Sigma^{a_1a_2}\{\Sigma^{b_1b_2},\,\Sigma^{c_1c_2}\}\right)$. Therefore, it reproduces the first term at the RHS of (6.39) with V_μ replaced by ω_μ, which corresponds to the lowest term of (5.51). But now

$$\mathrm{tr}\left(\Sigma^{a_1a_2}\{\Sigma^{b_1b_2},\,\Sigma^{c_1c_2}\}\right) = 0, \tag{9.87}$$

as we have explained several times. By consistency we expect also the cubic term of the anomaly to vanish. But what about the terms corresponding to the ellipses? There is a large number of such terms which are of a different type with respect the one just discussed, because they contain contractions of the Lie algebra indices among themselves or with world indices (with a μ index of the ε tensor, of ∂_μ or of ω_μ). Therefore they do not have the form of the terms contained in (5.51). But, among them, there are terms that violate diffeomorphisms. On the other hand, we know from Sect. 7.4.1 that diffeomorphisms are conserved at the three-point correlator level. Therefore diffeomorphism and local Lorentz cross-consistency conditions are not satisfied, and, in particular, it is not possible to restore diffeomorphism invariance by subtracting suitable counterterms from the effective action. We seem to be at a dead end.

We are in fact in the same situation explained in Sects. 7.4.5 and 7.4.6. We face an exceptional case, like the one illustrated there. The reason is the same: the identical vanishing of the odd three-point function of the e.m. tensor. This means that a solution cannot be found at the lowest perturbative order. As explained there, the way out is provided either by the calculation of higher order correlators of the e.m. or by the recourse to non-perturbative methods. The higher order correlator way seems to be forbidding and, as will be pointed out in a dedicated remark in Sect. 10.2, the SDW method (at least in the form presented in this book) is not applicable to local Lorentz anomalies. What remains is the index theorem (Chap. 12) and the equivalence of diffeomorphism and local Lorentz anomalies by means of a Wess-Zumino term (Sect. 15.2). For the time being, we have to settle for as much.

Appendix 9A: Mode-Cutoff Regularization and Heat Kernel

Let us consider the formula for the effective action L, (8.49),

$$W = -\frac{1}{2}\int_0^\infty \frac{ds}{is}\langle x|e^{i\mathcal{F}s}|x\rangle + \text{const} \equiv L + \text{const} \tag{9.88}$$

and compare it with the formula for the projector P_Λ:

$$\langle x|P_\Lambda|x\rangle = \int_{-\infty}^{+\infty} \frac{d\alpha}{2\pi i\alpha} e^{i\alpha}\langle x|e^{-i\alpha\left(\frac{\not{D}^2}{\Lambda^2}\right)}|x\rangle \tag{9.89}$$

They are very similar, the main difference being the integration contour. In (9.88), the contour is the positive real axis. In the second case, it is the entire real axis with a small bend below the origin. In the first case, convergence at infinity is ensured by the $i\epsilon$ prescription. In the second case, the factor $e^{i\alpha}$ allows us to close the contour at infinity in the upper half plane. In the first case, the regulator is $e^{-im_2 s}$, in the second case it is Λ.

In fact in a flat metric background, the mode-cutoff regularization turns out to be a version of the heat kernel method, as we show hereafter.

To start with let us consider Eq. (9.89) and, in order to work with dimensionless quantities let us absorb Λ in $\not{D}$: $\not{D} \to \Lambda\not{D}$. We will rescale back $\not{D}$ later on. Then, let us define

$$\langle x, \alpha|x', 0\rangle = \langle x|e^{-i\alpha\not{D}^2}|x'\rangle \tag{9.90}$$

It is evident that this quantity satisfies the heat kernel equation

$$i\frac{\partial}{\partial\alpha}\langle x, \alpha|x', 0\rangle = \not{D}_x^2\langle x, \alpha|x', 0\rangle \tag{9.91}$$

Next, we rewrite $\langle x, \alpha|x', 0\rangle$ in the form of Eq. (9.40) and (8.58), i.e. as a fat covariant delta function times $\Phi(x, x'; \alpha)$. When the background metric is flat we have $\sqrt{\mathcal{D}} = 1$ and $\sigma(x, x') = \frac{1}{2}(x - x')^2$. Therefore, in $\mathsf{d} = 4$, we can write

$$\langle x, \alpha|x', 0\rangle = -\frac{1}{(4\pi\alpha)^2} e^{i\frac{(x-x')^2}{4\alpha}}\Phi(x, x'; \alpha) \tag{9.92}$$

with α near 0. This formula comes from the Euclidean delta function definition in 4d

$$\delta^{(4)}(x - x') = \lim_{t\to 0}\frac{1}{(4\pi t)^2}e^{-\frac{(x-x')^2}{4t}} \tag{9.93}$$

Keeping t small, without taking the $t \to 0$ limit, we obtain a fat delta function. We recover the form (9.92) by taking $t = i\alpha$.

To proceed further, we consider the Fourier transform of the fat delta function

$$\int \mathrm{d}^4 x \frac{\mathrm{e}^{ikx}}{(4\pi t)^2} \mathrm{e}^{-\frac{x^2}{4t}} = \mathrm{e}^{-tk^2} \tag{9.94}$$

Then, we can write (taking $t = i\alpha$ small)

$$\langle x, \alpha | x', 0 \rangle = -\int \frac{\mathrm{d}^4 k}{(2\pi)^4} \mathrm{e}^{-ik(x-x')} \mathrm{e}^{-i\alpha k^2} \Phi(x, x'; \alpha) \tag{9.95}$$

Now, we apply the heat kernel equation (9.91) to (9.92). Dropping for simplicity the symbols of limit as well as the x and k integration, the LHS of the latter takes the form

$$i \frac{\partial}{\partial \alpha} \left(\mathrm{e}^{-ik(x-x')} \mathrm{e}^{-i\alpha k^2} \Phi \right) = \mathrm{e}^{-ik(x-x')} \mathrm{e}^{-i\alpha k^2} \left(i \frac{\partial \Phi}{\partial \alpha} + k^2 \Phi \right) \tag{9.96}$$

while the RHS is

$$\mathrm{e}^{-ik(x-x')} \mathrm{e}^{-i\alpha k^2} \left(k^2 + \not{k} \not{\mathcal{D}} + \not{\mathcal{D}} \not{k} + \not{\mathcal{D}}^2 \right) \Phi \tag{9.97}$$

Thus,

$$i \frac{\partial \Phi}{\partial \alpha}(x, x'; \alpha) = \left(\not{k} \not{\mathcal{D}} + \not{\mathcal{D}} \not{k} + \not{\mathcal{D}}^2 \right) \Phi(x, x'; \alpha) \tag{9.98}$$

Now, we reinsert the Λ dependence by rescaling $\not{\mathcal{D}}$ and $\not{k}$, and get

$$i \frac{\partial \Phi}{\partial \alpha}(x, x'; \alpha) = \left(\frac{\not{k} \not{\mathcal{D}} + \not{\mathcal{D}} \not{k}}{\Lambda^2} + \frac{\not{\mathcal{D}}^2}{\Lambda^2} \right) \Phi(x, x'; \alpha) \tag{9.99}$$

Assuming $\Phi(x, x'; 0) = 1$, we can integrate and get

$$\Phi(x, \alpha, \Lambda) = \mathrm{e}^{i\alpha \left(\frac{\not{\mathcal{D}}_x \not{k} + \not{k} \not{\mathcal{D}}_x}{\Lambda^2} + \frac{\not{\mathcal{D}}_x^2}{\Lambda^2} \right)} \tag{9.100}$$

Therefore, after rescaling back k: $k_\mu \to \Lambda k_\mu$, we can rewrite $\langle x | P_\Lambda | x \rangle$ as

$$\langle x | P_\Lambda | x \rangle = \Lambda^4 \int_{-\infty}^{+\infty} \frac{\mathrm{d}\alpha}{2\pi i\alpha} \int \frac{\mathrm{d}^4 k}{(2\pi)^4} \mathrm{e}^{i\alpha(1-k^2)} \, \mathrm{e}^{-i\alpha \left(\frac{\not{\mathcal{D}}_x \not{k} + \not{k} \not{\mathcal{D}}_x}{\Lambda} + \frac{\not{\mathcal{D}}_x^2}{\Lambda^2} \right)} \tag{9.101}$$

which coincides with (9.19, 9.20). The important thing we have shown here is that the mode-cutoff regularization is one of the various versions of the heat kernel approach.

Appendix 9B: Auxiliary Formulas

Here, we report explicit formulas for $[\partial_\lambda \partial_\mu a_0]$, $[\partial_\lambda \Box a_0]$ and $[\Box\Box a_0]$ needed in Sect. 9.3:

$$\begin{aligned}[\partial_\lambda \partial_\mu a_0] &= -\frac{1}{4}\left(\partial_\lambda V_\mu + \partial_\mu V_\lambda\right) + \frac{1}{8}\left(V_\lambda V_\mu + V_\mu V_\lambda\right) + \frac{1}{4}\gamma_5\left(\Sigma_\lambda{}^\nu\{V_\mu, V_\nu\} + \Sigma_\mu{}^\nu\{V_\lambda, V_\nu\}\right)\\ &-\frac{1}{2}\gamma_5(\Sigma_\lambda{}^\nu \partial_\mu V_\nu + \Sigma_\mu{}^\nu \partial_\lambda V_\nu) + \frac{1}{2}\left(\Sigma_\lambda{}^\nu \Sigma_\mu{}^\rho + \Sigma_\mu{}^\nu \Sigma_\lambda{}^\rho\right) V_\nu V_\rho \end{aligned} \tag{9.102}$$

$$\begin{aligned}[\partial_\lambda \Box a_0] = -\frac{1}{3}\Bigg\{ &\partial_\lambda \partial\cdot V - \frac{1}{2}\partial\cdot V V_\lambda - \gamma_5 \Sigma_\lambda{}^\nu \partial\cdot V V_\nu - \frac{1}{2}\partial_\lambda V^\nu V_\nu - \gamma_5 \partial_\lambda V^\mu \Sigma_\mu{}^\nu V_\nu\\
&-\frac{1}{4}(V^\mu \partial_\mu V_\lambda + V^\mu \partial_\lambda V_\mu) + \frac{1}{8}(V^2 V_\lambda + V^\mu V_\lambda V_\mu) + \frac{1}{4}\gamma_5(\Sigma_\lambda{}^\nu V^2 V_\nu + \Sigma_\mu{}^\nu V^\mu V_\lambda V_\nu)\\
&-\frac{1}{2}\gamma_5\left(\Sigma_\mu{}^\nu V^\mu \partial_\lambda V_\nu + \Sigma_\lambda{}^\nu V^\mu \partial_\mu V_\nu\right) + \frac{1}{4}\gamma_5\left(\Sigma_\mu{}^\nu V^\mu V_\nu V_\lambda + \Sigma_\lambda{}^\nu V^\mu V_\nu V_\mu\right)\\
&+\frac{1}{2}(\Sigma_\mu{}^\nu V^\mu V_\nu \Sigma_\lambda{}^\rho V_\rho + \Sigma_\lambda{}^\nu V^\mu V_\nu \Sigma_\mu{}^\rho V_\rho) + 2\gamma_5 \Sigma^{\mu\nu}\partial_\lambda \partial_\mu V_\nu\\
&-\gamma_5 \Sigma^{\mu\nu}\partial_\mu V_\nu V_\lambda - 2\Sigma^{\mu\nu}\partial_\mu V_\nu \Sigma_\lambda{}^\rho V_\rho - \gamma_5 \Sigma^{\mu\nu}\partial_\lambda V_\nu V_\mu - 2\Sigma^{\mu\nu}\partial_\lambda V_\nu \Sigma_\mu{}^\rho V_\rho\\
&+2\gamma_5 \Sigma^{\mu\nu} V_\nu\Big[-\frac{1}{4}\left(\partial_\lambda V_\mu + \partial_\mu V_\lambda\right) + \frac{1}{8}\left(V_\lambda V_\mu + V_\mu V_\lambda\right) - \frac{1}{2}\gamma_5(\Sigma_\lambda{}^\nu \partial_\mu V_\nu + \Sigma_\mu{}^\nu \partial_\lambda V_\nu)\\
&+\frac{1}{4}\gamma_5\left(\Sigma_\lambda{}^\nu\{V_\mu, V_\nu\} + \Sigma_\mu{}^\nu\{V_\lambda, V_\nu\}\right) + \frac{1}{2}\left(\Sigma_\lambda{}^\nu \Sigma_\mu{}^\rho + \Sigma_\mu{}^\nu \Sigma_\lambda{}^\rho\right) V_\nu V_\rho\Big]\\
&+\frac{1}{2}\Box V_\lambda - \frac{1}{2}\partial^\mu V_\lambda V_\mu - \gamma_5 \Sigma^{\mu\nu}\partial_\mu V_\lambda V_\nu - \frac{1}{4}V_\lambda \partial\cdot V - \frac{3}{16}V_\lambda V^2 - \frac{1}{4}V_\lambda \Sigma^{\mu\nu} C_{\mu\nu}\\
&-\frac{1}{2}\Sigma^{\mu\nu} V_\lambda V_\mu V_\nu + \gamma_5 \Sigma_\lambda{}^\nu \Box V_\nu - \gamma_5 \Sigma_\lambda{}^\nu \partial^\mu V_\nu V_\mu - 2\Sigma_\lambda{}^\nu \partial^\mu V_\nu \Sigma_\mu{}^\rho V_\rho\\
&-\frac{1}{2}\gamma_5 \Sigma_\lambda{}^\nu V_\nu \partial\cdot V - \frac{3}{8}\gamma_5 \Sigma_\lambda{}^\nu V_\nu V^2 - \frac{2}{3}\Sigma_\lambda{}^\nu V_\nu \Sigma^{\sigma\rho} C_{\sigma\rho} - \gamma_5 \Sigma_\lambda{}^\nu V_\nu \Sigma^{\sigma\rho} V_\sigma V_\rho \Bigg\}\end{aligned} \tag{9.103}$$

Finally,

$$\begin{aligned}[\Box\Box a_0] &= -\frac{1}{2}\Box\partial\cdot V - \partial^\lambda \partial\cdot V[\partial_\lambda a_0] - \frac{1}{2}\partial\cdot V[\Box a_0] - \frac{1}{2}\Box V^\lambda[\partial_\lambda a_0] - \partial^\lambda V^\rho[\partial_\lambda \partial_\rho a_0]\\
&-\frac{1}{2}V^\lambda[\partial_\lambda \Box a_0] - \gamma_5\Big(\Sigma^{\mu\nu}\partial_\mu \Box V_\nu + 2\Sigma^{\mu\nu}\partial^\lambda \partial_\mu V_\nu[\partial_\lambda a_0] + \Sigma^{\mu\nu}\partial_\mu V_\nu[\Box a_0]\\
&+\Sigma^{\mu\nu}\Box V_\nu[\partial_\mu a_0] + 2\Sigma^{\mu\nu}\partial^\lambda V_\nu[\partial_\lambda \partial_\mu a_0] + \Sigma^{\mu\nu} V_\nu[\partial_\mu \Box a_0]\Big)\end{aligned} \tag{9.104}$$

References

1. W.A. Bardeen, Anomalous ward identities in spinor field theories. Phys. Rev. **184**, 1848 (1969)
2. A.P. Balachandran, G. Marmo, V.P. Nair, C.G. Trahern, A nonperturbative proof of the nonabelian anomalies. Phys. Rev. D **25**, 2713 (1982)
3. A. Andrianov, L. Bonora, R. Gamboa-Saravi, Regularized functional integral for fermions and anomalies. Phys. Rev. D **26**, 2821 (1982)
4. A. Andrianov, L. Bonora, Finite-mode regularization of the fermion functional integral. I. Nucl. Phys. **B233**, 232 (1984)
5. A. Andrianov, L. Bonora, Finite-mode regularization of the fermion functional integral. II. Nucl. Phys. **B233**, 247 (1984)
6. A.A. Andrianov, L. Bonora, P. Pasti, Anomalies, cohomology and finite-mode regularization in higher dimensions. Ann. Phys. **158**, 374 (1984)
7. B.S. DeWitt, *Global Approach to Quantum Field Theory*, vol. I and II (Oxford University Press, Oxford, 2003)

Chapter 10
Metric-Axial-Tensor (MAT) Background

In this chapter, we introduce a generalization of the theory of gravity coupled to fermions. The reason is that we wish to formulate the anomaly problem in such a way as to be able to answer all possible questions about (local) anomalies in fermionic theories and get explicit formulas for all of them (gauge, trace and diffeomorphism anomalies). In this regard, let us recall the distinction between *split* and *non-split* anomalies. Split anomalies have opposite sign for opposite fermion chiralities. Non-split anomalies have the same sign for opposite chiralities. An example of the first is the consistent chiral gauge or gravity anomalies. They may of course arise only in the presence of a chiral asymmetry. The gauge-induced odd parity trace anomaly is split, while the even parity ones are non-split. Consistent chiral anomalies are split. In the previous chapters we have calculated them with several methods, perturbative and non-perturbative. The most general and fruitful setting is the one introduced first by W.A.Bardeen, which we have already used before and repeat here for clarity. One considers a theory of Dirac fermions coupled to two external non-Abelian (a vector V_μ and an axial A_μ) gauge potentials. Eventually, one takes the chiral limit $V \to \frac{V}{2}$ and $A \to \frac{V}{2}$ and verifies that, in such a limit, the anomaly becomes the expected consistent chiral gauge anomaly. Another output of this setup is that by choosing $A_\mu = 0$ one obtains the so-called covariant (or ABJ) anomaly.

Our program in this chapter is to generalize Bardeen's setup of non-Abelian gauge theories to general theories in a metric background. That is, beside the usual metric, the model is endowed with an additional symmetric tensor that interacts axially with fermions. We refer to it as the metric-axial-tensor (MAT) gravity. We can analyze anomalies in such a theory both with perturbative and non-perturbative methods. But, no doubt, in this context the latter is much more powerful and far-reaching. Therefore, we will apply the Schwinger-DeWitt (and the Seeley-DeWitt) method to calculate anomalies in a metric-axial-tensor (MAT) background (the references are [1–3], see also [4–10]).

Let us dwell a bit on the motivation for this move. As was pointed out in Chap. 5, in order to reproduce split anomalies, one should be careful to preserve the definite fermion chirality throughout the calculation. The difficulty stems from the fact that we need the inverse of the kinetic operator (the propagator) and there is no direct

L. Bonora, *Fermions and Anomalies in Quantum Field Theories*, Theoretical and Mathematical Physics, https://doi.org/10.1007/978-3-031-21928-3_10

way to derive it, basically because the Dirac operator for a Weyl fermion contains a chiral projector. Therefore, one has to resort to some indirect method. Like in many other cases in physics, the best way to circumvent similar problems is to embed the system in a larger one containing more variables and/or parameters. The metric-axial-tensor (MAT) gravity is designed for this purpose. It is formulated for Dirac fermions coupled to the usual metric plus an axial symmetric tensor. In this case, the operator involved is the usual Dirac operator and its inverse is naturally defined. The situation appropriate for Weyl fermions turns out to be a particular case of this construction, which is recovered in a specific limit, the chiral limit.

To be more explicit, we introduce, beside the usual metric $g_{\mu\nu}$, an axial symmetric 2-tensor $f_{\mu\nu}$, and couple them to a Dirac fermion. Then, we compute the trace of the energy-momentum tensor and of its axial companion and eventually take the limit $g \to \frac{g}{2}$ and $f \to \frac{g}{2}$. Using this extended metric, we enact the Schwinger-DeWitt method, that is we combine the latter with Bardeen's approach. The Schwinger-DeWitt method is based on point-splitting; thus, we need a regularization in order to avoid divergences, but since the point-splitting is along a geodesic, covariance under diffeomorphisms is guaranteed. The MAT theory requires a new basic formalism we call axial-complex analysis (which is similar to the hyper-complex or pseudo-complex analysis already introduced in the literature, [11, 12]). We will spend the first part of the chapter to illustrate it and dig out the formulas we need in the sequel, without any pretension of completeness (see also Appendix 10A).

10.1 Axial-Complex Analysis and Geometry

Axial-complex numbers are defined by

$$\hat{a} = a_1 + \gamma_5 a_2 \tag{10.1}$$

where a_1 and a_2 are real numbers. Arithmetic is defined in the obvious way. We can define a conjugation operation

$$\overline{\hat{a}} = a_1 - \gamma_5 a_2 \tag{10.2}$$

We will denote by $\mathcal{AC}$ the set of axial-complex numbers, by $\mathcal{AR}$ the set of axial-complex numbers with $a_2 = 0$ (the axial-real numbers) and by $\mathcal{AI}$ the set of axial-complex numbers with $a_1 = 0$ (the axial-imaginary numbers). We can define a (pseudo)norm

$$(a, a) = \hat{a}\overline{\hat{a}} = a_1^2 - a_2^2 \tag{10.3}$$

This determines an axial-light-cone with all the related problems. In general we will keep away from it by considering the case $|a_1| > |a_2|$.

Introducing the chiral projectors $P_\pm = \frac{1\pm\gamma_5}{2}$, we can also write

$$\hat{a} = a_+ P_+ + a_- P_-, \qquad a_\pm = a_1 \pm a_2 \tag{10.4}$$

We will consider functions $\hat{f}(\hat{x})$ of the axial-complex variable

$$\widehat{x} = x_1 + \gamma_5 x_2 \tag{10.5}$$

from $\mathcal{AC}$ to $\mathcal{AC}$, which are axial-analytic, i.e. admit a Taylor expansion, and we actually identify the functions with their expansions. Using the property of the projectors, it is easy to see that

$$\hat{f}(\hat{x}) = P_+ \hat{f}(x_+) + P_- \hat{f}(x_-) = \frac{1}{2}\left(\hat{f}(x_+) + \hat{f}(x_-)\right) + \frac{\gamma_5}{2}\left(\hat{f}(x_+) - \hat{f}(x_-)\right) \tag{10.6}$$

In the same, way we will consider functions from $\mathcal{AC}^4$ to $\mathcal{AC}$, with analogous properties.

$$\hat{f}(\hat{x}^\mu) = P_+ \hat{f}(x_+^\mu) + P_- \hat{f}(x_-^\mu) = \frac{1}{2}\left(\hat{f}(x_+^\mu) + \hat{f}(x_-^\mu)\right) + \frac{\gamma_5}{2}\left(\hat{f}(x_+^\mu) - \hat{f}(x_-^\mu)\right) \tag{10.7}$$

with $\mu = 0, 1, 2, 3$, and

$$\widehat{x}^\mu = x_1^\mu + \gamma_5 x_2^\mu \tag{10.8}$$

are the axial-complex coordinates.

Derivatives are defined in the obvious way:

$$\frac{\partial}{\partial \hat{x}^\mu} = \frac{1}{2}\left(\frac{\partial}{\partial x_1^\mu} + \gamma_5 \frac{\partial}{\partial x_2^\mu}\right), \qquad \frac{\partial}{\partial \overline{\hat{x}}^\mu} = \frac{1}{2}\left(\frac{\partial}{\partial x_1^\mu} - \gamma_5 \frac{\partial}{\partial x_2^\mu}\right) \tag{10.9}$$

Notice that for axial-analytic functions

$$\frac{d}{d\hat{x}} = \frac{\partial}{\partial x_1} \equiv \frac{\partial}{\partial \hat{x}}, \tag{10.10}$$

whereas $\frac{\partial}{\partial \overline{\hat{x}}} \widehat{f}(\hat{x}) = 0$.

As for integrals, since we will always deal with rapidly decreasing functions at infinity, we define

$$\int d\hat{x}\, \widehat{f}(\hat{x})$$

as the rapidly decreasing primitive $\widehat{g}(\hat{x})$ of $\widehat{f}(\hat{x})$. Therefore, the property

$$\int \mathrm{d}\hat{x}\, \frac{\partial}{\partial \hat{x}} \hat{f}(\hat{x}) = 0 \tag{10.11}$$

follows immediately. As a consequence of (10.10) it follows that, for an axial-analytic function,

$$\int \mathrm{d}\hat{x}\, \widehat{f}(\hat{x}) = \int \mathrm{d}x_1\, \widehat{f}(\hat{x}) \tag{10.12}$$

We can define definite integrals such as

$$\int_{\hat{a}}^{\hat{b}} \mathrm{d}\hat{x}\, \widehat{f}(\hat{x}) = \widehat{g}(\hat{b}) - \widehat{g}(\hat{a}) \tag{10.13}$$

In this axial-spacetime, we introduce an axial-Riemannian geometry as follows. Starting from a metric $\widehat{g}_{\mu\nu} = g_{\mu\nu} + \gamma_5 f_{\mu\nu}$, the Christoffel symbols (see Appendix 10A) are defined by

$$\widehat{\Gamma}^{\lambda}_{\mu\nu} = \frac{1}{2}\widehat{g}^{\lambda\rho}\left(\frac{\partial}{\partial \widehat{x}^{\mu}}\widehat{g}_{\rho\nu} + \frac{\partial}{\partial \widehat{x}^{\nu}}\widehat{g}_{\mu\rho} - \frac{\partial}{\partial \widehat{x}^{\rho}}\widehat{g}_{\mu\nu}\right) \tag{10.14}$$

They split as follows

$$\widehat{\Gamma}^{\mu}_{\nu\lambda} = \Gamma^{(1)\mu}_{\nu\lambda} + \gamma_5 \Gamma^{(2)\mu}_{\nu\lambda} \tag{10.15}$$

and are such that the metricity condition is satisfied

$$\frac{\partial}{\partial \hat{x}^{\mu}}\widehat{g}_{\nu\lambda} = \widehat{\Gamma}^{\rho}_{\mu\nu}\,\widehat{g}_{\rho\lambda} + \widehat{\Gamma}^{\rho}_{\mu\lambda}\,\widehat{g}_{\nu\rho}, \tag{10.16}$$

which in components takes the form

$$\frac{\partial}{\partial \hat{x}^{\mu}} g_{\nu\lambda} = \Gamma^{(1)\rho}_{\mu\nu}\, g_{\rho\lambda} + \Gamma^{(1)\rho}_{\mu\lambda}\, g_{\nu\rho} + \Gamma^{(2)\rho}_{\mu\nu}\, f_{\rho\lambda} + \Gamma^{(2)\rho}_{\mu\lambda}\, f_{\nu\rho} \tag{10.17}$$

$$\frac{\partial}{\partial \hat{x}^{\mu}} f_{\nu\lambda} = \Gamma^{(1)\rho}_{\mu\nu}\, f_{\rho\lambda} + \Gamma^{(1)\rho}_{\mu\lambda}\, f_{\nu\rho} + \Gamma^{(2)\rho}_{\mu\nu}\, g_{\rho\lambda} + \Gamma^{(2)\rho}_{\mu\lambda}\, g_{\nu\rho} \tag{10.18}$$

10.1.1 MAT Geodesics

The SDW method is based on point-splitting along a geodesic. Therefore, it is of upmost importance to define geodesics in a MAT background.

The equation for MAT geodesics is

$$\ddot{\widehat{x}}^{\mu} + \widehat{\Gamma}^{\mu}_{\nu\lambda}\dot{\widehat{x}}^{\nu}\dot{\widehat{x}}^{\lambda} = 0 \tag{10.19}$$

where a dot denotes derivation with respect to an axial-affine parameter $\widehat{t} = t_1 + \gamma_5 t_2$. For axial-real and axial-imaginary components, this means

$$\ddot{x}_1^{\mu} + \Gamma^{(1)\mu}_{\nu\lambda}(\dot{x}_1^{\nu}\dot{x}_1^{\lambda} + \dot{x}_2^{\nu}\dot{x}_2^{\lambda}) + \Gamma^{(2)\mu}_{\nu\lambda}(\dot{x}_1^{\nu}\dot{x}_2^{\lambda} + \dot{x}_2^{\nu}\dot{x}_1^{\lambda}) = 0 \tag{10.20}$$

$$\ddot{x}_2^{\mu} + \Gamma^{(1)\mu}_{\nu\lambda}(\dot{x}_1^{\nu}\dot{x}_2^{\lambda} + \dot{x}_2^{\nu}\dot{x}_1^{\lambda}) + \Gamma^{(2)\mu}_{\nu\lambda}(\dot{x}_1^{\nu}\dot{x}_1^{\lambda} + \dot{x}_2^{\nu}\dot{x}_2^{\lambda}) = 0 \tag{10.21}$$

These geodesic equations can be obtained as equations of motion from the action

$$\widehat{S} = \int \mathrm{d}\hat{t}\sqrt{\widehat{g}_{\mu\nu}\dot{\widehat{x}}^{\mu}\dot{\widehat{x}}^{\nu}} = S_1 + \gamma_5 S_2 \tag{10.22}$$

where $\widehat{g}_{\mu\nu} = g_{\mu\nu} + \gamma_5 f_{\mu\nu}$.

The action takes values in $\mathcal{AC}$. We require the action principle to be specified by $\delta\widehat{S}[\widehat{x}] = 0$. Taking the variation of $S[\widehat{x}]$ with respect to $\delta\widehat{x} = \delta x_1 + \gamma_5\delta x_2$, with

$$\begin{aligned}
\delta\widehat{g}_{\mu\nu} &= \frac{\partial\widehat{g}_{\mu\nu}}{\partial\widehat{x}^{\lambda}}\delta\widehat{x}^{\lambda}, \qquad \text{i.e.} \\
\delta g_{\mu\nu} &= \frac{1}{2}\left(\frac{\partial g_{\mu\nu}}{\partial x_1^{\lambda}} + \frac{\partial f_{\mu\nu}}{\partial x_2^{\lambda}}\right)\delta x_1^{\lambda} + \left(\frac{\partial f_{\mu\nu}}{\partial x_1^{\lambda}} + \frac{\partial g_{\mu\nu}}{\partial x_2^{\lambda}}\right)\delta x_2^{\lambda} \\
&= \frac{\partial g_{\mu\nu}}{\partial x_1^{\lambda}}\delta x_1^{\lambda} + \frac{\partial f_{\mu\nu}}{\partial x_1^{\lambda}}\delta x_2^{\lambda} \\
\delta f_{\mu\nu} &= \frac{1}{2}\left(\frac{\partial g_{\mu\nu}}{\partial x_1^{\lambda}} + \frac{\partial f_{\mu\nu}}{\partial x_2^{\lambda}}\right)\delta x_2^{\lambda} + \left(\frac{\partial f_{\mu\nu}}{\partial x_1^{\lambda}} + \frac{\partial g_{\mu\nu}}{\partial x_2^{\lambda}}\right)\delta x_1^{\lambda} \\
&= \frac{\partial g_{\mu\nu}}{\partial x_1^{\lambda}}\delta x_2^{\lambda} + \frac{\partial f_{\mu\nu}}{\partial x_1^{\lambda}}\delta x_1^{\lambda}
\end{aligned} \tag{10.23}$$

we get the eom

$$\widehat{g}_{\mu\rho}\ddot{\widehat{x}}^{\rho} + \widehat{\Gamma}^{\rho}_{\nu\lambda}\widehat{g}_{\mu\rho}\dot{\widehat{x}}^{\mu}\dot{\widehat{x}}^{\nu} = 0, \qquad \text{i.e.} \qquad \ddot{\widehat{x}}^{\mu} + \widehat{\Gamma}^{\mu}_{\nu\lambda}\dot{\widehat{x}}^{\nu}\dot{\widehat{x}}^{\lambda} = 0 \tag{10.24}$$

Let us rewrite

$$\begin{aligned}
\sqrt{\widehat{g}_{\mu\nu}\dot{\widehat{x}}^{\mu}\dot{\widehat{x}}^{\nu}} &= \sqrt{A + \gamma_5 B}, \\
A &= g_{\mu\nu}\left(\dot{x}_1^{\mu}\dot{x}_1^{\nu} + \dot{x}_2^{\mu}\dot{x}_2^{\nu}\right) + 2 f_{\mu\nu}\dot{x}_1^{\mu}\dot{x}_2^{\nu}, \\
B &= f_{\mu\nu}\left(\dot{x}_1^{\mu}\dot{x}_1^{\nu} + \dot{x}_2^{\mu}\dot{x}_2^{\nu}\right) + 2 g_{\mu\nu}\dot{x}_1^{\mu}\dot{x}_2^{\nu},
\end{aligned} \tag{10.25}$$

so that we have ($\hat{\tau} = \tau_1 + \gamma_5 \tau_2$ is the proper time)

$$\widehat{S}[\widehat{x}] = \int d\hat{\tau} \sqrt{\widehat{g}_{\mu\nu} \dot{\widehat{x}}^\mu \dot{\widehat{x}}^\nu}$$
$$= \frac{1}{2} \left[\int \mathrm{d}\tau_1 \left(\sqrt{A+B} + \sqrt{A-B} \right) + \int \mathrm{d}\tau_2 \left(\sqrt{A+B} - \sqrt{A-B} \right) \right]$$
$$+ \frac{\gamma_5}{2} \left[\int \mathrm{d}\tau_1 \left(\sqrt{A+B} - \sqrt{A-B} \right) + \int \mathrm{d}\tau_2 \left(\sqrt{A+B} + \sqrt{A-B} \right) \right] \quad (10.26)$$

Varying this action with respect to δx^λ, we obtain the same eom (10.24). This is due to (10.12) and to the fact that, the action is an analytic function of $\widehat{x}$, so that the variation with respect to $\delta \widehat{x}^\lambda$ is the same as the variation with respect to δx_1^λ.

In our final results, we shall always set $x_2 = 0$, but it is very convenient to keep the axial-analytic notation as far as possible.

10.1.2 Geodetic Interval and Distance

The quantity

$$\widehat{E} = E_1 + \gamma_5 E_2 = \frac{1}{2} \widehat{g}_{\mu\nu} \dot{\widehat{x}}^\mu \dot{\widehat{x}}^\nu \quad (10.27)$$

is conserved as a function of $\hat{t}$. Since $\widehat{g}_{\mu\nu} \dot{\widehat{x}}^\mu \dot{\widehat{x}}^\nu$ is constant for geodesics, we can write for the arc length parameter $\widehat{s}$

$$\frac{\mathrm{d}\widehat{s}}{\mathrm{d}\hat{t}} = \sqrt{\widehat{g}_{\mu\nu} \dot{\widehat{x}}^\mu \dot{\widehat{x}}^\nu}, \quad (10.28)$$

and

$$\widehat{s} - \widehat{s'} = \int_{\hat{t}'}^{\hat{t}} \mathrm{d}\hat{\tau} \sqrt{2\widehat{E}} = \sqrt{2\widehat{E}} \, (\hat{t} - \hat{t}'). \quad (10.29)$$

$\widehat{s} - \widehat{s'}$ is the axial arc length along the geodesic between $\widehat{x}$ and $\widehat{x'}$. The half square of it is called the *world function*, and it is denoted

$$\widehat{\sigma}(\widehat{x}, \widehat{x'}) = \frac{1}{2} (\widehat{s} - \widehat{s'})^2 = \widehat{E} (\hat{t} - \hat{t}')^2 = (\hat{t} - \hat{t}') \int_{\hat{t}'}^{\hat{t}} \widehat{E} \mathrm{d}\hat{\tau} \quad (10.30)$$

The main properties are

$$\widehat{\sigma}_{;\mu} = \widehat{\partial}_\mu \widehat{\sigma} = (\hat{t} - \hat{t}')\widehat{g}_{\mu\nu}\dot{\widehat{x}}^\nu \equiv -\widehat{g}_{\mu\nu}\widehat{y}^\nu \tag{10.31}$$

$\widehat{y}^\mu$ are the *normal coordinates* based at $\widehat{x}$.

Using (10.30 and 10.31), one can see that

$$\frac{1}{2}\widehat{\sigma}_{;\mu}\widehat{\sigma}_{;}{}^{\mu} = \widehat{\sigma} \tag{10.32}$$

The subscript $_{;\mu}$ means the covariant derivative with respect to $\widehat{x}^\mu$, while $_{;\mu'}$ means the covariant derivative with respect to $\widehat{x'}^{\mu'}$.

Remark. $\widehat{\sigma} = \sigma_1 + \gamma_5\sigma_2$, but notice that, even when we set $x_2 = 0$, we cannot infer that $\sigma_2 = 0$. This descends from Eq. (10.28). Looking at (10.26), we see that B does not vanish even when $x_2^\nu = 0$. As a consequence, the axial-imaginary part of (10.25) does not vanish, so the axial-imaginary part of Eq. (10.28) will not automatically vanish either.

10.1.3 Normal Coordinates

The tangent vector $\frac{\mathrm{d}\widehat{x}^\mu}{\mathrm{d}\hat{t}}$ to the geodesic at $\hat{x}$ satisfies

$$\frac{D}{\mathrm{d}\hat{t}}\frac{\mathrm{d}\widehat{x}^\mu}{\mathrm{d}\hat{t}} \equiv \frac{d^2\widehat{x}^\mu}{\mathrm{d}\hat{t}^2} + \widehat{\Gamma}^\mu_{\nu\lambda}\frac{\mathrm{d}\widehat{x}^\nu}{\mathrm{d}\hat{t}}\frac{\mathrm{d}\widehat{x}^\lambda}{\mathrm{d}\hat{t}} = 0 \tag{10.33}$$

and an analogous equation at $\hat{x}'$. Now, we can write

$$\begin{aligned} \widehat{y}^{\mu'}{}_{;\nu}(\hat{x}', \hat{x})\widehat{y}^\nu(\hat{x}, \hat{x}') &= (\hat{t}' - \hat{t})\widehat{y}^{\mu'}{}_{;\nu}(\widehat{x}', \widehat{x})\frac{\mathrm{d}\widehat{x}^\nu(\hat{t})}{\mathrm{d}\hat{t}} \\ &= (\hat{t}' - \hat{t})\frac{d}{\mathrm{d}\hat{t}}\widehat{y}^{\mu'}(\widehat{x}', \widehat{x}) = (\hat{t}' - \hat{t})\frac{\mathrm{d}\widehat{x}^{\mu'}(\hat{t}')}{\mathrm{d}\hat{t}'} \\ &= -\widehat{y}^{\mu'}(\widehat{x'}, \widehat{x}) \end{aligned} \tag{10.34}$$

Dividing by $\hat{t} - \hat{t}'$ the second and fourth terms and taking the coincidence limit $\widehat{x'} \to \widehat{x}$, one gets

$$[\widehat{y}^{\mu'}{}_{;\nu}]\frac{\mathrm{d}\widehat{x}^\nu}{\mathrm{d}\hat{t}} = \frac{\mathrm{d}\widehat{x}^\mu}{\mathrm{d}\hat{t}} \qquad \to \qquad [\widehat{y}^{\mu'}{}_{;\nu}] = \delta^\mu_\nu \tag{10.35}$$

where $[X]$ denotes the result of the coincidence limit on the quantity X. In a similar way, one can prove

$$[\widehat{y}^{\mu'}{}_{;\nu'}]\frac{\mathrm{d}\widehat{x}^{\nu}}{\mathrm{d}\hat{t}} = -\frac{\mathrm{d}\widehat{x}^{\mu}}{\mathrm{d}\hat{t}} \qquad \rightarrow \qquad [\widehat{y}^{\mu'}{}_{;\nu'}] = -\delta^{\mu}_{\nu} \tag{10.36}$$

$$[\widehat{y}^{\mu}{}_{;\nu}]\frac{\mathrm{d}\widehat{x}^{\nu}}{\mathrm{d}\hat{t}} = -\frac{\mathrm{d}\widehat{x}^{\mu}}{\mathrm{d}\hat{t}} \qquad \rightarrow \qquad [\widehat{y}^{\mu}{}_{;\nu}] = -\delta^{\mu}_{\nu} \tag{10.37}$$

$$[\widehat{y}^{\mu}{}_{;\nu'}]\frac{\mathrm{d}\widehat{x}^{\nu}}{\mathrm{d}t} = \frac{\mathrm{d}\widehat{x}^{\mu}}{\mathrm{d}t} \qquad \rightarrow \qquad [\widehat{y}^{\mu}{}_{;\nu'}] = \delta^{\mu}_{\nu} \tag{10.38}$$

10.1.4 Coincidence Limits of $\widehat{\sigma}$

Covariantly differentiating (10.32), we get

$$\widehat{\sigma}_{;\nu} = \widehat{\sigma}_{;\mu\nu}\widehat{\sigma}_{;}{}^{\mu} \tag{10.39}$$

In the coincidence limit $[\widehat{\sigma}_{;\nu}] = 0$. Therefore, (10.39) is trivial in the coincidence limit. Differentiating the first and last member of (10.31), we get

$$\widehat{\sigma}_{;\mu\lambda} = -\widehat{g}_{\mu\nu}\,\widehat{y}^{\nu}{}_{;\lambda} \tag{10.40}$$

Using (10.37), one gets

$$[\widehat{\sigma}_{;\mu\lambda}] = \widehat{g}_{\mu\lambda} \tag{10.41}$$

Similarly,

$$[\widehat{\sigma}_{;\mu\lambda'}] = -\widehat{g}_{\mu\lambda} \tag{10.42}$$

Differentiating (10.39) once more, one gets

$$\widehat{\sigma}_{;\nu\lambda} = \widehat{\sigma}_{;\mu\nu\lambda}\,\widehat{\sigma}_{;}{}^{\mu} + \widehat{\sigma}_{;\mu\nu}\,\widehat{\sigma}^{\mu}_{;\lambda}$$

which, in the coincidence limit, using the previous results, yields an identity. Differentiating it again

$$\widehat{\sigma}_{;\nu\lambda\rho} = \widehat{\sigma}_{;\mu\nu\lambda\rho}\,\widehat{\sigma}_{;}{}^{\mu} + \widehat{\sigma}_{;\mu\nu\lambda}\,\widehat{\sigma}^{\mu}_{;\rho} + \widehat{\sigma}_{;\mu\nu\rho}\,\widehat{\sigma}^{\mu}_{;\lambda} + \widehat{\sigma}_{;\mu\nu}\,\widehat{\sigma}_{;}{}^{\mu}{}_{\lambda\rho} \tag{10.43}$$

In the coincidence limit, this becomes

$$[\widehat{\sigma}_{;\nu\lambda\rho}] = [\widehat{\sigma}_{;\rho\nu\lambda}] + [\widehat{\sigma}_{;\lambda\nu\rho}] + [\widehat{\sigma}_{;\nu\lambda\rho}] \tag{10.44}$$

Since $\widehat{\sigma}$ is a biscalar, we have

$$[\widehat{\sigma}_{;\nu\lambda\rho}] = [\widehat{\sigma}_{;\nu\rho\lambda}] + \widehat{R}_{\rho\lambda\nu}{}^{\tau}[\widehat{\sigma}_{;\tau}] = [\widehat{\sigma}_{;\rho\nu\lambda}] \tag{10.45}$$

Therefore,

$$[\widehat{\sigma}_{;\rho\nu\lambda}] = [\widehat{\sigma}_{;\lambda\nu\rho}] = [\widehat{\sigma}_{;\nu\lambda\rho}] = 0 \tag{10.46}$$

Differentiating (10.43) once more and taking the coincidence limit, one gets

$$[\widehat{\sigma}_{;\nu\lambda\rho\tau}] = -\frac{1}{3}\left(\widehat{R}_{\nu\tau\lambda\rho} + \widehat{R}_{\nu\rho\lambda\tau}\right) \equiv \widehat{S}_{\nu\lambda\rho\tau} \tag{10.47}$$

where $\widehat{R}_{\nu\tau\lambda\rho} = \widehat{g}_{\nu\mu}\widehat{R}^{\mu}{}_{\tau\lambda\rho}$. Differentiating once more,

$$[\widehat{\sigma}_{;\nu\lambda\rho\sigma\tau}] = \frac{3}{4}\left(\widehat{S}_{\nu\lambda\sigma\tau;\rho} + \widehat{S}_{\nu\lambda\sigma\rho;\tau} + \widehat{S}_{\nu\lambda\tau\rho;\sigma}\right) \tag{10.48}$$

We will need also the coincidence limits of tensors covariantly differentiated with respect to a primed index ν'. In general,

$$[t_{\mu_1\ldots\mu_k;\nu'}] = [t_{\mu_1\ldots\mu_k}]_{;\nu} - [t_{\mu_1\ldots\mu_k;\nu}] \tag{10.49}$$

So

$$[\widehat{\sigma}_{;\mu\nu'}] = [\widehat{\sigma}_{;\mu}]_{;\nu} - [\widehat{\sigma}_{;\mu\nu}] = -\widehat{g}_{\mu\nu} \tag{10.50}$$

$$[\widehat{\sigma}_{;\mu\nu'\lambda}] = [\widehat{\sigma}_{;\mu\lambda\nu'}] = [\widehat{\sigma}_{;\mu\lambda}]_{;\nu} - [\widehat{\sigma}_{;\mu\lambda\nu}] = 0 \tag{10.51}$$

$$[\widehat{\sigma}_{;\mu\nu'\lambda\rho}] = [\widehat{\sigma}_{;\mu\lambda\rho\nu'}] = [\widehat{\sigma}_{;\mu\lambda\rho}]_{;\nu} - [\widehat{\sigma}_{;\mu\lambda\rho\nu}] = -[\widehat{\sigma}_{;\mu\lambda\rho\nu}] = -\widehat{S}_{\mu\lambda\rho\nu} \quad (10.52)$$

and

$$\begin{aligned}[\widehat{\sigma}_{;\mu\nu'\lambda\rho\sigma}] &= [\widehat{\sigma}_{;\mu\lambda\rho\sigma\nu'}] = [\widehat{\sigma}_{;\mu\lambda\rho\sigma}]_{;\nu} - [\widehat{\sigma}_{;\mu\lambda\rho\sigma\nu}] \\ &= \frac{1}{4}\widehat{S}_{\mu\lambda\rho\sigma;\nu} - \frac{3}{4}\left(\widehat{S}_{\mu\lambda\nu\rho;\sigma} + \widehat{S}_{\mu\lambda\sigma\nu;\rho}\right)\end{aligned} \quad (10.53)$$

Similarly, one obtains

$$[\widehat{\sigma}_{;\mu}{}^{\mu}{}_{\nu}{}^{\nu}{}_{\rho}{}^{\rho}] = -\frac{8}{5}R_{;\mu}{}^{\mu} + \frac{4}{15}\widehat{R}_{\mu\nu}\widehat{R}^{\mu\nu} - \frac{4}{15}\widehat{R}_{\mu\nu\lambda\rho}\widehat{R}^{\mu\nu\lambda\rho} \quad (10.54)$$

$$[\widehat{\sigma}_{;\mu\nu}{}^{\nu}{}_{\rho}{}^{\rho\mu}] = -[\widehat{\sigma}_{;\mu}{}^{\mu'}{}_{\nu}{}^{\nu}{}_{\rho}{}^{\rho}] = \frac{2}{5}R_{;\mu}{}^{\mu} - \frac{1}{15}\widehat{R}_{\mu\nu}\widehat{R}^{\mu\nu} - \frac{4}{15}\widehat{R}_{\mu\nu\lambda\rho}\widehat{R}^{\mu\nu\lambda\rho} \quad (10.55)$$

10.1.5 Van Vleck-Morette Determinant

The Van Vleck-Morette determinant in MAT is defined by

$$\widehat{\mathcal{D}}(\widehat{x}, \widehat{x'}) = \det(-\widehat{\sigma}_{;\mu\nu'}) \quad (10.56)$$

$\widehat{\mathcal{D}}(\widehat{x}, \widehat{x'})$ is a bi-density of weight 1 both at $\widehat{x}$ and $\widehat{x'}$. Later on, we will need a bi-density of weight 0:

$$\widehat{\Delta}(\widehat{x}, \widehat{x'}) = \frac{1}{\sqrt{\widehat{g}(\widehat{x})}}\widehat{\mathcal{D}}(\widehat{x}, \widehat{x'})\frac{1}{\sqrt{\widehat{g}(\widehat{x'})}} \quad (10.57)$$

The VVM determinant also satisfies (in four dimensions)

$$(\widehat{\mathcal{D}}(\widehat{x}, \widehat{x'})\widehat{\sigma}^{;\mu})_{;\mu} = 4\widehat{\mathcal{D}}(\widehat{x}, \widehat{x'}) \quad (10.58)$$

In the coincidence limit

$$[\widehat{\Delta}^{\frac{1}{2}}_{;\lambda}] = [\widehat{g}^{-\frac{1}{4}}(\widehat{x})\sqrt{\widehat{\mathcal{D}}(\widehat{x},\widehat{x}')}\frac{1}{2}\left(\widehat{\sigma}^{-1\mu\nu'}\widehat{\sigma}_{;\mu\nu'\lambda}\right)\widehat{g}^{-\frac{1}{4}}(\widehat{x}')] = \frac{1}{2}[\widehat{\sigma}^{\mu}_{;\mu\lambda}] = 0 \quad (10.59)$$

We need to compute the covariant derivatives of $\widehat{\sigma}^{-1\mu\nu'} \equiv \{\widehat{\sigma}^{-1}_{;\mu\nu'}\}$. The latter is defined as

$$\widehat{\sigma}^{-1\mu\nu'}\widehat{\sigma}_{;\nu'\lambda} = \delta^{\mu}_{\lambda} \quad (10.60)$$

Differentiating this relation once, twice and thrice, one gets

$$\begin{aligned}
&[\widehat{\sigma}^{-1\mu\nu'}{}_{;\lambda}] = 0, \\
&[\widehat{\sigma}^{-1}{}_{\mu\lambda';\rho\sigma}] = -[\widehat{\sigma}_{;\mu'\lambda\rho\sigma}] = [\widehat{\sigma}_{;\lambda\rho\sigma\mu}] = \widehat{S}_{\lambda\rho\sigma\mu}
\end{aligned} \quad (10.61)$$

and

$$[\widehat{\sigma}^{-1}{}_{\mu\lambda';\rho\sigma\tau}] = -[\widehat{\sigma}_{;\lambda\mu'\rho\sigma\tau}] = \frac{1}{4}\widehat{S}_{\mu\rho\sigma\tau;\lambda} - \frac{3}{4}\left(\widehat{S}_{\mu\rho\lambda\sigma;\tau} + \widehat{S}_{\mu\rho\tau\lambda;\sigma}\right) \quad (10.62)$$

Differentiating once more, one gets

$$\begin{aligned}
[\widehat{\Delta}^{\frac{1}{2}}_{;\lambda\rho}] &= \frac{1}{6}\widehat{g}^{\mu\nu}\left(\widehat{R}_{\mu\nu\lambda\rho} + \widehat{R}_{\mu\lambda\nu\rho}\right) = \frac{1}{6}\widehat{g}^{\mu\nu}\widehat{g}_{\mu\sigma}\widehat{R}^{\sigma}{}_{\lambda\nu\rho} \\
&= \frac{1}{6}\left(R^{(1)}_{\lambda\rho} + \gamma_5 R^{(2)}_{\lambda\rho}\right)
\end{aligned} \quad (10.63)$$

and

$$[\widehat{\Delta}^{\frac{1}{2}}_{;\lambda\rho\sigma}] = \frac{1}{12}\left(\widehat{R}_{\lambda\rho;\sigma} + \widehat{R}_{\rho\sigma;\lambda} + \widehat{R}_{\sigma\lambda;\rho}\right) \quad (10.64)$$

Finally,

$$[\widehat{\Delta}^{\frac{1}{2}}{}_{;\mu}{}^{\mu}{}_{\nu}{}^{\nu}] = +\frac{1}{5}\widehat{R}_{;\mu}{}^{\mu} + \frac{1}{36}\widehat{R}^{2} - \frac{1}{30}\widehat{R}_{\mu\nu}\widehat{R}^{\mu\nu} + \frac{1}{30}\widehat{R}_{\mu\nu\lambda\rho}\widehat{R}^{\mu\nu\lambda\rho} \quad (10.65)$$

10.1.6 The Geodetic Parallel Displacement Matrix

The geodetic parallel displacement matrix $\widehat{G}^{\mu}{}_{\nu'}(\widehat{x}, \widehat{x'})$ is needed in order to parallel displace vectors from one end to the other of the geodetic interval. It is defined by

$$[\widehat{G}^{\mu}{}_{\nu'}] = \delta^{\mu}_{\nu}, \qquad \widehat{G}^{\mu}{}_{\nu';\lambda}\widehat{\sigma}^{;\lambda} = 0 \tag{10.66}$$

The second condition means that the covariant derivative of $\widehat{G}^{\mu}{}_{\nu'}$ vanishes in directions parallel to the geodesic. Since tangents to the geodesics are self-parallel, it follows that

$$\begin{aligned} &\widehat{G}_{\mu}{}^{\nu'}\widehat{\sigma}_{;\nu'} = -\sigma_{;\mu}, \qquad \widehat{\sigma}_{;\mu}\widehat{G}^{\mu}{}_{\nu'} = -\widehat{\sigma}_{;\nu'} \\ &\widehat{G}_{\mu\nu'} = \widehat{G}_{\nu'\mu}, \qquad \widehat{\sigma}_{;}{}^{\lambda'}\widehat{G}^{\mu}{}_{\nu';\lambda'} = 0 \\ &\widehat{G}_{\mu}{}^{\nu'}\widehat{G}_{\nu'}{}^{\lambda} = \delta^{\lambda}_{\mu} \end{aligned} \tag{10.67}$$

The analogous parallel displacer for spinors is denoted $\widehat{I}(\widehat{x}, \widehat{x'})$: the object $\widehat{I}(\widehat{x}, \widehat{x'})\widehat{\psi}(\widehat{x'})$ is the spinor $\psi(\widehat{x})$ obtained by parallel displacing $\widehat{\psi}(\widehat{x'})$ along the geodesic from $\widehat{x'}$ to $\widehat{x}$. It is a bispinor quantity satisfying

$$\widehat{\sigma}_{;}{}^{\mu}\widehat{I}_{;\mu} = 0, \qquad [\widehat{I}] = \mathbf{1} \tag{10.68}$$

and $\mathbf{1}$ is the identity matrix in the spinor space. Differentiating (10.68) once, we get $[\widehat{I}_{;\mu}] = 0$. Differentiating twice, we get

$$[\widehat{I}_{;(\mu\nu)}] = 0, \tag{10.69}$$

while

$$\widehat{I}(x, x')_{;\mu\nu} - \widehat{I}(x, x')_{;\nu\mu} = -\frac{1}{2}\left(\mathrm{d}\widehat{\Omega} + \widehat{\Omega}\widehat{\Omega}\right)_{\mu\nu}\widehat{I}(x, x') = -\frac{1}{2}\widehat{\mathcal{R}}_{\mu\nu}I(x, x') \tag{10.70}$$

where $\widehat{\mathcal{R}}_{\mu\nu} = \widehat{R}_{\mu\nu}{}^{ab}\Sigma_{ab}$. So

$$[\widehat{I}(x, x')_{;[\mu,\nu]}] = [\widehat{I}(x, x')_{;\mu\nu}] = -\frac{1}{4}\widehat{\mathcal{R}}_{\mu\nu} \tag{10.71}$$

Proceeding with the differentiation of (10.68), we find

$$[\widehat{I}_{;\nu\lambda\rho}] + [\widehat{I}_{;\lambda\nu\rho}] + [\widehat{I}_{;\rho\lambda\nu}] = 0 \tag{10.72}$$

Now,

$$[\widehat{I}_{;\nu\lambda\rho}] - [\widehat{I}_{;\nu\rho\lambda}] = \frac{1}{2}\widehat{\mathcal{R}}_{\rho\lambda}[\widehat{I}_{;\nu}] = 0 \tag{10.73}$$

and

$$3[\widehat{I}_{;\nu\lambda\rho}] = \frac{1}{2}\widehat{\nabla}_{\rho}\widehat{\mathcal{R}}_{\lambda\nu} + \frac{1}{2}\widehat{\nabla}_{\lambda}\widehat{\mathcal{R}}_{\rho\nu} \tag{10.74}$$

In particular,

$$[\widehat{I}_{;\nu}{}^{\nu}{}_{\rho}] = \frac{1}{6}\widehat{\nabla}^{\nu}\widehat{\mathcal{R}}_{\rho\nu} \tag{10.75}$$

Differentiating (10.68) once more with respect to x^{σ}, using (10.47) and then contracting with $\widehat{g}^{\nu\lambda}\widehat{g}^{\sigma\rho}$, we find, after simplifying,

$$[\widehat{I}_{;\mu}{}^{\mu}{}_{\nu}{}^{\nu}] + [\widehat{I}_{;\mu\nu}{}^{\nu\mu}] = 0 \tag{10.76}$$

A contraction with $\widehat{g}^{\nu\sigma}\widehat{g}^{\lambda\rho}$ gives:

$$[\widehat{I}_{;\mu\nu}{}^{\nu\mu}] + 2[\widehat{I}_{;\mu\nu}{}^{\mu\nu}] + [\widehat{I}_{;\mu}{}^{\mu}{}_{\nu}{}^{\nu}] = 0 \tag{10.77}$$

Using (10.70), we get

$$[\widehat{I}_{;\sigma\rho\mu\nu}] = [\widehat{\nabla}_{\nu}\widehat{\nabla}_{\mu}(\widehat{I}_{;\sigma\rho})] = -\frac{1}{2}\widehat{\mathcal{R}}_{\sigma\rho;\mu\nu} + \frac{1}{8}\widehat{\mathcal{R}}_{\sigma\rho}\widehat{\mathcal{R}}_{\mu\nu} + [\widehat{I}_{;\rho\sigma\mu\nu}] \tag{10.78}$$

Contracting with $\widehat{g}^{\mu\sigma}\widehat{g}^{\nu\rho}$ gives

$$[\widehat{I}_{;\mu\nu}{}^{\mu\nu}] = 0 + \frac{1}{8}\widehat{\mathcal{R}}_{\mu\nu}\widehat{\mathcal{R}}^{\mu\nu} + [\widehat{I}_{;\mu\nu}{}^{\nu\mu}] \tag{10.79}$$

since by Walker's identity

$$\widehat{\nabla}_{\rho}\widehat{\nabla}_{\lambda}\widehat{\mathcal{R}}^{\rho\lambda} = 0 \tag{10.80}$$

Finally, by using (10.76) and (10.77), one gets

$$[\widehat{I}_{;\nu}{}^{\nu}{}_{\rho}{}^{\rho}] = \frac{1}{8}\widehat{\mathcal{R}}_{\rho\lambda}\widehat{\mathcal{R}}^{\rho\lambda} \tag{10.81}$$

10.2 Fermions in MAT Background

The action of a fermion interacting with a metric and an axial tensor is

$$\begin{aligned}\widehat{S} &= \int d^4\widehat{x}\left(i\overline{\psi}\sqrt{\overline{\widehat{g}}}\gamma^a\widehat{e}_a^\mu\left(\partial_\mu+\frac{1}{2}\widehat{\Omega}_\mu\right)\psi\right)(\widehat{x}) \\ &= \int d^4\widehat{x}\left(i\overline{\psi}\sqrt{\overline{\widehat{g}}}\gamma^a(\tilde{e}_a^\mu+\gamma_5\tilde{c}_a^\mu)\left(\partial_\mu+\frac{1}{2}\left(\Omega_\mu^{(1)}+\gamma_5\Omega_\mu^{(2)}\right)\right)\psi\right)(\widehat{x}) \\ &= \int d^4\widehat{x}\left(i\overline{\psi}\sqrt{\widehat{g}}(\tilde{e}_a^\mu-\gamma_5\tilde{c}_a^\mu)\left[\frac{1}{2}\gamma^a\overset{\leftrightarrow}{\partial}_\mu+\frac{1}{4}\left(\gamma^a\widehat{\Omega}_\mu+\overline{\widehat{\Omega}}_\mu\gamma^a\right)\right]\psi\right)(\widehat{x}) \\ &= \int d^4\widehat{x}\left(i\overline{\psi}\sqrt{\widehat{g}}(\tilde{e}_a^\mu-\gamma_5\tilde{c}_a^\mu)\left[\frac{1}{2}\gamma^a\overset{\leftrightarrow}{\partial}_\mu\psi+\frac{i}{4}\gamma_d\varepsilon^{dabc}\widehat{\Omega}_{\mu bc}\gamma_5\right]\psi\right)(\widehat{x})\end{aligned} \tag{10.82}$$

It must be noticed that this action takes axial-real values.[1] The field $\psi(\widehat{x})$ can be understood, classically, as a series of powers of $\widehat{x}$ applied to constant spinors on their right, and the symmetry transformations act on it from the left. The analogous definitions for $\psi^\dagger$ are obtained via hermitean conjugation. In the second line, it is stressed that the action contains also an axial part. It is understood that $\partial_\mu = \frac{\partial}{\partial\widehat{x}^\mu}$ applies only to ψ or $\overline{\psi}$, as indicated, and $\overline{\widehat{g}}$ denotes, as usual, the axial-complex conjugate of $\widehat{g}$ (for the other symbols see Appendix 10A).

A few comments are in order. The density $\sqrt{\overline{\widehat{g}}}$ must be inserted between $\overline{\psi}$ and ψ, due to the presence in it of the γ_5 matrix. Moreover one has to take into account that the kinetic operator contains a γ matrix that anticommutes with γ_5. Thus, for instance, using $\widehat{D}_\lambda\widehat{g}_{\mu\nu}=0$ and $(\widehat{D}_\lambda+\frac{1}{2}\widehat{\Omega}_\lambda)\widehat{e}=0$, where $\widehat{D}=\partial+\widehat{\Gamma}$, one gets

$$\overline{\psi}\gamma^a\widehat{e}_a^\mu\left(\partial_\mu+\frac{1}{2}\Omega_\mu\right)\psi = \overline{\psi}(\overline{\widehat{D}}_\mu+\frac{1}{2}\overline{\widehat{\Omega}}_\mu)\gamma^a\widehat{e}_a^\mu\psi \tag{10.83}$$

We recall again that a bar denotes axial-complex conjugation, i.e. a sign reversal in front of each γ_5 contained in the expression, for instance $\overline{\widehat{\Omega}}_\mu = \Omega_\mu^{(1)} - \gamma_5\Omega_\mu^{(2)}$.

To obtain the two last lines in (10.82), one must use (10.260) and (10.83).

10.2.1 Classical Ward Identities

Let us consider AE (axially extended) diffeomorphisms first, (10.237). It is not hard to prove that the action (10.82) is invariant under these transformations. Now, define the full MAT e.m. tensor by means of

[1] One could consider also an axial complex action, but for our purposes this is a useless complication. That is why we use the notation $\psi(\widehat{x})$ instead of $\widehat{\psi}(\widehat{x})$.

$$\mathbf{T}_{\mu\nu} = \frac{2}{\sqrt{g}} \frac{\overleftarrow{\delta}\, \widehat{S}}{\delta \widehat{g}^{\mu\nu}} \tag{10.84}$$

This formula needs a comment, since $\sqrt{g}$ contains γ_5. To give a meaning to it, we understand that the operator $\frac{2}{\sqrt{g}} \frac{\overleftarrow{\delta}}{\delta \widehat{g}^{\mu\nu}}$ in the RHS acts on the operatorial expression, say $\mathcal{O}\sqrt{g}$, which is inside the spinor scalar product, $\overline{\psi}\mathcal{O}\sqrt{g}\psi$. Moreover, the functional derivative acts from the right of the action. Now, the conservation law under diffeomorphisms is

$$\begin{aligned} 0 = \delta_{\widehat{\xi}} S &= \int \overline{\psi} \frac{\overleftarrow{\delta}\, \mathcal{O}}{\delta \widehat{g}^{\mu\nu}} \delta \widehat{g}^{\mu\nu} \psi = - \int \overline{\psi} \frac{\overleftarrow{\delta}\, \mathcal{O}}{\delta \widehat{g}^{\mu\nu}} \left(\widehat{D}^{\mu} \widehat{\xi}^{\nu} + \widehat{D}^{\nu} \widehat{\xi}^{\mu} \right) \psi \\ &= 2 \int \overline{\psi} \frac{\overleftarrow{\delta}\, \mathcal{O}\overleftarrow{\ }}{\delta \widehat{g}_{\mu\nu}} \widehat{D}_{\mu} \widehat{\xi}_{\nu} \psi \end{aligned} \tag{10.85}$$

where $\widehat{D}$ acts (from the right) on everything except the parameter $\widehat{\xi}_{\nu} = \xi_{\nu} + \gamma_5 \zeta_{\nu}$. Differentiating with respect to the arbitrary parameters ξ^{μ} and ζ^{ν}, we obtain two conservation laws involving the two tensors

$$T_{\mu\nu} = 2 \overline{\psi} \frac{\overleftarrow{\delta}\, \mathcal{O}}{\delta \widehat{g}^{\mu\nu}} \psi \tag{10.86}$$

$$T_{5\mu\nu} = 2 \overline{\psi} \frac{\overleftarrow{\delta}\, \mathcal{O}}{\delta \widehat{g}^{\mu\nu}} \gamma_5 \psi \tag{10.87}$$

To give a less abstract idea of these tensors, at the lowest order (flat background) and setting $x_2^{\mu} = 0$, they are given on shell by

$$T_{\mu\nu} \approx T_{\mu\nu}^{(flat)} = -\frac{i}{4} \left(\overline{\psi} \gamma_{\mu} \overleftrightarrow{\partial}_{\nu} \psi + \mu \leftrightarrow \nu \right), \tag{10.88}$$

and

$$T_{5\mu\nu} \approx T_{5\mu\nu}^{(flat)} = \frac{i}{4} \left(\overline{\psi} \gamma_5 \gamma_{\mu} \overleftrightarrow{\partial}_{\nu} \psi + \mu \leftrightarrow \nu \right), \tag{10.89}$$

Repeating the same derivation for the axial complex Weyl transformation, one can prove that assuming for the fermion field the transformation rule

$$\psi \rightarrow \mathrm{e}^{-\frac{3}{2}(\omega + \gamma_5 \eta)} \psi, \tag{10.90}$$

Equation (10.82) is invariant, and obtain the Ward identity

$$0 = \int \overline{\psi} \frac{\overleftarrow{\delta}\, \mathcal{O}}{\delta \widehat{g}^{\mu\nu}} \widehat{g}^{\mu\nu} (\omega + \gamma_5 \eta) \psi \tag{10.91}$$

One gets in this way two WI's

$$\mathcal{T}(x) \equiv T_{\mu\nu} g^{\mu\nu} + T_{5\mu\nu} f^{\mu\nu} = 0, \tag{10.92}$$

$$\mathcal{T}_5(x) \equiv T_{\mu\nu} f^{\mu\nu} + T_{5\mu\nu} g^{\mu\nu} = 0, \tag{10.93}$$

10.2.2 A More Explicit Formula for the e.m. Tensor

In our calculation, a more explicit formula of the e.m. tensor is needed. The e.m. tensor is defined by

$$\mathbf{T}_{\mu\nu} = \frac{2}{\sqrt{g}} \frac{\overleftarrow{\delta}\, \widehat{S}}{\delta \widehat{g}^{\mu\nu}} = \frac{1}{2} \left(\mathbf{T}^a_\mu \widehat{e}_{a\nu} + \mathbf{T}^a_\nu \widehat{e}_{a\mu} \right) \tag{10.94}$$

where

$$\mathbf{T}^a_\mu = \frac{1}{\sqrt{|g|}} \frac{\overleftarrow{\delta}\, \widehat{S}}{\delta \widehat{e}^\mu_a} \tag{10.95}$$

In (10.94), it is understood that, for instance, $\widehat{e}^\nu_a$ is inserted in the rightmost position past ψ (inserting it from the right) inside $\mathbf{T}^a_\mu$. Let us prove first that the functional derivative of $\widehat{\Omega}_m$ does not contribute to the e.m. tensor. Consider the general variational formula

$$\begin{aligned} \delta \widehat{\Omega}^{bc}_\mu &= \frac{1}{2} \widehat{e}^{b\nu} \left(\widehat{\nabla}_\mu (\delta \widehat{e}^c_\nu) - \widehat{\nabla}_\nu (\delta \widehat{e}^c_\mu) \right) - \frac{1}{2} \widehat{e}^{c\nu} \left(\widehat{\nabla}_\mu (\delta \widehat{e}^b_\nu) - \widehat{\nabla}_\nu (\delta \widehat{e}^b_\mu) \right) \\ &+ \frac{1}{2} \widehat{e}^{b\nu} \widehat{e}^{c\lambda} \left(\widehat{\nabla}_\lambda (\delta \widehat{e}^e_\nu) - \widehat{\nabla}_\nu (\delta \widehat{e}^e_\lambda) \right) \widehat{e}_{e\mu} \end{aligned} \tag{10.96}$$

where $\widehat{\nabla}$ denotes the covariant derivative such that $\widehat{\nabla}_\mu \widehat{e}^a_\lambda = 0$. After some algebra, one gets

$$\gamma_d\, \varepsilon^{dabc} \widehat{e}^\mu_a\, \delta \widehat{\Omega}_{\mu bc} = \gamma_d\, \varepsilon^{dabc} \widehat{e}^\mu_a \widehat{e}^\nu_b \nabla_\mu \delta e_{c\nu} \tag{10.97}$$

Now use this and

$$\frac{\delta \widehat{e}^a_\mu(\widehat{x})}{\delta \widehat{e}^b_\nu(\widehat{y})} = \delta^a_b \delta^\nu_\mu \delta(\widehat{x}, \widehat{y})$$

and insert them into the definition (10.94). The relevant contribution (labeled with the subscript $_\Omega$) is

$$\begin{aligned}
\mathbf{T}^{\lambda\rho}_{\Omega} &= \frac{1}{2}\left(\mathbf{T}^{\lambda}_{a}\widehat{e}^{a\rho} + \mathbf{T}^{\rho}_{a}\widehat{e}^{a\lambda}\right)_{\Omega} \qquad (10.98)\\
&\equiv \frac{1}{8}\int \overline{\psi}\gamma_d \varepsilon^{dabc}\widehat{e}^{\mu}_{a}\left(\frac{\delta\widehat{\Omega}_{\mu bc}}{\delta\widehat{e}^{e}_{\lambda}}\widehat{e}^{e\rho} + \frac{\delta\widehat{\Omega}_{\mu bc}}{\delta\widehat{e}^{e}_{\rho}}\widehat{e}^{e\lambda}\right)\gamma_5\psi\\
&= \frac{1}{8}\int \overline{\psi}\gamma_d \varepsilon^{dabc}\widehat{e}^{\mu}_{a}\left(\widehat{e}^{\lambda}_{b}\widehat{e}^{\rho}_{c}\widehat{\nabla}_{\mu}\delta(\widehat{x},\widehat{y}) + \widehat{e}^{\rho}_{b}\widehat{e}^{\lambda}_{c}\widehat{\nabla}_{\mu}\delta(\widehat{x},\widehat{y})\right)\gamma_5\psi = 0
\end{aligned}$$

Therefore, the only contribution to the e.m. tensor comes from the variation of the first $\widehat{e}^{m}_{a}$ factor in (10.82). The result (on shell) is

$$\mathbf{T}_{\lambda\rho} = -\frac{i}{2}\overline{\psi}\widehat{\gamma}_{\lambda}\left(\partial_{\rho} + \frac{1}{2}\widehat{\Omega}_{\rho}\right)\psi + (\lambda \leftrightarrow \rho) = -\frac{i}{2}\overline{\psi}\widehat{\gamma}_{\lambda}\widehat{\nabla}_{\rho}\psi + (\lambda \leftrightarrow \rho) \qquad (10.99)$$

where $\widehat{\gamma}_{\lambda} = \gamma_a \widehat{e}^{a}_{\lambda}$.

10.2.3 The Dirac Operator and Its Inverse

In the action (10.82), the Dirac operator is

$$\widehat{F} = i\widehat{\gamma}\cdot\widehat{\nabla} = i\widehat{\gamma}^{\mu}\widehat{\nabla}_{\mu} = i\gamma^{a}\widehat{e}^{\mu}_{a}\widehat{\nabla}_{\mu} \equiv \gamma^{a}\,\widehat{F}_{a} \qquad (10.100)$$

where the $\widehat{\nabla}$ operator is, schematically, $\widehat{D} + \frac{1}{2}\widehat{\Omega}$ and satisfies $\widehat{\nabla}_{\mu}\widehat{e}^{a}_{\nu} = 0$.

Under AE diffeomorphisms ψ transforms as: $\delta_{\widehat{\xi}}\psi = \widehat{\xi}\cdot\partial\psi$, while

$$\delta_{\widehat{\xi}}\left(i\widehat{\gamma}\cdot\widehat{\nabla}\psi\right) = \overline{\widehat{\xi}}\cdot\partial\left(i\widehat{\gamma}\cdot\widehat{\nabla}\psi\right) \qquad (10.101)$$

Under AE Weyl transformation $\widehat{F}$ transform as

$$\delta_{\widehat{\omega}}\widehat{F} = -\frac{1}{2}\gamma^{a}\{\widehat{F}_{a}, \widehat{\omega}\} \qquad (10.102)$$

and it has the following hermiticity property

$$\widehat{F}^{\dagger} = \beta\widehat{F}\beta \qquad (10.103)$$

where $\beta = \gamma_0$ and γ_0 is the non-dynamical (flat) gamma matrix.

Integrating out the fermion field in (10.82) is understood to correspond to evaluating the determinant of the Dirac operator $\widehat{F}$. This is, however, not yet the starting point for the SDW method. $\widehat{F}$ is a linear operator while we need a quadratic operator, which becomes elliptic and self-adjoint in the Euclidean. This quantity can be constructed as follows.

One starts from

$$\widehat{G}(\widehat{x}, \widehat{x'}) = \langle 0|T\psi(\widehat{x})\psi^\dagger(\widehat{x'})|0\rangle \tag{10.104}$$

which satisfies

$$i\sqrt{\widehat{g}}\,\beta\,\widehat{\gamma}^\mu\widehat{\nabla}_\mu\widehat{G}(\widehat{x}, \widehat{x'}) = -\mathbf{1}\delta(\widehat{x}, \widehat{x'}), \tag{10.105}$$

where $\mathbf{1}$ is the unit matrix in the spinor space. Next, we make the ansatz

$$\widehat{G}(x, x') = i\overline{\widehat{\gamma}}^\mu\overline{\widehat{\nabla}}_\mu\overline{\widehat{\mathcal{G}}}(x, x')\beta^{-1} \tag{10.106}$$

Inserting (10.106) in (10.105), we construct a quadratic operator, which after an axial conjugation becomes

$$\widehat{\mathcal{F}} = \overline{\widehat{\gamma}}^\mu\overline{\widehat{\nabla}}_\mu\widehat{\gamma}^\nu\widehat{\nabla}_\nu \tag{10.107}$$

This is the quadratic operator we need.

10.2.3.1 Why the Ansatz (10.107)

As we have explained in Sect. 8.3.1, the SDW method can be applied only to systems where diffeomorphisms are conserved. In a MAT background, we must generalize this to include axial diffeomorphisms, the reason being that we use throughout procedures which are covariant in both ordinary and axial diffeomorphisms. On the other hand, we know from the perturbative calculations of Chap. 7 that in 4d for Weyl fermions diffeomorphisms are conserved—and, in any case, consistent chiral diffeomorphisms anomalies do not find place in 4d. Since a Dirac fermion can be viewed as a couple of Weyl fermion of opposite chiralities, we deduce that diffeomorphism anomalies cannot exist for Dirac fermion in a MAT background. Therefore, our choice for the square Dirac operator must conform to it. Thus, the choice of (10.107). Another motivation for this choice is that it leads to a self-adjoint elliptic operator in a Euclidean metric background. Let us see the two motivations in turn.

In ordinary gravity, from the diffeomorphism invariance of the fermion action, we can extract the transformation rule

$$\delta_\xi\left(i\gamma^\mu\nabla_\mu\psi\right) = \xi\cdot\partial\left(i\gamma\cdot\nabla\psi\right) \tag{10.108}$$

while $\delta_\xi\psi = \xi\cdot\partial\psi$. Therefore it makes sense to apply $\gamma\cdot\nabla$ to $\gamma\cdot\nabla\psi$, because the latter transforms as ψ. This allows us to define the square of the Dirac operator:

$$F^2\psi = (i\gamma\cdot\nabla)^2\,\psi \tag{10.109}$$

If we want to preserve general covariance, it is not possible to repeat the same for MAT because of (10.101), from which we see that $\left(i\widehat{\gamma}\cdot\widehat{\nabla}\psi\right)$ does not transform like ψ, and an expression like $\left(i\widehat{\gamma}\cdot\widehat{\nabla}\right)^2\psi$ would break general covariance.[2] Noting that

$$\delta_{\widehat{\xi}}\left(i\overline{\widehat{\gamma}}\cdot\overline{\widehat{\nabla}}\psi\right) = \widehat{\xi}\cdot\partial\left(i\overline{\widehat{\gamma}}\cdot\overline{\widehat{\nabla}}\psi\right) \tag{10.110}$$

when $\delta_{\overline{\widehat{\xi}}}\psi = \overline{\widehat{\xi}}\cdot\partial\psi$, we are led to consider, instead, the *covariant* quadratic operator

$$\left(i\overline{\widehat{\gamma}}\cdot\overline{\widehat{\nabla}}\right)\left(i\widehat{\gamma}\cdot\widehat{\nabla}\right)\psi \tag{10.111}$$

This is the first reason that justifies the use of $\left(i\overline{\widehat{\gamma}}\cdot\overline{\widehat{\nabla}}\right)\left(i\widehat{\gamma}\cdot\widehat{\nabla}\right)$. The second reason has to do with his Euclidean version and will be explained in a moment. In preparation, let us quote a few useful identities.

$$\overline{\widehat{\nabla}}_\mu\widehat{\gamma}_\nu - \widehat{\gamma}_\nu\widehat{\nabla}_\mu = \gamma^a\left(\partial_\mu\,\widehat{e}_{a\nu} - \widehat{\Gamma}^\lambda_{\mu\nu}\widehat{e}_{a\lambda} + \frac{1}{2}\widehat{\Omega}_{\mu ab}\,\widehat{e}^b_\nu\right) = 0 \tag{10.112}$$

because of metricity, and

$$\overline{\widehat{\nabla}}_\mu\gamma^a - \gamma^a\widehat{\nabla}_\mu = 0 \tag{10.113}$$

The axial conjugate relation holds as well. Therefore,

$$\widehat{\gamma}^\mu\widehat{\nabla}_\mu\overline{\widehat{\gamma}}^\nu\overline{\widehat{\nabla}}_\nu = \gamma^a\gamma^b\overline{\widehat{e}}^\mu_a\overline{\widehat{e}}^\nu_b\overline{\widehat{\nabla}}_\mu\overline{\widehat{\nabla}}_\nu = \eta^{ab}\overline{\widehat{e}}^\mu_a\overline{\widehat{e}}^\nu_b\overline{\widehat{\nabla}}_\mu\overline{\widehat{\nabla}}_\nu + \Sigma^{ab}\overline{\widehat{e}}^\mu_a\overline{\widehat{e}}^\nu_b[\overline{\widehat{\nabla}}_\mu, \overline{\widehat{\nabla}}_\nu] \tag{10.114}$$

On the other hand, when acting on a (bi-)spinor quantity,

$$\Sigma^{ab}\overline{\widehat{e}}^\mu_a\overline{\widehat{e}}^\nu_b[\overline{\widehat{\nabla}}_\mu, \overline{\widehat{\nabla}}_\nu] = \frac{1}{8}\gamma^a\gamma^b\gamma^c\gamma^d\widehat{R}_{abcd} = -\frac{1}{4}\widehat{R}_{\mu\nu\lambda\rho}\widehat{g}^{\mu\lambda}\widehat{g}^{\nu\rho} = -\frac{1}{4}\widehat{R} \tag{10.115}$$

where use is made of

$$\widehat{R}_{abcd} = \widehat{e}^\mu_a\widehat{e}^\nu_b\widehat{e}^\lambda_c\widehat{e}^\rho_d\,\widehat{R}_{\mu\nu\lambda\rho}. \tag{10.116}$$

---

[2] One could object that the choice of $\left(i\widehat{\gamma}\cdot\widehat{\nabla}\right)^2$ breaks only the axial diffeomorphisms, while eventually we are interested in ordinary diffeomorphisms. This remark, however, is misleading, because in the chiral limit a violation of the axial diffeomorphisms would precisely affect also the ordinary diffeomorphism conservation.

Now replacing (10.106) into (10.105) and using the above, we get

$$\sqrt{\overline{\widehat{g}}}\left(\overline{\widehat{\nabla}}_{\mu}\overline{\widehat{g}}^{\mu\nu}\overline{\widehat{\nabla}}_{\nu}-\frac{1}{4}\overline{\widehat{R}}\right)\overline{\widehat{\mathfrak{G}}}(\widehat{x},\widehat{x}')=\mathbf{1}\delta(\widehat{x},\widehat{x}') \tag{10.117}$$

The differential operator acting on $\overline{\widehat{\mathfrak{G}}}$ in this equation will be denoted by $\overline{\widehat{\mathfrak{F}}}_{\hat{g}}$. In compact operator notation

$$\overline{\widehat{\mathfrak{F}}}_{\hat{g}}\overline{\widehat{\mathfrak{G}}}_{\hat{g}}=-\mathbf{1}, \tag{10.118}$$

Of course also the axial-conjugated relation holds. We set $\langle\widehat{x}|\widehat{\mathfrak{G}}|\widehat{x}'\rangle=\widehat{\mathfrak{G}}_{\hat{g}}(\widehat{x},\widehat{x}')$.

As a consequence of (10.103), we have

$$\left[\sqrt{\widehat{g}}\left(\widehat{\nabla}_{\mu}\widehat{g}^{\mu\nu}\widehat{\nabla}_{\nu}-\frac{1}{4}\widehat{R}\right)\right]^{\dagger}=\beta\left[\sqrt{\widehat{g}}\left(\widehat{\nabla}_{\mu}\widehat{g}^{\mu\nu}\widehat{\nabla}_{\nu}-\frac{1}{4}\widehat{R}\right)\right]\beta \tag{10.119}$$

or

$$\left(\widehat{\mathfrak{F}}_{\hat{g}}\right)^{\dagger}=\beta\,\widehat{\mathfrak{F}}_{\hat{g}}\,\beta \tag{10.120}$$

We shall be using mostly the related operator (10.107)

$$\widehat{\mathfrak{F}}=\frac{1}{\sqrt{\widehat{g}}}\,\widehat{\mathfrak{F}}_{\hat{g}},\qquad \widehat{\mathfrak{F}}^{\dagger}=\eta\,\widehat{\mathfrak{F}}\eta \tag{10.121}$$

and its inverse $\widehat{\mathfrak{G}}$: $\widehat{\mathfrak{F}}\widehat{\mathfrak{G}}=-\mathbf{1}$.

Let us come finally to the Euclidean version of $\widehat{\mathfrak{F}}$. To deal with it it is easier to remark that

$$\left(i\widetilde{\widehat{\gamma}}\widetilde{\widehat{\nabla}}\right)^{\dagger}=-i\widetilde{\overline{\widehat{\gamma}}}\widetilde{\overline{\widehat{\nabla}}} \tag{10.122}$$

Therefore,

$$\left(\widetilde{\widehat{\mathfrak{F}}}\right)^{\dagger}=\widetilde{\widehat{\mathfrak{F}}} \tag{10.123}$$

This is the second fundamental reason for using $\widehat{\mathfrak{F}}$.

Summarizing. The operator $\widehat{\mathfrak{F}}$ is the main intermediate result we need. It is natural to assume that its inverse $\widehat{\mathfrak{G}}$ exists. The differential operator $\widehat{\mathfrak{F}}$ (after a Wick rotation) can be defined as an axial-elliptic operator, at least under reasonable conditions on the axial tensor $f_{\mu\nu}$. In fact, its quadratic part can be cast in the form $\partial_i A_{ij}(x)\partial_j$, where A_{ij} is an invertible matrix and its leading term is symmetric and positive definite. Moreover $\widetilde{\widehat{\mathfrak{F}}}$ is self-adjoint. Since in the SDW method the Euclidean version

$\widehat{\widehat{\mathfrak{F}}}$ of $\widehat{\mathfrak{F}}$ is to be used, all the requirements are satisfied. However, as we have done before, we will always work with Minkowski quantities, paying attention, however, that no relation is used that cannot be mapped from the Minkowski to the Euclidean by a Wick rotation and back by an inverse Wick rotation.

Remark. The action (10.82) is invariant not only under the AE diffeomorphisms, but also under the AE local Lorentz transformations (for relevant definitions, see Appendix 10A)

$$\delta_{\widehat{\Lambda}}\psi = -\frac{1}{2}\widehat{\Lambda}\,\psi, \qquad \delta_{\widehat{\Lambda}}\Omega_\mu = \partial_\mu\widehat{\Lambda} + [\Omega_\mu, \widehat{\Lambda}] \tag{10.124}$$

As a consequence,

$$\delta_{\widehat{\Lambda}}\left(i\overline{\overline{\gamma}}\cdot\overline{\overline{\nabla}}\psi\right) = \frac{1}{2}\widehat{\Lambda}\left(i\overline{\overline{\gamma}}\cdot\overline{\overline{\nabla}}\psi\right) \tag{10.125}$$

It follows that the quadratic operator $\left(i\overline{\overline{\gamma}}\cdot\overline{\overline{\nabla}}\right)\,(i\widehat{\gamma}\cdot\widehat{\nabla})$ behaves covariantly not only for AE diffeomorphisms but also under AE local Lorentz transformations. As a consequence of this, one can anticipate that diffeomorphism and local Lorentz anomalies cannot be produced with the SDW method as long as we use this quadratic operator.

10.3 The Schwinger Proper Time Method

Hereafter, we mostly repeat, with adapted symbols, much of the content of Sect. 8.3

Let us define the amplitude

$$\langle\widehat{x}, \widehat{s}|\widehat{x}', 0\rangle = \langle\widehat{x}|e^{i\widehat{\mathfrak{F}}\widehat{s}}|\widehat{x}'\rangle \tag{10.126}$$

which satisfies the (heat kernel) differential equation

$$i\frac{\partial}{\partial\widehat{s}}\langle\widehat{x}, \widehat{s}|\widehat{x}', 0\rangle = -\widehat{\mathfrak{F}}_{\hat{x}}\langle\widehat{x}, \widehat{s}|\widehat{x}', 0\rangle \equiv K(\widehat{x}, \widehat{x}', \widehat{s}) \tag{10.127}$$

where $\widehat{\mathfrak{F}}_{\hat{x}}$ is the differential operator

$$\widehat{\mathfrak{F}}_{\widehat{x}} = \widehat{\nabla}_\mu\widehat{g}^{\mu\nu}\widehat{\nabla}_\nu - \frac{1}{4}\widehat{R} \tag{10.128}$$

In formula (10.126), as usual, we understand the $i\epsilon$ prescription.

Then, we make the ansatz

$$\langle\widehat{x}, \widehat{s}|\widehat{x}', 0\rangle = -\lim_{m\to 0}\frac{i}{16\pi^2}\frac{\sqrt{\widehat{D}(\widehat{x}, \widehat{x}')}}{\widehat{s}^2}e^{i\left(\frac{\widehat{\sigma}(\widehat{x}, \widehat{x}')}{2\widehat{s}} - m^2\widehat{s}\right)}\widehat{\Phi}(\widehat{x}, \widehat{x}', \widehat{s}) \tag{10.129}$$

where $\widehat{D}(\widehat{x}, \widehat{x}')$ is the VVM determinant and $\widehat{\sigma}$ is the world function (see above). $\widehat{\Phi}(\widehat{x}, \widehat{x}', \widehat{s})$ is a function to be determined. It is useful to introduce also the mass parameter m, which we will eventually set to zero. In the limit $\widehat{s} \to 0$, the RHS of (10.129) becomes the definition of a delta function multiplied by $\widehat{\Phi}$. More precisely, since it must be $\langle \widehat{x}, 0|\widehat{x}', 0\rangle = \delta(\widehat{x}, \widehat{x}')$, and

$$\lim_{\widehat{s}\to 0} \frac{i}{(4\pi)^2} \frac{\sqrt{\widehat{D}(\widehat{x}, \widehat{x}')}}{\widehat{s}^2} e^{i\left(\frac{\widehat{\sigma}(\widehat{x},\widehat{x}')}{2\widehat{s}} - m^2\widehat{s}\right)} = \sqrt{\widehat{g}(\widehat{x})}\ \delta(\widehat{x}, \widehat{x}'), \tag{10.130}$$

we must have

$$\lim_{\widehat{s}\to 0} \widehat{\Phi}(\widehat{x}, \widehat{x}', \widehat{s}) = \mathbf{1} \tag{10.131}$$

Equation (10.127) becomes an equation for $\widehat{\Phi}(\widehat{x}, \widehat{x}', \widehat{s})$. Using (8.61) and (10.58), after some algebra, one gets

$$i\frac{\partial \widehat{\Phi}}{\partial \widehat{s}} + \frac{i}{\widehat{s}} \widehat{\nabla}^\mu \widehat{\Phi} \widehat{\nabla}_\mu \widehat{\sigma} + \frac{1}{\sqrt{\widehat{D}}} \widehat{\nabla}^\mu \widehat{\nabla}_\mu \left(\sqrt{\widehat{D}}\widehat{\Phi}\right) - \left(\frac{1}{4}\widehat{R} - m^2\right)\widehat{\Phi} = 0 \tag{10.132}$$

Now, we expand

$$\widehat{\Phi}(\widehat{x}, \widehat{x}', \widehat{s}) = \sum_{n=0}^{\infty} \widehat{a}_n(\widehat{x}, \widehat{x}')(i\widehat{s})^n \tag{10.133}$$

with the boundary condition $[\widehat{a}_0] = 1$. The $\widehat{a}_n$ must satisfy the recursive relations:

$$\begin{aligned}(n+1)\widehat{a}_{n+1} + \widehat{\nabla}^\mu \widehat{a}_{n+1} \widehat{\nabla}_\mu \widehat{\sigma} - \frac{1}{\sqrt{\widehat{D}}} \widehat{\nabla}^\mu \widehat{\nabla}_\mu \left(\sqrt{\widehat{D}}\widehat{a}_n\right) \\ + \left(\frac{1}{4}\widehat{R} - m^2\right)\widehat{a}_n = 0\end{aligned} \tag{10.134}$$

Using these relations and the coincidence results of Sects. 10.1.4, 10.1.5 and 10.1.6, it is possible to compute each coefficient a_n at the coincidence limit.

10.3.1 Computing $[\widehat{a}_n]$

In this subsection, we wish to compute $[\widehat{a}_1]$ and $[\widehat{a}_2]$, which will be needed later on. We start from (10.134) for $n = -1$.:

$$\widehat{\nabla}^\mu \widehat{a}_0\, \sigma_{;\mu} = 0, \qquad \text{with} \qquad [\widehat{a}_0] = \mathbf{1}, \tag{10.135}$$

which implies that

$$\widehat{a}_0(\widehat{x}, \widehat{x}') = \widehat{I}(\widehat{x}, \widehat{x}'). \tag{10.136}$$

Replacing this inside (10.134) for $n = 0$, one gets

$$\widehat{a}_1(\widehat{x}, \widehat{x}') + \widehat{\nabla}^\mu \widehat{\sigma} \widehat{\nabla}_\mu \widehat{a}_1(\widehat{x}, \widehat{x}') - \frac{1}{\sqrt{\widehat{\Delta}}} \widehat{\nabla}^\mu \widehat{\nabla}_\mu \left(\sqrt{\widehat{\Delta}}\, \widehat{I}(\widehat{x}, \widehat{x}') \right) + \left(\frac{1}{4}\widehat{R} - m^2 \right) \widehat{I}(\widehat{x}, \widehat{x}') = 0, \tag{10.137}$$

which implies

$$[\widehat{a}_1] = \left(-\frac{1}{12}\widehat{R} + m^2 \right) \mathbf{1} \tag{10.138}$$

Next, differentiating (10.137) with respect to ∇_λ and taking the coincidence limit:

$$2[\widehat{\nabla}_\lambda \widehat{a}_1] = \frac{1}{4}\widehat{R}_{;\lambda}\mathbf{1} - [\sqrt{\widehat{\Delta}}_{;\mu}{}^{\mu}{}_{\lambda}\widehat{I} + \widehat{\nabla}_\lambda \widehat{\nabla}^\mu \widehat{\nabla}_\mu \widehat{I}]$$

so

$$[\widehat{\nabla}_\lambda \widehat{a}_1] = \left(\frac{1}{12}\widehat{\mathcal{R}}_{\lambda\nu;}{}^{\nu} - \frac{1}{24}\widehat{R}_{;\lambda} \right) \mathbf{1}. \tag{10.139}$$

Likewise, we have

$$[\widehat{\nabla}^\lambda \widehat{\nabla}_\lambda \left(\widehat{a}_1 + \widehat{\nabla}^\mu \widehat{\sigma}\, \widehat{\nabla}_\mu \widehat{a}_1 \right)] = 3[\widehat{\nabla}^\lambda \widehat{\nabla}_\lambda \widehat{a}_1]$$

so that

$$\begin{aligned} [\widehat{\nabla}^\lambda \widehat{\nabla}_\lambda \widehat{a}_1] &= \frac{1}{3}[\widehat{\nabla}^\lambda \widehat{\nabla}_\lambda \left(\frac{1}{\sqrt{\widehat{\Delta}}} \widehat{\nabla}^\mu \widehat{\nabla}_\mu \left(\sqrt{\widehat{\Delta}}\, \widehat{I} \right) - \left(\frac{1}{4}\widehat{R} - m^2 \right) \widehat{I} \right)] \\ &= \frac{1}{3}\left(-\frac{1}{20}\widehat{R}_{;\mu}{}^{\mu} - \frac{1}{30}\widehat{R}_{\mu\nu}\widehat{R}^{\mu\nu} + \frac{1}{30}\widehat{R}_{\mu\nu\lambda\rho}\widehat{R}^{\mu\nu\lambda\rho} + \frac{1}{8}\widehat{\mathcal{R}}_{\mu\nu}\widehat{\mathcal{R}}^{\mu\nu} \right) \end{aligned} \tag{10.140}$$

Finally,

$$\begin{aligned} [\widehat{a}_2] &= \frac{1}{2}[\widehat{\nabla}^\lambda \widehat{\nabla}_\lambda \widehat{a}_1 - \left(\frac{1}{12}\widehat{R} - m^2 \right) \widehat{a}_1] \\ &= \frac{1}{2}m^4 - \frac{1}{12}m^2\widehat{R} + \frac{1}{288}\widehat{R}^2 - \frac{1}{120}\widehat{R}_{;\mu}{}^{\mu} - \frac{1}{180}\widehat{R}_{\mu\nu}\widehat{R}^{\mu\nu} + \frac{1}{180}\widehat{R}_{\mu\nu\lambda\rho}\widehat{R}^{\mu\nu\lambda\rho} \\ &\quad + \frac{1}{48}\widehat{\mathcal{R}}_{\mu\nu}\widehat{\mathcal{R}}^{\mu\nu} \end{aligned} \tag{10.141}$$

We recall that $\widehat{\mathcal{R}}_{\mu\nu} = \widehat{R}_{\mu\nu}{}^{ab}\Sigma_{ab}$.

10.4 The Odd Trace Anomaly

We are now ready to compute even and odd parity trace anomaly. Beside the point-splitting, which we have used above, we need a regulator to get rid of the infinities at coincident point. We will use two regularizations: the dimensional and zeta function ones.

10.4.1 Schwinger-DeWitt and Dimensional Regularization

We start again from the Dirac operator (10.100). We have defined above, Eq. (10.107), the covariant quadratic differential operator $\widehat{\mathfrak{F}}$. We identify the effective action for Dirac fermions with

$$\widehat{W} = -\frac{i}{2}\mathsf{Tr}\,(\ln\,\widehat{\mathfrak{F}}) \tag{10.142}$$

Tr includes also the spacetime integration.

So we can write

$$\delta_{\widehat{\omega}}\widehat{W} = \delta_{\widehat{\omega}}\left(-\frac{1}{2}\int_0^\infty \frac{\mathrm{d}\widehat{s}}{i\widehat{s}}\mathrm{e}^{i\widehat{\mathfrak{F}}\widehat{s}}\right) = -\frac{1}{2}\mathsf{Tr}\left(\int_0^\infty \mathrm{d}\widehat{s}\,\mathrm{e}^{i\widehat{\mathfrak{F}}\widehat{s}}\delta_{\widehat{\omega}}\widehat{\mathfrak{F}}\right). \tag{10.143}$$

where an $i\epsilon$ prescription is understood; i.e. $\widehat{\mathfrak{F}}$ must be understood as $\widehat{\mathfrak{F}} + i\epsilon$ in order to guarantee the convergence of the integral. It follows that, as far as the variation with respect to axial-Weyl transform is concerned, the effective action can be represented as

$$\widehat{W} = -\frac{1}{2}\int_0^\infty \frac{\mathrm{d}\widehat{s}}{i\widehat{s}}\mathrm{e}^{i\widehat{\mathfrak{F}}\widehat{s}} + \mathrm{const} \equiv \widehat{L} + \mathrm{const} \tag{10.144}$$

where $\widehat{L}$ is the effective action in configuration space:

$$\widehat{L} = \int d^{\mathrm{d}}\widehat{x}\,\widehat{L}(\widehat{x}) \tag{10.145}$$

which can be written as

$$\widehat{L}(\widehat{x}) = -\frac{1}{2}\lim_{\widehat{x}'\to\widehat{x}}\mathrm{tr}\int_0^\infty \frac{\mathrm{d}\widehat{s}}{i\widehat{s}}\widehat{K}(\widehat{x},\widehat{x}',\widehat{s}) \tag{10.146}$$

where the kernel $\widehat{K}$ is defined by

$$\widehat{K}(\widehat{x}, \widehat{x}', \widehat{s}) = \langle x|\mathrm{e}^{i\widehat{\mathcal{F}}\widehat{s}}|x'\rangle \tag{10.147}$$

Inserted in $\delta_{\widehat{\omega}}\widehat{W}$, under the symbol Tr, it means integrating over x after taking the limit $x' \to x$. So, looking at (10.129), in dimension d, we define

$$\widehat{K}(\widehat{x}, \widehat{x}, \widehat{s}) \equiv \lim_{\widehat{x}' \to \widehat{x}} \widehat{K}(\widehat{x}, \widehat{x}', \widehat{s}) = \frac{i}{(4\pi i\widehat{s})^{\frac{\mathsf{d}}{2}}} \sqrt{\widehat{g}}\,\mathrm{e}^{-im^2\widehat{s}}[\widehat{\Phi}(\widehat{x}, \widehat{x}, \widehat{s})] \tag{10.148}$$

10.4.2 Analytic Continuation in d

The purpose now is to analytically continue in d. But we can do this only for dimensionless quantities. We therefore multiply $\widehat{L}$ by $\mu^{-\mathsf{d}}$, where μ is a mass parameter. We have for a Dirac fermion

$$\frac{\widehat{L}(x)}{\mu^{\mathsf{d}}} = -\frac{i}{2}(4\pi\mu^2)\mathrm{tr}\int_0^\infty \mathrm{d}\widehat{s}\,(4\pi i\mu^2\widehat{s})^{-\frac{\mathsf{d}}{2}-1}\sqrt{\widehat{g}}\mathrm{e}^{-im^2\widehat{s}}[\widehat{\Phi}(\widehat{x}, \widehat{x}, \widehat{s})] \tag{10.149}$$

where tr denotes the trace over gamma matrices

Now, we make the assumption that

$$\lim_{s\to\infty} \mathrm{e}^{-im^2\widehat{s}}[\widehat{\Phi}(\widehat{x}, \widehat{x}, \widehat{s})] = 0 \tag{10.150}$$

As a consequence, we can integrate by parts

$$\begin{aligned}
\frac{\widehat{L}(x)}{\mu^{\mathsf{d}}} &= \frac{i}{\mathsf{d}}\mathrm{tr}\int_0^\infty \mathrm{d}\widehat{s}\frac{\partial}{\partial(i\widehat{s})}(4\pi i\mu^2\widehat{s})^{-\frac{\mathsf{d}}{2}}\sqrt{\widehat{g}}\mathrm{e}^{-im^2\widehat{s}}[\widehat{\Phi}(\widehat{x}, \widehat{x}, \widehat{s})] \\
&= -\frac{i}{\mathsf{d}}\mathrm{tr}\int_0^\infty \mathrm{d}\widehat{s}\,(4\pi i\mu^2\widehat{s})^{-\frac{\mathsf{d}}{2}}\sqrt{\widehat{g}}\frac{\partial}{\partial(i\widehat{s})}\left(\mathrm{e}^{-im^2\widehat{s}}[\widehat{\Phi}(\widehat{x}, \widehat{x}, \widehat{s})]\right) \\
&= \frac{2i}{\mathsf{d}(2-\mathsf{d})4\pi\mu^2}\mathrm{tr}\int_0^\infty \mathrm{d}\widehat{s}\,(4\pi i\mu^2\widehat{s})^{1-\frac{\mathsf{d}}{2}}\sqrt{\widehat{g}}\frac{\partial^2}{\partial(i\widehat{s})^2}\left(\mathrm{e}^{-im^2\widehat{s}}[\widehat{\Phi}(\widehat{x}, \widehat{x}, \widehat{s})]\right) \\
&= -\frac{4i}{\mathsf{d}(2-\mathsf{d})(4-\mathsf{d})}\frac{1}{(4\pi\mu^2)^2}\mathrm{tr}\int_0^\infty d\widehat{s}\,(2\pi i\mu^2\widehat{s})^{2-\frac{\mathsf{d}}{2}}\sqrt{\widehat{g}}\frac{\partial^3}{\partial(i\widehat{s})^3}\left(\mathrm{e}^{-im^2\widehat{s}}[\widehat{\Phi}(\widehat{x}, \widehat{x}, \widehat{s})]\right)
\end{aligned} \tag{10.151}$$

Now, we use

$$[\widehat{\Phi}(\widehat{x}, \widehat{x}, \widehat{s})] = 1 + [\widehat{a}_1]i\widehat{s} + [\widehat{a}_2](i\widehat{s})^2 + \cdots \tag{10.152}$$

Around $\mathsf{d} = 2$, $\frac{1}{\mathsf{d}(2-\mathsf{d})} = \frac{1}{2}\left(\frac{1}{\mathsf{d}-2} - \frac{1}{\mathsf{d}}\right)$ and in the third line of (10.151), we write

$$(4\pi i \mu^2 \widehat{s})^{1-\frac{\mathsf{d}}{2}} = 1 - \frac{\mathsf{d}-2}{2}\ln(4\pi i \mu^2 \widehat{s}) + \cdots$$

Then, we differentiate once $[\widehat{\Phi}(\widehat{x}, \widehat{x}, \widehat{s})]$, and the remaining derivation we get rid of by integrating by parts. Finally, one gets

$$\begin{aligned}\widehat{L}(\widehat{x}) = & \frac{1}{4\pi}\left(\frac{1}{\mathsf{d}-2} - \frac{1}{2}\right)\mathrm{tr}\left(([\widehat{a}_1] - m^2)\sqrt{\widehat{g}}\right) \\ & -\frac{i}{8\pi}\mathrm{tr}\int_0^\infty d\widehat{s}\,\ln(4\pi i \mu^2 \widehat{s})\sqrt{\widehat{g}}\frac{\partial^2}{\partial (i\widehat{s})^2}\left(\mathrm{e}^{-im^2\widehat{s}}[\widehat{\Phi}(\widehat{x}, \widehat{x}, \widehat{s})]\right)\end{aligned} \tag{10.153}$$

Around $\mathsf{d} = 4$, we use $\frac{1}{\mathsf{d}(\mathsf{d}-2)(\mathsf{d}-4)} \approx \frac{1}{8}\left(\frac{1}{\mathsf{d}-4} - \frac{3}{4}\right)$. With reference to the last line of (10.151), we differentiate twice $[\widehat{\Phi}(x, x, s)]$ and integrate by parts the third derivative. The result is

$$\begin{aligned}\widehat{L}(\widehat{x}) \approx & \frac{1}{32\pi^2}\left(\frac{1}{\mathsf{d}-4} - \frac{3}{4}\right)\mathrm{tr}\left(m^4 - 2m^2[\widehat{a}_1] + 2[\widehat{a}_2]\right)\sqrt{\widehat{g}} \\ & +\frac{i}{64\pi^2}\mathrm{tr}\int_0^\infty d\widehat{s}\,\ln(4\pi i \mu^2 \widehat{s})\sqrt{\widehat{g}}\frac{\partial^3}{\partial (i\widehat{s})^3}\left(\mathrm{e}^{-im^2\widehat{s}}[\widehat{\Phi}(\widehat{x}, \widehat{x}, \widehat{s})]\right)\end{aligned} \tag{10.154}$$

The last line depends explicitly on the parameter μ and represents a part which cannot contribute to the anomaly for dimensional reasons.

10.4.3 The Anomaly

Let us take the variation of (10.154) with respect to $\widehat{\omega} = \omega + \gamma_5\eta$.

Recall that

$$\delta_{\widehat{\omega}}\sqrt{\widehat{g}} = \mathsf{d}\,\widehat{\omega}\sqrt{\widehat{g}} \tag{10.155}$$

$$\delta_{\widehat{\omega}}\widehat{R} = -2\widehat{\omega}\,\widehat{R} - 2(\mathsf{d}-1)\widehat{\square}\widehat{\omega} \tag{10.156}$$

$$\begin{aligned}\delta_{\widehat{\omega}}\widehat{R}_{\mu\nu\lambda}{}^{\rho} = & -\delta_\nu^\rho\widehat{D}_\mu\widehat{D}_\lambda\widehat{\omega} + \delta_\mu^\rho\widehat{D}_\nu\widehat{D}_\lambda\widehat{\omega} \\ & + \widehat{D}_\mu\widehat{D}_\sigma\widehat{\omega}\,\widehat{g}^{\rho\sigma}\widehat{g}_{\nu\lambda} - \widehat{D}_\nu\widehat{D}_\sigma\widehat{\omega}\,\widehat{g}^{\rho\sigma}\widehat{g}_{\mu\lambda}\end{aligned} \tag{10.157}$$

From these follows, for instance,

$$\delta_{\widehat{\omega}}\left(\sqrt{\widehat{g}}\widehat{R}^2\right) = (\mathsf{d}-4)\sqrt{\widehat{g}}\,\widehat{\omega}\,\widehat{R}^2 - 4(\mathsf{d}-1)\widehat{R}\,\sqrt{\widehat{g}}\,\widehat{\square}\widehat{\omega} \tag{10.158}$$

$$\begin{aligned}\delta_{\widehat{\omega}}\left(\sqrt{\widehat{g}}\widehat{R}_{\mu\nu}\widehat{R}^{\mu\nu}\right) &= (\mathsf{d}-4)\widehat{\omega}\,\sqrt{\widehat{g}}\,\widehat{R}_{\mu\nu}\widehat{R}^{\mu\nu} + 2(2-\mathsf{d})\sqrt{\widehat{g}}\,\widehat{R}^{\mu\nu}\widehat{D}_\mu\widehat{D}_\nu\widehat{\omega} - 2\sqrt{\widehat{g}}\,\widehat{R}\widehat{\square}\widehat{\omega}\\ &= (\mathsf{d}-4)\widehat{\omega}\,\sqrt{\widehat{g}}\,\widehat{R}_{\mu\nu}\widehat{R}^{\mu\nu} - \mathsf{d}\sqrt{\widehat{g}}\,\widehat{R}\widehat{\square}\widehat{\omega}\end{aligned} \tag{10.159}$$

$$\begin{aligned}\delta_{\widehat{\omega}}\left(\sqrt{\widehat{g}}\widehat{R}_{\mu\nu\lambda\rho}\widehat{R}^{\mu\nu\lambda\rho}\right) &= (\mathsf{d}-4)\widehat{\omega}\,\sqrt{\widehat{g}}\,\widehat{R}_{\mu\nu\lambda\rho}\widehat{R}^{\mu\nu\lambda\rho} - 8\sqrt{\widehat{g}}\,\widehat{R}^{\mu\nu}\widehat{D}_\mu\widehat{D}_\nu\widehat{\omega}\\ &= (\mathsf{d}-4)\widehat{\omega}\,\sqrt{\widehat{g}}\,\widehat{R}_{\mu\nu\lambda\rho}\widehat{R}^{\mu\nu\lambda\rho} - 4\sqrt{\widehat{g}}\,\widehat{R}\widehat{\square}\widehat{\omega}\end{aligned} \tag{10.160}$$

$$\begin{aligned}\delta_{\widehat{\omega}}\left(\sqrt{\widehat{g}}\widehat{\square}\widehat{R}\right) &= (\mathsf{d}-4)\widehat{\omega}\,\sqrt{\widehat{g}}\,\widehat{\square}\widehat{R} + (\mathsf{d}-6)\sqrt{\widehat{g}}\,\partial_\mu\widehat{\omega}\,\partial^\mu\widehat{R}\\ &\quad -2\sqrt{\widehat{g}}\,\widehat{R}\,\widehat{\square}\,\widehat{\omega} - 2(\mathsf{d}-1)\sqrt{\widehat{g}}\,\widehat{\square}^2\,\widehat{\omega}\end{aligned}$$

and

$$\begin{aligned}\delta_{\widehat{\omega}}\mathrm{tr}\left(\sqrt{\widehat{g}}\,\widehat{\mathcal{R}}_{\mu\nu}\widehat{\mathcal{R}}^{\mu\nu}\right) &= (\mathsf{d}-4)\mathrm{tr}\left(\widehat{\omega}\,\sqrt{\widehat{g}}\,\widehat{\mathcal{R}}_{\mu\nu}\widehat{\mathcal{R}}^{\mu\nu}\right) + 4\,\mathrm{tr}\left(\sqrt{\widehat{g}}\,\widehat{R}^{\mu\nu}\widehat{D}_\mu\widehat{D}_\nu\widehat{\omega}\right)\\ &= (\mathsf{d}-4)\mathrm{tr}\left(\widehat{\omega}\,\sqrt{\widehat{g}}\,\widehat{\mathcal{R}}_{\mu\nu}\widehat{\mathcal{R}}^{\mu\nu}\right) + 2\,\mathrm{tr}\left(\sqrt{\widehat{g}}\,\widehat{R}\widehat{\square}\widehat{\omega}\right)\end{aligned} \tag{10.161}$$

In the first line of (10.154), one can ignore m^2 or m^4 terms (they can be subtracted away because they are trivial). As we said before, the second line (10.154) can be disregarded in this calculation, and we can limit ourselves to

$$\widehat{L} = \frac{1}{16\pi^2}\left(\frac{1}{\mathsf{d}-4} - \frac{3}{4}\right)\int d^{\mathsf{d}}\widehat{x}\,\mathrm{tr}\left([\widehat{a}_2]|_{m=0}\sqrt{\widehat{g}}\right) \tag{10.162}$$

We now act with $\delta_{\widehat{\omega}} = \int d^{\mathsf{d}}\widehat{x}\,2\mathrm{tr}\left(\widehat{\omega}\,\widehat{g}_{\mu\nu}\frac{\delta}{\delta\widehat{g}_{\mu\nu}}\right)$.

From (10.155)-(10.159), it follows that

$$\delta_{\widehat{\omega}}\mathrm{tr}\left(\sqrt{\widehat{g}}\,[\widehat{a}_2]|_{m=0}\right) = (\mathsf{d}-4)\mathrm{tr}\left(\sqrt{\widehat{g}}\,\widehat{\omega}\,[\widehat{a}_2]|_{m=0}\right) - \frac{\mathsf{d}-4}{120}\mathrm{tr}\left(\sqrt{\widehat{g}}\,\widehat{R}\widehat{\square}\widehat{\omega}\right) \tag{10.163}$$

The second piece can be canceled, e.g. by a counterterm proportional to tr $\left(\sqrt{\widehat{g}}\widehat{R}^2\right)$.

Now, the variation of the effective action under the $\widehat{\omega}$ transformations defines the integrated anomaly. Therefore, defining $\frac{2}{\sqrt{\widehat{g}}}\frac{\delta}{\delta\widehat{g}^{\mu\nu}}\widehat{L} = \widehat{\Theta}_{\mu\nu}$ and recalling the definition (5.8), and using the fact that the second line of (10.154) is Weyl invariant, we get

$$\int d^{\mathsf{d}}\widehat{x}\,\mathrm{tr}\left(\widehat{\omega}\sqrt{\widehat{g}}\widehat{g}^{\mu\nu}\widehat{\Theta}_{\mu\nu}\right) = \frac{1}{16\pi^2}\int d^{\mathsf{d}}\widehat{x}\,\mathrm{tr}\left(\sqrt{\widehat{g}}\,\widehat{\omega}\,[\widehat{a}_2]|_{m=0}\right) \quad (10.164)$$

where the $\mathsf{d}-4$ factor in (10.163) has canceled the pole $\frac{1}{\mathsf{d}-4}$ in (10.162).

Clearly, the odd parity anomaly can come only from the term $\widehat{\mathcal{R}}_{\mu\nu}\widehat{\mathcal{R}}^{\mu\nu}$ contained in $[\widehat{a}_2]$. For the odd part, we have

$$\int d^{\mathsf{d}}\widehat{x}\,\mathrm{tr}\sqrt{\widehat{g}}\,\widehat{\omega}\,\widehat{\mathfrak{T}} = \frac{1}{768\pi^2}\int d^4x\,\mathrm{tr}\sqrt{\widehat{g}}\,\widehat{\omega}\,\widehat{\mathcal{R}}_{\mu\nu}\widehat{\mathcal{R}}^{\mu\nu}\Big|_{\mathrm{odd}} \quad (10.165)$$

where we denoted $\widehat{\mathfrak{T}} = \widehat{g}^{\mu\nu}\widehat{\Theta}_{\mu\nu} = \widehat{g}^{\mu\nu}\langle\langle\widehat{T}_{\mu\nu}\rangle\rangle$.

10.4.4 ζ function Regularization

The zeta function regularization has been introduced in Sect. 8.3. Here, we wish to apply it to the MAT case.

Given a differential operator A in analogy with the Riemann ζ function, the expression A^{-z}, for complex z, is called ζ function regularization of A:

$$\zeta(z, A) = A^{-z} = \frac{1}{\Gamma(z)}\int_0^\infty \mathrm{d}t\, t^{z-1}\,\mathrm{e}^{-tA} \quad (10.166)$$

We will apply this representation to the operator $\widehat{\mathfrak{F}}(\widehat{x}, \widehat{x})$, :

$$(\widehat{\mathfrak{F}}(\widehat{x}))^{-z} = \frac{1}{\Gamma(z)}\int_0^\infty \mathrm{d}t\, t^{z-1}\,\langle\widehat{x}|\mathrm{e}^{-t\widehat{\mathfrak{F}}}|\widehat{x}\rangle \quad (10.167)$$

where $\langle\widehat{x}|\mathrm{e}^{-t\widehat{\mathfrak{F}}}|\widehat{x}\rangle$ means the coincidence limit of $\langle\widehat{x}|\mathrm{e}^{-t\widehat{\mathfrak{F}}}|\widehat{x'}\rangle$. Equation (10.167) is not accurate because only dimensionless quantities can be raised to an arbitrary power. Moreover, the object of interest will be $\widehat{\mathcal{G}}$, rather than $\widehat{\mathfrak{F}}$. Thus, we introduce again the mass parameter μ and shift from t to $i\widehat{s}\mu$.

$$\zeta(\widehat{x,z}) \equiv (\mu^2\widehat{\mathcal{G}}(\widehat{x},\widehat{x}))^z = \frac{1}{\Gamma(z)}\int_0^\infty (i\mu^2)d\widehat{s}\,(i\widehat{s}\mu^2)^{z-1}\,\langle x|\mathrm{e}^{i\widehat{s}\widehat{\mathfrak{F}}}|\widehat{x}\rangle \quad (10.168)$$

Finally we replace $\langle\widehat{x}|\mathrm{e}^{i\widehat{s}\widehat{\mathfrak{F}}}|\widehat{x}\rangle$ with $\widehat{K}(\widehat{x},\widehat{x},\widehat{s})$ in Eq. (10.148). The result is

$$\zeta(\widehat{x}, z) = (\mu^2\widehat{\mathcal{G}}(\widehat{x},\widehat{x}))^z = \frac{i}{\Gamma(z)}\frac{\mu^{\mathsf{d}}}{(4\pi)^{\frac{d}{2}}}\sqrt{\widehat{g}}\int_0^\infty (i\mu^2)\mathrm{d}\widehat{s}\,(i\widehat{s}\mu^2)^{z-1-\frac{\mathsf{d}}{2}}\,\mathrm{e}^{-im^2\widehat{s}}[\widehat{\Phi}(\widehat{x},\widehat{x},\widehat{s})]$$

(10.169)

which can be rewritten as

$$\zeta(\widehat{x}, z) = (\mu^2 \widehat{\mathcal{G}}(\widehat{x}, \widehat{x}))^z = -\frac{i}{\Gamma(z)} \frac{\mu^{\mathsf{d}-4}}{(4\pi)^{\frac{\mathsf{d}}{2}}} \frac{\sqrt{g}}{(z - \frac{\mathsf{d}}{2})(z - \frac{\mathsf{d}}{2} + 1)(z - \frac{\mathsf{d}}{2} + 2)} \times \int_0^\infty d(i\widehat{s})\, (i\widehat{s}\mu^2)^{z - \frac{\mathsf{d}}{2} + 2} \frac{\partial^3}{\partial (i\widehat{s})^3} \left(\mathrm{e}^{-im^2\widehat{s}} [\widehat{\Phi}(\widehat{x}, \widehat{x}, \widehat{s})] \right) \tag{10.170}$$

This is well defined for $\mathsf{d} = 4$ at $z = 0$.

$$\zeta(\widehat{x}, 0) = \frac{i\sqrt{g}}{2(4\pi)^2} \left[\frac{\partial^2}{\partial (i\widehat{s})^2} \left(\mathrm{e}^{-im^2\widehat{s}} [\widehat{\Phi}(\widehat{x}, \widehat{x}, \widehat{s})] \right) \right]_{\widehat{s}=0} \tag{10.171}$$

Now, differentiating (10.166) with respect to z and evaluating at $z = 0$, we get formally

$$\frac{d}{dz} \zeta(z, A)|_{z=0} = -\ln A \tag{10.172}$$

This suggests the procedure to regularize $\widehat{W}$ (which is the trace of a log). More precisely

$$\widehat{W} \to \widehat{W}_\zeta = \frac{i}{2} \zeta'(0), \qquad \text{where} \qquad \zeta(z) = \int \mathrm{tr}\, \zeta(\widehat{x}, z) d^{\mathsf{d}}\widehat{x} \tag{10.173}$$

At this point, we have two possibilities. One is to expand the factor in front of the integral in the RHS of (10.170) first around $z = 0$ and then around $\mathsf{d} = 4$. The result is

$$\frac{1}{\Gamma(z)} \frac{\mu^{\mathsf{d}-4}}{(4\pi)^{\frac{\mathsf{d}}{2}}} \frac{\sqrt{g}}{(z - \frac{\mathsf{d}}{2})(z - \frac{\mathsf{d}}{2} + 1)(z - \frac{\mathsf{d}}{2} + 2)} = -\frac{1}{(4\pi)^2} \frac{1}{\mathsf{d} - 4} z + O(z, \mathsf{d} - 4) \tag{10.174}$$

where $O(z, \mathsf{d} - 4)$ means terms of order at least linear in z and/or in $\mathsf{d} - 4$. Similarly we have

$$\frac{d}{dz} \left(\frac{1}{\Gamma(z)} \frac{\mu^{\mathsf{d}-4}}{(4\pi)^{\frac{\mathsf{d}}{2}}} \frac{\sqrt{g}}{(z - \frac{\mathsf{d}}{2})(z - \frac{\mathsf{d}}{2} + 1)} \right) = \frac{1}{2(4\pi)^2} + O(z, \mathsf{d} - 4) \tag{10.175}$$

Inserting these expansions in (10.170) and further expanding

$$(i\widehat{s}\mu^2)^{z - \frac{\mathsf{d}}{2} + 2} = 1 + (z - \frac{\mathsf{d}}{2} + 2) \log(i\widehat{s}\mu^2) + \cdots,$$

then differentiating twice $\left(\mathrm{e}^{-im^2\widehat{s}}[\widehat{\Phi}(\widehat{x},\widehat{x},\widehat{s})]\right)$ and integrating by parts the remaining derivative, (10.173) can be written as

$$L_\zeta = \frac{1}{32\pi^2}\frac{1}{\mathsf{d}-4}\mathrm{tr}\left(\sqrt{\widehat{g}}\left(m^4 - 2m^2[\widehat{a}_1] + 2[\widehat{a}_2]\right)\right) \\ + \frac{i}{64\pi^2}\mathrm{tr}\int\limits_0^\infty d\widehat{s}\,\ln(i\mu^2\widehat{s})\sqrt{\widehat{g}}\frac{\partial^3}{\partial(i\widehat{s})^3}\left(\mathrm{e}^{-im^2\widehat{s}}[\widehat{\Phi}(\widehat{x},\widehat{x},\widehat{s})]\right) \tag{10.176}$$

This reproduces (10.154). Then, one can proceed in the same way as above and arrive at the anomaly (10.165).

There is an alternative procedure. Suppose that the operator A, under a symmetry transformation with parameter ϵ, transforms as

$$\delta_\epsilon A = \{A, \epsilon\}. \tag{10.177}$$

Then

$$\delta_\epsilon \mathrm{Tr} A^{-z} = -2z\mathrm{Tr}\left(A^{-z}\epsilon\right) = -2z\mathrm{Tr}\left(\zeta(z, A)\epsilon\right) \tag{10.178}$$

Since the relevant result is obtained by differentiating with respect to z and setting $z = 0$, once the functional is regularized, the anomalous part of the effective action is extremely easy to derive:

$$L_{\mathcal{A}} = \mathrm{Tr}\left(\zeta(0, A)\epsilon\right) \tag{10.179}$$

This elegant formula must, however, be employed with care. For example, for the case considered in Sect. 9.1 where the quadratic operator is the square of $\not{D}$, it can be applied straight away. In the present case, where the quadratic operator is $\mathcal{F} = -\widehat{\overline{\not\nabla}}\widehat{\not\nabla}$ and the transformation property under an extended Weyl transformation is

$$\delta_\omega \widehat{\not\nabla} = -\frac{1}{2}\left(\widehat{\overline{\omega}}\widehat{\not\nabla} + \widehat{\not\nabla}\widehat{\omega}\right), \tag{10.180}$$

a naive application would lead to the result $\widehat{L}_{\mathcal{A}} = -\mathrm{Tr}\left(\zeta(\widehat{x}, 0,)\omega\right)$ and the dependence on η would disappear. This is a false result, the reason being that we cannot forget in the amplitude the regularizing factor given by the exponential of the quadratic elliptic operator. In the case of section 9.1, we can freely move $\not{D}$ past this factor. This is particularly clear if we use the language of that section, where we introduced the projector P_Λ, which projects to the eigenspace spanned by the eigenvectors of $\not{D}^2$ whose eigenvalues are in modulus $\leq \Lambda$. In that case $\not{D}$ commutes with P_Λ and we can freely move it around inside the trace. A similar situation was remarked after Eq. (9.38). In the present case, we cannot do the same: $\widehat{\mathcal{F}}$ does not commute neither with $\widehat{\not\nabla}$ nor with $\widehat{\overline{\not\nabla}}$ and, as a consequence, we have to take into account their commutators. Therefore, in the present case we can use a formula like (8.78) provided we do not forget the commutators whenever we need to move an operator past another.

Let us return to our problem. The operator to be regulated is $\widehat{\mathcal{F}} = \widehat{\mathcal{F}}_{\hat{x}}$. Taking into account the previous remark, its AE Weyl transformation is

$$\begin{aligned}
\delta_{\hat{\omega}}\widehat{\mathcal{F}} &= -\widehat{\omega}\widehat{\mathcal{F}} + \frac{1}{2}\left(\left(\overline{\widehat{\gamma}}^{\mu}\widehat{\gamma}^{\nu} + \widehat{g}^{\mu\nu}\right)\partial_\nu\widehat{\omega}\widehat{\nabla}_\mu + \frac{3}{2}\widehat{\Box}\widehat{\omega}\right)\\
&= -\widehat{\omega}\widehat{\mathcal{F}} + \frac{1}{2}\widehat{\mathcal{F}}\left[\frac{1}{\widehat{\mathcal{F}}}\left(\left(\overline{\widehat{\gamma}}^{\mu}\widehat{\gamma}^{\nu} + \widehat{g}^{\mu\nu}\right)\partial_\nu\widehat{\omega}\widehat{\nabla}_\mu + \frac{3}{2}\widehat{\Box}\widehat{\omega}\right)\right]
\end{aligned}$$

$\widehat{\mathcal{G}}(\widehat{x}, \widehat{x})$ is the inverse of $\widehat{\mathcal{F}}$ and its transformation is similar:

$$\delta_{\hat{\omega}}\widehat{\mathcal{G}} = \widehat{\mathcal{G}}\widehat{\omega} + \frac{1}{2}\widehat{\mathcal{G}}\left[\left(\left(\overline{\widehat{\gamma}}^{\mu}\widehat{\gamma}^{\nu} + \widehat{g}^{\mu\nu}\right)\partial_\nu\widehat{\omega}\widehat{\nabla}_\mu + \frac{3}{2}\widehat{\Box}\widehat{\omega}\right)\widehat{\mathcal{G}}\right]$$

The first piece in the RHS reproduces exactly the mechanism in (8.77). The second is a non-local term of the effective action; we simply drop it counting on the fact that the non-local part of the effective action is *bona fide* invariant. As noticed above, this procedure gives the anomalous part of the effective action, i.e. the anomaly integrated with the insertion of $\sqrt{\widehat{g}}$:

$$\begin{aligned}
\widehat{L}_{\rm an}(\widehat{\omega}) &= -i\,{\rm Tr}\left(\widehat{\omega}\,\zeta(\widehat{x}, 0)\right) \qquad (10.181)\\
&= {\rm Tr}\left(\frac{\sqrt{\widehat{g}}}{2(4\pi)^2}\left[\frac{\partial^2}{\partial(i\widehat{s})^2}\left(e^{-im^2\widehat{s}}[\widehat{\Phi}(\widehat{x}, \widehat{x}, \widehat{s})]\right)\right]_{s=0}\widehat{\omega}\right)\\
&= {\rm Tr}\left(\frac{\sqrt{\widehat{g}}}{2(4\pi)^2}\left(2[\widehat{a}_2(\widehat{x})] - 2m^2[\widehat{a}_1(\widehat{x})] + m^4\right)\widehat{\omega}\right)
\end{aligned}$$

This leads to the same results as above.

10.4.5 The Chiral Limit

We are now interested in returning to the original problem, that is the trace anomaly of a Weyl tensor in a chiral fermion theory coupled to ordinary gravity. To this end, we need to take the chiral limit. But before we rewrite the anomaly (10.165) by splitting, it in the chiral and anti-chiral parts. We use (10.257) below, written in another form,

$$\sqrt{\widehat{g}} = P_+\sqrt{\det g_+} + P_-\sqrt{\det g_-}, \qquad g_{\pm\mu\nu} = g_{\mu\nu} \pm f_{\mu\nu} \qquad (10.182)$$

and split

$$\widehat{\omega} = P_+\omega_+ + P_-\omega_-, \qquad \widehat{\mathcal{R}}_{\mu\nu} = P_+\mathcal{R}^{(+)}_{\mu\nu} + P_-\mathcal{R}^{(-)}_{\mu\nu} \qquad (10.183)$$

where

$$\omega_\pm = \omega \pm \eta, \qquad \mathcal{R}^\pm_{\mu\nu} = \mathcal{R}^{(1)}_{\mu\nu} \pm \mathcal{R}^{(2)}_{\mu\nu} \tag{10.184}$$

Then, we can rewrite (10.165) as follows

$$\int d^d\widehat{x}\,\mathrm{tr}\sqrt{\widehat{g}}\,\widehat{\omega}\,\widehat{\mathcal{T}}\Big|_{\mathrm{odd}} = \frac{1}{1536\pi^2}\int d^4x\,\mathrm{tr}\Big(P_+\sqrt{\det g_+}\,\omega_+\,\mathcal{R}^{(+)}_{\mu\nu}\mathcal{R}^{(+)\mu\nu} \tag{10.185}$$
$$+P_-\sqrt{\det g_-}\,\omega_-\,\mathcal{R}^{(-)}_{\mu\nu}\mathcal{R}^{(-)\mu\nu}\Big)\Big|_{\mathrm{odd}}$$

The chiral limit is defined by making the replacements

$$g_{\mu\nu} \to \eta_{\mu\nu} + \frac{h_{\mu\nu}}{2}, \qquad f_{\mu\nu} \to \frac{h_{\mu\nu}}{2} \tag{10.186}$$

in the previous formulas, with finite $h_{\mu\nu}$. With this choice, one has

$$\hat{g}_{\mu\nu} = P_-\,\eta_{\mu\nu} + P_+\,\mathsf{g}_{\mu\nu}, \qquad \mathsf{g}_{\mu\nu} \equiv \eta_{\mu\nu} + h_{\mu\nu} \tag{10.187}$$

From this, we see that the left-handed part couples to the flat metric, while the right-handed part couples to the (generic) metric $\mathsf{g}_{\mu\nu}$. As a consequence, we have also

$$\widehat{\mathrm{e}}^a_m \to \delta^a_m P_- + \mathrm{e}^a_m\,P_+, \qquad \widehat{\mathrm{e}}^m_a \to \delta^m_a P_- + \mathrm{e}^m_a\,P_+ \tag{10.188}$$

as well as

$$\sqrt{\widehat{g}} \to P_- + P_+\sqrt{\mathsf{g}}, \tag{10.189}$$

Similarly, for the Christoffel symbols,

$$\Gamma^{(1)\lambda}_{\mu\nu} \to \frac{1}{2}\Gamma^\lambda_{\mu\nu}, \qquad \Gamma^{(2)\lambda}_{\mu\nu} \to \frac{1}{2}\Gamma^\lambda_{\mu\nu}, \tag{10.190}$$

for the spin connections

$$\Omega^{(1)ab}_\mu \to \frac{1}{2}\omega^{ab}_\mu, \qquad \Omega^{(2)ab}_\mu \to \frac{1}{2}\omega^{ab}_\mu, \tag{10.191}$$

and for the curvatures

$$R^{(1)}_{\mu\nu\lambda}{}^\rho \to \frac{1}{2}R_{\mu\nu\lambda}{}^\rho, \qquad R^{(2)}_{\mu\nu\lambda}{}^\rho \to \frac{1}{2}R_{\mu\nu\lambda}{}^\rho, \tag{10.192}$$

where all the quantities on the RHS of these limits are built with the metric $\mathsf{g}_{\mu\nu}$.

As a consequence, the action (10.82) becomes

$$\widehat{S} \to S' \tag{10.193}$$
$$S' = \int d^4x\, i\overline{\psi}\gamma^a P_- \partial_a \psi + \int d^4x\, \sqrt{\mathsf{g}}\, i\overline{\psi}\gamma^a e_a^\mu \left(\partial_\mu + \frac{1}{2}\omega_\mu\right) P_+ \psi$$

where γ^a is the flat (non-dynamical) gamma matrix, while the vierbein e_a^μ and the connection ω_μ are compatible with the metric $\mathsf{g}_{\mu\nu}$. Up to the term that represents a decoupled left-handed fermion in the flat spacetime, the action S' is the action of a right-handed Weyl fermion coupled to ordinary gravity.

Moreover, in the chiral limit, we have $\mathcal{R}_{\mu\nu}^{(-)} \to 0$, $\mathcal{R}_{\mu\nu}^{(+)} \to 2\mathcal{R}_{\mu\nu}$. Therefore, (10.185) becomes

$$\int d^d\hat{x}\, \mathrm{tr}\sqrt{\hat{g}}\,\hat{\omega}\hat{\mathcal{T}} \to -\frac{1}{768\pi^2}\int d^4x\, \mathrm{tr}\left(P_+\sqrt{g}\,\omega_+\mathcal{R}_{\mu\nu}\mathcal{R}^{\mu\nu}\right)$$
$$= \frac{i}{2\times 768\pi^2}\int d^4x\sqrt{g}\,\omega_+\varepsilon^{\mu\nu\lambda\rho}R_{\mu\nu\alpha\beta}R_{\lambda\rho}{}^{\alpha\beta} \tag{10.194}$$

Now notice that an extended Weyl trnsformation split as follows

$$e^{2\hat{\omega}}\hat{g} = P_+ e^{2\omega_+} g_+ + P_- e^{2\omega_-} g_- \tag{10.195}$$

therefore, in the chiral limit, for consistency with (10.193), we have to require that η coincides with ω, more precisely $\omega \to \omega/2$, $\eta \to \omega/2$. We conclude that in the chiral limit the e.m. trace is

$$\mathcal{T}(x) = \frac{i}{1536\pi^2}\varepsilon^{\mu\nu\lambda\rho}R_{\mu\nu\alpha\beta}R_{\lambda\rho}{}^{\alpha\beta} \equiv \mathcal{T}_R(x) \tag{10.196}$$

If, instead of (10.186), we take the following chiral limit

$$g_{\mu\nu} \to \eta_{\mu\nu} + \frac{h_{\mu\nu}}{2}, \qquad f_{\mu\nu} \to -\frac{h_{\mu\nu}}{2} \tag{10.197}$$

then one obtains

$$\hat{g}_{\mu\nu} = P_-\,\mathsf{g}_{\mu\nu} + P_+\eta_{\mu\nu}, \qquad \mathsf{g}_{\mu\nu} \equiv \eta_{\mu\nu} + h_{\mu\nu} \tag{10.198}$$

Now, the right-handed part is coupled to the flat metric and left-handed part to generic curved metric. We can now repeat the arguments from above and obtain the Pontryagin Weyl anomaly for left-handed Weyl fermion

$$\mathcal{T}_L(x) = -\frac{i}{1536\pi^2}\varepsilon^{\mu\nu\lambda\rho}R_{\mu\nu\alpha\beta}R_{\lambda\rho}{}^{\alpha\beta}. \tag{10.199}$$

The relative minus sign with respect to the right-handed case is because of the opposite sign in front of γ_5 matrix in the defining relation for projectors $P_\pm$.

Remark. Contrary to what may seem at first sight the anomaly (10.196) violates parity, but, due to the imaginary coefficient in front, does not violate time reversal. Therefore, assuming CPT invariance, it does not violate CP either.

The Full MAT Trace Anomalies

From Eq. (10.185), it is easy to compute the $\mathcal{T}$ and $\mathcal{T}_5$ anomalies in the general MAT background. They are the coefficients of ω and η, respectively.:

$$\mathcal{T}(x) = \frac{i}{1536\pi^2}\varepsilon^{\mu\nu\lambda\rho}\left(\sqrt{g_+}R^{(+)}_{\mu\nu\alpha\beta}R^{(+)\alpha\beta}_{\lambda\rho} + \sqrt{g_-}R^{(-)}_{\mu\nu\alpha\beta}R^{(-)\alpha\beta}_{\lambda\rho}\right) \tag{10.200}$$

$$\mathcal{T}_5(x) = \frac{i}{1536\pi^2}\varepsilon^{\mu\nu\lambda\rho}\left(\sqrt{g_+}R^{(+)}_{\mu\nu\alpha\beta}R^{(+)\alpha\beta}_{\lambda\rho} - \sqrt{g_-}R^{(-)}_{\mu\nu\alpha\beta}R^{(-)\alpha\beta}_{\lambda\rho}\right) \tag{10.201}$$

We have Wick-rotated back the result: this is the origin of the i in the anomaly coefficient. At this point we can safely set $x_2^\mu = 0$ everywhere.

10.4.6 The Even Parity Trace Anomaly

As a byproduct of the previous calculation, we can derive the even parity trace anomaly of a Dirac fermion coupled to an ordinary metric background. This is obtained by simply taking the limit $g_{\mu\nu} \to g_{\mu\nu}$, $f_{\mu\nu} \to 0$ in (10.164). Therefore,

$$\sqrt{\hat{g}}\,\hat{\omega}\,[\hat{a}_2]|_{m=0} \to \omega\sqrt{g}\left(\frac{1}{72}R^2 - \frac{1}{45}R_{\mu\nu}R^{\mu\nu} - \frac{7}{360}R_{\mu\nu\lambda\rho}R^{\mu\nu\lambda\rho}\right) \tag{10.202}$$

from which we can extract the even parity trace anomaly

$$\mathcal{T}(x) = \frac{1}{16\pi^2}\left(\frac{1}{72}R^2 - \frac{1}{45}R_{\mu\nu}R^{\mu\nu} - \frac{7}{360}R_{\mu\nu\lambda\rho}R^{\mu\nu\lambda\rho}\right) \tag{10.203}$$

for a Dirac fermion.

10.4.7 The ABJ-Like Trace Anomaly

The odd parity trace anomalies (10.196) and (10.199) are the analogs of the consistent chiral gauge anomalies for Weyl fermions. We wish now to compute the analog of the ABJ anomaly, that is the response of the effective action of a Dirac fermion ψ coupled to ordinary metric $g_{\mu\nu}$ under the transformation

$$\delta_\eta g_{\mu\nu} = 2\gamma_5 \eta\, g_{\mu\nu}, \qquad \delta_\eta \psi = -\frac{3}{2}\gamma_5 \eta\, \psi \tag{10.204}$$

This response can be extracted from (10.185) in the limit $g_{\mu\nu} \to g_{\mu\nu}$, $f_{\mu\nu} \to 0$. In this limit, $\Gamma^{(1)\lambda}_{\mu\nu} \to \Gamma^{\lambda}_{\mu\nu}$ and $\Gamma^{(2)\lambda}_{\mu\nu} \to 0$, $R^{(1)}_{\mu\nu\lambda}{}^{\rho} \to R_{\mu\nu\lambda}{}^{\rho}$, $R^{(2)}_{\mu\nu\lambda}{}^{\rho} \to 0$. Therefore, $g_{\pm\mu\nu} \to g_{\mu\nu}$ and $R^{\pm}_{\mu\nu\lambda\rho} \to R_{\mu\nu\lambda\rho}$. As a consequence the odd parity (ABJ-like) trace anomaly is

$$\mathcal{T}_5(x) = \frac{i}{768\pi^2}\varepsilon^{\mu\nu\lambda\rho} R_{\mu\nu\alpha\beta} R_{\lambda\rho}{}^{\alpha\beta}. \tag{10.205}$$

Contrary to the anomalies (10.196 and 10.199), the anomaly (10.205) is not a risk for the consistency of a theory of Dirac fermions coupled to gravity. Rather, its absorbitive nature suggests a possible phenomenological application to a decay of a neutral bound state into two gravitons, similar to the one of the ABJ anomalies for a double photon decay of π^0.

10.4.8 The Gauge-Induced Odd Trace Anomaly via SDW

For completeness, we wish to compute the odd trace anomaly for a Weyl fermion coupled to a vector potential V_μ. We use Sect. 9.3 as a shortcut, but in this case, we need to introduce an axial partner also for the metric; that is, we couple the Weyl fermion to a MAT (metric-axial-tensor) background. The kinetic operator in this case is $i\not{\mathbb{D}}$ with

$$\mathbb{D}_\mu = \widehat{D}_\mu + \frac{1}{2}\widehat{\Omega}_\mu + V_\mu \tag{10.206}$$

where $\widehat{D}_\mu$ is the covariant derivative and $\widehat{\Omega}_\mu = \Omega^{(1)}_\mu + \gamma_5 \Omega^{(2)}_\mu$, the spin connection with respect to the metric $\widehat{g}_{\mu\nu} = g_{\mu\nu} + \gamma_5 f_{\mu\nu}$. Here, V_μ is an imaginary vector potential. In reality, for our present calculation, we will not need the full SDW derivation in a MAT background. A shortcut is viable and is as follows. The density of the anomaly we are after is $\varepsilon^{\mu\nu\lambda\rho}\partial_\mu V_\nu \partial_\lambda V_\rho$ which 'occupies' the dimensionally available slots in 4d. The only change that MAT can (and does) bring about is the multiplication by the factor $\sqrt{g}\,\widehat{\omega}$ (instead of $\gamma_5\rho$), where $\widehat{\omega} = \omega + \gamma_5\eta$ is the extended Weyl parameter. Therefore, using the ζ function regularization as in Sect. 10.4.4, we can simply replace the symbols in the final formula. The variation of the effective action under an extended infinitesimal Weyl transformation is

$$\begin{aligned} \delta_{\widehat{\omega}} \widehat{L} &= -i\mathrm{Tr}\,(\widehat{\omega}\,\zeta(x,0))_c \\ &= i\,\mathrm{Tr}\left(\frac{\sqrt{g}}{2(4\pi)^2}\widehat{\omega}\left(2[a_2(x)] - 2m^2[a_1(x)] + m^4\right)\right)_c \end{aligned} \tag{10.207}$$

where Tr denotes spacetime integration and gamma matrix trace, and $[a_1]$, $[a_2]$ are those of the previous subsection. The subscript $_c$ denotes the chiral limit. The terms proportional to m^2 drop out because of the limit $m \to 0$. It remains for us to compute the chiral limit of $\sqrt{\widehat{g}}\,\widehat{\omega}$.

In the right-handed chiral limit, we have not only $V_\mu \to V_\mu/2$, $A_\mu \to V_\mu/2$ but also $h_{\mu\nu} \to h_{\mu\nu}/2$, $f_{\mu\nu} \to h_{\mu\nu}/2$, and $\eta \to \omega$. Therefore, $\sqrt{\widehat{g}}\,\widehat{\omega} \to P_+\sqrt{g}\omega$. As a consequence, we obtain

$$\delta_\omega L = -\frac{i}{32\pi^2}\int d^4x\,\sqrt{g}\,\omega\,\mathrm{tr}\,(P_+[a_2](x)) \tag{10.208}$$

Arguing as in last paragraph of Sect. 9.3, the odd part of tr $(P_+[a_2](x))$ produces precisely the density $-\varepsilon^{\mu\nu\lambda\rho}\partial_\mu V_\nu \partial_\lambda V_\rho$. Therefore, (10.208), together with the definition (5.18), yields

$$\mathcal{A}_\omega = -\frac{1}{96\pi^2}\int d^4x\,\sqrt{g}\,\omega\,\varepsilon^{\mu\nu\lambda\rho}\partial_\mu V_\nu \partial_\lambda V_\rho, \tag{10.209}$$

wich coincides with Eq. (7.136).

ABJ-Like Gauge-Induced Trace Anomaly

The trace analog of the ABJ anomaly induced by a $U(1)$ gauge field for Dirac fermions is obtained in the same way by taking the limit $A \to 0$ and $f_{\mu\nu} \to 0$. Therefore, $\sqrt{\widehat{g}}\,\widehat{\omega} \to \sqrt{g}\eta$. This leads to

$$\mathcal{A}^{ABJ}_\omega = -\frac{1}{48\pi^2}\int d^4x\,\sqrt{g}\,\omega\,\varepsilon^{\mu\nu\lambda\rho}\partial_\mu V_\nu \partial_\lambda V_\rho, \tag{10.210}$$

10.4.9 A Gauge-Induced Even Trace Anomalies with SDW

To complete the panorama of trace anomalies, let us show an example of how to compute an even trace anomaly originated from a background gauge field. To this end, we do not need the full MAT equipment, but only the even part of it, which can be anyhow derived from the previous general formulas. We consider the theory of Dirac fermions coupled to a vector potential V_μ, (7.2), where the covariant operator is

$$\nabla_\mu = D_\mu + \frac{1}{2}\omega_\mu + V_\mu \tag{10.211}$$

To apply the SDW method, we need the square of the Dirac operator, which is not self-adjoint

$$\left(\nabla\!\!\!/^2\right)^\dagger = \gamma_0 \nabla\!\!\!/^2 \gamma_0 \tag{10.212}$$

To obtain a self-adjoint operator, we Wick-rotate it, see Sect. 8.1. Denoting, as usual, with a tilde a Wick rotated object, we have

$$\left(\widetilde{\nabla}^2\right)^\dagger = \widetilde{\nabla}^2 \tag{10.213}$$

Therefore, we can use $\widetilde{\nabla}^2$ for the SDW approach and return to the Minkowski metric with a reverse Wick rotation after applying this method. However, as before, we will avoid clogging the formulas with Euclidean symbols. For the same reasons explained before, we will use all the time Minkowski symbols and obtain the results directly in Minkowski background.

Without repeating all the steps of SDW, we remark that in this case $\mathcal{F}_x$ is given by

$$\mathcal{F}_x = \nabla_\mu g^{\mu\nu}\nabla_\nu - \frac{1}{4}R + \mathcal{V} \tag{10.214}$$

where $\mathcal{V} = \Sigma^{ab}\mathcal{V}_{ab} = \Sigma^{ab}\, e_a^\mu e_b^\nu\, V_{\mu\nu}$, and $V_{\mu\nu}$ is the curvature of V_μ. We proceed directly to the calculation of the coefficients $[a_n]$ and notice that they are a particular case of the coefficients already computed in Sect. 10.3.1:

$$[a_1] = -\frac{1}{12}R + \mathcal{V} \tag{10.215}$$

and

$$\begin{aligned}[a_2] &= \frac{1}{2}m^4 - \frac{1}{12}m^2 R + \frac{1}{288}R^2 - \frac{1}{120}R_{;\mu}{}^{\mu} - \frac{1}{180}R_{\mu\nu}R^{\mu\nu} + \frac{1}{180}R_{\mu\nu\lambda\rho}R^{\mu\nu\lambda\rho}\\ &-\frac{1}{2}\mathcal{V}^2 + \frac{1}{6}\nabla^\lambda\nabla_\lambda\,(\mathcal{V}) + \frac{1}{12}\left(\frac{1}{2}\mathcal{R}_{\mu\nu} + V_{\mu\nu}\right)\left(\frac{1}{2}\mathcal{R}^{\mu\nu} + V^{\mu\nu}\right)\end{aligned} \tag{10.216}$$

where $\mathcal{R}_{\mu\nu} = R_{\mu\nu}{}^{ab}\Sigma_{ab}$.

As explained before, after a double partial integration in s, the effective action turns out

$$\begin{aligned}L(x) &= \frac{1}{32\pi^2}\left(\frac{1}{\mathsf{d}-4} - \frac{3}{4}\right)\mathrm{tr}\left(m^4 - 2m^2[a_1] + 2[a_2]\right)\sqrt{g}\\ &+ \frac{i}{64\pi^2}\mathrm{tr}\int_0^\infty ds\, \ln(4\pi i\mu^2 s)\sqrt{g}\frac{\partial^3}{\partial(is)^3}\left(\mathrm{e}^{-im^2 s}[\Phi(x,x,s)]\right)\end{aligned} \tag{10.217}$$

where tr refers to the gamma matrix trace. The last line does not have the right dimensions to contribute to an anomaly (it is a non-local part of the effective action). Let us take the variation under a Weyl transformation

$$\delta_\omega g_{\mu\nu}(x) = 2\,\omega(x)\,g_{\mu\nu}(x), \quad \delta_\omega\sqrt{g} = \mathsf{d}\,\omega\,\sqrt{g}, \quad \delta_\omega\left(V_{\mu\nu}V^{\mu\nu}\right) = -4\,\omega\,V_{\mu\nu}V^{\mu\nu} \tag{10.218}$$

Focusing on the V_μ dependence in the $\mathsf{d} \to 4$ limit, we find

$$\delta_\omega\,L(x) = -\frac{1}{32\pi^2}\int d^4x\,\sqrt{g}\,\frac{4}{3}\omega\,V_{\mu\nu}V^{\mu\nu} \tag{10.219}$$

and then we take the limit $g_{\mu\nu} \to \eta_{\mu\nu}$. This defines the anomaly

$$\mathcal{T}(x) = \frac{1}{24\pi^2}V_{\mu\nu}V^{\mu\nu} \tag{10.220}$$

which coincides with (7.120).

We have already pointed out that the result (7.120) is the correct one provided the invariance under diffeomorphisms is preserved. In the perturbative case, we have verified it by computing the divergence of the e.m. tensor at least to second order in V_μ. In the case of the SDW method, this invariance is embedded in the method itself because the latter is designed to respect the diffeomorphisms, thanks to the rules for point-splitting regularization, which is always applied to covariant objects, such as $\sigma, \mathfrak{D}$ and so on.

Appendix 10A. The Axial-Riemannian Geometry

In this Appendix, we collect the formulas of axial-Riemannian geometry.

Axial Metric

We use the symbols $g_{\mu\nu}$, $g^{\mu\nu}$ and e_μ^a, e_a^μ in the usual sense of metric and vierbein and their inverses, except for the fact that they are functions of $\widehat{x}^\mu$. Then, we introduce the MAT metric

$$\widehat{g}_{\mu\nu} = g_{\mu\nu} + \gamma_5 f_{\mu\nu} \tag{10.221}$$

where f is a symmetric tensor. Their background values are $\eta_{\mu\nu}$ and 0, respectively. So, we write as usual $g_{\mu\nu} = \eta_{\mu\nu} + h_{\mu\nu}$.

In matrix notation, the inverse of $\widehat{g}$, $\widehat{g}^{-1}$, is defined by

$$\widehat{g}^{-1} = \tilde{g} + \gamma_5\tilde{f}, \qquad \widehat{g}^{-1}\widehat{g} = 1, \qquad \widehat{g}^{\mu\lambda}\widehat{g}_{\lambda\nu} = \delta^\mu_\nu \tag{10.222}$$

which implies

$$\tilde{g}f + \tilde{f}g = 0, \qquad \tilde{g}g + \tilde{f}f = 1. \tag{10.223}$$

So

$$\tilde{g} = (1 - g^{-1}fg^{-1}f)^{-1}g^{-1}, \qquad \tilde{f} = -(1 - g^{-1}fg^{-1}f)^{-1}g^{-1}fg^{-1} \tag{10.224}$$

MAT Vierbein

Likewise for the vierbein one writes

$$\widehat{e}^a_\mu = e^a_\mu + \gamma_5 c^a_\mu, \qquad \widehat{e}^\mu_a = \tilde{e}^\mu_a + \gamma_5 \tilde{c}^\mu_a \tag{10.225}$$

This implies

$$\eta_{ab}\left(e^a_\mu e^b_\nu + c^a_\mu c^b_\nu\right) = g_{\mu\nu}, \qquad \eta_{ab}\left(e^a_\mu c^b_\nu + e^a_\nu c^b_\mu\right) = f_{\mu\nu} \tag{10.226}$$

Moreover, from $\widehat{e}^\mu_a \widehat{e}^a_\nu = \delta^\mu_\nu$,

$$\tilde{e}^\mu_a c^a_\nu + \tilde{c}^\mu_a e^a_\nu = 0, \qquad \tilde{e}^\mu_a e^a_\nu + \tilde{c}^\mu_a c^a_\nu = \delta^\mu_\nu, \tag{10.227}$$

one gets

$$\tilde{e}^\mu_a = \left(\frac{1}{1 - e^{-1}c\,e^{-1}c} e^{-1}\right)^\mu_a \tag{10.228}$$

and

$$\tilde{c}^\mu_a = -\left(\frac{1}{1 - e^{-1}c\,e^{-1}c} e^{-1}ce^{-1}\right)^\mu_a \tag{10.229}$$

Christoffel and Riemann

The ordinary Christoffel symbols are

$$\Gamma^\lambda_{\mu\nu} = \frac{1}{2}g^{\lambda\rho}\left(\partial_\mu g_{\rho\nu} + \partial_\nu g_{\rho\mu} - \partial_\rho g_{\mu\nu}\right) \tag{10.230}$$

The MAT Christoffel symbols are defined in a similar way

$$\begin{aligned}
\widehat{\Gamma}^{\lambda}_{\mu\nu} &= \frac{1}{2}\widehat{g}^{\lambda\rho}\left(\partial_\mu \widehat{g}_{\rho\nu} + \partial_\nu \widehat{g}_{\rho\mu} - \partial_\rho \widehat{g}_{\mu\nu}\right) \qquad (10.231)\\
&= \frac{1}{2}\left(\tilde{g}^{\lambda\rho}\left(\partial_\mu g_{\rho\nu} + \partial_\nu g_{\rho\mu} - \partial_\rho g_{\mu\nu}\right) + \tilde{f}^{\lambda\rho}\left(\partial_\mu f_{\rho\nu} + \partial_\nu f_{\rho\mu} - \partial_\rho f_{\mu\nu}\right)\right)\\
&\quad + \frac{1}{2}\gamma_5\left(\tilde{g}^{\lambda\rho}\left(\partial_\mu f_{\rho\nu} + \partial_\nu f_{\rho\mu} - \partial_\rho f_{\mu\nu}\right) + \tilde{f}^{\lambda\rho}\left(\partial_\mu g_{\rho\nu} + \partial_\nu g_{\rho\mu} - \partial_\rho g_{\mu\nu}\right)\right)\\
&\equiv \Gamma^{(1)\lambda}_{\mu\nu} + \gamma_5 \Gamma^{(2)\lambda}_{\mu\nu}
\end{aligned}$$

where it is understood that $\partial_\mu = \frac{\partial}{\partial \hat{x}^\mu}$, etc.

Proceeding the same way, one can define the MAT Riemann tensor via $\widehat{R}_{\mu\nu\lambda}{}^{\rho}$:

$$\begin{aligned}
\widehat{R}_{\mu\nu\lambda}{}^{\rho} &= -\partial_\mu \widehat{\Gamma}^{\rho}_{\nu\lambda} + \partial_\nu \widehat{\Gamma}^{\rho}_{\mu\lambda} - \widehat{\Gamma}^{\rho}_{\mu\sigma}\widehat{\Gamma}^{\sigma}_{\nu\lambda} + \widehat{\Gamma}^{\rho}_{\nu\sigma}\widehat{\Gamma}^{\sigma}_{\mu\lambda} \qquad (10.232)\\
&= -\partial_\mu \Gamma^{(1)\rho}_{\nu\lambda} + \partial_\nu \Gamma^{(1)\rho}_{\mu\lambda} - \Gamma^{(1)\rho}_{\mu\sigma}\Gamma^{(1)\sigma}_{\nu\lambda} + \Gamma^{(1)\rho}_{\nu\sigma}\Gamma^{(1)\sigma}_{\mu\lambda} - \Gamma^{(2)\rho}_{\mu\sigma}\Gamma^{(2)\sigma}_{\nu\lambda} + \Gamma^{(2)\rho}_{\nu\sigma}\Gamma^{(2)\sigma}_{\mu\lambda}\\
&+\gamma_5\Big(-\partial_\mu \Gamma^{(2)\rho}_{\nu\lambda} + \partial_\nu \Gamma^{(2)\rho}_{\mu\lambda} - \Gamma^{(1)\rho}_{\mu\sigma}\Gamma^{(2)\sigma}_{\nu\lambda} + \Gamma^{(1)\rho}_{\nu\sigma}\Gamma^{(2)\sigma}_{\mu\lambda} - \Gamma^{(2)\rho}_{\mu\sigma}\Gamma^{(1)\sigma}_{\nu\lambda} + \Gamma^{(2)\rho}_{\nu\sigma}\Gamma^{(1)\sigma}_{\mu\lambda}\Big)\\
&\equiv \widehat{R}^{(1)}_{\mu\nu\lambda}{}^{\rho} + \gamma_5 \widehat{R}^{(2)}_{\mu\nu\lambda}{}^{\rho}
\end{aligned}$$

The MAT spin connection is introduced in analogy

$$\widehat{\Omega}^{ab}_{\mu} = \widehat{e}^{a}_{\nu}\left(\partial_\mu \widehat{e}^{\nu b} + \widehat{e}^{\sigma b}\widehat{\Gamma}^{\nu}_{\sigma\mu}\right) = \Omega^{(1)ab}_{\mu} + \gamma_5 \Omega^{(2)ab}_{\mu} \qquad (10.233)$$

where

$$\Omega^{(1)ab}_{\mu} = e^{a}_{\nu}\left(\partial_\mu \tilde{e}^{\nu b} + \tilde{e}^{\sigma b}\Gamma^{(1)\nu}_{\sigma\mu} + \tilde{c}^{b\sigma}\Gamma^{(2)\nu}_{\sigma\mu}\right) + c^{a}_{\nu}\left(\partial_\mu \tilde{c}^{\nu b} + \tilde{e}^{\sigma b}\Gamma^{(2)\nu}_{\sigma\mu} + \tilde{c}^{b\sigma}\Gamma^{(1)\nu}_{\sigma\mu}\right) \qquad (10.234)$$

$$\Omega^{(2)ab}_{\mu} = e^{a}_{\nu}\left(\partial_\mu \tilde{c}^{\nu b} + \tilde{e}^{\sigma b}\Gamma^{(2)\nu}_{\sigma\mu} + \tilde{c}^{b\sigma}\Gamma^{(1)\nu}_{\sigma\mu}\right) + c^{a}_{\nu}\left(\partial_\mu \tilde{e}^{\nu b} + \tilde{e}^{\sigma b}\Gamma^{(1)\nu}_{\sigma\mu} + \tilde{c}^{b\sigma}\Gamma^{(2)\nu}_{\sigma\mu}\right) \qquad (10.235)$$

Transformations. Diffeomorphisms

We recall that under a diffeomorphism, $\delta x^\mu = \xi^\mu$, the ordinary Christoffel symbols transform as tensors except for one non-covariant piece

$$\delta^{(n.c.)}_{\xi}\Gamma^{\lambda}_{\mu\nu} = \partial_\mu \partial_\nu \xi^\lambda \qquad (10.236)$$

In the MAT context, one must introduce also axially extended (AE) diffeomorphisms. They are defined by

$$\widehat{x}^\mu \to \widehat{x}^\mu + \widehat{\xi}^\mu(\widehat{x}^\mu), \qquad \widehat{\xi}^\mu = \xi^\mu + \gamma_5 \zeta^\mu \qquad (10.237)$$

Since operationally these transformations act in the same way as the usual diffeomorphisms, it is easy to obtain for the non-covariant part

$$\delta^{(n.c.)}\widehat{\Gamma}^{\lambda}_{\mu\nu} = \partial_\mu \partial_\nu \widehat{\xi}^{\lambda} \tag{10.238}$$

where the derivatives are understood with respect to $\widehat{x}^{\mu}$ and $\widehat{x}^{\nu}$.

We have also

$$\delta_{\widehat{\xi}}\widehat{g}_{\mu\nu} = \widehat{D}_\mu \widehat{\xi}_\nu + \widehat{D}_\nu \widehat{\xi}_\mu \tag{10.239}$$

where $\widehat{\xi}_\mu = \widehat{g}_{\mu\nu}\widehat{\xi}^{\nu}$ and $\widehat{D}_\mu$ is the covariant derivative with respect to $\widehat{\Gamma}$.

In components, one easily finds

$$\delta_\xi g_{\mu\nu} = \xi^\lambda \partial_\lambda g_{\mu\nu} + \partial_\mu \xi^\lambda g_{\lambda\nu} + \partial_\nu \xi^\lambda g_{\lambda\mu} \tag{10.240}$$
$$\delta_\xi f_{\mu\nu} = \xi^\lambda \partial_\lambda f_{\mu\nu} + \partial_\mu \xi^\lambda f_{\lambda\nu} + \partial_\nu \xi^\lambda f_{\lambda\mu}$$
$$\delta_\zeta g_{\mu\nu} = \zeta^\lambda \partial_\lambda f_{\mu\nu} + \partial_\mu \zeta^\lambda f_{\lambda\nu} + \partial_\nu \zeta^\lambda f_{\lambda\mu} \tag{10.241}$$
$$\delta_\zeta f_{\mu\nu} = \zeta^\lambda \partial_\lambda g_{\mu\nu} + \partial_\mu \zeta^\lambda g_{\lambda\nu} + \partial_\nu \zeta^\lambda g_{\lambda\mu}$$

Summarizing

$$\delta_\xi^{(n.c.)}\Gamma^{(1)\lambda}_{\mu\nu} = \partial_\mu\partial_\nu\xi^\lambda, \qquad \delta_\xi^{(n.c.)}\Gamma^{(2)\lambda}_{\mu\nu} = 0 \tag{10.242}$$
$$\delta_\zeta^{(n.c.)}\Gamma^{(1)\lambda}_{\mu\nu} = 0, \qquad \delta_\zeta^{(n.c.)}\Gamma^{(2)\lambda}_{\mu\nu} = \partial_\mu\partial_\nu\zeta^\lambda$$

and the overall Riemann and Ricci tensors are tensor, and the Ricci scalar $\widehat{R}$ is a scalar. But also $\widehat{R}^{(1)}$ and $\widehat{R}^{(2)}$, separately, have the same tensorial properties.

Transformations. Local Lorentz Transformations

There are two types of AE local Lorentz transformations. The corresponding AE parameter is

$$\widehat{\Lambda} = \Lambda + \gamma_5 \mathsf{H}, \qquad \widehat{\Lambda} = \widehat{\Lambda}_{ab}\Sigma^{ab},$$

and the induced transformations are

$$\delta_{\widehat{\Lambda}}\widehat{e}^{a}_{\mu} = \widehat{e}^{b}_{\mu}\widehat{\Lambda}_b{}^a, \tag{10.243}$$

which splits into

$$\delta_{\widehat{\Lambda}} e^a_\mu = e^b_\mu \Lambda_b{}^a + c^b_\mu \mathsf{H}_b{}^a, \qquad \delta_{\widehat{\Lambda}} c^a_\mu = e^b_\mu \mathsf{H}_b{}^a + c^b_\mu \Lambda_b{}^a.$$

As a consequence, the transformation of the spin connection is

$$\delta_{\widehat{\Lambda}}\widehat{\Omega} = d\widehat{\Lambda} + [\widehat{\Omega}, \widehat{\Lambda}], \tag{10.244}$$

with the corresponding obvious splitting for $\Omega^{(1)}_\mu$ and $\Omega^{(2)}_\mu$.

Transformations. Weyl Transformations

There are two types of Weyl transformations. The first is the obvious one

$$\widehat{g}_{\mu\nu} \longrightarrow \mathrm{e}^{2\omega}\widehat{g}_{\mu\nu}, \qquad \widehat{g}^{\mu\nu} \to \mathrm{e}^{-2\omega}\widehat{g}^{\mu\nu} \tag{10.245}$$

and

$$\widehat{e}^{a}_{\mu} \longrightarrow \mathrm{e}^{\omega}\widehat{e}^{a}_{\mu}, \qquad \widehat{\mathrm{e}}^{\mu}_{a} \to \mathrm{e}^{-\omega}\widehat{\mathrm{e}}^{\mu}_{a} \tag{10.246}$$

This leads to the usual relations

$$\widehat{\Gamma}^{\lambda}_{\mu\nu} \longrightarrow \widehat{\Gamma}^{\lambda}_{\mu\nu} + \partial_\mu\omega\,\delta^{\lambda}_{\nu} + \partial_\nu\omega\,\delta^{\lambda}_{\mu} - \partial_\rho\omega\,\widehat{g}^{\lambda\rho}\widehat{g}_{\mu\nu} \tag{10.247}$$

and

$$\widehat{\Omega}^{ab}_{\mu} \longrightarrow \widehat{\Omega}^{ab}_{\mu} + \left(\widehat{e}^{a}_{\mu}\widehat{\mathrm{e}}^{\sigma b} - \widehat{e}^{b}_{\mu}\widehat{\mathrm{e}}^{\sigma a}\right)\partial_\sigma\omega \tag{10.248}$$

The second type of Weyl transformation is the axial one

$$\widehat{g}_{\mu\nu} \longrightarrow \mathrm{e}^{2\gamma_5\eta}\widehat{g}_{\mu\nu}, \qquad \widehat{g}^{\mu\nu} \to \mathrm{e}^{-2\gamma_5\eta}\widehat{g}^{\mu\nu} \tag{10.249}$$

and

$$\widehat{e}^{a}_{\mu} \longrightarrow \mathrm{e}^{\gamma_5\eta}\widehat{e}^{a}_{\mu}, \qquad \widehat{\mathrm{e}}^{\mu}_{a} \to \mathrm{e}^{-\gamma_5\eta}\widehat{\mathrm{e}}^{\mu}_{a} \tag{10.250}$$

This leads to

$$\widehat{\Gamma}^{\lambda}_{\mu\nu} \longrightarrow \widehat{\Gamma}^{\lambda}_{\mu\nu} + \gamma_5\left(\partial_\mu\eta\,\delta^{\lambda}_{\nu} + \partial_\nu\eta\,\delta^{\lambda}_{\mu} - \partial_\rho\eta\,\widehat{g}^{\lambda\rho}\widehat{g}_{\mu\nu}\right) \tag{10.251}$$

and

$$\widehat{\Omega}^{ab}_{\mu} \longrightarrow \widehat{\Omega}^{ab}_{\mu} + \gamma_5\left(\widehat{e}^{a}_{\mu}\widehat{\mathrm{e}}^{\sigma b} - \widehat{e}^{b}_{\mu}\widehat{\mathrm{e}}^{\sigma a}\right)\partial_\sigma\eta \tag{10.252}$$

Equation (10.249) implies

$$g_{\mu\nu} \longrightarrow \cosh(2\eta)\,g_{\mu\nu} + \sinh(2\eta)\,f_{\mu\nu}, \quad f_{\mu\nu} \longrightarrow \cosh(2\eta)\,f_{\mu\nu} + \sinh(2\eta)\,g_{\mu\nu} \tag{10.253}$$

We can write the axially extended (AE) Weyl transformation in compact form using the parameter $\widehat{\omega} = \omega + \gamma_5\eta$

$$\widehat{g}_{\mu\nu} \longrightarrow \mathrm{e}^{2\widehat{\omega}}\widehat{g}_{\mu\nu}, \tag{10.254}$$

etc.

Volume Density

The ordinary density $\sqrt{g}$ is replaced by

$$\sqrt{\widehat{g}} = \sqrt{\det(\widehat{g})} = \sqrt{\det(g + \gamma_5 f)} \tag{10.255}$$

The expression in the RHS has to be understood as a formal Taylor expansion in terms of the axial-complex variable $g + \gamma_5 f$. This means

$$\mathrm{tr}\ln(g + \gamma_5 f) = \frac{1+\gamma_5}{2}\,\mathrm{tr}\ln(g+f) + \frac{1-\gamma_5}{2}\,\mathrm{tr}\ln(g-f) \tag{10.256}$$

It follows that

$$\sqrt{\widehat{g}} = \frac{1}{2}\left(\sqrt{\det(g+f)} + \sqrt{\det(g-f)}\right) + \frac{\gamma_5}{2}\left(\sqrt{\det(g+f)} - \sqrt{\det(g-f)}\right) \tag{10.257}$$

$\sqrt{\widehat{g}}$ has the basic property that, under AE diffeomorphisms,

$$\delta_{\hat{\xi}}\sqrt{\widehat{g}} = \widehat{\xi}^\lambda \partial_\lambda \sqrt{\widehat{g}} + \sqrt{\widehat{g}}\,\partial_\lambda \widehat{\xi}^\lambda \tag{10.258}$$

This is a volume density, with the following rule

$$\sqrt{\widehat{g}} \to \mathrm{e}^{4\widehat{\omega}}\sqrt{\widehat{g}}, \tag{10.259}$$

under an axial-Weyl transformations. Moreover,

$$\frac{1}{\sqrt{\widehat{g}}}\partial_\nu \sqrt{\widehat{g}} = \frac{1}{2}\widehat{g}^{\mu\lambda}\partial_\nu \widehat{g}_{\mu\lambda} = \widehat{\Gamma}^\mu_{\mu\nu} \tag{10.260}$$

References

1. B.S. DeWitt, *Global Approach to Quantum Field Theory*, vols. I and II (Oxford University Press, Oxford, 2003)
2. L. Bonora, M. Cvitan, P. Dominis Prester, A. Duarte Pereira, S. Giaccari, T. Štemberga, Axial gravity, massless fermions and trace anomalies. Eur. Phys. J. C **77**, 511 (2017) [arXiv:1703.10473 [hep-th]]
3. L. Bonora, M. Cvitan, P. Dominis Prester, S. Giaccari, M. Paulisic, T. Stemberga, Axial gravity: a non-perturbative approach to split anomalies. Eur. Phys. J. C **78**, 652 (2018) [arXiv:1807.01249]
4. D.M. Capper, M.J. Duff, Trace anomalies in dimensional regularization. Nuovo Cim. **23 A**, 173 (1974)
5. S.M. Christensen, Vacuum expectation value of the stress tensor in an arbitrary curved background: the covariant point-separation method. Phys. Rev. D **14**, 2490 (1976)

6. S.M. Christensen, S.A. Fulling, Trace anomalies and the hawking effect. Phys. Rev. D **15**, 2088–2104 (1977)
7. J.S. Dowker, R. Critchley, Stress-tensor conformal anomaly for scalar and spinor fields. Phys. Rev. D **16**, 3390 (1977)
8. S.M. Christensen. Regularization, renormalization and covariant geodesic point separation. Phys. Rev. **D17**, 946 (1978)
9. S.M. Christensen, M.J. Duff Axial and conformal anomalies for arbitrary spin in gravity and supergravity. Phys. Lett. **B 76**, 571 (1978)
10. S.M. Christensen, M.J. Duff, New gravitational index theorems and super theorems. Nucl. Phys. B **154**, 301 (1979)
11. A . Crumeyrolle, Variétés différentiables à coordonnées hypercomplexes. Application à une géométrisation et à une généralisation de la théorie d'Einstein-Schrodinger. Ann. de la Fac. des Sciences de Toulouse, 4e série, **26**, 105 (1962)
12. R.-L. Clerc, Résolution des équations aux connexions du cas antisymétrique de la théorie unitaire hypercomplexe. Application à un principe variationnel. Ann. de l'I.H.P. Section A **12**(4), 343 (1970)

Part V
Nonperturbative Methods. (B) Index Theorem

In this sector (B) of Part V of the book we illustrate a different non-perturbative method to deal with anomalies: the method based on the family's index theorem of Atiyah and Singer, where anomalies play a different role. So far anomalies have meant breaking a conservation law, i.e. the corresponding Ward identity. With the index theorem they change nature and appear as obstructions to the existence of the fermion propagator. In simple words we could phrase this issue in the following way. In Sect. 6.1 we have introduced the crucial problem of whether the inverse of the Dirac-Weyl operator exists, we dubbed it the Weyl fermion catastrophe. There, we presented it in the form of the inverse of $\not{D}_L = \not{D} P_L$, where $\not{D}$ is the Dirac operator and P_L is the chiral left-projector. This is a naive, although suggestive, way of formulating the problem. The more sophisticated and rigorous way is through the index theorem, which tells us when this operation of inversion is possible and when it is not. We may insist in defining an inverse by somewhat deforming the Dirac-Weyl operator by adding a free part with opposite chirality, as we have done in Eq. (6.8). However, even so, it is impossible to avoid the obstruction, because it reappears as a violation of the relevant WI. Here is where index theorem and algorithmic methods meet.

Before dealing with the family's index theorem it is necessary to prepare the ground by analyzing the connection between anomalies and geometry, notably the geometry of principal fiber bundles and their automorphisms. We shall see that when applying such general geometric constructs to quantum field theory a modification is necessary with respect to standard geometry: the relevant cohomology needed in the study of anomalies is not a topological one (the singular one or even the familiar de Rham cohomology) but a somewhat different version, the *local cohomology*, also known as BRST cohomology. In turn this is related to the concept of universality, a characteristic of quantum field theory, according to which for local anomalies the topology of space-time is irrelevant.

Chapter 11
Geometry of Anomalies

In Chap. 4, we have introduced the BRST (or local) cohomology and constructed corresponding non-trivial cocycles. Although the formulas used there are formally correct, one should remark that in order to describe a consistent anomaly in a $\mathsf{d} = 2n - 2$ dimensional spacetime, we need two forms $P_n(F, \ldots, F)$ and $\Delta^{(0)}_{2n-1}(A)$, which are identically vanishing. This is an unsatisfactory aspect of the purely algebraic approach adopted there. In order to make sense of it, we must resort to geometry, more precisely to the geometry of principal fiber bundles. The purpose of the next section is to offer an interpretation of the BRST transformations, the BRST cohomology and its cocycles in the framework of that geometry. The basic idea is to use the evaluation map $ev : \mathsf{P} \times \mathcal{G} \to \mathsf{P}$, where P is the principal fiber bundle where the gauge theory is defined and $\mathcal{G}$ is the relevant group of gauge transformations. It turns out that this simple setup contains all the information concerning the BRST transformations. Exploiting this new tool, we rewrite the formulas of Chap. 4 in a geometrical language. Next, we extend this treatment to diffeomorphisms and local Lorentz transformations. This language is essential to understand the origin of anomalies, which lies in the geometry of the classifying space and classifying bundle. More precisely, local anomalies of field theory are rooted in the cohomology of the classifying space. The classifying space is a universal construction, because, given a spacetime, any principal fiber bundle and any connection in it can be obtained as the pull-back of the classifying bundle and relative universal connection. For this reason, we claim that locality in quantum field theory is linked to universality. This is the main result of this chapter and a fundamental result for the comprehension of the anomaly problem. It is also a crucial preliminary for the application of the family's index theorem in the next chapter. Main references for this chapter are [1–7] as well as [8–11].

L. Bonora, *Fermions and Anomalies in Quantum Field Theories*, Theoretical and Mathematical Physics, https://doi.org/10.1007/978-3-031-21928-3_11

The framework for the geometry of anomalies analyzed in this chapter as well as in the following one is based on a Euclidean spacetime. It would be necessary to formulate the same geometry in Minkowski spacetime but this is out of our reach, for the time being. The understanding is that properties and formulas are translated from the Euclidean to the Minkowskian counterpart via an inverse Wick rotation and that the geometrical meaning we are not able to describe in the original Lorentzian language is *bona fide* provided by the corresponding Euclidean geometry.

11.1 Evaluation Map and BRST

Let $\mathsf{P}(\mathsf{M}, \mathsf{G})$ be a principal bundle with base a d-dimensional Euclidean manifold M, structure group G, which we suppose to be compact, and total space P. π will denote the projection $\pi : \mathsf{P} \to \mathsf{M}$.

$$\begin{array}{ccc} \mathsf{G} & \hookrightarrow & \mathsf{P} \\ & & \downarrow{\scriptstyle \pi} \\ & & \mathsf{M} \end{array}$$

An automorphism is a diffeomorphisms of P, $\psi : \mathsf{P} \to \mathsf{P}$, such that $\psi(pg) = \psi(p)g$, for any $p \in \mathsf{P}$ and any $g \in \mathsf{G}$. Automorphisms form a group which will be denoted $\mathsf{Aut}(\mathsf{P})$. For any automorphism ψ, the map $\pi \circ \psi$ is a diffeomorphism of the base space, i.e. $\in \mathsf{Diff}(\mathsf{M})$. Vertical automorphisms are those that do not move the base point, $\pi(\psi(p)) = \pi(p)$. Vertical automorphisms form a subgroup of $\mathsf{Aut}(\mathsf{P})$ denoted by $\mathsf{Aut}_\mathsf{v}(\mathsf{P})$. The corresponding Lie algebras will be denoted by $\mathsf{aut}(\mathsf{P})$ and $\mathsf{aut}_\mathsf{v}(\mathsf{P})$, respectively; the latter are spaces of vector fields in P generated by one-parameters subgroups of $\mathsf{Aut}(\mathsf{P})$ and $\mathsf{Aut}_\mathsf{v}(\mathsf{P})$, respectively. Finally, a basic form ω in P is the pullback by the projection π of a form $\bar{\omega}$ on M: $\omega = \pi^*\bar{\omega}$.

For the time being, we will disregard diffeomorphisms of the base space and focus on $\mathsf{Aut}_\mathsf{v}(\mathsf{P})$. The latter is to be identified with the group $\mathcal{G}$ of gauge transformations, whose Lie algebra $Lie\,\mathcal{G}$ is, in turn, identified with $\mathsf{aut}_\mathsf{v}(\mathsf{P})$, the Lie algebra of $\mathsf{Aut}_\mathsf{v}(\mathsf{P})$. The reason of this identification is clear from the way a connection transforms under vertical automorphisms. Let A be a connection with curvature $F = dA + \frac{1}{2}[A, A]$. In local coordinates, it takes the form $A = A^a_\mu T^a \mathrm{d}x^\mu$, where T^a are the generators of $\mathfrak{g} \equiv Lie(\mathsf{G})$. Let ψ be a vertical automorphism: we can associate to it a map $\gamma : \mathsf{P} \to \mathsf{G}$ defined by $\psi(p) = p\,\gamma(p)$ satisfying $\gamma(pg) = g^{-1}\gamma(u)g$. Then, one can show that

$$\psi^* A = \gamma^{-1} A \gamma + \gamma^{-1} d\gamma, \quad \psi^* F = \gamma^{-1} F \gamma$$

Thus, we set $\mathsf{Aut_v(P)} \equiv \mathcal{G}$ and introduce the evaluation map

$$ev : \mathsf{P} \times \mathcal{G} \to \mathsf{P}, \quad ev(p, \psi) = \psi(p) \tag{11.1}$$

We suppose that G acts trivially on $\mathcal{G}$, so that $\mathsf{P} \times \mathcal{G}$ is a principal fiber bundle over $\mathsf{M} \times \mathcal{G}$, with group G.

$$\begin{array}{ccc} \mathsf{G} & \hookrightarrow & \mathsf{P} \times \mathcal{G} \\ & & \downarrow{\scriptstyle \pi \times \mathrm{id}} \\ & & \mathsf{M} \times \mathcal{G} \end{array}$$

This means that by pulling back a connection A in P we obtain a connection $\mathcal{A} = ev^*A$ in $\mathsf{P} \times \mathcal{G}$. This connection contains all the information about the propertie of the FP ghosts and BRST transformations. Let us see how this comes about.

We evaluate ev^*A over a couple (X, Y), where $\mathrm{X} \in T_p\mathsf{P}$ and $Y \in T_\psi\mathcal{G}$. Here, $T_p\mathsf{P}$, $T_\psi\mathcal{G}$ denote the tangent space of P at p and of $\mathcal{G}$ at ψ, respectively. Since $\mathcal{G}$ is a Lie group, there exists a $Z \in T_{id}\mathcal{G}$, such that $\psi_* Z = Y$. Let ψ_t ($t \in \mathbb{R}$) with $\psi_0 = \psi$, generate Y, i.e. given $f \in C^\infty(\mathcal{G})$ let

$$Yf = \frac{\mathrm{d}}{\mathrm{d}t} f\left(\psi_t\right)\Big|_{t=0} \tag{11.2}$$

Then, $\mathring{\psi}_t = \psi^{-1}\psi_t$ generates Z:

$$Zf = \frac{\mathrm{d}}{\mathrm{d}t} f\left(\mathring{\psi}_t\right)\Big|_{t=0} = (\psi_*^{-1} Y) f \tag{11.3}$$

Now, it is useful to introduce two auxiliary maps:

$$\begin{aligned} ev_p &: \mathcal{G} \to \mathsf{P}, \quad ev_p(\psi) = \psi(p) \\ ev_\psi &: P \to \mathsf{P}, \quad ev_\psi(p) = \psi(p) \end{aligned}$$

For any $h \in C^\infty(\mathsf{P})$, we have

$$(ev_{p*}Y)h = \frac{\mathrm{d}}{\mathrm{d}t} h(ev_p \circ \psi_t)\Big|_{t=0} = \frac{\mathrm{d}}{\mathrm{d}t} h(\psi_t(p))\Big|_{t=0} = Y_{\psi(p)}h \tag{11.4}$$

Given this, we have

$$\begin{aligned} (ev^*A)_{p,\psi}(\mathrm{X}, Y) &= A_{\psi(p)}\left(ev_{\psi*}\mathrm{X}\right) + A_{\psi(p)}\left(ev_{p*}Y\right) = A_{\psi(p)}\left(\psi_*\mathrm{X}\right) + A_{\psi(p)}\left(Y_{\psi(p)}\right) \\ &= \left(\psi^*A\right)_p(\mathrm{X}) + \left(\psi^*A\right)_p\left(Z_p\right) = \left(\psi^*A\right)_p(\mathrm{X}) + \left(i_{\psi_*^{-1}(Y)}\psi^*A\right)_{\psi(p)} \\ &= \left(\psi^*A\right)_p(\mathrm{X}) + \left(i_{\psi_*^{-1}(\cdot)}\psi^*A\right)_{\psi(p)}(Y) \end{aligned} \tag{11.5}$$

At $\psi = id$, the identity of $\mathcal{G}$, this formula can be written in the compact form

$$\mathbf{A} \equiv ev^* A = A + i_{(\cdot)} A \tag{11.6}$$

where i is the interior product and $i_{(\cdot)} A$ denotes the map $Z \to i_Z A$, that associates to every $Z \in Lie(\mathcal{G})$ the map $\xi_Z = A(Z) : \mathsf{P} \to Lie(\mathsf{G})$. ξ_Z is an infinitesimal gauge transformation, for let us recall that the action of Z over the connection A is given by the Lie derivative L_Z, which takes the form:

$$\begin{aligned} L_Z A &= (di_Z + i_Z d)\, A = \mathrm{d}\,(i_Z A) + i_Z\,(\mathrm{d}A) = \mathrm{d}\,(i_Z A) + i_Z \left(F - \frac{1}{2}[A, A] \right) \\ &= i_Z F + \mathrm{d}\,(i_Z A) + [A, i_Z A] = \mathrm{d}\xi_Z + [A, \xi_Z] \end{aligned} \tag{11.7}$$

because $i_Z F = 0$, being Z a vertical vector, while F is a horizontal two-form.

The above means in particular that $i_{(\cdot)} A$ behaves like the Maurer-Cartan form on the group G.

We record for later use the relation

$$i_Z \mathrm{d}A = -[A, i_Z A], \quad \forall Z \in Lie(\mathcal{G}) \tag{11.8}$$

Formula (11.6) is the geometrical transcription of the writing

$$\mathsf{A} = A + c \tag{11.9}$$

where c is the FP anticommuting ghost field, we have introduced in Chap. 2. An alternative way of writing (11.9) is

$$\mathsf{A} = A + c\, \mathrm{d}\vartheta \tag{11.10}$$

where ϑ is an anticommuting variable (see Appendix 15A). We will use (11.10) later on. Equation (11.6) is a poly-form (the sum of a 1-form and a 0-form with values in the Lie algebra $\mathfrak{g} \equiv Lie(\mathsf{G})$), which is the basis of our geometrical interpretation of BRST. From now on, our purpose is to justify the fact that $i_{(\cdot)} A$ can play the role of the ghost field c.

If $Q(A)$ is a polynomial of $A, \mathrm{d}A$ and $[A, A]$, the product being the exterior product of forms, the Formula (11.6) is generalized to

$$\begin{aligned} ev^* Q(A) &= Q(A) + i_{(\cdot)} Q(A) - i_{(\cdot)} i_{(\cdot)} Q(A) - \cdots \\ &\quad + (-1)^{\frac{k(k-1)}{2}} \underbrace{i_{(\cdot)} \ldots i_{(\cdot)}}_{k \text{ terms}} Q(A) + \cdots \end{aligned} \tag{11.11}$$

where the interior products are understood with respect to vectors in $Lie(\mathcal{G})$. Of course we have also

$$ev^* Q(A) = Q(ev^* A) = Q(A + i_{(\cdot)} A) \tag{11.12}$$

because the pull-back 'passes through' exterior products and differentials. So the RHS of this equation equals the RHS of (11.11). This allows us to read off the meaning of the expression $Q(A + i_{(\cdot)} A)$, which may not be immediately transparent.

Let us consider an example. A remarkable consequence of (11.6) is

$$\mathsf{F} = ev^* F = F \tag{11.13}$$

because $i_Z F = 0$ for any $Z \in Lie(\mathcal{G})$. This formula is sometimes called in the literature the *russian formula*. Let us write down the explicit form of $ev^* F$

$$\begin{aligned} ev^* F = F + i_{(\cdot)} F + i_{(\cdot)} i_{(\cdot)} F &= F + i_{(\cdot)} \mathrm{d}A \\ &+ [i_{(\cdot)} A, A] + i_{(\cdot)} i_{(\cdot)} \mathrm{d}A + \frac{1}{2} [i_{(\cdot)} A, i_{(\cdot)} A] \end{aligned} \tag{11.14}$$

On the other hand, if $\hat{\delta}$ is the exterior differential in $\mathcal{G}$, we have

$$\begin{aligned} F(ev^* A) &= (d + \hat{\delta}) \mathbf{A} + \frac{1}{2} [\mathbf{A}, \mathbf{A}] \\ &= F + \hat{\delta} A + d i_{(\cdot)} A + \frac{1}{2} [A, i_{(\cdot)} A] + \frac{1}{2} [i_{(\cdot)} A, A] \\ &+ \hat{\delta} i_{(\cdot)} A + \frac{1}{2} [i_{(\cdot)} A, i_{(\cdot)} A] \end{aligned} \tag{11.15}$$

which means, splitting it according to the form degree,

$$\hat{\delta} A = -d i_{(\cdot)} A - \frac{1}{2} [A, i_{(\cdot)} A] - \frac{1}{2} [i_{(\cdot)} A, A] \tag{11.16}$$

$$\hat{\delta} i_{(\cdot)} A = \frac{1}{2} [i_{(\cdot)} A, i_{(\cdot)} A] \tag{11.17}$$

The first equation must correspond to the term $i_{(\cdot)} F$ in Eq. (11.14). But

$$i_{(\cdot)} F = i_{(\cdot)} d A + [i_{(\cdot)} A, A] = L_{(\cdot)} A - d i_{(\cdot)} A - [A, i_{(\cdot)} A] = 0 \tag{11.18}$$

Therefore, we must understand that $\hat{\delta} = (-1)^k \delta$, where k is the degree of the form in P it acts on, and δ is the ordinary BRST variation. Moreover, $[i_{(\cdot)} A, A] = [A, i_{(\cdot)} A]$, i.e. A and $i_{(\cdot)} A$ behave like one-forms. Finally, we have

$$\delta A = di_{(.)}A + [A, i_{(.)}A] \tag{11.19}$$

$$\delta i_{(.)}A = -\frac{1}{2}[i_{(.)}A, i_{(.)}A] \tag{11.20}$$

On the other hand, Eq. (11.17) must correspond to the last two terms in Eq. (11.14). To see that this is the case, one must recall some basic formulas in differential geometry, where for any one-form ω and any two vector fields X, Y, we have

$$d\omega(\mathsf{X}, \mathsf{Y}) = \frac{1}{2}\big(\mathsf{X}\,\omega(\mathsf{Y}) - \mathsf{Y}\,\omega(\mathsf{X})\big) - \omega([\mathsf{X}, \mathsf{Y}]) \tag{11.21}$$

and

$$L_{\mathsf{X}}\omega(\mathsf{Y}) = \mathsf{X}\,\omega(\mathsf{Y}) - \omega([\mathsf{X}, \mathsf{Y}]) \tag{11.22}$$

The skew double interior product $i_{(.)}i_{(.)}dA$ in (11.14) must be understood as follows

$$i_{Z_2}i_{Z_1}dA = dA(Z_1, Z_2) = \frac{1}{2}\big(Z_1 A(Z_2) - Z_2\, A(Z_1)\big) - A([Z_1, Z_2])$$
$$= \frac{1}{2}L_{Z_1}A(Z_2) - \frac{1}{2}L_{Z_2}A(Z_1)$$

In other words, $i_{(.)}i_{(.)}dA$ is to be understood as the Lie derivative of $i_{(.)}A$, or its BRST transform once we interpret $i_{(.)}A$ as the FP ghost, in agreement with (11.20).

From the above formulas, we see that the evaluation map provides a geometrical interpretation of the BRST transformations.

11.2 Evaluation Map and Anomalies

Let us consider an ad-invariant polynomial P_n, with n entries as in Chap. 5. The expression

$$TP_n(A) = n \int_0^1 \mathrm{d}t\, P_n(A, F_t, \ldots, F_t) \tag{11.23}$$

introduced above is called *transgression formula*. Let $n = \frac{\mathsf{d}}{2} + 1$, then

$$\mathrm{d}\,TP_n(A) = P_n(F, \ldots, F) = 0 \tag{11.24}$$

because of dimensional reasons, for $P_n(F, \ldots, F)$ is a $\mathsf{d}+2$ basic form in a d dimensional base spacetime M. Let us pull back (11.24) through the evaluation map to $\mathsf{P} \times \mathcal{G}$. We get

$$(d + \hat{\delta}) ev^* TP_n(A) = (d + \hat{\delta}) TP_n(ev^* A) = 0 \tag{11.25}$$

Now, using (11.11), we can decompose this equation according to the form degree

$$\hat{\delta}\, TP_n(A) + d\, i_{(\cdot)} TP_n(A) = 0 \tag{11.26}$$
$$\hat{\delta}\, i_{(\cdot)} TP_n(A) + d\, i_{(\cdot)} i_{(\cdot)} TP_n(A) = 0 \tag{11.27}$$
$$\ldots\ldots$$
$$\ldots\ldots$$

These are the descent equations written in geometrical language. In order to justify this identification, we have to show how $i_{(\cdot)} TP_n(A)$ relates to the expression of the anomaly (5.40). In order to do so, we use the following formulas

$$F_t = t\mathrm{d}A + \frac{t^2}{2}[A, A], \quad \frac{\mathrm{d}}{\mathrm{d}t} F_t = \mathrm{d}A + t[A, A] = D_{tA} A, \quad D_{tA} F_t = \mathrm{d}F_t + [tA, F_t] = 0 \tag{11.28}$$

where D_{tA} denotes the covariant derivative with respect to the connection tA: $D_{tA} = d + [tA, \cdot\ ]$. We have

$$i_Z TP_n(A) = n \int_0^1 \mathrm{d}t\, \left(P_n(i_Z A, F_t, \ldots, F_t) - (n-1) P_n(A, i_Z F_t, \ldots, F_t)\right) \tag{11.29}$$

Now, $i_Z F_t = (t - t^2) i_z \mathrm{d}A$; therefore, using also (11.8), one gets

$$\begin{aligned} & i_Z TP_n(A) \\ &= n \int_0^1 \mathrm{d}t \left(\frac{\mathrm{d}}{\mathrm{d}t}(t-1) P_n(i_Z A, F_t, \ldots, F_t) - (n-1)(t - t^2) P_n(A, i_Z \mathrm{d}A, \ldots, F_t) \right) \\ &= -n(n-1) \int_0^1 \mathrm{d}t \left((t-1) P_n(i_Z A, \frac{\mathrm{d}}{\mathrm{d}t} F_t, \ldots, F_t) + (1-t) P_n(A, [tA, i_Z A], F_t, \ldots, F_t) \right) \end{aligned} \tag{11.30}$$

Now replacing $\frac{\mathrm{d}}{\mathrm{d}t}F_t$ with (11.28) and using the ad-invariance of P_n, we find

$$
\begin{aligned}
&i_Z TP_n(A) &(11.31)\\
&=-n(n-1)\int_0^1 \mathrm{d}t(t-1)\Big(P_n(di_Z A, A, F_t, \dots, F_t) - \mathrm{d}P_n(A, i_Z A, F_t, \dots, F_t)\Big)
\end{aligned}
$$

The first term on the RHS is precisely, up to a global sign, the anomaly (5.40):

$$
\Delta^1_{\mathsf{d}}(i_{(\cdot)}A, A) = n(n-1)\int_0^1 \mathrm{d}t(t-1)P_n(di_{(\cdot)}A, A, F_t, \dots, F_t). \quad (11.32)
$$

The second piece is a total differential which drops out upon spacetime integration.

Another possibility is to start, instead, from $TP_n(ev^*A) = TP_n(A + i_{(\cdot)}A)$ and dig out the term linear in $i_{(\cdot)}A$. The end result is the same, i.e. the RHS of (11.31).

In sum, the evaluation map $ev : \mathsf{P} \times \mathcal{G} \to \mathsf{P}$ contains the information about the BRST symmetry and the consistent chiral anomalies.

11.3 Background Connection

The transgression formula $TP_n(A)$ is defined in the total space of the principal fiber bundle $\mathsf{P}(\mathsf{M}, \mathsf{G})$ and its pull-back via a local section to the base manifold vanishes for dimensional reasons. Also, the expression for the anomalies are generally defined only in the total space P . It is, however, possible, with some modest changes, to express them as forms in M. The space $\mathcal{A}$ of connections in a principal fiber bundle is an affine space constructed over a fixed (background) connection A_0 by adding to it any member of the vector space of ad-covariant 1-forms $\Omega^1(\mathsf{M}, \mathsf{ad}\,\mathsf{P})$. If A_1, A_2 are connections also $tA_1 + (1-t)A_2$, for $0 \le t \le 1$, is a connection. In this section, we would like to exploit this property in order to express the anomaly as a basic form. To this end, we will fix a connection A_0 (the background connection) and use the relevant connection A to construct $\mathsf{A}_t = tA + (1-t)A_0$, which, as just mentioned, is also a connection. The related curvatures will be denoted by F_0, F and F_t, respectively.

Next, let us start from

$$
\begin{aligned}
P_n(F, \dots, F) - P_n(F_0, \dots, F_0) &= \mathrm{d}\left(n\int_0^1 \mathrm{d}t\, P_n(A - A_0, \mathsf{F}_t, \dots, \mathsf{F}_t)\right)\\
&\equiv d\Delta_{2n-1}(A, A_0) \qquad (11.33)
\end{aligned}
$$

This is easily proven by noting that $\frac{d}{dt}\mathsf{F}_t = d_{\mathsf{A}_t}(A - A_0)$ and $d_{\mathsf{A}_t}\mathsf{F}_t = 0$. Next, we take the BRST variation of $\Delta_{2n-1}(A, A_0)$ with the rule:

$$\mathfrak{s}A = d_A c, \qquad \mathfrak{s}A_0 = 0 \tag{11.34}$$

It follows in particular that

$$\mathfrak{s}\mathsf{F}_t = [\mathsf{F}_t, c] - (1-t)\left(d_{\mathsf{A}_t}([A_0, c]) + [\mathsf{A}_t, dc]\right) \tag{11.35}$$

Notice also that $[\mathsf{A}_t, dc] = d_{\mathsf{A}_t} dc$. Now

$$\begin{aligned}
\mathfrak{s}\Delta_{2n-1}(A, A_0) &= n\int_0^1 \mathrm{d}t\Big[P_n(dc + [A, c], \mathsf{F}_t, \ldots, \mathsf{F}_t)\\
&\quad + (n-1)P_n(A - A_0, [\mathsf{F}_t, c] - (1-t)\left(d_{\mathsf{A}_t}([A_0, c]) + [\mathsf{A}_t, dc]\right), \mathsf{F}_t, \ldots, \mathsf{F}_t)\Big]\\
&= n\int_0^1 \mathrm{d}t\Big[P_n(dc + [A_0, c], \mathsf{F}_t, \ldots, \mathsf{F}_t)\\
&\quad - (n-1)(1-t)P_n(A - A_0, \left(d_{\mathsf{A}_t}([A_0, c]) + d_{\mathsf{A}_t} dc\right), \mathsf{F}_t, \ldots, \mathsf{F}_t)\Big]
\end{aligned}$$

using ad-invariance. Next,

$$\begin{aligned}
\mathfrak{s}\Delta_{2n-1}(A, A_0) &= n\int_0^1 \mathrm{d}t\Big[P_n(dc + [A_0, c], \mathsf{F}_t, \ldots, \mathsf{F}_t)\\
&\quad + (n-1)(1-t)\mathrm{d}\big(P_n(A - A_0, d_{A_0}c, \mathsf{F}_t, \ldots, \mathsf{F}_t)\big)\\
&\quad - (n-1)(1-t)P_n(d_{\mathsf{A}_t}(A - A_0), d_{A_0}c, \mathsf{F}_t, \ldots, \mathsf{F}_t)\Big]\\
&= n\int_0^1 \mathrm{d}t\Big[P_n(dc + [A_0, c], \mathsf{F}_t, \ldots, \mathsf{F}_t)\\
&\quad + (n-1)(1-t)\mathrm{d}\big(P_n(A - A_0, d_{A_0}c, \mathsf{F}_t, \ldots, \mathsf{F}_t)\big)\\
&\quad - (1-t)\frac{\mathrm{d}}{\mathrm{d}t}P_n(d_{A_0}c, \mathsf{F}_t, \ldots, \mathsf{F}_t)\Big]
\end{aligned}$$

Finally, integrating by parts,

$$\begin{aligned}
&\mathfrak{s}\Delta_{2n-1}(A, A_0) \\
&= \mathrm{d}\left(-n(n-1)\int_0^1 \mathrm{d}t(1-t)\, P_n(d_{A_0}c, A - A_0, \mathsf{F}_t, \ldots, \mathsf{F}_t) + nP_n(c, F_0, \ldots F_0)\right)
\end{aligned} \tag{11.36}$$

This leads us to identify the basic cocycle

$$\begin{aligned}&\Delta^{(1)}_{2n-2}(A, A_0, c) \\ &= n(n-1)\int_0^1 dt(1-t)\, P_n(d_{A_0}c, A - A_0, \mathsf{F}_t, \dots, \mathsf{F}_t) - n\, P_n(c, F_0, \dots F_0)\end{aligned} \tag{11.37}$$

This basic form in M is the anomaly with background connection. Needless to say, the procedure is somewhat formal because the RHS of (11.36) is actually 0. The evaluation map provides a less formal derivation of the same formulas.

11.3.1 Background Connection and Evaluation Map

Here, we would like to derive the previous cocycles via the evaluation map. In such a case, we need connections defined in the principal fiber bundle $\mathsf{P} \times \mathcal{G}$. As we have seen ev^*A is one. As for A_0, we pull it back via the projection map $pr_1 : \mathsf{P} \times \mathcal{G} \to \mathsf{P}$. We obtain in this way a connection $pr_1^* A_0$ in $\mathsf{P} \times \mathcal{G}$, which we denote simply A_0. So, we can define a new connection in $\mathsf{P} \times \mathcal{G}$ given by $\mathcal{A}_t = t\, ev^*A + (1-t)A_0$, whose curvature will be denoted $\mathcal{F}_t$. Finally, let the curvature of A_0 be F_0. Then, we have

$$\begin{aligned}0 &= P_n(ev^*F, \dots, ev^*F) - P_n(F_0, \dots, F_0) \\ &= (d+\hat{\delta})\left(n\int_0^1 dt\, P_n(ev^*A - A_0, \mathcal{F}_t, \dots, \mathcal{F}_t)\right) \equiv (d+\hat{\delta})\, \mathcal{T}P_n(ev^*A, A_0)\end{aligned} \tag{11.38}$$

because $\frac{d}{dt}\mathcal{F}_t = (d+\hat{\delta})_{\mathcal{A}_t}(ev^*A - A_0)$. Proceeding as above, we can show that

$$\begin{aligned}\mathcal{T}P_n(ev^*A, A_0) &= \mathcal{T}P_n(\psi^*A, A_0) + j_{(\cdot)}\mathcal{T}P_n(\psi^*A, A_0) - j_{(\cdot)}j_{(\cdot)}\mathcal{T}P_n(\psi^*A, A_0) \\ &+ \dots + (-1)^{\frac{k(k-1)}{2}} \underbrace{j_{(\cdot)} \dots j_{(\cdot)}}_{k \text{ terms}} \mathcal{T}P_n(\psi^*A, A_0) + \cdots\end{aligned} \tag{11.39}$$

where $j_{(\cdot)} = i_{(\cdot)}$ when applied to ψ^*A and $j_{(\cdot)} = 0$ when applied to A_0. So, in particular, $j_{(\cdot)}A_0 = 0$, $j_{(\cdot)}dA_0 = 0$.

We remark that even though $\mathcal{T}P_n(\psi^*A, A_0)$ vanishes for dimensional reasons (it is a basic $\mathsf{d}+1$ form), $j_{(\cdot)}\mathcal{T}P_n(\psi^*A, A_0)$ does not and, since the consistency condition

$$\delta j_{(\cdot)}\mathcal{T}P_n(\psi^*A, A_0) + d j_{(\cdot)}j_{(\cdot)}\mathcal{T}P_n(\psi^*A, A_0) = 0 \tag{11.40}$$

is satisfied at $\psi = id$, it represents the expression of the anomaly with a background connection, and is a basic form on M. It obviously coincides with Eq. (11.37).

We would like now to enlighten the relation between the anomalies with form $i_{(\cdot)}TP_n(A)$ of the previous section and the anomalies of type $j_{(\cdot)}\mathcal{T}P_n(\psi^* A, A_0)$. To this end, given two connections A_1 and A_2, with curvature F_1 and F_2, let us consider the expression

$$SP_n(A_1, A_2) = -n(n-1)\int_0^1 \mathrm{d}s \int_0^1 \mathrm{d}t\, t\, P_n(A_1 - A_2, A_s, \mathcal{F}_t^s, \ldots, \mathcal{F}_t^s) \tag{11.41}$$

where $A_s = sA_1 + (1-s)A_2$ and $\mathcal{F}_t^s = t dA_s + \frac{t^2}{2}[A_s, A_s]$. Let us notice first that

$$\frac{\mathrm{d}}{\mathrm{d}t}\mathcal{F}_t^s = d_{tA_s}A_s, \quad \frac{\mathrm{d}}{\mathrm{d}s}\mathcal{F}_t^s = t d_{tA_s}(A_1 - A_2), \quad d_{tA_s}\mathcal{F}_t^s = 0 \tag{11.42}$$

Taking the exterior derivative of (11.41) and using ad-invariance of P_n, we get

$$\begin{aligned} dSP_n(A_1, A_2) &= -n(n-1)\int_0^1 \mathrm{d}s \int_0^1 \mathrm{d}t \Big[P_n(td_{tA_s}(A_1 - A_2), A_s, \mathcal{F}_t^s, \ldots, \mathcal{F}_t^s) \\ &\quad - tP_n(A_1 - A_2, d_{tA_s}A_s, \mathcal{F}_t^s, \ldots, \mathcal{F}_t^s)\Big] \\ &= -n(n-1)\int_0^1 \mathrm{d}s \int_0^1 \mathrm{d}t \Big[P_n(\frac{d}{ds}\mathcal{F}_t^s, A_s, \mathcal{F}_t^s, \ldots, \mathcal{F}_t^s) \\ &\quad - tP_n(A_1 - A_2, \frac{\mathrm{d}}{\mathrm{d}t}\mathcal{F}_t^s, \mathcal{F}_t^s, \ldots, \mathcal{F}_t^s)\Big] \end{aligned} \tag{11.43}$$

Integrating by parts in s and t, we get

$$\begin{aligned} dSP_n(A_1, A_2) &= -n\int_0^1 \mathrm{d}t \Big[P_n(A_1, F_{1t}, \ldots, F_{1t}) - P_n(A_2, F_{2t}, \ldots, F_{2t}) \\ &\quad - \int_0^s \mathrm{d}s\, P_n(A_1 - A_2, \mathcal{F}_1^s, \ldots, \mathcal{F}_1^s)\Big] \\ &= -TP_n(A_1) + TP_n(A_2) + \mathcal{T}P_n(A_1, A_2) \end{aligned} \tag{11.44}$$

Applying this result to $A_1 = ev^*A$ and $A_2 = A_0$, we get

$$\mathcal{T}P_n(ev^*A, A_0) = TP_n(ev^*A) - TP_n(A_0) - (d + \hat{\delta})SP_n(ev^*A, A_0) \tag{11.45}$$

The highest order component of this relation, (the $(\mathsf{d}+1, 0)$ one), is

$$0 = TP_n(A) - TP_n(A_0) + d\, SP_n(A, A_0) \tag{11.46}$$

which means that $TP_n(A)$ and $TP_n(A_0)$ belongs to the same de Rham cohomology class of P . The $(n, 1)$ component is

$$\begin{aligned} j_{(\cdot)}\mathcal{T}P_n(\psi^* A, A_0) = i_{(\cdot)}TP_n(\psi^* A) + \hat{\delta} S P_n(\psi^* A, A_0) \\ + d\, j_{(\cdot)} S P_n(\psi^* A, A_0) \end{aligned} \tag{11.47}$$

This establishes the relation between the two anomalies, with and without background. If the principal fiber bundle $\mathsf{P}(\mathsf{M}, \mathsf{G})$ is trivial, there exists a global section σ_0 and a connection A_0 such that $\sigma_0^* A_0 = 0$. In this case, pulling back via σ_0 the two expressions $\sigma_0^* (\mathcal{T} P_n(ev^* A, A_0))$ and $\sigma_0^* (TP_n(ev^* A))$, one can see that they collapse to the same function of $(ev \circ \sigma_0)^* A$, and the corresponding anomaly can be identified with (5.40). This is the situation we meet in perturbative field theory, where the background connection is in general assumed to vanish. The same happens in any local patch of M, because locally any bundle is trivial. But if the principal bundle is not trivial, in passing from one local patch to another, one must take into account possible boundary contributions.

Remark The form $j_{(\cdot)}\mathcal{T} P_n(\psi^* A, A_0)$ is defined on the base manifold M. Therefore, we can integrate it. The object we obtain

$$\int_M j_{(\cdot)}\mathcal{T} P_n(\psi^* A, A_0) \tag{11.48}$$

is a 1-form in $\mathcal{G}$, closed under $\hat{\delta}$, see Eq. (11.40). Since $\hat{\delta}$ is the exterior differential in $\mathcal{G}$, (11.48) defines an element of the cohomology group $H^1(\mathcal{G}_0)$, where $\mathcal{G}_0$ is the component of $\mathcal{G}$ connected to the identity. From this, we see that an anomaly can be regarded not only as a non-trivial class in the BRST cohomology, but also as an element of the cohomology of $\mathcal{G}_0$. The difference between the two is that, while (11.48) is always a non-trivial anomaly, it may be a trivial element of $H^1(\mathcal{G}_0)$ depending on the topology of M.

11.4 Trivial and Non-trivial Cocycles

A cocycle $\Delta^{(1)}_{2n-2}(A, c)$ is trivial when there are local cochains $C^{(0)}_{2n-2}(A, c)$ and $C^{(1)}_{2n-3}(A, c)$ such that

$$\Delta^1_{\mathsf{d}}(A, c) = \mathfrak{s}\, C^{(0)}_{2n-2}(A, c) + dC^{(1)}_{2n-3}(A, c) \tag{11.49}$$

where $C^{(p)}_k(A, c)$ denotes a polynomial k-form of ghost number p, built with A, c, their commutators, wedge products and exterior differentials. If such local cochains do not exist, the cocycle is said to be non-trivial and defines a non-trivial class of the BRST cohomology. What we would like to do now is translate these ideas in the geometrical language of principal fiber bundles and evaluation map.

Let us return to the descent Eq. (11.25) and, in particular, to the consistency conditions (11.27). Since $d\, TP_n(A) = 0$, the anomalies are to be found among the representatives of the de Rham cohomology group of order $\mathsf{d}+1$ in P . Now, the de Rham cohomology group of order $\mathsf{d}+1$ in P , $\mathsf{H}^{\mathsf{d}+1}_{\mathsf{de\,Rham}}(\mathsf{P})$, is spanned by forms

$$TP_i(A) \wedge \beta_i \tag{11.50}$$

where β_i are basic and closed forms of order $h_i = 2(n-i)$, such that $2i-1+h_i = \mathsf{d}+1$. The β_i may be forms of the type $P_{n-i}(F, \ldots, F)$ (this is the case of a reducible polynomial $P_n(F) = P_i(F)P_{n-i}(F)$), or of the type $P'_{n-i}(R, \ldots, R)$, where R is the Riemann (or spin) curvature 2-form on M, and P'_i are the corresponding ad-invariant symmetric polynomials. But the set of β_i may also contain forms that cannot be written in local form like in the previous examples. Such non-local β_i do not give rise to anomalies in local field theory.

Let us suppose next that $\chi = TP_n(A)$ is cohomologically trivial, i.e. $TP_n(A) = \mathrm{d}\eta$. Then, repeating the steps that lead from (11.25) to the descent equations, we get, in particular,

$$i_{(\cdot)}TP_n(A) = i_{(\cdot)}\mathrm{d}\eta = -di_{(\cdot)}\eta + \hat{\delta}\eta \tag{11.51}$$

Since $\hat{\delta}\eta = -\delta\eta$, comparing with (11.49), we see that this is a triviality condition for the anomaly $i_{(\cdot)}TP_n(A)$. However, this does not correspond automatically to a trivial field theory anomaly. The condition $TP_n(A) = \mathrm{d}\eta$ means that $TP_n(A)$ is exact in the de Rham cohomology of P . But this does not automatically mean that it is trivial with respect to the local BRST cohomology. For this to be the case, it must be that η is a local expression of A. If such a local η does not exist, we are in the presence of a true non-trivial anomaly originated from a cocycle $d\eta$ in P which is trivial (a coboundary) in the de Rham cohomology. When $TP_n(A) = \mathrm{d}\eta$ only for a non-local η, the anomaly is sometime called *non-topological*. In the same tune, if χ and χ' are two closed $\mathsf{d}+1$ forms in P such that $\chi - \chi' = \mathrm{d}\eta$, we can say the corresponding anomalies are the same only if η is a local form.

Consider, as an example, P_n being reducible to the product of two symmetric ad-invariant polynomials: $P_n(F) = P_{n_1}(F)P_{n_2}(F)$, with $n_1 + n_2 = n$. Then

$$\begin{aligned}
P_n(F) = P_{n_1}(F)P_{n_2}(F) &= \mathrm{d}\left(n_1 \int_0^1 \mathrm{d}t\, P_{n_1}(A, F_t, \ldots, F_t)P_{n_2}(F, \ldots, F)\right)\\
&= \mathrm{d}\left(n_2 \int_0^1 \mathrm{d}t\, P_{n_2}(A, F_t, \ldots, F_t)P_{n_1}(F, \ldots, F)\right)
\end{aligned} \tag{11.52}$$

We have

$$TP_{n_2}(A)P_{n_1}(F) - TP_{n_1}(A)P_{n_2}(F) = \mathrm{d}\left(TP_{n_1}(A)TP_{n_2}(A)\right)$$

This is a case in which the anomaly ensuing from $TP_{n_1}(A)P_{n_2}(F)$ is the same as the anomaly originated from $TP_{n_2}(A)P_{n_1}(F)$.

11.5 Locality and Universality

In, conclusion, there may be a difference between topology (i.e. the cohomology group $\mathsf{H}^{\mathsf{d}+1}_{\mathsf{de\,Rham}}(\mathsf{P})$) and the BRST cohomology of local field theory. Coboundaries of the former can give rise to non-trivial cocycles in the latter. It is interesting that geometry and local field theory can be reconciled in such a way that anomalies in all cases correspond to non-trivial geometrical object. This is the geometry of the universal bundle. Given a compact Lie group G, any principal fiber bundle $\mathsf{P}(\mathsf{M}, \mathsf{G})$ can be obtained as the pull-back of the so-called *universal principal bundle* $\mathsf{EG}(\mathsf{BG}, \mathsf{G})$, where EG, the total space, is contractible

$$\begin{array}{ccc} \mathsf{G} & \hookrightarrow & \mathsf{EG} \\ & & \big\downarrow{\scriptstyle \pi} \\ & & \mathsf{BG} \end{array}$$

BG, the basis, is called *classifying space.*

For any principal bundle $\mathsf{P}(\mathsf{M}, \mathsf{G})$, there exists a bundle morphisms $(\hat{f}, f)$ such that the following diagram is commutative

$$\begin{array}{ccc} \mathsf{P} & \xrightarrow{\hat{f}} & \mathsf{EG} \\ \big\downarrow{\scriptstyle \pi} & & \big\downarrow{\scriptstyle \pi} \\ \mathsf{M} & \xrightarrow{f} & \mathsf{BG} \end{array}$$

f is unique up to homotopy[1] and is called classifying map. Both EG and BG are generally infinite dimensional. For instance, if $\mathsf{G} = \mathsf{SU(N)}$, they are defined as the limits

$$\mathsf{EG} = \lim_{\mathsf{n}\to\infty} \frac{\mathsf{U(n)}}{\mathsf{U(n-N)}}, \qquad \mathsf{BG} = \lim_{\mathsf{n}\to\infty} \frac{\mathsf{U(n)}}{\mathsf{U(n-N)} \times \mathsf{SU(N)}}. \tag{11.53}$$

[1] Two maps f_1, f_2 are homotopic if there exist a continuous map $F : \mathsf{M} \times I \to \mathsf{BG}$ such that $F(x, 0) = f_1(x)$ and $F(x, 1) = f_2(x)$, $\forall x \in \mathsf{m}$.

In many practical applications, one can take n finite provided it is large enough. In this case, BG is called n-classifying space.

The important fact for our present purposes is the existence of a universal connection $\mathfrak{a}$ from which any connection on P can be derived via pullback [12, 13] : for any A in P there exists a bundle morphisms $(\hat{f}, f)$ such that

$$A = \hat{f}^* \mathfrak{a} \tag{11.54}$$

A form χ in P constructed with the connection form A will be called, with some straining of language, *universal* if it can be derived by pulling back an analogous form in EG constructed with $\mathfrak{a}$.

$$\chi(A) = \chi(\hat{f}^* \mathfrak{a}) = \hat{f}^*(\chi(\mathfrak{a})), \tag{11.55}$$

for a bundle map $\hat{f}$. Therefore, forms like $TP_n(A)$ and $P_n(F)$ are universal. But, of course, in P there may be non-universal forms. In particular, the forms β_i in Formula (11.50) may not be universal. As already remarked, not all the forms (11.50), which span the non-trivial cohomology classes of $\mathsf{H}^{\mathsf{d+1}}_{\mathsf{de\,Rham}}(\mathsf{P})$, necessarily generate local anomalies.

We are now ready to connect universal geometry and anomalies in field theory. We can pull-back $\mathfrak{a}$ and relevant expressions to $\mathsf{P} \times \mathcal{G}$ by combining the evaluation map and bundle maps

$$\mathsf{P} \times \mathcal{G} \xrightarrow{ev} \mathsf{P} \xrightarrow{\hat{f}} \mathsf{EG} \tag{11.56}$$

Starting with local expressions of $\mathfrak{a}$ in EG and pulling them back with $ev \circ \hat{f}$, we create universal expressions in $\mathsf{P} \times \mathcal{G}$. We focus in particular on the forms of the type (11.50) of order $\mathsf{d}+1$ that are universal, and state that *non-trivial local field theory anomalies are identified by the quotient*

$$\frac{closed\ universal\ (\mathsf{d}{+}1){-}forms\ in\ P}{differentials\ of\ universal\ \mathsf{d}{-}forms\ in\ P}$$

For instance, let us consider $TP_{n_1}(A) P_{n_2}(F)$ and suppose that $P_{n_2}(F) = d\eta$ with η a basic form in some manifold M of dimension $\mathsf{d} = 2(n_1+n_2)–2$. Then $TP_{n_1}(A) P_{n_2}(F) = d\left(TP_{n_1}(A)\,\eta\right)$ (because $P_{n_1}(F)\,\eta$ is basic and therefore vanishes on M for dimensional reasons). Nevertheless, the anomaly corresponding to $TP_{n_1}(A) P_{n_2}(F)$ is universal, because, even though $P_{n_2}(F) = d\eta$ in M, the form η cannot be universal. To explain the reason for this, we need to introduce the concept of Weil homomorphism.

Chern-Weil homomorphism. One of the basic results of bundle cohomology is the Weil (or Chern-Weil) homomorphism. Given a principal fiber bundle $\mathsf{P}(\mathsf{M}, \mathsf{G})$, there is a homomorphism

$$w : \mathbb{I}(\mathsf{G}) \longrightarrow \mathsf{H}^*(\mathsf{M})$$

$$P_n \longmapsto w(P_n)$$

where $w(P_n) = P_n(F, \dots, F)$, between the algebra of ad-invariant symmetric polynomials and the de Rham cohomology of M. The homomorphism is constructed by filling in the curvature F of P(M, G) in all the entries of P_n, so as to build the forms $P_n(F, \dots, F)$, which are closed forms on M. They may or may not be exact according to the topology of M. If the bundle is the universal bundle, then M = BG, the classifying space. In this case, the Weil homomorphism is an isomorphism; i.e. no form $P_n(F, \dots, F)$ is exact.

Let us now return to the previous example and explain why the anomaly corresponding to $TP_{n_1}(A)P_{n_2}(F)$ is universal. This is so because, even if $P_{n_2}(F) = d\eta$ in M, the form η cannot be universal. The reason is that for BG the Weil homomorphism is an isomorphism when the group G is compact. This means that $P_{n_2}(F)$ is not the differential of any universal form, and thus, $TP_{n_1}(A)P_{n_2}(F)$ cannot be written itself as a differential of any universal form. From the field theory point of view, this means that $P_{n_2}(F)$ cannot be written as the total differential of a local expression in the fields. In other words, the anomaly originated from $TP_{n_1}(A)P_{n_2}(F)$ is a true local field theory anomaly.

We can summarize the previous discussion with a dense sentence:

- *locality* in perturbative QFT is *universality* in geometry,

at least for gauge theories.

A comment is in order to explain this at first sight surprising correspondence. The origin of it lies in the circumstance that standard perturbative field theory is developed in a local geometry. Propagators and vertices are evaluated on plane wave configurations of fields, thus in a unique local patch isomorphic to a Minkowski (or Euclidean) spacetime. In this way, the cohomological properties of the base space M, reflected also in the topology of P, are inevitably ignored. Therefore, as far as anomalies are concerned, the results of perturbative field theories are independent of the topological features of the base manifold M, the only thing that matters being its dimension. In this sense, they are universal. One can express this also in another way: considering the previous example, even though the class $P_{n_2}(F, \dots, F)$ may be trivial for many manifolds M, there will certainly exist other manifolds for which this class is non-trivial. This is the meaning of the classifying space BG. In turn, perturbative field theories, while ignoring the topology of M, are able to detect non-trivial cohomology classes of the space of fields or of the group $\mathcal{G}$ of gauge transformations. The reason for this is the connection we have illustrated with the universal bundle and classifying space, where such classes acquire a topological significance.

So far, in this section, we have considered the case where the background connection can be set to 0. We can extend all we have said so far to the cases with a non-vanishing background connection A_0. Beside (11.56) we consider the composite map

$$\mathsf{P} \times \mathcal{G} \xrightarrow{pr_1} \mathsf{P} \xrightarrow{\hat{f}_0} \mathsf{EG} \tag{11.57}$$

where $\hat{f}_0$ is the map that generates A_0 from the universal connection $\mathfrak{a}$, $A_0 = \hat{f}_0^* \mathfrak{a}$. Universal forms will be basic forms on $\mathsf{P} \times \mathcal{G}$, i.e. forms in $\mathsf{M} \times \mathcal{G}$, constructed out of $(\hat{f} \circ ev)^* \mathfrak{a}$ and $(\hat{f}_0 \circ pr_1)^* \mathfrak{a}$.

11.6 Diffeomorphisms and Linear Frame Bundles

In theories including gravity symmetry, under diffeomorphisms is a fundamental invariance. If the theory involves fermions also, the local Lorentz symmetry enters the game. They are both fundamental symmetries and are not allowed to be broken by anomalies, the price being the inconsistency of the theory. The study of the corresponding anomalies is thus of utmost importance. In Sect. 5.2.2, we have introduced cocycles of both symmetries. Local Lorentz anomalies are expressed in terms of the connection

$$\boldsymbol{\omega} = \omega_\mu \mathrm{d}x^\mu, \quad \omega_\mu = \omega_\mu^{ab} \Sigma_{ab} \tag{11.58}$$

whose transformation law is

$$\delta_\Lambda \boldsymbol{\omega} = d\,\Lambda + [\boldsymbol{\omega}, \Lambda], \qquad \Lambda = \Lambda^{ab} \Sigma_{ab}, \tag{11.59}$$

where $\Sigma_{ab} = \frac{1}{4}[\gamma_a, \gamma_b]$ are the generators of the Lorentz Lie algebra and Λ is the ghost field. The diffeomorphism cocycles are expressed in terms of the connection

$$\Gamma \equiv \{\Gamma_\nu{}^\lambda\}, \quad \Gamma_\nu{}^\lambda = \mathrm{d}x^\mu \, \Gamma_{\mu\nu}{}^\lambda \tag{11.60}$$

and the ghost ξ^μ. We refer to Sect. 5.2.2 for other relevant formulas.

Let us consider next the geometry where these objects belong. Given a principal bundle $\mathsf{P}(\mathsf{M}, \mathsf{G})$, we have already noticed that the projection of an automorphism in P is a diffeomorphism in M; therefore, we have the exact sequence

$$id \longrightarrow \mathsf{Aut_v(P)} \longrightarrow \mathsf{Aut(P)} \xrightarrow{j} \mathsf{Diff(M)} \tag{11.61}$$

where j is the homomorphism defined by $j(\psi)(x) = \pi(\psi(p))$. We recall that the Lie algebra of $\mathsf{Diff(M)}$, $\mathsf{diff(M)}$, is the Lie algebra of vector fields on M (with the sign of vector field Lie bracket reversed). In order to be able to exploit the fiber bundle formalism for gravitational anomalies, we must embed diffeomorphisms in $\mathsf{Aut(P)}$, i.e. lift $\mathsf{Diff(M)}$ to $\mathsf{Aut(P)}$. To do so, we must invert the map j, that is define a map l such that $j \circ l = id$. This is possible, for instance, when there exists a global section $\sigma : \pi \circ \sigma(x) = x$, i.e. if P is a trivial bundle; in particular when $\mathsf{P} = \mathsf{M} \times \mathsf{G}$, in which case we can trivially set for any diffeomorphism φ:

$$l(\varphi)(x, g) = (\varphi(x), g), \quad \forall x \in \mathsf{M} \quad \text{and} \quad g \in \mathsf{G}$$

Another remarkable case is when P is the frame bundle LM, i.e the principal bundle of linear frames whose structure group is the general linear group $\mathsf{GL(d}, \mathbb{R})$. A frame u is a set of d linearly independent tangent vectors $u = (X_1, \dots, X_{\mathsf{d}}) \in T_x\mathsf{M}$, associated to any point $x \in \mathsf{M}$. In this case, for a diffeomorphism φ of M, we set:

$$l(\varphi)(p) = (\varphi(x); \varphi_* X_1, \dots, \varphi_* X_{\mathsf{d}}), \quad \forall p = (x; X_1, \dots X_{\mathsf{d}})$$

where $\varphi_* X_i$ denotes the push-forward of the vector X_i at x. The automorphism $l(\varphi)$ is called the *natural lift* of φ and $\mathsf{Aut(P)}$ can be thought of as the semi-direct product of $\mathsf{Aut_v(P)}$ and $\mathsf{Diff(M)}$.

A linear frame bundle is characterized by a mathematical object which is not present in a generic principal bundle: the canonical form $\boldsymbol{\theta}$. It is an $\mathbb{R}^{\mathsf{d}}$-valued form in LM such that $\boldsymbol{\theta}(Y) = u^{-1}(\pi(Y))$ for any $Y \in T_u(\mathsf{LM})$, where u is interpreted here as the map that associates to a vector in $\mathbb{R}^n$ a corresponding frame in $T_{\pi(u)}\mathsf{M}$.

The Lie algebra $\mathsf{diff(M)}$ can be thought of as a subalgebra of $\mathsf{aut(P)}$. For any $X \in \mathsf{Diff(M)}$, the natural lift $l(X)$ is the unique vector field in P with the following properties:

- it is invariant under the right multiplication $\mathsf{P} \times \mathsf{G} \to \mathsf{P}$;
- it leaves the canonical form $\boldsymbol{\theta}$ invariant: $L_{l(X)}\boldsymbol{\theta} = 0$, where L is the Lie derivative;
- $\pi_* l(X) = X$.

We would like now to describe the previous concept with the language of field theory, that is in local coordinates. We first recall that $\boldsymbol{\theta}$ maps vector fields to $\mathbb{R}^{\mathsf{d}}$; thus, we introduce its components $\boldsymbol{\theta}^\mu$, ($\mu = 1, \dots, \mathsf{d}$ in the Euclidean case, $\mu = 0, \dots, \mathsf{d} - 1$ in the Minkowski case). We also introduce the components ϖ^λ_ρ of the connection ϖ and write

$$\mathbf{d} = \boldsymbol{\theta}^\mu \partial_\mu, \quad \varpi^\lambda_\rho = \boldsymbol{\theta}^\mu \Gamma^\lambda_{\mu\rho} \tag{11.62}$$

They form a basis of differential forms in P . We introduce a dual basis of vector fields formed by P_μ and Q^λ_ρ. The former has components only in the horizontal direction and the latter only in the vertical direction. The duality relation $\langle \cdot, \cdot \rangle$ between forms and vector fields are

$$\langle \boldsymbol{\theta}^\mu, P_\nu \rangle = \delta^\mu_\nu, \quad \langle \boldsymbol{\theta}^\mu, Q^\lambda_\rho \rangle = 0, \quad \langle \varpi^\lambda_\rho, P_\mu \rangle = \Gamma^\lambda_{\mu\rho}, \quad \langle \varpi^\lambda_\rho, Q^\sigma_\tau \rangle = \delta^\lambda_\tau \delta^\sigma_\rho$$

Any vector field Z in P can be decomposed according to this basis of vector fields

$$Z = \xi^\mu P_\mu + Y^\rho_\lambda\, Q^\lambda_\rho, \quad \xi^\mu = \langle \theta^\mu, Z\rangle = i_Z\theta^\mu, \quad \langle \varpi^\rho_\lambda, Z\rangle = i_Z\varpi^\rho_\lambda = Y^\rho_\lambda + \xi^\mu \Gamma^\rho_{\mu\lambda} \tag{11.63}$$

Since $\boldsymbol{\theta}$ is invariant under the natural lift of a vector field in M, if $Z = l(X)$ is such a natural lift, we have

$$0 = L_Z\theta^\mu = (\mathbf{d}i_Z + i_Z\mathbf{d})\theta^\mu = i_Z\Theta^\mu + D(i_Z\theta^\mu) - (i_Z\varpi^\mu{}_\nu)\,\theta^\nu$$

where $\boldsymbol{\Theta} = d\boldsymbol{\theta} + \varpi \wedge \boldsymbol{\theta}$ is the torsion. If the torsion vanishes we have

$$(i_Z\varpi^\mu_\nu)\,\theta^\nu = D\xi^\mu, \quad \text{i.e.} \quad Y^\mu_\nu + \xi^\lambda\Gamma^\mu_{\lambda\nu} = \langle P_\nu, D\xi^\mu\rangle = \partial_\nu\xi^\mu + \Gamma^\mu_{\nu\lambda}\xi^\lambda, \tag{11.64}$$

which means

$$Y^\mu_\nu = \partial_\nu \xi^\mu \tag{11.65}$$

Y^μ_ν are the vertical components of the lift $l(X)$, while its horizontal components are X^μ.

11.6.1 Diffeomorphism Anomalies

Diffeormorphism anomalies in the fiber bundle framework can be described by adapting the previous geometrical methods to the case in which $\mathcal{G}$ is the lift of diffeomorphisms, i.e. the semi-direct product of $\mathsf{Diff(M)}$ and $\mathsf{Aut_v(P)}$. The diffeomorphism anomalies are determined by

$$i_{(\cdot)}TP_n(\psi^*\Gamma) \tag{11.66}$$

or

$$j_{(\cdot)}\mathcal{T}P_n(\psi^*\Gamma, \Gamma_0) \tag{11.67}$$

where Γ_0 is a linear background connection, at $\psi = id$. In order to make a link with the field theory language let us consider a local chart U in M and a local section $\sigma_U \equiv \sigma$, and let us pull-back (11.67) by it, in which case we can always choose $\sigma^*\Gamma_0 = 0$[2]. Then, the anomaly takes the simplified form (11.32):

$$\sigma^* i_Z TP_n(\Gamma). \tag{11.68}$$

[2] This is the case globally when LM is trivial, so that a global cross section exists.

Now comes an important remark. Only the vertical part of Z contributes to (11.68), for if Z_h is its horizontal component,

$$\sigma^* i_{Z_h} TP_n(\Gamma) = i_{\pi_* Z}\, \sigma^* TP_n(\Gamma) = 0 \tag{11.69}$$

because $\sigma^* TP_n(\Gamma) = 0$ for dimensional reasons. Now, for simplicity, we drop σ^* and replace $\sigma^*\Gamma$ with $\Gamma^\lambda_\rho = \theta^\mu \Gamma^\lambda_{\mu\rho}$ and $i_{Z_v}\Gamma^\lambda_\rho$ with $Y^\lambda_\rho = \partial_\rho \xi^\lambda \equiv \Xi^\lambda_\rho$. Moreover, we factor out $\sigma^*\theta^{\mu_1} \wedge \ldots \wedge \sigma^*\theta^{\mu_{\mathsf{d}}} = \mathfrak{v}\, \mathrm{d}x^{\mu_1} \wedge \ldots \wedge \mathrm{d}x^{\mu_{\mathsf{d}}}$. In the metric case $\mathfrak{v}$ becomes $\sqrt{g}$. Finally, we recover the expression (5.61) and (5.57) of the diffeomorphism anomaly.

The relation between universality and locality can be analyzed in a way similar to gauge anomalies. One has to resort to the universal frame bundle $\mathsf{EGL(d}, \mathbb{R})$ with structure group $\mathsf{GL(d}, \mathbb{R})$ over the classifying space $\mathsf{BGL(d}, \mathbb{R})$.

$$\mathsf{LM} \times \mathcal{G}^{gl} \xrightarrow{ev} \mathsf{LM} \xrightarrow{\hat{\mathsf{f}}} \mathsf{EGL(d}, \mathbb{R}) \tag{11.70}$$

where $\mathcal{G}^{gl}$ is the Lie group generated by the lift of vector fields of M, i.e. the semi-direct product of $\mathsf{Diff(M)}$ and $\mathsf{Aut_v(LM)}$, and $(\mathsf{f}, \hat{\mathsf{f}})$ is the bundle map between LM and $\mathsf{EGL(d}, \mathbb{R})$, such that $\mathsf{M} = \mathsf{f}^{-1}(\mathsf{BGL(d}, \mathbb{R}))$. Now, as above, starting with local expressions of $\mathfrak{a}^{gl}$, the universal connection, its exterior differential and its commutators in $\mathsf{EGL(d}, \mathbb{R})$, and pulling them back via $ev \circ \hat{\mathsf{f}}$ one generates universal expressions in $\mathsf{LM} \times \mathcal{G}^{gl}$. Diffeomorphism anomalies in LM are generated by $\mathsf{d}+1$ forms of this type that are closed but not exact in $\mathsf{EGL(d}, \mathbb{R})$. The latter originate from irreducible or reducible ad-invariant polynomials $P_n(\mathfrak{F}^{gl}, \ldots, \mathfrak{F}^{gl})$ of the Lie algebra $\mathfrak{gl}(d)$, with $n = \frac{\mathsf{d}}{2}+1$, where $\mathfrak{F}^{gl}$ is the curvature of $\mathfrak{a}^{gl}$. As we know, every such polynomial expression is closed in $\mathsf{BGL(d}, \mathbb{R})$, but not exact, because the Weil homomorphism is an isomorphism in the universal bundle. Therefore, pulling back $TP_n(\mathfrak{a}^{gl})$ to LM, and applying the operation $i_{(\cdot)}$ to it we get all the possible local anomalies of diffeomorphisms: $i_{(\cdot)}TP_n(\Gamma)$. In a similar way, we can derive diffeomorphism anomalies with a background connection.

For later use we notice that, since the group $\mathsf{GL(d}, \mathbb{R})$ is non-compact and its maximal compact subgroup is $\mathsf{O(d}, \mathbb{R})$, the ad-invariant polynomials are determined by the latter. In the case of an oriented manifold the relevant groups are $\mathsf{GL}^+\mathsf{(d}, \mathbb{R})$, the symmetry group of oriented frames, and $\mathsf{SO(d}, \mathbb{R})$, respectively. But the ad-invariant polynomials are the same.

11.6.2 Orthonormal Frame Bundles and Lorentz Anomalies

The spacetime manifold M is supposed to be Riemannian, i.e. is endowed with a Riemannian metric g. In this case one can define over M a principal fiber bundle $\mathsf{O(M)}$ of orthonormal frames. Its structure group is the orthogonal group $\mathsf{O(d}, \mathbb{R})$, which is the maximal compact subgroup of $\mathsf{GL(d}, \mathbb{R})$. $\mathsf{O(M)}$ is a reduced bundle of

LM, i.e. there exist a bundle morphism which maps M to M, maps fiber into fiber, and is a monomorphism of $\mathsf{O}(\mathsf{d}, \mathbb{R})$ into $\mathsf{GL}(\mathsf{d}, \mathbb{R})$. Any distinct reduction of LM into $\mathsf{O}(\mathsf{M})$ defines a metric and vice versa. When we wish to specify this fact we will denote the latter by $\mathsf{O}_g(\mathsf{M})$. Every connection in $\mathsf{O}(\mathsf{M})$ defines a linear connection in LM, which is then called *metric connection*. Vice versa, a linear connection Γ of LM with a metric g is a metric connection if g is parallel with respect Γ. Every Riemannian manifold admits a metric connection with vanishing torsion. Such a connection will be referred to henceforth as Levi-Civita (LC) connection.

In the sequel we will consider orientable Riemannian manifolds M, so that the appropriate linear frame bundle is LM^+ with structure group $\mathsf{G}(\mathsf{d}, \mathbb{R})^+$, the subgroup of $\mathsf{G}(\mathsf{d}, \mathbb{R})$ with positive determinant, and the orthonormal frame bundle has structure $\mathsf{SO}(\mathsf{d}, \mathbb{R})$, rather than $\mathsf{O}(\mathsf{d}, \mathbb{R})$. When fermions are involved we will need to consider a spin bundle SpinM with structure group $\mathsf{Spin}(\mathsf{d})$, a connected double covering of $\mathsf{SO}(\mathsf{d}, \mathbb{R})$. For simplicity, for the time being, we refer to $\mathsf{O}(\mathsf{M})$ as the orthonormal frame bundle with structure group $\mathsf{SO}(\mathsf{d}, \mathbb{R})$. In this bundle the group $\mathcal{G}$ of gauge transformation are the vertical automorphisms $\mathsf{Aut_V O}(\mathsf{M})$, which in field theory are called local Lorentz transformations. Therefore the analysis of anomalies is the same as in the generic gauge case. Given an ad-invariant polynomial $P_n(\cdot, \dots, \cdot)$ in the Lie algebra $\mathfrak{so}(\mathsf{d})$, with $n = \frac{\mathsf{d}}{2} + 1$, we construct the transgression $TP_n(\omega)$ where ω is the spin connection (11.58), whose transformation law is given by (11.59).

Then anomalies are given by

$$i_{(\cdot)} TP_n(\omega) \tag{11.71}$$

or

$$j_{(\cdot)} \mathcal{T} P_n(\omega, \omega_0) \tag{11.72}$$

when a background connection ω_0 is needed. In particular, the explicit Minkowski form of (11.71) is given in Eq. (5.51), after an inverse Wick rotation. An explicit expression for (11.72) is given by Eq. (11.37) with the obvious replacements.

The relation between universality and locality can be analyzed in the same way as for gauge anomalies. One starts from the universal frame bundle $\mathsf{ESO}(\mathsf{d}, \mathbb{R})$ with structure group $\mathsf{SO}(\mathsf{d}, \mathbb{R})$ over the classifying space $\mathsf{BSO}(\mathsf{d}, \mathbb{R})$ and define the evaluation map

$$\mathsf{OM} \times \mathcal{G}^{so} \xrightarrow{ev} \mathsf{OM} \xrightarrow{\hat{f}} \mathsf{ESO}(\mathsf{d}, \mathbb{R}), \tag{11.73}$$

where $\mathcal{G}^{so}$ is the just mentioned Lie group of vertical automorphisms, and $(f, \hat{f})$ is the bundle map between OM and $\mathsf{ESO}(\mathsf{d}, \mathbb{R})$. Now, as above, starting with local expressions of the universal connection, $\mathfrak{a}^{so}$, its exterior differential and its commutators, and pulling them back via $ev \circ \hat{f}$ one generates universal expressions in $\mathsf{OM} \times \mathcal{G}^{so}$. Anomalies in OM are generated by $\mathsf{d} + 1$ forms of this type that are closed but not exact in $\mathsf{ESO}(\mathsf{d}, \mathbb{R})$. They originate from irreducible or reducible ad-invariant symmetric polynomials $P_n(\mathfrak{F}^{so}, \ldots, \mathfrak{F}^{so})$ of the Lie algebra $\mathfrak{so}(d)$, with $n = \frac{\mathsf{d}}{2} + 1$, where $\mathfrak{F}^{so}$ is the curvature of $\mathfrak{a}^{so}$. Every such reducible or irreducible polynomial expression, is closed in $\mathsf{BSO}(\mathsf{d}, \mathbb{R})$, but not exact. Therefore, pulling back $TP_n(\mathfrak{a}^{so})$ to OM, and applying the operation $i_{(\cdot)}$ to it we get all the local anomalies in the form $i_{(\cdot)}TP_n(\omega)$. In a similar way we can derive local Lorentz anomalies with a background connection.

Here we have obtained the Lorentz anomalies, while just above we got the diffeomorphism anomalies. They are connected to one another in a one-to-one way. The reason is that they originate from expressions in the relevant universal bundles of the form $\mathrm{Tr}\big((\mathfrak{F}^{so})^n\big)$ and $\mathrm{Tr}\big((\mathfrak{F}^{gl})^n\big)$, respectively. The relation between $\mathfrak{F}^{gl}$ and $\mathfrak{F}^{so}$ is the same as the relation between the Riemann curvature two-form R and the spin curvature two-form $\mathcal{R}$, see Eq. (2.40).[3] Therefore we can make the identification

$$\mathrm{Tr}\big((\mathfrak{F}^{gl})^n\big) = \mathrm{Tr}\big((\mathfrak{F}^{so})^n\big) \tag{11.74}$$

This can be justified also as follows. If H is a closed subgroup of a Lie group G, then one can take $\mathsf{EG} = \mathsf{EH}$ and $\mathsf{BH} = \mathsf{EG} \times_{\mathsf{G}} (\mathsf{G}/\mathsf{H})$. Applying this to $\mathsf{GL}(\mathsf{d}, \mathbb{R})$ and $\mathsf{O}(\mathsf{d})$, or to $\mathsf{GL}^+(\mathsf{d}, \mathbb{R})$ and $\mathsf{SO}(\mathsf{d})$, we see that the identification (11.74) is justified.

In a more general way one can notice that $\mathsf{BO}(\mathsf{n})$ is homotopically equivalent to $\mathsf{BGL}(\mathsf{n})$, and $\mathsf{BSO}(\mathsf{n})$ to $\mathsf{BGL}^+(\mathsf{n})$. Moreover, as remarked above, the ad-invariant polynomials are the same in the two cases. Therefore Lorentz and diffeomorphism anomalies originate from identical closed forms in their classifying spaces. They give rise to two related transgression formulas, which are then pulled back via the evaluation maps. This is summarized in the diagram

$$\mathsf{LM} \times \mathcal{G}^{gl} \xrightarrow{ev} \mathsf{LM} \xrightarrow{\hat{\mathsf{f}}} \mathsf{EGL}(\mathsf{d}, \mathbb{R}) \hookleftarrow \mathsf{ESO}(\mathsf{d}, \mathbb{R}) \xleftarrow{\hat{f}} \mathsf{OM} \xleftarrow{ev} \mathsf{OM} \times \mathcal{G}^{so} \tag{11.75}$$

These operations do not modify the form and coefficient of the relevant polynomials. Thus we can conclude that *Lorentz and diffeomorphism anomalies are uniquely related.* A local field theory argument for this identification, by means of a Wess-Zumino term, is given below in Sect. 15.2.

11.6.3 Mixed Anomalies

Consider two distinct principal fiber bundles $\mathsf{P}(\mathsf{M}, \mathsf{G})$ and $\mathsf{Q}(\mathsf{M}, \mathsf{H})$ over M, with structure groups G and H, respectively. Then we can define a principal fiber bundle $\mathsf{P} \times \mathsf{Q}$ over $\mathsf{M} \times \mathsf{M}$, with group $\mathsf{G} \times \mathsf{H}$ (direct product of groups). We can restrict our consideration to the diagonal $\Delta\mathsf{M}$ of $\mathsf{M} \times \mathsf{M}$. We obtain in this way a principal

[3] A prerequisite for this argument is that BG can be given a Riemannian structure. Now, every classifying space like BG can be given a CW complex structure; every CW complex is paracompact; every paracompact space admit a Riemannian metric. Therefore any BG can be given a Riemannian structure.

fiber bundle, which we denote $\mathsf{P}+\mathsf{Q}$ with structure group $\mathsf{G}\times\mathsf{H}$. Let us denote by π_{P} the restriction of the projection of $\mathsf{P}+\mathsf{Q}$ to P, and π_{Q} the analogous restriction to Q. Given two connections $\mathbf{A}_{\mathsf{P}}$ in P with curvature $\mathbf{F}_{\mathsf{P}}$ and $\mathbf{A}_{\mathsf{Q}}$ in Q with curvature $\mathbf{F}_{\mathsf{Q}}$, there is a unique connection $\mathbf{A}$ in $\mathsf{P}+\mathsf{Q}$, with curvature $\mathbf{F}$, such that

$$\mathbf{A}=\pi_{\mathsf{P}}^{*}\mathbf{A}_{\mathsf{P}}+\pi_{\mathsf{Q}}^{*}\mathbf{A}_{\mathsf{Q}},\qquad \mathbf{F}=\pi_{\mathsf{P}}^{*}F_{\mathsf{Q}}+\pi_{\mathsf{Q}}^{*}\mathbf{F}_{\mathsf{Q}} \tag{11.76}$$

We can also adopt a simplified notation: $\mathbf{A}$ and $\mathbf{F}$ can be thought of as a connection and curvature on a principal fiber bundle over M with values in the direct sum $\mathfrak{g}+\mathfrak{h}$, the Lie algebra of $\mathsf{G}\times\mathsf{H}$, and write

$$\mathbf{A}=\mathbf{A}^{\mathfrak{g}}+\mathbf{A}^{\mathfrak{h}},\quad \mathbf{F}=\mathbf{F}^{\mathfrak{g}}+\mathbf{F}^{\mathfrak{h}}$$

The same can be done also for the corresponding universal fiber bundles. So the local field theory anomalies will originate from the ad-invariant polynomials of the Lie algebra $\mathfrak{g}+\mathfrak{h}$. These, in turn, are decomposed in products of reducible or irreducible polynomials of $\mathfrak{g}$ or $\mathfrak{h}$. For instance, if $\mathfrak{a}^{\mathfrak{g}}$ and $\mathfrak{a}^{\mathfrak{h}}$ are the universal connections with curvature $\mathfrak{F}^{\mathfrak{g}}$ and $\mathfrak{F}^{\mathfrak{h}}$, respectively, a typical example is $P_n(\mathfrak{F},\ldots,\mathfrak{F})=P_k(\mathfrak{F}^{\mathfrak{g}},\ldots,\mathfrak{F}^{\mathfrak{g}})P_{n-k}(\mathfrak{F}^{\mathfrak{h}},\ldots,\mathfrak{F}^{\mathfrak{h}})$. Then the relevant transgression formula in the total space can be either $TP_k(\mathfrak{a}^{\mathfrak{g}})P_{n-k}(\mathfrak{F}^{\mathfrak{h}},\ldots,\mathfrak{F}^{\mathfrak{h}})$ or $TP_{n-k}(\mathfrak{a}^{\mathfrak{h}})P_k(\mathfrak{F}^{\mathfrak{g}},\ldots,\mathfrak{F}^{\mathfrak{g}})$ or a combination thereof.

Now, pulling back the universal transgression formulas to $\mathsf{P}+\mathsf{Q}$, we get, for instance, $TP_k(\mathbf{A}^{\mathfrak{g}})P_{n-k}(\mathbf{F}^{\mathfrak{h}},\ldots,\mathbf{F}^{\mathfrak{h}})$ or $TP_{n-k}(\mathbf{A}^{\mathfrak{h}})P_k(\mathbf{F}^{\mathfrak{g}},\ldots,\mathbf{F}^{\mathfrak{g}})$. From the former we obtain the cocycle

$$\Delta_d^{(1)}(\mathbf{A}^{\mathfrak{g}},c^{\mathfrak{g}},\mathbf{F}^{\mathfrak{h}})=\Delta_{2k-2}^{(1)}(\mathbf{A}^{\mathfrak{g}},c^{\mathfrak{g}})P_{n-k}(\mathbf{F}^{\mathfrak{h}},\ldots,\mathbf{F}^{\mathfrak{h}}) \tag{11.77}$$

and from the latter

$$\Delta_d^{(1)}(\mathbf{A}^{\mathfrak{h}},c^{\mathfrak{h}},\mathbf{F}^{\mathfrak{g}})=\Delta_{2n-2k-2}^{(1)}(\mathbf{A}^{\mathfrak{h}},c^{\mathfrak{h}})P_k(\mathbf{F}^{\mathfrak{g}},\ldots,\mathbf{F}^{\mathfrak{g}}) \tag{11.78}$$

where $c^{\mathfrak{g}}, c^{\mathfrak{h}}$ are the ghosts valued in $\mathfrak{g}$, $\mathfrak{h}$, respectively. These are example of mixed $\mathfrak{g}$-$\mathfrak{h}$ anomalies. However, they are not independent. For let us consider the following chain: $TP_k(\mathbf{A}^{\mathfrak{g}})TP_{n-k}(\mathbf{A}^{\mathfrak{h}})$. Its gauge or BRST variation is easily computed

$$\begin{aligned}&\mathfrak{s}\left(TP_k(\mathbf{A}^{\mathfrak{g}})TP_{n-k}(\mathbf{A}^{\mathfrak{h}})\right)\\&=\Delta_{2k-2}^{(1)}(\mathbf{A}^{\mathfrak{g}},c^{\mathfrak{g}})P_{n-k}(\mathbf{F}^{\mathfrak{h}},\ldots,\mathbf{F}^{\mathfrak{h}})-\Delta_{2n-2k-2}^{(1)}(\mathbf{A}^{\mathfrak{h}},c^{\mathfrak{h}})P_k(\mathbf{F}^{\mathfrak{g}},\ldots,\mathbf{F}^{\mathfrak{g}})\\&-d\Big(\Delta_{2k-2}^{(1)}(\mathbf{A}^{\mathfrak{g}},c^{\mathfrak{g}})TP_{n-k}(\mathbf{A}^{\mathfrak{h}})-TP_k(\mathbf{A}^{\mathfrak{g}})\Delta_{2n-2k-2}^{(1)}(\mathbf{A}^{\mathfrak{h}},c^{\mathfrak{h}})\Big)\end{aligned} \tag{11.79}$$

Integrating over M, we see that the difference between(11.77) and (11.78) is a coboundary; i.e. they are representatives of the same cohomology class. From the field theory point of view, this means that we can get rid of either (11.77) or (11.78) by subtracting from or adding to the effective action the term $\int_{\mathsf{M}} TP_k(\mathbf{A}^{\mathfrak{g}})TP_{n-k}(\mathbf{A}^{\mathfrak{h}})$.

All the above can be repeated in the presence of a background connection.

U(1) Anomalies

So far we have considered simple gauge groups G or H. The extension to semisimple groups is straightforward. Let us consider the U(1) case, the corresponding complex valued connection **A** and curvature **F**. The BRST transformation is: $\mathfrak{s}\mathbf{A} = dc$, $\mathfrak{s}\mathbf{F} = 0$. The ad-invariant polynomials are replaced in this case by (exterior) products of **F**. The anomaly analysis is a simple subcase of the previous discussion. The U(1) non-trivial cocycles have the form

$$\Delta_d^{(1)}(\mathbf{A}, c) = dc\,\mathbf{A}\,\mathbf{F}^{n-2} \tag{11.80}$$

A typical and important case in field theory is when we have the product of U(1) times a simple group G. The non-trivial mixed cocycles take the form

$$\Delta_d^{(1)}(\mathbf{A}, c, \mathbf{F}^{\mathfrak{g}}) = dc\,\mathbf{A}\,\mathbf{F}^{k-2}P_{n-k}(\mathbf{F}^{\mathfrak{g}}, \dots, \mathbf{F}^{\mathfrak{g}}) \tag{11.81}$$

This cocycle, for the reason previously explained, belongs to the same cohomology class (in the mixed U(1) × G BRST cohomology) as

$$\Delta_d^{(1)}(\mathbf{A}^{\mathfrak{g}}, c^{\mathfrak{g}}, \mathbf{F}) = \Delta_{2n-2k-1}^{(1)}(\mathbf{A}^{\mathfrak{g}}, c^{\mathfrak{g}})\,\mathbf{F}^k \tag{11.82}$$

An example of this kind of anomalies has been calculated in Sect. 7.4.4.

Appendix 11A. Comment on Covariant and Consistent Anomalies

The consistent anomaly (5.40) is the RHS of a Ward identity which is the covariant divergence of a, say left-handed, current $J_\mu^L(x) = J_\mu^{La}(x)T^a$, i.e. $D^\mu J_\mu^L = \partial^\mu J_\mu^L + [A^\mu, J_\mu^L]$. The Ward identity, written in integrated form, is

$$\int_x c^a (D^\mu J_\mu^L)^a$$

$$= -\int_x \mathrm{tr}(J_\mu^L D^\mu c) = \alpha_n n(n-1)\int_x\int_0^1 \mathrm{d}t(t-1)P_n(dc, A, F_t, \dots F_t) \tag{11.83}$$

where α_n is a suitable numerical coefficient $\sim \frac{1}{(2\pi)^{n-1}}$. There exists also a covariant anomaly whose WI for a chiral current $J_\mu(x) = J_\mu^a(x)T^a$ can be written as

$$\int_x \mathrm{tr}(cD^\mu J_\mu) = -\int_x \mathrm{tr}(J_\mu D^\mu c) = \beta_n \int_x P_n(dc, F, F, \dots F) \qquad (11.84)$$

where β_n is another numerical coefficient $\sim \frac{1}{(2\pi)^{n-1}}$. There is an algebraic relation between the two formulas. Let us rewrite the integral in the RHS of (11.83) as follows

$$\int_x \int_0^1 \mathrm{d}t(1-t)P_n(dc, A, F_t, \dots F_t) \qquad (11.85)$$

$$= \int_x \int_0^1 \mathrm{d}t\, P_n(dc + [tA, c], A, F_t, \dots F_t) - \int_x \int_0^1 \mathrm{d}t\, t\, P_n(dc + [A, c], A, F_t, \dots F_t)$$

The first term on the RHS leads to

$$(n-1)\int_x \int_0^1 \mathrm{d}t\, P_n(dc + [tA, c], A, F_t, \dots F_t) = (n-1)\int_x \int_0^1 \mathrm{d}t\, P_n(d_{tA}c, A, F_t, \dots F_t)$$

$$= (n-1)\int_x \int_0^1 \mathrm{d}t\, P_n(c, A, F_t, \dots F_t) - (n-1)\int_x \int_0^1 \mathrm{d}t\, P_n(c, \frac{\mathrm{d}F_t}{\mathrm{d}t}, F_t, \dots F_t)$$

$$= -\int_x \int_0^1 \frac{\mathrm{d}}{\mathrm{d}t} P_n(c, F_t, F_t, \dots F_t) = -\int_x P_n(c, F, \dots, F) \qquad (11.86)$$

which is proportional to the covariant anomaly. On the other hand, the second piece in the RHS of (11.85) is

$$-\int_x \int_0^1 \mathrm{d}t\, t\, P_n(dc + [A, c], A, F_t, \dots F_t) \qquad (11.87)$$

$$= -\int_x \int_0^1 \mathrm{d}t\, t (Dc)^a P_n(T^a, A, F_t, \dots F_t) = \int_x \int_0^1 \mathrm{d}t\, t\, c^a D P_n(T^a, A, F_t, \dots F_t)$$

Therefore if we look at the LHS of (11.83) and compare it with the RHS of (11.87) we see that we can redefine J^L in the following way

$$\tilde{J}^a_\mu = J^{La}_\mu - B^a_\mu \qquad (11.88)$$

where

$$B^a_\mu = (n-1)\int_0^1 t\,\mathrm{d}t\, D_\mu P_n(T^a, A, F_t, \dots F_t) \tag{11.89}$$

then $\tilde{J}^a_\mu$ has the same WI as J^a_μ, (11.84), with coefficient $\beta_n = n\alpha_n$. However, the ratio of coefficients of the consistent and covariant anomaly do not correspond to the computed ones: for instance, in 4d the covariant anomaly has coefficient $\frac{1}{16\pi^2}$ and the consistent one has coefficient $\frac{1}{24\pi^2}$, while here their ratio is 1. Moreover, consistent anomalies have opposite sign for opposite chiralities, while for Dirac fermion the covariant anomaly has a definite sign. In fact the relation (11.88) is not surprising: while manipulating the descent equations, covariant anomalies pop up at various stages, but the above relation between consistent and covariant anomalies is purely formal and does not clarify the nature of either.

Covariant anomalies have remarkable field theory properties. There are in fact several non-renormalization theorem concerning them. On the other hand covariant anomalies are the relevant ones in relation to the fixed background Atiyah-Singer index theorem, see Sect. 12.11. Understandably, there has been in the literature a lot of emphasis on covariant anomalies and on how to derive them. We think the most effective link between the two species has been illustrated in several examples above. For instance, the covariant chiral anomaly for Dirac fermions coupled to a vector potential V_μ is obtained as follows: couple the theory to both the vector potential and an axial potential A_μ and derive an anomaly in terms of both potentials; this anomaly satisfies the appropriate consistency condition; then set $A_\mu = 0$; what survives is called covariant anomaly. This scheme applies to all covariant anomalies, and, as we see, it is obtained via consistency. It accounts not only for the forms of the two anomalies but also for their coefficients and signs.

References

1. W.A. Bardeen, B. Zumino, Consistent and covariant anomalies in gauge and gravitational theories. Nucl. Phys. B **244**, 421 (1984)
2. P. Mitter, C. Viallet, On the bundle of connections and the gauge orbits manifolds in Yang-Mills theory. Commun. Math. Phys. **79**, 457 (1981)
3. L. Bonora, P. Cotta-Ramusino, Some remarks on BRS transformations, anomalies and the cohomology of the Lie algebra of the group of gauge transformations. Commun. Math. Phys. **87**, 589 (1982)
4. L. Bonora, P. Cotta-Ramusino, M. Rinaldi, J. Stasheff, The evaluation map in field theory, sigma-models and strings. I. Commun. Math. Phys. **112**, 237 (1987)
5. R. Stora, Algebraic structure and topological origin of anomalies, in *Progress in Gauge Field Theory*, NATO ASI, Series B, Vol. 115, ed. by G. 't Hooft, A. Jaffe, G. Lehmann, P. K. Mitter, I. M. Singer (Plenum Press, 1984)
6. A. Trautman, Geometrical aspects of gauge configurations. Acta Phys. Austriaca[Supp] **XXIII**, 401 (1981)

7. A. Trautman, Einstein-Cartan theory, in *Encyclopedia of Mathematical Physics* ed. by J.-P. Francoise, G. L. Naber, S. T. Tsou (Oxford, 2006). ArXiv: gr-qc/0606062
8. D. Husemoller, *Fiber Bundles* (Springer, New York-Heidelberg-Berlin, 1966)
9. S.S. Chern, *Complex Manifolds without Potential Theory* (Springer, Berlin, 1969)
10. S. Kobayashi, K. Nomizu, *Foundations of Differential Geometry*, Vol. I–II (Interscience Pub., New York, 1969)
11. W. Greub, S. Halperin, R. Vanstone, *Connections, Curvature and Cohomology*, vol. II (Academic Press, New York, London, 1973)
12. M. Narasimhan, S. Ramanan, Existence of universal connections. Am. J. Math. **83**, 563 (1961)
13. R. Schlafly, Universal connections. Inv. Math. **59**, 59 (1980)

Chapter 12
Anomalies as Obstructions: The Atiyah-Singer Family's Index Theorem

The family's index theorem is the relevant index theorem in relation to consistent chiral anomalies. In general, an (ordinary) index theorem counts the difference between the graded zero modes of an elliptic operator, zero modes being labeled by 0 and 1, or by + and −, like in the chiral case. In the latter case, it counts the difference between the zero modes of opposite chiralities of a Dirac operator. The family's index theorem does the same, except that the operator depends on a continuous parameter and, in the case of interest to us, it varies on a Hausdorff space, the moduli space of connections (or, the space of orbits under the action of the gauge group). The index is formally represented by the space of zero mode eigenvectors (kernel of the operator) of one chirality *minus* the space of zero mode eigenvectors of the opposite chirality (cokernel). The *difference* of two vector spaces varying from point to point of the parameter space is formalized by the K-theory of vector bundles. What really matters to us in relation to the anomaly problem is not as much the number of zero modes, but the existence or non-existence of the inverse of the Dirac operator. This is the same problem we met in the perturbative approach to anomalies, as well as in the non-perturbative approach à la Seeley-Schwinger-DeWitt. There, the inverse of the relevant elliptic operator (or, the existence of an effective action for Weyl fermions) was required, and we were obliged, for instance, to replace the Dirac-Weyl operator

$$\mathcal{D} = i(\partial\!\!\!/ + V\!\!\!\!/)P_+, \qquad P_\pm = \frac{1 \pm \gamma_5}{2} \tag{12.1}$$

by

$$\mathcal{D} = i(\partial\!\!\!/ + V\!\!\!\!/ P_+) \tag{12.2}$$

The original version of the chapter has been revised: Equations have been corrected. A correction to this chapter can be found at https://doi.org/10.1007/978-3-031-21928-3_19

L. Bonora, *Fermions and Anomalies in Quantum Field Theories*, Theoretical and Mathematical Physics,
https://doi.org/10.1007/978-3-031-21928-3_12

in order to guarantee the invertibility of the kinetic operator, i.e. the existence of the fermion propagator. In that approach, anomalies show up as non-conservation of the chiral gauge currents. The family's index theorem, instead, calculates for us the obstructions to the invertibility of the Dirac-Weyl operator. In other words, it tells us what are the topological conditions for the Weyl fermion propagator to exist.

Although the just mentioned obstructions manifest themselves in different ways in the two types of approach, their origin is the same and can be traced to the geometry of the universal bundle and classifying space. The surplus value of the family's index theorem approach is that it furnishes explicit formulas for the obstructions for all dimensions in terms of ad-invariant polynomials, gauge curvatures and/or Riemann curvatures; from which one can derive the anomalies by the transgression formulas and the descent equations. The important feature is that the index theorem also provides the numerical coefficients (depending on the fermion field representations) in front of the anomalies, based on which one can study in general the conditions for their cancelation.

This chapter is organized as follows. The first part is a synthesis of the mathematical material needed to introduce index theorems: elliptic operators, spin bundles, Dirac operators. Then we formulate the ordinary (fixed background) index theorem. Next we explain why in quantum field theory we need the family's index theorem and why in that case we need K-theory. Next we formulate the Atiyah-Singer family's index theorem [1–6]. To present all this basic material, we follow in particular the textbook by H.B. Lawson jr and M.-L. Michelsohn, *Spin geometry* [7]. We then introduce the Quillen determinant bundle [8, 9], which is an intermediate step for the application of the family's index theorem to anomalies. Finally, we apply it to the anomaly problem [10], that is we analyze the obstructions to the existence of the fermion propagators and relate them to the geometry of the classifying space, where they share the same origin as the chiral anomalies studied in the previous chapter.

In the following sections, until further notice, the background metric is understood to be Euclidean.

12.1 Elliptic Operators

Given two fiber bundles E, F over a Riemannian manifold X, we call $\Gamma(E), \Gamma(F)$ their spaces of smooth sections. Let D be a differential operator mapping sections of E to sections of F, $D : \Gamma(E) \to \Gamma(F)$. In local coordinates, D takes the form

$$D = \sum_{|\alpha| \le m} M_\alpha(x) \frac{\partial^{|\alpha|}}{\partial x^\alpha}, \qquad \text{where} \quad \frac{\partial^{|\alpha|}}{\partial x^\alpha} = \frac{\partial^{|\alpha|}}{\partial x^{\alpha_1} \dots \partial x^{\alpha_k}}, \quad |\alpha| = k \tag{12.3}$$

where x^{α_i} are the coordinates of the point $x \in X$ and $M_{\alpha_1 \dots \alpha_i}(x)$ are smooth linear operators mapping $E_x \to F_x$, the local expressions of a bundle map $E \to F$. Then saturating the local expression of D with a cotangent vector $\xi = \xi_k dx^k \in T_x^*(X)$ we obtain the symbol of D: $\sigma_\xi(D) = \sum_{|\alpha|\le m} M_\alpha(x)\xi^\alpha$. The *leading* or *principal symbol* is

$$\sigma_\xi(D) = i^m \sum_{|\alpha|=m} M_\alpha(x)\xi^\alpha \tag{12.4}$$

This is a linear map from fiber to fiber $E_x \to F_x$.

The operator D is *elliptic* if its symbol is an isomorphism $E_x \to F_x$ for any $\xi \neq 0$.

If X is a compact manifold the kernel and cokernel (i.e. $\Gamma(F)/\overline{\text{Im}\, D}$, where $\overline{\text{Im}\, D}$ is the closure of the image of D) of the elliptic operator D are finite dimensional. Therefore, we can define its (analytic) index

$$\text{ind}\, D = \dim(\ker D) - \dim(\text{coker}\, D) \tag{12.5}$$

The properties of an elliptic operator and the stability of its index follow from a series of well-known theorems, valid for a compact Riemannian space X, which we briefly recall:

- an elliptic operator P of order m in a vector bundle E has a Fredholm extension $P : L^2_s(E) \to L^2_{s-m}(E)$, where the latter are Sobolev spaces (a Fredholm operator has just finite dimensional kernel and cokernel), and the index of P equals the index of its extensions;
- a self-adjoint elliptic operator $P : \Gamma(E) \to \Gamma(E)$ has real discrete eigenvalues with finite eigenspaces consisting of smooth sections;
- the index of an elliptic operator depends only on its leading symbol;
- the index of an elliptic operator depends only on its homotopy class (two operators P_0 and P_1 are homotopic if they can be joined by a continuous family of operators $P_t, 0 \le t \le 1$).

The Dirac operator is the central operator in our study of anomalies. In physics, it is usually (but not exclusively) defined on a Minkowski spacetime (a pseudo-Riemannian manifold). After a Wick rotation, i.e. on a Euclidean background, it is an example of linear elliptic operator. In order to be able to properly define a Dirac operator, we need to introduce spinors, spin structures and spinor bundles.

12.2 Spin Structures and Spinor Bundles

To define, Dirac operators the first requirement is that X be a spin manifold. X is spin if it is an oriented Riemannian manifold with a spin structure. A spin structure relative to a vector bundle E over X is defined as follows. Let us define first the principal fiber bundle $\mathsf{P}_{SO}(E)$ of orthonormal frames of E. Then a spin structure on E is a principal fiber bundle $\mathsf{P}_{\text{Spin}}(E)$ together with a two-sheeted covering

$$\xi : \mathsf{P}_{\text{Spin}}(E) \longrightarrow \mathsf{P}_{SO}(E)$$

such that $\xi(p, g) = \xi(p)\xi_0(g)$, for any $p \in \mathsf{P}_{\text{Spin}}(E)$ and any $g \in \text{Spin}_n, n = \dim(E)$, where $\xi_0 : \text{Spin}_n \longrightarrow SO(n)$ is the universal covering map.

A spin structure exists for E provided the second Stiefel-Withney class of E vanishes, $w_2(E) = 0$.

A manifold is *spin* if its tangent bundle admits a spin structure, i.e. iff $w_2(TX) = 0$.

The next step consists in defining spinor bundles, which will take the place of the E and F bundles above. Let $\mathsf{P}(X, \mathsf{G})$ be a principal fiber bundle with structure group G and let us consider a vector space F with an action of G in it given by a representation $\rho : \mathsf{G} \to \mathsf{GL}(F)$, where $\mathsf{GL}(F)$ is the group of invertible linear transformation of F. Then we can define the action of G on the product $\mathsf{P} \times F$ as follows

$$\varphi_g(p, v) = (pg^{-1}, \rho(g)v)$$

The quotient of $\mathsf{P} \times F$ by this action is a fiber bundle, the *associated fiber bundle* to P by ρ, $\mathsf{P} \times_\rho F$, with projection π_ρ induced by $\pi(p, f) = \pi(p) = x$.

Spinor bundles are constructed in an analogous way, but in order for their sections to represent the spinor fields of quantum field theory, it must be that the fibers are representation spaces for both the γ-matrix algebra and the relevant $Spin_n$ group, the connected double covering of $SO(n)$.

To arrive at a spinor bundle definition, we need first the concept of *Clifford bundle*. We have seen in Sect. 1.3 that $Spin_n$, and so also $SO(n)$, acts on the Clifford algebra $C\ell(\mathbb{R}^n)$. A Clifford bundle is a bundle associated to $\mathsf{P}_{SO}(E)$

$$C\ell(E) = \mathsf{P}_{SO}(E) \times_{c\ell_n} C\ell(\mathbb{R}^n) \tag{12.6}$$

where $c\ell_n$ is the representation given by the just mentioned left action of $SO(n)$ on $C\ell(\mathbb{R}^n)$.

We are now ready to define spinor bundles. Let E be a Riemannian vector bundle with a spin structure $\xi : \mathsf{P}_{\text{Spin}}(E) \longrightarrow \mathsf{P}_{SO}(E)$, then a *real spinor bundle* $S(E)$ is the associated bundle

$$S(E) = \mathsf{P}_{\text{Spin}}(E) \times_\mu M \tag{12.7}$$

where M is a left module of $C\ell(\mathbb{R}^n)$ and $\mu : Spin_n \longrightarrow SO(M)$ is the representation defined by left multiplication of elements of $Spin_n \subset C\ell^0(\mathbb{R}^n)$.

We will mostly be interested in complex spinor bundles. A *complex spinor bundle* $S_{\mathbb{C}}(E)$ is an associated bundle

$$S_{\mathbb{C}}(E) = \mathsf{P}_{\text{Spin}}(E) \times_\mu M_{\mathbb{C}} \tag{12.8}$$

where $M_{\mathbb{C}}$ is a left module of $C\ell(\mathbb{R}^n) \otimes \mathbb{C}$.

Now consider a complex spinor bundle $S_{\mathbb{C}}(E)$ and assume $n = 2m$. Consider

$$\omega_{\mathbb{C}} = i^m e_1 \cdot \ldots \cdot e_{2m} \tag{12.9}$$

where $e_1, \ldots, e_{2m}$ is an oriented orthonormal basis of E_x. $\omega_{\mathbb{C}}$ is a global section of the Clifford bundle $C\ell(E)$. It satisfies

$$\omega_{\mathbb{C}}^2 = 1, \quad \omega_{\mathbb{C}}\, e = -e\, \omega_{\mathbb{C}} \tag{12.10}$$

for any $e \in C\ell^1(E) \otimes \mathbb{C}$. Left multiplication of $S_{\mathbb{C}}(E)$ by $\omega_{\mathbb{C}}$ is an operation with ± 1 eigenvalues, which splits it into the corresponding eigenbundles

$$S_{\mathbb{C}}(E) = S_{\mathbb{C}}^{+}(E) \oplus S_{\mathbb{C}}^{-}(E) \tag{12.11}$$

These two eigenbundles can be written as associated bundles

$$S_{\mathbb{C}}^{\pm}(E) \cong P_{\text{Spin}}(E) \times_{\varrho^{\pm}} \mathbb{C}^{2^{m-1}} \tag{12.12}$$

where $\varrho^{\pm}$ denote the fundamental complex representations of $Spin_{2m}$ in the vector space $\mathbb{C}^{2^{m-1}}$.

The meaning of $\omega_{\mathbb{C}}$ is transparent as an alias of the chirality matrix in field theory. The latter is in fact a representative of it.

12.3 Dirac Operators

In this subsection, $\mathsf{P}_{SO}(TX)$ will be simply written $\mathsf{P}_{SO}(X)$ and $C\ell(TX)$ will be denoted $C\ell(X)$. $\mathsf{P}_{SO}(X)$ is the *orthonormal tangent frame bundle* over X, and $C\ell(X)$ the *Clifford bundle* over X.

Let us denote simply by S a spinor bundle associated with $\mathsf{P}_{Spin}(X)$ (as above, i.e. S_x is a left module of the algebra $C\ell(X)_x$) and assume that it has a Riemannian connection with connection form ω and curvature $\Omega = \mathrm{d}\omega + [\omega, \omega]$. We will denote by $\Gamma(S)$ the space of section of S. Given a basis of orthonormal frames $(e_1, \ldots, e_n)$ of TX, i.e. a set of orthonormal sections of $\mathsf{P}_{SO}(X)$, we can define a covariant derivative ∇ satisfying

$$\nabla e_i = \sum_{j=1}^{n} \omega_{ji} \otimes e_j, \quad \omega = -\omega_{ij} e_i \wedge e_j \tag{12.13}$$

and acting on any section $\sigma \in \Gamma(S)$

$$\nabla\sigma = \sum_{j=1}^{n} e_j \cdot \nabla_{e_j}\sigma, \quad \nabla(f\sigma) = \mathrm{d}f\,\sigma + f\nabla\sigma \tag{12.14}$$

where $\cdot$ denotes, as usual, the Clifford product.

There is an embedding $\mathsf{P}_{SO}(X) \subset \mathsf{P}_{SO}(S)$. Let us call $\sigma_1, \ldots, \sigma_n$ the image of $e_1, \ldots, e_n$ under this embedding. They are local sections of $\mathsf{P}_{SO}(S)$. Then

$$\nabla \sigma_\alpha = \frac{1}{2} \sum_{i<j} \omega_{ij} \otimes e_i e_j \cdot \sigma_\alpha \tag{12.15}$$

Moreover, we shall require the existence in $\Gamma(S)$ of a local inner product $\langle \cdot, \cdot \rangle$ with the property

$$\langle e\sigma_1, \sigma_2 \rangle + \langle \sigma_1, e\sigma_2 \rangle = 0 \tag{12.16}$$

for any $e \in T_x(X)$ with $e \cdot e = -1$.

Under these condition we call S a *Dirac bundle* and the covariant derivative $D\sigma = \sum_i e_i \cdot \nabla_{e_i} \sigma$ a *Dirac operator.*

Let us define an inner product in $\Gamma(S)$:

$$(\sigma_2, \sigma_2) = \int_X \langle \sigma_1, \sigma_2 \rangle \tag{12.17}$$

Then the Dirac operator satisfies

$$(D\sigma_1, \sigma_2) = (\sigma_1, D\sigma_2) \tag{12.18}$$

for sections σ_1, σ_2 with compact support. Therefore, in a compact space, D is a self-adjoint operator.

We easily recognize in the above formulas the familiar ones in field theory. A section of S is a spinor field ψ. The Clifford multiplication by e_i is the multiplication of gamma matrices γ_i. The inner product $\langle \sigma_1, \sigma_2 \rangle$ is the field theory bilinear product $\bar{\psi}_1 \psi_2$, and, of course, the Dirac operator D is $\not{D}$, the Euclidean version of

$$\not{D} = i\gamma^\mu (\partial_\mu + \frac{1}{2}\omega_\mu) \tag{12.19}$$

where $\omega_\mu = \omega_\mu^{ab} \Sigma_{ab}$ is the spin connection. The property (12.18) is a well-known property of the Dirac operator in field theory.

At this stage, in order to avoid a too cumbersome notation, we choose to simplify it, considering this simplification preferable to the cost of a possible confusion, which we hope the reader will easily avoid. In this and the following section, and until further notice, we use for a Dirac operator the notation $\not{D}$, even though the metric is Euclidean and the spacetime Riemannian. An important remark to be added is that, while in mathematics spinors are real or complex-valued (c-number) sections, in QFT they are a-numbers; i.e. they take anticommuting values. However, since this fact becomes relevant only when quadratic or higher order polynomials of the spinor fields are involved, it will not affect index theorems, which have to do with the Dirac operator in isolation and the relative Dirac equation. Therefore, we hold *bona fide* valid the index theorem results below also for the a-number spinors of QFT.

The leading symbol of $\not{D}$ is

$$\sigma_\xi(\not{D}) = -\tilde{\gamma}^\mu \xi_\mu \tag{12.20}$$

which is invertible for any $\xi \neq 0$. Therefore the Dirac operator is elliptic. So is the square $\not{D}^2$, also called the *Dirac Laplacian*.

Since the Dirac operator $\not{D}$ is elliptic on a compact Riemannian manifold X, it has, in particular, finite dimensional kernel and cokernel, and, thus, a well-defined index. But one can say more.

In an even dimensional ($n = 2m$) oriented manifold, we can introduce the complex volume element (12.9) and obtain a $\mathbb{Z}_2$ grading on the spin bundle S.

$$S_\mathbb{C} = S_\mathbb{C}^+ \oplus S_\mathbb{C}^-, \quad S_\mathbb{C}^\pm = \frac{1}{2}(1 \pm \omega_\mathbb{C}) S_\mathbb{C} \tag{12.21}$$

and, for $e \in TX$

$$e \cdot S_\mathbb{C}^\pm \subseteq S_\mathbb{C}^\mp$$

Therefore $\not{D}$ splits according to

$$\not{D} = \begin{pmatrix} 0 & \not{D}^- \\ \not{D}^+ & 0 \end{pmatrix} \tag{12.22}$$

Therefore, we have $\ker \not{D} = \ker \not{D}^+ \oplus \ker \not{D}^-$. Now since $\not{D}$ is self-adjoint, it follows that $(\not{D}^\pm)^\dagger = \not{D}^\mp$. And since for any elliptic operator T, $\operatorname{coker} T = \ker T^\dagger$, it follows that

$$\operatorname{Ind} \not{D}^+ = \dim(\ker \not{D}^+) - \dim(\ker \not{D}^-) \tag{12.23}$$

12.4 Field Theory Families of Operators

The field theory path integral is a functional of the fields, in particular of the connections or the metrics or both. Let us focus on a gauge theory with a space $\mathcal{A}$ of connections and a symmetry group $\mathcal{G}$ of gauge transformations. Let us denote by $\mathcal{Z}[A]$ the fermion determinant, i.e. the partial path integral in which we suppose only the quadratic part of the action is considered, and fermions have been integrated over. $\mathcal{Z}[A]$ can be regarded as a function of the gauge potential alone: by means of perturbative and non-perturbative methods, we can define it as a functional of A. But we have to get rid of the gauge degrees of freedom. In other words, the information relevant to physics is stored in the orbit space $\mathcal{A}/\mathcal{G}$. The full path integral is obtained by integrating over the orbit space

$$\mathcal{Z} = \int_{\mathcal{A}/\mathcal{G}} \mathrm{d}[A]\, \mathcal{Z}[A] \tag{12.24}$$

Here $[A]$ denotes a point in the orbit space. Therefore, $\mathcal{Z}[A]$ must be a well-defined function, integrable over $\mathcal{A}/\mathcal{G}$.

Before proceeding further we need to clarify a technical point. The gauge transformations $\in \mathcal{G}$ is a subgroup of $\mathsf{Aut(P)}$, therefore they act on connections on the right

$$\mathcal{A} \times \mathcal{G} \to \mathcal{A}, \quad (A, \psi) \mapsto \psi^* A$$

Let us focus on $\mathcal{G} = \mathsf{Aut_v(P)}$. The action of $\mathcal{G}$ in general is not free. Since we need a free action for a well-defined quotient $\mathcal{A}/\mathcal{G}$, we will work with the group

$$\mathsf{Aut}_\mathsf{v}^\mathsf{m}(\mathsf{P}) = \left\{\psi \in \mathsf{Aut_v(P)}, \quad \text{s.t.} \quad \psi(p) = p, \quad \forall p \in \pi^{-1}(\mathsf{m})\right\} \tag{12.25}$$

where m is a preferred point in X (the point at infinity, in field theory). When speaking of orbit space $\mathcal{A}/\mathcal{G}$, we shall refer to this group of gauge transformations.

Now we can introduce the bundles which will play a role in the sequel

$$\begin{array}{ccc} \mathcal{G} & \hookrightarrow & \mathcal{A} \\ & & \downarrow \\ & & \mathcal{A}/\mathcal{G} \end{array} \tag{12.26}$$

It is a principal fiber bundle with structure group $\mathcal{G}$. Since the actions of G and $\mathcal{G}$ commute, another relevant principal fiber bundle is

$$\begin{array}{ccc} \mathsf{G} & \hookrightarrow & \frac{\mathsf{P}\times\mathcal{A}}{\mathcal{G}} \\ & & \downarrow \\ & & \frac{\mathsf{M}\times\mathcal{A}}{\mathcal{G}} \end{array} \tag{12.27}$$

with structure group G. In turn, $\frac{\mathsf{M}\times\mathcal{A}}{\mathcal{G}}$ is a fiber bundle over $\frac{\mathcal{A}}{\mathcal{G}}$, with fiber M (and trivial structure group, for the time being). Given a complex vector space V, which is a representation space of G, we have also two associated fiber bundles

- $V = \mathsf{P} \times_\mathsf{G} \mathsf{V}$

- $\mathcal{V} = \frac{\mathsf{P} \times \mathcal{A}}{\mathcal{G}} \times_{\mathsf{G}} \mathsf{V}$

The index theorem (12.47) involves a fixed gauge potential. But what is required in a gauge theory is an index theorem over the entire orbit space, because we have to deal with an operator that varies from point to point in it. The dimensions of its kernel and cokernel may jump from point to point in the parameter space provided by the orbit space. The object that makes sense in this case is the formal difference between the kernel and cokernel as they vary from point to point, i.e. the difference between two vector bundles over the parameter space. There is a fitting mathematical theory in this situation, it is K-theory [11, 12]. Before proceeding, we need a rapid recall of this theory.

12.5 K-Theory of Vector Bundles. A Lightning Introduction

As mentioned above, the index of an elliptic operator is the formal difference of two vector bundles. To understand this concept, one must resort to K-theory. This would require a specific mathematical course. Here we limit ourselves to a sketchy summary. Let X be a compact space. Given any vector bundle $E \xrightarrow{\pi} X$ let us denote by $[E]$ the class of vector bundles isomorphic to it. Now consider the set $Vect\ X$ of classes of isomorphic vector bundles E over X. We can sum two bundles E and F by considering the direct sum $E \oplus F$, in which the fiber at $x \in X$ is the direct sum of the fibers E_x and F_x. We shall denote by $[E \oplus F]$ the corresponding class. Therefore $Vect\ X$ is an additive semi-group, but a difference operation is not defined in it. This requires a further step. A construction due to Grothendieck allows us to define an associated Abelian group. This construction is in fact general and applies to any Abelian semi-group (or monoid). In the particular case of $Vect\ X$ let us consider the free Abelian group $F(X)$ generated by all elements (words) $[E] \in Vect\ X$ and quotient it by the ideal $I(X)$ generated by all words of the type $[E \oplus F] - [E] - [F]$. The result is automatically an Abelian group. This quotient is denoted $K(X)$:

$$K(X) = F(X)/I(X) \tag{12.28}$$

The quotient tells us of course that the sum of two elements $[E]$ and $[F]$ is represented by the coset $[E \oplus F]$, but now the difference $[E] - [F]$ is well-defined. Any elements of $K(X)$ can be written as a difference $[E] - [F]$. Such objects are called *virtual bundles*. Moreover, one can show that $[E] - [F] = [E'] - [F']$ if there is $G \in Vect\ X$ such that $E \oplus F' \oplus G \cong E' \oplus F \oplus G$, where '$\cong$' stands for 'isomorphic.'

There are even more stringent results. Let us call ε_n the trivial bundle of rank n over X. Then any element of $K(X)$ can be represented as $[E] - [\varepsilon_n]$, and $[E] = [F]$ if and only if there is n such that $E \oplus \varepsilon_n \cong F \oplus \varepsilon_n$.

If X is a point pt, then any vector bundle over pt is simply a vector space, and the only distinguishing feature is its rank n. Therefore, $K(pt) \cong \mathbb{Z}$. There is an exact sequence

$$0 \longrightarrow K(pt) \longrightarrow K(X) \longrightarrow \tilde{K}(X) \longrightarrow 0 \tag{12.29}$$

so that

$$K(X) \cong \mathbb{Z} + \tilde{K}(X) \tag{12.30}$$

$\tilde{K}(X)$ is called the *reduced* K-group.

If we have two spaces X and Y and two vector bundles $E \to X$ and $F \to Y$, we can define the tensor product bundle $E \otimes F$ over $X \times Y$, whose transition functions are the products of transition functions of E and F. Moreover, if f is a map: $f : X \to Y$, we have the pullback bundle f^*F over X, which is defined by pulling back to $x \in X$ the fiber $F_{f(x)}$.

When $X = Y$ we have the diagonal map $\Delta : X \to X \times X$ and we can define the product

$$[E] \cdot [F] \equiv \Delta^*[E \otimes F] \tag{12.31}$$

by means of the pullback of the class $[E \otimes F]$ in $X \times X$ to X. Therefore, $K(X)$ is a *ring*.

So far we have not specified the field of scalars. If we deal with complex vector bundles the K group is denoted simply $K(X) \equiv K_{\mathbb{C}}(X)$, if with real vector bundles, it is denoted $KO(X)$.

Next let us denote by $X \vee Y = (X \times pt_Y) \cup (pt_X \times Y) \subset X \times Y$ the *wedge product* and by $X \wedge Y = X \times Y / X \vee Y$ the *smash product*. Then we can define the *reduced suspension*: $\Sigma(X) = S^1 \wedge X$ and the iterated i-fold suspension $\Sigma^i(X)$. Finally, we define the higher $\tilde{K}^{-i}$ groups

$$\tilde{K}^{-i}(X) = \tilde{K}(\Sigma^i(X)), \qquad \tilde{K}^0 X \equiv K(X) \tag{12.32}$$

The set of K-theory groups are collectively denoted $K^{-*}(X)$

A celebrated result in K-theory is the *Bott periodicity*. In the complex case, we have the isomorphism

$$K^{-i}(X) \cong K^{-i-2}(X), \tag{12.33}$$

and in the real case

$$KO^{-i}(X) \cong KO^{-i-8}(X). \tag{12.34}$$

So far we have considered the case of a compact space X. The K-theory can be defined also for non-compact spaces. It is K_{cpt}, that is the K-theory with compact support:

$$K_{cpt}(X) = \tilde{K}(X^{\bullet}), \quad X^{\bullet} = X \cup pt. \tag{12.35}$$

$X^{\bullet}$ is the compactification of X, where pt denotes the compactification point, or point at infinity. Higher groups are defined

$$K^{-i}_{\rm cpt}(X) = K_{\rm cpt}(X \times \mathbb{R}^i) \tag{12.36}$$

For locally compact spaces, there is another description of $K^0(X)$. Let $[E] - [F]$ be an element of $K^0(X)$, and let the vector bundle E and F be isomorphic and trivial outside a compact set (a neighborhood of infinity), then $[E] - [F]$ can be represented by the triple $[E, F, \alpha]$ where $\alpha : E \to F$ is a smooth bundle map which is an isomorphism outside a compact set. Two such classes $[E, F, \alpha]$ and $[E', F', \alpha']$ are said to be equivalent if and only if there are bundles G and H such that

$$(E, F, \alpha) \oplus (G, G, id_G) \cong (E', F', \alpha') \oplus (H, H, id_H)$$

that is if $E \oplus G \cong E' \oplus H$ and $F \oplus G \cong F' \oplus H$ and the maps $\alpha \oplus id_G$ and $\alpha' \oplus id_H$ agree with these isomorphisms. The set of equivalence classes of triples is a group under the direct sum and the correspondence

$$[E] - [F] \longleftrightarrow [E, F, \alpha]$$

is an isomorphism.

12.6 The Atiyah-Singer Index Theorems

Let us return now to the index theorem. We shall consider first the fixed background case. The relevant configuration is given by two vector bundles E and F on a compact space X, the corresponding spaces of section $\Gamma(\mathsf{E})$ and $\Gamma(\mathsf{F})$, and an elliptic operator $P : \Gamma(\mathsf{E}) \to \Gamma(\mathsf{F})$. The case of interest to us will be of course when P is the Dirac operator $\not{D}$ and $\mathsf{E} = S^+_{\mathbb{C}} \otimes V$ while $\mathsf{F} = S^-_{\mathbb{C}} \otimes V$, where V is a vector bundle associated with $\mathsf{P}(M, \mathsf{G})$ (see above). Looking at this example, it is not hard to see that the principal symbol $\sigma(P)$ of $P : \Gamma(\mathsf{E}) \to \Gamma(\mathsf{F})$, if $\xi \in T^*_x(X)$, defines a map between E_x and F_x:

$$\sigma_\xi(P) : \mathsf{E}_x \longrightarrow \mathsf{F}_x \tag{12.37}$$

which, for an elliptic operator, is an isomorphism. Given the projection $\pi : TX \to X$ we can pullback the vector bundles E and F over TX and define the element

$$\boldsymbol{\sigma}(P) = [\pi^*\mathsf{E}, \pi^*\mathsf{F}, \sigma(P)] \in K_{\rm cpt}(TX) \tag{12.38}$$

Now we choose an embedding $f : X \hookrightarrow \mathbb{R}^n$ ($n = \dim(X)$). This induces a map

$$f_! : K_{\rm cpt}(TX) \longrightarrow K_{\rm cpt}(T\mathbb{R}^n) \tag{12.39}$$

Finally, the *scrunch* map to a point, $q : T\mathbb{R}^n \to pt$, induces a homomorphism

$$q_! : K_{\rm cpt}(T\mathbb{R}^n) \longrightarrow K(pt) \cong \mathbb{Z} \tag{12.40}$$

Now we can define the *topological index* of P:

$$\text{top}-\text{ind}\, P = q_! f_! \,\boldsymbol{\sigma}(P) \in \mathbb{Z} \tag{12.41}$$

which takes values in $\mathbb{Z}$. On the other hand, the *analytic index* is (see above)

$$\text{ind}\, P = \dim(\ker P) - \dim(\text{coker}\, P) \tag{12.42}$$

The *Atiyah-Singer index theorem* proves that if P is an elliptic operators on a compact space X, then

$$\text{ind}\, P = \text{top}-\text{ind}\, P \tag{12.43}$$

This fundamental theorem opens the way to express the index in terms of characteristic classes. For instance, in the case of interest to us, i.e. when P is the Dirac operator $\mathfrak{D}^+$ and $\mathsf{E} = S^+_{\mathbb{C}} \otimes E$ while $\mathsf{F} = S^-_{\mathbb{C}} \otimes E$, we have

$$\text{ind}\, \mathfrak{D}^+ = \left(\text{ch}(E) \cdot \hat{A}(X)\right)[X] \tag{12.44}$$

where $[X]$ means evaluated on the fundamental cycle $[X]$.

To be more explicit let us consider the tensor product of a spinor bundle $S^{\pm}_{\mathbb{C}}$ with a vector bundle E corresponding to a representation ρ of the structure group G of $\mathsf{P}(X, \mathsf{G})$: $S^{\pm}_{\mathbb{C}} \otimes E$. The relevant connection will be the spin connection plus a gauge connection $V = V_\mu \text{d}x^\mu$ valued in some representation ρ of the Lie algebra of G with anti-hermitean generators. The corresponding Dirac operator

$$\mathfrak{D} = \not{D} + i \not{V} \tag{12.45}$$

acts on the space of sections $\Gamma(S^{\pm}_{\mathbb{C}} \otimes E)$ mapping it to the same space. Then the *Atiyah-Singer index theorem* is:

$$\text{Ind}(\mathfrak{D}^+) = \int_X \text{ch}(E)\, \hat{A}(X) \tag{12.46}$$

The symbol $ch(E)$ indicates the Chern character of the E bundle, i.e. the rational characteristic class given in terms of the curvature F of V by

$$\mathrm{ch}(E) = r + \frac{i}{2\pi}\mathrm{tr}\, F + \frac{i^2}{2(2\pi)^2}\mathrm{tr}\, F^2 + \cdots + \frac{i^n}{n!(2\pi)^n}\mathrm{tr}\, F^n + \cdots \quad (12.47)$$

where r is the dimension of the representation ρ. The symbol $\widehat{A}(X)$ denotes the $\widehat{A}$-*genus*, which is the (rational) Pontryagin characteristic class of X. It can be expressed in terms of the Riemann curvature R as follows

$$\begin{aligned}\widehat{A}(X) =& 1 + \frac{1}{(4\pi)^2}\frac{1}{12}\,\mathrm{tr}\, R^2 + \frac{1}{(4\pi)^4}\left[\frac{1}{288}\,\mathrm{tr}\,(R^2)^2 + \frac{1}{360}\,\mathrm{tr}\, R^4\right] \\ &+ \frac{1}{(4\pi)^6}\left[\frac{1}{10368}\,\mathrm{tr}\,(R^2)^3 + \frac{1}{4320}\,\mathrm{tr}\, R^2\,\mathrm{tr}\, R^4 + \frac{1}{5670}\,\mathrm{tr}\, R^6\right] + \cdots\end{aligned} \quad (12.48)$$

However, this theorem, which we shall refer to as the *fixed background* index theorem, is not what we need in relation to anomalies. For anomalies a richer mathematical structure is involved and, to take it into account, the family's index theorem is necessary.

12.7 The Family's Index Theorem

Let us define first the index for families. The geometrical environment is the one corresponding to (12.26, 12.27). We have a Hausdorff space $\mathcal{B}$ which is the basis of a principal fiber bundle (similar to $\mathcal{A}/\mathcal{G}$ in (12.26)). A continuous family $\mathcal{E}$ of smooth vector bundles over X, parametrized by $\mathcal{B}$, is a fiber bundle $\mathcal{E} \to \mathcal{B}$ whose fiber is a smooth vector bundle E over X, associated with a principal fiber bundle $\mathsf{P}(X, \mathsf{G})$, on which the group $\mathsf{Aut_v(P)}$ acts ($\mathcal{E}$ is similar to $\mathcal{V}$, defined above). Now, let us consider two families $\mathcal{E}$ and $\mathcal{F}$ with the relative spaces of sections $\Gamma(\mathcal{E})$ and $\Gamma(\mathcal{F})$, and the differential operators $P : \Gamma(\mathcal{E}) \to \Gamma(\mathcal{F})$ of order $\leq m$, with the transformation property $P \to g_1 \cdot P \cdot g_2^{-1}$, where $g_1 \in \mathsf{Aut_v}(\mathcal{E})$ and $g_2 \in \mathsf{Aut_v}(\mathcal{F})$ (the latter are the groups $\mathsf{Aut_v}(P)$ in the representation appropriate to $\mathcal{E}$ and $\mathcal{F}$, respectively).

Therefore, if we have two continuous families $\mathcal{E}$ and $\mathcal{F}$, we can consider the family of elliptic operators of order $\leq m$ from $\mathcal{E}$ to $\mathcal{F}$ on each point of $\mathcal{B}$. They form a family $\mathsf{Op^m}(\mathcal{E}, \mathcal{F})$ which is a bundle over $\mathcal{B}$ with structure group $\mathsf{Aut_v}(\mathcal{E}) \times \mathsf{Aut_v}(\mathcal{F})$. Given one such operator P the kernel and cokernel are vector spaces that vary from point to point in $\mathcal{B}$. Therefore, their difference defines an element of $K(\mathcal{B})$:

$$\mathrm{ind}\, P = [\ker P] - [\mathrm{coker}\, P] \in K(\mathcal{B}) \quad (12.49)$$

This is called the *analytic index* of P. Even though the dimensions of these vector bundles may vary in $\mathcal{B}$, the definition makes sense.

The next step consists in defining the topological index. For that purpose let us introduce the bundle $\pi : \mathcal{X} \to \mathcal{B}$ with fiber X (this is the analog of $\frac{\mathsf{M}\times\mathcal{A}}{\mathcal{G}} \to \frac{\mathcal{A}}{\mathcal{G}}$ with fiber M introduced above). For s large enough one can introduce a map $f : \mathcal{X} \to \mathcal{B} \times \mathbb{R}^s$, which induces a map in K-theory

$$f_! : K_{\text{cpt}}(T\mathcal{X}) \longrightarrow K_{\text{cpt}}(\mathcal{B} \times \mathbb{R}^s) \tag{12.50}$$

Moreover, the projection $q : \mathcal{B} \times \mathbb{R}^s \to \mathcal{B}$ induces an isomorphism (the Thom isomorphism)

$$q_! : K_{\text{cpt}}(\mathcal{B} \times \mathbb{R}^s) \xrightarrow{\cong} K(\mathcal{B}) \tag{12.51}$$

The composed map $q_! \circ f_! : K_{\text{cpt}}(T\mathcal{X}) \to K(\mathbb{B})$ does not depends on the choice of f. Under the same hypotheses as for the analytic index, the *topological index for families* is defined by

$$\text{top}-\text{ind}(P) \equiv q_! f_!\big(\sigma(P)\big) \in K(\mathcal{B}) \tag{12.52}$$

Again, the *Atiyah-Singer index theorem for families* asserts that[1]

$$\text{ind}(P) = \text{top}-\text{ind}(P) \tag{12.53}$$

This allows to express the index in terms of characteristic classes. In the case outlined above in which P is the Dirac operator $\not{D}^+$ between two families of sections $\Gamma(S^+_{\mathbb{C}}) \otimes \mathcal{V}$ and $\Gamma(S^-_{\mathbb{C}} \otimes \mathcal{V})$ the index takes the form

$$\text{ind}(\not{D}^+) = \pi_! \left(\text{ch}(\mathcal{V}) \cdot \hat{A}(T\mathcal{X})\right) \tag{12.54}$$

where $\pi : \mathcal{X} \to \mathcal{B}$ and ch denotes the Chern character. Where this passage from K theory to cohomology comes from is suggested by the following: if X is compact we have the isomorphism: $\text{ch} : K^*(X) \xrightarrow{\cong} H^*(X)$ given by

$$[E] - [F] \longrightarrow \text{ch}(E) - \text{ch}(F). \tag{12.55}$$

A more ready-to-use formula for gauge field theories is

$$\text{ch}\left(\text{ind}(\not{D}^+)\right) = \int_M \text{ch}(\mathcal{V}) \cdot \hat{A}(T\mathcal{Q}) \tag{12.56}$$

where $X = \mathsf{M}$ is the spacetime manifold, $\mathcal{Q} \equiv \frac{\mathcal{A}}{\mathcal{G}}$ is the orbit space of connections and $\mathcal{V}$ is the gauge bundle. The Chern character of the index measures the extent to which $\ker \not{D}^+$ differs from $\text{coker}\, \not{D}^+ = \ker \not{D}^-$. It is intuitive that, as long as, $\text{ind}(\not{D}^+)$ differs from 0, the inverse of $\not{D}^+$ does not exist. Therefore, *the only way to*

[1] After the proof of Atiyah and Singer, several other proofs of this theorem were produced in the 80s of the last century, see the main bibliography.

ensure the existence of this inverse (and, so, the existence of the fermion propagator and, in turn, the existence of the fermion path integral) is that the RHS of (12.56) correspond to the 0 class. This is particularly visible with the first Chern class, i.e. the first nontrivial term of the Chern character. Let us recall that, for a vector bundle V

$$\mathrm{ch}(V) = \mathrm{rank}(V) + c_1(V) + \frac{1}{2}\left(c_1(V)^2 - 2c_2(V)\right) + \cdots \tag{12.57}$$

Therefore,

$$c_1\left(\mathrm{ind}(\not{D}^+)\right) = \int_M \mathrm{ch}(\mathcal{V}) \cdot \hat{A}(T\mathcal{Q})\Big|_{\mathsf{d},2} \tag{12.58}$$

where $\mathsf{d} = \dim(M)$ and in the RHS is the $(\mathsf{d}, 2)$ component in $M \times \mathcal{Q}$. The RHS is a two-form in $\mathcal{Q}$, therefore it belongs to a class in $H^2(\mathcal{Q}, \mathbb{Z})$. It represents the first Chern class of the so-called Quillen determinant bundle. Given the importance of the latter, it is worth spending some time to introduce it in more detail.

12.8 The Quillen Determinant Bundle

The field theory path integral we are considering is a functional of the fields, in particular of the connections or the metrics. Let us focus on a gauge theory with a space $\mathcal{A}$ of connections and a symmetry group $\mathcal{G}$ of gauge transformations. The information relevant to physics is stored in the orbit space $\mathcal{A}/\mathcal{G}$. Thus, we must be able to properly define the path integral over this space. In other words, a crucial step in quantization is the elimination of the gauge degrees of freedom. In perturbative field theory, one proceeds by fixing the gauge by hand. This however does not eliminate the symmetry of the theory, which reappears, at the quantum level, as a FP ghost-dependent symmetry, the BRST symmetry. The price is that in the path integral, while integrating over the gauge potentials, we meet the Gribov ambiguity and we have to integrate also over the ghost fields. The BRST symmetry can be iterated order by order in the perturbative approach and is crucial in the quantization process because it guarantees the existence of Ward-Takahashi identities, which provide, so to speak, the road map for renormalization. This procedure however works only if the BRST symmetry is preserved throughout and is not broken at some stage of the quantization by anomalies. In non-perturbative methods of type (B), at least for the anomaly analysis, we can bypass the FP ghost and Gribov problem and can directly deal with the moduli space $\mathcal{A}/\mathcal{G}$. It should be added, however, that, at least at present, such non-perturbative methods cannot be extended to the full theory, for instance to include matter field interactions.

Let us denote by $\mathcal{Z}[A]$ the partial path integral in which we suppose only matter quadratic terms, in particular fermionic quadratic terms, have been integrated over. $\mathcal{Z}[A]$ can be regarded as a function of the gauge potential alone: as already pointed out, by means of perturbative and non-perturbative methods we can define it as a functional of A. In order to get the full path integral, we have to integrate also over the moduli space (or orbit space)

$$\mathcal{Z} = \int_{\mathcal{A}/\mathcal{G}} d[A]\,\mathcal{Z}[A] \tag{12.59}$$

where $[A]$ represents the orbit through A. Since the measure $\mathrm{d}[A]$ is gauge invariant, also $\mathcal{Z}[A]$ must be so. Therefore, $\mathcal{Z}[A]$ must be a well-defined non-vanishing function integrable over $\mathcal{A}/\mathcal{G}$.

The case of interest to us now is a twisted Dirac operator, for instance $\not{D}$ considered above (we are still considering the case of a Euclidean background metric). It is an elliptic operator on a compact space, and so a Fredholm operator, which splits according to Eq. (12.22), and we wish to define the path integral of either $\not{D}^+$ or $\not{D}^-$. We can use, for instance, the definition of the (partially integrated) path integral for $\not{D}^+$ as the product of its eigenvalues. But there are obstacles. As we have pointed out in correspondence with Eq. (6.7), the first obstacle is the very definition of eigenvalues of a Dirac-Weyl operator. In perturbative and SDW approaches, we have bypassed this problem by modifying the Dirac-Weyl operator like in (6.8). In relation with the index theorem, it is more opportune to consider instead the quadratic operator $(\not{D}^+)^\dagger\not{D}^+$. It has the same eigenvalues as $(\not{D}^-)^\dagger\not{D}^-$, which are the square of the eigenvalues of the Dirac operator $\not{D}$, but of course the eigenvectors are different. More precisely, the nonzero eigenvalues of $(\not{D}^+)^\dagger\not{D}^+$ are paired with the nonzero eigenvalues of $(\not{D}^-)^\dagger\not{D}^-$, in the sense that to each common eigenvalue there correspond two distinct eigenvectors of the two operators with the same multiplicity. But this may not be true anymore for the zero eigenvalues. It goes without saying that this is the crucial point for the index theorem. But it is crucial also for defining the path integrals of the two Dirac-Weyl operators. If we wish to define the relevant path integrals as the square root of the product of their eigenvalues, the nonzero eigenvalues do not pose a problem because they are gauge invariant, but for the zero eigenvalues this may not be true. Therefore, we have to settle the question of the zero modes in such a way as to produce a well-defined integrable $\mathcal{Z}[A]$.

Let us denote by $\psi_1, \ldots, \psi_r$ a basis for the zero modes of $\not{D}^+$. In field theory, due to the Pauli principle, it is natural to study the wedge product (Slater determinant) $\psi_1 \wedge \ldots \wedge \psi_r$. It is a complex number (a determinant) that varies as $\not{D}^+$ varies over $\mathcal{X} \equiv \frac{\mathsf{M}\times\mathcal{A}}{\mathcal{G}}$, but, in general, does not form a complex line bundle over $\mathcal{X}$. In the same way, one can construct an analogous wedge product for coker $\not{D}^+ = \ker \not{D}^-$, which also form a complex line, but in general not a line bundle over $\mathcal{Q}$. However, the formal difference of the two, i.e.

$$\coprod_{[A]\in\mathcal{X}} \bigwedge^{\dim\ker \not{D}_A^+} \ker(\not{D}_A^+)^\dagger \otimes \bigwedge^{\dim\operatorname{coker} \not{D}_A^+} \operatorname{coker}(\not{D}_A^+) \tag{12.60}$$

where we have made explicit the dependence on the connection A, constitutes an element of $K(\mathcal{X})$, which is called the Quillen determinant line bundle. This complex line bundle has remarkable topological properties. For us it is a fundamental object because it is trivial, i.e. it has a global non-vanishing section, precisely when $\not{D}^+$ is invertible.[2]

In the case of the Dirac-Weyl operator $\not{D}^+$, the Quillen determinant bundle is a bundle over the base space $\mathcal{X} \equiv \frac{M\times\mathcal{A}}{\mathcal{G}}$. It is a non-trivial bundle as long as $H^2(\mathcal{X},\mathbb{Z}) \neq 0$. Upon integrating along the fiber M, the relevant cohomology group is $H^2(\mathcal{Q},\mathbb{Z})$ and coincides with $c_1\left(\mathrm{ind}(\not{D}^+)\right)$. If this class vanishes the Quillen bundle is trivial, that is it has a global section, which is a non-vanishing function over $\mathcal{A}/\mathcal{G}$ and can be integrated over. In this case, $\mathcal{Z}[A]$ is well-defined.

The formal Eq. (6.3) implies that, in order for $\mathcal{Z}[A]$ to be well-defined, the inverse of $\not{D}^+$, i.e. the fermion propagator, must be well-defined. The inverse operator is represented by the inverse eigenvalues. For the nonzero eigenvalues, invertibility is obvious. For the 0 eigenvalue, the inverse is represented by the inverse of the global section of the Quillen determinant bundle. We expect, therefore, that, if $H^2(\mathcal{Q},\mathbb{Z})$ is trivial, this global section is invertible. On the contrary a nonzero $c_1\left(\mathrm{ind}(\not{D}^+)\right)$, given by (12.58), is an obstruction to defining the inverse of $\not{D}^+$, i.e. the Weyl fermion propagator. The vanishing of $c_1\left(\mathrm{ind}(\not{D}^+)\right)$ is therefore a necessary condition for the existence of a well-defined partition function.

A comment. The previous discussion based on the Quillen determinant bundle is the mathematical way of investigating the existence of a fermion propagator. It is different from the physical approach, which is of a constructive type. The mathematical approach highlights the obstructions to its existence. As long as there are obstructions, there is no way of defining a propagator and no way of defining a path integral. Therefore, there is no way of defining a current, and a current conservation via a Ward identity. The only thing we can do is to verify the conditions under which there are no obstructions to the existence of a propagator.

The physical approach is different: we start from $i\not{\partial}$, which is invertible, and add (left or right-handed) stuff of order 0, the potentials; provided the potentials are small enough that there is always an inverse of the kinetic operator, and we can define a path integral and a Ward identity; the process does not stop even if the current is not conserved and the Ward identity violated, i.e. if an anomaly appears; but, at that

[2] The theorem has been proven by D. Quillen for a Cauchy-Riemann operator over a Riemann surface. Here we are extending its scope to a general framework with the same basic characteristics [10]

point, physical considerations take over, because, in the presence of an anomaly, we cannot ensure unitarity.

This said the question is: where is the connection, if any, between the just outlined mathematical and the physical approaches? The answer is: the connection is in the topology of the classifying space (for applications of the family's index theorem to anomalies see [10, 13–15]).

12.9 Obstructions. The Gauge Case

In the previous section, we have seen that, in a Euclidean d-dimensional spacetime M, the obstructions to the existence of a Weyl fermion propagator (i.e. the existence of the inverse of $\not{D}^+$, for instance), is contained in the class

$$\mathrm{ch}(\mathcal{V}) \cdot \hat{A}(T\mathcal{X}) \tag{12.61}$$

where $\mathcal{X} \equiv \frac{\mathsf{M}\times\mathcal{A}}{\mathcal{G}}$ ($\mathcal{A}$ is the space of gauge connections and $\mathcal{G}$ the group of gauge transformations). In this section, we would like to establish the relation of this class with the topology of the classifying space. We start with the (simpler) pure gauge case, which corresponds to selecting the part of (12.61) with $\hat{A}(T\mathcal{X}) = 1$. Then we recall that $\mathcal{G} = \mathsf{Aut_v(P)}$ in reference to the principal fiber bundle $\mathsf{P(M, G)}$, that $\mathcal{A} \to \mathcal{A}/\mathcal{G}$ is a principal fiber bundle with structure group $\mathcal{G}$, (12.26), that $\frac{\mathsf{P}\times\mathcal{A}}{\mathcal{G}} \to \frac{\mathsf{M}\times\mathcal{A}}{\mathcal{G}}$ is a principal fiber bundle with group G, (12.27), and, finally, that $\frac{M\times\mathcal{A}}{\mathcal{G}}$ is a fiber bundle over $\mathcal{Q} = \frac{\mathcal{A}}{\mathcal{G}}$ with fiber M (in fact in the case of $\mathcal{G} = \mathsf{Aut_v(P)}$, the bundle $\frac{M\times\mathcal{A}}{\mathcal{G}}$ reduces to $M \times \frac{\mathcal{A}}{\mathcal{G}}$ because the basepoint is not moved by a vertical automorphisms). Let us endow $\mathcal{A}$ with a connection ω in the fiber bundle $\mathcal{A} \to \mathcal{A}/\mathcal{G}$. Then we can define a connection η on the G bundle $\mathsf{P} \times \mathcal{A}$ over $\mathsf{M} \times \mathcal{A}$ as follows:

$$\eta_{p,A}(X, Y) = A_p(X) + A(\omega(Y))_p, \quad X \in T_p\mathsf{P}, \quad Y \in T_A\mathcal{A} \tag{12.62}$$

where A is a connection in P and ω a connection in $\mathcal{A} \to \mathcal{A}/\mathcal{G}$ (see Appendix 12A for further details). This connection descends to a connection η' on the fiber bundle (12.27)

$$\begin{array}{ccc} \mathsf{G} & \hookrightarrow & \frac{\mathsf{P}\times\mathcal{A}}{\mathcal{G}} \\ & & \downarrow \\ & & \frac{\mathsf{M}\times\mathcal{A}}{\mathcal{G}} \end{array} \tag{12.63}$$

We can therefore introduce the curvature $\mathfrak{F}_{\eta'}$ of η' and express $\mathrm{ch}(\mathcal{V})$ in terms of it.[3] Since the Chern character can be expressed as a polynomial in the curvature, the relevant piece for dimension d will take the form $P_n(\mathfrak{F}_{\eta'}, \ldots, \mathfrak{F}_{\eta'})$, where $n = \frac{\mathsf{d}}{2} + 1$ and P_n is an ad-invariant (reducible or irreducible) polynomial for the group G. Now let us recall that, in the definition (12.62) of the connection η, the connection A in P, can be obtained from the universal connection $\mathfrak{a}$ (with curvature $F_{\mathfrak{a}}$) via a map $\hat{f} : \mathsf{P} \to \mathsf{EG}$, $A = \hat{f}^* \mathfrak{a}$. It can be shown that, in the same way also ω, the connection in $\mathcal{A}$, can be pulled back from the same universal connection via a map $\hat{\mathfrak{f}} : \mathcal{A} \to \mathsf{EG}$: $\omega = \hat{\mathfrak{f}}^* \mathfrak{a}$. Therefore, there exists an overall map $\hat{f}_\times = (\hat{f}, \hat{\mathfrak{f}}) : \mathsf{P} \times \mathcal{A} \to \mathsf{EG}$, such that $\eta = \hat{f}^*_\times \mathfrak{a}$ (for a more detailed discussion of η and η', see Appendix 12A[4]). Therefore,

$$P_n(\mathfrak{F}_\eta, \ldots, \mathfrak{F}_\eta) = f^*_\times P_n(F_{\mathfrak{a}}, \ldots, F_{\mathfrak{a}}) \tag{12.64}$$

The LHS then descends to $P_n(\mathfrak{F}_{\eta'}, \ldots, \mathfrak{F}_{\eta'})$ on (12.63). Remember that $P_n(F_{\mathfrak{a}}, \ldots, F_{\mathfrak{a}})$ is a closed form in EG that descends to a closed form in BG, which represents a nontrivial cohomology class of the classifying space.

Now, we know that, via the evaluation map and the transgression formula, we can extract from this class a corresponding anomaly, which is given by $i_{(\cdot)} T P_n(A)$. This can be seen directly from the transgression formula ensuing from (12.64)

$$T P_n(\eta) = n \int_0^1 \mathrm{d}t \; P_n(\eta, \mathfrak{F}_\eta, \ldots, \mathfrak{F}_\eta), \tag{12.65}$$

which splits in components of order $(2n{-}p{-}1, p)$ in P and $\mathcal{A}$. The $(2n{-}1, 0)$ is precisely $T P_n(A)$. We also know that any local anomaly takes this form and originates from a class in BG. Therefore, we see that anomalies and obstructions to the existence of a Weyl fermion propagator coming from the index theorem, are rigidly connected. The added value of the index theorem is that it determines precisely the numerical coefficients of the irreducible polynomials and the product of reducible ones (see (12.47) for the case of spin 1/2). This allows us to cope with the problem of anomaly cancelation in general. In a specific model, anomalies are absent/canceled if the sum of coefficients in front of each independent polynomial P_n, of the various component fields involved, vanishes. The important remark here is that the index theorem allows us to solve this problem in complete generality once we know the component fields of a theory at the level of classifying space (i.e. no need to go to the real spacetime, only the datum of its dimension is necessary, see the superstring examples in Chap. 17).

---

[3] The curvature $\mathfrak{F}_{\eta'}$ has been explicitly computed in [10]. It has components (i, j) in $\mathsf{M} \times \mathcal{A}$, with $i = 0, 1, 2$, $j = 2 - i$. In local coordinates, the component (2,0) is $F_{\mu\nu}$, the component (1,1) is δA_μ and the one (0,2) is $\left(D_A^\dagger D_A\right)^{-1} [\delta A_\mu, \delta B^\mu]$, where D_A is the covariant derivative and $\delta A_\mu, \delta B_\mu$ are infinitesimal variation with respect to the background connection.

[4] In the language of Appendix 12A $\eta' = \widehat{\mathsf{Ev}}^* \mathfrak{a}$ and $\eta = \widehat{\mathsf{ev}}^* \mathfrak{a}$.

12.9.1 Diffeomorphism Obstructions

The application of the family's index theorem to diffeomorphisms requires some preparation. First of all, the relevant principal fiber bundle is LM^+, the bundle of oriented frames, whose structure group is $\mathsf{GL}^+(\mathsf{d}, \mathbb{R})$, the general linear group with positive determinant. As explained elsewhere the relevant group of gauge transformations depends on what we consider the primary mathematical object we choose to work with. The latter can be a linear connection Γ, a metric connection, a Levi-Civita connection or simply a metric. If we consider the space of linear connections, the relevant gauge group is $l\,(\mathsf{Diff}_*(\mathsf{M}))$, i.e. the group of natural lifts of orientation preserving diffeomorphisms of M. This is a subgroup of $\mathsf{Aut}(\mathsf{LM}^+)$, which is the semidirect product of the group of vertical authomorphisms, $\mathsf{Aut}_\mathsf{v}(\mathsf{LM}^+)$, and the orientation preserving diffeomorphisms of the base M, $\mathsf{Diff}_*(\mathsf{M})$ (a subscript $_*$ signals orientation preserving diffeomorphisms). Now, if we wish a free action of diffeomorphisms we must consider the subgroup $\mathsf{Diff}_*^{\mathsf{m},1}(\mathsf{M})$ of $\mathsf{Diff}_*(\mathsf{M})$, formed by the diffeomorphisms φ of M satisfying the following conditions:

$$\varphi(\mathsf{m}) = \mathsf{m}, \quad \text{and} \quad \varphi_*\Big|_{\mathsf{m}} = id \quad \text{in} \quad T_\mathsf{m}\mathsf{M}$$

that is, φ does not move m and does not change the tangent vectors at m. Therefore, to guarantee a free action we will consider the subgroup $l\big(\mathsf{Diff}_*^{\mathsf{m},1}(\mathsf{M})\big)$, i.e. the lift of $\mathsf{Diff}_*^{\mathsf{m},1}(\mathsf{M})$.

If we consider the space $\mathcal{M}$ of all metrics on M, the relevant gauge transformations are $\mathsf{Diff}_*(\mathsf{M})$, and, if we want a free action, $\mathsf{Diff}_*^{\mathsf{m},1}(\mathsf{M})$. To any metric there corresponds a Levi-Civita (LC) connection and, again, $l\big(\mathsf{Diff}_*^{\mathsf{m},1}(\mathsf{M})\big)$ acts freely on the space of LC connections. If we call $\mathcal{A}^{LC}$ this space, we have a principal fiber bundle

$$\mathcal{A}^{LC} \to \frac{\mathcal{A}^{LC}}{l\big(\mathsf{Diff}_*^{\mathsf{m},1}(\mathsf{M})\big)} \tag{12.66}$$

But we can also consider the space $\mathcal{A}^{\text{metric}}$ of linear connections whose holonomy group is $\mathsf{SO}(\mathsf{d})$. This space is the product of $\mathcal{A}^{LC}$ times the space of torsion tensors and is contractible. It is also the total space of a principal fiber bundle

$$\mathcal{A}^{\text{metric}} \to \frac{\mathcal{A}^{\text{metric}}}{l\big(\mathsf{Diff}_*^{\mathsf{m},1}(\mathsf{M})\big)} \tag{12.67}$$

While applying (12.61) to diffeomorphisms, we choose this last bundle. We suppose, for the time being, that there is no gauge field coupled to our system, so we disregard $ch(\mathcal{V})$ and focus on $\widehat{A}(T\mathcal{X})$. Unlike the gauge case, where $\mathcal{X}$ reduces to $\mathsf{M} \times \mathcal{A}/\mathcal{G}$, here we have to consider the full quotient $\mathcal{X} = \frac{\mathsf{M}\times\mathcal{A}^{\text{metric}}}{l\left(\mathsf{Diff}_*^{\mathsf{m},1}(\mathsf{M})\right)}$ and its tangent space, because, obviously, the diffeomorphisms act simultaneously on M and on the connections. Therefore, the relevant bundle is

$$\mathcal{X} = \frac{\mathsf{M} \times \mathcal{A}^{\text{metric}}}{\mathcal{I}\big(\mathsf{Diff}_*^{\mathsf{m},1}(\mathsf{M})\big)} \to \frac{\mathcal{A}^{\text{metric}}}{\mathcal{I}\big(\mathsf{Diff}_*^{\mathsf{m},1}(\mathsf{M})\big)} \tag{12.68}$$

It is along the fiber of this bundle that we have to integrate. The tangent $T\mathcal{X}$ is the bundle

$$T\mathcal{X} = \frac{T\mathsf{M} \times \mathcal{A}^{\text{metric}}}{\mathcal{I}\big(\mathsf{Diff}_*^{\mathsf{m},1}(\mathsf{M})\big)} \to \frac{\mathsf{M} \times \mathcal{A}^{\text{metric}}}{\mathcal{I}\big(\mathsf{Diff}_*^{\mathsf{m},1}(\mathsf{M})\big)} \tag{12.69}$$

This is the bundle whose $\widehat{A}$-genus appears in (12.61).

What remains for us to do is work out the characteristic class corresponding to $\widehat{A}(T\mathcal{X})$. To do so we construct the corresponding universal polynomial poly-form. To this end, we consider the diagram

$$\begin{array}{ccc} \dfrac{\mathsf{LM}^+ \times \mathcal{A}^{\text{metric}}}{\mathcal{I}\big(\mathsf{Diff}_*^{\mathsf{m},1}(\mathsf{M})\big)} & \xrightarrow{\ \mathsf{Ev}\ } & \mathsf{EGL(d,\mathbb{R})^+} \\ \Big\downarrow \pi & & \Big\downarrow \pi \\ \dfrac{\mathsf{M} \times \mathcal{A}^{\text{metric}}}{\mathcal{I}\big(\mathsf{Diff}_*^{\mathsf{m},1}(\mathsf{M})\big)} & \xrightarrow{\ \mathsf{Ev}\ } & \mathsf{BGL(d,\mathbb{R})^+} \end{array} \tag{12.70}$$

(for details see Appendix 12A) and starting from the universal connection $\mathfrak{a}^{gl}$ with curvature $F(\mathfrak{a}^{gl})$ we pull it back to the corresponding η^{gl}: $\eta^{gl} = \mathsf{Ev}^*\mathfrak{a}^{gl}$, with curvature $\mathcal{F}_{\eta^{gl}}$. With these, we form the characteristic classes

$$P_n(\mathcal{F}_{\eta^{gl}}, \ldots, \mathcal{F}_{\eta^{gl}}) = \mathsf{Ev}^* P_n\big(F(\mathfrak{a}^{gl}), \ldots, F(\mathfrak{a}^{gl})\big) \tag{12.71}$$

which are closed forms in $\mathsf{EGL(d,\mathbb{R})^+}$ that descend to closed non-trivial forms in $\mathsf{BGL(d,\mathbb{R})^+}$.

At this stage, we should pause a moment in order to generalize this construction to spinor fields. Spinors are section of spinor bundles, which are associated bundles to the bundle $\mathsf{LM_{Spin}}$ of spin frames. The relevant connections are spin connections, which form a space $\mathcal{A}^{\text{spin}}$. Since $\mathcal{A}^{\text{spin}} \cong \mathcal{A}^{\text{metric}}$, and the Lie algebras of the structure groups are isomorphic to the previous ones, nothing changes concerning the classes (12.71). Therefore, we assume the identification between spin and metric connections and postpone a dedicated comment to Appendix 12B.

Now we can repeat what we already said for gauge anomalies. Via the evaluation map and the transgression formula, we can extract from these classes the corresponding anomalies, which are given by $i_{(\cdot)}TP_n(\Gamma)$ for any metric connection Γ. We can extract them from the transgression formulas derived from (12.71)

$$TP_n(\eta^{gl}) = n\int_0^1 \mathrm{d}t\ P_n(\eta, \mathcal{F}_{\eta^{gl}}, \ldots, \mathcal{F}_{\eta^{gl}}) \tag{12.72}$$

This splits in various components of order $(p, 2n-p-1)$ in LM^+ and $\mathcal{A}^{\text{metric}}$. The $(2n-1, 0)$ is $TP_n(\Gamma)$. Any local diffeomorphism anomaly takes this form and originates from a class in $\mathsf{BGL}(\mathsf{d}, \mathbb{R})$. Therefore, we see that anomalies and obstructions to the existence of fermion propagators coming from the index theorem are bijectively connected. As already stressed, the added value of the index theorem is that it determines precisely the numerical coefficients of the obstructions and allows us to analyze anomaly cancelation in an exact way.

12.9.2 *Mixed Gauge-Diffeomorphism Obstructions*

The next to be considered is the case of mixed anomalies, where we have to deal simultaneously with two distinct principal fiber bundles $\mathsf{P}(\mathsf{M}, \mathsf{G})$ and LM^+ over M, with structure groups G and $\mathsf{GL}^+(\mathsf{d}, \mathbb{R})$, respectively. Then, see Sect. 11.6.3, we can define a principal fiber bundle $\mathsf{P} \times \mathsf{LM}^+$ over $\mathsf{M} \times \mathsf{M}$, with group $\mathsf{G} \times \mathsf{GL}^+(\mathsf{d}, \mathbb{R})$ (direct product of groups), and restrict our consideration to the diagonal $\Delta\mathsf{M}$ of $\mathsf{M} \times \mathsf{M}$. We obtain in this way a principal fiber bundle, which we denote PLM with structure group $\mathsf{G} \times \mathsf{GL}^+(\mathsf{d}, \mathbb{R})$. Let us denote by π_1 the restriction of the projection PLM to P , and π_2 the analogous projection to LM^+. Given two connections, A_1 in P with curvature F_1 and A_2 in LM^+ with curvature F_2, there is a unique connection A in PLM, with curvature F, such that

$$A = \pi_1^* A_1 + \pi_2^* A_2, \quad F = \pi_1^* F_1 + \pi_2^* F_2 \tag{12.73}$$

On the other hand, if A is a connection on a principal fiber bundle $\mathsf{PLM}\big(\mathsf{M}, \mathsf{G} \times \mathsf{GL}^+(\mathsf{d}, \mathbb{R})\big)$, its reduction to the subgroup G gives a unique connection A_1 valued in the Lie algebra $\mathfrak{g}$ of G, and its reduction to the subgroup $\mathsf{GL}^+(\mathsf{d}, \mathbb{R})$ yields a unique metric connection A_2. Therefore, there is a one-to-one correspondence between connections in PLM and their components obtained by their reduction to $\mathfrak{g}$-valued connections and metric connections over M. We will simply write

$$A = A_1 \oplus A_2 \tag{12.74}$$

meaning that A_1 is a connection valued in $\mathfrak{g}$ and trivially extended to $\mathsf{GL}^+(\mathsf{d}, \mathbb{R})$, and likewise for A_2.

The same can be said about universal connections. So we will write the universal connection in $\mathsf{E}(\mathsf{G} \times \mathsf{GL})$ in the form $\mathfrak{b} = \mathfrak{a}^{\mathsf{G}} + \mathfrak{a}^{gl}$. In parallel with the previous formulas, we obtain $\eta = \widehat{Ev}^* \mathfrak{b}$ and construct the corresponding ad-invariant polynomials and transgression formulas similar to (12.65, 12.71, 12.72). Due to the direct sum structure of η, the polynomials P_n will break down to a sum of product of irreducible polynomials for the group G and $\mathsf{GL}(\mathsf{d}, \mathbb{R})$, respectively.

In order to obtain the relevant transgression formulas, we have to specify the framework in which the map $\widehat{Ev}$ is defined. There is nothing new with respect to the previous cases, but we have to adapt the diagrams, such as (12.63), to this more general framework. In the diagram (12.63) $\mathcal{A}$ will be replaced by $\mathcal{A}^{\mathsf{G}} \oplus \mathcal{A}^{\text{metric}}$, P will be replaced by $\mathsf{P} + \mathsf{LM}^+$ and G by $\mathsf{G} \times \mathsf{GL}(\mathsf{d}, \mathbb{R})$. A particular attention must be paid to the group of gauge transformations, because diffeomorphisms act also on connections in P. We have therefore to consider a substitute of the lift of $\mathsf{Diff(M)}$ to P, because the latter may not exist if P is non-trivial. To avoid this obstacle, we consider the projection of $J : \mathsf{Aut}^{\mathsf{m}}(\mathsf{P}) \to \mathsf{Diff(M)}$, which is a subgroup of $\mathsf{Diff(M)}$. This subgroup acts both on P via $\mathsf{Aut}^{\mathsf{m}}\mathsf{P}$ and on LM^+ via the lift l; let us call it $\mathsf{Diff}_J(\mathsf{M})$. Therefore, $\mathcal{G}$ will be replaced by $\mathcal{G} \times \mathsf{Diff}^{\mathsf{m},1}_{J*}(\mathsf{M})$, where $J : \mathsf{Aut}(\mathsf{P} + \mathsf{LM}^+) \to \mathsf{Diff(M)}$. For instance, the bundle $\mathcal{X}$ will be

$$\mathcal{X} = \frac{\mathsf{M} \times \left(\mathcal{A}^{\mathsf{G}} \oplus \mathcal{A}^{\text{metric}}\right)}{\mathcal{G} \times \mathsf{Diff}^{\mathsf{m},1}_{J*}(\mathsf{M})} \tag{12.75}$$

and so on.

12.9.3 Local Lorentz Obstructions

It is often said that local Lorentz anomalies are the gauge anomalies of the $\mathsf{SO(d)}$ group. There is something true in it, but it should not be taken too literally. For gauge anomalies come from the $ch(\mathcal{V})$ factor of the index theorem (12.61), while local Lorentz anomalies are rooted in the $\widehat{A}$-genus term. It is nevertheless true that given a metric g and the corresponding principal fiber bundle of special orthogonal frames $\mathsf{O}_g\mathsf{M}$, the group of the relevant gauge transformations is the group of vertical automorphisms $\mathsf{Aut_V O}_g\mathsf{M}$ ($\mathsf{Aut_V O}^{\mathsf{m}}_g\mathsf{M}$ for a free action). Now, as we have learnt from the above, the two crucial things to be identified are the bundle $\mathcal{X}$ and its tangent space on one side and the map $\widehat{Ev}$ on the other. Then, denoting by $\mathcal{A}$ the space of $\mathsf{O}_g\mathsf{M}$ connections, let us consider the $\mathsf{SO(d)}$ bundle

$$\frac{\mathsf{O}_g\mathsf{M} \times \mathcal{A}}{\mathsf{Aut}^{\mathsf{m}}_{\mathsf{V}}\mathsf{O}_g\mathsf{M}} \to \mathsf{M} \times \frac{\mathcal{A}}{\mathsf{Aut}^{\mathsf{m}}_{\mathsf{V}}\mathsf{O}_g\mathsf{M}} \tag{12.76}$$

which has a connection η' similar to (12.62). Now we can construct the diagram

$$\begin{array}{ccc} \frac{\mathsf{O}_g\mathsf{M} \times \mathcal{A}}{\mathsf{Aut}^{\mathsf{m}}_{\mathsf{V}}\mathsf{O}_g\mathsf{M}} & \xrightarrow{\ \mathsf{Ev}\ } & \mathsf{ESO(d)} \\ \Big\downarrow \pi & & \Big\downarrow \pi \\ \mathcal{X} = \mathsf{M} \times \frac{\mathcal{A}}{\mathsf{Aut}^{\mathsf{m}}\mathsf{O}_g\mathsf{M}} & \xrightarrow{\ \mathsf{Ev}\ } & \mathsf{BSO(d)} \end{array} \tag{12.77}$$

and proceed to define $\eta = \mathsf{Ev}^* \mathfrak{a}^{so}$ and the relevant transgression formula. As we have already pointed out any irreducible polynomial of $\mathsf{GL(d)}$ is identified with (reduces to) an irreducible polynomial of $\mathsf{SO(d)}$, and we have

$$P_n(\mathcal{F}_{\mathfrak{a}^{gl}}, \ldots, \mathcal{F}_{\mathfrak{a}^{gl}}) = P_n(\mathcal{F}_{\mathfrak{a}^{so}}, \ldots, \mathcal{F}_{\mathfrak{a}^{so}}) \tag{12.78}$$

Therefore, the obstructions related to the local Lorentz symmetry are in one-to-one correspondence with the obstructions of the diffeomorphisms. However, in the two cases the same ad-invariant polynomial forms are pulled back via different paths and lead to different expressions of the relevant anomalies.

12.10 The Absence of Consistent Chiral Anomalies

As explained earlier, consistent chiral anomalies are fatal anomalies for a quantum field theory. If they are present they break, irreparably, the BRST Ward identities, which are crucial in order to guarantee renormalizability together with unitarity. It is therefore important to single out the theories in which anomalies are absent.

The absence or cancelation of local anomalies takes place:

- (A) in gauge field theories or sigma models in which the gauge group G has vanishing ad-invariant tensors. Consistent anomalies are determined by ad-invariant polynomials P_n, or by the corresponding ad-invariant tensors $t^{a_1 \ldots a_n}$ with a specific coefficient depending on the representation of the matter fields involved. In many simple groups, some of these tensors vanish identically. This is the case, as far as the polynomial P_3 is concerned, for all simple groups except $SU(N)$, for $N \geq 3$. This cancelation happens at the very source, in the sense that the form $P_3(F(\mathfrak{a}), \ldots, F(\mathfrak{a}))$, where $\mathfrak{a}$ is the universal connection, identically vanishes in the classifying space. Even in groups where these tensors are non-trivial, it may happen that the coefficient in front of it vanishes, depending on the representation of the fundamental fields. A well-known example is the case of matter in the adjoint representation in 4d for which $\mathrm{tr}(T^a_{(ad)}\{T^b_{(ad)}, T^c_{(ad)}\}) = 0$, where $(T^a_{(ad)})^{bc} = f^{abc}$.
- (B) in gauge field theories or sigma models with different fermion fields, which are separately anomalous, but when put together in the same theory the coefficients of the various consistent anomalies sum up to 0. The standard model of particle physics is an example (see below).

There is also a third case, where cancelation of chiral anomalies due to elementary fermion fields (let us call them primitive) may take place thanks to other fields in the theory. Such fields are endowed with transformation properties that allow them to cancel the primitive anomalies of the theory:

- (C) in sigma models, see Chap. 16, if the form $P_n(F, \ldots, F)$ in the target space T is non-vanishing but trivial, i.e. $P_n(F, \ldots, F) = dK$, for some $2n$–1 form K in T, it is possible to construct a Wess-Zumino term that cancels the corresponding anomaly. Another similar example is the Green-Schwarz mechanism, vastly used in target space field theories derived from superstrings: the type (A) and (B) mechanisms may be unable to get rid completely of the chiral anomalies produced by fermions, so that some residual anomalies (originated from reducible polynomials) are left;

in this case specific fields may be endowed with *ad hoc* transformation properties that erase the residual anomalies (see Chap. 17).

12.10.1 Index Theorem And Trace Anomalies

The family's index theorem of an elliptic operator (like Dirac-Weyl operator) signals the obstructions to the existence of an inverse (the propagator). So far we have considered the obstruction related to the $(\mathsf{d}, 2)$ component in $P \times \mathcal{A}$, see (12.58). We have seen that this obstruction is linked to the non-triviality of the Quillen determinant. There are other components in the RHS of (12.58). What is their meaning? There is no literature on this subject, but at least in the simplest cases, an explanation is possible.

Let us focus on the $(\mathsf{d}, 0)$ component, for $\mathsf{d}=4$. There are components coming from the $\widehat{A}$-genus and from $ch(\mathcal{V})$. The first gives a term $\frac{1}{192\pi^2}\mathrm{tr}(R^2)$, the second (in the non-Abelian case) a term $\frac{1}{8\pi^2}\mathrm{tr}F^2$. What can be the meaning of these two terms? The geometrical meaning of the $(\mathsf{d}, 0)$ component in $P \times \mathcal{A}$ could be the 'virtual' rank of the index bundle, as one can deduce from Formulas (12.56) and (12.57). As for a physical interpretation, it is natural to suggest a connection with the corresponding *odd* trace anomalies we have calculated previously for Weyl fermions coupled to a metric and a gauge field. The corresponding trace anomalies are not obtained via antitransgression from the universal bundle like the chiral consistent anomalies, but are pulled back directly to the spacetime M via the classifying map. The first motivation for the suggested connection, the obvious one, is the coincidence between the densities (Pontryagin and Chern density, respectively) for both the obstructions and the trace anomaly densities. The second, more important, is that the trace anomaly calculation with nonperturbative (heat-kernel-like) methods requires the existence of the full propagator (i.e. the complete inverse of the kinetic operator), while the non-vanishing obstructions represented by $\frac{1}{192\pi^2}\mathrm{tr}(R^2)$ and $\frac{1}{8\pi^2}\mathrm{tr}F^2$ may represent an obstacle to its existence.

Let us elaborate a bit on this argument. The analytic index, (12.5), is a local mathematical object that represents an asymmetry between the kernel of $\not{D}^+$ and the kernel of $\not{D}^-$. As long as this asymmetry is non-vanishing it represents a threat to the existence of the propagator (and the theory). The above-mentioned densities (Chern and Pontryagin class ones) are an additional hazard beside the ones contained in the $(\mathsf{d}, 2)$ component (in $P \times \mathcal{A}$) of the index, which give rises to the consistent chiral anomalies. This is the mathematical side of the problem. It is to be expected that such a problem will show up also on the field theory side.

The imaginary coefficient in (7.137) and (10.196) implies, in particular, that the Hamiltonian density upon quantization becomes complex, consequently it may break unitarity. So one may wonder whether this anomaly may be considered on the same footing as consistent chiral gauge anomalies in a chiral gauge theory, which, when present, spoils its consistency. On the one hand, one can object that in the case of chiral gauge anomalies in gauge theories the gauge fields propagate and drag the inconsistency in the internal loops, while in the present book the metric is treated as a background field. So, first let us notice that there are background geometries where the very Pontryagin density vanishes. They include for instance the FRW, Schwarzschild and gravitational wave metrics. Therefore, in such backgrounds the odd trace anomaly simply vanishes. On the other hand, these are very special macroscopic geometries. For a generic geometry, the Pontryagin density does not vanish. For instance in a cosmological framework, we can imagine to go back to higher energies where gravity inevitably back-reacts; or we can consider a quantized gravity theory. In these cases, it does not seem to be possible to avoid the conclusion that the Pontryagin density does not vanish and unitarity may be affected due to the trace anomaly. These considerations on the Pontryagin density are even more cogent for the Chern class densities and the corresponding trace anomalies, especially because they involve quantized gauge fields which certainly propagate in the internal loops. Thus, seen in this more general context, the breakdown of unitarity due to a chiral unbalance in an asymptotically free matter theory should be taken seriously into account.

The mathematical way of analyzing consistent chiral gauge anomalies is to search for obstructions to the existence of the fermion propagator, while the field theory approach is constructive: as we have seen several times, it concocts a propagator (for instance, by adding to the theory a free Weyl fermion of opposite chirality) and pushes the analysis as far as possible. The remarkable thing is that this procedure comes across the same obstacles as the mathematical approach, in the form of consistent chiral anomalies. Now for trace anomalies, we face a similar situation. On one side we have an obstacle to the existence of the fermion propagator represented by the Pontryagin and/or the Chern classes, on the field theory side we have the calculation of the trace of the e.m. which requires the existence of the fermion propagator. Therefore the trace anomalies related to Pontryagin and Chern class densities represent a threat to a theory analogous to those of chiral consistent anomalies. Looking, at Eqs. (7.137) and (7.203), we see that they imply imaginary quantum corrections to the Hamiltonian. In the prospect of a quantum gravity renormalization program, this is, no doubt, a severe challenge.

The trace anomalies due to the Pontryagin class obstruction exist not only in $\mathsf{d} = 4$, but also in any dimension $\mathsf{d} = 4k$. The trace anomalies due to the Chern class obstruction exist in any $\mathsf{d} = 2k$.

The conditions for vanishing of odd parity trace anomalies in the context of the standard model of particle physics is analyzed in Sect. 12.10.3.

Note. The existence of odd parity trace anomalies, originated both from a gauge or a metric background, has been sometime called into question. It is worth spelling out any doubt about it. Of course, the principal and necessary evidence is provided by the direct calculation. Another argument is the rigid link between the odd parity trace anomaly and the chiral (ABJ) anomaly. In the previous chapters, we have seen that they both confirm the existence of odd parity trace anomalies. But, sometime, a more qualitative and heuristic argument may be more convincing than rigorous, but inevitably less intuitive, calculations.

The first necessary condition for the appearance of an anomaly is that its quantum numbers correspond to those of the current divergence or of the energy-momentum trace in question. Let us ask: when is the case that chiral anomalies, which are possible on the basis of their quantum numbers, do not make their appearance? In quantum field theory, there are two possibilities, which we have classified as cases (A) and (B) above (case (C) is not pertinent to this discussion). Case (A) has its origin in group and group representation theory, and for this reason, we can call it *group theoretical* cancelation mechanism. Case (B) is instead the result of a collection of elementary fields, which are anomalous when considered separately, but conjure up an overall cancelation when inserted in the same theory. We can call it a *collective* cancelation mechanism. A simple example of this kind is the vanishing of the consistent gauge anomaly for a Dirac fermion. A Dirac fermion can be viewed as the sum of a left-handed plus a right-handed Weyl fermion. The latter have each a consistent gauge anomaly with opposite coefficients so that the overall coefficient is 0. This example can be regarded also in another way. A Dirac fermion theory is parity invariant, while a non-vanishing consistent gauge anomaly would break parity. So the vanishing of the consistent anomaly can be viewed as due to the protection of parity symmetry.

If we now examine all the various cases in which a (single) Weyl fermion field is involved, we see that it can be anomaly free only due to a *group theoretical cancelation*, in the case of consistent gauge or diffeomorphism/local Lorentz anomalies. In all the other cases, including trace anomalies, Weyl fermions are anomalous. Coming now to the Chern and Pontryagin trace anomalies in 4d in this backdrop, we face a situation in which a density like the Chern and Pontryagin class ones have the right quantum number, dimensions and properties to couple to the trace of the energy-momentum tensor of a Weyl fermion in the Ward identity of conformal symmetry (in particular it is consistent). None of the cancelation mechanisms described above can operate: the group theoretical mechanism is absent, the collective mechanism is absent, and there is no extra symmetry that can prevent its appearance. The conclusion is that the Pontryagin and Chern trace anomalies must be there, because nothing prevents their appearance. Actually, had we found a vanishing coefficient of these anomalies we would face a serious problem, because, like in other similar problems, the question would be: how can we explain it?

12.10.2 The Standard Model Example

The standard model of particle physics is a good example of anomaly cancelation, where both mechanisms (A) and (B) are present. As is well known the gauge group is $SU(3) \times SU(2) \times U(1)$. The quark and lepton content is summarized in the following table.

$$\begin{array}{cccc} \text{G/fields} & SU(3) & SU(2) & U(1) \\ \begin{pmatrix} u \\ d \end{pmatrix}_L & 3 & 2 & \frac{1}{6} \\ \widehat{u_R} & \bar{3} & 1 & \frac{2}{3} \\ \widehat{d_R} & \bar{3} & 1 & -\frac{1}{3} \\ \begin{pmatrix} \nu_e \\ e \end{pmatrix}_L & 1 & 2 & -\frac{1}{2} \\ \widehat{e_R} & 1 & 1 & -1 \end{array} \tag{12.79}$$

where $\hat{\psi} = \gamma^0 C\psi$, C being the charge conjugation matrix; i.e. $\hat{\psi}$ represents the Lorentz covariant conjugate field. The second column assigns the relevant representation of $SU(3)$, the third of $SU(2)$ and the last of $U(1)$, whose representations are denoted by the corresponding hypercharge eigenvalues. In this section, we consider the high energy limit of the SM, which we believe corresponds to a conformal fixed point, where all the elementary particles are massless (here we are interested in fermions, but one can think of a model with all fields being massless).

As we know, the chiral consistent anomaly is determined by the tensor $d^{abc} = \frac{1}{2}\text{tr}\left(T^a\{T^b, T^c\}\right)$, where T^a denotes the total anti-hermitean generator of the Lie algebra $\mathfrak{su}(3) \oplus \mathfrak{su}(2) \oplus \mathfrak{u}(1)$: dropping labels we can generically write $T = T^{\mathfrak{su}(3)} \oplus T^{\mathfrak{su}(2)} \oplus T^{\mathfrak{u}(1)}$. The tensor d^{abc} decomposes into various independent components, which we list hereafter:

- $T^{\mathfrak{su}(3)} \times T^{\mathfrak{su}(3)} \times T^{\mathfrak{su}(3)}$: counting the component fields, the total $SU(3)$ representation is a $3 \oplus 3 \oplus \bar{3} \oplus \bar{3}$, which is real, because $\bar{3}$ is the complex conjugate of 3. If a representation is real $T^{a\dagger} = -T^a$. Taking the hermitean conjugate of d^{abc} we find $d^{abc} = -d^{abc}$, therefore $d^{abc} = 0$.
- $T^{\mathfrak{su}(2)} \times T^{\mathfrak{su}(2)} \times T^{\mathfrak{su}(2)}$, which vanishes because the tensor d^{abc} vanishes in general for the Lie algebra $\mathfrak{su}(2)$.
- $T^{\mathfrak{su}(2)} \times T^{\mathfrak{su}(2)} \times T^{\mathfrak{u}(1)}$, in which case we have the trace of two $\mathfrak{su}(2)$ generators in two doublet representations. These traces are non-vanishing because $\text{tr}(T^a T^b) \sim \delta^{ab}$, but they are multiplied by the corresponding $\mathfrak{u}(1)$ charges, whose total value is $6\left(\frac{1}{6}\right) - 2\left(\frac{1}{2}\right) = 0$.
- $T^{\mathfrak{su}(3)} \times T^{\mathfrak{su}(3)} \times T^{\mathfrak{u}(1)}$, in which case we have the trace of two $\mathfrak{su}(3)$ left triplet generators and two right triplet generators. These traces are again non-vanishing, but they are multiplied by the corresponding $\mathfrak{u}(1)$ charge, whose total value is $3\left(2\left(\frac{1}{6}\right) - \frac{2}{3} + \frac{1}{3}\right) = 0$.
- $T^{\mathfrak{u}(1)} \times T^{\mathfrak{u}(1)} \times T^{\mathfrak{u}(1)}$, in this case the tensor is proportional to the overall sum of the charge products: $6\left(\frac{1}{6}\right)^3 - 3\left(\frac{2}{3}\right)^3 - 3\left(-\frac{1}{3}\right)^3 + 2\left(-\frac{1}{2}\right)^3 - (-1)^3 = 0$

This completes the analysis of gauge anomalies in the standard model. There are no local anomalies. We should consider however also the coupling to gravity, as weak as it may be. The fermion sector of the massless standard model is conformal invariant, and therefore, it makes sense to consider not only diffeomorphism and mixed anomalies, but also trace anomalies generated by this interaction, in particular the odd parity trace anomalies (see below).

For the time being let us deal with chiral diffeomorphism anomalies. We have already pointed out several times that they are absent in 4d. However, there could be mixed gauge and gravitational anomalies. Adopting the notation Σ for the generic Lorentz generator, it is easy to see that the only a priori non-vanishing possibility is

- $\Sigma \times \Sigma \times T^{u(1)}$, the trace tr $\left(\Sigma^{ab}\Sigma^{cd}\right)$ is non-vanishing, but it is multiplied by the total $U(1)$ charge: $6\left(\frac{1}{6}\right) - 3\left(\frac{2}{3}\right) - 3\left(-\frac{1}{3}\right) + 2\left(-\frac{1}{2}\right) + 1 = 0$.

12.10.3 Cancelation of Odd Trace Anomalies in the SM

The formulation (12.79) of the standard model is the old one, in the preneutrino-mass-discovery era, because at present its complete formulation is still an open problem. All the fields are Weyl spinors and a hat represents Lorentz covariant conjugation. If a field is right-handed, its conjugate is left-handed. Thus, all the fields in (12.79) are left-handed. This is the well-known chiral formulation of the SM. So we could represent the entire family as a unique left-handed spinor ψ_L and write the kinetic part of the action as in (6.9). However, the coupling to gravity of a Lorentz covariant conjugate field is better described as follows. With the help of the C and γ_5 properties introduced in Chap. 1, for a generic spinor field ψ, we have

$$\widehat{\psi_R} = \gamma^0 C \psi_R^* = \gamma^0 C P_R^* \psi^* = P_L \gamma^0 C \psi^* = P_L \hat{\psi} = \hat{\psi}_L \tag{12.80}$$

Then

$$\begin{aligned}\sqrt{|g|}\,\overline{\hat{\psi}_L}\,\gamma^m(\nabla_m + \frac{1}{2}\omega_m)\hat{\psi}_L &= \sqrt{|g|}\,\overline{\widehat{\psi_R}}\,\gamma^m(\nabla_m + \frac{1}{2}\omega_m)\widehat{\psi_R}\\ &= \sqrt{|g|}\,\psi_R^T\, C^{-1}\gamma^m\gamma^0(\nabla_m + \frac{1}{2}\omega_m)C\psi_L^*\end{aligned}$$

which, after a partial integration and an overall transposition, becomes

$$\sqrt{|g|}\,\overline{\psi_R}\,\gamma^m(\nabla_m + \frac{1}{2}\omega_m)\psi_R \tag{12.81}$$

i.e. the right-handed version of the action (6.9). This follows in particular from the property $C^{-1}\Sigma_{ab}C = -\Sigma_{ab}^T$.

Now, as pointed in Chap. 1, a second quantized field ψ_L describes simultaneously the creation of left-handed particles and the annihilation of right-handed antiparticles. The conjugate $\widehat{\psi_R}$ does the same. We see therefore that cancelation of anomalies takes place for left-handed particles and right-handed antiparticles (while right-handed particles and left-handed antiparticles are absent).

Now let us come to the odd parity trace anomalies of the SM. First, we must clarify that they are relevant only when gravity is effectively coupled to the model. For, looking at the effective action (7.108) one can see that any quantum contribution to the trace of the e.m. tensor is multiplied by the metric fluctuation h. If the latter vanishes, no quantum contribution from the trace anomaly can affect the effective action. Therefore, the previous anomaly analysis is completely satisfactory if gravity can be disregarded. On the other hand, if gravity is assumed to interact with the standard model via the minimal covariant couplings we have been considering throughout this book, odd trace anomalies may have significant fallouts, as was pointed out before. Therefore, it makes sense to study the conditions under which also these anomalies cancel. Hereafter we broach this subject.

For odd parity trace anomalies, the cancelation takes place in any case if there is a perfect balance between opposite chiralities, between, say, left-handed particles and right-handed antiparticles. From the above, we see that in the multiplet (12.79) there is a balance between the left-handed and right-handed components except for the left-handed ν_e. Therefore the multiplet (12.79), when weakly coupled to gravity, will produce an overall non-vanishing (imaginary) coefficient for the Pontryagin density in the trace anomaly. This breakdown is naturally avoided if we add to the above SM multiplet other Weyl fermions (for instance a right-handed neutrino) so as to produce a chirally symmetric model without compromising the cancelation of the chiral gauge and gravity anomalies.

The analysis concerning gauge-induced odd trace anomalies, see (7.137, 10.209), is more complex. First of all, we have three types of such anomalies, constructed with $SU(3)$, $SU(2)$ and $U(1)$ gauge fields, respectively. We have six units of the anomaly (7.137) with curvature $F \equiv F^{\mathfrak{su}(3)}$ and six units with opposite sign. Therefore, the multiplet (12.79) is free of these anomalies. We have instead 8 units of the same anomaly with gauge field $F \equiv F^{\mathfrak{su}(2)}$ and positive sign; see (7.138). Finally, we have a '$U(1)$ gauge-induced trace anomaly with vanishing coecient'. The simplest way to cancel these anomalies is to have perfect left-right symmetry for the electroweak sector. But of course, there are many other possibilities.

12.11 Index Theorem and Covariant Anomalies

Besides the use of the family's index theorem in describing consistent anomalies, which has been the main focus of this chapter, there is in field theory another realm of applications of the index theorem: the ones that concern covariant anomalies. The classical example is the covariant (ABJ) anomaly (6.54), which we copy here,

$$\partial^\mu \langle\langle j_{5\mu}(x)\rangle\rangle = \frac{1}{4\pi^2}\varepsilon_{\mu\nu\lambda\rho}\,\partial^\mu V^\nu(x)\partial^\lambda V^\rho(x). \tag{12.82}$$

This anomaly makes its appearance in the theory of a Dirac fermion ψ coupled to an Abelian vector potential $V_\mu(x)$. The theory is classically invariant under the chiral transformation $\psi(x) \to e^{-i\gamma_5\alpha(x)}\psi(x)$ and $\not{V}(x) \to \not{V}(x) + i\not{\partial}\alpha(x)\gamma_5$, where $\alpha(x)$ is a generic function, which implies a conservation of the current $j_{5\mu} = i\bar{\psi}\gamma_\mu\gamma_5\psi$. Quantization leads to the violation (12.82). Generalizing it to the non-Abelian case is straightforward. Hereafter in this section, for simplicity, we limit ourselves to the Abelian case.

Let us take a Dirac operator $\not{D}$, twisted by an Abelian gauge connection V, as above, on a compact Euclidean spacetime M (we dispense with the tilde), and consider a fixed configuration of V. Then the spectrum of $\not{D}$ will be discrete. Now, let us apply the method of Sect. 9.1, but let us focus on the zero modes of $\not{D}$. Since we want to separate the zero modes from the nonzero ones, we define the surrogate functional integral $\overline{Z} = \prod_i \bar{\mu}_i$,[5] where $\bar{\mu}_i \neq 0$ are the non-vanishing eigenvalues of $\not{D}$. As a consequence, we introduce the projector P_0

$$P_0 = \sum_{i=1}^{n_0} |\psi_i^{(0)}\rangle\langle\psi_i^{(0)}| \tag{12.83}$$

of the zero modes of $\not{D}$ and, instead of (9.5), we use the new regularizing projector

$$\overline{P}_\Lambda = P_\Lambda - P_0 \tag{12.84}$$

The surrogate regularized fermion determinant is

$$\overline{Z}_\Lambda[\not{D}] = \sqrt{\det\left(1 - \overline{P}_\Lambda + \overline{P}_\Lambda \not{D}^2 \overline{P}_\Lambda\right)} \tag{12.85}$$

and the (surrogate) regularized chiral current is

$$\overline{j}_{5\mu}^{(\Lambda)} = i\,\mathrm{tr}\left(\gamma_\mu\gamma_5\langle x|\frac{\not{D}}{\overline{P}_\Lambda}|x\rangle\right) \tag{12.86}$$

where the trace tr refers to the gamma matrices. Proceeding as in Sect. 9.1, we find the anomalous conservation law

$$\partial^\mu \overline{j}_{5\mu}^{(\Lambda)} = -2i\left(\gamma_5\langle x|\overline{P}_\Lambda|x\rangle\right) \tag{12.87}$$

Now observe that $\overline{j}_{5\mu}^{(\Lambda)}$ does not contain the zero modes, and therefore, its divergence cannot contain singularities (singularities are related to zero modes). Therefore, we can integrate the LHS of (12.87) over M and get 0. Therefore,

$$0 = \int_{\mathsf{M}} \mathrm{tr}\left(\gamma_5\langle x|P_\Lambda|x\rangle\right) - \int_{\mathsf{M}} \mathrm{tr}\left(\gamma_5\langle x|P_0|x\rangle\right) \tag{12.88}$$

[5] An object like $\overline{Z}$ makes sense only for a fixed configuration of the potential V_μ. Should we consider different gauge-inequivalent configurations the zero modes and their multiplicity may change from one to another and, in that case, the only sensible treatment is by means of K-theory and the index theorem for families.

The first term in the RHS of this equation has already been calculated and equals the RHS of (12.82). It is universal and holds for any field configuration. To evaluate the second term, we can split the zero modes $\psi_i^{(0)}$ into eigenvactors of γ_5 with eigenvalue ± 1, $\psi_j^{(0+)}$ and $\psi_k^{(0-)}$, respectively. Then, due to orthonormality of these eigenfunctions, we have

$$\int_{\mathsf{M}} \mathrm{tr}\,(\gamma_5 \langle x | P_0 | x \rangle) = \int_{\mathsf{M}} \sum_{j=1}^{n_+} \psi_j^{(0+)*}(x)\psi_j^{(0+)}(x) - \int_{\mathsf{M}} \sum_{k=1}^{n_-} \psi_j^{(0-)*}(x)\psi_k^{(0-)}(x)$$
$$= n_+ - n_- \qquad (12.89)$$

where $n_\pm$ is the number of eigenvalues with ± 1 chirality. Finally, we have the index theorem formula

$$n_+ - n_- = \frac{1}{8\pi^2} \int_{\mathsf{M}} \varepsilon_{\mu\nu\lambda\rho}\, \partial^\mu V^\nu(x) \partial^\lambda V^\rho(x). \qquad (12.90)$$

The previous one is a homemade proof of the Atiyah-Singer index theorem of Sect. 12.6 (which we call *fixed background* to distinguish it from the family's index theorem), see Eqs. (12.23, 12.46) with reference in particular to the third term in the RHS of (12.47). If the index is nonvanishing the field $V_\mu(x)$ must have singularties at some points, thus we need to extend the validity of (12.82) beyond the perturbative configurations considered in Chap. 6.

A formula similar to (12.90) holds also in the non-Abelian singlet and non-singlet case, but in the latter case the relevant theorem is the equivariant index theorem. And an analogous formula holds also for the Kimura-Delbourgo-Salam anomaly (7.205).

It is important to stress the differences between the just illustrated index theorem and the family's index theorem. To start with, the first applies to the covariant anomalies of some chiral currents, the second to consistent anomalies. The first can be read off directly from the general Formulas (12.23, 12.46) and (12.47), the index is a number that can be calculated by integrating the covariant anomaly over spacetime, and Eq. (12.90) holds for a single configuration of $V_\mu(x)$. The family's index is not a number but an element of a suitable K-theory, it can be connected to the corresponding consistent anomaly only indirectly: on the one side via the transgression formula the consistent anomaly is connected to the cohomology of the classifying space, on the other side the latter is connected via the $\eta(\eta')$ connection to the family's index Formula (12.58) and the Quillen determinant bundle.

There is another important difference between the two index theorems' applications. As is evident that in Formula (12.90), the topology of the spacetime M plays a fundamental role. On the contrary in the case of application of the family's index theorem, due to locality, one has to abstract from the topology of spacetime.

The previous remarks are intended to highlight the fact that the above two index theorem applications should not be confused.

12.11.1 Even and Odd Trace Anomalies

Besides the family of odd parity trace anomalies discussed above, there is a second large family, the even parity anomalies. They include anomalies which appear in the trace of the energy-momentum tensor. They may have a partner in the divergence of the e.m. tensor (a second type diffeomorphism anomaly), but, in all known cases, this partner is trivial. We have calculated examples of this second family only occasionally, in Chaps. 7 and 10, see in particular Eq. (10.217) for Dirac fermions. In Sect. 17.1, we will compute the trace anomaly for a scalar field theory in 2d. Even parity trace anomalies have a different origin than the odd parity ones, and they affect differently the theory where they appear. In the just mentioned examples, it is evident that inverting the kinetic operator (the Dirac and the quadratic scalar, respectively) is no problem at all, and therefore those trace anomalies are no obstructions to the existence of the relevant propagators.

Mainly for this reason, even parity trace anomalies are not the main subject of this book. It is, nevertheless, worth delving into their role and significance in order to appreciate the difference with the odd parity ones. There is a vast literature on them. They have been computed mostly with heat kernel-like methods. Apart from the examples considered in this book, where the integrated out fields are either scalars or spin 1/2 fields, they appear in theories where the integrated out fields have spin ≥ 1. The simplest example is provided by the Maxwell field, which we have already considered in Sect. 6.1 in interaction with a Weyl fermion. There, in order to invert the Maxwell kinetic operator we introduced a gauge fixing, which transformed it into a D'Alambertian operator. The latter is invertible in the field theory sense and, after a Wick rotation, it becomes a Euclidean quadratic operator. In a nontrivial background metric, it reduces to the Maxwell operator: $\Box A_\mu - R_{\mu\nu} A^\nu = 0$. In any case, it is self-adjoint. Now, the family's index of a self-adjoint operator is (generically) zero [12]. Therefore, also in this case, we find an accord between quantum field theory and the family's index theorem.

For any other higher spin bosonic field, the treatment is similar [16]. One chooses the gauge in order to reduce the kinetic operator to an invertible quadratic self-adjoint operator. A more involved but similar treatment is devoted to fermion fields (such as the spin 3/2 Rarita-Schwinger field). The basic strategy for all these fields (except of course for the chiral ones, in which case one faces a non-trivial family's index), leads to an invertible quadratic self-adjoint operator, which guarantees both that the corresponding propagator exists and the family's index is trivial. In 4d, the trace anomaly for the Maxwell field and all the other mentioned higher spin fields is a superposition of the Euler density and the square Weyl tensor (plus a trivial $\Box R$ term with a regularization-dependent coefficient); see Sect. 5.3. As for the Euler density, it is the Pfaffian of the Riemann curvature and has a topological character. The square Weyl tensor is non-topological.

The previous considerations suggest that for field theory anomalies a somewhat subtler subdivision is needed than the simple separation between the two families of even and odd parity. The first set encompasses all odd parity anomalies (except the

ABJ ones). It includes all chirally split (i.e. with opposite sign for opposite chiralites) anomalies: consistent gauge and diffeomorphism/Lorentz anomalies which appear in the divergence of a current or of the e.m. tensor, as well as the odd parity trace anomalies. The origin of all these anomalies can be traced back to the index theorem for families and to the topology of the classifying space. This family includes also the chiral gauge anomalies of the $V-A$ background for Dirac fermions (except for the limiting case $A=0$). At first sight, it would seem that they are non-split, but, as we have seen, such systems do split into two subsystems of Weyl fermions of opposite chirality coupled to potentials $V_\mu - A_\mu$ and $V_\mu + A_\mu$, respectively, to each of which we can apply the family's index theorem and each of which may lack the fermion propagator. The same holds for the chiral gravitational anomalies in a MAT background.

The second set includes all the above-mentioned even trace anomalies. They do not have any bearing with the topology of the corresponding classifying space. The relevant (fermion or boson) propagators exist, and they are no threat to the consistency of the theory. Some of the corresponding densities may have a topological significance, in that they may appear in some fixed background index theorem (like the Gauss-Bonnet, for instance). But the connection between these trace anomalies and index theorems seems to be rather accidental, since (apart from the 2d case), neither a trace anomaly fully determines an index theorem nor vice versa.

But we need also another category, a subset of the odd parity trace anomalies. It is formed by the (odd) ABJ anomalies in gauge field theories and the ABJ-like anomalies, like the KDS anomaly, in non-trivial gravitational backgrounds. They are chiral, they appear for instance in theories of Dirac fermions as well as in all theories, such as the SM, where cancelation of local anomalies occur. In those theories, they feature as covariant anomalies. They do not bear a direct relation to the topology of the relevant classifying space, their family's index being 0. They are characterized by well-defined propagators and do not threaten the consistency of the theory. In the example of the previous subsection, they have a topological meaning, in that they correspond to the formulation of the fixed background Atiyah-Singer index theorem. Whether all the anomalies of this category, for instance (10.205), may have a similar interpretation is an open question.

Appendix 12A. The η Connection

The connection ω in the definition of the connection η, (12.62), is introduced as follows. The space $\mathcal{A}$ admit a Riemannian metric $\langle \delta A, \delta A\rangle = \int d^{\mathsf{d}}x \, \mathrm{tr}\left(\delta A_\mu \delta A^\mu\right)$, where δA_μ denotes the variation with respect to a background connection A_0. Therefore $\delta_{\mathcal{G}} \delta A_\mu = [\delta A_\mu, \lambda]$ for $\lambda \in \mathcal{G}$, so that the metric is $\mathcal{G}$ invariant. The orthogonal complements of the orbits of $\mathcal{G}$ define a connection in $\mathcal{A} \to \mathcal{A}/\mathcal{G}$, which we denote by ω.

Now we have also a Riemannian metric in P, which is determined by the metric in M, the metric in G and by a connection A. A determines a splitting of the tangent bundle $T\mathsf{P}$ of P into a vertical and a horizontal bundle. This metric satisfies

$k(l_{g*}X, l_{g*}Y) = k(X, Y)$ for all vertical vectors X, Y and $g \in \mathsf{G}$, where l_g is the left multiplication along the fiber and l_{g*} is the corresponding push-forward. Thus, it is invariant under the action of G. On the other hand, the metric $\langle \delta A, \delta A \rangle$ is also invariant under the action of G in P, therefore the overall metric in $\mathsf{P} \times \mathcal{A}$ descends to a metric in $\mathsf{P} \times \mathcal{A}/\mathcal{G}$ due to $\mathcal{G}$ invariance, and to a metric in $\mathsf{M} \times \mathcal{A}/\mathcal{G}$, due to G invariance. The orthogonal complements of the orbits of G define η.

As a side remark, let us notice that, looking at the definition (12.62), if we restrict η to the orbit of $\mathcal{G}$ we obtain $\eta = ev^* A$, where $ev : \mathsf{P} \times \mathcal{G} \to \mathsf{P}$.

The just defined η connection is universal, meaning that there exists a bundle map, which we call $(\mathsf{Ev}, \underline{\mathsf{Ev}}) \equiv (\hat{f}_\times, f_\times)$, between $\mathsf{P} \times \mathcal{A}/\mathcal{G}$ and $\mathsf{M} \times \mathcal{A}/\mathcal{G}$ such that η' is the pullback of the universal connection $\mathfrak{a}$: $\eta' = \mathsf{Ev}^*\mathfrak{a}$. In the sequel, we would like to motivate this.

Let us consider the following diagrams

$$\begin{array}{ccc} \mathsf{P} \times \mathcal{A} & \xrightarrow{\mathsf{ev}} & \mathsf{EG} \\ \downarrow \pi & & \downarrow \pi \\ \mathsf{M} \times \mathcal{A} & \xrightarrow{\underline{\mathsf{ev}}} & \mathsf{BG} \end{array} \tag{12.91}$$

and

$$\begin{array}{ccc} \frac{\mathsf{P} \times \mathcal{A}}{\mathcal{G}} & \xrightarrow{\mathsf{Ev}} & \mathsf{EG} \\ \downarrow \pi & & \downarrow \pi \\ \frac{\mathsf{M} \times \mathcal{A}}{\mathcal{G}} & \xrightarrow{\underline{\mathsf{Ev}}} & \mathsf{BG} \end{array} \tag{12.92}$$

In these commutative diagrams, the use of evaluation map symbols (although with a different font than the usual ev and Ev symbols) is a slight abuse of language which needs some explanation. These symbols would be appropriate if we could identify $\mathcal{A}$ with $Hom(\mathsf{P}, \mathsf{EG})$. Any element of $Hom(\mathsf{P}, \mathsf{EG})$ is a bundle map $(\hat{f}, f)$ for which from the universal connection $\mathfrak{a}$ we can obtain the connection $A = \hat{f}^*\mathfrak{a}$ in P. So such an identification is not unmotivated, but it is somewhat approximate. If this identification were correct ev would be the usual evaluation map $ev(u, \hat{f}) = \hat{f}(u)$. This is the reason for using the symbol ev in (12.91) as the map replacing ev when $\mathcal{A}$ replaces $Hom(\mathsf{P}, \mathsf{EG})$. Similarly, $\underline{\mathsf{ev}}$ replaces $\underline{ev}$, which is defined by $\underline{ev}(x, f) = f(x)$.

As for (12.92), Ev and $\underline{\mathsf{Ev}}$ take the place of Ev and $\underline{Ev}$, which descend from ev and $\underline{ev}$, respectively. For we define the action of $\mathcal{G}$ on $\mathsf{P} \times Hom(\mathsf{P}, \mathsf{EG})$ as

$$(u, \hat{f}) \to (u, \hat{f})\psi \equiv (\psi(u), \hat{f} \circ \psi^{-1}), \qquad \forall \psi \in \mathcal{G}$$

and on $\mathsf{M} \times Hom(\mathsf{P}, \mathsf{EG})$ as

$$(x, \hat{f}) \rightarrow (x, \hat{f})\psi \equiv (j(\psi)(x), \hat{f} \circ \psi^{-1})$$

Therefore, $Ev(u, \hat{f}) = \hat{f}(u)$ and $\underline{Ev}(x, f) = f(x)$.

The reason why we cannot simply identify the symbols ev, Ev with ev, Ev is because the map $\hat{f}$, which cover f, is not uniquely defined; i.e. one cannot just identify $\mathcal{A}$ with $Hom(\mathsf{P}, \mathsf{EG})$. However, it has been proved that this is essentially true if we take, for instance, n large enough in the approximate representation (11.53) of the principal fiber bundle EG. In that case, the map $\hat{f}$ is unique in the sense that there is a connection preserving homotopy between any two such maps. In other words, it does not matter what $\hat{f}$ we choose in any homotopy class, because the pulled back connection is always the same. Since we are interested not as much in the maps $\hat{f}$ but rather in the connections, taking n large enough for our needs, we are justified in using the above two diagrams (12.91) and (12.92).

Appendix 12B. The Bundle of Spin Frames

Appendix 12B. The Bundle of Spin Frames

The bundle of spin frames $\mathsf{LM}_{\mathsf{Spin}}$ has structure group $\widetilde{\mathsf{GL}}(\mathsf{d}, \mathbb{R}))$, which is a double covering of $\mathsf{GL}(\mathsf{d}, \mathbb{R})^+$, the subgroup of $\mathsf{GL}(\mathsf{d}, \mathbb{R})$ with positive determinants. It is defined by being the double covering of LM^+, the bundle of oriented frames, and by the existence of a bundle morphism $\mathsf{LM}_{\mathsf{Spin}} \rightarrow \mathsf{LM}^+$ which is induced by the standard double covering homomorphism $\widetilde{\mathsf{GL}}(\mathsf{d}, \mathbb{R}) \rightarrow \mathsf{GL}(\mathsf{d}, \mathbb{R})^+$. The metric connections in $\mathsf{LM}_{\mathsf{Spin}}$ are denoted by $\mathcal{A}^{\mathrm{spin}}$. One can prove that $\mathcal{A}^{\mathrm{spin}}$ is isomorphic to $\mathcal{A}^{\mathrm{metric}}$. The group of gauge transformations relevant in this case is denoted $\widetilde{\mathsf{Diff}}^{\mathsf{m},1}\mathsf{M}$ and corresponds to the diffeomorphisms that lift (in two ways) to $\mathsf{LM}_{\mathsf{Spin}}$. It follows that, instead of (12.70), we have

$$\begin{array}{ccc} \dfrac{\mathsf{LM}_{\mathsf{Spin}} \times \mathcal{A}^{\mathrm{spin}}}{\widetilde{\mathsf{Diff}}^{\mathsf{m},1}(\mathsf{M})} & \xrightarrow{\mathsf{Ev}} & \mathsf{EGL}(\mathsf{d}, \mathbb{R})^+ \\ \big\downarrow \pi & & \big\downarrow \pi \\ \dfrac{\mathsf{M} \times \mathcal{A}^{\mathrm{spin}}}{\widetilde{\mathsf{Diff}}^{\mathsf{m},1}(\mathsf{M})} & \xrightarrow{\mathsf{Ev}} & \mathsf{BGL}(\mathsf{d}, \mathbb{R})^+ \end{array} \tag{12.93}$$

References

1. M.F. Atiyah, I.M. Singer, The index of elliptic operators on compact manifolds. Bull. Am. Math. Soc. **69**, 422–433 (1963)

2. M.F. Atiyah, I.M. Singer, The index of elliptic operators I. Ann. Math. **87**, 484–530 (1968)
3. M.F. Atiyah, I.M. Singer, The index of elliptic operators III. Ann. Math. **87**, 484–530 (1968)
4. M.F. Atiyah, G.B. Segal, The index of elliptic operators II. Ann. Math. **87**, 531–545 (1968)
5. M.F. Atiyah, I.M. Singer, The index of elliptic operators IV. Ann. Math. **93**, 119–138 (1971)
6. M.F. Atiyah, I.M. Singer, The index of elliptic operators. V. Ann. Math. **93**, 139–149 (1971)
7. H.B. Lawson Jr, M.-L. Michelsohn, *Spin Geometry* (Princeton University Press, Princeton, 1989)
8. D. Quillen, Determinants of a Cauchy-Riemann operators over a Riemann surface. Funks. Anal. Prilozh. **19**, 37 (1985)
9. K. Furutani, On the Quillen determinant. J. Geom. Phys. **49**, 366–375 (2004)
10. M.F. Atiyah, I.M. Singer, Dirac operators coupled to vector potentials. Proc. Nat. Acad. Sci. **81**, 2597 (1984)
11. M. Karoubi, *K-Theory* (Springer, New York-Heidelberg-Berlin, An Introduction, 1970)
12. M.F. Atiyah, *K-Theory* (Addison-Wesley, 1989)
13. O. Alvarez, I.M. Singer, B. Zumino, Gravitational anomalies and the family's index theorem. Commun. Math. Phys. **96**, 409 (1984)
14. L. Alvarez-Gaumé, E. Witten, Gravitational anomalies. Nuclear Phys. B **234**, 269 (1984)
15. L. Alvarez-Gaumé, P. Ginsparg, The topological meaning of nonabelian anomalies. Nuclear Phys. B **243**, 449–474 (1984)
16. S.M. Christensen, M.J. Duff, New gravitational index theorems and super theorems. Nuclear Phys. B **154**, 301 (1979)

Chapter 13
Global Anomalies

Global anomalies in the field theory literature may have an exotic flavor, for they require an amount of mathematics which is not usually requested in ordinary quantum field theory. But the basic idea, once we are familiar with local anomalies, is very simple. Local anomalies are cocycles of the local cohomology defined by BRST transformations, which are infinitesimal gauge transformations. The latter are contained in a neighborhood of the identity of the group $\mathcal{G}$ of gauge transformations, that is they define the Lie algebra of $\mathcal{G}$. They are unable to probe the global structure of $\mathcal{G}$. Therefore, the question arise, first, as to whether it is possible to extend the local cocycles to the full group, second, whether this extension, if any, is uniquely defined. We shall see that the answer to the first question is yes, while the answer to the second depends on the topology of the spaces involved. More precisely for ordinary field theory, the crucial topology is the topology of the classifying space, while for sigma model, it is the topology of the target space.

When one tries to extend a local cocycles to the entire group $\mathcal{G}$, one finds out that there are in general different possibilities. The mathematical tools that detect these different possibilities are the *differential characters*, a somewhat more sophisticated mathematical structure than the ones considered so far in this book. Differential characters classify the different possible extensions of local cocycles to the full group of gauge transformations, i.e. their *indeterminacies*, which we call *global anomalies*. These are related, via differential characters, to the presence of torsion in the appropriate homology/cohomology groups of the classifying spaces or target spaces, which in turn may be related to the homotopy groups of $\mathcal{G}$. For global anomalies, there are, in general, no precise mechanisms of cancelation like for local anomalies. However, we can at least establish some sufficient conditions for their absence, which are, for instance, the absence of torsion in the just mentioned groups.

L. Bonora, *Fermions and Anomalies in Quantum Field Theories*, Theoretical and Mathematical Physics, https://doi.org/10.1007/978-3-031-21928-3_13

13.1 Introduction

So far, we have studied local field theory anomalies and their embedding in the geometry of fiber bundles. Throughout we have had in mind infinitesimal gauge transformations, as opposed to finite ones. The next step is to take into consideration also finite gauge transformations and analyze their relation to anomalies. To do so, it is evident that we need two ingredients which have not been considered so far, not explicitly at least: we need to consider the topology of the group $\mathcal{G}$ of gauge transformations, in particular, its cohomology; and we need a better knowledge of the space of connections; if $\mathcal{A}$ denotes this space in relation to a principal fiber bundle $\mathsf{P}(\mathsf{M}, \mathsf{G})$, we need to know in particular the structure of the moduli space $\frac{\mathcal{A}}{\mathcal{G}}$.

Let us return to the gauge anomaly with background connection $j_{(\cdot)}\mathcal{T}P_n(\psi^*A, A_0)$ at $\psi = id$. They are well-defined basic d-form in M; therefore, we can integrate them over M. The object we obtain is BRST invariant

$$\mathfrak{s}\int_M \Delta_d^{(1)}(A, A_0, c) = 0 \tag{13.1}$$

On the other hand $\mathfrak{s}$ is an alias of δ,[1] the differential in $\mathcal{G} = \mathsf{Aut_V(P)}$. Therefore, the expression $\int_M \Delta_d^{(1)}(A, A_0, c)$ is a cocycle in the cohomology of δ (or of $\mathfrak{s}$), that is in the cohomology of $\mathcal{G}$. But, in fact, so far, we are justified only to speak of cohomology of $\mathcal{G}_0$,[2] which is the group of gauge transformations connected to the identity. That is $\int_M \Delta_d^{(1)}(A, A_0, c)$ is an element of $\mathsf{H}^1(\mathcal{G}_0)$. Whether it is a trivial element or not depends on whether $\Delta_d^{(1)}(A, A_0, c)$ is trivial or not as an anomaly, with an important difference: what matters in $\mathsf{H}^1(\mathcal{G}_0)$ is topology; thus, local non-topological anomalies may not correspond to non-trivial classes of $\mathsf{H}^1(\mathcal{G}_0)$. The precise statement is: $j_{(\cdot)}\mathcal{T}P_n(\psi^*A, A_0)$ at $\psi = id$ generates a non-trivial class of $\mathsf{H}^1(\mathcal{G}_0)$ if and only if (1) P_n is irreducible, or (2) P_n is reducible and its polynomial factors, say Q_i for $i =, 1, \ldots, p$, correspond under the Weil homomorphism to non-trivial cohomology classes of $\mathsf{H}^*(\mathsf{M})$, except possibly one.

Let us recall a few notions we have already met previously. The gauge transformations $\in \mathcal{G}$ form a subgroup of $\mathsf{Aut(P)}$; therefore, they act on connections on the right

$$\mathcal{A} \times \mathcal{G} \to \mathcal{A}, \quad (A, \psi) \mapsto \psi^*A \equiv A\psi$$

This applies both to $\mathsf{Aut_V(P)}$ and to $\mathsf{Diff(M)}$, as long as via lift the latter can be seen as a subgroup of $\mathsf{Aut(P)}$. The action of $\mathcal{G}$ in general is not free. Since we need a free action for a well-defined quotient $\mathcal{A}/\mathcal{G}$, we will work as before with the group

---

[1] In fact $\mathfrak{s}$ corresponds to $(-1)^k\delta$, where k is the spacetime degree of the form it is applied to.

[2] Actually so far we are justified to speak only of an infinitesimal region around the identity of $\mathcal{G}$, but we shall see in Sect. 13.4 that this neighborhood of the identity can be extended to the whole of $\mathcal{G}_0$.

$$\mathsf{Aut}_{\mathsf{v}}^{\mathsf{m}}(\mathsf{P}) = \left\{\psi \in \mathsf{Aut}_{\mathsf{v}}(\mathsf{P}), \quad \text{s.t.} \quad \psi(p) = p, \quad \forall p \in \pi^{-1}(\mathsf{m})\right\} \tag{13.2}$$

where m is a preferred point in M (the point at infinity, in field theory: this means that we are considering gauge transformations that vanish at infinity). When speaking of moduli space $\mathcal{A}/\mathcal{G}$, we shall refer to this group of gauge transformations. As it turns out

$$\begin{array}{ccc} \mathcal{G} & \hookrightarrow & \mathcal{A} \\ & & \downarrow \\ & & \mathcal{A}/\mathcal{G} \end{array}$$

is a principal fiber bundle.

We need some general topological properties of these spaces. The space $\mathcal{A}$ is an affine space; therefore, its topology is trivial. All cohomology and homotopy groups are trivial. Using the long exact sequence for homotopy groups of a fibration $F \to E \to B$:

$$\ldots \to \pi_i(F) \to \pi_i(E) \to \pi_i(B) \to \pi_{i-1}(F) \to \ldots \tag{13.3}$$

one can conclude, in particular, that $\pi_{i+1}(\mathcal{A}/\mathcal{G}) = \pi_i(\mathcal{G})$, $i = 0, 1, \ldots$.

Concerning diffeomorphisms the situation is more complex. We start from $\mathsf{Diff}^{\mathsf{m}}(\mathsf{M})$, the subgroup of diffeomorphisms that leave the point m fixed. Then we restrict it by defining $\mathsf{Diff}^{\mathsf{m},1}(\mathsf{M})$, see above. The group $\mathsf{Diff}^{\mathsf{m},1}(\mathsf{M})$ acts freely on the space of linear connection in LM and acts freely in the space $\mathcal{M}$ of metrics. In particular

$$\begin{array}{ccc} \mathcal{G} & \hookrightarrow & \mathcal{M} \\ & & \downarrow \\ & & \mathcal{M}/\mathcal{G} \end{array}$$

is a principal fiber bundle with $\mathcal{G} = \mathsf{Diff}^{\mathsf{m},1}(\mathsf{M})$.

A more restricted space than $\mathcal{M}$ is the space of Levi-Civita connections, $\mathcal{A}_{LC}$, for which we have the commuting diagram

$$\begin{array}{ccc} \mathcal{M} & \xrightarrow{\psi} & \mathcal{M} \\ \downarrow{\scriptstyle lc} & & \downarrow{\scriptstyle lc} \\ \mathcal{A}_{LC} & \xrightarrow{l(\psi)} & \mathcal{A}_{LC} \end{array}$$

where lc is the map that associate to each metric its LC connection, and $l(\psi)$ is the lift of ψ. Then, again we can consider the principal fiber bundle

$$\begin{array}{ccc} \mathcal{G} & \hookrightarrow & \mathcal{A}_{LC} \\ & & \downarrow \\ & & \mathcal{A}_{LC}/\mathcal{G} \end{array}$$

where, again, $\mathcal{G} = \mathsf{Diff}^{\mathsf{m},1}(\mathsf{M})$.

We will consider also the space $\mathcal{A}^{\text{metric}}$ of metric connections. It is the sum $\mathcal{A}_{LC} + \mathcal{T}$, where $\mathcal{T}$ is the space of torsion tensor fields. We have another principal fiber bundle

$$\mathcal{A}^{\text{metric}} \longrightarrow \frac{\mathcal{A}^{\text{metric}}}{\mathsf{Diff}^{m,1}(M)}$$

The spaces $\mathcal{M}$, $\mathcal{A}_{LC}$ and $\mathcal{A}^{\text{metric}}$ are contractible, thus topologically trivial. We can therefore derive results analogous to the space of gauge connections.

13.2 Global Analysis and Global Anomalies

Let us return to the previous discussion about the QFT path integral and Formula (12.59). We have seen that $\mathcal{Z}[A]$ must be a well-defined function integrable over the moduli space $\mathcal{A}/\mathcal{G}$. The problem we are faced with concerning anomalies is therefore a two-step problem. First, find the obstructions to defining $\mathcal{Z}[A]$ locally (local anomalies); second, find the obstructions to extending $\mathcal{Z}[A]$ over the entire moduli space as an integrable functional (global anomalies).

To be more specific, we have to consider the action of $\mathcal{G}$ on $\mathcal{Z}[A]$ in general. Let us denote the right action of $\psi \in \mathcal{G}$ on A as $A\psi \equiv \psi^* A$ and represent its effect on $\mathcal{Z}[A]$ in the following way

$$U(\psi)\mathcal{Z}[A] = \mathcal{Z}[A\psi] = \rho(A, \psi)\mathcal{Z}[A] \tag{13.4}$$

where $U(\psi)$ denotes the representative of ψ, when acting on functionals $\mathcal{Z}[A]$. Therefore, we assume $U(\psi_1\psi_2) = U(\psi_1)U(\psi_2)$. So, by performing two transformations ψ_1, ψ_2 on a row, we find that the complex factor ρ must satisfy a consistency condition

$$\rho(A, \psi_1\psi_2) = \rho(A\psi_1, \psi_2)\rho(A, \psi_1) \tag{13.5}$$

If we assume, as is the case in field theory (see note below), that $\rho(A, \psi) = e^{2\pi i\alpha(A,\psi)}$, then the consistency condition becomes

$$\alpha(A, \psi_1\psi_2) - \alpha(A\psi_1, \psi_2) - \alpha(A, \psi_1) = 0 \mod \mathbb{Z} \tag{13.6}$$

If in Eq. (13.4) the factor ρ is not 1, we have a breaking of the gauge symmetry. However, if there exists a (complex) function σ over $\mathcal{A}$ such that

$$\rho(A, \psi) = \sigma(A\psi)\sigma^{-1}(A) \tag{13.7}$$

then we can define

$$\widehat{\mathcal{Z}}[A] = \sigma^{-1}(A)\mathcal{Z}[A] \tag{13.8}$$

It follows that

$$U(\psi)\widehat{\mathcal{Z}}[A] = \widehat{\mathcal{Z}}[A] \tag{13.9}$$

and gauge invariance is recovered. Therefore, (13.7) is a triviality condition, which, setting $\sigma(A) = e^{2\pi i\theta(A)}$, can be rewritten as

$$\alpha(A, \psi) = \theta(A\psi) - \theta(A), \qquad \text{mod } \mathbb{Z} \tag{13.10}$$

If ψ is an infinitesimal (i.e. infinitesimally close to the identity) gauge transformation, then the WI (13.4) takes the form of the infinitesimal one

$$\mathfrak{s} \log \mathcal{Z}[A] = i\mathcal{A}(A, c) \tag{13.11}$$

and we are back to the analysis of local anomalies.

Let us suppose now that a given theory is free of local anomalies, that is, $\widehat{\mathcal{Z}}[A]$ is locally well-defined. The problem of global anomalies may arise when we face the next step, which consists in integrating the local anomaly free $\widehat{\mathcal{Z}}[A]$ over the full moduli space $\mathcal{A}/\mathcal{G}$, Eq. (12.59). For the integral to make sense, the integrand must transform under a gauge transformation in a way complementary to the measure $d[A]$. Since the latter is BRST invariant, the same must be true for the integrand. The condidate integrand is therefore $\widehat{\mathcal{Z}}[A]$. But this is possible only if both $\mathcal{Z}[A]$ and $\sigma(A)$ are globally defined. This is the problem now.

In this sense, we face two possible obstacles. The first is that the factors $\alpha(A, \psi)$ corresponding to different species of fields in a theory cancel locally (cancelation scheme (B)) but not globally. This may happen due to the presence of torsion in the cohomology of $\mathcal{A}$, which renders indeterminate the global extension of local anomalies. We call these indeterminacies *global anomalies*.

The second obstacle can arise as follows: let us suppose that for a given theory $\rho(A, \psi)$ is locally trivial, i.e. can be written as $\rho(A, \psi) = \sigma(A\psi)\sigma^{-1}(A)$ for ψ in the component $\mathcal{G}_0$ connected to the identity of $\mathcal{G}$; the absence of any anomaly means the absence of any obstruction to extending $\sigma(A)$ for any A in $\mathcal{A}$ and any ψ in $\mathcal{G}$, so that $\rho(A, \psi) = \sigma(A\psi)\sigma^{-1}(A)$ globally. But the question is: is this guaranteed in the theory or are there non-trivial extensions of $\rho(A, \psi)$ in $\mathcal{G}$, which reduce to the previous factorized form in $\mathcal{G}_0$ but cannot be written in such a form for any ψ in $\mathcal{G}$?, i.e. do there exist non-trivial one cocycles of $\mathcal{G}$ that become trivial when restricted to

$\mathcal{G}_0$? Such possible non-trivial extensions are also a manifestation of *global anomalies* (references to global anomalies are, beside [1], [2–5].

To dress this problem in an appropriate mathematical form, we formulate it in the language of line bundles. In the presence of a cocycle $\rho(A, \psi)$, we can construct a (complex) line bundle over $\mathcal{A}/\mathcal{G}$, as follows. Consider the equivalence relation in $\mathcal{A} \times \mathbb{C}$

$$(A, c) \sim (A\psi, \rho(A, \psi)^{-1}c), \quad A \in \mathcal{A}, \ \psi \in \mathcal{G}, \ c \in \mathbb{C} \tag{13.12}$$

and denote this class as $[(A, c)]$. This relation defines a line bundle over $\mathcal{A}/\mathcal{G}$, which we call $\mathfrak{L}_\rho$. If $\rho(a, \psi)$ is trivial there exists a global section which associates to every class $[A] \in \mathcal{A}/\mathcal{G}$ the point in $\mathfrak{L}_\rho$ defined by $[(A, \sigma^{-1}(A))]$. Thus, $\mathfrak{L}_\rho$ is trivial. Vice-versa if $\mathfrak{L}_\rho$ is trivial there exists a global section $s([A]) = [(A, f(A))]$, where f is a globally defined non-vanishing complex function. Since, from the definition $(A\psi, f(A\psi)) \sim (A\psi, \rho(A, \psi)^{-1} f(A))$, it follows that $f(A\psi) = \rho(A, \psi)^{-1} f(A)$. i.e. $f(A)$ plays the role of a global section $\sigma(A)$.

The integration in (12.59) makes sense only if $\mathcal{Z}[A]$ is a non-vanishing global section of a trivial line bundle.

Note. In this note, our aim is to motivate the formula $\rho(A, \psi) = e^{2\pi i\alpha(A,\psi)}$. In a Minkowski spacetime, the partial path integral $\mathcal{Z}[A, g]$ (where we add also the dependence on the metric g) has the form

$$\mathcal{Z}[A, g] = \int \prod_i \mathrm{d}\phi_i \, \mathrm{e}^{iS(A,g,\phi_i)} \tag{13.13}$$

where ϕ_i generically denotes matter fields. $S(A, g, \phi_i)$ denotes the action, and the action is real. Classically after a Wick rotation, we end up with the Euclidean path integral

$$\mathcal{Z}_E[A, g] = \int \prod_i \mathrm{d}\phi_i \, \mathrm{e}^{-S_E(A,g,\phi_i)} \tag{13.14}$$

where the $-$ sign comes from Wick rotating the measure in the action. In both expressions the action is supposed to be invariant under gauge transformations. However, quantization may produce an extra factor $\rho(A, g, \psi)$ as a consequence of operating a gauge transformation represented by $U(\psi)$. This factor (an anomaly) appears in exponential form, because it effectively is a local modification of the action; therefore, in a Minkowski spacetime, it is natural to represent it as $\rho(A, g, \psi) = \mathrm{e}^{2\pi i\alpha(A,g,\psi)}$. However, we need to consider also its Euclidean version, because in the sequel, in order to analyze its properties, we will resort to the language of Euclidean geometry. The Euclidean form $\rho_E(A, g, \psi)$ depends crucially on the type of anomaly we are considering. For instance, for even parity trace anomalies, such as the Euler one, $\rho_E(g, \psi) = \mathrm{e}^{-2\pi\alpha_E(g,\psi)}$. Instead, for parity odd anomalies, like the chiral ones, we have $\rho_E(A, g, \psi) = \mathrm{e}^{2\pi i\alpha_E(A,g,\psi)}$. The reason is the presence of the completely antisymmetric tensor $\varepsilon_{\mu_1,\dots,\mu_d}$. A Wick rotation $x^0 \to it_E$, where t_E is the Euclidean time, acts on two factors, the measure $d^d x$ and a unique derivative ∂_{x^0} or unique time component of A or g. The result is that, contrary to what happens to the action, the i factor in the exponential survives. Therefore, for this kind of anomalies, the form $\rho(A, g, \psi) = e^{2\pi i \alpha(A,g,\psi)}$ is correct both for Lorentz or Euclidean background metric. For this reason, we drop the label $_E$ when dealing with them, which is the subject of the following sections.

13.3 Differential Characters

To analyze global anomalies with the appropriate mathematical language, we need a digression in order to introduce the concept and properties of differential characters [6].

From Eqs. (13.6) and (13.10), it is evident that in the global case, we need quantities that are defined up to integers, i.e. we must consider the $\mathbb{R} \to \mathbb{R}/\mathbb{Z}$ reduction of such quantities. So, for any real number r or form α let us denote by $\tilde{r}$ and $\tilde{\alpha}$ their reductions mod $\mathbb{Z}$. Under this reduction, Eqs. (13.6) and (13.10) become

$$\tilde{\alpha}(A, \psi_1\psi_2) - \tilde{\alpha}(A\psi_1, \psi_2) - \tilde{\alpha}(A, \psi_1) = 0 \tag{13.15}$$

$$\tilde{\alpha}(A, \psi) = \tilde{\theta}(A\psi) - \tilde{\theta}(A) \tag{13.16}$$

α (θ) are quantities defined in $\mathcal{A}$ and in $\mathcal{G}$. They define cocycles (coboundaries) which are relevant to the cohomology of $\mathcal{A}$, $\mathcal{A}/\mathcal{G}$ and $\mathcal{G}$. When $\tilde{\alpha}$ ($\tilde{\theta}$) enter into play we have to be more specific and consider cohomologies with values in $\mathbb{R}$, $\mathbb{Z}$, $\mathbb{R}/\mathbb{Z}$. So, let us consider a manifold X and its cohomology groups $H^*(\mathsf{X}, \mathbb{R})$, $H^*(\mathsf{X}, \mathbb{Z})$ and $H^*(\mathsf{X}, \mathbb{R}/\mathbb{Z})$. The first basic information about these groups comes from the short exact sequence

$$0 \longrightarrow \mathbb{Z} \longrightarrow \mathbb{R} \longrightarrow \mathbb{R}/\mathbb{Z} \longrightarrow 0$$

from which the Bockstein exact sequence follows

$$\dots \longrightarrow H^k(\mathsf{X}, \mathbb{Z}) \xrightarrow{r} H^k(\mathsf{X}, \mathbb{R}) \xrightarrow{\sim} H^k(\mathsf{X}, \mathbb{R}/\mathbb{Z}) \xrightarrow{\beta} H^{k+1}(\mathsf{X}, \mathbb{Z}) \longrightarrow \dots \tag{13.17}$$

Here r is the natural map between integral and real cohomology, $\tilde{}$ denotes mod $\mathbb{Z}$ reduction and β is the Bockstein homomorphism.

Having introduced these groups, let us come to differential characters. Let Z_k be the group of normalized smooth singular k-cycles and ∂, δ denote the boundary and coboundary operators. *A k-differential character is a homomorphism $u : Z_k \to \mathbb{R}/\mathbb{Z}$ subject to the condition that $u \circ \partial$ is the reduction mod $\mathbb{Z}$ of a $k+1$ differential form.* The space of differential characters is denoted $\widehat{H}^k(\mathsf{X}, \mathbb{R}/\mathbb{Z})$.

Given a k-differential character u, one can find a real k-cochain b such that

$$\tilde{b}\Big|_{Z_k} = u \tag{13.18}$$

If b_1, b_2 both satisfy this equation then

$$b_1 = b_2 + q + \text{exact} \tag{13.19}$$

where q is an integral k-cochain. In field theory, it is more useful to represent u in terms of differential forms. In terms of forms, b can be characterized by the following property

$$\delta b = \omega - s \tag{13.20}$$

where ω is a closed $k+1$ differential form with integral periods and s is an integral $k+1$ cocycle such that

$$r([s]) = [\omega] \tag{13.21}$$

where, as above, r is the map $H^{k+1}(\mathsf{X}, \mathbb{Z}) \to H^{k+1}(\mathsf{X}, \mathbb{R})$ and $[a]$ denotes the class of a. Next we define the operations

$$\delta_1(u) = \omega, \quad \delta_2(u) = [s] \tag{13.22}$$

These definitions do not depend on the real cochain b in (13.18).

Let us introduce $\Omega_0^*(\mathsf{X})$, the space of closed form with integral periods, and

$$R^k(\mathsf{X}, \mathbb{Z}) = \left\{(\omega, h) \in \Omega_0^k(\mathsf{X}) \times H^k(\mathsf{X}, \mathbb{Z}) \Big| r(h) = [\omega]\right\}. \tag{13.23}$$

Then one can prove the existence of the following exact sequences

$$0 \longrightarrow H^k(\mathsf{X}, \mathbb{R}/\mathbb{Z}) \longrightarrow \widehat{H}^k(\mathsf{X}, \mathbb{R}/\mathbb{Z}) \xrightarrow{\delta_1} \Omega_0^{k+1}(\mathsf{X}) \longrightarrow 0 \tag{13.24}$$

$$0 \longrightarrow \frac{\Omega^k(\mathsf{X})}{\Omega_0^k(\mathsf{X})} \longrightarrow \widehat{H}^k(\mathsf{X}, \mathbb{R}/\mathbb{Z}) \xrightarrow{\delta_2} H^{k+1}(\mathsf{X}, \mathbb{Z}) \longrightarrow 0 \tag{13.25}$$

from which one can derive

$$0 \longrightarrow \frac{H^k(\mathsf{X}, \mathbb{R})}{r(H^k(\mathsf{X}, \mathbb{Z}))} \longrightarrow \widehat{H}^k(\mathsf{X}, \mathbb{R}/\mathbb{Z}) \xrightarrow{(\delta_1, \delta_2)} R^{k+1}(\mathsf{X}, \mathbb{Z}) \longrightarrow 0 \tag{13.26}$$

Moreover, δ_1 restricted to $\frac{\Omega^k(\mathsf{X})}{\Omega_0^k(\mathsf{X})}$ is the exterior derivative d, and δ_2 restricted to $H^k(\mathsf{X}, \mathbb{R}/\mathbb{Z})$ coincides with $-\beta$.

From the previous exact sequences, we deduce that if $H^k(\mathsf{X}, \mathbb{R}) = 0$ then k-differential characters are completely identified by their images in $R^{k+1}(\mathsf{X}, \mathbb{Z})$. Moreover, we have the isomorphisms

$$\begin{aligned} \widehat{H}^0(\mathsf{X}, \mathbb{R}/\mathbb{Z}) &= C^\infty(\mathsf{X}, S^1) \\ \widehat{H}^d(\mathsf{X}, \mathbb{R}/\mathbb{Z}) &= H^d(\mathsf{X}, \mathbb{R}/\mathbb{Z}), \quad d = \dim(\mathsf{X}) \\ \widehat{H}^k(\mathsf{X}, \mathbb{R}/\mathbb{Z}) &= 0, \quad k > d \end{aligned} \tag{13.27}$$

13.3.1 Differential Characters and Classifying Space

Let us apply these ideas to the classifying space BG of a compact Lie group G. We know that the cohomology group $H^{2n-1}(\mathsf{BG}, \mathbb{R}) = 0$, while we have non-vanishing even order classes $\in H^{2n}(\mathsf{BG}, \mathbb{R})$ represented by forms $P_n\,(F(\mathfrak{a}), \ldots, F(\mathfrak{a}))$, where $\mathfrak{a}$ is the universal connection. The first element in the sequence (13.26) vanishes; thus, we have

$$0 \longrightarrow \widehat{H}^{2n-1}(\mathsf{BG}, \mathbb{R}/\mathbb{Z}) \overset{(\delta_1,\delta_2)}{\longrightarrow} R^{2n}(\mathsf{BG}, \mathbb{Z}) \longrightarrow 0 \tag{13.28}$$

Therefore we are in the condition outlined above: there is a one-to-one correspondence between $2n$–1–differential characters and elements of $R^{2n}(\mathsf{X}, \mathbb{Z})$. That is, for any integral class $[s] \in H^{2n}(\mathsf{BG}, \mathbb{Z})$ with $r([s]) = [P_n\,(F(\mathfrak{a}), \ldots, F(\mathfrak{a}))]$, there exists a unique differential character $\mathcal{U}_{P_n}([s])$ such that $\delta_1\left(\mathcal{U}_{P_n}([s])\right) = P_n\,(F(\mathfrak{a}), \ldots, F(\mathfrak{a}))$ and $\delta_2\left(\mathcal{U}_{P_n}([s])\right) = [s]$.

The trouble may come from the relation between the *form* $P_n\,(F(\mathfrak{a}), \ldots, F(\mathfrak{a}))$ and $\mathcal{U}_{P_n}([s])$. The reason is that while $H^{2n-1}(\mathsf{BG}, \mathbb{R}) = 0$, the homology group $H_{2n-1}(\mathsf{BG}, \mathbb{Z})$ might not vanish due to the torsion part[3]. If there is no torsion, the map $H^{2n}(\mathsf{BG}, \mathbb{Z}) \overset{r}{\longrightarrow} H^{2n}(\mathsf{BG}, \mathbb{R})$ is an injection. Then looking at the Bockstein sequence (13.17), one can see that $H^{2n-1}(\mathsf{BG}, \mathbb{R}/\mathbb{Z}) = 0$. Finally, using the exact sequence (13.24), one concludes that

$$0 \longrightarrow \widehat{H}^{2n-1}(\mathsf{BG}, \mathbb{R}/\mathbb{Z}) \overset{\delta_1}{\longrightarrow} \Omega_0^{2n}(\mathsf{BG}) \longrightarrow 0 \tag{13.29}$$

Therefore there is a unique possibility for which $\delta_1 \mathcal{U}_{P_n}([s]) = P_n\,(F(\mathfrak{a}), \ldots, F(\mathfrak{a}))$, that is $\mathcal{U}_{P_n}([s])$ is completely determined by P_n. On the contrary, in the presence of torsion there is an indeterminacy due to the different possible choices of the class $[s]$, which in turn are due to the torsion in $H^{2n}(\mathsf{BG}, \mathbb{Z})$.

Recall now that for any principal bundle $\mathsf{P}(\mathsf{M}, \mathsf{G})$, there exists a bundle morphism $(\hat{f}, f)$ such that the following diagram is commutative

---

[3] By the universal coefficient theorem, the free component of $H_{2n-1}(\mathsf{BG}, \mathbb{Z})$ vanishes, while a (finitely generated) non-vanishing torsion component implies that also $H^{2n}(\mathsf{BG}, \mathbb{Z})$ has torsion.

$$\begin{array}{ccc} \mathsf{P} & \xrightarrow{\hat{f}} & \mathsf{EG} \\ \downarrow{\scriptstyle \pi} & & \downarrow{\scriptstyle \pi} \\ \mathsf{M} & \xrightarrow{f} & \mathsf{BG} \end{array}$$

f is unique up to homotopy. Differential characters can be pulled back from BG to M via f. Once the class $[s]$ is chosen the differential character $f^*\mathcal{U}_{P_n}([s])$ is the unique one that satisfies the conditions

- $\delta_1 \left(f^*\mathcal{U}_{P_n}([s])\right) = f^* P_n \left(F(\mathfrak{a}), \dots, F(\mathfrak{a})\right)$,
- $\delta_2 \left(f^*\mathcal{U}_{P_n}([s])\right) = f^*[s]$,
- $f^*\mathcal{U}_{P_n}([s])$ is functorial with respect to bundle morphisms.

Important properties for the sequel are the following ones. Let $(\hat{f}_i, f_i)$ with $i = 1, 2$ be two classifying morphisms with the corresponding two connections $A_i = \hat{f}_i \mathfrak{a}$, $i = 1, 2$. Then

$$\pi^* f_i^*\mathcal{U}_{P_n}([s]) = \widetilde{TP}_n(A_i)\big|_{Z_{2n-1}(\mathsf{P})} \tag{13.30}$$

$$f_1^*\mathcal{U}_{P_n}([s]) - f_2^*\mathcal{U}_{P_n}([s]) = \widetilde{TP}_n(A_1, A_2)|_{Z_{2n-1}(\mathsf{M})} \tag{13.31}$$

where on the RHS we understand restrictions to cycles in P and M, respectively.

In the rest of this chapter, we will apply differential characters to the study of global anomalies. We have anticipated above that analysis of global anomalies becomes relevant after making sure that local anomalies are absent. Such situations of anomaly absence or cancelation have been classified in Sect. 12.10. The analysis in this chapter relies on that classification.

13.4 Cocycles in $\mathcal{G}_0$

So far by local anomalies, we have meant cocycles that satisfy the consistency conditions in a neighborhood of the identity of $\mathcal{G}$, or, even more precisely, for infinitesimal transformations of $\mathcal{G}$. Before facing global anomalies, we need an intermediate step, that is we need to consider cocycles that satisfy the consistency conditions for finite gauge transformations inside the component $\mathcal{G}_0$ connected to the identity.

Our aim is to find a cocycle that satisfies the consistency condition (13.15) for $\psi_1, \psi_2 \in \mathcal{G}_0$. To this end, let us recall the connection η in the principal fiber bundle $\mathsf{P} \times \mathcal{A} \to \mathsf{M} \times \mathcal{A}$ with group G, which is defined by

$$\eta_{p,A}(X_1, X_2) = A_p(X_1) + A(\omega(X_2))_p, \quad X_1 \in T_p\mathsf{P}, \quad X_2 \in T_A\mathcal{A} \tag{13.32}$$

where A is a connection in P and ω a connection in $\mathcal{A} \to \mathcal{A}/\mathcal{G}$. We also consider a background connection A_0 in M. We recall that all these connections can be pulled back from the universal connection $\mathfrak{a}$ in EG. Then we can construct the transgression $\mathcal{T}P_n(\eta, A_0)$ which is a $\mathsf{d}+1 = 2n-1$ form in $\mathsf{M} \times \mathcal{A}$. This in turn is derived by transgressing the $2n$ form $P_n(F(\mathfrak{a}, \ldots, \mathfrak{a}) \in \Omega_0^{2n}(\mathsf{BG})$. If we restrict $\mathcal{A}$ to the orbit of $\mathcal{G}$ passing through A, then, as pointed out above, $\eta = ev^* A$, where $ev : \mathsf{P} \times \mathcal{G} \to \mathsf{P}$ is the usual evaluation map.

Now pick an integral class $[s] \in H^{2n}(\mathsf{BG}, \mathbb{Z})$. As explained above, this defines a differential character $\mathcal{U}_{P_n}([s])$. Let $(\hat{f}_1, f_1)$ and $(\hat{f}_0, f_0)$ be the classifying maps relative to A and A_0, respectively. Due to (13.27), the pullback of $\mathcal{U}_{P_n}([s])$ through these maps vanishes:

$$f_i^* \mathcal{U}_{P_n}([s]) = 0, \quad \text{for} \quad i = 0, 1. \tag{13.33}$$

Now let us consider the following construct: choose a path in $\mathcal{A}$, that is a map $p : I \to \mathcal{A}$, i.e. every point of the path corresponds to a connection, in particular $p(0) = A$ and $p(1) = A_1$. Then we define the functional

$$\mathcal{T}P_n(A, A_1, p) = \int_{\mathsf{M} \times p(I)} \mathcal{T}P_n(\eta, A_0) \tag{13.34}$$

integrated over a $\mathsf{d}+1$ cycle in $\mathsf{M} \times \mathcal{A}$. We want to prove now that $\widetilde{\mathcal{T}P}_n(A, A_1, p)$ does not depend on the path p when A and $A_1 = \psi^* A \equiv A\psi$ lie on the same $\mathcal{G}_0$ orbit. To this end, let us pick two distinct paths p_1, p_2 with the same endpoints A and A_1 and opposite orientation. They form a loop $l : S^1 \to \mathcal{A}$. Then, using first (13.31) and then (13.33), we conclude that

$$\widetilde{\mathcal{T}P}_n(ev^* A, A_0)\Big|_{l(S^1) \times \mathsf{M}} = \left(\underline{ev}^* f_1^* \mathcal{U}_{P_n}([s]) - f_0^* \mathcal{U}_{P_n}([s])\right)\big|_{l(S^1) \times \mathsf{M}} = 0 \tag{13.35}$$

where $\underline{ev}$ is the map $\mathsf{M} \times \mathcal{G} \to \mathsf{M}$ induced by $ev : \mathsf{P} \times \mathcal{G} \to \mathsf{P}$. Therefore

$$\int_{l(S^1) \times \mathsf{M}} \widetilde{\mathcal{T}P}_n(ev^* A, A_0) = 0 \tag{13.36}$$

Thus, $\widetilde{\mathcal{T}P}_n(A, A\psi, p)$ does not depend on p and we will denote it simply $\widetilde{\mathcal{T}P}_n(A, A\psi)$. As a consequence it satisfies

$$\widetilde{\mathcal{T}P}_n(A, A\psi_1) + \widetilde{\mathcal{T}P}_n(A\psi_1, A\psi_1\psi_2) = \widetilde{\mathcal{T}P}_n(A, A\psi_1\psi_2) \tag{13.37}$$

which is (13.15). Therefore $\alpha(A; \psi) \equiv \widetilde{\mathcal{T}P}_n(A, A\psi)$ *is a group theoretical one-cocycle in* $\mathcal{G}_0$ *with values in* $\mathbb{R}/\mathbb{Z}$.

We remark that, contrary to $\widetilde{\mathcal{T}P}_n(A, A\psi, p)$, in general $\mathcal{T}P_n(A, A\psi, p)$ does depend on the path p. Moreover, let us pick an infinitesimal path $p(\epsilon) = A + \epsilon L_X A$ with $X \in T_{id}\mathcal{G}$ (remember that $L_X A = \mathrm{d}\lambda + [A, \lambda]$ where $\lambda = A(X)$), then

$$\lim_{\epsilon\to 0}\frac{\mathcal{T}P_n(A, A+\epsilon L_X A, p_\epsilon)}{\epsilon}(\mathsf{M}) = \int_{\mathsf{M}} j_X \mathcal{T}P_n(ev^* A, A_0)$$
$$\equiv \delta_{\mathcal{G}}\,\alpha(A;\psi)|_{\psi=id} \tag{13.38}$$

i.e. by considering an infinitesimal path along the fiber, we obtain the integrated local anomaly generated by $P_n(F(\mathfrak{a}), \ldots, F(\mathfrak{a}))$.

Due to the path-connectedness, the group theoretical one cocycle $\widetilde{\mathcal{T}P_n}(A, A\psi)$ in $\mathcal{G}_0$ is homotopically equivalent to the one cocycle $\int_{\mathsf{M}} j_X \mathcal{T}P_n(ev^* A, A_0)$ and represents the same class in $H^1(\mathcal{G}_0)$.

13.4.1 Coboundaries in $\mathcal{G}_0$

Having constructed a mod $\mathbb{Z}$ cocycle satisfying (13.6), one may wonder in what situations it is a coboundary, and, in that case, how to construct a term like $\widetilde{\theta}(A)$, in order to satisfy Eq. (13.10). This depends on the topology of $\mathcal{G}$. It would seem that $\widetilde{\mathcal{T}P_n}(A, A\psi)$ is a non-trivial element of $H^1(\mathcal{G}, \mathbb{R})$. But this is not always the case. As we have already warned the reader, the cohomology of $\mathcal{G}$ does not always coincide with the BRST cohomology of local field theory. It all depends on the topological or non-topological nature of the corresponding field theory anomaly. As we have already pointed out, this issue is not relevant for local field theory anomalies, which are universal objects and therefore insensitive to topology. It may be important in sigma models, depending on the topology of the target space T. However, in the rest of this subsection, we proceed on a general ground, without referring to a particular theory. Let us first state the problem more precisely.

Local anomalies in field theory come from polynomial forms $P_n(F(\eta'), \ldots, F(\eta'))$, of order $2n = \mathsf{d}+2$, where η' is a connection in $\frac{\mathsf{P}\times\mathcal{A}}{\mathcal{G}}$ induced by a connection η in $\mathsf{P}\times\mathcal{A}$, which is in turn derived, in the way explained earlier, from the universal connection $\mathfrak{a}$ in the universal bundle (in the sigma model it is derived from a connection in T). Now $\frac{\mathsf{P}\times\mathcal{A}}{\mathcal{G}}$ is a principal G-bundle over $\frac{\mathsf{M}\times\mathcal{A}}{\mathcal{G}}$, and the latter is a fiber bundle over $\frac{\mathcal{A}}{\mathcal{G}}$ with fiber M. Therefore, we can integrate $P_n(F(\eta'), \ldots, F(\eta'))$ along the fiber and obtain an element of $H^2\left(\frac{\mathcal{A}}{\mathcal{G}}, \mathbb{R}\right)$ which, upon transgression

$$H^2\left(\frac{\mathcal{A}}{\mathcal{G}}, \mathbb{R}\right) \longrightarrow H^1(\mathcal{G}, \mathbb{R}), \tag{13.39}$$

gives rise to the anomaly $\int_{\mathsf{M}} j_{(\cdot)}\mathcal{T}P_n(ev^* A, A_0)$. If $\mathcal{A}/\mathcal{G}$ is simply connected, which implies that $\pi_0(\mathcal{G}) = 0$, then the map (13.39) is an isomorphism. Let us see what happens if we suppose that $\int_{\mathsf{M}} j_{(\cdot)}\mathcal{T}P_n(ev^* A, A_0)$ (which comes from the cocycle $\alpha(A, \psi) = \mathcal{T}P_n(A, A\psi, p)$ with p a path joining A and $A\psi$) represents the 0 element of $H^1(\mathcal{G}, \mathbb{R})$. We wish to show that in this case, there exists $\theta(A)$ such that $\alpha(A; \psi) = \theta(A\psi) - \theta(A)$.

Let us call κ the closed 2-form that represents a class $[\kappa] \in H^2\left(\frac{\mathcal{A}}{\mathcal{G}}\right)$. Then a transgression is the restriction to the orbit of a form $\sigma \in \Omega^1(\mathcal{A})$ such that $d\sigma = \pi^*\kappa$. In our case, $\kappa = \int_{\mathsf{M}} P_n(F(\eta'), \ldots, F(\eta')$ and $\sigma = \int_{\mathsf{M}} \mathcal{T}P_n(\eta, A_0)$. If $[\kappa] = 0$ then $\kappa = d\gamma$ in $\mathcal{A}/\mathcal{G}$, and, so, $d(\pi^*\gamma - \sigma) = 0$ in $\mathcal{A}$. Since the topology of $\mathcal{A}$ is trivial, there exist $\theta \in C^\infty(\mathcal{A})$ such that $\sigma - \pi^*\gamma = d\theta$. Now, let $\theta_{\mathcal{G}}$ denotes the restriction of θ to the orbit and $\delta_{\mathcal{G}}$ the differential along the orbit. Then $\delta_{\mathcal{G}}\theta_{\mathcal{G}}$ coincides with the restriction of σ to the orbit. Therefore, integrating along the fiber from A to $A\psi$, we get $\alpha(A; \psi) = \theta(A\psi) - \theta(A)$.

Vice-versa, one can show that if there exists $\theta(A)$ such that $\alpha(A; \psi) = \theta(A\psi) - \theta(A)$ then this happens only if the anomaly is obtained by antitransgressing an exact 2-form in $\frac{\mathcal{A}}{\mathcal{G}}$.

To summarize, if $\mathcal{G}$ is connected and the anomaly in question corresponds to $0 \in H^1(\mathcal{G}, \mathbb{R})$, and only in this case, can one construct via transgression, and only in this way, a function $\theta(A)$ in $\mathcal{A}$, whose reduction mod $\mathbb{Z}$ trivializes the group 1-cocycle $\widetilde{\alpha}(A; \psi) = \widetilde{\mathcal{T}P_n}(A, A\psi)$, i.e. such that $\widetilde{\mathcal{T}P_n}(A, A\psi) = \widetilde{\theta}(A\psi) - \widetilde{\theta}(A)$.

The condition that $\mathcal{G}$ is simply connected is crucial for this result. If $\mathcal{A}/\mathcal{G}$ is not simply connected, which implies $\pi_0(\mathcal{G}) \neq 0$, then the transgression map (13.39) can have a non-trivial kernel, and the above conclusion is obstructed.

13.5 Cocycles in $\mathcal{G}$, Global Indeterminacy and Global Anomalies

It makes sense to consider global anomalies when local anomalies are absent (are canceled). The list of situations where this happens has been given above, and classified in case (A), (B) and (C). Case (A) is when the relevant polynomial P_n (or its coefficient) vanishes identically. In this case there is no room for corresponding global anomalies. Case (B) is characterized by different species of fundamental fields, each contributing a non-trivial anomaly in such a way that the overall anomaly is zero. Finally case (C) corresponds to non-topological local anomalies.

Excluding (A), in case (B), we are faced with one or more closed $2n$ forms of the type $P_n(F(\eta'), \ldots, F(\eta'))$ in $\mathsf{M} \times \mathcal{A}/\mathcal{G}$, with $2n = \mathsf{d} + 2$. Upon integrating over M we obtain for each one a closed 2-form

$$\mathcal{F}_{P_n} = \int_{\mathsf{M}} P_n(F(\eta'), \ldots, F(\eta')) \in \Omega_0^2\left(\frac{\mathcal{A}}{\mathcal{G}}\right) \tag{13.40}$$

which defines a class in $H^2\left(\frac{\mathcal{A}}{\mathcal{G}}, \mathbb{R}\right)$. By transgression we get a class in $H^1(\mathcal{G}, \mathbb{R})$, which we suppose gives rise to a non-trivial anomaly. The problem we would like to examine now is the relation between $H^2\left(\frac{\mathcal{A}}{\mathcal{G}}, \mathbb{R}\right)$ and $H^2\left(\frac{\mathcal{A}}{\mathcal{G}}, \mathbb{Z}\right)$ and its effect on the anomaly cancelation. As we have seen above this relation is mediated by differential characters. To make this explicit, we will construct a differential character $\mathcal{U}_{P_n} \in \widehat{H}^1\left(\frac{\mathcal{A}}{\mathcal{G}}, \frac{\mathbb{R}}{\mathbb{Z}}\right)$ such that $\delta_1 \mathcal{U}_{P_n} = \mathcal{F}_{P_n}$. Then acting with δ_2 will identify an integral 2-class in $\frac{\mathcal{A}}{\mathcal{G}}$.

$\mathcal{U}_{P_n}$ is constructed as follows. We start from the classifying space BG and pick a differential character $u_{P_n}([s]) \in \widehat{H}^{2n-1}(\mathsf{BG}, \mathbb{R}/\mathbb{Z})$ determined by P_n, such that $\delta_1 u_{P_n}([s]) = P_n(F(\mathfrak{a}), \ldots, F(\mathfrak{a}))$ and $\delta_2 u_{P_n}([s]) = [s]$. Then we pull it back through the map $\underline{\mathsf{Ev}} : \frac{\mathsf{M}\times\mathcal{A}}{\mathcal{G}} \to \mathsf{BG}$ (see Appendix 12A for the notation). Since differential characters are functorial, we obtain a differential character in $\widehat{H}^{2n-1}\left(\frac{\mathsf{M}\times\mathcal{A}}{\mathcal{G}}, \mathbb{R}/\mathbb{Z}\right)$: $\underline{\mathsf{Ev}}^* u_{P_n}([s])$. Now consider a loop l in $\frac{\mathcal{A}}{\mathcal{G}}$ and the induced bundle

$$\begin{array}{ccc} l^*\left(\frac{\mathsf{M}\times\mathcal{A}}{\mathcal{G}}\right) & \xrightarrow{\bar{l}} & \frac{\mathsf{M}\times\mathcal{A}}{\mathcal{G}} \\ \downarrow \pi & & \downarrow \pi \\ S^1 & \xrightarrow{l} & \frac{\mathcal{A}}{\mathcal{G}} \end{array} \tag{13.41}$$

where $\bar{l}$ is the canonical cover of l. In other words $l^*\left(\frac{\mathsf{M}\times\mathcal{A}}{\mathcal{G}}\right)$ is the bundle $\frac{\mathsf{M}\times\mathcal{A}}{\mathcal{G}} \to \frac{\mathcal{A}}{\mathcal{G}}$ restricted to the image of S^1 in $\frac{\mathcal{A}}{\mathcal{G}}$ by l.

The differential character $\mathcal{U}_{P_n}([s]) \in \widehat{H}^1\left(\frac{\mathcal{A}}{\mathcal{G}}, \frac{\mathbb{R}}{\mathbb{Z}}\right)$ is defined for any one-cycle $l(S^1)$ by

$$\mathcal{U}_{P_n}([s])[l(S^1)] = \bar{l}^* \underline{\mathsf{Ev}}^* u_{P_n}([s]) \left[l^*\left(\frac{\mathsf{M}\times\mathcal{A}}{\mathcal{G}}\right)\right] \tag{13.42}$$

Now, the integral class we were looking for is determined by:

$$\beta\left(\mathcal{U}_{P_n}([s])\right) \in H^2\left(\frac{\mathcal{A}}{\mathcal{G}}, \mathbb{Z}\right) \tag{13.43}$$

where β is the Bockstein homomorphism (remember that $\beta = -\delta_2$).

The global definition of the anomaly determined by P_n includes a specification of the element $\beta\left(\mathcal{U}_{P_n}([s])\right)$ inside $H^2\left(\frac{\mathcal{A}}{\mathcal{G}}, \mathbb{Z}\right)$. If the map $H^2\left(\frac{\mathcal{A}}{\mathcal{G}}, \mathbb{Z}\right) \xrightarrow{r} H^2\left(\frac{\mathcal{A}}{\mathcal{G}}, \mathbb{R}\right)$ is one-to-one, the identification of $\beta\left(\mathcal{U}_{P_n}([s])\right)$ is unambiguous. On the other hand, if it happens that the Abelian group $H^2\left(\frac{\mathcal{A}}{\mathcal{G}}, \mathbb{Z}\right)$ contains a torsion subgroup, then there is an indeterminacy in the identification of the differential character corresponding to the anomaly. As a consequence, the anomaly itself is globally ill-defined. Therefore, a non-trivial

$$Tor\, H^2\left(\frac{\mathcal{A}}{\mathcal{G}}, \mathbb{Z}\right) \tag{13.44}$$

where by *Tor* we denote the torsion subgroup of an Abelian group[4] , is an anomaly within the anomaly, which we call *global anomaly* or *global indeterminacy*. A first conclusion is that a sufficient condition for absence of this indeterminacy is

$$Tor\, H^2\left(\frac{\mathcal{A}}{\mathcal{G}}, \mathbb{Z}\right) = 0 \tag{13.45}$$

Now, from the universal coefficient theorems it follows that for a space X, if the groups $H_q(\mathsf{X})$ and $H_{q-1}(\mathsf{X})$ are finitely generated, then the torsion part of $H^q(\mathsf{X})$ is given by the torsion part of $H_{q-1}(\mathsf{X})$. Therefore, if the homology groups $H_1\left(\frac{\mathcal{A}}{\mathcal{G}}, \mathbb{Z}\right)$ and $H_2\left(\frac{\mathcal{A}}{\mathcal{G}}, \mathbb{Z}\right)$ are finitely generated, then

$$Tor\ H^2\left(\frac{\mathcal{A}}{\mathcal{G}}, \mathbb{Z}\right) \cong Tor\ H_1\left(\frac{\mathcal{A}}{\mathcal{G}}, \mathbb{Z}\right) \cong Tor\ \pi_1\left(\frac{\mathcal{A}}{\mathcal{G}}\right) \cong Tor\ \pi_0\left(\mathcal{G}\right) \tag{13.46}$$

where $\cong$ means isomorphism. The second step follows from the theorem by which, if X is path-connected, then $H_1(\mathsf{X})$ is the Abelianization of $\pi_1(\mathsf{X})$. The third passage follows from the long exact sequence (13.3).

Therefore, under mild conditions, we can conclude that if $\pi_0\left(\mathcal{G}\right)$ contains no torsion subgroup, there is no global indeterminacy.

Another sufficient condition comes from the choice of $[s] \in H^{\mathsf{d}+2}(\mathsf{BG}, \mathbb{Z})$, with $\mathsf{d} = 2n - 2$. If the latter group has no torsion

$$Tor\ H^{\mathsf{d}+2}(\mathsf{BG}, \mathbb{Z}) = 0, \tag{13.47}$$

there is no source of indeterminacy and the anomaly is globally well-defined. In this case as well, there cannot be global anomalies.

The condition (13.45) and (13.47) are drastic. It is not excluded that milder conditions may guarantee the absence of global anomalies, but this can be decided only on a case by case analysis.

13.5.1 Extension to $\mathcal{G}$

How do the just constructed differential characters intercept the components of $\mathcal{G}$ not connected to the identity? To answer this question we need a short digression.

For any $\psi \in \mathcal{G}$, let us define the space $\mathsf{Map}_\psi(I, \mathcal{A})$ of all smooth maps φ from the unit interval $\times \mathcal{A}$ to $\mathcal{A}$ such that

$$\varphi(0) = A \quad \text{and} \quad \varphi(1) = \psi^* A \equiv A\psi, \quad \forall A \in \mathcal{A}$$

[4] A typical torsion group is the direct sum of cyclic groups $\mathbb{Z}_{p_1} \oplus \ldots \oplus \mathbb{Z}_{p_n}$.

For any $\varphi \in \mathsf{Map}_\psi(I, \mathcal{A})$, if π is the projection $\mathcal{A} \to \frac{\mathcal{A}}{\mathcal{G}}$, we have that $\pi \circ \varphi$ is a loop in $\frac{\mathcal{A}}{\mathcal{G}}$, i.e.

$$\pi \circ \varphi \in \Omega\left(\frac{\mathcal{A}}{\mathcal{G}}\right)$$

If $\varphi \in \mathsf{Map}_\psi(I, \mathcal{A})$ and $\varphi' \in \mathsf{Map}_{\psi'}(I, \mathcal{A})$ and $\varphi(0) = \varphi'(0)$, it follows that $\pi \circ \varphi$ and $\pi \circ \varphi'$ are homotopic if and only if ψ and ψ' belong to the same connected component of $\mathcal{G}$, because $\pi_1\left(\frac{\mathcal{A}}{\mathcal{G}}\right) \cong \pi_0(\mathcal{G})$.

Moreover, given a loop $\ell \in \Omega\left(\frac{\mathcal{A}}{\mathcal{G}}\right)$, there exists a $\psi \in \mathcal{G}$ and a $\varphi \in \mathsf{Map}_\psi(I, \mathcal{A})$ such that $\ell = \pi \circ \varphi$. Now, for any ψ denote by $\Omega_\psi\left(\frac{\mathcal{A}}{\mathcal{G}}\right)$ the subset of $\Omega\left(\frac{\mathcal{A}}{\mathcal{G}}\right)$ made of loops that come from elements in $\mathsf{Map}_\psi(I, \mathcal{A})$. Then the union of $\Omega_\psi\left(\frac{\mathcal{A}}{\mathcal{G}}\right)$ for all ψ coincides with $\Omega\left(\frac{\mathcal{A}}{\mathcal{G}}\right)$. On the other hand if ψ and ψ' belong to the same connected component of $\mathcal{G}$, then $\Omega_\psi\left(\frac{\mathcal{A}}{\mathcal{G}}\right) = \Omega_{\psi'}\left(\frac{\mathcal{A}}{\mathcal{G}}\right)$. Therefore choosing (arbitrary) ψ_i, one for each connected component of $\mathcal{G}$, we conclude that $\Omega\left(\frac{\mathcal{A}}{\mathcal{G}}\right) = \amalg_i \Omega_{\psi_i}\left(\frac{\mathcal{A}}{\mathcal{G}}\right)$.

Now let us consider the manifold $(\mathsf{P} \times S^1)_\psi$ constructed from $\mathsf{P} \times I$ by identifying $(u, 0)$ with $(\psi^{-1}(u), 1)$ for any $u \in \mathsf{P}$. Then we can construct the diagram

$$\begin{array}{ccccc}
(\mathsf{P} \times S^1)_\psi \times \mathsf{Map}_\psi(I, \mathcal{A}) & \xrightarrow{ev_\psi} & \frac{\mathsf{P}\times\mathcal{A}}{\mathcal{G}} & \xrightarrow{\mathsf{Ev}} & \mathsf{EG} \\
\downarrow & & \downarrow & & \downarrow \\
(\mathsf{M} \times S^1) \times \mathsf{Map}_\psi(I, \mathcal{A}) & \xrightarrow{\underline{ev}_\psi} & \mathsf{M} \times \frac{\mathcal{A}}{\mathcal{G}} & \xrightarrow{\underline{\mathsf{Ev}}} & \mathsf{BG}
\end{array} \tag{13.48}$$

where ev_ψ is defined as follows: $ev_\psi(u, s, \varphi_\ell) = (id \times \pi)(u, \varphi_\ell(s))$, for $u \in \mathsf{P}$, $s \in I$ and $\varphi_\ell \in \mathsf{Map}_\psi(I, \mathcal{A})$ such that $\pi \circ \varphi_\ell = \ell$, and where π is the projection $\mathcal{A} \to \frac{\mathcal{A}}{\mathcal{G}}$. This is correct since

$$ev_\psi(u, 0, \varphi_\ell) = ev_\psi(\psi^{-1}(u), 1, \varphi_\ell)$$

The map $\underline{ev}_\psi$ is the product of the identity on M and the map

$$S^1 \times \mathsf{Map}_\psi(I, \mathcal{A}) \xrightarrow{id\times\pi} S^1 \times \Omega_\psi\left(\frac{\mathcal{A}}{\mathcal{G}}\right) \xrightarrow{ev} \frac{\mathcal{A}}{\mathcal{G}} \tag{13.49}$$

Now let us turn to differential characters. We can pull back $\underline{\mathsf{Ev}}^* \mathcal{U}_{P_n}(\tau)$ to $(\mathsf{M} \times S^1) \times \mathsf{Map}_\psi(I, \mathcal{A})$ via $\underline{ev}_\psi$, and extract the component corresponding to $\mathsf{M} \times S^1$, which we write simply

$$\underline{ev_{\psi}^{*}\mathsf{Ev}^{*}\mathcal{U}_{P_n}(\tau)}\Big|_{\mathsf{M}\times S^1} \in \widehat{H}^{2n-1}(M \times S^1) \tag{13.50}$$

This differential character clearly depends on the connected component of $\mathcal{G}$ determined by ψ. If it happens to be uniquely defined by the polynomial P_n, then it identifies the unique extension of the corresponding anomaly to the whole of $\mathcal{G}$.

A construction similar to the previous one can be carried out also in the case of the diffeomorphisms, with a diagram similar to (13.48) in which P is replaced by LM^+, $\mathcal{A}$ by $\mathcal{A}^{\text{metric}}$, $\mathcal{G}$ by $\mathsf{Diff}_*^{\mathsf{m},1}\mathsf{M}$ and the group G by $\mathsf{GL}(\mathsf{d}, \mathbb{R})^+$. The rest of the construction need not be repeated.

13.5.2 Cancelation of Global Indeterminacy and of Global Anomalies. Case (B)

Local anomalies in case (B) cancel out among different species of particles. Concerning global indeterminacy and global anomalies the safest situation corresponds to one of the conditions (13.45) or (13.47) being satisfied, in which case there is nothing to worry about. However we cannot always assume this. A simple example is the case of a Dirac fermion coupled to a gauge potential, which can be viewed as a system of two Weyl fermions with opposite chiralities. The local chiral anomalies for the two Weyl fermions come up with opposite signs, thus they cancel. The question one must pose next is: if there is a global indeterminacy in these anomalies, does the cancelation still occur? The answer is yes. In fact whatever class $[s] \in H^{n+2}(\mathsf{BG}, \mathbb{Z})$ or $\beta\left(\mathcal{U}_{P_n}([s])\right)$ in $H^2\left(\frac{\mathcal{A}}{\mathcal{G}}, \mathbb{Z}\right)$ we choose for one chirality, there is a canonical choice for the opposite chirality: we choose the same element with opposite sign. Therefore, the global indeterminacy is irrelevant for the Dirac fermion system, which is therefore globally anomaly free.

There are more complicated theories in which the local anomaly cancelation occurs between different fermion species, for instance different species of spin 1/2 and spin 3/2 fermions. Such models require a detailed case by case analysis; it does not seem to be possible to resolve their global indeterminacy with a general argument.

An important example, in this regard, is the standard model of particle physics, where the groups involved are $U(1)$, $SU(2)$ and $SU(3)$. The relevant local anomaly for $SU(2)$ is absent (it is an example of case (A)) because the ad-invariant polynomial $P_3^{SU(2)}$ identically vanishes. As for $U(1)$ and $SU(3)$ let us assume that $\mathsf{M} = S^4$, then $\mathcal{G}$ (understood as $\mathsf{Aut}_{\mathsf{V}}^{\mathsf{m}}(\mathsf{P})$) is weakly homotopic equivalent [7] to the pointed space of $\mathsf{Map}^{\mathsf{m}}(\mathsf{S}^4, \mathsf{G})$, where m is identified with the north pole of S^4. Therefore $\pi_0(\mathcal{G})$ is isomorphic to $\pi_4(\mathsf{G})$, which is 0 both for $U(1)$ and for $SU(3)$. Therefore, see Eqs. (13.45) and (13.46), there is no global indeterminacy and no global anomaly. Concerning global gravitational anomalies, we recall that in $\mathsf{d} = 4$ local gravitational anomalies are absent as in case A, thus there is no room for global gravitational anomalies.

In more generality, suppose $\mathsf{M} = S^4$, $\mathcal{G}$ is weakly homotopic equivalent to $Map^{\mathsf{m}}(S^4, \mathsf{G}) = \pi_4(\mathsf{G})$ (see [8]); therefore, $\pi_0(\mathcal{G})$ is isomorphic to $\pi_4(\mathsf{G})$. Now the fourth homotopy group is trivial for all the groups $SU(n)$ and $U(n)$ with $n \neq 2$, as well as for all exceptional groups (here and in the sequel we use data taken from [8]). It is non-trivial for $n = 2$, but, as we have seen, the local anomaly is absent for $SU(2)$; therefore, there cannot be any global indeterminacy. As for other gauge groups, of the $SO(n)$ and $Sp(n)$ series, there is no possibility of global anomalies, because, as we have already signaled, the corresponding local anomalies are absent.

13.6 Real Line Bundles

From the previous sections, we deduce that the path integral $\mathcal{Z}[A, g]$ is, in general, not a function on $\mathcal{A}/\mathcal{G}$ but a local section of a non-trivial complex line bundle over $\mathcal{A}/\mathcal{G}$. On the other hand, if the factor $\rho(A, g, \psi)$ does not appear after the action of $U(\psi)$ on $\mathcal{Z}[A, g]$, i.e. if

$$U(\psi)\mathcal{Z}[A, g] = \mathcal{Z}[A, g] \tag{13.51}$$

i.e. if there are no local or corresponding global anomalies, then $\mathcal{Z}[A, g]$ is represented by a global section of a trivial complex line bundle and can be safely integrated over $\mathcal{A}/\mathcal{G}$. This analysis applies to a gauge theory defined in a Minkowski spacetime.

Now the path integral $\mathcal{Z}[A, g]$ in Eq. (13.13) after a Wick rotation becomes (13.14). The relevant question is now whether the integrand in it is real or complex. In Sect. 1.5, we have considered the problem of what is the Euclidean field theory obtained via a Wick rotation from a Minkowski field theory. The conclusion is that, as long as we consider the prescription introduced there (essentially the Osterwalder-Schrader prescription), we end up with the doubling of the fermionic fields: the $\bar{\psi}$ in the action has to be replaced with a bispinor independent of ψ, moreover we have to abandon the expectation of a real action. There are however more refined prescriptions [9, 10], which, for instance, accompany the Wick rotation with a rotation of the spinor fields. In these new settings, one can impose hermiticity on the action for a Euclidean Dirac spinor, but it remains impossible to do so for a Weyl spinor. In the latter case, the action is not Hermitean and, as a consequence, the integrand in (13.14) is complex. In this situation, the integrand is a section of a complex bundle, and there is nothing else we can add to the previous analysis of local and global anomalies.

However, there is an additional possibility when a suitable Wick rotation drives us to Euclidean theory for a Dirac fermion, because in this case, the action is Hermitean and, as a consequence, the integrand in (13.14) is real. As shown in [2] for a theory of an $SU(2)$ doublet of Dirac fermions, we can extract the square root of the Dirac determinant. The object we obtain is a real Pfaffian. We can try to interpret it, at least locally, as a Euclidean field theory of chiral spin 1/2 fermion coupled to a gauge potential, that is a field theory defined in a Euclidean spacetime, where $\mathcal{Z}[A]$ is real. The path integral makes sense if it is a global section of a real bundle over $\mathcal{A}/\mathcal{G}$. Therefore, we should ask whether there are obstructions to the existence of a global section, i.e. to the triviality of the bundle. If $\mathcal{Z}[A]$ is the square root of a real determinant line bundle, that is a Pfaffian, there are possible obstructions. Such real line bundles are classified by $H^1(\mathcal{A}/\mathcal{G}, \mathbb{Z}_2)$. Therefore, a sufficient condition for absence of global anomalies is that this group vanishes. On the other hand $H^1(\mathcal{A}/\mathcal{G}, \mathbb{Z}_2) = Hom(\pi_1(\mathcal{A}/\mathcal{G}), \mathbb{Z}_2) = Hom(\pi_0(\mathcal{G}), \mathbb{Z}_2)$ (see above for the last passage). An example, of this obstruction is the $SU(2)$ gauge theory of a Weyl fermion defined on a Euclidean spacetime S^4. As we know already this field theory does not have local anomalies. But $\mathcal{G}$ in this case is weakly homotopic equivalent to $Map^{\mathrm{m}}(S^4, SU(2)) = \pi_4(SU(2)) = \mathbb{Z}_2$. It follows that $H^1(\mathcal{A}/\mathcal{G}, \mathbb{Z}_2) = Hom(\mathbb{Z}_2, \mathbb{Z}_2) = \mathbb{Z}_2$. Therefore this model has a global anomaly.

Similar obstructions exist, if one tries a similar construction on S^4, for a few groups $SO(N)$ because: $\pi_4(SO(3)) = \pi_4(SO(5)) = \mathbb{Z}_2$ and $\pi_4(SO(4)) = \mathbb{Z}_2 \times \mathbb{Z}_2$, while $\pi_4(SO(N))$ is 0 for all the other N. Analogous data for groups $Sp(n)$ are: $\pi_4(Sp(n)) = \mathbb{Z}_2$ for $n = 1, \ldots, 5$. The fourth homotopy group is instead trivial for all the other simple groups: $SU(n)$ and $U(n)$ with $n \neq 2$, as well as for all exceptional groups.

Comment. The case illustrated above for the $SU(2)$ Euclidean field theory should not be confused with the discussion of the global anomalies for the group $SU(2)$ of the standard model in Sect. 13.5.2. In that case, the Weyl fermion (in Minkowski background) is complex, and conjugate to the Weyl fermion of opposite chirality. The fermion determinant is complex and, according to all prescriptions, a Wick rotation transforms it into a complex Euclidean fermion determinant.

References

1. L. Bonora, P. Cotta-Ramusino, M. Rinaldi, J. Stasheff, The evaluation map in field theory, sigma-models and strings. II. Commun. Math. Phys. **114**, 381 (1988)
2. E. Witten, An SU(2) anomaly. Phys. Lett. **117B**, 324 (1982)
3. E. Witten, Global gravitational anomalies. Commun. Math. Phys. **100**, 197 (1985)
4. E. Witten, Global anomalies. Conf. Proc. C 850623 21–27. Contribution to: Nuffield Workshop on Supersymmetry and its Applications (1985)
5. R. Holman, T.W. Kephart, Global anomalies in higher dimensions. Phys. Lett. B **167**, 417 (1986)
6. J. Cheeger, J. Simons, Differential characters and geometric invariants, in *Geometry and Topology. Lectures Notes in Mathematics*, vol. 1167 (Springer, Berlin, Heidelberg, New York, 1985)
7. I.M. Singer, Some remarks on the Gribov ambiguity. Commun. Math. Phys. **60**, 7–12 (1978)
8. *The Encyclopedic Dictionary of Mathematics, by the Mathematical Society of Japan*, 2nd edn. Vol. I–IV, ed. by K. Itô (MIT Press, Cambridge and London, 1987)
9. M.R. Mehta, Euclidean continuation of the Dirac fermion. Phys. Rev. Lett. **65**, 1983 (1990)
10. P. van Nieuwenhuizen, A. Waldron, On euclidean spinors and wick rotations. Phys. Lett. **B389**, 29 (1996). e-print: 9608174 [hep-th]

Part VI
Special Topics

Chapter 14
MAT in 2d

In 2d, it is possible to make calculations which may be inaccessible, due to their complexity, in higher dimensions. In this chapter, we illustrate an example of complete anomaly calculation in a MAT background in 2d. We compute all the relevant diffeomorphism and trace anomalies and check that we get the same results with the SDW method. We start with the perturbative calculation of the diffeomorphism and trace anomalies of a Dirac fermion in a MAT background. We repeat this derivation in a different way, by splitting the Dirac fermion into two Weyl fermions of opposite chiralities, and perform the calculation for the two Weyl fermions separately. Finally, we use the SDW method and verify that all these methods yield the same results. The main reason for this chapter is that it permits a clear verification of the consistency of Formula (7.1) for the trace anomaly. At the same time, it offers a nice example of the interplay between diffeomorphisms and trace anomalies.

14.1 Anomalies of a Dirac Fermion in a Metric-Axial-Tensor (MAT) Background

The MAT background in 4d has been introduced above. Here, we limit ourselves to a summary for 2d. A MAT background is specified by a metric $\widehat{g}_{\mu\nu} = g_{\mu\nu} + \gamma_* f_{\mu\nu}$ or a frame $\widehat{e}^a_\mu = e^a_\mu + \gamma_* c^a_\mu$. They give rise, in particular, to a MAT Christoffel symbol $\widehat{\Gamma}^\lambda_{\mu\nu} = \Gamma^{(1)\lambda}_{\mu\nu} + \gamma_*\Gamma^{(2)\lambda}_{\mu\nu}$ and to MAT spin connection $\widehat{\Omega}^{ab}_\mu = \Omega^{(1)ab}_\mu + \gamma_*\Omega^{(2)ab}_\mu$. The action of a Dirac fermion in this background is

$$\widehat{S} = \int d^2\widehat{x}\,\left(i\overline{\psi}\sqrt{\widehat{g}}\gamma^a\widehat{e}^\mu_a\left(\partial_\mu + \frac{1}{2}\widehat{\Omega}_\mu\right)\psi\right)(\widehat{x}) \tag{14.1}$$

where $\widehat{x}^\mu = x^\mu_1 + \gamma_* x^\mu_2$. Now, we simplify this action as we did in Chap. 6 with the action (7.60) and for the same reason, by dropping the term containing the spin connection. Henceforth, we shall refer to the action

L. Bonora, *Fermions and Anomalies in Quantum Field Theories*, Theoretical and Mathematical Physics, https://doi.org/10.1007/978-3-031-21928-3_14

$$\widehat{S} = \int d^2\widehat{x} \left(i\overline{\psi}\gamma^a \widehat{e}_a^\mu \partial_\mu \psi\right)(\widehat{x}) \tag{14.2}$$

where we have also absorbed the $\sqrt{\overline{\overline{g}}}$ factor in ψ and $\bar{\psi}$.

We expand around $\widehat{g}_{\mu\nu} = \eta_{\mu\nu}$, $g_{\mu\nu} \approx \eta_{\mu\nu} + h_{\mu\nu}$, $f_{\mu\nu} \approx k_{\mu\nu}$. The Feynman rules for this simplified theory are as follows. The fermion propagator is

$$\frac{i}{\not{p} + i\epsilon} \tag{14.3}$$

The two-fermion-h-graviton vertex (V_{ffh}) is:

$$\frac{i}{8}\left[(p+p')_\mu \gamma_\nu + (p+p')_\nu \gamma_\mu\right] \tag{14.4}$$

The axial two-fermion-k-graviton vertex (V_{ffk}) is:

$$\frac{i}{8}\left[(p+p')_\mu \gamma_\nu + (p+p')_\nu \gamma_\mu\right]\gamma_* \tag{14.5}$$

(p incoming, p' outgoing). There are also three 2-fermion-2-graviton vertices and higher order vertices, but they do not enter the present calculation, being relevant to higher order approximations than those considered here. We recall that in this model we have conservation of diffeomorphisms, both ordinary and axial, which are defined by

$$\widehat{x}^\mu \to \widehat{x}^\mu + \widehat{\xi}^\mu(\widehat{x}^\mu), \qquad \widehat{\xi}^\mu = \xi^\mu + \gamma_* \zeta^\mu, \tag{14.6}$$

and under Weyl transformations (both ordinary and axial)

$$\widehat{g}_{\mu\nu} \longrightarrow e^{2\omega}\widehat{g}_{\mu\nu}, \qquad \widehat{g}_{\mu\nu} \to e^{2\gamma_*\eta}\widehat{g}_{\mu\nu} \tag{14.7}$$

and

$$\widehat{e}_\mu^a \longrightarrow e^{\omega}\widehat{e}_\mu^a, \qquad \widehat{e}_\mu^a \to e^{\gamma_*\eta}\widehat{e}_\mu^a \tag{14.8}$$

These invariances imply the conservation and tracelessness of two 'e.m. tensors', which at the lowest level of approximation take the form

$$T^{\mu\nu} \approx T^{\mu\nu}_{flat} = -\frac{i}{4}\left(\overline{\psi}\gamma^\mu \overleftrightarrow{\partial}^\nu \psi + \mu \leftrightarrow \nu\right), \tag{14.9}$$

and

$$T_*^{\mu\nu} \approx T^{\mu\nu}_{*flat} = -\frac{i}{4}\left(\overline{\psi}\gamma^\mu \overleftrightarrow{\partial}^\nu \gamma_* \psi + \mu \leftrightarrow \nu\right), \tag{14.10}$$

The classical conformal WI are:

$$T^{\mu\nu}g_{\mu\nu} + T_5^{\mu\nu}f_{\mu\nu} = 0, \tag{14.11}$$
$$T^{\mu\nu}f_{\mu\nu} + T_5^{\mu\nu}g_{\mu\nu} = 0, \tag{14.12}$$

The quantum effective action is defined by

$$W[h,k] = W[0] + \sum_{n,m=0}^{\infty} \frac{i^{m+n-1}}{2^{n+m}n!m!} \int \prod_{i=1}^{n} dx_i h_{\mu_i\nu_i}(x_i) \prod_{j=1}^{m} dy_j k_{\lambda_j\rho_j}(y_j)$$
$$\cdot\langle 0|\mathcal{T}T^{\mu_1\nu_1}(x_1)\dots T^{\mu_n\nu_n}(x_n)T_5^{\lambda_1\rho_1}(y_1)\dots T_5^{\lambda_m\rho_m}(y_m)|0\rangle \tag{14.13}$$

and the infinitesimal (0th order) Weyl transformations and diffeomorphism are

$$\delta_\omega^{(0)}h_{\mu\nu} = 2\omega\,\eta_{\mu\nu}, \qquad \delta_\omega^{(0)}k_{\mu\nu} = 0$$
$$\delta_\eta^{(0)}h_{\mu\nu} = 0, \qquad \delta_\eta^{(0)}k_{\mu\nu} = 2\eta\,\eta_{\mu\nu} \tag{14.14}$$

for ordinary and axial Weyl transformations, and

$$\delta_\xi^{(0)}h_{\mu\nu} = \partial_\mu\xi_\nu + \partial_\nu\xi_\mu, \qquad \delta_\xi^{(0)}k_{\mu\nu} = 0,$$
$$\delta_\zeta^{(0)}h_{\mu\nu} = 0, \qquad \delta_\zeta^{(0)}k_{\mu\nu} = \partial_\mu\zeta_\nu + \partial_\nu\zeta_\mu \tag{14.15}$$

for ordinary and axial diffeomorphisms.

Disregarding for simplicity one-point functions of the e.m. tensors, the conformal WIs for two-point correlators, corresponding to the parameter ω, are

$$\eta^{\mu\nu}\langle 0|\mathcal{T}T_{\mu\nu}(x)T_{\lambda\rho}(y)|0\rangle = 0 \tag{14.16}$$
$$\eta^{\mu\nu}\langle 0|\mathcal{T}T_{\mu\nu}(x)T_{*\lambda\rho}(y)|0\rangle = 0 \tag{14.17}$$

and corresponding to η

$$\eta^{\mu\nu}\langle 0|\mathcal{T}T_{*\mu\nu}(x)T_{\lambda\rho}(y)|0\rangle = 0 \tag{14.18}$$
$$\eta^{\mu\nu}\langle 0|\mathcal{T}T_{*\mu\nu}(x)T_{*\lambda\rho}(y)|0\rangle = 0 \tag{14.19}$$

The WI for diffeomorphisms with parameter ξ^μ are

$$\partial^\mu\langle 0|\mathcal{T}T_{\mu\nu}(x)T_{\lambda\rho}(y)|0\rangle = 0 \tag{14.20}$$
$$\partial^\mu\langle 0|\mathcal{T}T_{\mu\nu}(x)T_{*\lambda\rho}(y)|0\rangle = 0 \tag{14.21}$$

and with parameter ζ^μ

$$\partial^\mu \langle 0|\mathcal{T} T_{*\mu\nu}(x) T_{\lambda\rho}(y)|0\rangle = 0 \tag{14.22}$$

$$\partial^\mu \langle 0|\mathcal{T} T_{*\mu\nu}(x) T_{*\lambda\rho}(y)|0\rangle = 0 \tag{14.23}$$

We will follow two different perturbative approaches, and check they yield coincident results.

14.1.1 First Approach: The Trace Cocycles

The first-order contribution to trace anomalies comes from two point functions of the two e.m. tensors in all combinations with subsequent contraction of the indices of one of them, minus, following (7.1), the two point functions of the corresponding trace with the corresponding e.m. tensor. Let us start from the two-point correlators involving only T's. The amplitude $\eta^{\mu\nu}\langle 0|\mathcal{T} T_{\mu\nu}(x) T_{\lambda\rho}(y)|0\rangle$ in momentum space is given by

$$\begin{aligned}
\overset{\approx}{\mathfrak{F}}{}^{(VV)\mu}_{\mu\lambda\rho}(k) &= -\frac{1}{8}\int \frac{d^2p}{(2\pi)^2}\,\mathrm{tr}\left(\frac{\not p}{p^2}\,(2\not p - \not q)\,\frac{\not p - \not k}{(p-k)^2}(2p-k)_\lambda\gamma_\rho\right) + (\lambda \leftrightarrow \rho)\\
&= -\frac{1}{8}\int \frac{d^2p\, d^\delta\ell}{(2\pi)^{2+\delta}}\mathrm{tr}\left(\frac{\not p + \not \ell}{p^2-\ell^2}\,(2\not p - \not k)\,\frac{\not p + \not\ell - \not k}{(p-k)^2-\ell^2}(2p-k)_\lambda\gamma_\rho\right)\\
&\quad + (\lambda \leftrightarrow \rho)\\
&= \frac{i}{24\pi}\left(2\eta_{\lambda\rho}k^2 + 3k_\lambda k_\rho\right)
\end{aligned} \tag{14.24}$$

The amplitude $\langle 0|\mathcal{T} T^\mu_\mu(x) T_{\lambda\rho}(y)\rangle$ in momentum space is given by

$$\begin{aligned}
\tilde{\mathfrak{F}}{}^{(VV)\mu}_{\mu\lambda\rho}(k) &= -\frac{1}{8}\int \frac{d^2p\, d^\delta\ell}{(2\pi)^{2+\delta}}\mathrm{tr}\left(\frac{\not p + \not\ell}{p^2-\ell^2}\,(2\not p + 2\not\ell - \not k)\,\frac{\not p + \not\ell - \not k}{(p-k)^2-\ell^2}(2p-k)_\lambda\gamma_\rho\right)\\
&\quad + (\lambda \leftrightarrow \rho) = \frac{i}{24\pi}k_\lambda k_\rho
\end{aligned} \tag{14.25}$$

Therefore, according to Formula (7.8), we have

$$\overset{\approx}{\mathfrak{F}}{}^{(VV)\mu}_{\mu\lambda\rho}(k) - \tilde{\mathfrak{F}}{}^{(VV)\mu}_{\mu\lambda\rho}(k) = \frac{i}{12\pi}\left(\eta_{\lambda\rho}k^2 + k_\lambda k_\rho\right) \tag{14.26}$$

in Euclidean background, which corresponds to the integrated (Minkowski) cocycle

$$\Delta_\omega^{(VV)} = \frac{1}{48\pi}\int \mathrm{d}^2x\,\omega(x)\left[\partial_\lambda\partial_\rho h^{\lambda\rho}(x) - \Box h_\lambda^\lambda(x)\right] \tag{14.27}$$

The next even case is the same calculation with two T_*'s. $\eta^{\mu\nu}\langle 0|\mathcal{T}T_{*\mu\nu}(x)T_{*\lambda\rho}(y)\rangle$ in momentum space is given by

$$\begin{aligned}\widetilde{\widetilde{\mathfrak{F}}}^{(AA)\mu}_{\mu\lambda\rho}(k) &= -\frac{1}{8}\int\frac{\mathrm{d}^2p\,\mathrm{d}^\delta\ell}{(2\pi)^{2+\delta}}\mathrm{tr}\left(\frac{\not p+\not\ell}{p^2-\ell^2}\,(2\not p-\not k)\,\gamma_*\frac{\not p+\not\ell-\not k}{(p-k)^2-\ell^2}(2p-k)_\lambda\gamma_\rho\gamma_*\right)\\ &+(\lambda\leftrightarrow\rho) = \frac{i}{24\pi}\left(4\eta_{\lambda\rho}k^2+5k_\lambda k_\rho\right)\end{aligned} \tag{14.28}$$

The amplitude $\langle 0|\mathcal{T}T^\mu_{*\mu}(x)T_{*\lambda\rho}(y)\rangle$ in momentum space is given by

$$\begin{aligned}\widetilde{\mathfrak{F}}^{(AA)\mu}_{\mu\lambda\rho}(k) &= -\frac{1}{8}\int\frac{\mathrm{d}^2p\,\mathrm{d}^\delta\ell}{(2\pi)^{2+\delta}}\mathrm{tr}\left(\frac{\not p+\not\ell}{p^2-\ell^2}\,(2\not p+2\not\ell-\not k)\,\gamma_*\frac{\not p+\not\ell-\not k}{(p-k)^2-\ell^2}(2p-k)_\lambda\gamma_\rho\gamma_*\right)\\ &+(\lambda\leftrightarrow\rho) = \widetilde{\mathfrak{F}}^{(VV)\mu}_{\mu\lambda\rho}\end{aligned} \tag{14.29}$$

Therefore

$$\widetilde{\widetilde{\mathfrak{F}}}^{(AA)\mu}_{\mu\lambda\rho}(k) - \widetilde{\mathfrak{F}}^{(AA)\mu}_{\mu\lambda\rho}(k) = \frac{i}{6\pi}\left(\eta_{\lambda\rho}k^2+k_\lambda k_\rho\right) \tag{14.30}$$

so that

$$\Delta_\eta^{(AA)} = \frac{1}{24\pi}\int \mathrm{d}^2x\,\eta(x)\left[\partial_\lambda\partial_\rho k^{\lambda\rho}(x) - \Box k_\lambda^\lambda(x)\right] \tag{14.31}$$

The odd parity contributions come from the amplitudes with one T and one T_*. The amplitude $\eta^{\mu\nu}\langle 0|\mathcal{T}T_{\mu\nu}(x)T_{*\lambda\rho}(y)\rangle$ in momentum space is given by

$$\begin{aligned}\widetilde{\widetilde{\mathfrak{F}}}^{(VA)\mu}_{\mu\lambda\rho}(k) &= -\frac{1}{8}\int\frac{\mathrm{d}^2p\,\mathrm{d}^\delta\ell}{(2\pi)^{2+\delta}}\mathrm{tr}\left(\frac{\not p+\not\ell}{p^2-\ell^2}\,(2\not p-\not k)\,\frac{\not p+\not\ell-\not k}{(p-k)^2-\ell^2}(2p-k)_\lambda\gamma_\rho\gamma_*\right)\\ &+(\lambda\leftrightarrow\rho) = 0\end{aligned} \tag{14.32}$$

and $\langle 0|\mathcal{T}T^\mu_\mu(x)T_{*\lambda\rho}(y)\rangle$ in momentum space is given by

$$\begin{aligned}\widetilde{\mathfrak{F}}^{(VA)\mu}_{\mu\lambda\rho}(k) &= -\frac{1}{8}\int\frac{\mathrm{d}^2p\,\mathrm{d}^\delta\ell}{(2\pi)^{2+\delta}}\mathrm{tr}\left(\frac{\not p+\not\ell}{p^2-\ell^2}\,(2\not p+2\not\ell-\not k)\,\frac{\not p+\not\ell-\not k}{(p-k)^2-\ell^2}(2p-k)_\lambda\gamma_\rho\gamma_*\right)\\ &+(\lambda\leftrightarrow\rho) = -\frac{i}{24\pi}\left(k_\lambda\varepsilon_{\rho\sigma}+k_\rho\varepsilon_{\lambda\sigma}\right)k^\sigma\end{aligned} \tag{14.33}$$

Therefore,

$$\widetilde{\widetilde{\mathfrak{F}}}^{(VA)\mu}_{\mu\lambda\rho}(k) - \widetilde{\mathfrak{F}}^{(VA)\mu}_{\mu\lambda\rho}(k) = \frac{i}{24\pi}\left(k_\lambda\varepsilon_{\rho\sigma}+k_\rho\varepsilon_{\lambda\sigma}\right)k^\sigma \tag{14.34}$$

which corresponds to the integrated anomaly

$$\Delta_{\omega}^{(VA)} = \frac{1}{48\pi} \int d^2x\, \omega\, \varepsilon_{\lambda\sigma} \partial^{\sigma} \partial_{\rho} k^{\lambda\rho} \tag{14.35}$$

Finally, $\eta^{\mu\nu}\langle 0|\mathcal{T} T_{*\mu\nu}(x) T_{\lambda\rho}(y)\rangle$ in momentum space is given by

$$\begin{aligned}\widetilde{\widetilde{\mathfrak{F}}}^{(AV)\mu}_{\mu\lambda\rho}(k) &= -\frac{1}{8}\int \frac{d^2p\, d^{\delta}\ell}{(2\pi)^{2+\delta}} \mathrm{tr}\left(\frac{\not p + \not \ell}{p^2-\ell^2}\,(2\not p - \not k)\, \gamma_* \frac{\not p + \not \ell - \not k}{(p-k)^2-\ell^2} (2p-k)_{\lambda}\gamma_{\rho}\right)\\ &+ (\lambda \leftrightarrow \rho) = -\frac{i}{24\pi}\left(k_{\lambda}\varepsilon_{\rho\sigma} + k_{\rho}\varepsilon_{\lambda\sigma}\right) k^{\sigma}\end{aligned} \tag{14.36}$$

and

$$\begin{aligned}\widetilde{\mathfrak{F}}^{(AV)\mu}_{\mu\lambda\rho}(k) &= -\frac{1}{8}\int \frac{d^2p\, d^{\delta}\ell}{(2\pi)^{2+\delta}} \mathrm{tr}\left(\frac{\not p + \not \ell}{p^2-\ell^2}\,(2\not p + 2\not \ell - \not k)\, \gamma_* \frac{\not p + \not \ell - \not k}{(p-k)^2-\ell^2} (2p-k)_{\lambda}\gamma_{\rho}\right)\\ &+ (\lambda \leftrightarrow \rho) = \widetilde{\widetilde{\mathfrak{F}}}^{(AV)\mu}_{\mu\lambda\rho}(k) - \frac{i}{24\pi}\left(k_{\lambda}\varepsilon_{\rho\sigma} + k_{\rho}\varepsilon_{\lambda\sigma}\right) k^{\sigma}\end{aligned} \tag{14.37}$$

as a consequence

$$\widetilde{\widetilde{\mathfrak{F}}}^{(AV)\mu}_{\mu\lambda\rho}(k) - \widetilde{\mathfrak{F}}^{(VA)\mu}_{\mu\lambda\rho}(k) = 0 \tag{14.38}$$

which implies

$$\Delta_{\eta}^{(AV)} = 0 \tag{14.39}$$

14.1.2 First Approach: The Diffeomorphism Cocycles

As we have learned in the case of a right-handed fermion, the trace cocycles calculated so far may not be the ultimate expression of the trace anomalies. We have to verify that diffeomorphisms are preserved. There are three possibilities: either the corresponding WI's are respected, in which case, the above results for the trace anomalies are the ultimate ones, or there are anomalies; the latter may be trivial coboundaries or non-trivial cocycles. In the first case, we can subtract counterterms from the effective action so as to eliminate them and realize full symmetry under diffeomorphisms, in which case most probably the trace anomalies calculated previously will be modified. In the second case, the game is over and only trivial modifications of the diffeomorphism anomalies are possible.

The basic amplitudes in this context are the divergence of the two-point correlators of two e.m. tensors. Contrary to the previous trace calculations, there are no ambiguities in these type of calculations. The result will be the same if we compute the correlator of the divergence of an e.m. tensor with another e.m. tensor:

$$\partial_x^{\mu}\langle 0|\mathcal{T}\mathsf{T}_{\mu\nu}(x)\mathsf{T}_{\lambda\rho}(y)|0\rangle = \langle 0|\mathcal{T}\partial^{\mu}\mathsf{T}_{\mu\nu}(x)\mathsf{T}_{\lambda\rho}(y)|0\rangle \tag{14.40}$$

where T is either T or T_5. Let us start with $\mathsf{T} = T$. The amplitude to be evaluated is

$$k^\mu \tilde{\mathfrak{F}}^{(VV)}_{\mu\nu\lambda\rho}(k) = -\frac{1}{16}\int \frac{d^2p}{(2\pi)^2}\mathrm{tr}\Big(\frac{\not p}{p^2}(2p\cdot k - k^2)\gamma_\nu \frac{\not p - \not k}{(p-k)^2}(2p-k)_\lambda \gamma_\rho + \frac{\not p}{p^2}(2p-k)_\nu \not k \frac{\not p - \not k}{(p-k)^2}(2p-k)_\lambda\gamma_\rho\Big). \tag{14.41}$$

The regulated integral is

$$\begin{aligned} k^\mu \tilde{\mathfrak{F}}^{(VV)}_{\mu\nu\lambda\rho}(k) &= -\frac{1}{16}\int \frac{d^2p\, d^\delta\ell}{(2\pi)^{2+\delta}}\mathrm{tr}\Big(\frac{\not p + \not \ell}{p^2-\ell^2}(2p\cdot k - k^2)\gamma_\nu \frac{\not p + \not\ell - \not k}{(p-k)^2-\ell^2}(2p-k)_\lambda \gamma_\rho \\ &+ \frac{\not p + \not\ell}{p^2-\ell^2}(2p-k)_\nu \not k \frac{\not p + \not \ell - \not k}{(p-k)^2-\ell^2}\,(2p-k)_\lambda\gamma_\rho\Big) \\ &= 0 \end{aligned} \tag{14.42}$$

In other words, no anomaly: $\Delta^{(VV)}_\xi = 0$.

Let us consider next two T_*, whose regularized expression is:

$$\begin{aligned} k^\mu \tilde{\mathfrak{T}}^{(AA)}_{\mu\nu\lambda\rho}(k) &= -\frac{1}{16}\int \frac{d^2p\, d^\delta\ell}{(2\pi)^{2+\delta}}\mathrm{tr}\Big(\frac{\not p + \not \ell}{p^2-\ell^2}(2p\cdot k - k^2)\gamma_\nu\gamma_* \frac{\not p + \not\ell - \not k}{(p-k)^2-\ell^2}(2p-k)_\lambda \gamma_\rho\gamma_* \\ &+ \frac{\not p + \not\ell}{p^2-\ell^2}(2p-k)_\nu \not k\gamma_* \frac{\not p + \not \ell - \not k}{(p-k)^2-\ell^2}(2p-k)_\lambda\gamma_\rho\gamma_*\Big) \\ &= \frac{i}{48\pi}\left(\eta_{\nu\lambda}k^2k_\rho + \eta_{\nu\rho}k^2k_\lambda + 2k_\nu k_\lambda k_\rho\right) \end{aligned} \tag{14.43}$$

So

$$\Delta^{(AA)}_\zeta = -\frac{1}{96\pi}\int d^2x\, \zeta^\nu(x)\Big[\partial_\nu\partial_\lambda\partial_\rho k^{\lambda\rho}(x) - \partial_\lambda \Box k^\lambda_\nu(x)\Big] \tag{14.44}$$

Then comes the odd parity two point functions. Let us consider first $\partial^\mu\langle 0|\mathcal{T}T_{\mu\nu}(x)\, T_{*\lambda\rho}|0\rangle$:

$$\begin{aligned} k^\mu \tilde{\mathfrak{T}}^{(VA)}_{\mu\nu\lambda\rho}(k) &= -\frac{1}{16}\int \frac{d^2p\, d^\delta\ell}{(2\pi)^{2+\delta}}\mathrm{tr}\Big(\frac{\not p + \not \ell}{p^2-\ell^2}(2p\cdot k - k^2)\gamma_\nu \frac{\not p + \not\ell - \not k}{(p-k)^2-\ell^2}(2p-k)_\lambda \gamma_\rho\gamma_* \\ &+ \frac{\not p + \not\ell}{p^2-\ell^2}(2p-k)_\nu \not k \frac{\not p + \not \ell - \not k}{(p-k)^2-\ell^2}(2p-k)_\lambda\gamma_\rho\gamma_*\Big) \\ &= 0 \end{aligned} \tag{14.45}$$

i.e., $\Delta^{(VA)}_\xi = 0$.

On the other hand, for $\partial^\mu\langle 0|\mathcal{T}T_{*\mu\nu}(x)T_{\lambda\rho}|0\rangle$, we have:

$$\begin{aligned} k^\mu \tilde{\mathfrak{T}}^{(AV)}_{\mu\nu\lambda\rho}(k) &= -\frac{1}{16}\int \frac{d^2p\, d^\delta\ell}{(2\pi)^{2+\delta}}\mathrm{tr}\Big(\frac{\not p + \not \ell}{p^2-\ell^2}(2p\cdot k - k^2)\gamma_\nu\gamma_* \frac{\not p + \not\ell - \not k}{(p-k)^2-\ell^2}(2p-k)_\lambda \gamma_\rho \\ &+ \frac{\not p + \not\ell}{p^2-\ell^2}(2p-k)_\nu \not k\gamma_* \frac{\not p + \not \ell - \not k}{(p-k)^2-\ell^2}(2p-k)_\lambda\gamma_\rho\Big) \\ &= -\frac{i}{48\pi}\left(k_\nu\left(k_\lambda\varepsilon_{\rho\sigma} + k_\rho\varepsilon_{\lambda\sigma}\right)k^\sigma + k^2\left(\eta_{\nu\lambda}\varepsilon_{\rho\sigma} + \eta_{\nu\rho}\varepsilon_{\lambda\sigma}\right)k^\sigma\right), \end{aligned} \tag{14.46}$$

that is,

$$\Delta_{\zeta}^{(AV)} = \frac{1}{96\pi} \int d^2x\, \zeta^{\nu}(x) \Big[\varepsilon_{\rho\sigma} \partial^{\sigma} \big(\partial_{\nu}\partial_{\lambda} - \eta_{\nu\lambda}\Box \big) h^{\lambda\rho}(x) \Big] \tag{14.47}$$

14.1.3 First Approach: Trivial and Non-trivial Cocycles

Let us start from the (nonzero) even cocycles. They satisfy the consistency conditions

$$\delta_{\omega}^{(0)} \Delta_{\omega}^{(VV)} = 0, \qquad \delta_{\eta}^{(0)} \Delta_{\omega}^{(VV)} = 0 \tag{14.48}$$

and

$$\delta_{\zeta}^{(0)} \Delta_{\zeta}^{(AA)} = 0, \qquad \delta_{\xi}^{(0)} \Delta_{\zeta}^{(AA)} = 0 \tag{14.49}$$

Moreover

$$\delta_{\omega}^{(0)} \Delta_{\zeta}^{(AA)} = 0 = \delta_{\zeta}^{(0)} \Delta_{\omega}^{(VV)}, \qquad \delta_{\eta}^{(0)} \Delta_{\zeta}^{(AA)} = 0 \tag{14.50}$$

The cocycle $\Delta_{\zeta}^{(AA)}$ is trivial. In fact, let us define the chain

$$\mathcal{C}^{(ff)} = -\frac{1}{192\pi} \int d^2x \, \big(k_{\rho}^{\nu}(x) \partial_{\lambda}\partial_{\nu} k^{\lambda\rho}(x) - k_{\lambda\rho}(x)\Box k^{\lambda\rho}(x) \big) \tag{14.51}$$

It is easy to prove that $\delta_{\zeta}^{(0)} \mathcal{C}^{(ff)} = -\Delta_{\zeta}^{(AA)}$, so

$$\begin{aligned} \mathcal{A}_{\xi}^{(\text{even})} &\equiv \Delta_{\xi}^{(VV)} = 0 \\ \mathcal{A}_{\zeta}^{(\text{even})} &\equiv \Delta_{\zeta}^{(AA)} + \delta_{\zeta}^{(0)} \mathcal{C}^{(ff)} = 0 \end{aligned} \tag{14.52}$$

As a consequence of adding the counterterm to the effective action we have: since $\delta_{\omega}^{(0)} \mathcal{C}^{(ff)} = 0$, the overall even ω-trace anomaly is

$$\begin{aligned} \mathcal{A}_{\omega}^{(\text{even})} &\equiv \Delta_{\omega}^{(VV)} + \delta_{\omega}^{(0)} \mathcal{C}^{(ff)} = \frac{1}{48\pi} \int d^2x\, \omega \left[\partial_{\lambda}\partial_{\rho} h^{\lambda\rho} - \Box h_{\lambda}^{\lambda} \right] \\ &\approx \frac{1}{48\pi} \int d^2x\, \omega\, R^{(1)} + \cdots \end{aligned} \tag{14.53}$$

On the other hand, since $\Delta_{\eta}^{(AA)} = \frac{1}{24\pi} \int d^2x\, \eta(x) \left[\partial_{\lambda}\partial_{\rho} k^{\lambda\rho}(x) - \Box k_{\lambda}^{\lambda}(x) \right]$, we have

$$\begin{aligned} \mathcal{A}_{\eta}^{(\text{even})} &\equiv \Delta_{\eta}^{(AA)} + \delta_{\eta}^{(0)} \mathcal{C}^{(ff)} = \frac{1}{48\pi} \int d^2x\, \eta \left[\partial_{\lambda}\partial_{\rho} k^{\lambda\rho} - \Box k_{\lambda}^{\lambda} \right] \\ &\approx \frac{1}{48\pi} \int d^2x\, \eta\, R^{(2)} + \cdots \end{aligned} \tag{14.54}$$

The last terms in the RHS of (14.53, 14.54) are the covariantization of the (ordinary and axial) expressions corresponding to the lowest order approximation we have found, up to other possible covariant terms represented by the ellipses.

In the chiral limit $h_{\mu\nu} \to h_{\mu\nu}/2$, $k_{\mu\nu} \to h_{\mu\nu}/2$, the curvatures $R^{(1)}$ and $R^{(2)}$ tend both to the Ricci scalar divided by 2. Therefore, the previous even trace anomalies (14.53) and (14.54) reproduce both the even Weyl fermion anomaly (7.94).

Let us now consider the odd parity trace and diffeomorphism cocycles. We have two non-vanishing, $\Delta_\omega^{(VA)}$and $\Delta_\eta^{(AV)}$ of the former and one non-vanishing $\Delta_\zeta^{(AV)}$ of the latter given by Eqs. (14.35), (14.39) and (14.47), respectively. We have

$$\delta_\xi \Delta_\omega^{(VA)} = 0, \quad \delta_\eta \Delta_\omega^{(VA)} = 0, \quad \delta_\zeta \Delta_\omega^{(VA)} = -\frac{1}{48\pi} \int d^2x\, \omega\, \varepsilon_{\nu\sigma}\, \partial^\sigma \Box \zeta^\nu \tag{14.55}$$

and

$$\delta_\zeta \Delta_\zeta^{(AV)} = 0, \quad \delta_\xi \Delta_\zeta^{(AV)} = 0, \quad \delta_\omega \Delta_\zeta^{(AV)} = \frac{1}{48\pi} \int d^2x\, \zeta^\nu \varepsilon_{\nu\sigma}\, \partial^\sigma \Box \omega. \tag{14.56}$$

So the consistency condition

$$\delta_\zeta \Delta_\omega^{(VA)} + \delta_\omega \Delta_\zeta^{(AV)} = 0 \tag{14.57}$$

is satisfied. All the other consistency conditions are trivially satisfied.

Let us summarize the situation for odd parity cocycles. We have two cocycles

$$\Delta_\omega^{(VA)} = \frac{1}{48\pi} \int d^2x\, \omega\, \varepsilon_{\lambda\sigma}\, \partial^\sigma \partial_\rho k^{\lambda\rho}, \tag{14.58}$$

and

$$\Delta_\zeta^{(AV)} = \frac{1}{96\pi} \int d^2x\, \zeta^\nu \Big[\varepsilon_{\rho\sigma}\, \partial^\sigma \left(\partial_\nu \partial_\lambda - \eta_{\nu\lambda} \Box\right) h^{\lambda\rho}(x) \Big]. \tag{14.59}$$

The remaining ones vanish. Consistency conditions are satisfied.

14.1.4 Second Approach

The second approach consists in splitting the Dirac fermion into two Weyl fermions, one left handed and one right handed. Let us start from writing in the action (14.2)

$$\widehat{e}_a^\mu = e_a^\mu + \gamma_* c_a^\mu = \widehat{e}_a^\mu \left(P_R + P_L\right) = e_a^{+\mu} P_R + e_a^{-\mu} P_L \tag{14.60}$$

where $e_a^{\pm\mu} = e_a^\mu \pm c_a^\mu$. Then, (14.2) splits into a left-handed plus a right-handed action

$$\widehat{S} = \int d^2\widehat{x}\,(i\overline{\psi}_R\gamma^a e_a^{+\mu}\partial_\mu\psi_R)\,(\widehat{x}) + \int d^2\widehat{x}\,(i\overline{\psi}_L\gamma^a e_a^{-\mu}\partial_\mu\psi_L)\,(\widehat{x}) \quad (14.61)$$

In this way, we have the sum of two systems to which we can apply the results of Sect. 7.2.3. The frame $e_a^{\pm\mu}$ corresponds to inverse metrics $g^{\pm\mu\nu}$:

$$g^{\pm\mu\nu} = e_a^\mu e^{a\nu} + c_a^\mu c^{a\nu} \pm \left(e_a^\mu c^{a\nu} + c_a^\mu e^{a\nu}\right) = g^{\mu\nu} \pm f^{\mu\nu} \quad (14.62)$$

as can be deduced from

$$\widehat{g}^{\mu\nu} = \widehat{e}_a^\mu \widehat{e}^{a\nu} = g^{\mu\nu} + \gamma_* f^{\mu\nu}. \quad (14.63)$$

$g^{\pm\mu\nu}$ are the inverse of the metrics $g^\pm_{\mu\nu} = g_{\mu\nu} \pm \gamma_* f_{\mu\nu}$, respectively:

$$g^{\pm\mu\lambda} g^\pm_{\lambda\nu} = g^{\mu\lambda} g_{\lambda\nu} + f^{\mu\lambda} f_{\lambda\nu} \pm \left(g^{\mu\lambda} f_{\lambda\nu} + g^{\mu\lambda} g_{\lambda\nu}\right) = \delta^\mu_\nu \quad (14.64)$$

as can be deduced from the relation

$$\widehat{g}^{\mu\lambda}\widehat{g}_{\lambda\nu} = g^{\mu\lambda} g_{\lambda\nu} + f^{\mu\lambda} f_{\lambda\nu} + \gamma_* \left(g^{\mu\lambda} f_{\lambda\nu} + g^{\mu\lambda} g_{\lambda\nu}\right) = \delta^\mu_\nu. \quad (14.65)$$

From their very definition, we can deduce the Weyl and diffeomorphism transformation properties (both ordinary and axial).

$$\delta_W g^\pm_{\mu\nu} = 2(\omega \pm \eta)\eta_{\mu\nu} \quad (14.66)$$

$$\delta_D g^\pm_{\mu\nu} = \partial_\mu \xi^\pm_\nu + \partial_\nu \xi^\pm_\mu, \qquad \xi^\pm_\mu = \xi_\mu \pm \zeta_\mu \quad (14.67)$$

From the results of Sect. 7.2.3, we can deduce the anomalies of the system. They are the sum of the even parity anomalies and the difference of the odd parity anomalies of the two subsystems. Setting $g^\pm_{\mu\nu} = \eta_{\mu\nu} + h^\pm_{\mu\nu}$, we have

$$\Delta_{\omega\pm\eta} = \frac{1}{192\pi}\int d^2x\,(\omega(x) \pm \eta(x))\left[\partial_\lambda\partial_\rho h^{\lambda\rho}_\pm(x) - \Box h^\lambda_{\pm\lambda}(x) \pm \varepsilon_{\lambda\sigma}\partial^\sigma\partial_\rho h^{\lambda\rho}_\pm\right] \quad (14.68)$$

for the trace anomalies, and

$$\Delta_{\xi^\pm} = \frac{1}{384\pi}\int d^2x\,\xi^\nu_\pm(x)\left[\partial_\nu\partial_\lambda\partial_\rho h^{\lambda\rho}_\pm(x) - \partial_\lambda\Box h^\lambda_{\pm\nu}(x) \pm \varepsilon_{\rho\sigma}\partial^\sigma\left(\partial_\nu\partial_\lambda - \eta_{\nu\lambda}\Box\right)h^{\lambda\rho}_\pm(x)\right] \quad (14.69)$$

for the diffeomorphism ones.

Parity Even Cocycles

Let us focus on the even anomalies. By construction, they are all consistent, as we have already verified in Sect. 7.2.3. We can introduce the even counterterm

$$\begin{aligned}\mathcal{C}^{(e)} &= \frac{1}{768\pi}\int \mathrm{d}^2x \left(h^{\nu}_{+\rho}\partial_\lambda\partial_\nu h^{\lambda\rho}_{+} - h_{+\lambda\rho}\Box h^{\lambda\rho}_{+} + h^{\nu}_{-\rho}\partial_\lambda\partial_\nu h^{\lambda\rho}_{-} - h_{-\lambda\rho}\Box h^{\lambda\rho}_{-}\right)\\ &= \frac{1}{768\pi}\int \mathrm{d}^2x \left(h^{\nu}_{\rho}\partial_\lambda\partial_\nu h^{\lambda\rho} - h_{\lambda\rho}\Box h^{\lambda\rho} + k^{\nu}_{\rho}\partial_\lambda\partial_\nu k^{\lambda\rho} - k_{\lambda\rho}\Box k^{\lambda\rho}\right) \quad (14.70)\end{aligned}$$

It is easy to prove that

$$\begin{aligned}\mathcal{A}^{(\text{even})}_{\xi} &\equiv \Delta^{(\text{even})}_{\xi} + \delta^{(0)}_{\xi}\mathcal{C}^{(e)} = 0\\ \mathcal{A}^{(\text{even})}_{\zeta} &\equiv \Delta^{(\text{even})}_{\zeta} + \delta^{(0)}_{\zeta}\mathcal{C}^{(e)} = 0 \quad (14.71)\end{aligned}$$

As a consequence of adding the counterterm to the effective action we have:

$$\begin{aligned}\mathcal{A}^{(\text{even})}_{\omega} &\equiv \Delta^{(\text{even})}_{\omega} + \delta^{(0)}_{\omega}\mathcal{C}^{(e)} = \frac{1}{48\pi}\int \mathrm{d}^2x\, \omega\left[\partial_\lambda\partial_\rho h^{\lambda\rho} - \Box h^{\lambda}_{\lambda}\right]\\ &\approx \frac{1}{48\pi}\int \mathrm{d}^2x\sqrt{g}\,\omega\, R^{(1)} + \cdots \quad (14.72)\end{aligned}$$

On the other hand,

$$\begin{aligned}\mathcal{A}^{(\text{even})}_{\eta} &\equiv \Delta^{(\text{even})}_{\eta} + \delta^{(0)}_{\eta}\mathcal{C}^{(e)} = \frac{1}{48\pi}\int \mathrm{d}^2x\, \eta\left[\partial_\lambda\partial_\rho k^{\lambda\rho} - \Box k^{\lambda}_{\lambda}\right]\\ &\approx \frac{1}{48\pi}\int \mathrm{d}^2x\sqrt{g}\,\eta\, R^{(2)} + \cdots \quad (14.73)\end{aligned}$$

This results coincide with those found with the first approach, see subsection 14.1.3 and comment there.

Parity Odd Cocycles

The odd parity trace cocycles are given by

$$\Delta^{(\text{odd})}_{\omega} = \frac{1}{96\pi}\int \mathrm{d}^2x\, \omega\, \varepsilon_{\lambda\sigma}\partial^\sigma\partial_\rho k^{\lambda\rho} \quad (14.74)$$

and

$$\Delta^{(odd)}_{\eta} = \frac{1}{96\pi}\int \mathrm{d}^2x\, \eta\, \varepsilon_{\lambda\sigma}\partial^\sigma\partial_\rho h^{\lambda\rho}, \quad (14.75)$$

and the odd parity diffeomorphism cocycles are

$$\Delta^{(\text{odd})}_{\xi} = \frac{1}{192\pi}\int \mathrm{d}^2x\, \xi^\nu\varepsilon_{\rho\sigma}\partial^\sigma\left(\partial_\nu\partial_\lambda - \eta_{\nu\lambda}\Box\right)k^{\lambda\rho}(x)\big] \quad (14.76)$$

and

$$\Delta_{\zeta}^{(\text{odd})} = \frac{1}{192\pi} \int d^2x\, \zeta^{\nu} \varepsilon_{\rho\sigma} \partial^{\sigma} \left(\partial_{\nu}\partial_{\lambda} - \eta_{\nu\lambda}\Box\right) h^{\lambda\rho}(x)\Big]. \tag{14.77}$$

It is easy to verify that all the consistency conditions are verified:

$$\delta_{\omega}\Delta_{\omega}^{(\text{odd})} = 0, \quad \delta_{\eta}\Delta_{\omega}^{(\text{odd})} = 0, \quad \delta_{\omega}\Delta_{\eta}^{\text{odd})} = 0, \quad \delta_{\eta}\Delta_{\eta}^{(\text{odd})} = 0 \tag{14.78}$$

$$\delta_{\xi}\Delta_{\xi}^{(\text{odd})} = 0, \quad \delta_{\zeta}\Delta_{\xi}^{(\text{odd})} = 0, \quad \delta_{\xi}\Delta_{\zeta}^{(\text{odd})} = 0, \quad \delta_{\zeta}\Delta_{\zeta}^{(\text{odd})} = 0 \tag{14.79}$$

and

$$\delta_{\zeta}\Delta_{\omega}^{(\text{odd})} + \delta_{\omega}\Delta_{\zeta}^{(\text{odd})} = 0, \quad \delta_{\xi}\Delta_{\eta}^{(\text{odd})} + \delta_{\eta}\Delta_{\xi}^{(\text{odd})} = 0 \tag{14.80}$$

The counterterm

$$\mathcal{C}^{(o)} = \frac{1}{192\pi} \int d^2x\, k^{\sigma\nu} \varepsilon_{\rho\sigma} \left(\partial_{\nu}\partial_{\lambda} - \eta_{\nu\lambda}\Box\right) h^{\lambda\rho} \tag{14.81}$$

gives

$$\delta_{\omega}\mathcal{C}^{(o)} = -\Delta_{\omega}^{(\text{odd})}, \quad \delta_{\eta}\mathcal{C}^{(o)} = \Delta_{\eta}^{(\text{odd})}, \quad \delta_{\xi}\mathcal{C}^{(o)} = \Delta_{\xi}^{(\text{odd})}, \quad \delta_{\zeta}\mathcal{C}^{(o)} = -\Delta_{\zeta}^{(\text{odd})} \tag{14.82}$$

Therefore, by adding or subtracting $\mathcal{C}^{(o)}$ we can cancel $\Delta_{\omega}^{(\text{odd})}$ and $\Delta_{\zeta}^{(\text{odd})}$, and double $\Delta_{\eta}^{(\text{odd})}$ and $\Delta_{\xi}^{(\text{odd})}$ or vice versa.

14.2 MAT Anomalies from Seeley-DeWitt

Let us consider the background metric $\widehat{g}_{\mu\nu} = g_{\mu\nu} + \gamma_* f_{\mu\nu}$, where γ_* is the chirality matrix, and build the Dirac operator $\widehat{\nabla\!\!\!/} = i\widehat{\gamma}^{\mu}\left(\widehat{D}_{\mu} + \frac{1}{2}\widehat{\Omega}_{\mu}\right)$, where $\widehat{D}_{\mu}$ is the covariant differential operator corresponding to $\widehat{g}$, $\widehat{\gamma}_{\mu} = \gamma^{a}\widehat{e}_{a}^{\mu}$, $\widehat{e}_{a}^{\mu} = e_{a}^{\mu} + \gamma_* c_{a}^{\mu}$, and $\widehat{\Omega}_{\mu} = \widehat{\Omega}_{\mu}^{ab}\Sigma_{ab}$ where $\widehat{\Omega}_{\mu} = \Omega_{\mu}^{(1)} + \gamma_*\Omega_{\mu}^{(2)}$.

As usual, we define the amplitude

$$\langle \widehat{x}, \widehat{s} | \widehat{x}', 0 \rangle = \langle \widehat{x} | e^{i\widehat{\mathfrak{F}}s} | \widehat{x}' \rangle \tag{14.83}$$

which satisfies the (heat kernel) differential equation

$$i\frac{\partial}{\partial \widehat{s}}\langle \widehat{x}, \widehat{s}|\widehat{x}', 0\rangle = -\widehat{\mathfrak{F}}_{\widehat{x}}\langle \widehat{x}, \widehat{s}|\widehat{x}', 0\rangle \tag{14.84}$$

At this stage, we have to make a choice for the quadratic differential operator $\widehat{\mathfrak{F}}$. As explained in Sect. 8.3.1, it makes sense to apply the SDW method only to systems where diffeomorphisms are conserved. In Sect. 10.2.3, this remark has been extended also to axial diffeomorphisms. Moreover, the Euclidean version $\widetilde{\widetilde{\widehat{\mathfrak{F}}}}$ must be self-adjoint. Therefore, using the same argument as in 10.2.3, the choice must be

$$\widehat{\mathfrak{F}}_{\widehat{x}} = -\overline{\overline{\widehat{\nabla\!\!\!/}}}\widehat{\nabla\!\!\!/} = \widehat{D}_\mu \widehat{g}^{\mu\nu} \widehat{D}_\nu - \frac{1}{4}\widehat{R} \tag{14.85}$$

where $\widehat{R} = R^{(1)} + \gamma_* R^{(2)}$. At first sight, however, this choice seems to be improper, because we have just shown (see (14.76) and (14.77)) that diffeomorphisms are anomalous. Therefore, whatever results we find for trace anomalies, they are not reliable. In this case, however, there is a way out. In Sect. 17.1.6, we shall prove that these two-dimensional odd parity anomalies are gauge artifacts and, choosing a suitable gauge (the conformal one), they vanish identically. In view of this the choice (14.85) makes sense, with the proviso that the results we will find are valid in this specific gauge.

Then, we make the ansatz (in $\mathsf{d} = 2$)

$$\langle \widehat{x}, \widehat{s}|\widehat{x}', 0\rangle = \lim_{m\to 0} \frac{i}{4\pi\widehat{s}}\sqrt{\widehat{\mathfrak{D}}(\widehat{x}, \widehat{x}')}\, e^{i\left(\frac{\widehat{\sigma}(\widehat{x},\widehat{x}')}{2\widehat{s}} - m^2\widehat{s}\right)}\widehat{\Phi}(\widehat{x}, \widehat{x}', \widehat{s}) \tag{14.86}$$

where $\widehat{\mathfrak{D}}(\widehat{x}, \widehat{x}')$ is the VVM determinant and $\widehat{\sigma}$ is the world function (one half the square geodesic distance). $\widehat{\Phi}(\widehat{x}, \widehat{x}', \widehat{s})$ is a function to be determined. As usual, we have also introduced the mass parameter m, which we will eventually set to zero. In the limit $s \to 0$, the RHS of (14.86) becomes the definition of a delta function multiplied by $\widehat{\Phi}$. As usual, we must have

$$\lim_{s\to 0} \widehat{\Phi}(\widehat{x}, \mathbf{x}', \widehat{s}) = \mathbf{1} \tag{14.87}$$

Eq. (14.84) becomes an equation for $\widehat{\Phi}(\widehat{x}, \widehat{x}', \widehat{s})$. Then, we expand

$$\widehat{\Phi}(\widehat{x}, \widehat{x}', \widehat{s}) = \sum_{n=0}^{\infty} \widehat{a}_n(\widehat{x}, \widehat{x}')(i\widehat{s})^n \tag{14.88}$$

with the boundary condition $[\widehat{a}_0] = 1$. In particular, we have

$$[\widehat{a}_1] = \left(-\frac{1}{12}\widehat{R} + m^2\right)\mathbf{1} \tag{14.89}$$

The effective action near $d \approx 2$ is

$$\widehat{\mathcal{L}}(\widehat{x}) = \frac{1}{4\pi}\left(\frac{1}{\mathsf{d}-2} - \frac{1}{2}\right)\mathsf{Tr}\left(([\widehat{a}_1] - m^2)\sqrt{\widehat{g}}\right)$$
$$-\frac{i}{8\pi}\mathsf{Tr}\int_0^\infty \mathrm{d}s\ \ln(4\pi i\mu^2 s)\sqrt{\widehat{g}}\frac{\partial^2}{\partial(is)^2}\left(\mathrm{e}^{-im^2 s}[\widehat{\Phi}(\widehat{x},\widehat{x},s)]\right) \quad (14.90)$$

Tr includes also spacetime integration. The last line depends explicitly on the μ parameter (a mass parameter introduced in order to produce dimensionless quantities) and represents a non-local part, which we ignore in the sequel.

Next, let us consider the axially extended Weyl transformations

$$\delta_{\widehat{\omega}}\sqrt{\widehat{g}} = \mathsf{d}\,\widehat{\omega}\sqrt{\widehat{g}} \quad (14.91)$$
$$\delta_{\widehat{\omega}}\widehat{R} = -2\widehat{\omega}\,\widehat{R} - 2(\mathsf{d}-1)\widehat{\square}\widehat{\omega} \quad (14.92)$$

Thus, when $\mathsf{d} \to 2$, we find

$$\delta_{\widehat{\omega}}\widehat{\mathcal{L}}(\widehat{x}) = -\frac{1}{48\pi}\int \mathrm{d}^2\widehat{x}\,\mathrm{tr}\left(\widehat{\omega}\sqrt{\widehat{g}}\widehat{R}\right) \quad (14.93)$$

where tr denotes gamma matrix trace. To interpret this result, let us recall that

$$\sqrt{\widehat{g}} = \sqrt{\det(\widehat{g})} = \sqrt{\det(g+\gamma_5 f)} \quad (14.94)$$

Using the formula $\det = \mathrm{e}^{\mathrm{Tr}\log}$ this gives

$$\mathrm{Tr}\ln(g+\gamma_5 f) = P_R\mathrm{Tr}\ln(g+f) + P_L\,\mathrm{Tr}\ln(g-f) \quad (14.95)$$

or

$$\sqrt{\widehat{g}} = \frac{1}{2}\left(\sqrt{\det(g+f)} + \sqrt{\det(g-f)}\right) + \frac{\gamma_5}{2}\left(\sqrt{\det(g+f)} - \sqrt{\det(g-f)}\right) \quad (14.96)$$

The integrand in (14.93) is

$$\mathrm{tr}\left(\widehat{\omega}\sqrt{\widehat{g}}\widehat{R}\right) = \frac{1}{2}\sqrt{\det(g+f)}\left[(\omega R^{(1)} + \eta R^{(2)}) + (\omega R^{(2)} + \eta R^{(1)})\right]$$
$$+\frac{1}{2}\sqrt{\det(g-f)}\left[(\omega R^{(1)} + \eta R^{(2)}) - (\omega R^{(2)} + \eta R^{(1)})\right] \quad (14.97)$$

Let us compare this with our previous results for Weyl fermions. To this end, we have to take the chiral limit. In the right-handed chiral limit $h_{\mu\nu} \to h_{\mu\nu}/2$, $k_{\mu\nu} \to h_{\mu\nu}/2$ we have

$$\sqrt{\det(g+f)} \to \sqrt{\det(g)} \equiv \sqrt{g},\quad \sqrt{\det(g-f)} \to 1,\quad R^{(1)} \to \frac{1}{2}R,\quad R^{(2)} \to \frac{1}{2}R.$$

where R is the Ricci scalar of the metric $g_{\mu\nu} = \eta_{\mu\nu} + h_{\mu\nu}$. Moreover, the parameters $\omega \to \omega/2$, $\eta \to \omega/2$ (as can be deduced, for instance, from (14.7), see also the discussion in Sect. 10.4.5). Moreover, we must take into account: $\mathrm{tr}(1) = 2$, $\mathrm{tr}(\gamma_*) = 0$. Therefore, the second line of (14.97) tends to 0, and

$$\mathrm{tr}\left(\widehat{\omega}\sqrt{\widehat{g}}\widehat{R}\right) \to \sqrt{g}\,\omega\,R \tag{14.98}$$

Hence, recalling (5.18), (14.93) yields the even trace anomaly (7.94) for a right-handed complex Weyl fermion.

In the left-handed chiral limit $h_{\mu\nu} \to h_{\mu\nu}/2$, $k_{\mu\nu} \to -h_{\mu\nu}/2$, we have instead

$$\sqrt{\det(g+f)} \to 1, \quad \sqrt{\det(g-f)} \to \sqrt{g}, \quad R^{(1)} \to \frac{1}{2}R, \quad R^{(2)} \to -\frac{1}{2}R.$$

while $\omega \to \omega/2$, $\eta \to -\omega/2$. Therefore, the first line in the RHS of (14.97) vanishes, and

$$\mathrm{tr}\left(\widehat{\omega}\sqrt{\widehat{g}}\widehat{R}\right) \to \sqrt{g}\,\omega\,R \tag{14.99}$$

gives the even trace anomaly of a left-handed (complex) fermion.

If instead we set $f = 0$, $\eta = 0$ we obtain

$$\mathrm{tr}\left(\widehat{\omega}\sqrt{\widehat{g}}\widehat{R}\right) \to 2\sqrt{g}\,\omega\,R \tag{14.100}$$

which gives the even trace anomaly for a complex Dirac fermion.

Let us now compare with the perturbative MAT results. At the level of approximation considered in the perturbative calculations above, $\sqrt{\det(g+f)} \approx 1 \approx \sqrt{\det(g-f)}$. Therefore, $\mathrm{Tr}\left(\widehat{\omega}\sqrt{\widehat{g}}\,\widehat{R}\right)$ reduces to

$$\mathrm{tr}\left(\widehat{\omega}\sqrt{\widehat{g}}\,\widehat{R}\right) \approx 2\left(\omega R_0^{(1)} + \eta R_0^{(2)}\right)$$

where $R_0^{(i)}$ represents the lowest order approximation of $R^{(i)}$. The even trace anomaly is made of two pieces, one proportional to ω and the other to η: $\omega R_0^{(1)} + \eta R_0^{(2)}$.

Therefore, the anomalies derived with the SDW (Schwinger-Seeley-DeWitt) method are:

$$\mathcal{A}_\omega^{(\mathrm{even})} \approx \frac{1}{48\pi}\int d^2x\,\omega\,R_0^{(1)}, \quad \mathcal{A}_\eta^{(\mathrm{even})} \approx \frac{1}{48\pi}\int d^2x\,\eta\,R_0^{(2)} \tag{14.101}$$

This corresponds to the results of the perturbative calculations, see (14.53, 14.54, 14.72) and (14.73).

Of course, (14.101) is only the lowest order result, the complete one being given by (14.93) and (14.97). In order to see the additional terms in (14.97) (with respect to the mid terms of (14.53, 14.54, 14.72, 14.73)) with a perturbative calculation, one should go to the next orders of approximation.

To conclude, let us remark that this method does not produce odd parity trace anomalies, at variance with the perturbative results (14.74) and (14.75). This is expected since it is related with the fact that both ordinary and axial diffeomorphisms are conserved by the differential operator (14.85). On the other hand, we know from the perturbative results (14.76) and (14.77) that there are diffeomorphism anomalies in the system we are considering. However, we have also anticipated that the just mentioned odd perturbative results for diffeomorphism anomalies are a gauge artifact, which vanishes in the conformal gauge. The same is true for the odd parity trace anomalies. Therefore, the fact that in the derivation of this section the odd trace anomalies do not show up is a result consistent with this gauge choice.

Chapter 15
Wess-Zumino Terms

The Wess-Zumino consistency conditions for anomalies can be interpreted as integrability conditions. They can in fact be integrated and produce local terms whose gauge variation yields the corresponding anomaly. In this chapter, we will show how to derive them. Such Wess-Zumino (WZ) terms have been variously utilized in the literature. In local gauge field theories, their significance is presumably academical, because the integration is possible only at the price of introducing new scalar fields in the theory: if on one side they cancel the anomalies, on the other side they may create problems for renormalization, except possibly in 2d. It is in fact mainly having in mind field theory in 2d (in particular string theory) that we introduce WZ terms.

A different application of WZ terms consists in establishing a connection between chiral diffeomorphism and local Lorentz anomalies. We will show that there exists a term (both with and without background connection) which under a diffeomorphism transformation and under a local Lorentz transformation reproduces the corresponding anomalies.

Another effective application of WZ terms is to sigma models. The problem of sigma model anomalies, and their cancelation will be discussed in the next chapter. In preparation to it, however, in the last section of this one, we present a more geometrical derivation of WZ terms in field theory.

15.1 Wess-Zumino Terms in Field Theories

In a gauge field theory with connection A valued in a Lie algebra with anti-Hermitean generators T^a and structure constants f^{abc}, an anomaly $\mathcal{A}^a$ must satisfy the WZ consistency conditions

$$X^a(x)\mathcal{A}^b(y) - X^b(y)\mathcal{A}^a(x) + f^{abc}\mathcal{A}^c(x)\delta(x-y) = 0 \tag{15.1}$$

L. Bonora, *Fermions and Anomalies in Quantum Field Theories*, Theoretical and Mathematical Physics, https://doi.org/10.1007/978-3-031-21928-3_15

where

$$X^a(x) = \partial_\mu \frac{\delta}{\delta A^a_\mu(x)} + f^{abc} A^b_\mu(x) \frac{\delta}{\delta A^c_\mu(x)} \tag{15.2}$$

The conditions (15.1) are integrability conditions. This means that one can find a functional of the fields $\mathcal{B}_{\mathrm{WZ}}$ such that

$$X^a(x)\mathcal{B}_{\mathrm{WZ}} = \mathcal{A}^a(x) \tag{15.3}$$

In this section, we show how to construct a term $\mathcal{B}_{\mathrm{WZ}}$ which, upon BRST variation, generates the anomaly

$$\mathcal{A} = -\int_{\mathsf{M}} c^a(x)\mathcal{A}^a(x) = n(n-1)\int_{\mathsf{M}}\int_0^1 \mathrm{d}t\,(1-t)P_n(dc, A, F_t, \dots, F_t) \tag{15.4}$$

where M is the spacetime of dimension $\mathsf{d} = 2n-2$, and $F_t = tdA + \frac{t^2}{2}[A, A]$.

This is possible provided we enlarge the set of fields of the theory, by adding new fields as follows. We introduce a set of auxiliary fields $\sigma(x) = \sigma^a(x)T^a$, which under a gauge transformation with parameters $\lambda(x) = \lambda^a T^a$ transform as

$$\mathrm{e}^{\sigma(x)} \longrightarrow \mathrm{e}^{\sigma'(x)} = \mathrm{e}^{-\lambda(x)}\mathrm{e}^{\sigma(x)} \tag{15.5}$$

Using the Campbell-Hausdorff formula this means $\delta\sigma(x) = -\lambda(x) - \frac{1}{2}[\lambda(x), \sigma(x)] + \dots$. Passing to the infinitesimal transformations, we replace the infinitesimal parameter $\lambda(x)$ with anticommuting fields $c(x) = c^a(x)T^a$. The BRST transformations of σ is given by

$$\begin{aligned} \mathfrak{s}\mathrm{e}^{\sigma(x)} &= -c(x)\mathrm{e}^{\sigma(x)}, \\ \mathfrak{s}\sigma(x) &= -c(x) + \frac{1}{2}[\sigma(x), c(x)] - \frac{1}{12}[\sigma(x), [\sigma(x), c(x)]] + \cdots \end{aligned} \tag{15.6}$$

Now we use a superspace technique ([1], see Appendix 15A for a short introduction to this formalism), and add to the spacetime coordinates x^μ an anticommuting one ϑ, but simultaneously enlarging the spacetime with the addition of a commuting parameter s, $0 \le s \le 1$. So the local coordinates in the superspace are (x^μ, s, ϑ). In particular,

$$\mathrm{e}^{-s(\sigma+\vartheta\mathfrak{s}\sigma)} = \mathrm{e}^{-s\sigma} + \vartheta\,\mathfrak{s}\mathrm{e}^{-s\sigma} \tag{15.7}$$

On this superspace, the superconnection is

$$\widetilde{\mathcal{A}}(x,s,\vartheta) = \mathrm{e}^{-s(\sigma+\vartheta\mathfrak{s}\sigma)}\mathrm{e}^{-\vartheta c}\left(\tilde{d}+A\right)\mathrm{e}^{\vartheta c}\mathrm{e}^{s(\sigma+\vartheta\mathfrak{s}\sigma)}$$
$$= \mathrm{e}^{-s(\sigma+\vartheta\mathfrak{s}\sigma)}\left(\mathcal{A}+d+\frac{\partial}{\partial s}\mathrm{d}s\right)\mathrm{e}^{s(\sigma+\vartheta\mathfrak{s}\sigma)} \quad (15.8)$$

where $\tilde{d} = d + \frac{\partial}{\partial s}\mathrm{d}s + \frac{\partial}{\partial\vartheta}\mathrm{d}\vartheta$, and $\mathcal{A} = A + \vartheta\,(\mathrm{d}c + [A, c]) + (c - \vartheta cc)\,\mathrm{d}\vartheta$. We decompose $\widetilde{\mathcal{A}}$ as follows:

$$\begin{aligned}
\widetilde{\mathcal{A}}(x,s,\vartheta) &= \phi(x,s,\vartheta) + \phi_s(x,s,\vartheta)\mathrm{d}s + \phi_\vartheta(x,s,\vartheta)\mathrm{d}\vartheta \\
\phi(x,s,\vartheta) &= A_s + \vartheta\,(dc_s + [A_s, c_s]) \\
\phi_s(x,s,\vartheta) &= \sigma + \vartheta\mathfrak{s}\sigma \\
\phi_\vartheta(x,s,\vartheta) &= c_s - \vartheta c_s c_s
\end{aligned} \quad (15.9)$$

where

$$\begin{aligned}
A_s &= \mathrm{e}^{-s\sigma} A\,\mathrm{e}^{s\sigma} + \mathrm{e}^{-s\sigma} d\mathrm{e}^{s\sigma} \\
c_s &= \mathrm{e}^{-s\sigma} c\,\mathrm{e}^{s\sigma} + \mathrm{e}^{-s\sigma}\mathfrak{s}\mathrm{e}^{s\sigma}
\end{aligned} \quad (15.10)$$

This means, in particular, that $A_0 = A$ and c_s interpolates between c and 0. The derivative with respect to ϑ corresponds to the BRST transformation, so we deduce

$$\mathfrak{s}A_s = dc_s + [A_s, c_s], \qquad \mathfrak{s}c_s = -c_s c_s \quad (15.11)$$

Moreover, if we denote by $\widetilde{\mathfrak{F}}, \mathfrak{F}$ and F the curvatures of $\widetilde{\mathcal{A}}, \mathcal{A}$ and A, respectively, we have

$$\widetilde{\mathfrak{F}} = \mathrm{e}^{-s(\sigma+\vartheta\mathfrak{s}\sigma)}\mathfrak{F}\,\mathrm{e}^{s(\sigma+\vartheta\mathfrak{s}\sigma)} = \mathrm{e}^{-s(\sigma+\vartheta\mathfrak{s}\sigma)}\mathrm{e}^{-\vartheta c}F\,\mathrm{e}^{\vartheta c}\mathrm{e}^{s(\sigma+\vartheta\mathfrak{s}\sigma)} \quad (15.12)$$

Now choose any ad-invariant polynomial P_n with $n = \frac{\mathsf{d}}{2-1}$, where d is the spacetime dimension. We have

$$P_n(\widetilde{\mathfrak{F}},\ldots,\widetilde{\mathfrak{F}}) = P_n(\mathfrak{F},\ldots,\mathfrak{F}) = P_n(F,\ldots,F) = 0 \quad (15.13)$$

where the last equality holds for dimensional reasons. But now we can write

$$P_n(\widetilde{\mathfrak{F}},\ldots,\widetilde{\mathfrak{F}}) = \tilde{d}\left(n\int_0^1 \mathrm{d}t\, P_n(\widetilde{\mathcal{A}},\widetilde{\mathfrak{F}}_t,\ldots,\widetilde{\mathfrak{F}}_t)\right) \equiv \tilde{d}\left(\mathrm{TP}_n(\widetilde{\mathcal{A}})\right) \quad (15.14)$$

For simplicity of notation, let us set $\widetilde{\mathcal{Q}}(\widetilde{\mathcal{A}}) \equiv \mathrm{TP}_n(\widetilde{\mathcal{A}})$, and decompose it in the various components according to the form degree

$$\widetilde{\mathcal{Q}} = \sum_{\substack{k,i,j \\ k+i+j=2n-1}} \widetilde{\mathcal{Q}}^j_{(k,i)}, \qquad \widetilde{\mathcal{Q}}^{\cdot j}_{(k,i)} = \left(\mathcal{Q}^j_{(k,i)} + \vartheta\, \hat{\mathcal{Q}}^j_{(k,i)}\right) (\mathrm{d}\vartheta)^j (\mathrm{d}s)^i \quad (15.15)$$

where k denotes the form degree in spacetime, j is the ghost number and i is either 0 or 1. Next let us decompose the equation $\tilde{\mathrm{d}}\widetilde{\mathcal{Q}} = 0$ in components and select in particular the component $(2n-2, 1, 1)$, i.e.

$$0 = \mathrm{d}\widetilde{\mathcal{Q}}^1_{(2n-3,1)} + \frac{\partial}{\partial\vartheta}\widetilde{\mathcal{Q}}^0_{(2n-2,1)}\mathrm{d}\vartheta + \frac{\partial}{\partial s}\widetilde{\mathcal{Q}}^1_{(2n-2,0)}\mathrm{d}s \qquad (15.16)$$

and let us integrate it over M and s. We get

$$0 = \int_{\mathsf{M}}\int_0^1 \mathrm{d}s\, \mathfrak{s}\mathcal{Q}^0_{(2n-2,1)} + \int_{\mathsf{M}}\int_0^1 \mathrm{d}s\, \frac{\partial}{\partial s}\mathcal{Q}^1_{(2n-2,0)} \qquad (15.17)$$

Since $\mathcal{Q}^1_{(2n-2,0)}$ is linear in c_s and $c_0 = c$, $c_1 = 0$, we get finally

$$\int_{\mathsf{M}} \mathcal{Q}^1_{(2n-2,0)} = \mathfrak{s}\int_{\mathsf{M}}\int_0^1 \mathrm{d}s\, \mathcal{Q}^0_{(2n-2,1)} \qquad (15.18)$$

Now remark that $\int_{\mathsf{M}} \mathcal{Q}^{(1)}_{(2n-2,0)}$ is linear in c and coincides precisely with the anomaly. On the other hand, $\mathcal{Q}^{(0)}_{(2n-2,1)}$ has the same expression as the anomaly with c replaced by σ and A by A_s, i.e.

$$\mathcal{Q}^0_{(2n-2,1)}(\sigma, A_s) = n(n-1)\int_0^1 \mathrm{d}t\,(1-t)\,P_n(\mathrm{d}\sigma, A_s, F_{s,t}, \ldots, F_{s,t}) \quad (15.19)$$

where $F_{s,t} = t\mathrm{d}A_s + \frac{t^2}{2}[A_s, A_s]$. We call

$$\mathcal{B}_{\mathrm{WZ}} = \int_{\mathsf{M}}\int_0^1 \mathrm{d}s\, \mathcal{Q}^0_{(2n-2,1)} \qquad (15.20)$$

the *Wess-Zumino term.*

The existence of $\mathcal{B}_{\mathrm{WZ}}$ for any anomaly seems to contradict its non-triviality. This is not so, for the price we have to pay to construct the term (15.20) is the introduction of the new fields σ^a, which are not present in the initial theory. The proof of non-triviality is based on a definite differential space formed by c, A and their exterior derivatives and commutators, which constrains the anomaly to be a polynomial in these fields.

Of course, in principle, it is not forbidden to enlarge the theory by adding new fields, plus the WZ term. In this particular case, we remark that one should physically motivate the addition of the fields σ^a, and, in particular, that they have 0 canonical dimension. This may be acceptable in two dimensions. But in higher dimensions, it is possible to construct out of these fields new invariant quadratic action terms with more than two derivatives. Such terms are generically excited by renormalization even if they are not present in the initial action. Renormalization in these conditions becomes problematic. For this reason, when introducing WZ terms, we have in mind essentially 2d field theory and, in particular, string theory.

***

A different and very significant role plays the WZ terms when applied to chiral diffeomorphism and local Lorentz anomalies. The existence of such terms proves that these two type of anomalies are linked in a one-to-one way, as was argued above in Chap. 11 (see in particular Sect. (11.6.2). One can cancel a chiral diffeomorphism anomaly by a WZ terms but the same term creates the corresponding local Lorentz anomaly, or vice-versa. In the next two sections, we are are going to construct these terms

15.2 The Wess-Zumino Term for Einstein and Lorentz Anomalies

Let us recall once again the formulas for Einstein and Lorentz anomalies. Anomalies of the local Lorentz symmetry are formally the same as local gauge anomalies. The role of gauge connection is played by the spin connection (in matrix notation)

$$\omega_\mu = \{\omega_\mu^{ab}\}, \qquad \omega^{ab} = \omega_\mu^{ab} \mathrm{d}x^\mu \tag{15.21}$$

whose transformation law is

$$\delta_\Lambda \omega = \mathrm{d}\Lambda + [\omega, \Lambda], \qquad \Lambda = \{\Lambda^{ab}\} \tag{15.22}$$

The curvature of ω is the Riemann two form

$$\mathcal{R} = \mathrm{d}\omega + \frac{1}{2}[\omega, \omega], \qquad \mathcal{R} = \{\mathcal{R}^{ab}\}, \qquad \delta_\Lambda \mathcal{R} = [\mathcal{R}, \Lambda] \tag{15.23}$$

Thus, the general cocycle in dimension $\mathrm{d} = 2n - 2$ is obtained from the usual formula for consistent gauge anomalies by simply making the replacements $c \to \Lambda$, $A \to \omega$ and $F \to \mathcal{R}$:

$$\Delta^1_{\mathrm{d}}(\Lambda, \omega) = n(n-1) \int_0^1 \mathrm{d}t (1-t) P_n(\mathrm{d}\Lambda, \omega, \mathcal{R}_t, \ldots \mathcal{R}_t) \tag{15.24}$$

In this case the anomaly is (up to an overall coefficient)

$$\mathcal{A}_L[\Lambda, \omega] = \int d^{\mathrm{d}}x \,\sqrt{g}\Delta^1_{\mathrm{d}}(\Lambda, \omega), \qquad \delta_\Lambda \mathcal{A}_L[\Lambda, \omega] = 0 \tag{15.25}$$

The $\sqrt{g}$ is introduced in order to guarantee diffeomorphism invariance

$$\delta_\xi \mathcal{A}_L[\Lambda, \omega] = 0 \tag{15.26}$$

Let us see next the cocycles of diffeomorphisms. In order to exploit the parallelism with the gauge case, we have introduced for the Christoffel symbols the matrix-form notation

$$\Gamma \equiv \{\Gamma_\nu{}^\lambda\}, \qquad \Gamma_\nu{}^\lambda = \mathrm{d}x^\mu\, \Gamma_{\mu\nu}{}^\lambda \tag{15.27}$$

and for the Riemann curvature

$$R = \mathrm{d}\Gamma + \Gamma^2, \qquad \Gamma_t, \qquad R_t = \mathrm{d}\,\Gamma_t + \Gamma_t^2 = tR + (t^2 - t)\Gamma^2 \tag{15.28}$$

The product between adjacent entries is the matrix product: $(XY)^\rho{}_\lambda = X^\rho{}_\nu Y^\nu{}_\lambda$, and $R = \{R^\rho{}_\lambda\}$, $R^\rho{}_\lambda = \frac{1}{2}\mathrm{d}x^\mu \wedge \mathrm{d}x^\nu R_{\mu\nu}{}^\rho{}_\lambda$. Now we can adapt the previous cocycle formulas to this case

$$\Delta^1_{\mathrm{d}}(\xi, \Gamma) = n(n-1)\int_0^1 \mathrm{d}t(1-t)\mathrm{Str}(\mathrm{d}\Xi\, \Gamma\, R_t \,\ldots\, R_t) \tag{15.29}$$

where $\Xi^\rho{}_\lambda = \partial_\lambda \xi^\rho$, and Str denotes the symmetric trace of the matrix entries. The integrated anomaly (up to an overall coefficient) is:

$$\mathcal{A}_D[\xi, \Gamma] = \int d^{\mathrm{d}}x \,\sqrt{g}\Delta^1_{\mathrm{d}}(\xi, \Gamma) \tag{15.30}$$

which is obviously local Lorentz invariant.

Let us introduce next the vielbein e^a_μ, whose transformation properties under diffeomorphisms and local Lorentz transformations are

$$\delta_\xi e^a_\mu = \xi^\lambda \partial_\lambda e^a_\mu + \partial_\mu \xi^\lambda e^a_\lambda, \qquad \delta_\Lambda e^a_\mu = e^b_\mu \Lambda_b{}^a \tag{15.31}$$

The inverse vielbein e^μ_a transforms as

$$\delta_\xi \mathrm{e}^\mu_a = \xi^\lambda \partial_\lambda \mathrm{e}^\mu_a - \partial_\lambda \xi^\mu \mathrm{e}^\lambda_a, \qquad \delta_\Lambda \mathrm{e}^\mu_a = \Lambda_a{}^b \mathrm{e}^\mu_b \tag{15.32}$$

The vielbein determines a direct relation between R and $\mathcal{R}$ (already pointed out in Eq. (2.40))

$$R^{\alpha}{}_{\beta} = e^{\alpha}_{a}\,\mathcal{R}^{a}{}_{b}\,e^{b}_{\beta} \tag{15.33}$$

which implies that

$$\mathrm{tr}\left(R^{n}\right) = \mathrm{tr}\left(\mathcal{R}^{n}\right), \tag{15.34}$$

Let us recall that if we write the same quantities in the form (2.33 and 2.35), we have

$$\mathrm{tr}\left(\mathbf{R}^{n}\right) = \alpha_{\mathrm{d}}\,\mathrm{tr}\left(\mathcal{R}^{n}\right) \tag{15.35}$$

where α_{d} is a constant which depends on the dimension: $\alpha_d = 2^{\frac{d}{2}-1}$.

The anomalies (15.25) and (15.30) identify two apparently disconnected cocycles of the cohomology of the local Lorentz transformations and of the diffeomorphisms, respectively: (15.25) preserves diffeomorphism invariance, while (15.30) preserves local Lorentz invariance. In a regularized theory either the former or the latter may appear. However we have shown in Chap. 11 that they arise from the same cohomology class of their classifying space, so that they are uniquely related. One can give a direct proof of this fact by constructing a counterterm that connects them. This is possible thanks to the vielbein which plays the role of the fields σ's of the previous section.

Following Bardeen and Zumino, [2], let us construct first such a WZ term for the Lorentz anomaly (15.25). Next we will show that the same term generates the corresponding diffeomorphism anomaly. The simplest way is to proceed like in Sect. 15.1. Using a matrix notation E for the matrix e^{a}_{μ}, we introduce a field H by

$$E = \mathrm{e}^{H}, \qquad \delta_{\Lambda}\mathrm{e}^{H} = \mathrm{e}^{H}\Lambda \tag{15.36}$$

and promote the parameter Λ to anticommuting field with the relevant BRST transformation property $\mathfrak{s}_{\mathrm{L}}\Lambda = -\Lambda^{2}$. The BRST transformation of H is given by

$$\mathfrak{s}_{\mathrm{L}}\mathrm{e}^{H(x)} = \mathrm{e}^{H(x)}\Lambda, \qquad \mathfrak{s}_{\mathrm{L}}H(x) = H(x) + \frac{1}{2}[H(x), \Lambda(x)] + \cdots \tag{15.37}$$

Now we use again a superspace technique, by adding to the spacetime coordinates x^{μ} an anticommuting one ϑ, but simultaneously enlarging the spacetime with the addition of a commuting parameter s, $0 \le s \le 1$. So the local coordinates in the superspace are (x^{μ}, s, ϑ). On this superspace, we have

$$\mathrm{e}^{-s(H+\vartheta\,\mathfrak{s}_{\mathrm{L}}H)} = \mathrm{e}^{-sH} + \vartheta\,\mathfrak{s}_{\mathrm{L}}\mathrm{e}^{-sH} \tag{15.38}$$

The superconnection is

$$\begin{aligned}\widetilde{\omega}(x,s,\vartheta) &= e^{-s(H+\vartheta\mathfrak{s}_L H)} e^{-\vartheta\Lambda}\left(\tilde{d}+\omega\right) e^{\vartheta\Lambda} e^{s(H+\vartheta\mathfrak{s}_L H)} \\ &= e^{-s(H+\vartheta\mathfrak{s}_L H)}\left(\hat{\omega}+\tilde{d}\right) e^{s(H+\vartheta\mathfrak{s}_L H)}\end{aligned} \tag{15.39}$$

where $\tilde{d} = d + \frac{\partial}{\partial s}\mathrm{d}s + \frac{\partial}{\partial\vartheta}\mathrm{d}\vartheta$, and $\hat{\omega} = \omega + \vartheta\,(d\Lambda + [\omega,\Lambda]) + \left(\Lambda - \vartheta\Lambda^2\right)\mathrm{d}\vartheta$. We decompose $\widetilde{\omega}$ as follows:

$$\begin{aligned}\widetilde{\omega}(x,s,\vartheta) &= \chi(x,s,\vartheta) + \chi_s(x,s,\vartheta)\mathrm{d}s + \chi_\vartheta(x,s,\vartheta)\mathrm{d}\vartheta \\ \chi(x,s,\vartheta) &= \omega_s + \vartheta\,(\mathrm{d}\Lambda_s + [\omega_s,\Lambda_s]) \\ \chi_s(x,s,\vartheta) &= H + \vartheta\mathfrak{s}_L H \\ \chi_\vartheta(x,s,\vartheta) &= \Lambda_s - \vartheta\Lambda_s\Lambda_s\end{aligned} \tag{15.40}$$

where

$$\begin{aligned}\omega_s &= e^{-sH}\omega\, e^{sH} + e^{-sH} d e^{sH} \\ \Lambda_s &= e^{-sH}\Lambda\, e^{sH} + e^{-sH}\mathfrak{s}_L e^{sH}\end{aligned} \tag{15.41}$$

In particular $\omega_0 = \omega$ and Λ_s interpolates between Λ and 0. Since the derivative with respect to ϑ corresponds to the BRST transformation, we deduce

$$\mathfrak{s}_L\omega_s = d\Lambda_s + [\omega_s,\Lambda_s], \qquad \mathfrak{s}_L\Lambda_s = -\Lambda_s\Lambda_s \tag{15.42}$$

Let us denote by $\widetilde{\mathcal{R}}$, $\hat{\mathcal{R}}$ and $\mathcal{R}$ the curvatures of $\widetilde{\omega}$, $\hat{\omega}$ and ω, where $\hat{\omega} \equiv e^{-\vartheta\Lambda}(\widetilde{d} + \omega)\, e^{\vartheta\Lambda}$. Then

$$\widetilde{\mathcal{R}} = e^{-s(H+\vartheta\mathfrak{s}_L H)}\hat{\mathcal{R}}\, e^{s(H+\vartheta\mathfrak{s}_L H)} = e^{-s(H+\vartheta\mathfrak{s}_L H)} e^{-\vartheta\Lambda} R\, e^{\vartheta\Lambda} e^{s(H+\vartheta\mathfrak{s}_L H)} \tag{15.43}$$

Now suppose the spacetime M has dimension d and choose any ad-invariant polynomial P_n with $n = \frac{\mathsf{d}}{2} - 1$. We have

$$P_n(\widetilde{\mathcal{R}},\ldots,\widetilde{\mathcal{R}}) = P_n(\hat{\mathcal{R}},\ldots,\hat{\mathcal{R}}) = P_n(\mathcal{R},\ldots,\mathcal{R}) = 0 \tag{15.44}$$

where the last equality holds for dimensional reasons. But now we can write

$$P_n(\widetilde{\mathcal{R}},\ldots,\widetilde{\mathcal{R}}) = \tilde{d}\left(n\int_0^1 \mathrm{d}t\, P_n(\widetilde{\omega},\widetilde{\mathcal{R}}_t,\ldots,\widetilde{\mathcal{R}}_t)\right) \equiv \tilde{d}\,(\mathrm{TP}_n(\widetilde{\omega})) \tag{15.45}$$

For simplicity of notation, we set as before $\widetilde{\mathcal{K}}(\widetilde{\omega}) \equiv \mathrm{TP}_n(\widetilde{\omega})$ and proceed to decompose it

$$\widetilde{\mathcal{K}} = \sum_{\substack{k,i,j \\ k+i+j=2n-1}} \widetilde{\mathcal{K}}^j_{(k,i)}, \qquad \widetilde{\mathcal{K}}^j_{(k,i)} = \left(K^j_{(k,i)} + \vartheta \widehat{K}^j_{(k,i)}\right) (\mathrm{d}\vartheta)^j (\mathrm{d}s)^i \quad (15.46)$$

where k denotes the form degree in spacetime, j is the ghost number and i is either 0 or 1.

Proceeding as before, we get finally

$$\int_{\mathsf{M}} K^1_{(2n-2,0)}(\omega, \Lambda) = \mathfrak{s}_{\mathsf{L}} \int_{\mathsf{M}} \int_0^1 \mathrm{d}s\, K^0_{(2n-2,1)}(\omega, H) \quad (15.47)$$

The integration over M is understood with the measure $\sqrt{g}\, d^{\mathsf{d}}x$. Now remark that $\int_{\mathsf{M}} K_{(2n-2,0,1)}$ is linear in Λ and coincides precisely with the anomaly (15.25). On the other hand, $K^{(0)}_{(2n-2,1)}$ has the same expression as the anomaly with Λ replaced by H and ω by ω_s, i.e.

$$K^0_{(2n-2,1)}(H, \omega) = n(n-1) \int_0^1 \mathrm{d}t\, (1-t)\, P_n(dH, \omega_s, \mathcal{R}_{s,t}, \ldots, \mathcal{R}_{s,t}) \quad (15.48)$$

where $\mathcal{R}_{s,t} = t\, d\, \omega_s + \frac{t^2}{2}[\omega_s, \omega_s]$.

$$\mathcal{L}_{\mathrm{WZ}}(\omega, H) = \int_{\mathsf{M}} \int_0^1 \mathrm{d}s\, K^0_{(2n-2,1)}(H, \omega) \quad (15.49)$$

is the *Wess-Zumino term* that cancels the Lorentz anomaly.

However, this term is not invariant under diffeomorphisms. To find the variation under a diffeomorphism transformation of $\mathcal{L}_{\mathrm{WZ}}$, let us recall that the spin connection is

$$\omega = E\mathrm{d}E^{-1} + E\,\Gamma E^{-1} = \mathrm{e}^H\,(d + \Gamma)\,\mathrm{e}^{-H} \quad (15.50)$$

so that

$$\omega_s = \mathrm{e}^{-(s-1)H}\,(d + \Gamma)\,\mathrm{e}^{(s-1)H} = \mathrm{e}^{-\tau H}\,(d + \Gamma)\,\mathrm{e}^{\tau H} \equiv \Gamma_\tau \quad (15.51)$$

with $\tau = s - 1$. Moreover

$$\mathcal{R}_s = d\omega_s + \omega_s\omega_s = \mathrm{e}^{-\tau H}\,(d\Gamma + \Gamma\,\Gamma)\,\mathrm{e}^{-\tau H} = \mathrm{d}\Gamma_\tau + \Gamma_\tau\Gamma_\tau \equiv R_\tau \quad (15.52)$$

Now

$$\delta_\xi E = \xi^\lambda \partial_\lambda E + \Xi\, E = \mathcal{L}_\xi E + \Xi E \tag{15.53}$$

The first part of this transformation, $\mathcal{L}_\xi E$, is the Lie derivative part of the δ_ξ transformation. The Lie derivative leaves $\mathcal{L}_{WZ}$ unchanged. Therefore, from now on, we disregard it and focus on the second part, $\Xi\, E$ where Ξ means the matrix $\partial_\lambda \xi^\nu$. We make ξ anticommuting and define the BRST transform

$$\mathfrak{s}_\xi E = \Xi E, \qquad \mathfrak{s}_\xi H = \Xi - \frac{1}{2}[H, \Xi] + \cdots, \qquad \mathfrak{s}_\xi \Xi = \Xi^2 \tag{15.54}$$

A comment is in order concerning the last transformation. The transformation of ξ^μ is $\delta_\xi \xi^\mu = \xi \cdot \partial \xi^\mu$ this implies

$$\delta_\xi \Xi_\mu{}^\nu = \xi \cdot \partial \Xi_\mu{}^\nu + (\Xi^2)_\mu{}^\nu, \qquad \delta_\xi \Xi = \mathcal{L}_\xi \Xi + \mathfrak{s}_\xi \Xi, \tag{15.55}$$

where Ξ features as a scalar. According to the previous prescription, we disregard $\mathcal{L}_\xi \Xi$ in this analysis.

Next, we introduce the superconnection

$$\begin{aligned}\widetilde{\Gamma}(x, s, \vartheta) &= \mathrm{e}^{-\tau(H+\vartheta \mathfrak{s}_\xi H)} \mathrm{e}^{-\vartheta \Xi} \left(\tilde{d} + \Gamma\right) \mathrm{e}^{\vartheta \Xi} \mathrm{e}^{\tau(H+\vartheta \mathfrak{s}_\Xi H)} \\ &= \mathrm{e}^{-\tau(H+\vartheta \mathfrak{s}_\xi H)} \left(\hat{\Gamma} + \tilde{d}\right) \mathrm{e}^{s(H+\vartheta \mathfrak{s}_\xi H)}\end{aligned} \tag{15.56}$$

where $\tilde{d} = d + \frac{\partial}{\partial s} ds + \frac{\partial}{\partial \vartheta} d\vartheta$ and $\hat{\Gamma} = \Gamma + \vartheta\,(d\Xi + [\Gamma, \Xi]) + \left(\Xi - \vartheta\, \Xi^2\right) d\vartheta$. The rest is very much like in the Lorentz case with: ω_s replaced by Γ_τ and Λ_s replaced by

$$\Xi_\tau = \mathrm{e}^{-\tau H}\, \Xi\, \mathrm{e}^{\tau H} + \mathrm{e}^{-\tau H} \mathfrak{s}_\Xi \mathrm{e}^{\tau H} \tag{15.57}$$

and

$$\mathfrak{s}_\xi \Gamma_\tau = \mathrm{d}\, \Xi_\tau + [\omega_\tau, \Xi_\tau], \qquad \mathfrak{s}_\xi \Xi_\tau = -\Xi_\tau \Xi_\tau \tag{15.58}$$

Now we denote by $\widetilde{R}$, $\hat{R}$ and R the curvatures of $\widetilde{\Gamma}$, $\hat{\Gamma}$ and Γ. Then

$$\widetilde{R} = \mathrm{e}^{-\tau(H+\vartheta \mathfrak{s}_\xi H)}\, \hat{R}\, \mathrm{e}^{s(H+\vartheta \mathfrak{s}_\xi H)} = \mathrm{e}^{-\tau(H+\vartheta \mathfrak{s}_\xi H)} \mathrm{e}^{-\vartheta \Xi} R\, \mathrm{e}^{\vartheta \Xi} \mathrm{e}^{\tau(H+\vartheta \mathfrak{s}_\xi H)} \tag{15.59}$$

Now in a spacetime M of dimension d choosing an ad-invariant polynomial P_n with $n = \frac{\mathsf{d}}{2} - 1$, we have

$$P_n(\widetilde{R}, \ldots, \widetilde{R}) = P_n(\hat{R}, \ldots, \hat{R}) = P_n(R, \ldots, R) = 0 \tag{15.60}$$

for dimensional reasons. Finally, we write

$$P_n(\widetilde{R}, \ldots, \widetilde{R}) = \tilde{d}\left(n \int_0^1 dt\, P_n(\widetilde{R}, \widetilde{R}_t, \ldots, \widetilde{R}_t)\right) \equiv \tilde{d}\left(\mathrm{TP}_n(\widetilde{R})\right) \tag{15.61}$$

Now, we proceed as before to decompose this equation and use the same symbols $K^j_{(k,i)}$ and $\widehat{K}^j_{(k,i)}$, but replace ω, ω_s with Γ, Γ_τ, respectively, and $\mathfrak{s}_L H$ with $\mathfrak{s}_\xi H$. Finally, upon integrating over M and τ from 0 to 1, we get

$$\int_{\mathsf{M}} K^1_{(2n-2,0)}(\Gamma, \Xi) = \mathfrak{s}_\xi \int_{\mathsf{M}} \int_0^1 d\tau\, K^0_{(2n-2,1)}(\Gamma, H) \tag{15.62}$$

Now remark that $\int_{\mathsf{M}} K^1_{(2n-2,0)}$ is linear in Ξ and coincides precisely with the anomaly (15.30) and (15.29). On the other hand, $K^{(0)}{}_{(2n-2,1)}$ has the same expression as the anomaly with Ξ replaced by H and Γ by Γ_τ, i.e.

$$K^0_{(2n-2,1)}(H, \Gamma) = n(n-1) \int_0^1 dt\, (1-t)\, P_n(dH, \Gamma_\tau, R_{\tau,t}, \ldots, R_{\tau,t}) \tag{15.63}$$

where $R_{\tau,t} = t\, d\Gamma_\tau + \frac{t^2}{2}[\Gamma_\tau, \Gamma_\tau]$.

$$\mathcal{L}_{\mathrm{WZ}}(\Gamma, H) = \int_{\mathsf{M}} \int_0^1 ds\, K^0_{(2n-2,1)}(H, \Gamma) \tag{15.64}$$

is the *Wess-Zumino term* that reproduces the diffeomorphism anomaly, but of course, by construction, it is the same as (15.49).

15.2.1 Lorentz and Gravitational WZ Term with Background Connection

The just found link between Lorentz and diffeomorphism anomalies via a Wess-Zumino term holds in a flat spacetime, or in any local patch, but does not hold in a general curved spacetime. To establish this link in general, we have to introduce background connections, one, $\overline{\omega}_0$, for the Lorentz side and one, $\overline{\Gamma}_0$, for the gravitational side, and link them in a suitable way. To this end, we pick a fixed vierbein $\overline{E}_0 = e^{\overline{H}_0}$, which generates $\overline{\omega}_0$. The BRST transformations are the same as before on E and ω, while we assume

$$\mathfrak{s}_L \overline{E}_0 = 0, \qquad \mathfrak{s}_L \overline{\omega}_0 = 0. \tag{15.65}$$

Then we define

$$\overset{\circ}{\overline{\omega}} = \mathrm{e}^{-s\overline{H}_0} \overline{\omega}_0\, \mathrm{e}^{s\overline{H}_0} + \mathrm{e}^{-s\overline{H}_0} d_\circ \mathrm{e}^{s\overline{H}_0} \equiv \overline{\omega}_{0s} + \overline{H}_0 \mathrm{d}s \tag{15.66}$$

where $d_\circ = d + \frac{\partial}{\partial s}\mathrm{d}s$, and notice that $\overline{\omega}_{0s}$ reduces to $\overline{\omega}_0$ at $s = 0$. Finally, we introduce the t-dependent superconnection and supercurvature

$$\widetilde{\boldsymbol{\varpi}}_t = t\,\widetilde{\omega} - (1-t)\,\overset{\circ}{\overline{\omega}}, \qquad \widetilde{\mathfrak{R}}_t = \tilde{d}\,\widetilde{\boldsymbol{\varpi}}_t + \widetilde{\boldsymbol{\varpi}}_t \widetilde{\boldsymbol{\varpi}}_t \tag{15.67}$$

where $\widetilde{\omega}$ is defined in (15.40).

Now, beside the definitions (15.43), let us consider also $\overset{\circ}{\overline{\mathfrak{R}}}$ and $\overline{\mathfrak{R}}_0$, the curvatures of $\overset{\circ}{\overline{\omega}}$ and $\overline{\omega}_0$, respectively. We have

$$\overset{\circ}{\overline{\mathfrak{R}}} = \mathrm{e}^{-sH_0} \overline{\mathfrak{R}}_0 \mathrm{e}^{sH_0} \tag{15.68}$$

With the same polynomial P_n as above, we have not only (15.44), but also

$$P_n(\overset{\circ}{\overline{\mathfrak{R}}}, \ldots, \overset{\circ}{\overline{\mathfrak{R}}}) = P_n(\overline{\mathfrak{R}}_0, \ldots, \overline{\mathfrak{R}}_0) = 0 \tag{15.69}$$

Now using the universal formula (11.33), with the appropriate substitutions, we have

$$\begin{aligned} 0 &= P_n(\mathfrak{R}, \ldots, \mathfrak{R}) - P_n(\overline{\mathfrak{R}}_0, \ldots, \overline{\mathfrak{R}}_0) = P_n(\widetilde{\mathfrak{R}}, \ldots, \widetilde{\mathfrak{R}}) - P_n(\overset{\circ}{\overline{\mathfrak{R}}}, \ldots, \overset{\circ}{\overline{\mathfrak{R}}}) \\ &= \tilde{d}\left(n \int_0^1 \mathrm{d}t\; P_n(\widetilde{\omega} - \overset{\circ}{\overline{\omega}}, \widetilde{\mathfrak{R}}_t, \ldots, \widetilde{\mathfrak{R}}_t) \right) = \tilde{d}\left(\mathrm{TP}_n(\widetilde{\omega}, \overset{\circ}{\overline{\omega}}) \right) \end{aligned} \tag{15.70}$$

We proceed as before. Set for simplicity $\widetilde{\mathcal{K}}(\widetilde{\omega}, \overset{\circ}{\overline{\omega}}) \equiv \mathrm{TP}_n(\widetilde{\omega}, \overset{\circ}{\overline{\omega}})$, and decompose it

$$\widetilde{\mathcal{K}} = \sum_{\substack{k,i,j \\ k+i+j=2n-1}} \widetilde{\mathcal{K}}^j_{(k,i)}, \qquad \widetilde{\mathcal{K}}^j_{(k,i)} = \left(\boldsymbol{K}^j_{(k,i)} + \vartheta \widehat{\boldsymbol{K}}^j_{(k,i)} \right) (\mathrm{d}\vartheta)^j (\mathrm{d}s)^i \tag{15.71}$$

In Eq. (15.70), we pick the component $(k, i, j) = (2n-2, 1, 1)$ and integrate it over M and s from 0 to 1. We get

$$\int_{\mathsf{M}} \boldsymbol{K}^1_{(2n-2,0)}(\omega, \overline{\omega}_0, \Lambda) = \mathfrak{s}_\mathsf{L} \int_{\mathsf{M}} \int_0^1 \mathrm{d}s\; \boldsymbol{K}^0_{(2n-2,1)}(\omega, \overline{\omega}_0, H, \overline{H}_0) \tag{15.72}$$

The LHS is linear in Λ_s and is obtained by evaluating $\boldsymbol{K}^1_{(2n-2,0)}(\widetilde{\omega}, \overset{\circ}{\overline{\omega}})$ at $s = 0$. Now recall that $\Lambda_0 = \Lambda$, $\Lambda_1 = 0$ and $\overline{\omega}_{0s} = \overline{\omega}_0$ at $s = 0$. Therefore, $\int_{\mathsf{M}} \boldsymbol{K}^1_{(2n-2,0)}$ is linear in Λ and coincides precisely with the anomaly (11.37) with the appropriate substitutions:

$$\Delta^{(1)}_{2n-2}(\omega, \overline{\omega}_0, \Lambda) = n(n-1) \int_0^1 \mathrm{d}t (1-t)\, P_n(d_{\overline{\omega}_0}\Lambda, \omega - \overline{\omega}_0, \mathcal{R}_t, \ldots, \mathcal{R}_t)$$
$$- n\, P_n(\Lambda, \overline{\mathcal{R}}_0, \ldots, \overline{\mathcal{R}}_0) \tag{15.73}$$

This is the Lorentz anomaly with background connection $\overline{\omega}_0$.

On the other hand, $\mathbf{K}^{(0)}_{(2n-2,1)}$ has the same expression as the anomaly with Λ replaced by $H - \overline{H}_0$, ω by ω_s and $\overline{\omega}_0$ by $\overline{\omega}_{0s}$, i.e. explicitly,

$$\mathbf{K}^0_{(2n-2,1)}(H - \overline{H}_0, \omega_s, \overline{\omega}_{0s})$$
$$= n(n-1) \int_0^1 \mathrm{d}t\, P_n(d_{\overline{\omega}_{0s}}(H - \overline{H}_0), \omega_s - \overline{\omega}_{0s}, \mathcal{R}_{t,s}, \ldots, \mathcal{R}_{t,s})$$
$$-n\, P_n(H - \overline{H}_0, \overline{\mathcal{R}}_{0s}, \ldots, \overline{\mathcal{R}}_{0s}), \tag{15.74}$$

where $\mathcal{R}_{t,s} = t d\omega_s + t^2 \omega_s \omega_s$ and $\overline{\mathcal{R}}_{0s}$ is the curvature of $\overline{\omega}_{0s}$. Therefore (15.74) is a basic form on M. We call

$$\mathcal{L}_{\mathrm{WZ}}(\omega, \omega_0, H, \overline{H}_0) = \int_{\mathsf{M}} \int_0^1 \mathrm{d}s\, \mathbf{K}^0_{(2n-2,1)}(H - \overline{H}_0, \omega_s, \overline{\omega}_{0s}) \tag{15.75}$$

the *Wess-Zumino term* that cancels the Lorentz anomaly with background connection.

As for the diffeomorphism side, we choose the metric $\overline{g}_0$ induced by $\overline{E}_0$ and define

$$\overline{\Gamma}_0 = \mathrm{e}^{-\overline{H}_0}(-d + \overline{\omega}_0)\mathrm{e}^{\overline{H}_0} \tag{15.76}$$

Then we use again (15.51) and define

$$\overline{\omega}_{0s} = \mathrm{e}^{-(s-1)\overline{H}_0}\left(d + \overline{\Gamma}_0\right)\mathrm{e}^{(s-1)\overline{H}_0} = \mathrm{e}^{-\tau\overline{H}_0}\left(d + \overline{\Gamma}_0\right)\mathrm{e}^{\tau\overline{H}_0} \equiv \overline{\Gamma}_{0\tau} \tag{15.77}$$

where $\tau = s - 1$. So

$$\overline{\mathcal{R}}_{0s} = \mathrm{d}\overline{\omega}_{0s} + \overline{\omega}_{0s}\overline{\omega}_{0s} = \mathrm{e}^{-\tau\overline{H}_0}\left(\mathrm{d}\overline{\Gamma}_0 + \overline{\Gamma}_0\overline{\Gamma}_0\right)\mathrm{e}^{-\tau\overline{H}_0}$$
$$= \mathrm{d}\overline{\Gamma}_{0\tau} + \overline{\Gamma}_{0\tau}\overline{\Gamma}_{0\tau} \equiv \overline{R}_{0\tau} \tag{15.78}$$

We assume the following transformation properties

$$\mathfrak{s}_\xi \overline{E}_0 = \xi^\mu \partial_\mu \overline{E}_0, \qquad \mathfrak{s}_\xi \overline{g}_0 = \xi^\mu \partial_\mu \overline{g}_0 = \mathcal{L}_\xi \overline{g}_0 \tag{15.79}$$

which are necessary but will be disregarded for the reason explained after Eq. (15.54). Next we define

$$\overset{\circ}{\overline{\Gamma}} = \mathrm{e}^{-\tau \overline{H}_0} \overline{\Gamma}_0 \, \mathrm{e}^{\tau \overline{H}_0} + \mathrm{e}^{-\tau \overline{H}_0} d_\circ \mathrm{e}^{\tau \overline{H}_0} \equiv \overline{\Gamma}_{0\tau} + \overline{H}_0 \mathrm{d}\tau \tag{15.80}$$

Notice that $\overline{\Gamma}_{0\tau}$ reduces to $\overline{\Gamma}_0$ at $\tau = 0$. Finally we introduce the superconnection and supercurvature

$$\widetilde{\boldsymbol{\Gamma}}_t = t\,\widetilde{\Gamma} - (1-t)\,\overset{\circ}{\overline{\Gamma}}, \qquad \widetilde{\boldsymbol{R}}_t = \tilde{d}\,\widetilde{\boldsymbol{\Gamma}}_t + \widetilde{\boldsymbol{\Gamma}}_t \widetilde{\boldsymbol{\Gamma}}_t \tag{15.81}$$

where $\widetilde{\Gamma}$ is defined by (15.56). Now beside the definitions (15.59) we consider also $\overset{\circ}{\overline{R}}$ and $\overline{R}_0$, the curvatures of $\overset{\circ}{\overline{\Gamma}}$ and $\overline{\Gamma}_0$, respectively. We have

$$\overset{\circ}{\overline{R}} = \mathrm{e}^{-\tau H_0} \overline{R}_0 \mathrm{e}^{\tau H_0} \tag{15.82}$$

With the same polynomial P_n as above, we have not only (15.60), but also

$$P_n(\overset{\circ}{\overline{R}}, \dots, \overset{\circ}{\overline{R}}) = P_n(\overline{R}_0, \dots, \overline{R}_0) = 0 \tag{15.83}$$

Now making the appropriate substitutions in the universal formula (11.33), we have

$$\begin{aligned} 0 &= P_n(R, \dots, R) - P_n(\overline{R}_0, \dots, \overline{R}_0) = P_n(\widetilde{R}, \dots, \widetilde{R}) - P_n(\overset{\circ}{\overline{R}}, \dots, \overset{\circ}{\overline{R}}) \\ &= \tilde{d}\left(n \int_0^1 \mathrm{d}t\; P_n(\widetilde{\Gamma} - \overset{\circ}{\overline{\Gamma}}, \widetilde{\boldsymbol{R}}_t, \dots, \widetilde{\boldsymbol{R}}_t) \right) = \tilde{d}\left(\mathrm{TP}_n(\widetilde{\Gamma}, \overset{\circ}{\overline{\Gamma}}) \right) \end{aligned} \tag{15.84}$$

and we proceed to decompose this equation in components. Then we pick the component $(k, i, j) = (2n-2, 1, 1)$ and integrate it over M and τ from 0 to 1. We get in particular

$$\int_{\mathsf{M}} \boldsymbol{K}^1_{(2n-2,0)}(\Gamma, \overline{\Gamma}_0, \Xi) = \mathfrak{s}_\xi \int_{\mathsf{M}} \int_0^1 \mathrm{d}\tau\; \boldsymbol{K}^0_{(2n-2,1)}(\Gamma, \overline{\Gamma}_0, H, \overline{H}_0) \tag{15.85}$$

The LHS is linear in Ξ_s and is obtained by evaluating $\boldsymbol{K}^1_{(2n-2,0)}(\widetilde{\Gamma}, \overset{\circ}{\overline{\omega}})$ at $\tau = 0$. Now recall that $\Xi_0 = \Xi$, $\Xi_1 = 0$ and $\overline{\Gamma}_{0\tau} = \overline{\Gamma}_0$ at $\tau = 0$. Therefore $\int_{\mathsf{M}} \boldsymbol{K}^1_{(2n-2,0)}$ is linear in Ξ and coincides precisely with the anomaly (11.37) with the appropriate substitutions:

$$\Delta^{(1)}_{2n-2}(\Gamma, \overline{\Gamma}_0, \Xi) = n(n-1) \int_0^1 \mathrm{d}t(1-t)\, P_n(d_{\overline{\Gamma}_0}\Xi, \Gamma - \overline{\Gamma}_0, R_t, \ldots, R_t) \\ - n\, P_n(\Xi, \overline{R}_0, \ldots, \overline{R}_0) \tag{15.86}$$

This is the diffeomorphism anomaly with background connection $\overline{\Gamma}_0$. On the other hand $\mathbf{K}^{(0)}_{(2n-2,1)}$ has the same expression as the anomaly with Ξ replaced by $H - \overline{H}_0$, Γ by Γ_τ and $\overline{\Gamma}_0$ by $\overline{\Gamma}_{0\tau}$, i.e. explicitly,

$$\mathbf{K}^0_{(2n-2,1)}(H - \overline{H}_0, \Gamma_\tau, \overline{\Gamma}_{0\tau}) \\ = n(n-1) \int_0^1 \mathrm{d}t(1-t)\, P_n(d_{\overline{\Gamma}_{0\tau}}(H - \overline{H}_0), \Gamma_\tau - \overline{\Gamma}_{0\tau}, R_{t,\tau}, \ldots, R_{t,\tau}) \\ -n\, P_n(H - \overline{H}_0, \overline{R}_{0\tau}, \ldots, \overline{R}_{0\tau}), \tag{15.87}$$

where $R_{t,\tau} = t\, d\Gamma_\tau + t^2 \Gamma_\tau \Gamma_\tau$ and $\overline{R}_{0\tau}$ is the curvature of $\overline{\Gamma}_{0\tau}$. Equation (15.87) is the desired basic form in M.

15.3 Wess-Zumino Terms in Field Theories. Another Derivation

In view of the derivation of a WZ term for sigma models, it is useful to consider a more geometrical formulation of the WZ term for gauge field theories derived in the previous subsection, [3]. So, let us consider a principal fiber bundle $\mathsf{P}(\mathsf{M}, \mathsf{G})$ and the relevant group of gauge transformations $\mathcal{G} = \mathsf{Aut_v}(\mathsf{P})$, and let us consider the space of paths in $\mathcal{G}$. The latter is defined as follows. A path in $\mathcal{G}$ is a map from the unit interval $I : 0 \leq v \leq 1$ to $\mathcal{G}$, $p : I \to \mathcal{G}$. We will consider a space $\mathcal{P}_{\psi_0}(\mathcal{G})$ with a base element $\psi_0 \in \mathcal{G}$:

$$\mathcal{P}_{\psi_0}(\mathcal{G}) = \{p : I \to \mathcal{G}, \quad \text{s.t.} \quad p(0) = \psi_0\}$$

There is a natural projection $\pi_1 : \mathcal{P}_{\psi_0}(\mathcal{G}) \to \mathcal{G}$ defined by $\pi_1(p) = p(1)$ and, as a consequence, a fibration (not a principal one, however)

$$\begin{array}{ccc} \Omega_{\psi_0}\mathcal{G} & \hookrightarrow & \mathcal{P}_{\psi_0}\mathcal{G} \\ & & \downarrow{\scriptstyle \pi_1} \\ & & \mathcal{G} \end{array} \tag{15.88}$$

where $\Omega_{\psi_0}\mathcal{G}$ is the space of loops passing through ψ_0. We have also the composite evaluation map

$$\begin{array}{ccccc} \mathsf{P} \times \mathcal{P}_{\psi_0}\mathcal{G} & \xrightarrow{id\times\pi_1} & \mathsf{P} \times \mathcal{G} & \xrightarrow{ev} & \mathsf{P} \\ \downarrow & & \downarrow & & \downarrow \\ \mathsf{M} \times \mathcal{P}_{\psi_0}\mathcal{G} & \xrightarrow{id\times\pi_1} & \mathsf{M} \times \mathcal{G} & \xrightarrow{ev} & \mathsf{M} \end{array} \tag{15.89}$$

Now, using an ad-invariant polynomial P_n with $n = \frac{\mathsf{d}}{2} - 1$ and a connection A together with a background connection A_0 in $\mathsf{P}(\mathsf{M}, \mathsf{G})$, we can form the transgression formula $d\mathcal{T}P_n(A, A_0) = P_n(F, \ldots, F) - P_n(F_0, \ldots, F_0)$. Next we pull the connection A back to $\mathsf{P} \times \mathcal{P}_{\psi_0}\mathcal{G}$ through the sequence (15.89) and obtain $\pi_1^* ev^* A$ (where by π_1 we actually mean $id \times \pi_1$); similarly, we pull back A_0 through the sequence

$$\mathsf{P} \times \mathcal{P}_{\psi_0}\mathcal{G} \xrightarrow{pr_1} \mathsf{P} \times \mathcal{G} \xrightarrow{pr_1} \mathsf{P} \tag{15.90}$$

where pr_1 is the projection to the first factor, to $pr_1^* pr_1^* A_0$, which we denote simply by A_0. Therefore we construct

$$\mathcal{T}P_n(\pi_1^* ev^* A, A_0) = n \int_0^1 dt\, P_n(\pi_1^* ev^* A - A_0, \mathcal{F}_t, \ldots, \mathcal{F}_t) \tag{15.91}$$

where $\mathcal{F}_t$ is the curvature of the connection $t\,\pi_1^* ev^* A + (1-t)A_0$. We have

$$(d + \widehat{\delta})\mathcal{T}P_n(\pi_1^* ev^* A, A_0) = \pi_1^* ev^* P_n(F, \ldots, F) - P_n(F_0, \ldots, F_0) = 0 \tag{15.92}$$

for dimensional reasons. Here, $\widehat{\delta}$ is the exterior differential in $\mathcal{P}_{\psi_0}\mathcal{G}$. So $\mathcal{T}P_n(\pi_1^* ev^* A, A_0)$ is a closed and basic form in $\mathsf{M} \times \mathcal{P}_{\psi_0}\mathcal{G}$.

Now, we recall that $\mathcal{P}_{\psi_0}\mathcal{G}$ is a contractible space, for we can define an operation h:

$$h : I \times \mathcal{P}_{\psi_0}\mathcal{G} \longrightarrow \mathcal{P}_{\psi_0}\mathcal{G}$$

$$(s, p) \rightsquigarrow p^s$$

where p^s is the path defined by $p^s(t) = p(st)$, which shrinks every path to ψ_0. In a contractible space, any closed form is exact. To show this explicitly, let us consider the sequence

$$\mathsf{M} \times \mathcal{P}_{\psi_0}\mathcal{G} \xrightarrow{r} \mathsf{M} \xrightarrow{i} \mathsf{M} \times \mathcal{P}_{\psi_0}\mathcal{G}$$

where r is the retraction and i is the inclusion. Then for forms on $\mathsf{M} \times \mathcal{P}_{\psi_0}\mathcal{G}$, one gets the identity

$$1^* - (i \circ r)^* = (d + \widehat{\delta})\mathcal{H} + \mathcal{H}(d + \widehat{\delta}) \tag{15.93}$$

where

$$\mathcal{H} = \int_0^1 \mathrm{d}s\, h^* \tag{15.94}$$

and, by abuse of notation, h^* represents the pullback by the map $id \times h : \mathsf{M} \times I \times \mathcal{P}_{\psi_0}\mathcal{G} \longrightarrow \mathsf{M} \times \mathcal{P}_{\psi_0}\mathcal{G}$. Now we apply both sides of (15.93) to $\mathcal{T}P_n(\pi_1^* ev^* A, A_0)$ and assume that ψ_0 is the identity automorphism, then

$$\mathcal{T}P_n(\pi_1^* ev^* A, A_0) = (d + \widehat{\delta}) \int_0^1 \mathrm{d}s\, h^* \mathcal{T}P_n(\pi_1^* ev^* A, A_0) \tag{15.95}$$

because of (15.92) and because i^* pulls back the form $\mathcal{T}P_n(\pi_1^* ev^* A, A_0)$ to M, where it vanishes for dimensional reasons. Therefore we have a d-form in $\mathsf{M} \times \mathcal{P}_{\psi_0}\mathcal{G}$

$$\beta = \int_0^1 \mathrm{d}s\, h^* \mathcal{T}P_n(\pi_1^* ev^* A, A_0) \tag{15.96}$$

that trivializes $\mathcal{T}P_n(\pi_1^* ev^* A, A_0)$.

Now we decompose (15.95) according to the different sectors and pick out the component $(\mathsf{d}, 1)$ in $\mathsf{M} \times \mathcal{P}_{\psi_0}\mathcal{G}$, i.e.

$$\mathcal{T}P_n(\pi_1^* ev^* A, A_0)\big|_{(\mathsf{d},1)} = \mathrm{d}\beta_{(\mathsf{d}-1,1)} + \widehat{\delta}\beta_{(\mathsf{d},0)} \tag{15.97}$$

Integrating over M, we get

$$\int_{\mathsf{M}} \mathcal{T}P_n(\pi_1^* ev^* A, A_0)\Bigg|_{(\mathsf{d},1)} = \widehat{\delta} \int_{\mathsf{M}} \beta_{(\mathsf{d},0)} \tag{15.98}$$

To render this formula more comprehensible, we have to work out the meaning of the maps $\pi_1^* ev^*$ and $h^* \pi_1^* ev^*$. For any $u \in \mathsf{P}$ and any path $p \in \mathcal{P}_{\psi_0}\mathcal{G}$, we have

$$(ev \circ \pi_1)(u, p) = ev(u, p(1)) = p(u, 1) \tag{15.99}$$

$$(ev \circ \pi_1 \circ h)(u, s, p) = ev(u, p^s) = p(u, s) \tag{15.100}$$

where we have used the fact that $p \in \mathcal{P}_{\psi_0}\mathcal{G}$ can be interpreted as a map $p : \mathsf{P} \times I \to \mathsf{P}$. Then, setting $p(\cdot, 1) = \psi(\cdot)$, we can see that

$$\mathcal{T}P_n(\pi_1^* ev^* A, A_0)\big|_{(\mathsf{d},1)} = j_{(\cdot)}\mathcal{T}P_n(\psi^* A, A_0) \tag{15.101}$$

where $j_{(\cdot)}$ coincides with $i_{(\cdot)}$ when applied to $\psi^* A$ and is 0 when applied to A_0, dA_0 etc. Therefore the RHS of (15.101) is the expression of the anomaly with a background connection A_0.

Next we pick the component of the integrand in (15.96) of degree $_{(\mathsf{d},0,1)}$ in $\mathsf{M} \times \mathcal{P}_{\psi_0}\mathcal{G} \times I$, and write it in the form

$$h^*\pi_1^* ev^* \mathcal{T}P_n(A, A_0)\big|_{(\mathsf{d},0,1)} = ds\, i_{\frac{\partial}{\partial s}}\mathcal{T}P_n(\psi_s^* A, A_0) \tag{15.102}$$

where $\psi_s(\cdot) = p(\cdot, s)$ and $\frac{\partial}{\partial s}$ is seen as an element of the tangent vector $T_s I$. Therefore (15.98) becomes

$$\int_{\mathsf{M}} j_{(\cdot)}\mathcal{T}P_n(\psi^* A, A_0) = \delta \int_{\mathsf{M}} \int_0^1 ds\, i_{\frac{\partial}{\partial s}}\mathcal{T}P_n(\psi_s^* A, A_0) \tag{15.103}$$

where δ is the BRST transform. Setting $\psi = p(1) = id \in \mathcal{G}$ the LHS is easily recognizable as the anomaly, the RHS needs some more elaboration. Let us write $p(u, s) = (u, s)\gamma(u, s) = (u\gamma(u, s), s)$ and the map $\gamma : \mathsf{P} \times I \to \mathsf{G}$ as $\gamma(u, s) = \gamma_s(u)$. The connection $\psi_s^* A$ takes the form

$$\widehat{A}_s = \gamma_s^{-1}(A + d)\gamma_s + \gamma_s^{-1}\frac{\partial}{\partial s}\gamma_s\, \mathrm{ds} \tag{15.104}$$

If we perform a gauge transformation on A: $A \to \Lambda^{-1}(A + d)\Lambda$, where $\Lambda = exp[\lambda^a(x)T^a]$, we see that

$$\widehat{A}_s \to \widehat{A}'_s = \gamma_s'^{-1}(A + d)\gamma'_s + \gamma_s'^{-1}\frac{\partial}{\partial s}\gamma'_s\, \mathrm{ds}, \qquad \gamma'_s = \Lambda\gamma_s \tag{15.105}$$

Now, the interior product $i_{\frac{\partial}{\partial s}}$, applied to $\widehat{A}_s$ selects the last term of (15.105). Thus, we have

$$\int_{\mathsf{M}} j_{(\cdot)}\mathcal{T}P_n(A, A_0) = \delta \int_{\mathsf{M}} \int_0^1 ds\, i_{\frac{\partial}{\partial s}}\mathcal{T}P_n(\widehat{A}_s, A_0) \equiv \delta\mathcal{B}_{\mathrm{WZ}} \tag{15.106}$$

which defines the WZ term, $\mathcal{B}_{\mathrm{WZ}}$. The latter has the same form as the anomaly with $i_{(\cdot)}A$ replaced by $\gamma_s^{-1}\frac{\partial}{\partial s}\gamma_s$ and A replaced by $A_s = \gamma_s^{-1}(A + d)\gamma_s$ and is integrated over s from 0 to 1. If we choose $\gamma_s = \exp[s\sigma]$ and set $A_0 = 0$, we reproduce the WZ term of Sect. 15.1.

Appendix 15A. The Superfield Formalism in Gauge Field Theories

This is a short introduction to the superfield formalism in gauge theories. Let us return to Chap. 2, Sect. 2.1.1, and consider a generic gauge theory in a d dimensional Minkowski spacetime M, with connection $A^a_\mu T^a$ ($\mu = 0, 1, \ldots, \mathsf{d}-1$), valued in a Lie algebra $\mathfrak{g}$ with anti-Hermitean generators T^a, such that $[T^a, T^b] = f^{abc}T^c$. In the sequel, we use the compact form notation $A = A^a_\mu T^a \mathrm{d}x^\mu$. The curvature and gauge transformation are

$$F = \mathrm{d}A + \frac{1}{2}[A, A] \quad \text{and} \quad \delta_\lambda A = \mathrm{d}\lambda + [A, \lambda], \tag{15.107}$$

with $\lambda(x) = \lambda^a(x)T^a$ and $d = \mathrm{d}x^\mu \frac{\partial}{\partial x^\mu}$. As we know, the infinite dimensional Lie algebra of gauge transformations and its cohomology can be formulated in a simpler way if we promote the gauge parameter λ to an anticommuting ghost field $c = c^a T^a$ and define the BRST transform as[1]

$$\mathfrak{s}A \equiv dc + [A, c], \qquad \mathfrak{s}\,c = -\frac{1}{2}[c, c]. \tag{15.108}$$

A very simple way to reproduce these formulas and properties is by enlarging the space to a superspace with coordinates (x^μ, ϑ), where ϑ is anticommuting, and promoting the connection A to a one-form superconnection $\widetilde{A} = \phi(x, \vartheta) + \phi_\vartheta(x, \vartheta)\mathrm{d}\vartheta$ with expansions

$$\phi(x, \vartheta) = \mathsf{A}(x) + \vartheta\Gamma(x), \qquad \phi_\vartheta(x, \vartheta) = \mathsf{c}(x) + \vartheta G(x), \tag{15.109}$$

and two-form supercurvature

$$\widetilde{F} = \widetilde{d}\widetilde{A} + \frac{1}{2}[\widetilde{A}, \widetilde{A}],$$
$$\widetilde{F} = \Phi(x, \vartheta) + \Phi_\vartheta(x, \vartheta)\mathrm{d}\vartheta + \Phi_{\vartheta\vartheta}(x, \vartheta)\mathrm{d}\vartheta \wedge \mathrm{d}\vartheta, \tag{15.110}$$

with $\Phi(x, \vartheta) = \mathsf{F}(x) + \vartheta\Lambda(x)$ and $\widetilde{d} = d + \frac{\partial}{\partial\vartheta}\mathrm{d}\vartheta$. Notice that since $\vartheta^2 = 0$, $\mathrm{d}\vartheta \wedge \mathrm{d}\vartheta \neq 0$, while $\mathrm{d}x^\mu \wedge \mathrm{d}\vartheta = -\mathrm{d}\vartheta \wedge \mathrm{d}x^\mu$. Then we impose the 'horizontality' condition

$$\widetilde{F} = \Phi(x, \vartheta), \quad \text{i.e.} \quad \Phi_\vartheta(x, \vartheta) = 0 = \Phi_{\vartheta\vartheta}(x, \vartheta). \tag{15.111}$$

The last two conditions imply

[1] The symbol $[\cdot, \cdot]$ denotes an ordinary commutator when both entries are non-anticommuting, and an anticommutator when both entries are anticommuting.

$$\Gamma(x) = d\mathsf{c}(x) + [\mathsf{A}(x), \mathsf{c}(x)], \qquad G(x) = -\frac{1}{2}[\mathsf{c}(x), \mathsf{c}(x)].$$

Moreover $\Lambda(x) = [\mathsf{F}(x), \mathsf{c}(x)]$.

This means that we can identify $\mathsf{c}(x) \equiv c(x)$, $\mathsf{A} \equiv A$, $\mathsf{F} \equiv F$, and the ϑ translation with the BRST transformation $\mathfrak{s}$, i.e. $\mathfrak{s} \equiv \frac{\partial}{\partial\vartheta}$. In this way all the previous transformations are naturally explained. It is also possible to push further the use of the superfield formalism by noting that, after imposing the horizontality condition, we have

$$\widetilde{A} = \mathrm{e}^{-\vartheta c} A\, \mathrm{e}^{\vartheta c} + \mathrm{e}^{-\vartheta c} \widetilde{d} \mathrm{e}^{\vartheta c}, \qquad \widetilde{F} = \mathrm{e}^{-\vartheta c} F\, \mathrm{e}^{\vartheta c}. \tag{15.112}$$

A comment is in order concerning the horizontality condition (HC). This condition is suggested by the analogy with the principal fiber bundle geometry. In the total space of a principal fiber bundle, one can define horizontal (or basic) forms. These are forms with no components in the vertical direction: for instance, given a connection, its curvature is horizontal. In our superfield approach, the ϑ coordinate mimics the vertical direction, for the curvature $\widetilde{F}$ does not have components in that direction.

References

1. L. Bonora, M. Bregola, L. Lucaroni, *Anomalies and Cohomology, in Anomalies, Phases, Defects* (Bibliopolis, Naples, 1990)
2. W.A. Bardeen, B. Zumino, Consistent and covariant anomalies in gauge and gravitational theories. Nucl. Phys. B **244**, 421 (1984)
3. L. Bonora, P. Cotta-Ramusino, M. Rinaldi, J. Stasheff, The evaluation map in field theory, sigma-models and strings. II. Comm. Math. Phys. **114**, 381 (1988)

Chapter 16
Sigma Model Anomalies

In this chapter, the analysis of anomalies extends to sigma models, i.e. to field theories in which the basic fields are maps from spacetime to a target space. Apart from some technicalities, it does not differ much from ordinary gauge field theories. The important realization is that in many respects ordinary gauge field theories can be regarded as sigma model in which the target space is the classifying space. But while in the classifying space the Weil homomorphism is an isomorphism, in sigma models this is in general not the case, a fact that opens extra possibilities to cancel anomalies by means of WZ terms. In this chapter, we analyze the anomalies of sigma models and construct the relevant WZ terms when this is possible. Finally, the last part of the chapter is devoted to the analysis of global anomalies and their cancelation. References for this chapter are [1–4].

16.1 Introduction

So far we have dealt with gauge field theories and their anomalies. These theories are characterized by elementary fields, mainly low spin fermion fields coupled to vector potentials or metrics. There are other models of field theory, the sigma models, whose fields are maps from a base spacetime to another space, the target space. To be more precise, let M denote the base spacetime (or world volume) and T the target space, with dimension d and D, with $\mathsf{D} > \mathsf{d}$, and local coordinates $x^\alpha, \alpha = 0, \ldots, \mathsf{d}-1$, and X^μ, $\mu = 0, \ldots, \mathsf{D}-1$, respectively. We assume both to be Riemannian, endowed with metrics $h_{\alpha\beta}(x)$ and $G_{\mu\nu}(X)$, respectively. Then we introduce the space $\mathsf{Map}(\mathsf{M}, \mathsf{T})$ of smooth maps $\phi : \mathsf{M} \to \mathsf{T}$. These maps will be also represented locally by means of the coordinates of the image points: $X^\mu(x) \equiv X^\mu(\phi(x))$. Actually, the latter will play the role of elementary fields of the sigma model. For instance, the kinetic term for them takes the form

$$\int_{\mathsf{M}} d^{\mathsf{d}}x\, h^{\alpha\beta}(x)\partial_\alpha X^\mu(x)\partial_\beta X^\nu(x)G_{\mu\nu}(\phi(x)) \tag{16.1}$$

L. Bonora, *Fermions and Anomalies in Quantum Field Theories*, Theoretical and Mathematical Physics, https://doi.org/10.1007/978-3-031-21928-3_16

In a similar way, we can have also kinetic terms for fermions. In the most general case, they take the form

$$\int_{\mathsf{M}} d^{\mathrm{d}}x\, i\bar{\psi}(x)\rho^{\alpha}\left(\nabla_{\alpha} + (\phi^{*}\mathbf{a})_{\alpha}(x) + \frac{1}{2}(\phi^{*}\boldsymbol{\omega})_{\alpha}(x)\right)\psi(x) \tag{16.2}$$

where ρ^{α} are the γ-matrices in M, $\mathbf{a}$ and $\boldsymbol{\omega}$ are the gauge and spin connections in T, and $\phi^{*}\mathbf{a}$ and $\phi^{*}\boldsymbol{\omega}$ the respective pullbacks to M. ∇ is the covariant derivative in M (so it may contain itself a gauge and spin connection living in a principal fiber bundle over M). In local coordinates, for instance, $(\phi^{*}\mathbf{a})_{\alpha}(x) = \partial_{\alpha}X^{\mu}(x)\mathbf{a}_{\mu}(\phi(x))$.

We disregard other possible interaction terms and focus on (16.1) and (16.2), which are the relevant ones for the study of anomalies. Let us start from the analysis of their symmetries. We have of course the usual symmetries of an ordinary field theory on M, i.e. $\mathsf{Diff}(\mathsf{M})$, and a possible gauge and local Lorentz symmetry accounted for by the covariant derivative ∇_{α}. Here however our focus will be on the induced symmetries from the target space to the world volume. Let us start from the induced gauge symmetry. We suppose that $\mathbf{a}$ is a connection in a principal fiber bundle $\mathsf{P}(\mathsf{T}, \mathsf{G})$ over the target space, valued in the Lie algebra $\mathfrak{g}$ of a compact Lie group G. Then we have the pulled back principal fiber bundle $\phi^{*}\mathsf{P}$:

$$\begin{array}{ccc} \phi^{*}\mathsf{P} & \xrightarrow{\hat{\phi}} & \mathsf{P} \\ \downarrow{\scriptstyle \pi} & & \downarrow{\scriptstyle \pi} \\ \mathsf{M} & \xrightarrow{\phi} & \mathsf{T} \end{array} \tag{16.3}$$

where $\hat{\phi}$ is the bundle map that covers ϕ. We recall that the fiber of $\phi^{*}\mathsf{P}$ at x is the fiber of P at $\phi(x)$ and $\hat{\phi}$ is the identification map between these two fibers. From any connection $\mathbf{a}$ in P, we obtain by pullback a connection $A = \hat{\phi}^{*}\mathbf{a}$ in $\phi^{*}\mathsf{P}$.

This reminds us of the analogous commutative diagram we have met in ordinary field theory

$$\begin{array}{ccc} \mathsf{P}(\mathsf{M}, \mathsf{G}) & \xrightarrow{\hat{f}} & \mathsf{EG} \\ \downarrow{\scriptstyle \pi} & & \downarrow{\scriptstyle \pi} \\ \mathsf{M} & \xrightarrow{f} & \mathsf{BG} \end{array}$$

f is a classifying map, unique up to homotopy, such that $(\hat{f}, f)$ constitute a bundle morphisms, by which we can obtain any connection A in P from a universal connection $\mathfrak{a}$ in EG: $A = \hat{f}^{*}\mathfrak{a}$. The group of gauge transformations relevant to gauged sigma model is, of course, $\mathcal{G}_{\phi} \equiv \mathsf{Aut}_{\mathsf{V}}(\phi^{*}\mathsf{P})$.

To summarize, the situation in gauged sigma models is similar to ordinary field theory with the following correspondences

	gauge theory over M	sigma model over M
target space	BG	T
induced bundle	$\mathsf{P} = f^*\mathsf{EG}$	$\phi^*\mathsf{P}$
group of gauge transformations	$\mathcal{G} \equiv \mathsf{Aut_v}(P)$	$\mathcal{G}_\phi \equiv \mathsf{Aut_v}(\phi^*\mathsf{P})$
space of maps	$\mathsf{Map(M,BG)}_f$	Map(M,T)

where $\mathsf{Map(M, BG)}_f$ denotes the space of maps[1] that induce a bundle equivalent to P. We can say that an ordinary gauge theory is a sigma model in which the target space is the classifying space BG, with an important difference, which we will highlight below. But, up to that point, the description of anomalies proceeds in a complete parallel way in the two cases.

16.2 Sigma Model Anomalies

We remark first that the perturbative and non-perturbative derivation of local anomalies in sigma models does not differ from the analogous gauge field theory calculations. What may be different in sigma models is the possibility of anomaly cancelation by means of Wess-Zumino terms, which is impossible in ordinary gauge theories. This is allowed by the fact that in sigma models the embedding coordinates X^μ are dynamical degrees of freedom.

Let us briefly describe anomalies in sigma models, adapting the language of gauge field theories to the latter.

In analogy with the evaluation map in gauge field theory $ev : \mathsf{P} \times \mathcal{G} \to \mathsf{P}$, we define the evaluation map for sigma models for a fixed ϕ:

$$\phi^*\mathsf{P} \times \mathsf{Aut_v}(\phi^*\mathsf{P}) \xrightarrow{ev_\phi} \phi^*\mathsf{P}$$
$$(u, \psi_\phi) \rightsquigarrow ev(u, \psi_\phi) = \psi_\phi(u) \tag{16.4}$$

For any $\psi \in \mathsf{Aut_v P}$, there is a ψ_ϕ such that $\hat{\phi} \circ \psi_\phi = \psi \circ \hat{\phi}$ (see (16.3)). Next, given a connection $\mathbf{a}$ in P(T, G), see (16.3), with curvature $F(\mathbf{a})$, we construct ad-invariant polynomials $P_n(F(\mathbf{a}), \ldots, F(\mathbf{a}))$, with $n = \frac{\mathsf{d}}{2} + 1$, which are basic and closed in T. We know that the source of anomalies in gauge field theories are these polynomials. The same is true in sigma models. From them, we can derive the transgression formulas

$$TP_n(\mathbf{a}) = n \int_0^1 \mathrm{d}t \, P_n(\mathbf{a}, F_t(\mathbf{a}), \ldots, F_t(\mathbf{a})) \tag{16.5}$$

where $F_t(\mathbf{a}) = t\, d\mathbf{a} + \frac{t^2}{2}[\mathbf{a}, \mathbf{a}]$. Pulling back (16.5) via the composed map

[1] For simplicity here we consider injective mappings, i.e. embeddings.

$$\phi^*\mathsf{P} \times \mathsf{Aut_v}(\phi^*\mathsf{P}) \xrightarrow{ev_\phi} \phi^*\mathsf{P} \xrightarrow{\hat{\phi}} \mathsf{P} \tag{16.6}$$

where $\hat{\phi}$ is the canonical covering of ϕ, we get $(\hat{\phi} \circ ev_\phi)^* T P_n(\mathbf{a}) = ev_\phi^* T P_n(A)$. The anomalies of our sigma model are given by

$$\Delta^1_{\mathsf{d}}(i_{(\cdot)}A, A) = n(n-1) \int_0^1 \mathrm{d}t(1-t) P_n(di_{(\cdot)}A, A, F_t, \dots, F_t) \tag{16.7}$$

where $A = \phi^*\mathbf{a}$ and $i_{(\cdot)}A$ denotes the map $Z \to i_Z A$, that associates to every $Z \in Lie(\mathcal{G})_\phi$ the map $\xi_Z = A(Z) : \phi^*\mathsf{P} \to Lie(\mathsf{G})$; $i_{(\cdot)}A$ plays the role of ghost field.

As already pointed out, the anomaly (16.7) is defined in $\phi^*\mathsf{P}$, not in M. We can pull it back to M via a local section, and this matches the result of perturbative field theory, where the calculation is carried out in a local patch. If we want a global expression of the anomaly in M, we must introduce a background connection A_0. Then we can construct the transgression

$$\mathcal{T} P_n((\hat{\phi} \circ ev_\phi)^*\mathbf{a}, A_0) = n \int_0^1 \mathrm{d}t\, P_n\left(ev_\phi^* A - A_0, \mathfrak{F}_t, \dots, \mathfrak{F}_t\right)) \tag{16.8}$$

where $\mathfrak{F}_t$ is the curvature of the connection $\mathcal{A}_t = t\, ev_\phi^* A + (1-t)A_0$, from which we can derive the basic formula for the anomaly in M:

$$\Delta^{(1)}_{\mathsf{d}}(A, A_0, c) = n(n-1) \int_0^1 \mathrm{d}t(1-t)\, P_n(d_{A_0}\mathsf{c}, A - A_0, \mathsf{F}_t, \dots, \mathsf{F}_t) \tag{16.9}$$

where F_t is the curvature of the connection $tA + (1-t)A_0$ and $\mathsf{c} = i_{(\cdot)}A$.

As anticipated above, up to now the description of gauge anomalies in sigma models parallels the analogous description for gauge field theories. We have simply replaced symbols when necessary. But now it is time to highlight the important difference between the two cases. The difference arises in relation to non-topological anomalies. In field theory, we called non-topological anomalies those that arise from reducible polynomials P_n, when a non-trivial factor, say P_k in $P_n = P_k\, P_{n-k}$, is such that the closed form $P_k(\mathsf{F}, \dots, \mathsf{F})$ is also exact: i.e. $P_k(\mathsf{F}, \dots, \mathsf{F}) = \mathrm{d}X_{2k-1}$, where X_{2k-1} is a $2k-1$-form in $\mathbf{M}$. When this happens $P_k(\mathsf{F}, \dots, \mathsf{F})$ is topologically trivial: we say that $P_k(\mathsf{F}, \dots, \mathsf{F})$ is in the kernel of the Weil homomorphism. In the case of gauge field theories, this does not imply any practical difference with the other anomalies (of topological origin) because of locality. In field theory, locality imposes that the only topology that matters be that of the classifying space BG, and, in BG, $P_k(\mathsf{F}, \dots, \mathsf{F})$, like any other polynomial of the same kind, is not exact.

The situation is different in sigma models because the role of classifying space BG is played by the manifold T. Suppose $P_k(\mathsf{F}, \dots, \mathsf{F})$ is trivial in T, what happens of the corresponding anomaly arising from $P_n = P_k\, P_{n-k}$? In this case, there is a possibility of cancelation via a Wess-Zumino term.

16.3 Wess-Zumino Terms in Sigma Models

Wess-Zumino terms in sigma models may take on a different meaning, not purely academical like the one mentioned at the end of Sect. 15.1 for chiral anomalies in gauge field theories. This is due to the fact that, under certain conditions, the role of the fields σ^a can be played by the basic fields of the sigma models, i.e. the maps $\in \mathsf{Map}(\mathsf{M}, \mathsf{T})$. The interesting cases we have in mind correspond to the so-called non-topological anomalies in field theory. To anticipate the content of this section let us consider a connection $\mathbf{a}$ in $\mathsf{P}(\mathsf{T}, \mathsf{G})$, see (16.3), with curvature $F(\mathbf{a})$, and let us construct an ad-invariant polynomial $P_n(F(\mathbf{a}), \ldots, F(\mathbf{a}))$, with $n = \frac{\mathsf{d}}{2} + 1$, which is basic and closed in T. We will focus on the case in which

$$P_n(F(\mathbf{a}), \ldots, F(\mathbf{a})) = \mathrm{d}H \tag{16.10}$$

where H is a *basic* $\mathsf{d}+1$ form in T. The strategy to construct the WZ terms suggested by the previous examples in field theory is to pull back the relation (16.10) to a contractible space, to extract from it a transgression form, which will be closed and therefore exact and, thus, automatically leads to a WZ term. The contractible space, much like in the previous section, will consist of a space of paths, but, differently from that case, the space of paths will not be a space of paths of gauge transformations, but a space of paths of maps from the world volume M to the target space T. Finally, we will pull back the so obtained WZ form to the space of loops, which induces the sought for gauge transformations in the sigma model.

To be more precise, let us define $\mathsf{Map}(\mathsf{M}, \mathsf{T})_{\phi_0}$, the space of maps which are path-connected to ϕ_0, and the space of paths in it with basis element ϕ_0, and call it $\mathcal{P}_{\phi_0}(\mathsf{Map}(\mathsf{M}, \mathsf{T}))$. This means that $p \in \mathcal{P}_{\phi_0}(\mathsf{Map}(\mathsf{M}, \mathsf{T}))$ is a map $p : \mathsf{M} \times I \to \mathsf{T}$, where $I : 0 \leq s \leq 1$, such that $p(0, x) = \phi_0(x)$. The space of loops with base element ϕ_0 will be denoted $\Omega_{\phi_0}((\mathsf{Mat}(\mathsf{M}, \mathsf{T}))$. Instead of (15.88), we have

$$\begin{array}{ccc} \Omega_{\phi_0}(\mathsf{Map}(\mathsf{M}, \mathsf{T})) & \xhookrightarrow{J} & \mathcal{P}_{\phi_0}(\mathsf{Map}(\mathsf{M}, \mathsf{T})) \\ & & \downarrow{\scriptstyle \pi_1} \\ & & \mathsf{Map}(\mathsf{M}, \mathsf{T})_{\phi_0} \end{array} \tag{16.11}$$

where π_1 is, as above, the projection of p to the endpoint $\pi_1(p)(s, x) = p(1, x) \equiv p_1(x) = \phi(x)$. J is the injection map. This is not a principal fiber bundle. However, since $\mathcal{P}_{\phi_0}(\mathsf{Map}(\mathsf{M}, \mathsf{T}))$ is contractible, (16.11) plays the same role as the universal bundle.

The construction of the WZ term relies on the following commutative diagram

$$
\begin{array}{ccccc}
\pi_1^* ev^* \mathsf{P} & \xrightarrow{\hat{\pi}_1} & ev^* \mathsf{P} & \xrightarrow{\hat{ev}} & \mathsf{P} \\
\downarrow{\scriptstyle \pi''} & & \downarrow{\scriptstyle \pi'} & & \downarrow{\scriptstyle \pi} \\
\mathsf{M} \times \mathcal{P}_{\phi_0}(\mathsf{Map}(\mathsf{M},\mathsf{T})) & \xrightarrow{id \times \pi_1} & \mathsf{M} \times \mathsf{Map}(\mathsf{M},\mathsf{T})_{\phi_0} & \xrightarrow{ev} & \mathsf{T}
\end{array}
\tag{16.12}
$$

The principal fiber bundle $ev^*\mathsf{P}$ is defined as follows

$$
\begin{aligned}
ev^*\mathsf{P} = \big\{(x, \phi, u) : ev(x, \phi) \equiv \phi(x) = \pi(u), \ \ \forall u \in \mathsf{P}, \\
\forall x \in \mathsf{M}, \quad \forall \phi \in \mathsf{Map}(\mathsf{M},\mathsf{T})_{\phi_0}\big\}
\end{aligned}
\tag{16.13}
$$

with $\pi'(x, \phi, u) = (x, \phi)$. $ev^*\mathsf{P}$ can be thought of as the union of the induced bundles $\phi^* P$: $\bigsqcup_\phi \phi^* P$, for all $\phi \in \mathsf{Map}(\mathsf{M},\mathsf{T})_{\phi_0}$.

In turn $\pi_1^* ev^*\mathsf{P}$ is defined as

$$
\begin{aligned}
\pi_1^* ev^*\mathsf{P} = \big\{(x, p, u, \phi) : \pi_1(x, p) = \pi'(x, \phi, u), \\
\forall p \in \mathcal{P}_{\phi_0}(\mathsf{Map}(\mathsf{M},\mathsf{T})), \quad \forall u \in \mathsf{P}, \quad \forall x \in \mathsf{M}, \quad \forall \phi \in \mathsf{Map}(\mathsf{M},\mathsf{T})_{\phi_0}\big\},
\end{aligned}
\tag{16.14}
$$

where $\pi_1(x, p) = (x, p_1)$ and $\pi'(x, \phi, u) = (x, \phi)$, i.e. $p_1 = \phi$. Moreover, $\pi''(x, p, u, \phi) = (x, p)$. Finally $\hat{ev}(x, \phi, u) = u$ and $\hat{\pi}_1(x, p, u, \phi) = (x, \phi, u)$. Therefore, $(\hat{ev} \circ \hat{\pi}_1)(x, p, u, \phi) = u$ such that $\pi(u) = \phi(x)$, with $\phi = p_1$. In the previous diagram for simplicity of notation, we have dropped the id map whenever it is self-evident. The bundle $\pi_1^* ev^*\mathsf{P}$ can be regarded as the union of all the induced bundles $p^*\mathsf{P}$: $\bigsqcup_p p^*\mathsf{P}$ for all p such that $p_1 = \phi \in \mathsf{Map}(\mathsf{M},\mathsf{T})_{\phi_0}$.

As it turns out, $\pi_1^* ev^*\mathsf{P}$ is isomorphic to $\phi_0^*\mathsf{P} \times \mathcal{P}_{\phi_0}(\mathsf{Map}(\mathsf{M},\mathsf{T}))$, for the latter is

$$
\phi_0^*\mathsf{P} \times \mathcal{P}_{\phi_0}(\mathsf{Map}(\mathsf{M},\mathsf{T})) = \big\{(x, u, p) : \phi_0(x) = \pi(u), \quad p_0 = \phi_0\big\}
\tag{16.15}
$$

with the same symbol as above. Also, in $\pi_1^* ev^*\mathsf{P}$ we have the constraint $p_0 = \phi_0$, while p_1 is free. But the condition $p_1 = \phi$ in $\pi_1^* ev^*\mathsf{P}$ does not restrict the set of p's. Moreover, we set $\pi''(x, u, p) = (x, p)$. Therefore, $\pi_1^* ev^*\mathsf{P}$ and $\phi_0^*\mathsf{P} \times \mathcal{P}_{\phi_0}(\mathsf{Map}(\mathsf{M},\mathsf{T}))$ can be identified. The explicit correspondence is given by $(x, p, u, \phi) \leftrightarrow (x, u, p)$ (there are no loose ends because in the LHS of this correspondence p and ϕ are related by the condition $p_1 = \phi$).

Now, starting with a connection $\mathbf{a}$ in P we can pull it back and get a connection $\hat{\pi}^* \hat{ev}^* \mathbf{a}$ in $\pi_1^* ev^*\mathsf{P}$. We can also define a background connection A_0 in $\phi_0^*\mathsf{P}$ and extend it trivially to $\phi_0^*\mathsf{P} \times \mathcal{P}_{\phi_0}(\mathsf{Map}(\mathsf{M},\mathsf{T}))$ and then define a background connection in $\pi_1^* ev^*\mathsf{P}$ via the isomorphism, which we denote with the same symbol A_0 (A_0 could be, for instance, $\phi_0^*(\mathbf{a}_0)$ where $\mathbf{a}_0$ is again a fixed connection in P). We can now construct the polynomials

$$\pi_1^* ev^* P_n(F(\mathbf{a}), \ldots, F(\mathbf{a})) = d\,(\pi_1^* ev^* H),$$
$$P_n(F(A_0), \ldots, F(A_0)) = 0 \tag{16.16}$$

The first relation follows from (16.10), the second because of dimensional reasons. From there, we can construct

$$(d + \tilde{\delta})\,\big(\mathcal{T} P_n(\hat{\pi}_1^* \hat{ev}^* \mathbf{a}, A_0) - \pi_1^* ev^* H\big) = 0 \tag{16.17}$$

The $\mathsf{d}+1$ form $\big(\mathcal{T} P_n(\hat{\pi}_1^* \hat{ev}^* \mathbf{a}, A_0) - \pi_1^* ev^* H\big)$ is basic in $\mathsf{M} \times \mathcal{P}_{\phi_0}(\mathsf{Map}(\mathsf{M}, \mathsf{T}))$. In (16.17) d and $\tilde{\delta}$ are the exterior differentials in M and $\mathcal{P}_{\phi_0}(\mathsf{Map}(\mathsf{M}, \mathsf{T}))$, respectively.

Now recall that the space $\mathcal{P}_{\phi_0}(\mathsf{Map}(\mathsf{M}, \mathsf{T}))$ is contractible, and therefore, its cohomology is that of a point. This in turn implies that, due to the Künneth theorem for the product of two spaces, also the (reduced) cohomology of $\mathsf{M} \times \mathcal{P}_{\phi_0}(\mathsf{Map}(\mathsf{M}, \mathsf{T}))$ is trivial. So there must exist a d-form β such that

$$(d + \tilde{\delta})\beta = \mathcal{T} P_n(\hat{\pi}_1^* \hat{ev}^* \mathbf{a}, A_0) - \pi_1^* ev^* H \tag{16.18}$$

To construct β we consider the sequence

$$\mathsf{M} \times \mathcal{P}_{\phi_0}(\mathsf{Map}(\mathsf{M}, \mathsf{T})) \xrightarrow{\;r\;} \mathsf{M} \xrightarrow{\;i\;} \mathsf{M} \times \mathcal{P}_{\phi_0}(\mathsf{Map}(\mathsf{M}, \mathsf{T}))$$

where r denotes the retraction and i the inclusion. The map $i \circ r$ is homotopic to the identity, so, by a standard argument, there exists a homotopy operator $\mathcal{H}$ such that

$$1^* - (i \circ r)^* = (d + \tilde{\delta})\mathcal{H} + \mathcal{H}(d + \tilde{\delta}) \tag{16.19}$$

The operator $\mathcal{H}$ is in fact given by $\int_0^1 dv\, h^*$, where

$$h : \mathsf{M} \times I \times \mathcal{P}_{\phi_0}(\mathsf{Map}(\mathsf{M}, \mathsf{T})) \longrightarrow \mathsf{M} \times \mathcal{P}_{\phi_0}(\mathsf{Map}(\mathsf{M}, \mathsf{T}))$$
$$(x, v, p) \rightsquigarrow (x, p^v) \tag{16.20}$$

where $p^v(x, s) = p(x, sv)$,i.e. h shrinks the paths.

Next, the fact that i^* pulls back forms to M, implies that, when applied to a $\mathsf{d}+1$ form, it gives 0. Therefore, applying (16.19) to $\mathcal{T} P_n(\hat{\pi}_1^* \hat{ev}^* \mathbf{a}, A_0) - \pi_1^* ev^* H$, and using (16.17), we get

$$\mathcal{T} P_n(\hat{\pi}_1^* \hat{ev}^* \mathbf{a}, A_0) - \pi_1^* ev^* H = (d + \tilde{\delta})\mathcal{H}\big(\mathcal{T} P_n(\hat{\pi}_1^* \hat{ev}^* \mathbf{a}, A_0) - \pi_1^* ev^* H\big) \tag{16.21}$$

Therefore

$$\beta = \mathcal{H}\big(\mathcal{T}P_n(\hat{\pi}_1^* \hat{ev}^* \mathbf{a}, A_0) - \pi_1^* ev^* H\big)$$

$$= \int_0^1 dv\, i_{\frac{\partial}{\partial v}} h^*\big(\mathcal{T}P_n(\hat{\pi}_1^* \hat{ev}^* \mathbf{a}, A_0) - \pi_1^* ev^* H\big) \qquad (16.22)$$

Before proceeding further we need an analysis and interpretation of the loop space $\Omega_{\phi_0}(\mathsf{Map}(\mathsf{M}, \mathsf{T}))$.

16.3.1 Space of Loops and Induced Gauge Transformations

Let us consider the space of loops $\Omega_{\phi_0}(\mathsf{Map}(\mathsf{M}, \mathsf{T}))$ and the diagram (16.11). For any path $p \in \mathcal{P}_{\phi_0}(\mathsf{Map}(\mathsf{M}, \mathsf{T}))$, we can consider $p_0^*\mathsf{P} \equiv \phi_0^*\mathsf{P}$ and $p_1^*\mathsf{P}$, where $p_0(x) = p(x, 0)$ and $p_1(x) = p(x, 1)$. The bundles $\phi_0^*\mathsf{P}$ and $p_1^*\mathsf{P}$ are isomorphic in many ways. Now pick a connection $\boldsymbol{\alpha}$ in P and pull it back via the canonical covering $\hat{p}$ of p: $\hat{p}^*\boldsymbol{\alpha}$. Now $\hat{p}^*\boldsymbol{\alpha}$ is a connection in $p^*\mathsf{P}$ over $I \times \mathsf{M}$. Next consider a trivial path (not to be confused with the paths in $\mathcal{P}_{\phi_0}(\mathsf{Map}(\mathsf{M}, \mathsf{T}))$) in $I \times \mathsf{M}$, specified by (s, x) with fixed x and lift it horizontally to a path in $p^*\mathsf{P}$, which starts at $\hat{u}_0 \in \pi^{-1}(x) \subset \phi_0^*\mathsf{P}$ and ends at $\hat{u}_1 \in \pi^{-1}(x) \subset p_1^*\mathsf{P}$. This establishes a (non-canonical) isomorphisms $\tau_{\boldsymbol{\alpha}} : \phi_0^*\mathsf{P} \to \phi_1^*\mathsf{P}$.

Next, if instead of a path p we consider a loop $\ell \in \Omega_{\phi_0}\big(\mathsf{Map}(\mathsf{M}, \mathsf{T})\big)$, the lifted loop becomes a path p_ℓ in $\ell^*\mathsf{P}$ which starts at $\hat{u}_0 \in \pi^{-1}(x) \subset \phi_0^*\mathsf{P}$ and ends at a point $\hat{u}_1$ of the same fiber over x. Therefore, $\hat{u}_1 = \hat{u}_0 g$, for some $g \in \mathsf{G}$. If the connection $\hat{p}^*\boldsymbol{\alpha}$ is irreducible these group elements g span the whole of G. Therefore, p_ℓ defines a vertical automorphism of $\phi_0^*\mathsf{P}$: $\hat{\psi}(\hat{u}) = \hat{u}\hat{\varphi}(\hat{u})$. All together they define a subgroup of $\mathsf{Aut}_\mathsf{V}(\phi_0^*\mathsf{P})$, which we call *the subgroup of induced gauge transformations* and, for simplicity, denote with the same symbol.

Therefore, we have an embedding (depending on $\boldsymbol{\alpha}$, any different connection may lead to a different embedding)

$$\begin{aligned} \tau : \Omega_{\phi_0}\big(\mathsf{Map}(\mathsf{M}, \mathsf{T})\big) &\hookrightarrow \mathsf{Aut}_\mathsf{V}(\phi_0^*\mathsf{P}) \\ \ell &\rightsquigarrow p_\ell \end{aligned} \qquad (16.23)$$

This map is many to one. Any loop in $\Omega_{\phi_0}\big(\mathsf{Map}(\mathsf{M}, \mathsf{T})\big)$ defines a vertical automorphism of $\phi_0^*\mathsf{P}$: $\psi(\hat{u}) = \hat{u}\hat{\varphi}(\hat{u})$. Once $\boldsymbol{\alpha}$ is kept fixed, $\hat{\varphi}$ depends only on ℓ: $\hat{\varphi} = \hat{\varphi}_\ell$. The image of τ is the subgroup of induced gauge transformations. The space $\Omega_{\phi_0}\big(\mathsf{Map}(\mathsf{M}, \mathsf{T})\big)$ does not coincide with this subgroup. The differential $\tilde{\delta}$ in $\Omega_{\phi_0}\big(\mathsf{Map}(\mathsf{M}, \mathsf{T})\big)$ does not coincide with the differential $\hat{\delta}$ in the subgroup, but is related to it by $\tilde{\delta} = \tau^*\hat{\delta}$. However, any variation in $\Omega_{\phi_0}\big(\mathsf{Map}(\mathsf{M}, \mathsf{T})\big)$ induces a gauge transformation (possibly the null transformation). Therefore, we shall use $\tilde{\delta}$ everywhere.

16.3.2 The WZ Term Made More Explicit

$\Omega_{\phi_0}(\mathsf{Map}(\mathsf{M},\mathsf{T})$ is a subspace of $\mathcal{P}_{\phi_0}(\mathsf{Map}(\mathsf{M},\mathsf{T})$ with inclusion map J. So let us restrict (16.21) to the loop space by pulling it back through J. We get

$$(d+\tilde{\delta})J^*\beta = J^*\mathcal{T}P_n(\hat{\pi}_1^*\hat{ev}^*\mathbf{a}, A_0) \tag{16.24}$$

where $\tilde{\delta}$ is the exterior differential in $\Omega_{\phi_0}(\mathsf{Map}(\mathsf{M},\mathsf{T}))$. Notice that the term $J^*\pi_1^*ev^*H$ vanishes for dimensional reasons because it is a $\mathsf{d}+1$ poly-form in $\mathsf{M}\times\Omega_{\phi_0}$ $(\mathsf{Map}(\mathsf{M},\mathsf{T}))$ with the only component $(\mathsf{d}+1,0)$. This is explained by the fact that $ev\circ\pi_1(x,\ell)=\phi_0(x)$, therefore ϕ_0^*H is a $\mathsf{d}+1$ form in M alone.

Next we restrict (16.24) to the $(\mathsf{d},0)$ component of β, $\beta_{(\mathsf{d},0)}$. We can write

$$\tilde{\delta}J^*\beta_{(\mathsf{d},0)} = J^*\mathcal{T}P_n(\hat{\pi}_1^*\hat{ev}^*\mathbf{a}, A_0)_{(\mathsf{d},0)} + \text{exact} \tag{16.25}$$

where $\tilde{\delta}$ is interpreted as the differential in the group of induced gauge transformations. Now $J^*\hat{\pi}_1^*\hat{ev}^*\mathbf{a}$ is, by construction, the same as $\left(\hat{\phi}_0\circ ev_{\phi_0}\right)^*\mathbf{a}$ in (16.9), since $ev\circ\pi_1(x,\ell)=\phi_0(x)$.

Therefore, looking back at Eq. (16.8), which gives rise to the anomaly (16.9), we find that $\mathcal{T}P_n((\hat{\phi}_0\circ ev_{\phi_0})^*\mathbf{a}, A_0)$ corresponds by construction to the RHS of Eq. (16.24) and, in particular, its $(\mathsf{d},0)$ component equals the first term in the RHS of (16.25). Therefore, we can write

$$\tilde{\delta}J^*\beta_{(\mathsf{d},0)} = \text{Anomaly} + \text{exact} \tag{16.26}$$

where by *Anomaly* we mean (16.9). We shall refer to $\mathcal{B}_{\mathsf{WZ}}\equiv\int_{\mathsf{M}}J^*\beta_{(n,0)}$ as generalized Wess-Zumino term.

Let us elaborate on it

$$\mathcal{B}_{\mathsf{WZ}} = \int_{\mathsf{M}}\left(J^*\int_0^1 \mathrm{d}v\, i_{\frac{\partial}{\partial v}}h^*\big(\mathcal{T}P_n(\hat{\pi}_1^*\hat{ev}^*\mathbf{a}, A_0)-\pi_1^*ev^*H\big)\Big|_{(\mathsf{d},0)}\right) \tag{16.27}$$

First, we have

$$ev\circ\pi_1\circ h\circ J(x,\hat{\varphi}_\ell) = ev\circ\pi_1\circ h(x,\ell) = ev\circ\pi_1(x,\ell^v) = ev(x,\ell_1^v) = \ell(x,v)$$

Let us denote by $H_{\mu_1\ldots\mu_m}(y_1,\ldots,y_m)\mathrm{d}y^{\mu_1}\wedge\ldots\wedge\mathrm{d}y^{\mu_m}$ the local expression of H in T, where $m=\mathsf{d}+1$. Then the local expression of ℓ^*H in M is

$$(\ell^*H)_{\nu_1\ldots\nu_{\mathsf{d}}v}\mathrm{d}x^{\nu_1}\ldots\mathrm{d}x^{\nu_{\mathsf{d}}}\mathrm{d}v = H_{\mu_1\ldots\mu_m}\mathcal{J}^{y^{\mu_1}\ldots y^{\mu_m}}_{x^{\nu_1}\ldots x^{\nu_{\mathsf{d}}}v}\mathrm{d}x^{\nu_1}\ldots\mathrm{d}x^{\nu_{\mathsf{d}}}\mathrm{d}v \tag{16.28}$$

where $\mathcal{J}$ is the Jacobian of the change of coordinates $(y^{\mu_1}, \ldots, y^{\mu_m})$ to $(x^{\nu_1}, \ldots, x^{\nu_d}, v)$.

Next $J^* \circ h^* A_0 = A_0$.

As far as the connection $J^* \circ h^* \circ \hat{\pi}_1^* \circ \hat{ev}^* \mathbf{a}$ is concerned the part relevant to $\beta_{(d,0)}$ is

$$J^* \circ h^* \circ \hat{\pi}_1^* \circ \hat{ev}^* \mathbf{a} = \hat{\varphi}_\ell^{-1} \big(\mathbf{a} + d + \frac{\partial}{\partial v} dv\big) \hat{\varphi}_\ell \tag{16.29}$$

where, now, d is the exterior differential in P. Thus, we can write

$$\begin{aligned} \mathcal{B}_{WZ} = \int_{\mathsf{M}} \Big[& n(n-1) \int_0^1 dt (1-t) \int_0^1 dv \, P_n(d_{A_0} \mathbf{c}, \mathbf{A} - A_0, \mathbf{F}_t, \ldots, \mathbf{F}_t) \\ & - \int_0^1 dv H_{\mu_1 \ldots \mu_m} \mathcal{J}^{y^{\mu_1} \ldots y^{\mu_m}}_{x^{\nu_1} \ldots x^{\nu_d} v} dx^{\nu_1} \ldots dx^{\nu_d} \Big] \end{aligned} \tag{16.30}$$

where $\mathbf{c} = \hat{\varphi}_\ell^{-1} \frac{\partial}{\partial v} \hat{\varphi}_\ell$,

$$\mathbf{A} = \hat{\varphi}_\ell^{-1} (\mathbf{a} + d) \hat{\varphi}_\ell \tag{16.31}$$

and $\mathbf{F}_t$ is the curvature of $t\mathbf{A} + (1-t)A_0$.

Comment. The Wess-Zumino term $\mathcal{B}_{WZ}$ is a complicated (usually transcendental) formula. It is expressed in terms of paths of the basic scalar fields, the maps ϕ from M to T. An example, when the target space is a group, is the exponential map $\gamma_s = e^{s\sigma}$ used in Chap. 15. Therefore, the anomaly is canceled by adding $\mathcal{B}_{WZ}$ to the effective action assuming the loops of maps are acceptable as dynamical variables in the field theory model one is considering.

16.4 Global Anomalies. Case (C)

As explained in Chap. 12 the cocycle $\rho(A, \psi)$ defines a complex line bundle $\mathcal{L}_\rho$ over $\frac{\mathcal{A}}{\mathcal{G}}$. The overall path integral $\mathcal{Z}$, Eq. (12.59), makes sense only if there is a global section of $\mathcal{A} \to \frac{\mathcal{A}}{\mathcal{G}}$, which, multiplied by $\mathcal{Z}[A]$, will form the integrand of (12.59). This means that $\mathcal{L}_\rho$ must be trivial. Complex line bundles over $\frac{\mathcal{A}}{\mathcal{G}}$ are classified by their Chern class, which is an element of $H^2\left(\frac{\mathcal{A}}{\mathcal{G}}, \mathbb{R}\right)$. This means, in particular, that the path integral (12.59) makes sense only if our line bundle corresponds to the 0 element of $H^2\left(\frac{\mathcal{A}}{\mathcal{G}}, \mathbb{R}\right)$. If the anomaly that gives rise to the cocycle $\rho(A, \psi)$ is generated by a polynomial P_n, we must identify the class of $\mathcal{L}_\rho$ with $[\mathcal{F}_{P_n}]$, i.e. with the class of the closed two-form in $\frac{\mathcal{A}}{\mathcal{G}}$ defined by $\mathcal{F}_{P_n}$ in (13.40).

The case we are interested in here is when the line bundle $\mathcal{L}_\rho$ is trivial. Therefore, its Chern class is trivial, which means that

$$\mathcal{F}_{P_n} = \mathrm{d}\mathcal{K}, \qquad \mathcal{K} \in \Omega^1\left(\frac{\mathcal{A}}{\mathcal{G}}\right). \tag{16.32}$$

In Sect. 13.4.1, we have shown that in this case $\widetilde{\alpha}(A, \psi)$ is a coboundary in $\mathcal{G}_0$. Here we want to investigate its extension to $\mathcal{G}$. To this end, we will construct a differential character $\mathcal{U}_{P_n} \in \widehat{H}^1\left(\frac{\mathcal{A}}{\mathcal{G}}, \frac{\mathbb{R}}{\mathbb{Z}}\right)$ that incorporates the information contained in (16.32) and provides a global definition of the corresponding (trivial local) anomaly via the image of δ_2.

Above, in Sect. 13.5, we have shown how, by picking an integral class $[s] \in H^{2n}(\mathsf{BG}, \mathbb{Z})$ we can construct a differential character $\mathcal{U}_{P_n}([s])$ in $\widehat{H}^1\left(\frac{\mathcal{A}}{\mathcal{G}}, \frac{\mathbb{R}}{\mathbb{Z}}\right)$ such that

$$\delta_1 \mathcal{U}_{P_n} = \mathcal{F}_{P_n} \tag{16.33}$$

and determines an integral class

$$\beta\left(\mathcal{U}_{P_n}([s])\right) \in H^2\left(\frac{\mathcal{A}}{\mathcal{G}}, \mathbb{Z}\right) \tag{16.34}$$

via the Bockstein homomorphism β. When this operation is unambiguous, the specification of $\mathcal{U}_{P_n}([s])$ completes the extension of the corresponding local anomaly to the full gauge group $\mathcal{G}$. Now we have to incorporate in these formulas the information contained in (16.32). It is evident that (16.33) must be replaced by

$$\delta_1\left(\mathcal{U}_{P_n} - \widetilde{\mathcal{K}}\right) = \mathcal{F}_{P_n} - d\mathcal{K} = 0 \tag{16.35}$$

As a consequence the associated integral class is

$$\beta\left(\mathcal{U}_{P_n} - \widetilde{\mathcal{K}}\right) \in H^2\left(\frac{\mathcal{A}}{\mathcal{G}}, \mathbb{Z}\right) \tag{16.36}$$

If this class is uniquely defined (and equals 0), we say that the anomaly is uniquely defined and trivial. This provides a global definition of the trivial anomaly and of its cancelation by the (global) section $\widetilde{\theta}(A)$. For there is no obstructions to its extension to the full $\frac{\mathcal{A}}{\mathcal{G}}$, or no obstruction to extending $\widetilde{\theta}(A\psi) - \widetilde{\theta}(A)$ to $\mathcal{G}$ (the proof of Sect. 13.4.1 holds for the whole of $\mathcal{G}$). To guarantee uniqueness, a sufficient condition is again

$$Tor\, H^2\left(\frac{\mathcal{A}}{\mathcal{G}}, \mathbb{Z}\right) = 0 \tag{16.37}$$

As already explained the previous conclusions do not affect anomalies in field theory because locality prevents any triviality relation such as (16.32). However, they may be relevant in sigma models because the topology of target manifolds is more flexible and may admit triviality conditions that are not possible in the universal bundle. Indeed the formulas and procedures to find or cancel anomalies are in sigma models the same as in local field theories, provided we replace the classifying space BG with the target space T and the universal bundle EG with the principal bundle P(T, G). But, of course, the topology of BG is in general different from that of a generic (although compact Riemannian) target space T. In particular, what matters most for anomalies, the Weil homorphism in the universal bundle is an isomorphism; in simpler words every ad-invariant polynomial P_n gives rise to a non-trivial de Rham class. While, for sigma models, depending on the topology of T, some of these polynomials may correspond to the trivial de Rham class.

Let us recall first the geometric setting for a sigma model, based on the diagram (16.3), where **a** is a connection in the principal fiber bundle P(T, G) over the target space, valued in the Lie algebra $\mathfrak{g}$ of a compact Lie group G, and the evaluation map (16.4) and the transgression formula formula (16.8) where $\mathfrak{F}_t$ is the curvature of the connection $\mathcal{A}_t = t\, ev_\phi^* A + (1-t)A_0$, A_0 being a fixed connection in $\phi^*\mathsf{P}$, and $n = \frac{\mathsf{d}+2}{2}$.

The case we want to consider is when $P_k(F(\mathbf{a}), \ldots, F(\mathbf{a})) = \mathrm{d}X_{2k-1}$, where X_{2k-1} is a $2k-1$−form in **T**, with $k \le n$. Then also $\bar{P}_n = P_k P_{n-k}$ is trivial and $= \mathrm{d}X_{2n-1}$ in T, so that we are in a situation similar to (16.32), where $\mathcal{A}$ is replaced by $\overline{\mathcal{A}}$, the space of induced connections $A = \phi^*\mathbf{a}$, and $\mathcal{G}$ by $\overline{\mathcal{G}}$, the space of induced gauge transformations $\mathsf{Aut}_{\mathsf{V}}(\phi^*\mathsf{P})$. Moreover,

$$\overline{\mathcal{F}}_{\bar{P}_n} = \int_{\mathsf{M}} \bar{P}_n(F(ev_\phi^*(\mathbf{a}), \ldots, F(ev_\phi^*(\mathbf{a}) = d\overline{\mathcal{K}} \tag{16.38}$$

Then we can repeat the same argument above and construct a differential character $\overline{\mathcal{U}}_{\bar{P}_n}(\tau)$ in $\widehat{H}^1\left(\frac{\bar{\mathcal{A}}}{\bar{\mathcal{G}}}, \frac{\mathbb{R}}{\mathbb{Z}}\right)$ such that $\delta_1\overline{\mathcal{U}_{P_n}} = \overline{\mathcal{F}}_{\bar{P}_n}$ holds, and it determines an integral class

$$\beta\left(\overline{\mathcal{U}}_{\bar{P}_n}(\tau) - \widetilde{\overline{\mathcal{K}}}\right) \in H^2\left(\frac{\overline{\mathcal{A}}}{\overline{\overline{\mathcal{G}}}}, \mathbb{Z}\right) \tag{16.39}$$

via the Bockstein homomorphism β. If this class is uniquely defined (and equals 0), we say that the anomaly is uniquely defined and trivial. This provides a global definition of the trivial anomaly and of its cancelation by a (global) section $\widetilde{\overline{\theta}}(A)$. To guarantee uniqueness, a sufficient condition is again

$$\operatorname{Tor} H^2\left(\frac{\overline{\mathcal{A}}}{\overline{\overline{\mathcal{G}}}}, \mathbb{Z}\right) = 0 \tag{16.40}$$

In the field theory case, we have seen above that the original source of ambiguities is the torsion subgroup of $H^{n+2}(\mathsf{BG}, \mathbb{Z})$ and that this gives rise to the sufficient condition (13.47) for the absence of ambiguities (and global anomalies). In the same way for sigma models, a sufficient condition for the absence of indeterminacies (and global anomalies), alternative to (16.40), is

$$T \text{ or } H^{\mathsf{d}+2}(\mathsf{T}, \mathbb{Z}) \cong T \text{ or } H_{\mathsf{d}+1}(\mathsf{T}, \mathbb{Z}) = 0 \qquad (16.41)$$

References

1. L. Bonora, P. Cotta-Ramusino, M. Rinaldi, J. Stasheff, The evaluation map in field theory, sigma-models and strings. II. Comm. Math. Phys. **114**, 381 (1988)
2. J. Cheeger, J. Simons, Differential characters and geometric invariants, in *Geometry and Topology. Lectures Notes in Mathematics*, vol. 1167 (Springer, Berlin, Heidelberg, New York, 1985)
3. G. Moore, P. Nelson, Anomalies in nonlinear σ models. Phys. Rev. Lett. **53**, 1519 (1984)
4. G. Moore, P. Nelson, The aetiology of sigma model anomalies. Comm. Math Phys. **100**, 83 (1985)

Chapter 17
Anomalies and (Super)String Theories

If QFTs are theories under construction, this is even more true for (super)string theories. These theories have multiple descriptions. They can be represented by 2d (worldsheet) field theories of scalars and fermions, as 2d sigma models with target space represented by the spacetimes where strings propagate, or as 26d (10d supersymmetric) field theories in the just mentioned target spacetime, where the fields are the (super)string excitations and, from the anomaly point of view, the relevant excitations are the massless ones. The quantum S-matrix can be defined by perturbative series whose terms are amplitudes of on shell vertex operators, each one of them representing a (super)string excitation, inserted at (internal or boundary) points of Riemann surfaces and integrated over the relevant moduli spaces. Second quantized (off-shell) superstring theories have been constructed, but they are not yet wieldy tools for calculations.

In any case, whatever description we have in mind, all different incarnations of string theory must be anomaly-free; that is, for them to be consistent, any type of anomalies, local or global, worldsheet or target space, must be absent. In particular, what is new with respect to ordinary field theories, not only spacetime anomalies but also worldsheet ones must be absent. This last fact is crucial for the critical dimension of spacetime.

A complete account of the (super)string anomalies and their cancelation would require an entire book. Especially their calculation in the framework of perturbative quantum S-matrix is out of question, because it would require a long dedicated preparation. As for target space local and, especially, global anomalies, they are still at present an active research terrain. In this chapter, we will limit ourselves to worldsheet, and some examples of sigma model and target space anomalies of the (super)string. In the first section, we deal with worldsheet anomalies in both perturbative and not perturbative approaches. The second is devoted to anomalies generated in spacetime by the massless modes of the (super)strings; the third to the sigma model anomalies of the string. Finally, in the fourth we deal with some aspects of global anomalies. General references for this chapter are [1, 2].

L. Bonora, *Fermions and Anomalies in Quantum Field Theories*, Theoretical and Mathematical Physics, https://doi.org/10.1007/978-3-031-21928-3_17

17.1 Worldsheet Anomalies of the String

String theory is defined by scalar and fermions fields in a 2d worldsheet. Their actions are

$$S_b = \frac{T}{2} \int_\Sigma d^2x \, \sqrt{h} h^{\alpha\beta} \partial_\alpha X^\mu \partial_\beta X^\nu \eta_{\mu\nu} \tag{17.1}$$

$$S_f = \frac{T}{2} \int_\Sigma \sqrt{h} \bar{\psi}^\mu \gamma^\alpha \nabla_\alpha \psi^\nu \eta_{\mu\nu} \tag{17.2}$$

and, in the case of heterotic superstrings,

$$S_f = \frac{T}{2} \int_\Sigma \sqrt{h} \bar{\lambda}^A \gamma^\alpha \nabla_\alpha \lambda^A \tag{17.3}$$

summed over $A = 1, \ldots, 32$.

X^μ, $\mu = 0, \ldots, \mathsf{d} - 1$ are the spacetime coordinates; ψ^μ and λ^A are Majorana-Weyl spinors of opposite chirality; Σ is the 2d worldsheet; $h_{\alpha\beta}$ is a metric on Σ $(\alpha, \beta = 0, 1)$; $\gamma^\alpha = e^\alpha_a \gamma^a$, where γ^a are the gamma matrices in 2d; ∇ is the covariant derivative on Σ; T is the string tension, $T = \frac{1}{2\pi\alpha'}$, and α' is the Regge slope, which coincides with the square of the characteristic length of the string. We shall set $\alpha' = 1/2$ throughout; $\eta_{\mu\nu}$ is the spacetime Minkowski metric; both worldsheet and spacetime metrics have negative spatial signatures.

A more geometrical description of both actions will be given in the next section. For the time being, we suppose that Σ is a flat Minkowski space, which becomes a complex $\mathbb{C}$ plane after a Wick rotation. In this section, we study the possible diffeomorphism and trace anomalies of scalar fields and fermion fields in 2d. In this book, we have treated so far only fermions, in particular in 2d. Concerning the latter, we will use the results of Chap. 7. Hereafter, we analyze first the case of a scalar field in 2d.

17.1.1 2d *Scalar*

The 2d (real) scalar model is the simplest possible field theory. Its action is

$$S_b = \frac{1}{2} \int d^2x \, \partial_\mu \phi(x) \partial^\mu \phi(x) \tag{17.4}$$

It is easy to covariantize it with respect to diffeomorphisms

$$S_b[\phi, g] = \frac{1}{2}\int d^2x\,\sqrt{g}\,g^{\mu\nu}\partial_\mu\phi(x)\partial_\nu\phi(x) \tag{17.5}$$

The eom is $\Box\phi = 0$. As for the e.m. tensor,

$$T^{(b)}_{\mu\nu} = \frac{2}{\sqrt{g}}\frac{\delta S}{\delta g^{\mu\nu}} = \partial_\mu\phi\,\partial_\nu\phi - \frac{1}{2}g_{\mu\nu}\,\partial_\lambda\phi\partial^\lambda\phi \tag{17.6}$$

We have both $D^\mu T^{(b)}_{\mu\nu} = 0$ on shell, and $T^{(b)\mu}_\mu = 0$ *identically*, i.e. on- and off-shell. This is to be compared with the 2d fermion model, whose covariant action is

$$S_f = \frac{i}{2}\int d^2x\,\sqrt{g}\,\bar\psi\gamma^\mu \overleftrightarrow{\partial}_\mu\,\psi \tag{17.7}$$

where $\gamma^\mu = \rho^a e^\mu_a$. Motivated by the previous analyses of the fermion e.m. tensor we will use the reduced form

$$T^{(f)}_{\mu\nu} = \frac{i}{4}\left(\bar\psi\gamma_\mu\overleftrightarrow{\partial}_\nu\,\psi + \mu\leftrightarrow\nu\right) \tag{17.8}$$

It satisfies both $D^\mu T^{(f)}_{\mu\nu} = 0$ and $T^{(f)\mu}_\mu = 0$ on shell, while $T^{(f)\mu}_\mu \neq 0$ off-shell.

By contrast $T^{(b)\mu}_\mu = 0$ identically, meaning that in QFT $T^{(b)\mu}_\mu$ is a *null quantum operator*. We cannot expect to find a trace anomaly by inserting $T^{(b)\mu}_\mu$ in a correlator. In fact, we are bound to find no trace anomalies as long as we use the definition (17.6). This is the case in the perturbative approach. The way a non-vanishing trace anomaly shows up is surprising and instructive.

17.1.2 Perturbative Approach

We refer to (17.5). The propagator is $\frac{i}{p^2}$. There are two scalar-scalar-graviton vertices:

$$V_{ssh} : \frac{i}{4}\left(p_\mu p'_\nu + p_\nu p'_\mu\right) \tag{17.9}$$

$$V'_{ssh} : -\frac{i}{4}\eta_{\mu\nu}\,p\cdot p' \tag{17.10}$$

where p is an entering scalar particle momentum and p' an exiting one. The relevant diagram is the bubble one: two external legs, one entering and one exiting graviton with momentum q, together with a fermion loop formed by a fermion of momentum p and $p-q$. There are two propagators and two vertices. We consider first two vertices V_{ssh}. The unregulated Fourier transformed amplitude is

$$\widetilde{B}^{(1)}_{\mu\nu\lambda\rho}(q) = \frac{1}{16}\int\frac{d^2p}{(2\pi)^2}\frac{(p_\mu(p-q)_\nu + p_\nu(p-q)_\mu)(p_\lambda(p-q)_\rho + p_\rho(p-q)_\lambda)}{p^2(p-q)^2} \tag{17.11}$$

Inserting also V'_{ssh} we get three more amplitudes

$$\widetilde{B}^{(2)}_{\mu\nu\lambda\rho}(q) = -\frac{1}{32}\int \frac{d^2p}{(2\pi)^2}\frac{\eta_{\mu\nu}\, p\cdot(p-q)\,(p_\lambda(p-q)_\rho + p_\rho(p-q)_\lambda)}{p^2(p-q)^2} \tag{17.12}$$

$$\widetilde{B}^{(2')}_{\mu\nu\lambda\rho}(q) = -\frac{1}{32}\int \frac{d^2p}{(2\pi)^2}\frac{(p_\mu(p-q)_\nu + p_\nu(p-q)_\mu)\,\eta_{\lambda\rho}\, p\cdot(p-q)}{p^2(p-q)^2} \tag{17.13}$$

$$\widetilde{B}^{(3)}_{\mu\nu\lambda\rho}(q) = \frac{1}{64}\int \frac{d^2p}{(2\pi)^2}\frac{\eta_{\mu\nu}\, p\cdot(p-q)\,\eta_{\lambda\rho}\, p\cdot(p-q)}{p^2(p-q)^2} \tag{17.14}$$

The total relevant amplitude is the sum of these four.

Now we regularize[1] these four expressions by adding extra Euclidean dimensions, $p_\mu \to p_\mu + \ell_{\bar\mu}$, $\bar\mu = 2, 3, \ldots, 1+\delta$, so that $p^2 \to p^2 - \ell^2$, $(p-q)^2 \to (p-q)^2 - \ell^2$, etc. We start by considering the non-local part of the two-point function. The latter comes from forgetting for the time being the pieces containing ℓ^2 powers in the numerator. So we consider first, the above expressions in which the denominators and the integration measure are replaced as follows

$$\int \frac{d^2p}{(2\pi)^2} \longrightarrow \int \frac{d^2p\,d^\delta\ell}{(2\pi)^{2+\delta}} \qquad p^2(p-q)^2 \longrightarrow (p^2-\ell^2)((p-q)^2-\ell^2)$$

A standard calculation, after a Wick rotation (meaning that $q^\mu q_\mu = -q^2$), yields the following total

$$\begin{aligned}
B_{\mu\nu\lambda\rho}(q) = \frac{i}{64\pi}\Bigg[&\frac{1}{3}q^2\left(\eta_{\mu\nu}\eta_{\lambda\rho}+\eta_{\mu\lambda}\eta_{\nu\rho}+\eta_{\mu\rho}\eta_{\nu\lambda}\right)\left(\frac{2}{\delta}-\frac{8}{3}+\gamma+\log\frac{q^2}{2\pi\mu^2}\right)\\
&+\frac{1}{6}\left(\eta_{\mu\nu}q_\nu q_\rho+\eta_{\mu\rho}q_\nu q_\lambda+\eta_{\nu\lambda}q_\mu q_\rho+\eta_{\nu\rho}q_\mu q_\lambda\right)\left(\frac{2}{\delta}-\frac{8}{3}+\gamma+\log\frac{q^2}{2\pi\mu^2}\right)\\
&+\frac{2}{3}\frac{1}{q^2}q_\mu q_\nu q_\lambda q_\rho-\frac{1}{3}\left(\eta_{\mu\nu}q_\lambda q_\rho+\eta_{\lambda\rho}q_\mu q_\nu\right)\left(\frac{2}{\delta}-\frac{5}{3}+\gamma+\log\frac{q^2}{2\pi\mu^2}\right)\\
&-\frac{1}{2}\eta_{\mu\nu}\Bigg(\left(\frac{4}{3}q^2\eta_{\lambda\rho}+\frac{2}{3}q_\lambda q_\rho\right)\left(\frac{2}{\delta}-\frac{8}{3}+\gamma+\log\frac{q^2}{2\pi\mu^2}\right)\\
&-\frac{2}{3}q_\lambda q_\rho-\frac{1}{3}\left(2q_\lambda q_\rho-q^2\eta_{\lambda\rho}\right)\left(\frac{2}{\delta}-\frac{5}{3}+\gamma+\log\frac{q^2}{2\pi\mu^2}\right)\Bigg)\\
&-\frac{1}{2}\eta_{\lambda\rho}\Bigg(\left(\frac{4}{3}q^2\eta_{\lambda\rho}+\frac{2}{3}q_\mu q_\nu\right)\left(\frac{2}{\delta}-\frac{8}{3}+\gamma+\log\frac{q^2}{2\pi\mu^2}\right)\\
&-\frac{2}{3}q_\mu q_\nu-\frac{1}{3}\left(2q_\mu q_\nu-q^2\eta_{\mu\nu}\right)\left(\frac{2}{\delta}-\frac{5}{3}+\gamma+\log\frac{q^2}{2\pi\mu^2}\right)\Bigg)\\
&+\frac{1}{4}\eta_{\mu\nu}\eta_{\lambda\rho}\Bigg(2q^2\left(\frac{2}{\delta}-\frac{8}{3}+\gamma+\log\frac{q^2}{2\pi\mu^2}\right)+\frac{2}{3}q^2\\
&+\frac{4}{3}q^2\left(\frac{2}{\delta}-\frac{5}{3}+\gamma+\log\frac{q^2}{2\pi\mu^2}\right)\Bigg)
\end{aligned} \tag{17.15}$$

[1] Before simplifying numerator with denominator one has to regularize.

Saturating with q^μ (conservation), we get

$$q^\mu \widetilde{B}_{\mu\nu\lambda\rho}(q) = \frac{i}{64\pi}\Big[\frac{1}{6}q^2\left(-q_\nu\eta_{\lambda\rho} + q_\lambda\eta_{\nu\rho} + q_\rho\eta_{\nu\lambda}\right)\left(\frac{2}{\delta} - \frac{8}{3} + \gamma + \log\frac{q^2}{2\pi\mu^2}\right)$$
$$-\frac{1}{3}q_\nu q_\lambda q_\rho - \frac{1}{6}q^2 q_\nu \eta_{\lambda\rho}\Big] \tag{17.16}$$

Saturating with $\eta^{\mu\nu}$ (trace), we get

$$\eta^{\mu\nu}\widetilde{B}_{\mu\nu\lambda\rho}(q) = 0 \tag{17.17}$$

The surprise is that the e.m. tensor is not conserved by a non-local term. However, let us consider first the last line in (17.16). After an inverse Wick rotation, i.e. after returning to Lorentz with an inverse Wick rotation, it corresponds to the following anomaly

$$\Delta_\xi = -\frac{1}{192\pi}\int d^2x\, \partial\cdot\xi\left(\partial_\lambda\partial_\rho h^{\lambda\rho} - \frac{1}{2}\Box h\right) \tag{17.18}$$

where $h = h^\lambda_\lambda$. Let us consider the counterterm

$$\mathcal{C}_l = \frac{1}{384\pi}\int d^2x\left(h\,\partial_\lambda\partial_\rho h^{\lambda\rho} - \frac{3}{4}h\Box h\right) \tag{17.19}$$

We have

$$\delta_\xi \mathcal{C}_l = -\Delta_\xi, \quad \delta_\omega \mathcal{C}_l = \frac{1}{96\pi}\int d^2x\,\omega\left(\partial_\lambda\partial_\rho h^{\lambda\rho} - \Box h\right) \equiv \Delta_\omega \tag{17.20}$$

Therefore, the counterterm $\mathcal{C}_l$ cancels the anomaly corresponding to the last line of (17.16) and produces a trace anomaly, which is exactly the expected one (if there are no more corrections).

Let us next consider the first line of (17.16). It corresponds to a non-local cocycle

$$\Delta_\xi^{(nl)} \sim \int d^2x\,\xi^\nu\left(2f(\Box)\partial_\lambda h^\lambda_\nu - \partial_\nu f(\Box)h\right) \tag{17.21}$$

where $f(\Box)$ could be any regular function of $\Box$, in our case it is a constant plus a logarithm. First we verify that

$$\delta_\xi \Delta_\xi^{(nl)} = 0 \tag{17.22}$$

In fact each of the two terms in the RHS of (17.21) is consistent due to the anticommutativity of ξ. We can cancel $\Delta_\xi^{(nl)}$ with a counterterm

$$\mathcal{C}^{(nl)} \sim \int d^2x \left(h^{\lambda\nu} f(\Box) h_{\lambda\nu} - \frac{1}{2} h f(\Box) h \right) \tag{17.23}$$

With a suitable coefficient this counterterm cancels $\Delta_\xi^{(nl)}$, but, in addition,

$$\delta_\omega \mathcal{C}^{(nl)} = 0 \tag{17.24}$$

so it does not modify the anomaly (17.20).

We have now to consider the local terms, which originate from regularizing the numerators of (17.12, 17.13) and (17.14). For instance, $p\cdot(p-q)$ becomes $p\cdot(p-q)-\ell^2$. The term proportional to ℓ^2 is a local term and has not been considered so far. For definiteness, the additional term for $\widetilde{B}^{(2)}_{\mu\nu\lambda\rho}(q)$ is

$$\begin{aligned}\Delta \widetilde{B}^{(2)}_{\mu\nu\lambda\rho}(q) &= -\frac{1}{32} \int \frac{d^2p\, d^\delta \ell}{(2\pi)^{2+\delta}} \ell^2\, \eta_{\mu\nu}\, \frac{p\cdot(p-q)\,(p_\lambda(p-q)_\rho + p_\rho(p-q)_\lambda)}{(p^2-\ell^2)((p-q)^2-\ell^2)} \\ &= \frac{i}{384\pi} \eta_{\mu\nu} \left(\frac{1}{2} q^2 \eta_{\lambda\rho} - q_\lambda q_\rho \right)\end{aligned} \tag{17.25}$$

Similarly,

$$\Delta \widetilde{B}^{(2')}_{\mu\nu\lambda\rho}(q) = \frac{i}{384\pi} \eta_{\lambda\rho} \left(q_\mu q_\nu - \frac{1}{2} q^2 \eta_{\mu\nu} \right) \tag{17.26}$$

and

$$\begin{aligned}\Delta \widetilde{B}^{(3)}_{\mu\nu\lambda\rho}(q) &= \frac{1}{64} \int \frac{d^2p\, d^\delta \ell}{(2\pi)^{2+\delta}}\, \eta_{\mu\nu}\, \eta_{\lambda\rho}\, \frac{\ell^4 - \ell^2 (p\cdot(p-q))^2}{(p^2-\ell^2)((p-q)^2-\ell^2)} \\ &= -\frac{i}{768\pi} q^2 \eta_{\mu\nu}\eta_{\lambda\rho}\end{aligned} \tag{17.27}$$

The total is

$$\Delta \widetilde{B}_{\mu\nu\lambda\rho}(q) = \frac{i}{384\pi} \left(\eta_{\mu\nu} q_\lambda q_\rho + \eta_{\lambda\rho} q_\mu q_\nu - \frac{3}{2} q^2 \eta_{\mu\nu}\eta_{\lambda\rho} \right) \tag{17.28}$$

Saturating with q^μ one gets

$$q^\mu \Delta \widetilde{B}_{\mu\nu\lambda\rho}(q) = \frac{i}{384\pi} \left(q_\nu q_\lambda q_\rho - \frac{5}{2} q^2 \eta_{\lambda\rho} \right) \tag{17.29}$$

which corresponds to the cocycle (after an inverse Wick rotation)

$$\Delta'_\xi = -\frac{1}{384\pi} \int d^2x\, \partial\cdot\xi \left(\partial_\lambda \partial_\rho h^{\lambda\rho} + \frac{5}{2} \Box h \right) \tag{17.30}$$

On the other hand, contracting (17.28) with $\eta^{\mu\nu}$ we get the additional term

$$\eta^{\mu\nu}\Delta\widetilde{B}_{\mu\nu\lambda\rho}(q) = \frac{i}{192\pi}\left(q_\lambda q_\rho - 2q^2\eta_{\lambda\rho}\right) \tag{17.31}$$

which corresponds to the additional cocycle

$$\Delta'_\omega = -\frac{1}{192\pi}\int d^2x\,\omega\left(\partial_\lambda\partial_\rho h^{\lambda\rho} + 2\Box h\right) \tag{17.32}$$

It is easy to prove that these cocycles are consistent

$$\delta_\xi\Delta'_\xi = 0, \quad \delta_\omega\Delta'_\omega = 0, \quad \delta_\xi\Delta'_\omega + \delta_\omega\Delta'_\xi = 0 \tag{17.33}$$

Next consider the counterterm

$$C' = \frac{1}{768\pi}\int d^2x\left(h\partial_\lambda\partial_\rho h^{\lambda\rho} + \frac{3}{4}h\Box h\right) \tag{17.34}$$

Its variations are

$$\delta_\xi C' = -\Delta'_\xi, \quad \delta_\omega C' = -\Delta'_\omega \tag{17.35}$$

Therefore, $\Delta\widetilde{B}$ does not modify the already obtained result. In conclusion, the net outcome is

$$\Delta_\xi^{(\text{total})} = 0, \quad \Delta_\omega^{(\text{total})} = \frac{1}{96\pi}\int d^2x\,\omega\left(\partial_\lambda\partial_\rho h^{\lambda\rho} - \Box h\right) \tag{17.36}$$

which is the leading term of the anomaly

$$\mathcal{A}_\omega^{(r)} = \frac{1}{96\pi}\int d^2x\,\sqrt{g}\,\omega\,R \tag{17.37}$$

This is the trace anomaly for a *real* scalar field in 2d. For a complex scalar boson, the action is equal to two copies of the action of a real scalar, and thus, the anomaly is twice that of a real scalar:

$$\mathcal{A}_\omega^{(c)} = \frac{1}{48\pi}\int d^2x\,\sqrt{g}\,\omega\,R \tag{17.38}$$

(17.37) is the same as the trace anomaly of a complex Weyl fermion.

Remark. With reference to the definition $g^{\mu\nu}\langle\langle T_{\mu\nu}\rangle\rangle - \langle\langle T_\mu^\mu\rangle\rangle$ of the trace anomaly, notice that, as already remarked, T_μ^μ is an identically null field operator. Therefore, the second term in this definition is absent. The first term is precisely what has just been calculated.

Comment. It is not surprising if our subtracting the non-local term (17.23) provokes the reader's reluctance. It is almost an axiom that, in order to preserve unitarity, we can modify the effective action only with local terms. Leaving aside the ongoing debate on locality, let us remark that the 2d worldsheet theory is not a physical theory, physics lives in the target space. The requirements on the 2d theory are that it preserves the two basic symmetries: diffeormorphism and conformal symmetry. Their mutual interplay and exact preservation are the basis of string theory and the origin of critical dimensions. This is well known, and we will come back to it below. However, let us remark that the 2d boson example unveils the origin and source of the trace anomaly. In fact, the (even parity) anomaly (17.37) or (17.38) does not come from a direct calculation of the trace of a correlator with insertions of $T_{\mu\nu}$, but it is generated as a diffeomorphism anomaly. This identifies a non-trivial cocycle of the joint conformal and diffeomorphism coboundary operator, which can be transformed into a pure conformal anomaly by subtracting a suitable counterterm. Therefore, the refrain that the vanishing of the trace anomaly is the root of critical dimensions is somewhat imprecise: critical dimensions stem from the vanishing of the trace anomaly *and* the diffeomorphism anomaly.

17.1.3 The Trace Anomaly via SDW

The quadratic kinetic operator in the scalar case is $\frac{1}{2}g^{\mu\nu}D_\mu D_\nu$. The application of DSW to it is straightforward. One can absorb the factor $\frac{1}{2}$ by a redefinition of s, $s \to s' = \frac{s}{2}$ and reduce this case to one already computed. The equation for the a_n coefficients is simplified, and there is no spinor covariant displacer. The only non-trivial contribution for $[a_1]$ comes from the double derivative of the VVM determinant. The coincidence limit gives

$$[a_1] = \frac{1}{6}R$$

and the effective Lagrangian is

$$L(x) = \frac{1}{4\pi}\sqrt{g}\left(\frac{1}{d-2} - \frac{1}{2}\right)\frac{[a_1(x)]}{2} \tag{17.39}$$

Since

$$\delta_\omega R = -2\omega R - (d-1)\Box\omega$$

this implies an anomaly

$$\mathcal{A}_\omega = \frac{1}{48\pi}\int d^2x\,\sqrt{g}\,\omega\,R \tag{17.40}$$

which coincides with (17.38).

The various cases of spin 1/2 fermions have been computed previously with the SDW method in Sect. 14.2.

17.1.4 Conformal e.m. Tensor Correlators and Anomalies

The conformal e.m. tensor 2-point function in Euclidean 2d is given by

$$\left\langle T_{\mu\nu}(x)\, T_{\rho\sigma}(0)\right\rangle = \frac{c}{2x^4}\left(I_{\mu\rho}(x)\, I_{\nu\sigma}(x) + I_{\nu\rho}(x)\, I_{\mu\sigma}(x) - \eta_{\mu\nu}\eta_{\rho\sigma}\right) \tag{17.41}$$

where c is the (unnormalized) central charge of the theory. Let us compute this correlator for the free scalar field theory using the propagator

$$\langle \phi(x)\phi(y)\rangle = \frac{1}{4\pi}\log(x-y)^2 \tag{17.42}$$

and the Wick theorem. The result is immediate

$$\left\langle T_{\mu\nu}(x)\, T_{\rho\sigma}(0)\right\rangle = \frac{1}{8\pi^2 x^4}\left(I_{\mu\rho}(x)\, I_{\nu\sigma}(x) + I_{\nu\rho}(x)\, I_{\mu\sigma}(x) - \eta_{\mu\nu}\eta_{\rho\sigma}\right) \tag{17.43}$$

The normalized central charge is $\mathbf{c} = 4\pi^2 c = 1$.

Let us next introduce the complex notation

$$\begin{aligned} z &= x^1 + ix^2, \quad \bar{z} = x^1 - ix^2, \quad z\bar{z} = (x^1)^2 + (x^2)^2 \equiv x^2 \\ w &= y^1 + iy^2, \quad \bar{w} = y^1 - iy^2, \quad w\bar{w} = (y^1)^2 + (y^2)^2 \equiv y^2 \end{aligned}$$

In this new coordinate system, the components of the e.m. tensor are

$$\begin{aligned} T_{zz} &= \frac{1}{4}\left(T_{11} - T_{22} - 2iT_{12}\right), \quad T_{\bar{z}\bar{z}} = \frac{1}{4}\left(T_{11} - T_{22} + 2iT_{12}\right), \\ T_{z\bar{z}} &= \frac{1}{4}\left(T_{11} + T_{22}\right) \end{aligned}$$

and (17.41) becomes simply

$$\langle T_{zz}(z)T_{ww}(w)\rangle = \frac{c}{2}\frac{1}{(z-w)^4}, \quad \langle T_{\bar{z}\bar{z}}(\bar{z})T_{\bar{w}\bar{w}}(\bar{w})\rangle = \frac{c}{2}\frac{1}{(\bar{z}-\bar{w})^4} \tag{17.44}$$

while the correlators with an insertion of $T_{z\bar{z}}$ vanish.

Let us apply these definitions to a bosonic field with a flat background metric. We have

$$T_{zz} = \partial_z\phi\,\partial_z\phi, \quad T_{\bar{z}\bar{z}} = \partial_{\bar{z}}\phi\,\partial_{\bar{z}}\phi, \quad T_{z\bar{z}} = 0. \tag{17.45}$$

T_{zz} is the e.m. tensor of a chiral scalar, defined by $\partial_{\bar{z}}\phi = 0$. The Lagrangian formulation of a chiral scalar is a longstanding problem, and we will make do without entering into it.

Let us consider next the odd parity correlator of the e.m. tensor. As we have seen in Chap. 7, the most general expression $\mathcal{T}^{\rm odd}_{\mu\nu\rho\sigma}(x)$ linear in the antisymmetric tensor $\varepsilon_{\alpha\beta}$ with the right dimensions, which is symmetric and traceless in μ,ν and ρ,σ separately, is symmetric in the exchange $(\mu,\nu) \leftrightarrow (\rho,\sigma)$, and is conserved, is

$$\mathcal{T}^{\rm odd}_{\mu\nu\rho\sigma}(x) = \frac{e}{2}\left(\varepsilon_{\alpha\mu}T^{\alpha}{}_{\nu\rho\sigma}(x) + \varepsilon_{\alpha\nu}T_{\mu}{}^{\alpha}{}_{\rho\sigma}(x) + \varepsilon_{\alpha\rho}T_{\mu\nu}{}^{\alpha}{}_{\sigma}(x) + \varepsilon_{\alpha\rho}T_{\mu\nu\rho}{}^{\alpha}(x)\right) \tag{17.46}$$

where e is a constant to be determined, and

$$T_{\mu\nu\rho\sigma} = \frac{1}{x^4}\left(I_{\mu\rho}(x)I_{\nu\sigma}(x) + I_{\mu\sigma}(x)I_{\nu\rho}(x) - \eta_{\mu\nu}\eta_{\rho\sigma}\right), \tag{17.47}$$

is proportional to the parity even 2-point function. The correlators (17.41) and (17.46) can be regularized, without reference to a specific model, by means of the differential regularization. For (17.41), the result is the conservation law

$$\partial_x^\mu \langle T_{\mu\nu}(x)T_{\lambda\rho}(y)\rangle = 0 \tag{17.48}$$

and the anomalous conformal Ward identity

$$\left\langle 0|T_\mu^\mu(x)\,T_{\rho\sigma}(y)\,|0\right\rangle = c\frac{\pi}{6}\left(\partial_\rho\partial_\sigma - \eta_{\rho\sigma}\Box\right)\delta^2(x-y), \tag{17.49}$$

This corresponds to the trace anomaly

$$\langle\!\langle T_\mu^\mu \rangle\!\rangle = c\frac{\pi}{12}(\partial_\mu\partial_\nu - \eta_{\mu\nu}\Box)h^{\mu\nu}, \tag{17.50}$$

which coincides with the lowest contribution of the expansion in h of the Ricci scalar, i.e.

$$\langle\!\langle T_\mu^\mu \rangle\!\rangle = c\frac{\pi}{12}R = \frac{\mathbf{c}}{48\pi}R. \tag{17.51}$$

For $\mathbf{c} = 1$ this is the anomaly of a boson, see (17.38).

On the other hand, the regularization of (17.46) leads to anomalous Ward identities, with

$$\mathcal{A}^{(0)}_\xi = -\frac{\pi e}{24}\int d^2x\,\xi^\nu\varepsilon_{\nu\alpha}\partial^\alpha\left(\eta_{\rho\sigma}\Box - \partial_\rho\partial_\sigma\right)h^{\sigma\rho} \tag{17.52}$$

and

$$T_{\omega}^{(0)} = -\frac{\pi e}{12} \int d^2x\, \omega\, \varepsilon_{\lambda\alpha} \partial^\alpha \partial_\rho h^{\lambda\rho} \tag{17.53}$$

The first corresponds to the consistent diffeomorphism anomaly

$$\langle\!\langle \nabla^\mu T_{\mu\nu}(x) \rangle\!\rangle = \frac{\pi e}{12} \frac{\varepsilon^{\nu\alpha}}{\sqrt{g}} \partial_\mu \partial_\alpha \Gamma^\alpha_{\rho\nu} \approx \frac{\pi e}{24} \varepsilon_{\alpha\rho} \partial^\alpha (\Box h^\rho_\nu - \partial_\nu \partial_\sigma h^{\rho\sigma}) + \mathcal{O}\left(h^2\right), \tag{17.54}$$

whose integrated version in compact form can be written

$$\frac{\pi e}{12} \int d^2x \sqrt{g}\, \mathrm{Tr}\big(d\,\Xi\, \Gamma\big) \tag{17.55}$$

where Ξ is the matrix with components $\Xi_\rho{}^\sigma = \partial_\rho \xi^\sigma$ and Γ is the matrix one-form $\Gamma^\rho_{\mu\sigma} dx^\mu$.

We have seen in Chap. 7 that the anomaly (17.54) is generated by Weyl fermions coupled to a metric. Now the important question is: what are the elementary fields that can generate this anomaly beside the Weyl spinors? and are they present in worldsheet formulation of string theories?

Equation (17.46) can be regarded as the two-point function of the odd e.m. tensor

$$T^{\mathrm{odd}}_{\mu\nu} = \frac{1}{2}\left(\varepsilon_{\lambda\mu} T^\lambda{}_\nu + \varepsilon_{\lambda\nu} T_\mu{}^\lambda\right) \tag{17.56}$$

and the even one. Equation (17.56) implies

$$T^{\mathrm{odd}}_{11} = -2T_{12}, \quad T^{\mathrm{odd}}_{22} = 2T_{12}, \quad T^{\mathrm{odd}}_{12} = T_{11} - T_{22} \tag{17.57}$$

and

$$T^{\mathrm{odd}}_{zz} = -iT_{zz}, \quad T^{\mathrm{odd}}_{\bar{z}\bar{z}} = iT_{\bar{z}\bar{z}}, \quad T^{\mathrm{odd}}_{z\bar{z}} = 0 \tag{17.58}$$

Let us apply these definitions to the scalar boson, Eq. (17.6). We find

$$T^{\mathrm{odd}}_{11} = -\partial_1\phi\partial_2\phi, \quad T^{\mathrm{odd}}_{22} = \partial_1\phi\partial_2\phi, \quad T^{\mathrm{odd}}_{12} = \frac{1}{2}\left(\partial_1\phi\partial_1\phi - \partial_2\phi\partial_2\phi\right) \tag{17.59}$$

which gives again (17.58). This can be thought of as the e.m. tensor of a 'pseudo-scalar' field. To see this consider that under a parity operation, $x^2 \to -x^2$ a scalar field does not transform, so that $T_{11} \to T_{11}$, $T_{22} \to T_{22}$ and $T_{12} \to -T_{12}$, while T^{odd} has an opposite transformation law. There are no such pseudo-scalar fields in the formulation of string theories.

The gist of this discussion is that the anomaly (17.54) cannot appear in bosonic string theory or in the bosonic sector of a superstring theory, because this would require the presence of fields with exotic properties such as those of (17.57) or

(17.58). Below we shall see that this anomaly cannot come from the FP ghost fields. The anomaly (17.54) can appear only in the fermion sector, as we shall see hereafter.

17.1.5 The Case of Fermions

The interpretation of the odd parity anomaly obtained above by regularizing the odd parity two-point correlator is evident if we consider the two-point correlator of the e.m. tensor of a Weyl fermion. The anomaly calculation is straightforward (see Sect. 7.2.3). In particular, the even trace anomaly is

$$\langle\!\langle T_\mu^{(R)\mu} \rangle\!\rangle = \frac{1}{48\pi} R. \tag{17.60}$$

while the odd diffeomorphism anomaly is

$$\mathcal{A}_\xi^{(R)} = \frac{1}{96\pi} \int d^2x \sqrt{g}\, \mathrm{Tr}\big(d\,\Xi\,\Gamma\big) \tag{17.61}$$

which implies $e = \frac{1}{4\pi^2}$ for a complex Weyl fermion. We recall that the latter is accompanied by a trace partner.

17.1.6 2d String Anomalies

In this section, we apply the previous results to string theory. Worldsheet string theory is different from most field theories in that it is required to be free of any anomaly, including the trace anomalies. Worldsheet conformal invariance is in fact a basic symmetry of any string theory.

Bosonic String Theory

The open bosonic string theory is a 2d theory of D scalars, each representing a coordinate in $(1, D-1)$ spacetime. They can be chosen to be holomorphic (they are defined in the upper half plane). As we have seen, scalars can be freed from diffeomorphism anomalies. However, their trace anomaly is still there with D units of central charge. This trace anomaly is canceled in $D = 26$ dimensions by the Faddeev-Popov ghosts b and c, whose central charge is -26. The FP ghosts are anticommuting scalars, and they do not carry diffeomorphism anomalies (see below).

Closed bosonic string theory corresponds to two copies of open string theories, one holomorphic and the other antiholomorphic. There are no diffeomorphism anomalies but there are trace anomalies that can be canceled by two sets of holomorphic and antiholomorphic b, c ghosts.

Superstring Theories
In the open superstring case, beside the D holomorphic scalars, we have in addition D Majorana-Weyl (holomorphic) fermions, which carry central charge 1/2 each. Therefore, in total we have $\frac{3}{2}D$ units of central charge **c**. If $D = 10$, these are canceled by the anticommuting b, c ghosts with central charge -26, and the commuting β, γ ghosts with central charge 11. Now we have to verify that also the diffeomorphism anomalies are absent. The scalars do not pose any problem as we have seen above. Fermions in principle carry diffeomorphism anomalies. Therefore, the next task is to prove that the latter vanishes. To prove it, we show that they are pure gauge effects and vanish in the conformal gauge.

Conformal Tensor Calculus
In string theory, it is customary to make a partial gauge fixing by choosing a fiducial metric, the conformal one:

$$\hat{h}_{\alpha\beta} = \frac{e^{\phi}}{2}\eta_{\alpha\beta} \tag{17.62}$$

where $\phi(z, \bar{z})$ is an arbitrary function; e^{ϕ} is often referred to as the conformal factor. The complex coordinates z and $\bar{z}$ are defined by $z = \frac{1}{\sqrt{2}}(x^1 + ix^2), \bar{z} = \frac{1}{\sqrt{2}}(x^1 - ix^2)$, where x^1, x^2 are flat Euclidean coordinates.

The conformal gauge (17.62) allows for a remarkable simplification of notation. Let us set

$$\hat{h}_{\alpha\beta} = \frac{e^{\phi}}{2}\begin{pmatrix} 0 & 1 \\ 1 & 0 \end{pmatrix}, \quad \text{i.e.} \quad \hat{h}_{zz} = \hat{h}_{\bar{z}\bar{z}} = 0, \quad \hat{h}_{z\bar{z}} = h_{\bar{z}z} = \frac{e^{\phi}}{2} \tag{17.63}$$

where α and β can be both z and $\bar{z}$. Then

$$ds^2 = \hat{h}_{\alpha\beta}dz^{\alpha}dz^{\beta} = e^{\phi}\,dzd\bar{z}$$

We remark that under a holomorphic transformation $z \to f(z)$, we have $e^{\phi} \to |f'(z)|^2 e^{\phi}$, so that the gauge is preserved. Lowering and raising of the indices take the following form. Let V_{α} be the intrinsic component of a 1-differential:

$$V_{\alpha} = \hat{h}_{\alpha\beta}V^{\beta}, \quad V_z = \frac{e^{\phi}}{2}V^{\bar{z}}, \quad V_{\bar{z}} = \frac{e^{\phi}}{2}V^z$$

Covariant derivatives take a simplified form. Let us consider $\hat{\nabla}_{\alpha}V_{\beta} = \partial_{\alpha}V_{\beta} - \hat{\Gamma}^{\gamma}_{\alpha\beta}V_{\gamma}$. It is easy to prove that the only non-trivial components of the Christoffel symbols $\hat{\Gamma}^{\gamma}_{\alpha\beta}$ are

$$\hat{\Gamma}^{z}_{zz} = \partial_z\phi, \quad \hat{\Gamma}^{\bar{z}}_{\bar{z}\bar{z}} = \partial_{\bar{z}}\phi \tag{17.64}$$

Using this result, the following formulas for the components of an n-th order symmetric tensor V are easy to prove

$$\hat{\nabla}_z V_{zz\ldots z} = (\partial_z - n\partial_z\phi)V_{zz\ldots z}$$
$$\hat{\nabla}_{\bar{z}} V_{zz\ldots z} = \partial_{\bar{z}} V_{zz\ldots z}$$
$$\hat{\nabla}_z V^{zz\ldots z} = (\partial_z + n\partial_z\phi)V^{zz\ldots z}$$
$$\hat{\nabla}_{\bar{z}} V^{zz\ldots z} = \partial_{\bar{z}} V^{zz\ldots z}$$

Moreover, we can derive

$$[\hat{\nabla}_{\bar{z}}, \hat{\nabla}_z]V_{zz\ldots z} = -4n\,\mathrm{e}^{\phi}\,R^{(2)}V_{zz\ldots z}, \quad R^{(2)} = 4\mathrm{e}^{-\phi}\partial_z\partial_{\bar{z}}\phi \tag{17.65}$$

where $R^{(2)}$ is the scalar curvature of the conformal metric.

We have also $\varepsilon_{ab}dx^a \wedge dx^b = \varepsilon_{\alpha\beta}dz^\alpha \wedge dz^\beta$, with $1 = \varepsilon_{12} = -i\varepsilon_{z\bar{z}} = i\varepsilon_{\bar{z}z}$.

Anomalies in Conformal Gauge

Let us consider the anomaly (17.61) in conformal gauge. The integrand is proportional to

$$\mathrm{tr}\,(\Xi d\Gamma) = \partial_\alpha\xi^\beta\,\varepsilon^{\mu\nu}\partial_\mu\Gamma^\alpha_{\nu\beta}$$

Since the only two non-vanishing components of the Christoffel symbols are (17.64), we have

$$\varepsilon^{\mu\nu}\partial_\mu\Gamma^\alpha_{\nu\beta} = \varepsilon^{z\bar{z}}\partial_z\Gamma^{\bar{z}}_{\bar{z}\bar{z}} + \varepsilon^{\bar{z}z}\partial_{\bar{z}}\Gamma^z_{zz} = \varepsilon^{z\bar{z}}\partial_z\partial_{\bar{z}}\phi + \varepsilon^{\bar{z}z}\partial_{\bar{z}}\partial_z\phi = 0 \tag{17.66}$$

Therefore, in conformal gauge there are no diffeomorphism anomalies. In other words, the diffeomorphism anomaly (17.61) is a pure gauge artifact.

This is not the end of the story, because we have recalled several times that the diffeomorphism anomaly (17.61) is accompanied by a Weyl partner T_ω, see (17.53), with opposite sign and half coefficient. However, this is also a pure gauge construct

$$T_\omega \sim \int d^2x\,\omega\,\varepsilon_{\lambda\sigma}\partial^\sigma\partial_\rho h^{\lambda\rho} = \int d^2x\,\omega\,\left(\varepsilon_{z\bar{z}}\partial^{\bar{z}}\partial_{\bar{z}}h^{z\bar{z}} + \varepsilon_{\bar{z}z}\partial^z\partial_z h^{\bar{z}z}\right) \tag{17.67}$$
$$= \frac{1}{2}\int d^2x\,\omega\,\mathrm{e}^{-\phi}\left(\varepsilon_{z\bar{z}}\partial_z\partial_{\bar{z}}h^{z\bar{z}} + \varepsilon_{\bar{z}z}\partial_{\bar{z}}\partial_z h^{\bar{z}z}\right) = 0$$

Once this is clarified let us proceed with the discussion of the anomalies in superstring theories. The b, c and β, γ ghosts arise in worldsheet string theory as a consequence of the gauge fixing, i.e. the choice of the conformal gauge and the analogous choice in superstring theory. Therefore, they cannot carry diffeomorphism anomalies because the only diffeomorphism anomaly in 2d is the one given by formula (17.61), which vanishes in conformal gauge together with its trace partner. Therefore, the overall diffeomorphism anomaly of the open superstring formulated in terms of FP ghost in the conformal gauge is 0, and so the open superstring is free of worldsheet anomalies.

The IIA and IIB theories split into two (holomorphic and antiholomorphic) sectors. In each of them, we have the same situation just described for the open superstring, thus not only holomorphic and antiholomorphic diffeomorphism anomalies are separately gauge artifacts, but they cancel exactly because they have opposite signs.

In the case of heterotic superstrings, the splitting is in two different sectors: one antiholomorphic is a copy of the open superstring (with b, c and β, γ ghosts), the other contains 10 chiral boson plus 32 holomorphic Weyl-Majorana fermions, or 16 compactified bosons (with b, c ghosts). Therefore, not only the total central charge is 0 in $D = 10$, but also the diffeomorphism anomalies are absent (being gauge artifacts).

What remains for us to do is prove that ghosts have the above-declared central charges.

b, c and β, γ Systems

Let us start with the b, c ghosts. The action is

$$S_{bc} = -\frac{1}{2}\int d^2z\sqrt{h}\, h^{\alpha\beta} c^{\gamma}\nabla_{\alpha} b_{\beta\gamma} \tag{17.68}$$

which is more usefully written as

$$S_{bc} = \frac{1}{2}\int d^2z\sqrt{h}\, h^{\alpha\beta}\left(2\nabla_{\alpha} c^{\gamma} b_{\beta\gamma} + c^{\gamma}\nabla_{\alpha} b_{\beta\gamma}\right) \tag{17.69}$$

c has conformal weight -1 (it is vector field), b has conformal weight 2 (a symmetric quadratic differential). The eom's are

$$\nabla_{\alpha} c^{\gamma} = 0, \quad \nabla_{\alpha} b_{\beta\gamma} = 0 \tag{17.70}$$

Taking the variation of (17.69) with respect to the metric we can derive the ghost energy-momentum tensor.

$$T^{(bc)}_{\alpha\beta} = \frac{1}{2}\left(2\nabla_{\alpha} c^{\gamma} b_{\beta\gamma} + c^{\gamma}\nabla_{\alpha} b_{\beta\gamma} + (\alpha \leftrightarrow \beta)\right) \tag{17.71}$$

After fixing the conformal gauge the two holomorphic and antiholomorphic components are:

$$T^{(bc)}_{zz} = c^{z}\partial_{z} b_{zz} + 2\left(\partial_{z} c^{z}\right) b_{zz} \tag{17.72}$$
$$T^{(bc)}_{\bar{z}\bar{z}} = c^{\bar{z}}\partial_{\bar{z}} b_{\bar{z}\bar{z}} + 2\left(\partial_{\bar{z}} c^{\bar{z}}\right) b_{\bar{z}\bar{z}} \tag{17.73}$$

Moreover

$$T^{(bc)}_{z\bar{z}} = \frac{1}{2}\left(2\partial_z c^{\bar{z}} b_{\bar{z}\bar{z}} + c^{\bar{z}}\partial_z b_{\bar{z}\bar{z}} + 2\partial_{\bar{z}} c^z b_{zz} + c^z \partial_{\bar{z}} b_{zz}\right) \tag{17.74}$$

which vanishes on shell. The propagators are

$$\langle c(z)b(w)\rangle = \frac{1}{2\pi}\ln(z-w), \quad \langle c(z)c(w)\rangle = 0 = \langle b(z)b(w)\rangle \tag{17.75}$$

It is easy to compute the two-point function

$$\langle T^{(bc)}_{zz}(z)T^{(bc)}_{zz}(0)\rangle = -\frac{13}{4\pi^2}\frac{1}{z^4} \tag{17.76}$$

which means that the normalized central charge is $\mathbf{c} = -26$. Similarly,

$$\langle T^{(bc)}_{\bar{z}\bar{z}}(\bar{z})T^{(bc)}_{\bar{z}\bar{z}}(0)\rangle = -\frac{13}{4\pi^2}\frac{1}{\bar{z}^4} \tag{17.77}$$

while the correlators involving $T^{(bc)}_{z\bar{z}}$ vanish. The overall two-point correlator is

$$\langle T^{(bc)}_{\mu\nu}(x)T^{(bc)}_{\lambda\rho}(0)\rangle = -\frac{13}{x^4}\left(I_{\mu\rho}(x)\, I_{\nu\sigma}(x) + I_{\nu\rho}(x)\, I_{\mu\sigma}(x) - \eta_{\mu\nu}\eta_{\rho\sigma}\right) \tag{17.78}$$

which can be regularized with the differential regularization and leads to a trace anomaly

$$T_\omega = -\frac{13}{24\pi}R. \tag{17.79}$$

A β, γ system is similar to a b, c system, except that the conformal weights are $\frac{3}{2}, -\frac{1}{2}$ and are commuting spinors. The e.m. tensor is

$$T^{(\beta\gamma)}_{zz} = \frac{1}{2}\gamma\, \partial_z\beta_z + \frac{3}{2}(\partial_z\gamma)\,\beta_z \tag{17.80}$$

$$T^{(\beta\gamma)}_{\bar{z}\bar{z}} = \frac{1}{2}\gamma\, \partial_{\bar{z}}\beta_{\bar{z}} + \frac{3}{2}(\partial_{\bar{z}}\gamma)\,\beta_{\bar{z}} \tag{17.81}$$

Repeating the same calculation as above the central charge is 11.

We repeat again that b, c and β, γ ghosts cannot carry diffeomorphism anomalies, because the only odd diffeomorphism anomaly in 2d is of the type(17.61), which, in the conformal gauge, vanishes identically. Of course, in principle, one could choose a gauge different from the conformal one. This would require a different approach to anomalies and a different, presumably more complicated, verification that they vanish.

17.2 Target Space Anomalies of the Superstrings

The zero mass spectrum of superstring theories contain chiral fields. Therefore, the corresponding target space field theories may be anomalous. For a superstring theory to be consistent, we require not only that the worldsheet anomalies (trace and diffeomorphism) vanish, but also that the chiral consistent target space anomalies cancel. As requested by the vanishing of the worldsheet central charge, superstrings propagate consistently only in a ten-dimensional spacetime. Since the massless superstring spectrum contains gravity, one of the compulsory symmetries is covariance under diffeomorphisms. The relevant group to be considered is $\mathsf{GL}(10)$ or its maximal compact subgroup $\mathsf{SO}(10)$ and the relevant ad-invariant polynomial is of order 6 in R, where R is the Riemann or spin curvature. The corresponding anomaly is pulled back from this 12-form in the classifying space to the physical spacetime after transgressing it, in the usual way. Now an irreducible polynomial P_6 for $\mathsf{SO}(10)$ exists and is not 0. There are also reducible polynomials $P_2(R^2)P_4(R^4)$ and $(P_2(R^2))^3$. Therefore, diffeomorphism anomalies may exist in a ten-dimensional spacetime, and the absence of anomalies can be achieved only by cancelation between different species (case (B)).

Besides diffeomorphisms we have to take care also of possible gauge anomalies in superstring theories carrying a gauge symmetry. This is the case for heterotic and type I superstrings. The anomalies may come from the polynomial $P_6(F^6)$, where F is the gauge curvature. Such an irreducible polynomial exists and is non-vanishing for $\mathsf{SO}(32)$. But for the same group, there are also reducible polynomials $P_2(F^2)P_4(F^4)$ and $(P_2(F^2))^3$. The last one exists also for the group E_8.

Besides the pure gauge and diffeomorphism anomalies, we have to consider also mixed anomalies. Indeed we have the following 12-form $P_2(R^2)(P_2(F^2))^2$, $P_4(R^4)P_2(F^2)$ for both $\mathsf{SO}(32)$ and E_8, and also $P_2(R^2)P_4(F^4)$ for the former.

Hereafter we intend to show how the just mentioned anomalies cancel in the target space field theory of the superstring massless modes. We shall use the index theorem for families (Chap. 12) and show that for these theories all obstructions implied by this theorem cancel out, except for some mixed anomalies to be treated separately with the Green-Schwarz mechanism. Therefore, the cancelations will take place at the level of classifying spaces. These are well-known results. We report them here because they are beautiful examples of the cancelation between different species of chiral fields. For explicit index formulas, we refer to (12.47) and (12.48) and Appendix 17A.

17.3 Partial Anomaly Cancelation for SO(32)

Type I and Heterotic $\mathsf{SO}(32)$ have the same massless spectrum. As far as anomalies are concerned, they contain a left-handed gravitino (spin 3/2) and a right-handed dilatino (spin 1/2), the superpartner of the graviton and the dilaton, respectively.

They are spinors belonging to the 144 and 16 (or conjugate) representations of SO(10). Moreover, there are $n = 496$ left-handed gauginos (the superpartners of the gauge bosons) transforming according to the adjoint representation of SO(32). Let us denote by $P_{\frac{3}{2}}(R)$, $P_{\frac{1}{2}}(R)$ and $P_{\frac{1}{2}}(R, F)$ the corresponding ad-invariant polynomials. As we have pointed out in Appendix 5A, the overall coefficient in front of each polynomial depends on the corresponding representation. So, for instance, the components of $P_{\frac{3}{2}}(R)$ and $P_{\frac{1}{2}}(R)$ are the same, but with different coefficients. The precise expressions are

$$P_{\frac{3}{2}}(R) = \frac{1}{(2\pi)^6}\frac{1}{2^7}\left[-\frac{495}{5670}\mathrm{tr}R^6 + \frac{225}{4320}\mathrm{tr}R^2\,\mathrm{tr}R^4 - \frac{63}{10368}(\mathrm{tr}R^2)^3\right] \tag{17.82}$$

$$P_{\frac{1}{2}}(R) = \frac{1}{(2\pi)^6}\frac{1}{2^7}\left[\frac{1}{5670}\mathrm{tr}R^6 + \frac{1}{4320}\mathrm{tr}R^2\,\mathrm{tr}R^4 + \frac{1}{10368}(\mathrm{tr}R^2)^3\right] \tag{17.83}$$

$$\begin{aligned} P_{\frac{1}{2}}(R, F) = \frac{1}{(2\pi)^6}\Bigg\{&\frac{n}{2^7}\left[\frac{1}{5670}\mathrm{tr}R^6 + \frac{1}{4320}\mathrm{tr}R^2\,\mathrm{tr}R^4 + \frac{1}{10368}(\mathrm{tr}R^2)^3\right] \\ &+\frac{1}{2}\left(-\frac{1}{720}\mathrm{Tr}F^6 + \frac{1}{1152}\mathrm{Tr}F^4\,\mathrm{tr}R^2 - \frac{1}{2}\mathrm{Tr}F^2\left(\frac{1}{5760}\mathrm{tr}R^4 + \frac{1}{4608}(\mathrm{tr}R^2)^2\right)\right)\Bigg\} \end{aligned} \tag{17.84}$$

where tr denotes the trace in the fundamental representation and Tr the trace in the adjoint representation of the corresponding group. In the case of SO(N), we have

$$\begin{aligned} \mathrm{Tr}F^2 &= (\mathrm{N}-2)\mathrm{tr}F^2 \\ \mathrm{Tr}F^4 &= (\mathrm{N}-8)\mathrm{tr}F^4 + 3(\mathrm{tr}F^2)^2 \\ \mathrm{Tr}F^6 &= (\mathrm{N}-32)\mathrm{tr}F^6 + 15\mathrm{tr}F^4\,\mathrm{tr}F^2 \end{aligned} \tag{17.85}$$

Now, inserting this in (17.84), the relevant polynomial for Type I and Heterotic SO(32) superstring is

$$\begin{aligned} &P_{\frac{3}{2}}(R) - P_{\frac{1}{2}}(R) + P_{\frac{1}{2}}(R, F) \\ &= \frac{1}{(2\pi)^6}\left(\mathrm{tr}R^2 - \mathrm{tr}F^2\right)\left(\frac{1}{96}\mathrm{tr}F^4 + \frac{1}{768}\mathrm{tr}R^4 - \frac{1}{768}\mathrm{tr}F^2\,\mathrm{tr}R^2 + \frac{1}{3072}(\mathrm{tr}R^2)^2\right) \\ &= \frac{1}{(2\pi)^6}\frac{1}{3\times 2^8}\left(\mathrm{tr}R^2 - \frac{1}{30}\mathrm{Tr}F^2\right) \\ &\quad\times\left(\frac{1}{3}\mathrm{Tr}F^4 - \frac{1}{300}(\mathrm{Tr}F^2)^2 + \mathrm{tr}R^4 - \frac{1}{30}\mathrm{Tr}F^2\mathrm{tr}R^2 + \frac{1}{4}(\mathrm{tr}R^2)^2\right) \end{aligned} \tag{17.86}$$

This is the residual anomaly to be canceled with the Green-Schwarz mechanism, while the irreducible terms have been canceled.

17.4 Partial Anomaly Cancelation for $\mathsf{E}_8 \times \mathsf{E}_8$

In the heterotic superstring with gauge group $\mathsf{E}_8 \times \mathsf{E}_8$, the only change with respect to the previous case is in $P_{\frac{1}{2}}(R, F)$. The adjoint representation of E_8 is the 248, there are no irreducible polynomials P_4 and P_6 and we have

$$\mathrm{Tr}\, F^4 = \frac{1}{100} \mathrm{tr}\, F^2, \quad \text{and} \quad \mathrm{Tr}\, F^6 = \frac{1}{7200} (\mathrm{tr}\, F^2)^3 \tag{17.87}$$

Therefore, the relevant polynomial reduces to

$$\begin{aligned} & P_{\frac{3}{2}}(R) - P_{\frac{1}{2}}(R) + P_{\frac{1}{2}}(R, F) \\ &= \frac{1}{(2\pi)^6} \frac{1}{768} \left(\mathrm{tr} R^2 - \frac{1}{30} \mathrm{Tr} F^2 \right) \left(\mathrm{tr} R^4 + \frac{1}{450} (\mathrm{Tr} F^2)^2 - \frac{1}{30} \mathrm{Tr} F^2 \mathrm{tr} R^2 + \frac{1}{4} (\mathrm{tr} R^2)^2 \right) \end{aligned} \tag{17.88}$$

Also, this residual anomaly will be dealt with further on.

17.5 Anomaly Cancelation in Type II Theories

In type II superstring theories gauge symmetries are absent, we have to do only with diffeomorphisms. In what concerns anomalies the relevant massless spectrum is as follows. In type IIA theory fermions are two gravitinos (spin 3/2) and two dilatinos (spin 1/2). They are spinors belonging to the 144 and 16 ($\overline{144}$ and $\overline{16}$)representation of $\mathsf{SO}(10)$ with opposite chiralities. We know that under these conditions diffeomorphism anomalies cancel exactly.

In type IIB theory, the two gravitini and the two dilatini have the same chirality. They compose a left-handed complex gravitino and a right-handed complex dilatino, from which we expect non-trivial anomalies. But there is also another field that may give rise to an anomaly: a four-form field usually denoted by C_4, whose curvature $F_5 = dC_4$ is self-dual: $F_5 = *F_5$. This is not a fermion, but it is a very close relative of fermions. To see this one can saturate the indices of F_5 with five γ matrices and form $\not{F}_5$; then verify that, being F_5 self-dual, $\not{F}_5$ is an eigenstate of the chirality matrix γ_{11}. The kinetic operator of this field gives rise to an index theorem. The corresponding obstructions are encoded in the polynomial poly-form $P_{sd}(R)$, given by

$$P_{sd}(R) = \frac{1}{(2\pi)^6} \frac{1}{2^6} \left[-\frac{496}{5670} \mathrm{tr} R^6 + \frac{224}{4320} \mathrm{tr} R^2\, \mathrm{tr} R^4 - \frac{64}{10368} (\mathrm{tr} R^2)^3 \right] \tag{17.89}$$

Comparing with (17.82) and (17.83), we find

$$P_{sd}(R) - 2P_{\frac{3}{2}}(R) + 2P_{\frac{1}{2}}(R) = 0 \tag{17.90}$$

(the factors 2 is because the fermions are complex). This means that in type IIB superstring diffeomorphism anomalies cancel out exactly.

17.6 The Green-Schwarz Cancelation Mechanism

In type I and heterotic target space theories, there remain uncanceled obstructions represented by Eqs. (17.86) and (17.89). The uncanceled polynomials can be written in both cases (up to numerical coefficients) as

$$P_{\rm res}(R, F) = ({\rm tr}R^2 - {\rm tr}F^2)X_8(R, F) \tag{17.91}$$

where $X_8(R, F)$ is an invariant polynomial of R and F. Upon transgression $P_{res}(R, F)$ gives rise to the anomaly

$$\mathcal{A}_{\rm res} = \int_{\mathsf{T}} \left(\Delta^1_{2L} - \Delta^1_{2G}\right) X_8(R, F) \tag{17.92}$$

where

$$\Delta^1_{2L} = {\rm tr}(d\Lambda\,\omega), \quad \Delta^1_{2G} = {\rm tr}(dc\,A) \tag{17.93}$$

have the same form as the 2d Lorentz and gauge anomalies (Λ and c are the corresponding parameters, and ω and A the respective connections). This anomaly can be canceled by subtracting a counterterm, [3]

$$\mathcal{C} = \int_{\mathsf{T}} B\, X_8(R, F) \tag{17.94}$$

provided we assume that the two-form field B, under an overall gauge and Lorentz transformation, transforms as

$$\delta B = \Delta^1_{2L} - \Delta^1_{2G} \tag{17.95}$$

where $\delta \equiv \delta_\Lambda + \delta_c$. Δ^1_{2L} and Δ^1_{2G} have the form of the two-dimensional anomalies generated via transgression from the polynomials $K = P_2$ in the Lorentz and gauge groups, respectively. The field B is identified with the massless two-form field that is part of the gravity spectrum of any (closed) string theory. Its curvature is usually denoted by $H = dB$. It is clear that here we have to modify this definition as follows

$$H = dB + \Delta^0_{3L} - \Delta^0_{3G} \tag{17.96}$$

where $\Delta^0_{3L} = TK(\omega) = \mathrm{tr}\big(\omega(d\omega + \frac{1}{3}[\omega,\omega])\big)$ and $\Delta^0_{3G} = TK(A) = \mathrm{tr}\big(A(dA + \frac{1}{3}[A,A])\big)$ are the Chern-Simons terms that give rise to the two anomalies, and assume $\delta H = 0$.

But, then, differentiating (17.96) we get

$$dH = \mathrm{tr}R^2 - \mathrm{tr}F^2 \tag{17.97}$$

which implies that the Pontryagin class of T must equal the second Chern class of the gauge bundle over it.

The last condition is, however, only a necessary condition for the anomaly cancelation. From a geometrical point of view, there are more stringent constraints. The form $TK(A)$ in $\mathsf{P}(\mathsf{T},\mathsf{G})$ restricted to the fiber reduces to the non-trivial generator of the third cohomology of $\mathsf{G} = \mathsf{SO}(32)$ or $\mathsf{E}_8 \times \mathsf{E}_8$. Analogously the form $TK(\omega)$ restricted to the fiber of $\mathsf{O}_g\mathsf{T}$ reduces to the non-trivial generator of the third cohomology group of $SO(10)$. Therefore, $\mathrm{tr}R^2$ and $\mathrm{tr}F^2$ cannot be separately exact. The only possibility to satisfy (17.97) is the so-called embedding of the orthogonal bundle $\mathsf{O}_g\mathsf{T}$ into the gauge bundle $\mathsf{P}(\mathsf{T},\mathsf{G})$. This means that there exists a bundle morphism $\mathsf{O}_g\mathsf{T} \to \mathsf{P}(\mathsf{T},\mathsf{G})$, and that in $\mathsf{P}(\mathsf{T},\mathsf{G})$, there is a connection A_ω reducible to ω. A direct implication is that it should be possible to lift the diffeomorphisms of T to P. This imposes conditions on the geometry of the latter.

These constraints should be understood in the appropriate way. As pointed out in the introduction of this chapter the just considered target space approach to superstring anomalies is not the only one. On shell quantum (super)string theories can be defined as perturbative series whose terms are amplitudes of vertex operators each one of them representing a (super)string excitation. In this framework, the crucial amplitudes for type I (open) superstring are the exagon ones, with six external gauge or gravity lines, inserted at the worldsheet boundary. From the worldsheet point of view, the contributions come from the annulus (with two boundaries) and the Moebius strip (with one single boundary). The annulus amplitudes can be planar, with vertex insertions along the same boundary, or non-planar, with insertions on both boundaries. Green and Schwarz have proved, in the $\mathsf{SO(N)}$ gauge case, that the non-planar amplitudes are not anomalous, [4]. The planar and non-orientable (Moebius) amplitudes do carry anomalous contributions, but in the case of $\mathsf{N} = 32$ the latter cancel each other. Therefore, the constraints met in the target space approach must be understood not as impossibility to cancel the anomaly (17.92) but as limitations on the geometry of the admissible target spaces. Although it is too hard to explain in detail the Green-Schwarz mechanism and the rather surprising transformation property of the field B, (17.95), the worldsheet approach provides suggestions of how this field can enter the game. For, in the annulus geometry, not only open strings propagate but also closed strings do, and the B field is a closed string excitation. From this, one can see that string dynamics is much richer than field theory dynamics and that, in order to encode the former in the latter a sort of 'squeezing' of ordinary field theory concepts is inevitable.

Other limitations of the target space and a different understanding of the Green-Schwarz mechanism, we will meet by studying strings from the sigma model point of view.

17.7 Worldsheet Geometry of the String

In order to examine more in depth the problems raised by the previous sections we will now return to the actions (17.1), (17.2) and (17.3) and present them in a more geometrical language. String theories are theories of embeddings of the worldsheet in spacetime and their supersymmetric partners. The dynamical fields are X^μ, ψ^μ and λ^A, where X^μ are the local coordinates of the embeddings $\phi : \Sigma \to \mathsf{T}$, i.e. $X^\mu(x) \equiv X^\mu(\phi(x))$. The above-mentioned actions are invariant under the following group of transformations:

- The diffeomorphisms of Σ, which form the group $\mathsf{Diff}\,\Sigma$, as these actions are constructed in an evidently covariant way in the worldsheet;
- Weyl rescaling of the metric $h_{\alpha\beta}$, i.e. $h_{\alpha\beta} \to \mathrm{e}^{2\omega} h_{\alpha\beta}$;
- The vertical automorphisms $\mathsf{Aut_v C}\Sigma$ and $\mathsf{Aut_v O_h}\Sigma$.

The last point deserves a detailed explanation. $\mathsf{Aut_v C}\Sigma$ is the group of vertical automorphisms of the principal bundle $\mathsf{C}\Sigma$ of complex structures of Σ. $\mathsf{Aut_v O_h}\Sigma$ is the group of vertical automorphisms of the bundle of orthogonal frames $\mathsf{O_h}\Sigma$ over Σ, whose structure group is $SO(1,1)$ in the Minkowski case and $SO(2)$ when the metric h is Euclidean. As usual, when discussing the geometrical aspects of anomalies we shall stick to Euclidean metrics. The bundle of orthogonal frames has been introduced and discussed at length before, and, in this respect, there is nothing new to be said about $\mathsf{O_h}\Sigma$ and its automorphisms. The bundle $\mathsf{C}\Sigma$ has not yet been introduced.

The worldsheet Σ is an orientable two-dimensional manifold. Therefore, it admits complex structures. A complex structure is a linear map $J_x : T_x\Sigma \to T_x\Sigma$ such that $J_x^2 = -1$, varying smoothly from point to point $x \in \Sigma$. A complex frame is a frame defined at every $x \in \Sigma$ by two tangent vectors X and $J_x X$, where $X \in T_x\Sigma$. A complex structure is compatible with the metric h if

$$h(J_x X_1, J_x X_2) = h(X_1, X_2) \quad \forall X_1, X_2 \in T\Sigma$$

Therefore, a complex frame satisfies $h(X, JX) = 0$ and $h(X, X) = h(JX, JX)$; in other words, it is formed by vectors which are orthogonal and of fixed length. It follows that the orthonormal frame bundle $\mathsf{O_h}\Sigma$ is a subbundle of $\mathsf{C}\Sigma$. Moreover, if h' is another compatible metric, then h and h' are conformally related, i.e. $h' = \mathrm{e}^{\omega} h$, where $\omega : \Sigma \to \mathbb{R}$. Finally, $\mathsf{C}\Sigma$ has structure group $\mathbb{C}^*$, the Abelian multiplicative group of complex numbers with the exclusion of the origin, and is a reduced subbundle of the linear frame bundle $L\Sigma$.

The Abelian group $\mathsf{Aut_v C}\Sigma$ can be represented simply by the group $\mathsf{Map}(\Sigma, \mathbb{C}^*)$ which is the direct product of $\mathsf{Map}(\Sigma, \mathbb{R}^+)$ and $\mathsf{Map}(\Sigma, U(1)) \cong \mathsf{Map}(\Sigma, SO(2))$ (where $\cong$ means isomorphic). These two factor groups act in different ways on $\mathcal{A}_{CS}^{\text{metric}}$, the space of connections on $\mathsf{C}\Sigma$ which are reducible to connections on $O_h\Sigma$ for some metric h. A map $\psi \in \mathsf{Map}(\Sigma, SO(2))$ maps the space of connections reducible to connections in $\mathsf{O_h}\Sigma$ into itself; while a map $\mathrm{e}^{\omega} \in \mathsf{Map}(\Sigma, \mathbb{R}^+)$ transforms the same space of connections into the space of connections reducible to connections in $O_{\mathrm{e}^{\omega}h}\Sigma$.

We would like now to apply to the anomalies in $\mathsf{C}\Sigma$ the methods of Chap. 11 and the family's index theorem of Chap. 12. We know that the source of anomalies for any gauge theory is the cohomology of the classifying space. The classifying space of $\mathsf{C}\Sigma$, that is the base manifold of the universal fiber bundle $\mathsf{E}\mathbb{C}^*$ is $\mathsf{B}\mathbb{C}^* = \mathsf{BU}(1)$, that is it is the classifying space of the group $U(1) \cong SO(2)$. If $\mathbf{a}^{\mathfrak{so}(2)}$ is the universal connection and $F(\mathbf{a}^{\mathfrak{so}(2)})$ its curvature, the relevant characteristic class is specified by $P_2(F(\mathbf{a}^{\mathfrak{so}(2)}), F(\mathbf{a}^{\mathfrak{so}(2)}))$. From it, via the transgression formula, we can derive the corresponding anomaly, which is the Lorentz version of the one computed above, see (7.48, 17.55), i.e. $\sim \Lambda\, \varepsilon^{\mu\nu}\partial_\mu\omega_\nu$, where $\Lambda = \Lambda_{12} = -\Lambda_{21}$ is the $SO(2)$ gauge transformation parameter, and $\omega_\mu = \omega_\mu^{12} = -\omega_\mu^{21}$ is the spin connection. Of course, as we have explained in Chap. 11, if one uses the transgression passing through the bundle $\mathsf{L}\Sigma^+$ of oriented frames, i.e. using the diagram

$$\begin{array}{ccc} \dfrac{\mathsf{L}\Sigma^+\times\mathcal{A}^{\text{metric}}}{\mathcal{I}\left(\mathsf{Diff}_*^{\mathsf{m},1}(\Sigma)\right)} & \xrightarrow{\widehat{Ev}} & \mathsf{EGL}(2,\mathbb{R})^+ \\ \Big\downarrow \pi & & \Big\downarrow \pi \\ \dfrac{\Sigma\times\mathcal{A}^{\text{metric}}}{\mathcal{I}\left(\mathsf{Diff}_*^{\mathsf{m},1}(\Sigma)\right)} & \xrightarrow{\widehat{Ev}} & \mathsf{BGL}(2,\mathbb{R})^+ \end{array} \tag{17.98}$$

(where $\mathsf{EGL}(2,\mathbb{R})^+ \cong \mathsf{EU}(1)$, $\mathsf{BGL}(2,\mathbb{R})^+ \cong \mathsf{BU}(1)$) one obtains exactly the anomaly (7.48), the diffeomorphism or gravitational anomaly. It goes without saying, therefore, that we can connect this anomaly to the family's index theorem. On the other hand, this local Lorentz anomaly can be canceled by means of a WZ term and generate a corresponding diffeomorphism anomaly, as we have seen in Chap. 15. We have also seen that the latter vanishes in the conformal gauge.

To complete the analysis of the worldsheet anomalies let us inquire whether there is anything else to be said about the anomalies of $\mathsf{C}\Sigma$ and whether, in particular, the even trace anomaly (7.25) can be obtained in the same way as the gravitational anomaly via transgression and can feature as an obstruction thanks to the family's index theorem. This is motivated by the fact that the factor subgroup $\mathsf{Map}(\Sigma, \mathbb{R}^+)$ of $\mathsf{Aut_v C}\Sigma$ operates Weyl transformations on metrics and connections. This would seem to offer the opportunity to relate the 2d even trace anomaly to the index theorem. But this is not enough. The group $\mathbb{C}^*$ reduces to its maximal compact subgroup $SO(2)$, and the cohomology of the classifying space of $\mathsf{B}\mathbb{C}^*$ reduces to the cohomology of $\mathsf{BSO}(2) \cong \mathsf{BU}(1)$. We can view $\mathsf{C}\Sigma$ as the direct sum of $\mathsf{O_h}\Sigma$ and $\mathsf{L}\mathbb{R}$, the linear bundle with fiber $\mathbb{R}$ on which $\mathbb{R}^+$ acts. There is no class we can pull back from the universal bundle with structure group $\mathbb{R}^+$ and gauge transformations $\mathsf{Map}(\Sigma, \mathbb{R}^+)$. Therefore, no corresponding anomaly can be obtained via the transgression formula, and, accordingly, no obstruction from the family's index theorem exists.

In other words, the even trace anomaly (7.25) does not belong to the family of anomalies related to the family's index theorem and derivable via transgression from the cohomology of the classifying space. As pointed out before this holds for all even dimensions. On the other hand, in 2d there is no room for a diffeomorphism preserving odd trace anomaly, unlike in 4d. In conclusion, the anomaly panorama for the string theory worldsheet encompasses two anomalies: the gravitational anomaly (7.48) and the trace anomaly (7.25). They have both been discussed at length in the previous section.

17.8 Sigma Model Anomalies of the String

String theory can be represented as a sigma model on the background of its massless modes. Let us consider the example of closed bosonic string. Its massless modes are a metric $G_{\mu\nu}$ and an antisymmetric tensor field $B_{\mu\nu}$. The corresponding action can therefore be written as

$$S_b = \frac{1}{2\pi} \int_\Sigma d^2x \sqrt{h} \left[h^{\alpha\beta}(x) \partial_\alpha X^\mu(x) \partial_\beta X^\nu(x) G_{\mu\nu}(X) + \varepsilon^{\alpha\beta} \partial_\alpha X^\mu(x) \partial_\beta X^\nu(x) B_{\mu\nu}(X) \right] \tag{17.99}$$

Σ denotes the worldsheet, a Riemann surface, x^α are local coordinates in Σ, while X^μ ($\mu = 0, 1, \ldots, \mathsf{d}-1$) are the local coordinates in the target space T (the physical spacetime). $\varepsilon^{\alpha\beta}$ is the completely antisymmetric tensor in two dimensions. Let us remark that the action (17.99) is invariant not only under the diffeomorphisms of Σ, but also under the diffeomorphisms of T. This way of representing strings has the following remarkable property: the vanishing of its beta functions yields the equations of motion of the background fields. Here however we are interested in another aspect, the possible anomalies of the string sigma model. Therefore, we have to include also fermions and turn to superstring theories.

Superstring sigma models, [5, 6], are often more conveniently formulated in the light-cone formalism. This means that a gauge is fixed for the worldsheet diffeomorphisms so that the fields $X^\pm = X^0 \pm X^{\mathsf{d}-1}$ are expressed in terms of the transverse coordinate fields X^i, $i = 1, \ldots, \mathsf{d}-2$. This means that the temporal and longitudinal modes are eliminated, and there is no need to introduce FP ghosts. For sure, this breaks spacetime Lorentz covariance. However, the gratifying aspect is that Lorentz covariance is recovered after quantization only in the critical dimensions, $\mathsf{d} = 26$ for the bosonic string and $D = 10$ for the superstring. Below is the kinetic part of the sigma model action for the heterotic superstring:

$$S_h = \frac{1}{2\pi}\int d\tau \int_0^{2\pi} d\sigma \Big[G_{ij}(X)\partial_\alpha X^i \partial^\alpha X^j + \varepsilon^{\alpha\beta} B_{ij}\partial_\alpha X^i \partial_\beta X^j$$
$$+ G_{ij}(X)\left(\psi^i \partial_- \psi^j + \psi^i \partial_- X^k \left(\Gamma^j_{kl}(X) - H^j_{kl}(X)\right)\psi^l\right)$$
$$+ i\lambda^r \left(\delta_{rs}\partial_+ + \partial_+ X^i A^a_i T^a_{rs}\right)\lambda^s\Big] \tag{17.100}$$

$\partial_\pm$ are the 2d chiral derivatives. Here ψ^i are 8 left-handed (Majorana-Weyl) worldsheet fermions, while λ^r, $r = 1, \dots, 32$ are right-handed fermions transforming according to the fundamental representation of $SO(32)$ or to the $(1, 16) + (16, 1)$ basic representation of the $SO(16) \times SO(16)$ subgroup of $E_8 \times E_8$; T^a are the corresponding Lie algebra generators. Finally, H_{ijk} are the components of the exterior differential of the antisymmetric field B_{ij}. In (17.100), the pullback to the worldsheet of both gravitational and gauge connections (+torsion) is clearly visible.

In (17.100), only the physical degrees of freedom appear. In the calculus of anomalies preserving covariance is preferable. Therefore, we schematize the kinetic part of the previous action in covariant form as follows

$$S_f = \frac{1}{2\pi}\int_\Sigma d^2x\sqrt{h}\, i\bar{\psi}^\mu(X)\gamma^\alpha \left(\mathbf{1}\nabla_\alpha + (\phi^*\Gamma)_\alpha\right)^\lambda_\nu \psi^\nu(X) G_{\mu\lambda}(X) \tag{17.101}$$

where $\mathbf{1}^\lambda_\nu = \delta^\lambda_\nu$ and $\Gamma^\lambda_\nu = \Gamma^\lambda_{\mu\nu} dX^\mu$ is the gravitational connection in LT. In the case of the heterotic superstring we have also

$$S'_f = \frac{1}{2\pi}\int_\Sigma d^2x\sqrt{h}\, i\bar{\lambda}(x)\gamma^\alpha \left(\nabla_\alpha + (\phi^*\mathbf{a})_\alpha(X)\right)\lambda(x) \tag{17.102}$$

where γ^α are the γ-matrices in Σ, $\mathbf{a}$ is a gauge connection in P(G, T) (G is either $SO(32)$ or $E_8 \times E_8$) and $\phi^*\mathbf{a}$, $\phi^*\Gamma$ are their pullbacks to Σ. ∇ is the covariant derivative in Σ (so it may contain itself a spin connection living in the appropriate principal fiber bundle over Σ). In local coordinates, for instance, $(\phi^*\mathbf{a})_\alpha(x) = \partial_\alpha X^\mu(x)\mathbf{a}_\mu(\phi(x))$. The 2d fermion fields ψ^μ are Majorana-Weyl spinors defined not in Σ but in $\phi(\Sigma)$. Instead λ denotes a collection of 2d Majorana-Weyl fields, λ^A, living in Σ, the index A being a group representation index running from 1 to 32.

17.8.1 Symmetries of the String Sigma Models

The actions (17.99) and (17.101, 17.102) are symmetric under the same worldsheet transformations of the previous section. But there are also transformations inherited from T which leave the above actions invariant:

- In the case (17.102), the gauge transformations induced by the gauge transformations in the principal fiber bundle $\mathsf{P}(\mathsf{T}, \mathsf{G})$ over the spacetime T. Any embedding ϕ induces a bundle $\phi^*\mathsf{P}$ over Σ, whose relevant group of gauge transformations is $\mathsf{Aut_v}\phi^*\mathsf{P}$. A gauge transformation of $\mathbf{a}_\mu$: $\mathbf{a}_\mu \to \mathbf{a}_\mu + D_\mu\lambda$, where λ takes values in the Lie algebra of G, which is either $SO(32)$ or $E_8 \times E_8$, induces a gauge transformation of $\phi^*\mathbf{a}$ given by: $(\phi^*\mathbf{a})_\alpha \to \partial_\alpha X^\mu(\mathbf{a}_\mu + D_\mu\lambda)$. This transformation of the induced connection is compensated for by a corresponding transformation of the spinors λ^A (with opposite sign).
- The diffeomorphisms of T, DiffT; this is easy to see for instance in (17.99), where a deformation $X^\mu \to X^\mu + \xi^\mu(X)$ is compensated by a diffeomorphism transformation of $G_{\mu\nu}$ with opposite parameter. To see it, in general, one must replace X^μ with $X^\mu + \xi^\mu$ and Taylor expand to first order in ξ^μ all the expressions wherever it appears; simultaneously transform the background fields $G_{\mu\nu}, \Gamma^\lambda_{\mu\nu}, A_\mu$ in the usual way they transform under diffeomorphisms in T with parameter $-\xi^\mu$. In this trade, $\psi^\mu(X)$ transforms as a vector field in T.

17.8.2 String Sigma Model Anomalies and Their Cancelation

The worldsheet anomalies of the sigma models (17.99) and (17.101, 17.102) are produced by the same bosons and fermions we have analyzed in the previous sections, interacting with the worldsheet metric $h_{\alpha\beta}$ (or the worldsheet zweibein e_a^α contained in $\gamma^\alpha = \mathrm{e}_a^\alpha\gamma^a$). The addition of the induced metric and gauge connection does not change anything from the point of view of the worldsheet anomalies, and they are the same as in the case of flat background. We know that in the conformal gauge the worldsheet diffeomorphism anomalies vanish identically. As for the trace anomalies we can argue exactly as in the previous chapter and conclude that they vanish in critical dimensions. So much for the worldsheet anomalies of the (super)string sigma models.

We would like next to deal with the induced anomalies of the actions (17.101, 17.102). Let us start from induced gauge anomalies in the second. In the perturbative approach, what matters is the fermion propagator and the gauge vertex. The fermion propagator of the λ fermions is always the same $\frac{i}{\not p}$, also the gauge potential vertex is the same γ^α, except that the associated fluctuating field to be inserted in the expression of the effective action is $\partial_\alpha X^\mu \mathbf{a}_\mu$, i.e. the pulled back connection from $\mathsf{P}(\mathsf{M}, \mathsf{G})$. So the anomaly is $\sim \int_\Sigma d^2x\,\sqrt{h}\,\phi^*\mathrm{tr}(d\lambda\,\mathbf{a})$ which is obtained by transgressing and pulling back to Σ the four-form in P

$$K(F(\mathbf{a}), F(\mathbf{a})) \equiv P_2(F(\mathbf{a}), F(\mathbf{a})) \tag{17.103}$$

The same result can be of course obtained with the SDW method.

Next let us turn to the action (17.101). From the point of view of the perturbative approach, we need again the propagator for the ψ^μ fields, which is the same as before, and the fermion-fermion-graviton vertex. The latter is hidden is the term $\bar{\psi}(X)\gamma^\alpha\partial_\alpha\psi(X)$, where ψ is any of the ψ^μ. We expand this term as follows

$$\bar{\psi}(X)\gamma^\alpha\partial_\alpha\psi(X) = \bar{\psi}(X)\gamma^a e_a^\alpha\partial_\alpha X^\mu\partial_\mu\psi(X) \tag{17.104}$$

It follows that the fluctuating field is $e_a^\alpha\partial_\alpha X^\mu(x)$. The square of it summed over a gives rise to a metric in T

$$e_a^\alpha\partial_\alpha X^\mu e_a^\beta\partial_\beta X^\nu = h^{\alpha\beta}\partial_\alpha X^\mu\partial_\beta X^\nu \tag{17.105}$$

In order to compute the anomaly, we remark that the calculation is the same as for the diffeomorphism anomaly in 2d (the relevant perturbative calculation was carried out in section 7.2.3), except that the inverse zweibein e_a^α must be replaced by $e_a^\alpha\partial_\alpha X^\mu$ and the metric $h_{\alpha\beta}$ by the induced metric from T. Therefore, the corresponding anomaly is the 2d diffeomorphism one originated via transgression from the four form

$$K'(R(G), R(G)) \equiv P_2'(R(G), R(G)) \tag{17.106}$$

and pulled back to Σ. Here $R(G)$ is the two-form curvature corresponding to the metric $G_{\mu\nu}$ in T. The same result can be obtained from the index theorem starting from the classifying space of the group $\mathsf{GL(d)}$. It originates from the ad-invariant polynomials $P_2(\mathfrak{F}^{gl}, \mathfrak{F}^{gl})$ of the Lie algebra $\mathfrak{gl}(\mathsf{d})$, where $\mathfrak{F}^{gl}$ is the curvature of the universal connection $\mathfrak{a}^{gl}$. One has to transgress it and pull it back first to T and, then, again to Σ via the embedding ϕ. A similar argument holds also for the anomaly corresponding to (17.103). It can be obtained starting from the classifying space of the group G, pulling back the form $K(F(\mathfrak{a}), F(\mathfrak{a}))$ to T, to get (17.103) and transgressing it before finally pulling it back to Σ.

As we have explained several times, the four-forms $K(F(\mathfrak{a}), F(\mathfrak{a}))$ and $K'(\mathfrak{F}^{gl}, \mathfrak{F}^{gl})$ are closed but not exact. On the contrary, the four-forms $K(F(\mathbf{a}), F(\mathbf{a}))$ and $K'(R(G), R(G))$ may be exact. In this case, as explained in Section 16.3, there is the possibility to cancel the corresponding anomalies with the relevant Wess-Zumino terms. This is the case, for instance, if there exists a three-form H in T, such that

$$K(F(\mathbf{a}), F(\mathbf{a})) = dH \tag{17.107}$$

To construct the corresponding WZ term, we use the diagram

$$\begin{array}{ccccc} \pi_1^* ev^*\mathsf{P} & \xrightarrow{\hat{\pi}_1} & ev^*\mathsf{P} & \xrightarrow{\hat{ev}} & \mathsf{P} \\ \downarrow{\scriptstyle \pi''} & & \downarrow{\scriptstyle \pi'} & & \downarrow{\scriptstyle \pi} \\ \mathsf{M} \times \mathcal{P}_{\phi_0}(\mathsf{Emb}(\Sigma, \mathsf{T})) & \xrightarrow{id\times\pi_1} & \mathsf{M} \times \mathsf{Emb}(\Sigma, \mathsf{T})_{\phi_0} & \xrightarrow{ev} & \mathsf{T} \end{array} \tag{17.108}$$

where $\mathsf{Emb}(\Sigma, \mathsf{T})$ is the space of embeddings of the worldsheet into the target spacetime, and $\pi_1^* ev^*\mathsf{P} \cong \phi_0^*\mathsf{P} \times \mathcal{P}_{\phi_0}(\mathsf{Emb}(\Sigma, \mathsf{T})$. The WZ terms are

$$\mathcal{B}_{WZ}(\mathbf{a}, H) = \int_{\Sigma} \left(J_\tau^* \int_0^1 dv\, i_{\frac{\partial}{\partial v}}\, h^* \left(\mathcal{T}K(\hat{\pi}_1^* \hat{ev}^* \mathbf{a}, A_0) - \pi_1^* ev^* H \right) \Big|_{(2,0)} \right) \tag{17.109}$$

The symbols in these formulas are explained in Section 16.3.

The same can be done for the diffeomorphism anomaly originated from $K'(R(G), R(G))$ in the case this four-form (the first Pontryagin class of T) is trivial. However, the most interesting case for string theory is when we consider both actions (17.101) and (17.102) together, and

$$K(F(\mathbf{a}), F(\mathbf{a})) - K'(R(G), R(G)) = d\mathcal{H} \tag{17.110}$$

The minus sign on the LHS is due to the opposite chiralities of the fermions in (17.101) and (17.102). In this case, we have to consider both the gauge and orthogonal frame bundle together, $\mathsf{P}(\mathsf{T}, \mathsf{G})$ and $\mathsf{O}_G\mathsf{T}$, respectively. Then the convenient geometrical structure is the bundle sum $\mathsf{P}(\mathsf{T}, \mathsf{G}) + \mathsf{O}_G\mathsf{T}$ (the sum of two principal bundles has been introduced in Subsection 11.6.3). To construct the WZ term for it, the basic diagram is

$$\begin{array}{ccccc} \pi_1^* ev^*(\mathsf{P} + \mathsf{O}_G\mathsf{T}) & \xrightarrow{\hat{\pi}_1} & ev^*(\mathsf{P} + \mathsf{O}_G\mathsf{T}) & \xrightarrow{\hat{ev}} & \mathsf{P} + \mathsf{O}_G\mathsf{T} \\ \downarrow \pi'' & & \downarrow \pi' & & \downarrow \pi \\ \mathsf{M} \times \mathcal{P}_{\phi_0}(\mathsf{Emb}(\Sigma, \mathsf{T})) & \xrightarrow{id \times \pi_1} & \mathsf{M} \times \mathsf{Emb}(\Sigma, \mathsf{T})_{\phi_0} & \xrightarrow{ev} & \mathsf{T} \end{array} \tag{17.111}$$

and $\pi_1^* ev^*(\mathsf{P} + \mathsf{O}_G\mathsf{T}) \cong \phi_0^*(\mathsf{P} + \mathsf{O}_G\mathsf{T}) \times \mathcal{P}_{\phi_0}(\mathsf{Emb}(\Sigma, \mathsf{T}))$. From this, we can construct

$$\begin{aligned} &\mathcal{B}_{WZ}(\mathbf{a}, G, \mathcal{H}) \\ &= \int_{\Sigma} \left(J_\tau^* \int_0^1 dv\, i_{\frac{\partial}{\partial v}}\, h^* \left(\mathcal{T}K(\hat{\pi}_1^* \hat{ev}^* \mathbf{a}, A_0) - \mathcal{T}K(\hat{\pi}_1^* \hat{ev}^* \mathbf{\Gamma}, \Gamma_0) - \pi_1^* ev^* \mathcal{H} \right) \Big|_{(2,0)} \right) \end{aligned} \tag{17.112}$$

where $\mathbf{\Gamma}$ is the Levi-Civita connection corresponding to the metric G and Γ_0 is a background connection analogous to A_0. This term cancels the anomaly originated from the four-form in the LHS of (17.110). This sigma model cancelation mechanism is, so to speak, a refinement of the Green-Schwarz cancelation mechanism which, in the target space field theory of the zero mass modes of the string, cancels the residual anomalies (see above). A complete unfolding of formula (17.112) and an analysis of its consequences in a string sigma model is still lacking.

17.9 Global Anomalies of the String

We have seen that local worldsheet diffeomorphism anomalies in superstring theories are absent (they vanish in the conformal gauge). The corresponding global anomalies can be excluded as well by the same argument used for ordinary field theories by providing sufficient conditions. Equation (13.46) shows that global anomalies are absent if the homotopy group $\pi_0(\mathcal{G})$ is torsionless. Now, this group is isomorphic to the group $\mathsf{Map}(\Sigma, \mathsf{SO}(2))$. In the case Σ is a two-sphere, this group is trivial. In the case it is an infinite cylinder, it is $\mathbb{Z}$. In the case of a torus, it is $\mathbb{Z} \times \mathbb{Z}$. In general for any compact Riemann surface Σ, there is no torsion and thus global anomalies are absent[2].

Global anomalies in the target space field theories of the superstrings are a much more complex matter, [7, 8]. As we have seen in Chap. 13, it is not possible to give general rules (i.e. necessary and sufficient conditions) for global anomaly cancelation, but in some cases it is possible to find sufficient conditions. Let us consider first the gauge case with a principal fiber bundle $\mathsf{P}(\mathsf{T}, \mathsf{G})$, and let us suppose that the group of gauge transformations $\mathcal{G} = \mathsf{Aut_v P}$ is weakly homotopically equivalent to the pointed space of maps $\mathsf{Map^m}(\mathsf{T}, \mathsf{G})$. Weak homotopy equivalence of two topological spaces means, roughly speaking, that they have the same homotopy groups; i.e. 'they have the same shape'. This weak equivalence has been proved for $\mathsf{d} \leq 4$, [9], and it is fairly reasonable to assume it in the context of global anomalies because the origin of torsion is in the homotopy groups. If, for instance $\mathsf{T} = S^{10}$, then $\pi_0(\mathcal{G}) = \pi_{10}(\mathsf{G})$. In both cases, $\mathsf{SO}(32)$ or $\mathsf{E}_8 \times \mathsf{E}_8$, this group is trivial. We can apply the same argument to the gravitational anomalies, in which case the relevant group is $\mathsf{SO}(10)$ (or $\mathsf{Spin}(10)$): we find $\pi_{10}(\mathsf{SO}(10)) = \mathbb{Z}$ and no torsion (no global anomalies). The same conclusions hold also in the case T is the product of a four-dimensional flat spacetime times S^6. But, of course, these are somewhat trivial examples: the superstring compactification spaces may be topologically much more complicated and the analysis must be adequately refined. There is not a single way of proceeding. One possibility is based on the so-called Postnikov systems, which we are going to present below.

Suppose X is a topological space with $\pi_1(\mathsf{X}) = \pi_2(\mathsf{X}) = 0$. Then one can approximate X with a sequence of spaces X_n, $n \geq 3$ such that

---

[2] The group $\mathsf{Map}(\Sigma, \mathsf{SO}(2))$ is the first cohomotopy group of Σ, $\pi^1(\Sigma)$. The latter is isomorphic to $H^1(\Sigma) = \mathbb{Z}^{2g}$, where g is the genus of Σ.

$$
\begin{array}{ccccc}
 & & \downarrow & & \\
K(\pi_5, 5) & \xrightarrow{i_5} & \mathsf{X}_5 & \xrightarrow{k_5} & K(\pi_6, 7) \\
 & & \downarrow{\scriptstyle p_4} & & \\
K(\pi_4, 4) & \xrightarrow{i_4} & \mathsf{X}_4 & \xrightarrow{k_4} & K(\pi_5, 6) \\
 & & \downarrow{\scriptstyle p_3} & & \\
K(\pi_3, 3) & = & \mathsf{X}_3 & \xrightarrow{k_3} & K(\pi_4, 5)
\end{array}
\tag{17.113}
$$

where $K(\pi_n, p)$ are Eilenberg-Mac Lane spaces and $\pi_n = \pi_n(X)$. The sequence

$$
\begin{array}{ccccc}
K(\pi_n, n) & \xrightarrow{i_n} & \mathsf{X}_n & & \\
 & & \downarrow{\scriptstyle p_{n-1}} & & \\
 & & \mathsf{X}_{n-1} & \xrightarrow{k_{n-1}} & K(\pi_n, n+1)
\end{array}
\tag{17.114}
$$

is a fibration with fiber $K(\pi_n, n)$ and projection p_{n-1}; k_{n-1} is the inducing map. If M is a CW complex, then

$$
[\mathsf{M}, \mathsf{X}]_* = [\mathsf{M}, \mathsf{X}_n]_* \tag{17.115}
$$

where $[\mathsf{X}, \mathsf{Y}]_*$ represent the homotopy classes of maps from the pointed space X to the pointed space Y. This is the way X_n approximates X. This construction is the Postnikov system.

Let now T be the ten-dimensional target space of our interest and let us suppose it is a CW complex (a very mild assumption). The classes of maps $[\mathsf{T}, \mathsf{G}]_*$ will be described by the approximant $[\mathsf{T}, X_{10}]_*$. Information about the latter can be extracted from the exact sequences obtained from (17.113):

$$
\begin{array}{ccccccc}
H^{10}(\mathsf{T}, \pi_{10}) & \longrightarrow & [\mathsf{T}, X_{10}]_* & \longrightarrow & [\mathsf{T}, X_9]_* & \longrightarrow & H^{11}(\mathsf{T}, \pi_{10}) \\
 & & & & & & \\
H^{9}(\mathsf{T}, \pi_{9}) & \longrightarrow & [\mathsf{T}, X_{9}]_* & \longrightarrow & [\mathsf{T}, X_8]_* & \longrightarrow & H^{10}(\mathsf{T}, \pi_{9}) \\
 & & & & & & \\
\dots\dots & \longrightarrow & \dots\dots & \longrightarrow & \dots\dots & \longrightarrow & \dots\dots \\
 & & & & & & \\
H^{4}(\mathsf{T}, \pi_{4}) & \longrightarrow & [\mathsf{T}, X_{4}]_* & \longrightarrow & [\mathsf{T}, \pi_3]_* & \longrightarrow & H^{5}(\mathsf{T}, \pi_{4})
\end{array}
\tag{17.116}
$$

where use has been made of the identification $[\mathsf{M}, K(\pi, n)]_* = H^n(\mathsf{M}, \pi)$. We can apply this to the group E_8 since we have $\pi_1 = \pi_2 = 0$. The remaining homotopy groups are $\pi_3 = \mathbb{Z}$ and $\pi_4 = \ldots = \pi_{14} = 0$. Therefore,

$$[\mathsf{T}, \mathsf{E}_8]_* = [\mathsf{T}, \mathsf{X}_{10}]_* = \mathsf{H}^3(\mathsf{T}, \mathbb{Z}) \tag{17.117}$$

from which we can derive

$$[\mathsf{T}, \mathsf{E}_8 \times \mathsf{E}_8]_* = \mathsf{H}^3(\mathsf{T}, \mathbb{Z} \oplus \mathbb{Z}) = \mathsf{H}^3(\mathsf{T}, \mathbb{Z}) \oplus \mathsf{H}^3(\mathsf{T}, \mathbb{Z}) \tag{17.118}$$

Therefore, an $\mathsf{E}_8 \times \mathsf{E}_8$ heteroric string compactification on T is global anomaly free if $H^3(\mathsf{T}, \mathbb{Z})$ is torsionless.

Postnikov systems are applicable also to the groups $\mathsf{Spin(N)}$, because their first and second homotopy groups are trivial for $\mathsf{N} \geq 3$. In the case of $\mathsf{Spin(32)}$, we have $\pi_4 = \pi_5 = \pi_6 = \pi_{10} = 0$, $\pi_3 = \pi_7 = \mathbb{Z}$ and $\pi_8 = \pi_9 = \mathbb{Z}_2$. Inserting these groups in the Postnikov system (17.113) does not imply a simple conclusion like (17.117), but leads to a set of (sufficient) restrictions on the cohomology groups of T. Even more restrictive are the conclusions if we apply the same method to the group $\mathsf{Spin(10)}$ and to the global gravitational anomalies. For the first 10 homotopy groups of $\mathsf{Spin(10)}$ are the same as $\mathsf{Spin(32)}$ except for $\pi_9 = \mathbb{Z} \oplus \mathbb{Z}_2$ and $\pi_{10} = \mathbb{Z}_4$.

Finally, we consider the possibility of global anomalies in the sigma model representation of superstrings. To this end, we have simply to adapt the conclusions of Chap. 16, and in particular Section 16.4, to the context of Section 17.8.2. For instance, one of the problems is to guarantee the global definition of the WZ term (17.112). A sufficient condition for that is

$$Tor\, H^4(\mathsf{T}, \mathbb{Z}) \cong Tor\, H_3(\mathsf{T}, \mathbb{Z}) = 0 \tag{17.119}$$

Appendix 17A. Index Formulas

The right-hand side of the index theorem formula is given by ordinary field theory classes, in terms of the curvature F of the involved bundle and the Riemann curvature R of the basis spacetime M. For spin 1/2, the formulas are in (12.47) and (12.48). For spin 3/2, the index is given by

$$Ind(\not{D}^+_{3/2}) = \int_{\mathsf{M}} ch(E)\, \hat{A}(\mathsf{M}) \left(\mathrm{tr}\left(\mathrm{e}^{i\frac{R}{2\pi}} - 1\right)\right) \tag{17.120}$$

The index for a real antisymmetric self-dual tensor fields is

$$ind(D_{sd}) = \frac{1}{8}\int_{\mathsf{M}} L(\mathsf{M}) \tag{17.121}$$

where

$$\begin{aligned} L(M) = 1 &- \frac{1}{(2\pi)^2}\frac{1}{6}\mathrm{trR}^2 + \frac{1}{(2\pi)^4}\left[\frac{1}{72}\left(\mathrm{trR}^2\right)^2 - \frac{7}{780}\mathrm{trR}^4\right] \\ &+ \frac{1}{(2\pi)^6}\left[-\frac{1}{1296}\left(\mathrm{trR}^2\right)^3 + \frac{7}{1080}\mathrm{trR}^2\mathrm{trR}^4 - \frac{31}{2835}\mathrm{trR}^6\right] + \ldots\ldots \end{aligned} \tag{17.122}$$

References

1. M.B. Green, J.H. Schwarz, E. Witten *Superstring Theory*, vol. I and II (Cambridge University Press, Cambridge, 1987)
2. J. Polchinski, *String Theory* (Cambridge University Press, Cambridge, I-II, 1998)
3. M. Green, J. Schwarz, Anomaly cancellation in supersymmetric D = 10 gauge theory and superstrings. Phys. Lett. **149B**, 117 (1984)
4. M. Green, J. Schwarz, The exagon gauge anomaly in type I superstring theory. Nucl. Phys. B **255**, 93 (1985)
5. C.M. Hull, E. Witten, Supersymmetric sigma models and the heterotic strings **160B**, 398 (1985)
6. Sen, A. Non-linear σ-models and string theory, in *Summer Workshop on High Energy Physics and Cosmology* (SLAC-PUB-4136, Trieste, 1986)
7. E. Witten, Topological tools in ten dimensional physics. Int. J. Mod. Phys. A **1**, 39 (1986)
8. L. Bonora, P. Cotta-Ramusino, M. Rinaldi, J. Stasheff, The evaluation map in field theory, sigma-models and strings. II. Comm. Math. Phys. **114**, 381 (1988)
9. I.M. Singer, Some remarks on the Gribov ambiguity. Comm. Math. Phys. **60**, 7 (1978)

Chapter 18
Literature and Further Readings

A full bibliography on anomalies consists of thousands of papers and books and is impractical for editorial reasons. Therefore I have made the choice of a short bibliography for any chapter containing papers and books which are a direct source of each chapter or have been basic for my understanding of fermions and anomalies.

After the limited list of references cited at the end of each chapter I have prepared an extended one. The latter, although very large, is still not an exhaustive collection of all the articles concerning anomalies in QFT and strings. I believe it is pretty so for the 'pioneering' period, which runs from the beginning till the late 80s of the last century. For the subsequent period until mid-2022, it is less complete, as it focuses mostly on the methodological papers rather than in the applicative ones. The range of applications of anomalies is so vast that for completeness it would have required, roughly speaking, more than twice as many pages. Therefore, I have chosen to be as thorough as possible for the first period and to become more selective in the most recent periods, privileging the papers which bear some relevance to the contents of the book and sacrificing the others. With this choice, some sectors are not covered or the coverage is limited to a small number of most significant papers, notably: anomalies in lattice gauge theories, phenomenological applications of anomalies to particle or gravity theories, anomalies and quantum Hall effect, supersymmetric anomalies, anomalies and Weyl semimetals, anomalies on orbifolds and on manifolds with boundary, anomalies and the CFT/AdS correspondence, anomalies in non-commutative field theories, rigid (often called global) anomalies, applications of t'Hooft anomalies, anomalies and entanglement, holomorphic anomalies; also the number of papers on anomalies in superstring theories had to be downsized.

This extended bibliography can be found appended to this chapter as Electronic Supplementary Material ESM_1.pdf.

Supplementary Information The online version contains supplementary material available at https://doi.org/10.1007/978-3-031-21928-3_18.

L. Bonora, *Fermions and Anomalies in Quantum Field Theories*, Theoretical and Mathematical Physics, https://doi.org/10.1007/978-3-031-21928-3_18

Literature and Further Readings

Correction to: Anomalies as Obstructions: The Atiyah-Singer Family'sIndex Theorem

Correction to:
Chapter 12 in: L. Bonora, *Fermions and Anomalies in Quantum Field Theories*, Theoretical and Mathematical Physics, https://doi.org/10.1007/978-3-031-21928-3_12

The original version of the book was published with some incorrect equations in Chapter 12 (Anomalies as Obstructions: The Atiyah-Singer Family's Index Theorem), which have now been corrected. The book and the chapter have been updated with the changes.

The updated version of this chapter can be found at
https://doi.org/10.1007/978-3-031-21928-3_12

L. Bonora, *Fermions and Anomalies in Quantum Field Theories*, Theoretical and Mathematical Physics, https://doi.org/10.1007/978-3-031-21928-3_19

Index

L. Bonora, *Fermions and Anomalies in Quantum Field Theories*, Theoretical and Mathematical Physics, https://doi.org/10.1007/978-3-031-21928-3

D

U

V

W

Y

Z

MIX
Papier aus verantwortungsvollen Quellen
Paper from responsible sources
FSC® C105338

If you have any concerns about our products,
you can contact us on
ProductSafety@springernature.com

In case Publisher is established outside the EU,
the EU authorized representative is:
Springer Nature Customer Service Center GmbH
Europaplatz 3, 69115 Heidelberg, Germany

Printed by Libri Plureos GmbH
in Hamburg, Germany